BASIC FUNCTIONS

Linear Function
$f(x) = x$

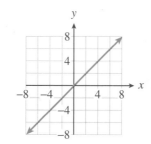

Absolute Value Function
$f(x) = |x|$

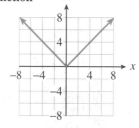

Square Function
$f(x) = x^2$

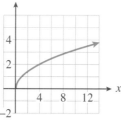

Cube Function
$f(x) = x^3$

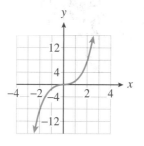

Square Root Function:
$f(x) = \sqrt{x}$

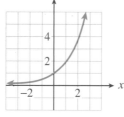

Cube Root Function
$f(x) = \sqrt[3]{x}$

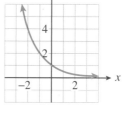

Reciprocal Function
$f(x) = \dfrac{1}{x}$

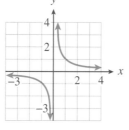

Reciprocal Square Function
$f(x) = \dfrac{1}{x^2}$

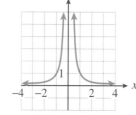

Exponential Function
$f(x) = 2^x$

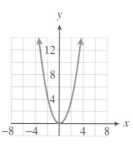

Exponential Function
$f(x) = \left(\dfrac{1}{2}\right)^x$

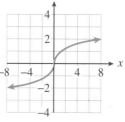

Common Logarithm Function
$f(x) = \log_{10} x$

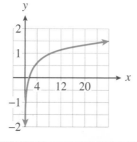

Natural Logarithm Function
$f(x) = \ln x$

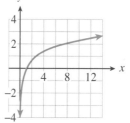

GEOMETRY

Perimeter and Area of Plane Figures

Triangle
$P = a + b + c$

$A = \dfrac{1}{2}bh$

Rectangle
$P = 2l + 2w$

$A = lw$

Parallelogram
$P = 2a + 2b$

$A = bh$

Trapezoid
$P = a + b + c + d$

$A = \dfrac{1}{2}h(a + b)$

Circle
$C = 2\pi r$

$A = \pi r^2$

Volume and Surface Area of Solid Figures

Rectangular Prism
$V = lwh$

$S = 2lw + 2lh + 2wh$

Right Circular Cylinder
$V = \pi r^2 h$

$S = 2\pi r^2 + 2\pi r h$

Sphere
$V = \dfrac{4}{3}\pi r^3$

$S = 4\pi r^2$

Right Circular Cone
$V = \dfrac{1}{3}\pi r^2 h$

$S = \pi r^2 + \pi r s$

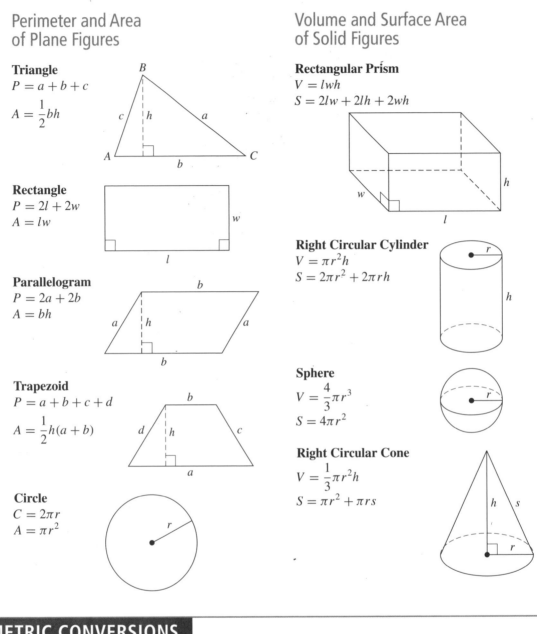

METRIC CONVERSIONS

Length
1 centimeter = 0.394 inch
1 meter = 3.281 feet
1 kilometer = 0.621 mile
1 inch = 2.54 centimeters
1 foot = 30.480 centimeters
1 mile = 1.609 kilometers

Volume
1 liter = 1.057 quarts
1 milliliter = 1 cubic centimeter
1 quart = 0.946 liter
1 gallon = 3.785 liters

Mass (Weight)
1 gram = 0.035 ounces
1 kilogram = 2.205 pounds
1 ounce = 28.350 grams
1 pound = 0.454 kilogram

MODELING, FUNCTIONS, AND GRAPHS

Algebra for College Students

THIS BOOK IS THE PROPERTY OF:

STATE_____

PROVINCE_____

COUNTY_____

PARISH_____

SCHOOL DISTRICT_____

OTHER_____

Book No. _____

Enter information
in spaces
to the left as
instructed

ISSUED TO	Year Used	CONDITION	
		ISSUED	RETURNED
Nathan Mosko 61181	2015	New	

PUPILS to whom this textbook is issued must not write on any page
or mark any part of it in any way, consumable textbooks excepted.

1. Teachers should see that the pupil's name is clearly written in ink in the spaces above in every book issued.
2. The following terms should be used in recording the condition of the book: New; Good; Fair; Poor; Bad.

FOURTH EDITION

MODELING, FUNCTIONS, AND GRAPHS
Algebra for College Students

Katherine Yoshiwara
Los Angeles Pierce College

Bruce Yoshiwara
Los Angeles Pierce College

THOMSON
™
BROOKS/COLE

Australia • Canada • Mexico • Singapore • Spain • United Kingdom • United States

THOMSON

BROOKS/COLE

Modeling, Functions, and Graphs: Algebra for College Students, Fourth Edition

Katherine Yoshiwara and Bruce Yoshiwara

Executive Editor: Charlie Van Wagner
Development Editor: Laura Stone
Assistant Editor: Rebecca Subity
Editorial Assistant: Christina Ho
Technology Project Manager: Sarah Woicicki
Marketing Manager: Greta Kleinert
Marketing Assistant: John Zents
Marketing Communications Manager: Darlene Amidon-Brent
Content Project Manager, Editorial Production: Cheryll Linthicum
Creative Director: Rob Hugel
Art Director: Lee Friedman
Print Buyer: Doreen Suruki

Permissions Editor: Roberta Broyer
Production Service: Interactive Composition Corporation
Text Designer: Diane Beasley
Photo Researcher: Roman A. Barnes
Copy Editor: Richard Camp, Randi Dubnick
Cover Designer: Roger Knox
Cover photo: Michael Trott/Wolfram Research. Infinite Windings was generated with Mathematica. www.wolfram.com/r/textbook.
Cover Printer: Courier Kendallville
Compositor: Interactive Composition Corporation
Printer: Courier Kendallville

Thomson Higher Education
10 Davis Drive
Belmont, CA 94002-3098
USA

Library of Congress Control Number: 2006925455

ISBN 0-534-41941-0

For more information about our products, contact us at:
Thomson Learning Academic Resource Center
1-800-423-0563

For permission to use material from this text or product, submit a request online at **http://www.thomsonrights.com**.

Any additional questions about permissions can be submitted by e-mail to **thomsonrights@thomson.com**.

BRIEF CONTENTS

CONTENTS

CHAPTER 2

Modeling with Functions 123

CHAPTER 4

Exponential Functions   321

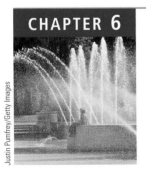

3D4Medical.com/Getty Images

Justin Pumfrey/Getty Images

CHAPTER 7

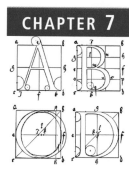

Polynomial and Rational Functions 561

CHAPTER 8

Adam Jones/Photo Researchers, Inc.

Linear Systems   637

APPENDIX A   Algebra Skills Refresher   707

APPENDIX B   Using a Graphing Calculator   797

PREFACE

$\mathbf{M}$athematics, as we all know, is the language of science, and fluency in algebraic skills has always been necessary for anyone aspiring to disciplines based on calculus. But in the information age, increasingly sophisticated mathematical methods are used in all fields of knowledge, from archaeology to zoology. Consequently, there is a new focus on the courses before calculus. The availability of calculators and computers allows students to tackle complex problems involving real data, but requires more attention to analysis and interpretation of results. All students, not just those headed for science and engineering, should develop a mathematical viewpoint, including critical thinking, problem-solving strategies, and estimation, in addition to computational skills. *Modeling, Functions and Graphs* employs a variety of applications to motivate mathematical thinking.

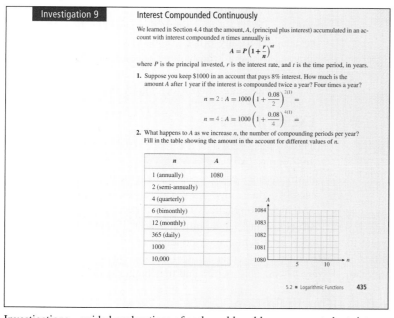

Investigations—guided explorations of real-world problems—ease students into intuitive problem-solving

MODELING

The ability to model problems or phenomena by algebraic expressions and equations is the ultimate goal of any algebra course. Through a variety of applications, we motivate students to develop the skills and techniques of algebra. Each chapter includes an interactive Investigation that gives students an opportunity to explore an open-ended modeling problem. These Investigations can be used in class as guided explorations or as projects for small groups. They are designed to show students how the mathematical techniques they are learning can be applied to study and understand new situations.

FUNCTIONS

The fundamental concept underlying calculus and related disciplines is the notion of function, and students should acquire a good understanding of functions before they embark on their study of college-level mathematics. While the formal study of functions is usually the content of precalculus, it is not too early to begin building an intuitive understanding of functional relationships in the preceding algebra courses. These ideas are useful not only in calculus but in practically any field students may pursue. We begin working with functions in Chapter 1 and explore the different families of functions in subsequent chapters.

In all our work with functions and modeling we employ the "Rule of Four," that all problems should be considered using algebraic, numerical, graphical, and verbal methods. It is the connections between these approaches that we have endeavored to establish in this course. At this level it is crucial that students learn to write an algebraic expression from a

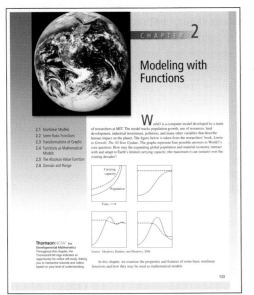

2.3 Transformations of Graphs

Models for real situations are often variations of the basic functions introduced in Section 2.2. In this section, we explore how certain changes in the formula for a function affect its graph. In particular, we will compare the graph of $y = f(x)$ with the graphs of $y = f(x) + k$, $y = f(x + h)$, and $y = af(x)$ for different values of the constants k, h, and a. Such variations are called **transformations** of the graph.

Vertical Translations

Figure 2.9 shows the graphs of $f(x) = x^2 + 4$, $g(x) = x^2 - 4$, and the basic parabola, $y = x^2$. By comparing tables of values, we can see exactly how the graphs of f and g are related to the basic parabola.

x	-2	-1	0	1	2
$y = x^2$	4	1	0	1	4
$f(x) = x^2 + 4$	8	5	4	5	8

x	-2	-1	0	1	2
$y = x^2$	4	1	0	1	4
$g(x) = x^2 - 4$	0	-3	-4	-3	0

FIGURE 2.9

Functions are studied in each chapter using algebraic expressions, tables of values, graphs, and verbal descriptions

Making a Table of Values

We can use a graphing calculator to make a table of values for a function defined by an equation. For the function in Example 4,

$$h = 1454 - 16t^2$$

we begin by entering the equation: Press the $\boxed{Y=}$ key, clear out any other equations, and define $Y_1 = 1454 - 16X^2$.

Next, we choose the x-values for the table. Press $\boxed{2nd}$ $\boxed{WINDOW}$ to access the **TblSet** (Table Setup) menu and set it to look like Figure 1.10a. This setting will give us an initial x-value of 0 (**TblStart = 0**) and an increment of one unit in the x-values, (**ΔTbl = 1**). It also fills in values of both variables automatically. Now press $\boxed{2nd}$ $\boxed{GRAPH}$ to see the table of values, as shown in Figure 1.10b. From this table, we can check the heights we found in Example 4.

Now try making a table of values with **TblStart = 0** and **ΔTbl = 0.5**. Use the $\boxed{\triangle}$ and $\boxed{\triangledown}$ keys to scroll up and down the table.

TABLE SETUP
TblMin=0
ΔTbl=1
Indpnt: **Auto** Ask
Depend: **Auto** Ask

(a)

(b)

FIGURE 1.10

Graphing Calculator boxes throughout the textbook explain when this tool can be appropriate to verify solutions or solve problems. An appendix provides more detailed assistance on using a calculator

CHAPTER 2

Modeling with Functions

2.1 Nonlinear Models
2.2 Some Basic Functions
2.3 Transformations of Graphs
2.4 Functions as Mathematical Models
2.5 The Absolute Value Function
2.6 Domain and Range

World3 is a computer model developed by a team of researchers at MIT. The model tracks population growth, use of resources, land development, industrial investment, pollution, and many other variables that describe human impact on the planet. The figure below is taken from the researchers' book, *Limits to Growth: The 30-Year Update*. The graphs represent four possible answers to World's core question: How may the expanding global population and material economy interact with and adapt to Earth's limited carrying capacity (the maximum it can sustain) over the coming decades?

ThomsonNOW™ for
Developmental Mathematics
Throughout this chapter, the
ThomsonNOW logo indicates an
opportunity for online self-study, linking
you to interactive tutorials and videos
based on your level of understanding.

In this chapter, we examine the properties and features of some basic nonlinear functions and how they may be used as mathematical models.

Source: Meadows, Randers, and Meadows, 2004

123

New chapter opener scenarios introduce the topic of the chapter through a real-life problem

verbal description, recognize trends in a table of data, and extract and interpret information from the graph of a function.

GRAPHS

No tool for conveying information about a system is more powerful than a graph. Yet many students have trouble progressing from a point-wise understanding of graphs to a more global view. By taking advantage of graphing calculators, we examine a large number of examples and study them in more detail than is possible when every graph is plotted by hand. We can consider more realistic models in which calculations by more traditional methods are difficult or impossible.

We have incorporated graphing calculators into the text wherever they can be used to enhance understanding. Calculator use is not simply an add-on, but in many ways shapes the organization of the material. The text includes instructions for the TI-84 graphing calculator, but these can easily be adapted to any other graphing utility. We have not attempted to use all the features of the calculator or to teach calculator use for its own sake, but in all cases have let the mathematics suggest how technology should be used.

EXERCISE SETS

The exercise sets reflect our focus on modeling, functions, and graphs. We have provided a wide range of open-ended problems that emphasize mathematical modeling using tables of values, algebraic expressions, and graphs. We have also included practice exercises for each new skill, so that instructors may decide how much drill or manipulation is appropriate for their classes. Each chapter concludes with a set of review exercises that students can use as a practice test for the chapter. Answers to the odd-numbered exercises in each section and the chapter reviews are provided.

About the Fourth Edition

This fourth edition of Modeling, Functions and Graphs has been revised to serve a modern Algebra for College Students course. There is a more stream-lined treatment of review topics from earlier algebra courses, and the text frequently refers students to the appropriate section of a new Appendix A, Algebra Skills Refresher. Function notation is introduced in Chapter 1 and is used consistently in subsequent chapters treating the various families of functions. In addition to studying mathematical topics from verbal, numerical, graphical, and algebraic viewpoints, we have added a selection of "proto-proofs": guided problems that require students to supply simple mathematical reasoning or arguments.

New to this edition, each chapter opens with a problem of historical or contemporary significance to introduce the material in the chapter. These scenarios encourage students to look for ways that real life can be described or predicted mathematically.

New Section Summaries have been added before each Homework set, and a Chapter Summary has been added before the Review problems. Each Section

SECTION 4.4 SUMMARY

VOCABULARY *Look up the definitions of new terms in the Glossary.*
Compounding period

CONCEPTS

1.

Properties of Logarithms

If x, y, and $b > 0$, then

1. $\log_b (xy) = \log_b x + \log_b y$
2. $\log_b \dfrac{x}{y} = \log_b x - \log_b y$
3. $\log_b x^k = k \log_b x$

2. We can use the properties of logarithms to solve exponential equations with any base.

Steps for Solving Exponential Equations

1. Isolate the power on one side of the equation.
2. Take the log base 10 of both sides.
3. Simplify by applying Log Property (3).
4. Solve for the variable.

3. The amount in an account earning interest compounded n times per year is an exponential function of time.

Compound Interest

The **amount** $A(t)$ accumulated (principal plus interest) in an account bearing interest compounded n times annually is

$$A(t) = P\left(1 + \frac{r}{n}\right)^{nt}$$

where P is the principal invested,
r is the interest rate,
t is the time period, in years.

STUDY QUESTIONS

1. The properties of logs are really another form of which familiar laws?
2. Which log property allows us to solve an exponential equation whose base is not 10?
3. Explain why $12 \cdot 10^{3t}$ is not the same as 120^{3t}.
4. Which of the following expressions are equivalent?

$\log\left(\dfrac{x}{4}\right) \qquad \dfrac{\log x}{\log 4} \qquad \log(x - 4) \qquad \log x - \log 4$

5. Which of the following expressions are equivalent?

$\log(x + 2) \qquad \log x + \log 2 \qquad \log(2x) \qquad (\log 2)(\log x)$

6. Which of the following expressions are equivalent?

$\log x^3 \qquad (\log 3)(\log x) \qquad 3 \log x \qquad \log 3^x$

SKILLS *Practice each skill in the Homework problems listed.*

1. Use the properties of logarithms to simplify expressions: #1–24, #45–52
2. Solve exponential equations using logs base 10: #25–36
3. Solve problems about exponential models: #37–44
4. Solve problems about compound interest: #53–58
5. Solve formulas involving exponential expressions: #59–64

372   Chapter 4 ■ Exponential Functions

HOMEWORK 4.4   Thomson

1. a. Simplify $10^2 \cdot 10^6$.
   b. Compute $\log 10^2$, $\log 10^6$, and $\log(10^2 \cdot 10^6)$. How are they related?
2. a. Simplify $\dfrac{10^9}{10^6}$.
   b. Compute $\log 10^9$, $\log 10^6$, and $\log\left(\dfrac{10^9}{10^6}\right)$. How are they related?
3. a. Simplify $\dfrac{b^8}{b^5}$.
   b. Compute $\log_b b^8$, $\log_b b^5$, and $\log_b\left(\dfrac{b^8}{b^5}\right)$. How are they related?
4. a. Simplify $b^4 \cdot b^3$.
   b. Compute $\log_b b^4$, $\log_b b^3$, and $\log(b^4 \cdot b^3)$. How are they related?
5. a. Simplify $(10^3)^5$.
   b. Compute $((10^3)^5)$ and $\log(10^3)$. How are they related?
6. a. Simplify $(b^2)^6$.
   b. Compute $\log_b(b^2)^6$ and $\log_b(b^2)$. How are they related?

Use the properties of logarithms to expand each expression in terms of simpler logarithms. Assume that all variable expressions denote positive numbers.

7. a. $\log_b 2x$ b. $\log_b \dfrac{x}{2}$ 8. a. $\log_b \dfrac{2x}{x-2}$ b. $\log_b x(2x + 3)$

9. a. $\log_3 3x^4$ b. $\log_5 1.1^{1/t}$ 10. a. $\log_b (4b)^t$ b. $\log_7 5(2^x)$

11. a. $\log_b \sqrt{bx}$ b. $\log_3 \sqrt{x^2 + 1}$ 12. a. $\log_{10} \sqrt{\dfrac{2L}{R^2}}$ b. $\log_{10} 2\pi \sqrt{\dfrac{l}{g}}$

13. a. $\log P_0 (1 - m)^t$ b. $\log_4\left(1 + \dfrac{r}{4}\right)^{4t}$ 14. a. $\log_3 \dfrac{a^2 - 2}{a^5}$ b. $\log \dfrac{a^3 b^2}{(a + b)^{3/2}}$

Combine into one logarithm and simplify. Assume all expressions are defined.

15. a. $\log_b 8 - \log_b 2$ b. $2 \log_4 x + 3 \log_4 y$ 16. a. $\log_b 5 + \log_b 2$ b. $\dfrac{1}{4} \log_5 x - \dfrac{3}{4} \log_5 y$

17. a. $\log 2x + 2 \log x - \log \sqrt{x}$ b. $\log(t^2 - 16) - \log(t + 4)$ 18. a. $\log x^2 + \log x^3 - 5 \log x$ b. $\log(x^2 - x) - \log \sqrt{x^3}$

19. a. $3 - 3 \log 30$ b. $\dfrac{1}{3} \log_6 (8w^6)$ 20. a. $2 - \log_4 (16z^2)$ b. $1 - 2 \log_3 x$

Use the three logs below to find the value of each expression.

$\log_b 2 = 0.6931$, $\log_b 3 = 1.0986$, $\log_b 5 = 1.6094$

(*Hint:* For example, $\log_b 15 = \log_b 3 + \log_b 5$.)

21. a. $\log_b 6$ b. $\log_b \dfrac{2}{5}$ 22. a. $\log_b 10$ b. $\log_b \dfrac{3}{2}$

23. a. $\log_b 9$ b. $\log_b \sqrt{50}$ 24. a. $\log_b 25$ b. $\log_b 75$

Solve each equation by using logarithms base 10.

25. $2^x = 7$ 26. $3^x = 4$ 27. $3^{x+1} = 8$ 28. $2^{x-1} = 9$

4.4 ■ Properties of Logarithms   373

New section summaries review key concepts. Homework sets provide plenty of skill practice as well as open-ended problems that emphasize mathematical modeling

Summary includes a list of new Vocabulary words that can be found in the Glossary, a brief review of new Concepts introduced in the section, a short set of Study Questions for students to test their understanding of the material, and a list of Mastery Skills. The Mastery Skills cite specific Homework Problem numbers for students to practice the listed skill, to help direct their studying.

In addition to the existing Investigations (guided explorations of an applied problem), five new Investigations have been added, with topics including vampire bats, compound interest, Bézier curves, polynomial interpolation, and the Leontief input-output model. New to this edition, Projects wrap up each chapter, to provide additional in-depth problems that challenge students with real-life data applications. In addition to the Graphing Calculator boxes that teach useful features of a graphing calculator as they arise in the exposition, we have included Appendix B, Using a Graphing Calculator, for easy reference.

As in the third edition, an in-text Exercise follows each Example in the reading material, allowing students to try out new concepts and skills as they are presented. Answers to Exercises appear at the end of the section. We have also included more boxed definitions and summaries of procedures to help organize the material for the reader.

The fourth edition reorganizes content from the third edition to flow more smoothly and to highlight functions earlier in the course:

- Chapter 1, Functions and their Graphs, includes introductory material on functions previously in Chapter 5, and a reworking of the treatment of linear functions from Chapter 1 of the third edition.

- Chapter 2, Modeling with Functions, covers the rest of the material on functions previously in Chapter 5 and includes a section on transformations of graphs.

- Chapter 3, Power Functions, contains the material on functions from Chapter 6 of the third edition, while the review material on radicals now appears in Appendix A.

- The treatment of exponential and logarithmic functions now appears in Chapters 4 and 5, including expanded coverage of inverse functions and the natural base. We have tried to delineate the difference between the simple operation of taking logarithms, which does not require an understanding of inverse functions, and logarithmic functions in themselves, which are introduced in Chapter 5. We have found that a clear understanding of these differences helps to demystify logarithms for students. We have also included a section on log scales and a brief introduction to log and log-log graphs. These tools constitute an enormously useful and widespread portion of the scientific literature, but have been neglected in the mathematics curriculum.

- Chapter 6, Quadratic Functions, contains the material of Chapters 3 and 4 from the third edition. Quadratic functions are actually more difficult for students than the power and exponential functions that precede them, and in a functions-based course provide a smoother introduction to the polynomial and rational functions of Chapter 7.

- Chapter 8, Linear Systems, considers both systems of equations and systems of inequalities.

The fourth edition includes new or expanded treatment of properties of functions, including concavity, transformations, inverse functions, and functions of two variables; more material on algebraic skills, including complex numbers, absolute value equations and inequalitites, equations quadratic in form, and the natural log; and a short section on annuities and amortization.

As with the third edition, we have endeavored to include authentic applications from a variety of disciplines, and to encourage some analysis and critical examination of the models chosen. We have avoided indiscriminate curve-fitting in favor of modeling phenomena that embody the properties of the functions they illustrate. In addition to the Projects, Investigations, and chapter openers designed to provide engaging examples of math in action, there are approximately 800 new problems, many of which draw on actual data, and some 6 dozen new Examples and Exercises in the reading.

Ancillaries

FOR INSTRUCTORS:

Test Bank (0-534-41944-5): The Test Bank includes multiple tests per chapter as well as final exams. The tests are made up of a combination of multiple-choice, free-response, true/false, and fill-in-the-blank questions.

Complete Solutions Manual (0-534-41943-7): The Complete Solutions Manual provides worked-out solutions to all of the problems in the text.

ThomsonNOW™

ThomsonNOW™ Instructor Version (0-534-41948-8): Providing instructors and students with unsurpassed control, variety, and all-in-one utility, ThomsonNOW is a powerful and fully integrated teaching and learning system that ties together tutorials, homework, and testing. Easy to use, ThomsonNOW offers instructors complete control when creating assessments in which they can draw from the wealth of exercises provided or create their own questions. A real timesaver for instructors, ThomsonNOW offers automatic grading of homework, quizzes, and tests with results flowing directly into the gradebook. Thomson-NOW provides seamless integration with Blackboard® and WebCT®. Contact your local Thomson sales representative for instructor access to ThomsonNOW.

JoinIn™ on Turning Point® (0-495-11437-5): JoinIn content for electronic response systems is tailored for *Modeling, Functions, and Graphs, Fourth Edition*. You can transform your classroom and assess your students' progress with instant in-class quizzes and polls. Turning Point software lets you pose book-specific questions and display studnets' answers seamlessly within Microsoft PowerPoint slides of your own lecture, in conjunction with the "clicker" hardware of your choice. Enhance the way your students interact with you, your lecture, and each other.

FOR STUDENTS:

Student Solutions Manual (0-534-41946-1): The Student Solutions Manual provides worked-out solutions to the odd-numbered problems in the text.

ThomsonNOW Student Version (0-534-41991-7): Featuring a variety of approaches that connect with all types of learners, ThomsonNOW offers text-specific tutorials that require no set-up by instructors. Students can begin exploring examples from the text by using the 1pass access code packaged with new books, or can purchase access by visiting www.thomsonedu.com. ThomsonNOW supports students with explanations from the text, step-by-step problem-solving help, unlimited practice, video lessons, quizzes, and more, ThomsonNOW is your key to better study and success.

Book Companion Website: Rich teaching and learning resources—including chapter-by-chapter online tutorial quizzes, a final exam, chapter outlines, chapter review, chapter-by-chapter weblinks, flashcards, and more.

Acknowledgments

We would like to thank the following reviewers for their valuable comments and suggestions:

Dr. Gisela Acosta, Valencia Community College

Peter J. Bloomsburg, Bellevue Community College

Lisa J. Carnell, High Point University

Steve E. Dostal, College of the Desert

James F. Dudley, University of New Mexico

Debbie Garrison, Valencia Community College

Lee Graubner, Valencia Community College

Judith Jones, Valencia Community College

Raimundo M. Kovac, Rhode Island College

Jolene Rhodes, Valencia Community College

John S. Robertson, Georgia Military College

Mark L. Sigfrids, Kalamazoo Valley Community College

Katherine Yoshiwara
Bruce Yoshiwara

Martin Ruegner/Getty Images

Functions and Their Graphs

Y ou may have heard that mathematics is the language of science. In fact, professionals in nearly every discipline take advantage of mathematical methods to analyze data, identify trends, and predict the effects of change. This process is called **mathematical modeling.** A model is a simplified representation of reality that helps us understand a process or phenomenon. Because it is a simplification, a model can never be completely accurate. Instead, it should focus on those aspects of the real situation that will help us answer specific questions. Here is an example.

The world's population is growing at different rates in different nations. Many factors, including economic and social forces, influence the birth rate. Is there a connection between birth rates and education levels? The figure shows the birth rate plotted against the female literacy rate in 148 countries. Although the data points do not all lie precisely on a line, we see a generally decreasing trend: The higher the literacy rate, the lower the birth rate. The **regression line** provides a model for this trend and a tool for analyzing the data. In this chapter, we study the properties of linear models and some techniques for fitting a linear model to data.

ThomsonNOW™

Throughout this chapter, the ThomsonNOW logo indicates an opportunity for online self-study, which:
- Links you to interactive tutorials and videos to help you study
- Helps to test your knowledge of material and prepare for exams

Visit ThomsonNOW at www.thomsonedu.com/login to access these resources.

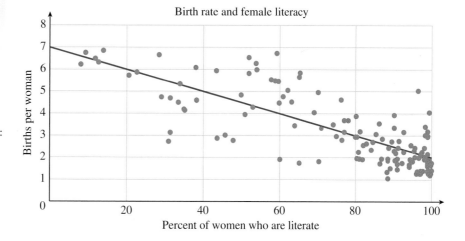

Birth rate and female literacy

Sales on Commission

Delbert is offered a part-time job selling restaurant equipment. He will be paid $1000 per month plus a 6% commission on his sales. The sales manager tells Delbert he can expect to sell about $8000 worth of equipment per month. To help him decide whether to accept the job, Delbert does a few calculations.

1. Based on the sales manager's estimate, what monthly income can Delbert expect from this job? What annual salary would that provide?

2. What would Delbert's monthly salary be if he sold only $5000 of equipment per month? What would his salary be if he sold $10,000 worth per month? Compute monthly incomes for each sales total shown in the table.

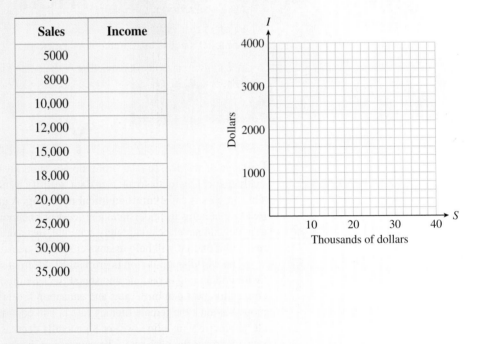

Sales	Income
5000	
8000	
10,000	
12,000	
15,000	
18,000	
20,000	
25,000	
30,000	
35,000	

3. Plot your data points on a graph, using sales, S, on the horizontal axis and income, I, on the vertical axis, as shown in the figure. Connect the data points to show Delbert's monthly income for all possible monthly sales totals.

4. Add two new data points to the table by reading values from your graph.

5. Write an algebraic expression for Delbert's monthly income, I, in terms of his monthly sales, S. Use the description in the problem to help you:

   He will be paid: $1000 . . . plus a 6% commission on his sales.

   $Income = $ _____.

6. Test your formula from part (5) to see if it gives the same results as those you recorded in the table.

7. Use your formula to find out what monthly sales total Delbert would need in order to have a monthly income of $2500.

8. Each increase of $1000 in monthly sales increases Delbert's monthly income by _____.

9. Summarize the results of your work: In your own words, describe the relationship between Delbert's monthly sales and his monthly income. Include in your discussion a description of your graph.

Tables, Graphs, and Equations

The first step in creating a model is to describe relationships between variables. In Investigation 1, we analyzed the relationship between Delbert's sales and his income. Starting from the verbal description of his income, we represented the relationship by a table of values, a graph, and an algebraic equation. Each of these mathematical tools is useful in a different way.

1. **A table of values** displays specific data points with precise numerical values.

2. **A graph** is a visual display of the data. It is easier to spot trends and describe the overall behavior of the variables from a graph.

3. **An algebraic equation** is a compact summary of the model. It can be used to analyze the model and to make predictions.

We begin our study of modeling with some examples of **linear models.** In the examples that follow, observe the interplay among the three modeling tools, and note how each contributes to the model.

EXAMPLE 1

Annelise is on vacation at a seaside resort. She can rent a bicycle from her hotel for $3 an hour, plus a $5 insurance fee. (A fraction of an hour is charged as the same fraction of $3.)

a. Make a table of values showing the cost, C, of renting a bike for various lengths of time, t.

b. Plot the points on a graph. Draw a curve through the data points.

c. Write an equation for C in terms of t.

Solutions
a. To find the cost, multiply the time by $3 and add the result to the $5 insurance fee. For example, the cost of a 1-hour bike ride is

$$\text{Cost} = (\$5 \text{ insurance fee}) + (\$3 \text{ per hour}) \times (\mathbf{1} \text{ hour})$$
$$C = 5 + 3(\mathbf{1}) = 8$$

A 1-hour bike ride costs $8. Record the results in a table, as shown here:

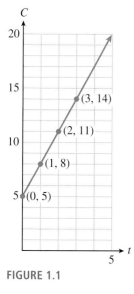

FIGURE 1.1

Length of rental (hours)	Cost of rental (dollars)		(t, C)
1	8	$C = 5 + 3(1)$	(1, 8)
2	11	$C = 5 + 3(2)$	(2, 11)
3	14	$C = 5 + 3(3)$	(3, 14)

b. Each pair of values represents a point on the graph. The first value gives the horizontal coordinate of the point, and the second value gives the vertical coordinate. The points lie on a straight line, as shown in Figure 1.1. The line extends infinitely in only one direction, because negative values of t do not make sense here.

(Solution continues on page 4.)

c. To find an equation, let C represent the cost of the rental, and use t for the number of hours:

$$\text{Cost} = (\$5 \text{ insurance fee}) + (\$3 \text{ per hour}) \times (\text{number of hours})$$
$$C = 5 + 3 \cdot t$$

EXAMPLE 2

Use the equation $C = 5 + 3t$ you found in Example 1 to answer the following questions. Then show how to find the answers by using the graph.

a. How much will it cost Annelise to rent a bicycle for 6 hours?

b. How long can Annelise bicycle for $18.50?

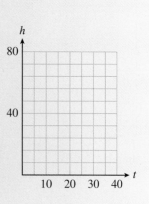

Solutions

a. Substitute $t = 6$ into the expression for C to find

$$C = 5 + 3(6) = 23$$

A 6-hour bike ride will cost $23. The point P on the graph in Figure 1.2 represents the cost of a 6-hour bike ride. The value on the C-axis at the same height as point P is 23, so a 6-hour bike ride costs $23.

b. Substitute $C = \mathbf{18.50}$ into the equation and solve for t.

$$\mathbf{18.50} = 5 + 3t$$
$$13.50 = 3t$$
$$t = 4.5$$

For $18.50, Annelise can bicycle for $4\frac{1}{2}$ hours. The point Q on the graph represents an $18.50 bike ride. The value on the t-axis below point Q is 4.5, so $18.50 will buy a 4.5-hour bike ride.

FIGURE 1.2

To review these algebraic techniques, see the Algebra Skills Refresher, A.2 and A.3.

In Example 2, notice the different algebraic techniques we used in parts (a) and (b). In part (a), we were given a value of t and we **evaluated the expression** $5 + 3t$ to find C. In part (b), we were given a value of C and we **solved the equation** $C = 5 + 3t$ to find t.

EXERCISE 1

Frank plants a dozen corn seedlings, each 6 inches tall. With plenty of water and sunlight they will grow approximately 2 inches per day. Complete the table of values for the height, h, of the seedlings after t days.

t	0	5	10	15	20
h					

a. Write an equation for the height, h, of the seedlings in terms of the number of days, t, since they were planted.

b. Graph the equation.

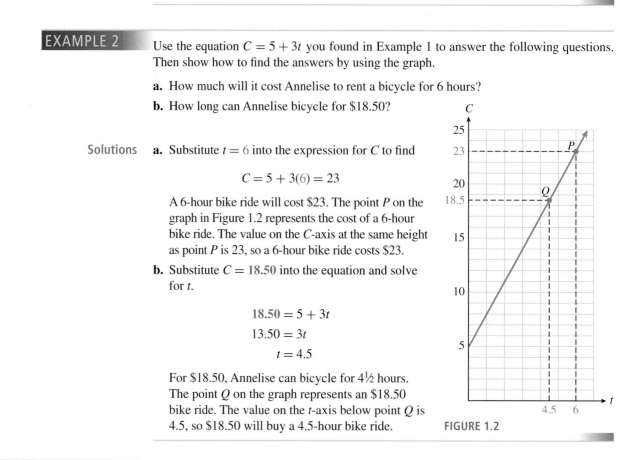

▶▶▶ Answers to Exercises appear at the end of each section, before the Summary.

Choosing Scales for the Axes

To draw a useful graph, we must choose appropriate scales for the axes. They must extend far enough to show the values of the variables, and the tick marks should be equally spaced. Usually no more than 10 or 15 tick marks are needed.

EXAMPLE 3

In 1990, the median home price in the United States was $92,000. The median price increased by about $4700 per year over the next decade. (Source: www. realestateabc.com)

▶▶▶ Consult the Glossary for the definition of *median*.

a. Make a table of values showing the median price of a house in 1990, 1994, 1998, and 2000.

b. Choose suitable scales for the axes and plot the values you found in part (a) on a graph. Use *t*, the number of years since 1990, on the horizontal axis and the price of the house, *P*, on the vertical axis. Draw a curve through the points.

c. Write an equation that expresses *P* in terms of *t*.

d. How much did the price of the house increase from 1990 to 1996? Illustrate the increase on your graph.

Solutions

a. In 1990, the median price was $92,000. Four years later, in 1994, the price had increased by $4(4700) = 18,800$ dollars, so

$$P = 92,000 + 4(4700) = 110,800$$

In 1998, the price had increased by $8(4700) = 37,600$ dollars, so

$$P = 92,000 + 8(4700) = 129,600$$

You can verify the price of the house in 2000 by a similar calculation.

Year	Price of house	(t, P)
1990	92,000	(0, 92,000)
1994	110,800	(4, 110,800)
1998	129,600	(8, 129,600)
2000	139,000	(10, 139,000)

b. Let *t* stand for the number of years since 1990, so that $t = 0$ in 1990, $t = 4$ in 1994, and so on. To choose scales for the axes, look at the values in the table. For this graph, we scale the horizontal axis, or *t*-axis, in 1-year intervals and the vertical axis, or *P*-axis, for $90,000 to $140,000 in intervals of $5000. The points in Figure 1.3 lie on a straight line.

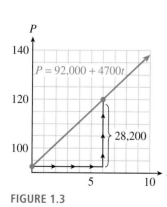

FIGURE 1.3

c. Look back at the calculations in part (a). The price of the house started at $92,000 in 1990 and increased by $t \times 4700$ dollars after t years. Thus,

$$P = 92{,}000 + 4700t \quad (t \geq 0)$$

d. Find the points on the graph corresponding to 1990 and 1996. These points lie above $t = 0$ and $t = 6$ on the t-axis. Now find the values on the P-axis corresponding to the two points. The values are $P = 92{,}000$ in 1990 and $P = 120{,}200$ in 1996. The increase in price is the difference of the two P-values.

$$\text{increase in price} = 120{,}200 - 92{,}000$$
$$= 28{,}200$$

The price of the home increased $28,200 between 1990 and 1996. This increase is indicated by the arrows in Figure 1.3.

The graphs in the preceding examples are **increasing graphs.** As we move along the graph from left to right (in the direction of increasing t), the second coordinate increases as well. Try Exercise 3, which illustrates a **decreasing graph.**

EXERCISE 3

Silver Lake has been polluted by industrial waste products. The concentration of toxic chemicals in the water is currently 285 parts per million (ppm). Local environmental officials would like to reduce the concentration by 15 ppm each year.

a. Complete the table of values showing the desired concentration, C, of toxic chemicals t years from now. For each t-value, calculate the corresponding value for C. Write your answers as ordered pairs.

t	C		(t, C)
0		$C = 285 - 15(0)$	$(0, \quad)$
5		$C = 285 - 15(5)$	$(5, \quad)$
10		$C = 285 - 15(10)$	$(10, \quad)$
15		$C = 285 - 15(15)$	$(15, \quad)$

b. To choose scales for the axes, notice that the value of C starts at 285 and decreases from there. We will scale the vertical axis up to 300 and use 10 tick marks at intervals of 30. Graph the ordered pairs on the grid and connect them with a straight line.

Extend the graph until it reaches the horizontal axis, but no farther. Points with negative C-coordinates have no meaning for the problem.

c. Write an equation for the concentration, C, of toxic chemicals t years from now. (*Hint*: The concentration is initially 285 ppm, and we *subtract* 15 ppm for each year that passes, or $15 \times t$.)

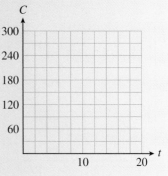

Graphing an Equation

We can use a graphing calculator to graph an equation. On most calculators, we follow three steps.

To Graph an Equation:

1. Press $\boxed{\textbf{Y=}}$ and enter the equation you wish to graph.
2. Press $\boxed{\textbf{WINDOW}}$ and select a suitable graphing window.
3. Press $\boxed{\textbf{GRAPH}}$.

EXAMPLE 4

In Example 3, we found the equation $P = 92{,}000 + 4700t$ for the median price of a house t years after 1990. Graph this equation on a calculator.

Solution To begin, we press $\boxed{\textbf{Y=}}$ and enter

$$Y_1 = 92{,}000 + 4700X$$

For this graph, we'll use the grid in Example 3 for our window settings, so we press $\boxed{\textbf{WINDOW}}$ and enter

$$\begin{array}{ll} \text{Xmin} = 0 & \text{Xmax} = 10 \\ \text{Ymin} = 90{,}000 & \text{Ymax} = 140{,}000 \end{array}$$

Finally, we press $\boxed{\textbf{GRAPH}}$. The calculator's graph is shown in Figure 1.4.

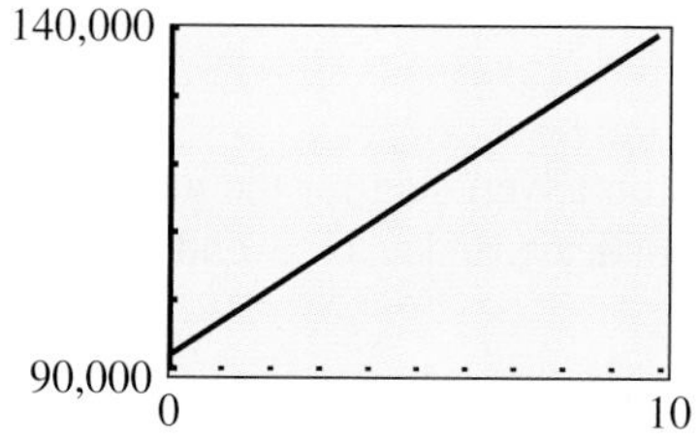

FIGURE 1.4

EXERCISE 4

a. Solve the equation $2y - 1575 = 45x$ for y in terms of x.
b. Graph the equation on a graphing calculator. Use the window

$$\begin{array}{lll} \text{Xmin} = -50 & \text{Xmax} = 50 & \text{Xscl} = 5 \\ \text{Ymin} = -500 & \text{Ymax} = 1000 & \text{Yscl} = 100 \end{array}$$

c. Sketch the graph on paper. Use the window settings to choose appropriate scales for the axes.

▶▶▶
See Appendix B for more about graphing with a calculator.

Linear Equations

All the models in these examples have equations with a similar form:

$$y = \textbf{(starting value)} + \textbf{(rate of change)} \cdot x$$

(We will talk more about rate of change in Section 1.4.) Their graphs were all portions of straight lines. For this reason, such equations are called **linear equations.** The order of the terms in the equation does not matter. For example, the equation in Example 1,

$$C = 5 + 3t$$

can be written equivalently as

$$-3t + C = 5$$

and the equation in Example 3,

$$P = 92{,}000 + 4700t$$

can be written as

$$-4700t + P = 92{,}000$$

This form of a linear equation,

$$Ax + By = C$$

is called the *general form*.

General Form for a Linear Equation

The graph of any equation

$$Ax + By = C$$

where A and B are not both equal to zero, is a straight line.

EXAMPLE 5   The manager at Albert's Appliances has $3000 to spend on advertising for the next fiscal quarter. A 30-second spot on television costs $150 per broadcast, and a 30-second radio ad costs $50.

   a. The manager decides to buy x television ads and y radio ads. Write an equation relating x and y.

   b. Make a table of values showing several choices for x and y.

   c. Plot the points from your table, and graph the equation.

Solutions   **a.** Each television ad costs $150, so x ads will cost $150x$. Similarly, y radio ads will cost $50y$. The manager has $3000 to spend, so the sum of the costs must be $3000. Thus,

$$150x + 50y = 3000$$

   b. Choose some values of x and solve the equation for the corresponding value of y. For example, if $x = 10$, then

$$150(10) + 50y = 3000$$
$$1500 + 50y = 3000$$
$$50y = 1500$$
$$y = 30$$

If the manager buys 10 television ads, she can also buy 30 radio ads. You can verify the other entries in the table.

x	8	10	12	14
y	36	30	24	18

   c. Plot the points from the table. All the solutions lie on a straight line, as shown in Figure 1.5.

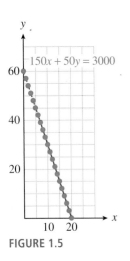

FIGURE 1.5

In central Nebraska, each acre of corn requires 25 acre-inches of water per year, and each acre of winter wheat requires 18 acre-inches of water. (An acre-inch is the amount of water needed to cover 1 acre of land to a depth of 1 inch.) A farmer can count on 9000 acre-inches of water for the coming year. (Source: Institute of Agriculture and Natural Resources, University of Nebraska)

a. Write an equation relating the number of acres of corn, x, and the number of acres of wheat, y, that the farmer can plant.

b. Complete the table.

x	50	100	150	200
y				

Intercepts

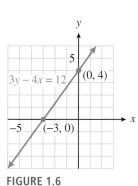

FIGURE 1.6

Consider the graph of the equation

$$3y - 4x = 12$$

shown in Figure 1.6. The points where the graph crosses the axes are called the **intercepts** of the graph.

The coordinates of these points are easy to find. The y-coordinate of the x-intercept is zero, so we set $y = 0$ in the equation to get

$$3(0) - 4x = 12,$$
$$x = -3$$

The x-intercept is the point $(-3, 0)$. Also, the x-coordinate of the y-intercept is zero, so we set $x = 0$ in the equation to get

$$3y - 4(0) = 12$$
$$y = 4$$

The y-intercept is $(0, 4)$.

Intercepts of a Graph

The points where a graph crosses the axes are called the **intercepts** of the graph.

1. To find the x-intercept, set $y = 0$ and solve for x.

2. To find the y-intercept, set $x = 0$ and solve for y.

The intercepts of a graph tell us something about the situation it models.

a. Find the intercepts of the graph in Exercise 3, about the pollution in Silver Lake.

b. What do the intercepts tell us about the problem?

Solutions **a.** An equation for the concentration of toxic chemicals is

$$C = 285 - 15t$$

To find the C-intercept, set t equal to zero.

$$C = 285 - 15(0) = 285$$

The C-intercept is the point $(0, 285)$, or simply 285. To find the t-intercept, set C equal to zero and solve for t.

$$0 = 285 - 15t \qquad \text{Add } 15t \text{ to both sides.}$$
$$15t = 285 \qquad \text{Divide both sides by 15.}$$
$$t = 19$$

The t-intercept is the point $(19, 0)$, or simply 19.

b. The C-intercept represents the concentration of toxic chemicals in Silver Lake now: When $t = 0$, $C = 285$, so the concentration is currently 285 ppm. The t-intercept represents the number of years it will take for the concentration of toxic chemicals to drop to zero: When $C = 0$, $t = 19$, so it will take 19 years for the pollution to be eliminated entirely.

EXERCISE 6

a. Find the intercepts of the graph in Example 5, about the advertising budget for Albert's Appliances: $150x + 50y = 3000$.

b. What do the intercepts tell us about the problem?

Intercept Method for Graphing Lines

Because we really only need two points to graph a linear equation, we might as well find the intercepts first and use them to draw the graph. The values of the intercepts will also help us choose suitable scales for the axes. It is always a good idea to find a third point as a check.

EXAMPLE 7 **a.** Find the x- and y-intercepts of the graph of $150x - 180y = 9000$.

b. Use the intercepts to graph the equation. Find a third point as a check.

Solutions **a.** To find the x-intercept, set $y = 0$.

$$150x - 180(0) = 9000 \qquad \text{Simplify.}$$
$$150x = 9000 \qquad \text{Dvide both sides by 150.}$$
$$x = 60$$

The x-intercept is the point $(60, 0)$. To find the y-intercept, set $x = 0$.

$$150(0) - 180y = 9000 \qquad \text{Simplify.}$$
$$-180y = 9000 \qquad \text{Dvide both sides by } -180.$$
$$y = -50$$

The y-intercept is the point $(0, -50)$.

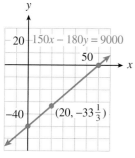

FIGURE 1.7

b. Scale both axes in intervals of 10 and then plot the two intercepts, $(60, 0)$ and $(0, -50)$. Draw the line through them, as shown in Figure 1.7. Now find another point and check that it lies on this line. We choose $x = 20$ and solve for y.

$$150(20) - 180y = 9000$$
$$3000 - 180y = 9000$$
$$-180y = 6000$$
$$y = -33.\overline{3}$$

Plot the point $(20, -33\frac{1}{3})$. Because this point lies on the line, we can be reasonably confident that our graph is correct.

Choosing a Graphing Window

Knowing the intercepts can also help us choose a suitable window on a graphing calculator. We would like the window to be large enough to show the intercepts. For the graph in Example 7, we can enter the equation

$$Y = (9000 - 150X)/{-}180$$

in the window

$$\begin{array}{ll} \text{Xmin} = -20 & \text{Xmax} = 70 \\ \text{Ymin} = -70 & \text{Ymax} = 30 \end{array}$$

We summarize the intercept graphing method as follows.

To Graph a Line Using the Intercept Method:

1. Find the intercepts of the line.
   a. To find the x-intercept, set $y = 0$ and solve for x.
   b. To find the y-intercept, set $x = 0$ and solve for y.
2. Plot the intercepts.
3. Choose a value for x and find a third point on the line.
4. Draw a line through the points.

EXERCISE 7

a. In Exercise 5, you wrote an equation about crops in Nebraska. Find the intercepts of the graph.
b. Use the intercepts to help you choose appropriate scales for the axes, and then graph the equation.
c. What do the intercepts tell us about the problem?

The examples in this section model simple linear relationships between two variables. Such relationships, in which the value of one variable is determined by the value of the other, are called *functions*. We will study various kinds of functions throughout the course.

ANSWERS TO 1.1 EXERCISES

1. a. $h = 6 + 2t$ **b.**

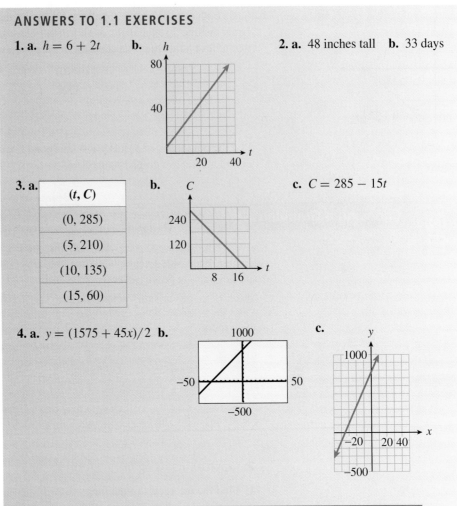

2. a. 48 inches tall   **b.** 33 days

3. a.

(t, C)
$(0, 285)$
$(5, 210)$
$(10, 135)$
$(15, 60)$

b.

c. $C = 285 - 15t$

4. a. $y = (1575 + 45x)/2$ **b.**

c.

5. a. $25x + 18y = 9000$ **b.**

x	50	100	150	200
y	430.6	361.1	291.7	222.2

6. (20, 0) The manager can buy 20 television ads if she buys no radio ads.
(0, 60) The manager can buy 60 radio ads if she buys no television ads.

7. a., c. (360, 0) If he plants no wheat, the
farmer can plant 360 acres of corn.
(0, 500) If he plants no corn, the farmer
can plant 500 acres of wheat.

b.

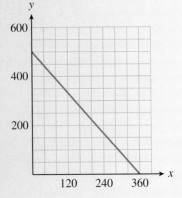

VOCABULARY   *Look up the definitions of new terms in the Glossary.*

Variable	Mathematical model	Increasing graph	Solve an equation
Linear equation	Evaluate an expression	Intercept	Decreasing graph

CONCEPTS

1. We can describe a relationship between variables with a table of values, a graph, or an equation.

2. Linear models have equations of the following form:

$$y = \text{(starting value)} + \text{(rate of change)} \cdot x$$

3. To make a useful graph, we must choose appropriate scales for the axes.

4.
> **General Form for a Linear Equation**
>
> The graph of any equation
>
> $$Ax + By = C$$
>
> where A and B are not both equal to zero, is a straight line.

5. The **intercepts** of a graph are the points where the graph crosses the axes.

6. We can use the intercepts to graph a line.

> **To Graph a Line Using the Intercept Method:**
>
> 1. Find the intercepts of the line.
>   a. To find the x-intercept, set $y = 0$ and solve for x.
>   b. To find the y-intercept, set $x = 0$ and solve for y.
> 2. Plot the intercepts.
> 3. Choose a value for x and find a third point on the line.
> 4. Draw a line through the points.

7. The intercepts are also useful for interpreting a model.

STUDY QUESTIONS

1. Name three ways to represent a relationship between two variables.
2. If C is expressed in terms of H, which variable goes on the horizontal axis?
3. Explain the difference between evaluating an expression and solving an equation.
4. How many points do you need to graph a linear equation?
5. Explain how the words *intercept* and *intersect* are related; explain how they are different.
6. Delbert says that the intercepts of the line $3x + 5y = 30$ are $(10, 6)$. What is wrong with his answer?

SKILLS   *Practice each skill in the Homework problems listed.*

1. Make a table of values: #1–4, 7 and 8
2. Plot points and draw a graph: #1–4, 7 and 8
3. Choose appropriate scales for the axes: #5–12
4. Write a linear model of the form
   $y = \text{(starting value)} + \text{(rate of change)} \cdot x$: #1–8
5. Write a linear model in general form: #25–28, 33–36
6. Evaluate a linear expression, algebraically and graphically: #1–4

7. Solve a linear equation, algebraically and graphically: #1–4
8. Find the intercepts of a graph: #5 and 6, 13–24, 45–52
9. Graph a line by the intercept method: #5 and 6, 13–24
10. Interpret the meaning of the intercepts: #5 and 6, 25–28
11. Use a graphing calculator to graph a line: #37–52
12. Sketch on paper a graph obtained on a calculator: #37–44

1. The temperature in the desert at 6 a.m., just before sunrise, was 65°F. The temperature rose 5 degrees every hour until it reached its maximum value at about 5 p.m. Complete the table of values for the temperature, T, at h hours after 6 a.m.

h	0	3	6	9	10
T					

   a. Write an equation for the temperature, T, in terms of h.
   b. Graph the equation.
   c. How hot is it at noon? Illustrate the answer on your graph.
   d. When will the temperature be 110°F? Illustrate the answer on your graph.

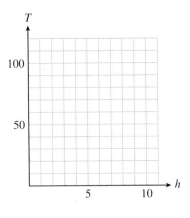

2. The taxi out of Dulles Airport charges a traveler with one suitcase an initial fee of $2.00, plus $1.50 for each mile traveled. Complete the table of values showing the charge, C, for a trip of n miles.

n	0	5	10	15	20	25
C						

   a. Write an equation for the charge, C, in terms of the number of miles traveled, n.
   b. Graph the equation.
   c. What is the charge for a trip to Mount Vernon, 40 miles from the airport? Illustrate the answer on your graph.
   d. If a ride to the National Institutes of Health (NIH) costs $39.50, how far is it from the airport to the NIH? Illustrate the answer on your graph.

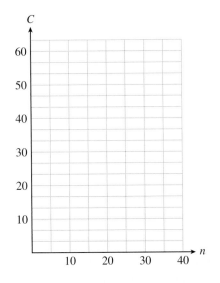

3. On October 31, Betty and Paul fill their 250-gallon oil tank for their heater. Beginning in November, they use an average of 15 gallons of oil per week. Complete the table of values for the amount of oil, A, left in the tank after w weeks.

w	0	4	8	12	16
A					

   a. Write an equation that expresses the amount of oil, A, in the tank in terms of the number of weeks, w, since October 31.
   b. Graph the equation.

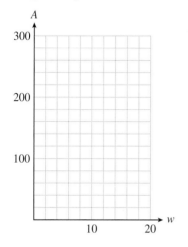

   c. How much did the amount of fuel oil in the tank decrease between the third week and the eighth week? Illustrate this amount on the graph.
   d. When will the tank contain more than 175 gallons of fuel oil? Illustrate on the graph.

4. Leon's camper has a 20-gallon gas tank, and he gets 12 miles to the gallon. (That is, he uses $\frac{1}{12}$ gallon per mile.) Complete the table of values for the amount of gas, g, left in Leon's tank after driving m miles.

m	0	48	96	144	192
g					

   a. Write an equation that expresses the amount of gas, g, in Leon's fuel tank in terms of the number of miles, m, he has driven.
   b. Graph the equation.

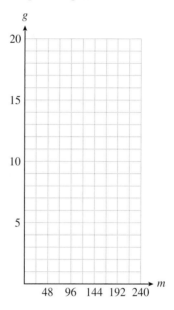

   c. How much gas will Leon use between 8 a.m., when his odometer reads 96 miles, and 9 a.m., when the odometer reads 144 miles? Illustrate on the graph.
   d. If Leon has less than 5 gallons of gas left, how many miles has he driven? Illustrate on the graph.

5. Phil and Ernie buy a used photocopier for $800 and set up a copy service on their campus. For each hour that the copier runs, Phil and Ernie make $40.
   a. Write an equation that expresses Phil and Ernie's profit (or loss), P, in terms of the number of hours, t, they run the copier.
   b. Find the intercepts and sketch the graph. (*Suggestion:* Scale the horizontal axis from 0 to 40 in increments of 5, and scale the vertical axis from −1000 to 400 in increments of 100.)
   c. What do the intercepts tell us about the profit?

6. A deep-sea diver is taking some readings at a depth of 400 feet. He begins rising at 20 feet per minute.
   a. Write an equation that expresses the diver's altitude, h, in terms of the number of minutes, m, elapsed. (Consider a depth of 400 feet as an altitude of −400 feet.)
   b. Find the intercepts and sketch the graph. (*Suggestion:* Scale the horizontal axis from 0 to 24 in increments of 2, and scale the vertical axis from −500 to 100 in increments of 50.)
   c. What do the intercepts tell us about the diver's depth?

7. There are many formulas for estimating the annual cost of driving. The Automobile Club estimates that fixed costs for a small car—including insurance, registration, depreciation, and financing—total about $5000 per year. The operating costs for gasoline, oil, maintenance, tires, and so forth are about 12.5¢ per mile. (Source: Automobile Association of America)

   a. Write an equation for the annual driving cost, C, in terms of d, the number of miles driven.
   b. Complete the table of values.

Miles driven	4000	8000	12,000	16,000	20,000
Cost ($)					

   c. Choose scales for the axes and graph the equation.
   d. How much does the annual cost of driving increase when the mileage increases from 8000 to 12,000 miles? Illustrate this amount on the graph.
   e. How much mileage will cause the annual cost to exceed $7000? Illustrate on the graph.

8. The boiling point of water changes with altitude. At sea level, water boils at 212°F, and the boiling point diminishes by approximately 0.002°F for each 1-foot increase in altitude.

   a. Write an equation for the boiling point, B, in terms of a, the altitude in feet.
   b. Complete the table of values.

Altitude (ft)	−500	0	1000	2000	3000	4000	5000
Boiling point (°F)							

   c. Choose scales for the axes and graph the equation.
   d. How much does the boiling point decrease when the altitude increases from 1000 to 3000 feet? Illustrate this amount on the graph.
   e. At what altitudes is the boiling point less than 204°F? Illustrate on the graph.

For each table, choose appropriate scales for the axes and plot the given points.

9.

x	0	80	90	120
y	6	2	1.5	1

10.

x	300	500	800	1100
y	1.2	1.3	1.5	1.9

11.

x	0.01	0.03	0.06	0.07
y	−0.2	−1	−1.1	−2

12.

x	0.003	0.005	0.008	0.011
y	6	2	1.5	1

For Problems 13–18,
(a) Find the intercepts of the graph.
(b) Graph the equation by the intercept method.

13. $x + 2y = 8$

14. $2x - y = 6$

15. $3x - 4y = 12$

16. $2x + 6y = 6$

17. $\dfrac{x}{9} - \dfrac{y}{4} = 1$

18. $\dfrac{x}{5} + \dfrac{y}{8} = 1$

In Problems 19–24,
(a) Find the intercepts of the graph.
(b) Use the intercepts to choose scales for the axes, and then graph the equation by the intercept method.

19. $20x = 30y - 45,000$

20. $30x = 45y + 60,000$

21. $0.4x + 1.2y = 4.8$

22. $3.2x - 0.8y = 12.8$

23. $\dfrac{2x}{3} + \dfrac{3y}{11} = 1$

24. $\dfrac{8x}{7} - \dfrac{2y}{7} = 1$

25. The owner of a gas station has $19,200 to spend on unleaded gas this month. Regular unleaded costs him $2.40 per gallon, and premium unleaded costs $3.20 per gallon.

    a. How much do x gallons of regular cost? How much do y gallons of premium cost?

    b. Write an equation in general form that relates the amount of regular unleaded gasoline, x, the owner can buy and the amount of premium unleaded, y.

    c. Find the intercepts and sketch the graph.

    d. What do the intercepts tell us about the amount of gasoline the owner can purchase?

26. Five pounds of body fat is equivalent to 16,000 calories. Carol can burn 600 calories per hour bicycling and 400 calories per hour swimming.

    a. How many calories will Carol burn in x hours of cycling? How many calories will she burn in y hours of swimming?

    b. Write an equation in general form that relates the number of hours, x, of cycling and the number of hours, y, of swimming Carol needs to perform in order to lose 5 pounds.

    c. Find the intercepts and sketch the graph.

    d. What do the intercepts tell us about Carol's exercise program?

27. Delbert must increase his daily potassium intake by 1800 mg. He decides to eat a combination of figs and bananas, which are both low in sodium. There are 9 mg potassium per gram of fig, and 4 mg potassium per gram of banana.

    a. How much potassium is in x grams of fig? How much potassium is in y grams of banana?

    b. Write an equation in general form that relates the number of grams, x, of fig and the number of grams, y, of banana Delbert needs to get 1800 mg of potassium.

    c. Find the intercepts and sketch the graph.

    d. What do the intercepts tell us about Delbert's diet?

28. Leslie plans to invest some money in two CD accounts. The first account pays 3.6% interest per year, and the second account pays 2.8% interest per year. Leslie would like to earn $500 per year on her investment.

    a. If Leslie invests x dollars in the first account, how much interest will she earn? How much interest will she earn if she invests y dollars in the second account?

    b. Write an equation in general form that relates x and y if Leslie earns $500 interest.

    c. Find the intercepts and sketch the graph.

    d. What do the intercepts tell us about Leslie's investments?

29. Find the intercepts of the graph for each equation.

    a. $\dfrac{x}{3} + \dfrac{y}{5} = 1$ **b.** $2x - 4y = 1$

    c. $\dfrac{2x}{5} - \dfrac{2y}{3} = 1$ **d.** $\dfrac{x}{p} + \dfrac{y}{q} = 1$

    e. Why is the equation $\dfrac{x}{a} + \dfrac{y}{b} = 1$ called the *intercept form* for a line?

30. Write an equation in intercept form (see Problem 29) for the line with the given intercepts. Then write the equation in general form.

    a. $(6, 0), (0, 2)$ **b.** $(-3, 0), (0, 8)$

    c. $\left(\dfrac{3}{4}, 0\right), \left(0, \dfrac{-1}{4}\right)$ **d.** $(v, 0), (0, -w)$

    e. $\left(\dfrac{1}{H}, 0\right), \left(0, \dfrac{1}{T}\right)$

31. a. Find the y-intercept of the line $y = mx + b$.

    b. Find the x-intercept of the line $y = mx + b$.

32. a. Find the y-intercept of the line $Ax + By = C$.

    b. Find the x-intercept of the line $Ax + By = C$.

Write an equation in general form for each line.

33.

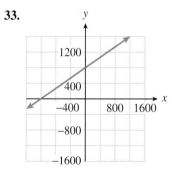

34.

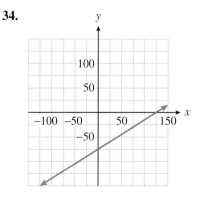

35.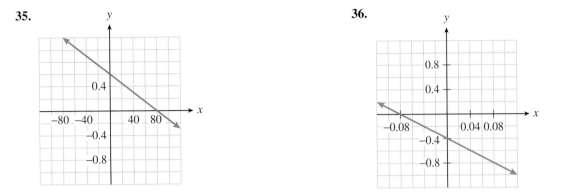

36.

For Problems 37–44,
(a) Solve each equation for *y* in terms of *x*. (See the Algebra Skills Refresher A.2 to review this skill.)
(b) Graph the equation on your calculator in the specified window. (See Appendix B for help with the graphing calculator.)
(c) Make a pencil and paper sketch of the graph. Label the scales on your axes, and the coordinates of the intercepts.

37. $2x + y = 6$
Xmin $= -10$ Ymin $= -10$
Xmax $= 10$ Ymax $= 10$
Xscl $= 1$ Yscl $= 1$

38. $8 - y + 3x = 0$
Xmin $= -10$ Ymin $= -10$
Xmax $= 10$ Ymax $= 10$
Xscl $= 1$ Yscl $= 1$

39. $3x - 4y = 1200$
Xmin $= -1000$ Ymin $= -1000$
Xmax $= 1000$ Ymax $= 1000$
Xscl $= 100$ Yscl $= 100$

40. $x + 2y = 500$
Xmin $= -1000$ Ymin $= -1000$
Xmax $= 1000$ Ymax $= 1000$
Xscl $= 100$ Yscl $= 100$

41. $0.2x + 5y = 0.1$
Xmin $= -1$ Ymin $= -0.1$
Xmax $= 1$ Ymax $= 0.1$
Xscl $= 0.1$ Yscl $= 0.01$

42. $1.2x - 4.2y = 3.6$
Xmin $= -1$ Ymin $= -1$
Xmax $= 4$ Ymax $= 1$
Xscl $= 0.2$ Yscl $= 0.1$

43. $70x + 3y = y + 420$
Xmin $= 0$ Ymin $= 0$
Xmax $= 10$ Ymax $= 250$
Xscl $= 1$ Yscl $= 25$

44. $40y - 5x = 780 - 20y$
Xmin $= -200$ Ymin $= 0$
Xmax $= 0$ Ymax $= 20$
Xscl $= 20$ Yscl $= 2$

For Problems 45–52,
(a) Find the *x*- and *y*-intercepts.
(b) Solve the equation for *y*.
(c) Choose a graphing window in which both intercepts are visible, and graph the equation on your calculator.

45. $x + 4y = 100$

46. $2x - 3y = -72$

47. $25x - 20y = 1$

48. $4x + 75y = 60,000$

49. $\dfrac{y}{12} - \dfrac{x}{60} = 1$

50. $\dfrac{x}{80} + \dfrac{y}{400} = 1$

51. $-2x = 3y + 84$

52. $7x = 91 - 13y$

Definition of Function

We often want to predict values of one variable from the values of a related variable. For example, when a physician prescribes a drug in a certain dosage, she needs to know how long the dose will remain in the bloodstream. A sales manager needs to know how the price of his product will affect its sales. A *function* is a special type of relationship between variables that allows us to make such predictions.

Suppose it costs \$800 for flying lessons, plus \$30 per hour to rent a plane. If we let C represent the total cost for t hours of flying lessons, then

$$C = 800 + 30t \qquad (t \geq 0)$$

Thus, for example,

$$\text{when} \quad t = 0, \quad C = 800 + 30(0) = 800$$
$$\text{when} \quad t = 4, \quad C = 800 + 30(4) = 920$$
$$\text{when} \quad t = 10, \quad C = 800 + 30(10) = 1100$$

The variable t is called the **input** or **independent** variable, and C is the **output** or **dependent** variable, because its values are determined by the value of t. We can display the relationship between two variables by a table or by ordered pairs. The input variable is the first component of the ordered pair, and the output variable is the second component.

t	C	(t, C)
0	800	(0, 800)
4	920	(4, 920)
10	1100	(10, 1100)

For this relationship, we can find the value C of associated with any given value of t. All we have to do is substitute the value of t into the equation and solve for C. The result has no ambiguity: Only *one* value for C corresponds to each value of t. This type of relationship between variables is called a **function.** In general, we make the following definition.

Definition of Function

A **function** is a relationship between two variables for which a *unique* value of the **output** variable can be determined from a value of the **input** variable.

What distinguishes functions from other variable relationships? The definition of a function calls for a *unique value*—that is, *exactly one value* of the output variable corresponding to each value of the input variable. This property makes functions useful in applications because they can often be used to make predictions.

EXAMPLE 1

a. The distance, *d*, traveled by a car in 2 hours is a function of its speed, *r*. If we know the speed of the car, we can determine the distance it travels by the formula $d = r \cdot 2$.

b. The cost of a fill-up with unleaded gasoline is a function of the number of gallons purchased. The gas pump represents the function by displaying the corresponding values of the input variable (number of gallons) and the output variable (cost).

c. Score on the Scholastic Aptitude Test (SAT) is *not* a function of score on an IQ test, because two people with the same score on an IQ test may score differently on the SAT; that is, a person's score on the SAT is not uniquely determined by his or her score on an IQ test.

EXERCISE 1

a. As part of a project to improve the success rate of freshmen, the counseling department studied the grades earned by a group of students in English and algebra. Do you think that a student's grade in algebra is a function of his or her grade in English? Explain why or why not.

b. Phatburger features a soda bar, where you can serve your own soft drinks in any size. Do you think that the number of calories in a serving of Zap Kola is a function of the number of fluid ounces? Explain why or why not.

A function can be described in several different ways. In the following examples, we consider functions defined by tables, by graphs, and by equations.

Functions Defined by Tables

When we use a table to describe a function, the first variable in the table (the left column of a vertical table or the top row of a horizontal table) is the input variable, and the second variable is the output. We say that the output variable *is a function of* the input.

EXAMPLE 2

a. Table 1.1 shows data on sales compiled over several years by the accounting office for Eau Claire Auto Parts, a division of Major Motors.

In this example, the year is the input variable, and total sales is the output. We say that total sales, *S*, *is a function of t*.

Year (*t*)	Total sales (*S*)
2000	$612,000
2001	$663,000
2002	$692,000
2003	$749,000
2004	$904,000

TABLE 1.1

Weight in ounces (*w*)	Postage (*P*)
$0 < w \le 1$	$0.39
$1 < w \le 2$	$0.63
$2 < w \le 3$	$0.87
$3 < w \le 4$	$1.11
$4 < w \le 5$	$1.35
$5 < w \le 6$	$1.59
$6 < w \le 7$	$1.83

TABLE 1.2

b. Table 1.2 gives the cost of sending printed material by first-class mail in 2006.

If we know the weight of the article being shipped, we can determine the required postage from Table 1.2. For instance, a catalog weighing 4.5 ounces would require $1.35 in postage. In this example, *w* is the input variable and *p* is the output variable. We say that *p is a function of w*.

c. Table 1.3 records the age and cholesterol count for 20 patients tested in a hospital survey.

Age	Cholesterol count	Age	Cholesterol count
53	217	51	209
48	232	53	241
55	198	49	186
56	238	51	216
51	227	57	208
52	264	52	248
53	195	50	214
47	203	56	271
48	212	53	193
50	234	48	172

TABLE 1.3

According to these data, cholesterol count is *not* a function of age, because several patients who are the same age have different cholesterol levels. For example, three different patients are 51 years old but have cholesterol counts of 227, 209, and 216, respectively. Thus, we cannot determine a *unique* value of the output variable (cholesterol count) from the value of the input variable (age). Other factors besides age must influence a person's cholesterol count.

EXERCISE 2

Decide whether each table describes *y* as a function of *x*. Explain your choice.

a.

x	3.5	2.0	2.5	3.5	2.5	4.0	2.5	3.0
y	2.5	3.0	2.5	4.0	3.5	4.0	2.0	2.5

b.

x	−3	−2	−1	0	1	2	3
y	17	3	0	−1	0	3	17

Functions Defined by Graphs

A graph may also be used to define one variable as a function of another. The input variable is displayed on the horizontal axis, and the output variable on the vertical axis.

EXAMPLE 3

Figure 1.8 shows the number of hours, H, that the sun is above the horizon in Peoria, Illinois, on day t, where January 1 corresponds to $t = 0$.

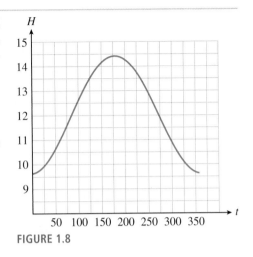

FIGURE 1.8

a. Which variable is the input, and which is the output?

b. Approximately how many hours of sunlight are there in Peoria on day 150?

c. On which days are there 12 hours of sunlight?

d. What are the maximum and minimum values of H, and when do these values occur?

Solutions

a. The input variable, t, appears on the horizontal axis. The number of daylight hours, H, is a function of the date. The output variable appears on the vertical axis.

b. The point on the curve where $t = 150$ has $H \approx 14.1$, so Peoria gets about 14.1 hours of daylight when $t = 150$, which is at the end of May.

c. $H = 12$ at the two points where $t \approx 85$ (in late March) and $t \approx 270$ (late September).

d. The maximum value of 14.4 hours occurs on the longest day of the year, when $t \approx 170$, about three weeks into June. The minimum of 9.6 hours occurs on the shortest day, when $t \approx 355$, about three weeks into December.

EXERCISE 3

Figure 1.9 shows the elevation in feet, a, of the Los Angeles Marathon course at a distance d miles into the race. (Source: *Los Angeles Times,* March 3, 2005)

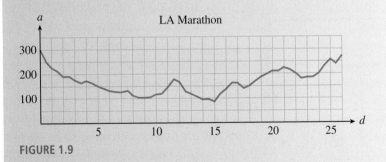

FIGURE 1.9

a. Which variable is the input, and which is the output?

b. What is the elevation at mile 20?

c. At what distances is the elevation 150 feet?

d. What are the maximum and minimum values of a, and when do these values occur?

e. The runners pass by the Los Angeles Coliseum at about 4.2 miles into the race. What is the elevation there?

Functions Defined by Equations

Example 4 illustrates a function defined by an equation.

EXAMPLE 4

As of 2006, the Sears Tower in Chicago is the nation's tallest building, at 1454 feet. If an algebra book is dropped from the top of the Sears Tower, its height above the ground after t seconds is given by the equation

$$h = 1454 - 16t^2$$

Thus, after **1** second the book's height is

$$h = 1454 - 16(1)^2 = 1438 \text{ feet}$$

After **2** seconds its height is

$$h = 1454 - 16(2)^2 = 1390 \text{ feet}$$

For this function, t is the input variable and h is the output variable. For any value of t, a unique value of h can be determined from the equation for h. We say that *h is a function of t*.

> A list of Geometric Formulas can be found inside the front cover of the book.

EXERCISE 4

Write an equation that gives the volume, V, of a sphere as a function of its radius, r.

Making a Table of Values

(a)

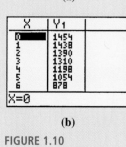

(b)

FIGURE 1.10

We can use a graphing calculator to make a table of values for a function defined by an equation. For the function in Example 4,

$$h = 1454 - 16t^2$$

we begin by entering the equation: Press the $\boxed{\text{Y=}}$ key, clear out any other equations, and define $Y_1 = 1454 - 16X^2$.

Next, we choose the x-values for the table. Press $\boxed{\text{2nd}}$ $\boxed{\text{WINDOW}}$ to access the **TblSet** (Table Setup) menu and set it to look like Figure 1.10a. This setting will give us an initial x-value of 0 (**TblStart = 0**) and an increment of one unit in the x-values, (**ΔTbl = 1**). It also fills in values of both variables automatically. Now press $\boxed{\text{2nd}}$ $\boxed{\text{GRAPH}}$ to see the table of values, as shown in Figure 1.10b. From this table, we can check the heights we found in Example 4.

Now try making a table of values with **TblStart = 0** and **ΔTbl = 0.5.** Use the $\boxed{\triangle}$ and $\boxed{\triangledown}$ keys to scroll up and down the table.

Function Notation

There is a convenient notation for discussing functions. First, we choose a letter, such as f, g, or h (or F, G, or H), to name a particular function. (We can use any letter, but these are the most common choices.) For instance, in Example 4, the height, h, of a falling algebra book is a function of the elapsed time, t. We might call this function f. In other words, f is the name of the relationship between the variables h and t. We write

$$h = f(t)$$

which means "h is a function of t, and f is the name of the function."

The new symbol $f(t)$, read "f of t," is another name for the height, h. The parentheses in the symbol $f(t)$ do *not* indicate multiplication. (It would not make sense to multiply the name of a function by a variable.) Think of the symbol $f(t)$ as a single variable that represents the output value of the function.

With this new notation we may write

$$h = f(t) = 1454 - 16t^2$$

or just

$$f(t) = 1454 - 16t^2$$

instead of

$$h = 1454 - 16t^2$$

to describe the function.

Perhaps it seems complicated to introduce a new symbol for h, but the notation $f(t)$ is very useful for showing the correspondence between specific values of the variables h and t.

EXAMPLE 5

In Example 4, the height of an algebra book dropped from the top of the Sears Tower is given by the equation

$$h = 1454 - 16t^2$$

We see that

$$\text{when } t = 1, \quad h = 1438$$
$$\text{when } t = 2, \quad h = 1390$$

Using function notation, these relationships can be expressed more concisely as

$$f(1) = 1438 \quad \text{and} \quad f(2) = 1390$$

which we read as "f of 1 equals 1438" and "f of 2 equals 1390." The values for the input variable, t, appear *inside* the parentheses, and the values for the output variable, h, appear on the other side of the equation.

Remember that when we write $y = f(x)$, the symbol $f(x)$ is just another name for the output variable.

Function Notation

Input variable ⟶ $f(x) = y$ ⟵ Output variable

Evaluating a Function

Finding the value of the output variable that corresponds to a particular value of the input variable is called **evaluating the function.**

EXAMPLE 6 Let g be the name of the postage function defined by Table 1.2 in Example 2b. Find $g(1)$, $g(3)$, and $g(6.75)$.

Solution   According to the table,

$$\text{when } w = 1, \qquad p = 0.39, \qquad \text{so} \quad g(1) = 0.39$$
$$\text{when } w = 3, \qquad p = 0.87, \qquad \text{so} \quad g(3) = 0.87$$
$$\text{when } w = 6.75, \quad p = 1.83, \qquad \text{so} \quad g(6.75) = 1.83$$

Thus, a letter weighing 1 ounce costs \$0.39 to mail, a letter weighing 3 ounces costs \$0.87, and a letter weighing 6.75 ounces costs \$1.83.

If a function is described by an equation, we simply substitute the given input value into the equation to find the corresponding output, or function value.

EXAMPLE 7 The function H is defined by $H = f(s) = \dfrac{\sqrt{s + 3}}{s}$. Evaluate the function at the following values.

a. $s = 6$ **b.** $s = -1$

Solution   Substitute the given values for s into the equation defining H.

a. $f(6) = \dfrac{\sqrt{6+3}}{6} = \dfrac{\sqrt{9}}{6} = \dfrac{3}{6} = \dfrac{1}{2}$. Thus, $f(6) = \dfrac{1}{2}$.

b. $f(-1) = \dfrac{\sqrt{-1+3}}{-1} = \dfrac{\sqrt{2}}{-1} = -\sqrt{2}$. Thus, $f(-1) = -\sqrt{2}$.

EXERCISE 7

Complete the table displaying ordered pairs for the function $f(x) = 5 - x^3$.
Evaluate the function to find the corresponding $f(x)$-value for each value of x.

x	$f(x)$
-2	
0	
1	
3	

$f(-2) = 5 - (-2)^3 =$

$f(0) = 5 - 0^3 =$

$f(1) = 5 - 1^3 =$

$f(3) = 5 - 3^3 =$

Evaluating a Function

We can use the table feature on a graphing calculator to evaluate functions. Consider the function of Exercise 7, $f(x) = 5 - x^3$.
Press $\boxed{Y=}$, clear any old functions, and enter

$$Y_1 = 5 - X\boxed{\wedge}3$$

Then press $\boxed{\text{TblSet}}$ ($\boxed{\text{2nd}}$ $\boxed{\text{WINDOW}}$) and choose **Ask** after **Indpnt,** as shown in Figure 1.11a, and press $\boxed{\text{ENTER}}$. This setting allows you to enter any x-values you like. Next, press $\boxed{\text{TABLE}}$ (using $\boxed{\text{2nd}}$ $\boxed{\text{GRAPH}}$).
  To follow Exercise 7, key in $\boxed{(-)}$ $\boxed{2}$ $\boxed{\text{ENTER}}$ for the x-value, and the calculator will fill in the y-value. Continue by entering 0, 1, 3, or any other x-values you choose. One such table is shown in Figure 1.11b.
  If you would like to evaluate a new function, you do not have to return to the $\boxed{Y=}$ screen. Use the $\boxed{\triangleright}$ and $\boxed{\triangle}$ keys to highlight Y_1 at the top of the second column. The definition of Y_1 will appear at the bottom of the display, as shown in Figure 1.11b. You can key in a new definition here, and the second column will be updated automatically to show the y-values of the new function.

```
TABLE SETUP
 TblMin=0
 ⌐Tbl=1
Indpnt: Auto Ask
Depend: Auto Ask
```

(a)

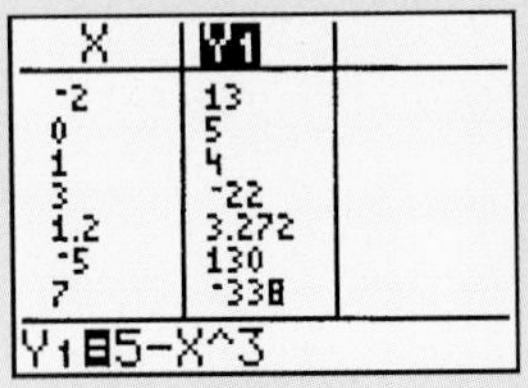

```
  X  │ Y1 │
 -2  │ 13    │
  0  │ 5     │
  1  │ 4     │
  3  │ -22   │
 1.2 │ 3.272 │
 -5  │ 130   │
  7  │ -338  │
Y1◻5-X^3
```

(b)

FIGURE 1.11

To simplify the notation, we sometimes use the same letter for the output variable and for the name of the function. In the next example, C is used in this way.

EXAMPLE 8

TrailGear decides to market a line of backpacks. The cost, C, of manufacturing backpacks is a function of the number, x, of backpacks produced, given by the equation

$$C(x) = 3000 + 20x$$

where $C(x)$ is measured in dollars. Find the cost of producing 500 backpacks.

Solution   To find the value of C that corresponds to $x = 500$, evaluate $C(500)$.

$$C(500) = 3000 + 20(500) = 13{,}000$$

The cost of producing 500 backpacks is \$13,000.

EXERCISE 8

The volume of a sphere of radius r centimeters is given by

$$V = V(r) = \frac{4}{3}\pi r^3$$

Evaluate $V(10)$ and explain what it means.

Operations with Function Notation

Sometimes we need to evaluate a function at an algebraic expression rather than at a specific number.

EXAMPLE 9   TrailGear manufactures backpacks at a cost of

$$C(x) = 3000 + 20x$$

for x backpacks. The company finds that the monthly demand for backpacks increases by 50% during the summer. The backpacks are produced at several small co-ops in different states.

a. If each co-op usually produces b backpacks per month, how many should it produce during the summer months?

b. What costs for producing backpacks should the company expect during the summer?

Solutions   a. An increase of 50% means an additional 50% of the current production level, b. Therefore, a co-op that produced b backpacks per month during the winter should increase production to $b + 0.5b$, or $1.5b$ backpacks per month in the summer.

b. The cost of producing $1.5b$ backpacks will be

$$C(1.5b) = 3000 + 20(1.5b) = 3000 + 30b$$

EXERCISE 9

A spherical balloon has a radius of 10 centimeters.

a. If we increase the radius by h centimeters, what will the new volume be?

b. If $h = 2$, how much did the volume increase?

EXAMPLE 10   Evaluate the function $f(x) = 4x^2 - x + 5$ for the following expressions.

    a. $x = 2h$ **b.** $x = a + 3$

Solution   **a.** $f(2h) = 4(2h)^2 - (2h) + 5$

$$= 4(4h^2) - 2h + 5 = 16h^2 - 2h + 5$$

    b. $(a + 3) = 4(a + 3)^2 - (a + 3) + 5$

$$= 4(a^2 + 6a + 9) - a - 3 + 5$$
$$= 4a^2 + 24a + 36 - a + 2$$
$$= 4a^2 + 23a + 38$$

CAUTION   In Example 10, notice that

$$f(2h) \neq 2f(h)$$

and

$$f(a + 3) \neq f(a) + f(3)$$

To compute $f(a) + f(3)$, we must first compute $f(a)$ and $f(3)$, then add them:

$$f(a) + f(3) = (4a^2 - a + 5) + (4 \cdot 3^2 - 3 + 5)$$
$$= 4a^2 - a + 43$$

In general, it is *not* true that $f(a + b) = f(a) + f(b)$. Remember that the parentheses in the expression $f(x)$ do not indicate multiplication, so the distributive law does not apply to the expression $f(a + b)$.

EXERCISE 10

Let $f(x) = x^3 - 1$ and evaluate each expression.

   a. $f(2) + f(3)$ **b.** $f(2 + 3)$ **c.** $2f(x) + 3$

ANSWERS TO 1.2 EXERCISES

  1. a. No, students with the same grade in English can have different grades in algebra.

     b. Yes, the number of calories is proportional to the number of fluid ounces.

  2. a. No, for example, $x = 3.5$ corresponds both to $y = 2.5$ and also to $y = 4.0$.

     b. Yes, each value of x has exactly one value of y associated with it.

  3. a. The input variable is d, and the output variable is a.

     b. Approximately 210 feet

     c. Approximately where $d \approx 5$, $d \approx 11$, $d \approx 12$, $d \approx 16$, $d \approx 17.5$, and $d \approx 18$

     d. The maximum value of 300 feet occurs at the start, when $d = 0$. The minimum of 85 feet occurs when $d \approx 15$.

     e. Approximately 165 feet

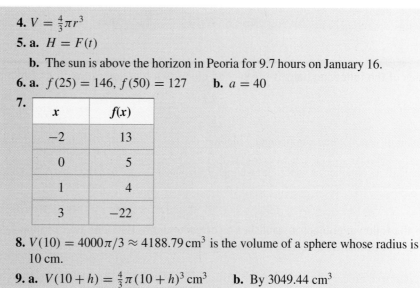

4. $V = \frac{4}{3}\pi r^3$

5. a. $H = F(t)$

   b. The sun is above the horizon in Peoria for 9.7 hours on January 16.

6. a. $f(25) = 146$, $f(50) = 127$ **b.** $a = 40$

7.

x	$f(x)$
-2	13
0	5
1	4
3	-22

8. $V(10) = 4000\pi/3 \approx 4188.79 \text{ cm}^3$ is the volume of a sphere whose radius is 10 cm.

9. a. $V(10 + h) = \frac{4}{3}\pi(10 + h)^3 \text{ cm}^3$ **b.** By 3049.44 cm^3

10. a. 33   **b.** 124   **c.** $2x^3 + 1$

SECTION 1.2 SUMMARY

VOCABULARY   *Look up the definitions of new terms in the Glossary.*

Function   Input variable   Independent variable   Dependent variable
Function value   Output variable

CONCEPTS

1. A function is a rule that assigns to each value of the input variable a unique value of the output variable.

2. Functions may be defined by words, tables, graphs, or equations.

3. Function notation: $y = f(x)$, where x is the input and y is the output.

STUDY QUESTIONS

1. What property makes a relation between two variables a function?

2. Name three ways to define a function.

3. Give an example of a function in which two distinct values of the input variable correspond to the same value of the output variable.

4. Use function notation to write the statement "G defines w as a function of p."

5. Give an example of a function for which $f(2 + 3) \neq f(2) + f(3)$.

SKILLS   *Practice each skill in the Homework problems listed.*

1. Decide whether a relationship between two variables is a function: #1–26

2. Evaluate a function defined by a table, a graph, or an equation: #27–54

3. Interpret function notation: #31–34, 49–54

4. Simplify expressions involving function notation: #59–76

ThomsonNOW™ Test yourself on key content at **www.thomsonedu.com/login.**

For which of the following pairs is the second quantity a function of the first? Explain your answers.

1. Price of an item; sales tax on the item at 4%

2. Time traveled at constant speed; distance traveled

3. Number of years of education; annual income

4. Distance flown in an airplane; price of the ticket

5. Volume of a container of water; the weight of the water

6. Amount of a paycheck; amount of Social Security tax withheld

Each of the following objects establishes a correspondence between two variables. Suggest appropriate input and output variables and decide whether the relationship is a function.

7. An itemized grocery receipt

8. An inventory list

9. An index

10. A will

11. An instructor's grade book

12. An address book

13. A bathroom scale

14. A radio dial

Which of the following tables define the second variable as a function of the first variable? Explain why or why not.

15.

x	t
-1	2
0	9
1	-2
0	-3
-1	5

16.

y	w
0	8
1	12
3	7
5	-3
7	4

17.

x	y
-3	8
-2	3
-1	0
0	-1
1	0
2	3
3	8

18.

s	t
2	5
4	10
6	15
8	20
6	25
4	30
2	35

19.

r	-4	-2	0	2	4
v	6	6	3	6	8

20.

p	-5	-4	-3	-2	-1
d	-5	-4	-3	-2	-1

21.

Pressure (p)	Volume (v)
15	100.0
20	75.0
25	60.0
30	50.0
35	42.8
40	37.5
45	33.3
50	30.0

22.

Frequency (f)	Wavelength (w)
5	60.0
10	30.0
20	15.0
30	10.0
40	7.5
50	6.0
60	5.0
70	4.3

23.

Temperature (T)	Humidity (h)
Jan. 1 34°F	42%
Jan. 2 36°F	44%
Jan. 3 35°F	47%
Jan. 4 29°F	50%
Jan. 5 31°F	52%
Jan. 6 35°F	51%
Jan. 7 34°F	49%

24.

Inflation rate (I)	Unemployment rate (U)
1972 5.6%	5.1%
1973 6.2%	4.5%
1974 10.1%	4.9%
1975 9.2%	7.4%
1976 5.8%	6.7%
1977 5.6%	6.8%
1978 6.7%	7.4%

25.

Adjusted gross income (I)	Tax bracket (T)
$0−2479	0%
$2480−3669	11%
$3670−4749	12%
$4750−7009	14%
$7010−9169	15%
$9170−11,649	16%
$11,650−13,919	18%

26.

Cost of merchandise (M)	Shipping charge (C)
$0.01−10.00	$2.50
10.01−20.00	3.75
20.01−35.00	4.85
30.01−50.00	5.95
50.01−75.00	6.95
75.01−100.00	7.95
Over 100.00	8.95

27. The function described in Problem 21 is called g, so that $v = g(p)$. Find the following:
   a. $g(25)$ **b.** $g(40)$ **c.** x so that $g(x) = 50$

28. The function described in Problem 22 is called h, so that $w = h(f)$. Find the following.
   a. $h(20)$ **b.** $h(60)$ **c.** x so that $h(x) = 10$

29. The function described in Problem 25 is called T, so that $T = T(I)$. Find the following:
  a. $T(8750)$ **b.** $T(6249)$ **c.** x so that $T(x) = 15\%$

30. The function described in Problem 26 is called C, so that $C = C(M)$. Find the following:
  a. $C(11.50)$ **b.** $C(47.24)$ **c.** x so that $C(x) = 7.95$

31. Data indicate that U.S. women are delaying having children longer than their counterparts 50 years ago. The table shows $f(t)$ the percent of 20–24-year-old women in year t who had not yet had children. (Source: U.S. Dept of Health and Human Services)

Year	1960	1965	1970	1975	1980
Percent of women	47.5	51.4	47.0	62.5	66.2

Year	1985	1990	1995	2000
Percent of women	67.7	68.3	65.5	66.0

  a. Evaluate $f(1985)$ and explain what it means.
  b. Estimate a solution to the equation $f(t) = 68$ and explain what it means.
  c. In 1997, 64.9% of 20–24-year-old women had not yet had children. Write an equation with function notation that states this fact.

32. The table shows $f(t)$, the death rate (per 100,000 people) from HIV among 15–24-year-olds, and $g(t)$, the death rate from HIV among 25–34-year-olds, for selected years from 1997 to 2002. (Source: U.S. Dept of Health and Human Services)

Year	1987	1988	1989	1990	1992
15–24-year-olds	1.3	1.4	1.6	1.5	1.6
25–34-year-olds	11.7	14.0	17.9	19.7	24.2

Year	1994	1996	1998	2000	2002
15–24-year-olds	1.8	1.1	0.6	0.5	0.4
25–34-year-olds	28.6	19.2	8.1	6.1	4.6

  a. Evaluate $f(1995)$ and explain what it means.
  b. Find a solution to the equation $g(t) = 28.6$ and explain what it means.
  c. In 1988, the death rate from HIV for 25–34-year-olds was 10 times the corresponding rate for 15–24-year-olds. Write an equation with function notation that states this fact.

33. When you exercise, your heart rate should increase until it reaches your target heart rate. The table shows target heart rate, $r = f(a)$, as a function of age.

a	20	25	30	35	40	45
r	150	146	142	139	135	131

a	50	55	60	65	70
r	127	124	120	116	112

  a. Does $f(50) = 2f(25)$?
  b. Find a value of a for which $f(a) = 2a$. Is $f(a) = 2a$ for all values of a?
  c. Is $r = f(a)$ an increasing function or a decreasing function?

34. The table shows $M = f(d)$, the men's Olympic record time, and $W = g(d)$, the women's Olympic record time, as a function of the length, d, of the race. For example, the women's record in the 100 meters is 10.62 seconds, and the men's record in the 800 meters is 1 minute, 42.58 seconds. (Source: www.hickoksports.com)

Distance (meters)	100	200	400	800
Men	9.84	19.32	43.49	1:42.58
Women	10.62	21.34	48.25	1:53.43

Distance (meters)	1500	5000	10,000
Men	3:32.07	13:05.59	27:07.34
Women	3:53.96	14:40.79	30:17.49

  a. Does $f(800) = 2f(400)$? Does $g(400) = 2g(200)$?
  b. Find a value of d for which $f(2d) < 2f(d)$. Is there a value of d for which $g(2d) < 2g(d)$?

In Problems 35–40, use the graph of the function to answer the questions.

35. The graph shows C as a function of t. C stands for the number of students (in thousands) at State University who consider themselves computer literate, and t represents time, measured in years since 1990.

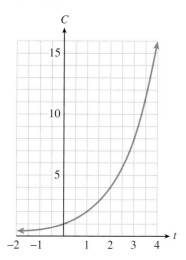

a. When did 2000 students consider themselves computer literate?
b. How long did it take that number to double?
c. How long did it take for the number to double again?
d. How many students became computer literate between January 1992 and June 1993?

36. The graph shows P as a function of t. P is the number of people in Cedar Grove who owned a portable DVD player t years after 2000.
a. When did 3500 people own portable DVD players?
b. How many people owned portable DVD players in 2005?

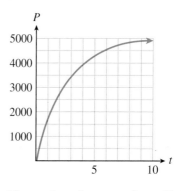

c. The number of owners of portable DVD players in Cedar Grove seems to be leveling off at what number?
d. How many people acquired portable DVD players between 2001 and 2004?

37. The graph shows the revenue, R, a movie theater collects as a function of the price, d, it charges for a ticket.

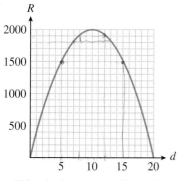

a. What is the revenue if the theater charges $12.00 for a ticket?
b. What should the theater charge for a ticket in order to collect $1500 in revenue?
c. For what values of d is $R > 1875$?

38. The graph shows S as a function of w. S represents the weekly sales of a best-selling book, in thousands of dollars, w weeks after it is released.

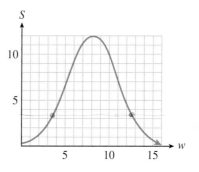

a. In which weeks were sales over $7000?
b. In which week did sales fall below $5000 on their way down?
c. For what values of w is $S > 3.4$?

39. The graph shows the federal minimum wage, M, as a function of time, t, adjusted for inflation to reflect its buying power in 2004 dollars. (Source: www.infoplease.com)

  a. When did the minimum wage reach its highest buying power, and what was it worth in 2004 dollars?

  b. When did the minimum wage fall to its lowest buying power after its peak, and what was its worth at that time?

  c. Give two years in which the minimum wage was worth $8 in 2004 dollars.

U.S. minimum wage

40. The graph shows the U.S. unemployment rate, U, as a function of time, t, for the years 1985–2004. (Source: U.S. Bureau of Labor Statistics)

  a. When did the unemployment rate reach its highest value, and what was its highest value?

  b. When did the unemployment rate fall to its lowest value, and what was its lowest value?

  c. Give two years in which the unemployment rate was 4.5%.

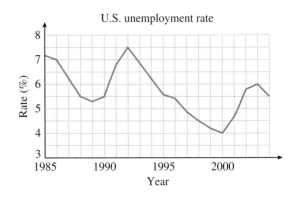

U.S. unemployment rate

In Problems 41–48, evaluate each function for the given values.

41. $f(x) = 6 - 2x$

  a. $f(3)$ **b.** $f(-2)$

  c. $f(12.7)$ **d.** $f\left(\dfrac{2}{3}\right)$

42. $g(t) = 5t - 3$

  a. $g(1)$ **b.** $g(-4)$

  c. $g(14.1)$ **d.** $g\left(\dfrac{3}{4}\right)$

43. $h(v) = 2v^2 - 3v + 1$

  a. $h(0)$ **b.** $h(-1)$

  c. $h\left(\dfrac{1}{4}\right)$ **d.** $h(-6.2)$

44. $r(s) = 2s - s^2$

  a. $r(2)$ **b.** $r(-4)$

  c. $r\left(\dfrac{1}{3}\right)$ **d.** $r(-1.3)$

45. $H(z) = \dfrac{2z - 3}{z + 2}$

  a. $H(4)$ **b.** $H(-3)$

  c. $H\left(\dfrac{4}{3}\right)$ **d.** $H(4.5)$

46. $F(x) = \dfrac{1 - x}{2x - 3}$

  a. $F(0)$ **b.** $F(-3)$

  c. $F\left(\dfrac{5}{2}\right)$ **d.** $F(9.8)$

47. $E(t) = \sqrt{t - 4}$

  a. $E(16)$ **b.** $E(4)$

  c. $E(7)$ **d.** $E(4.2)$

48. $D(r) = \sqrt{5 - r}$

  a. $D(4)$ **b.** $D(-3)$

  c. $D(-9)$ **d.** $D(4.6)$

49. A sport utility vehicle costs $28,000 and depreciates according to the formula $V(t) = 28,000\,(1 - 0.08t)$, where V is the value of the vehicle after t years.
  a. Evaluate $V(12)$ and explain what it means.
  b. Solve the equation $V(t) = 0$ and explain what it means.
  c. If this year is $t = n$, what does $V(n + 2)$ mean?

50. In a profit-sharing plan, an employee receives a salary of $S(x) = 20,000 + 0.01x$, where x represents the company's profit for the year.
  a. Evaluate $S(850,000)$ and explain what it means.
  b. Solve the equation $S(x) = 30,000$ and explain what it means.
  c. If the company made a profit of p dollars this year, what does $S(2p)$ mean?

51. The number of compact cars that a large dealership can sell at price p is given by $N(p) = \dfrac{12,000,000}{p}$.
  a. Evaluate $N(6000)$ and explain what it means.
  b. As p increases, does $N(p)$ increase or decrease? Why is this reasonable?
  c. If the current price for a compact car is D, what does $2N(D)$ mean?

52. A department store finds that the market value of its Christmas-related merchandise is given by $M(t) = \dfrac{600,000}{t}$, $t \le 30$, where t is the number of weeks after Christmas.
  a. Evaluate $M(2)$ and explain what it means.
  b. As t increases, does $M(t)$ increase or decrease? Why is this reasonable?
  c. If this week $t = n$, what does $M(n + 1)$ mean?

53. The velocity of a car that brakes suddenly can be determined from the length of its skid marks, d, by $v(d) = \sqrt{12d}$, where d is in feet and v is in miles per hour.
  a. Evaluate $v(250)$ and explain what it means.
  b. Estimate the length of the skid marks left by a car traveling at 100 miles per hour.
  c. Write your answer to part (b) with function notation.

54. The distance, d, in miles that a person can see on a clear day from a height, h, in feet is given by $d(h) = 1.22\sqrt{h}$.
  a. Evaluate $d(20, 320)$ and explain what it means.
  b. Estimate the height you need in order to see 100 miles.
  c. Write your answer to part (b) with function notation.

55. The figure gives data about snowfall, air temperature, and number of avalanches on the Mikka glacier in Sarek, Lapland, in 1957. (Source: Leopold, Wolman, Miller, 1992)
  a. During June and July, avalanches occurred over three separate time intervals. What were they?
  b. Over what three time intervals did snow fall?
  c. When was the temperature above freezing (0°C)?
  d. Using your answers to parts (a)–(c), make a conjecture about the conditions that encourage avalanches.

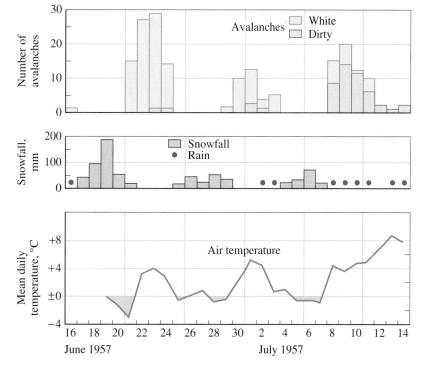

56. The bar graph shows the percent of Earth's surface that lies at various altitudes or depths below the surface of the oceans. (Depths are given as negative altitudes.) (Source: Open University)

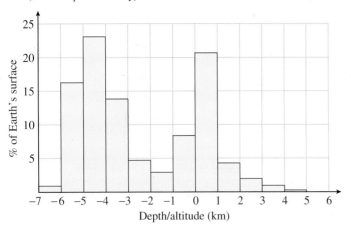

a. Read the graph and complete the table.

Altitude (km)	Percent of Earth's surface
−7 to −6	
−6 to −5	
−5 to −4	
−4 to −3	
−3 to −2	
−2 to −1	
−1 to 0	
0 to 1	
1 to 2	
2 to 3	
3 to 4	
4 to 5	

b. What is the most common altitude? What is the second most common altitude?

c. Approximately what percent of the Earth's surface is below sea level?

d. The height of Mt. Everest is 8.85 kilometers. Can you think of a reason why it is not included in the graph?

57. The graph shows the temperature of the ocean at various depths. (Source: Open University)

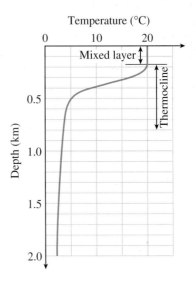

a. Is depth a function of temperature?

b. Is temperature a function of depth?

c. The axes are scaled in an unusual way. Why is it useful to present the graph in this way?

58. The graph shows the relationship between annual precipitation, p, in a region and the amount of erosion, measured in tons per square mile, s. (Source: Leopold, Wolman, Miller, 1992)

    a. Is the amount of erosion a function of the amount of precipitation?

    b. At what annual precipitation is erosion at a maximum, and what is that maximum?

    c. Over what interval of annual precipitation does erosion decrease?

    d. An increase in vegetation inhibits erosion, and precipitation encourages vegetation. What happens to the amount of erosion as precipitation increases in each of these three environments?

        desert-shrub $0 < p < 12$
        grassland $12 < p < 30$
        forest $30 < p < 60$

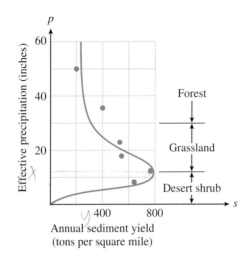

Annual sediment yield
(tons per square mile)

Evaluate the function and simplify.

59. $G(s) = 3s^2 - 6s$

    a. $G(3a)$ **b.** $G(a + 2)$

    c. $G(a) + 2$ **d.** $G(-a)$

60. $h(x) = 2x^2 + 6x - 3$

    a. $h(2a)$ **b.** $h(a + 3)$

    c. $h(a) + 3$ **d.** $h(-a)$

61. $g(x) = 8$

    a. $g(2)$ **b.** $g(8)$

    c. $g(a + 1)$ **d.** $g(-x)$

62. $f(t) = -3$

    a. $f(4)$ **b.** $f(-3)$

    c. $f(b - 2)$ **d.** $f(-t)$

63. $P(x) = x^3 - 1$

    a. $P(2x)$ **b.** $2P(x)$

    c. $P(x^2)$ **d.** $[P(x)]^2$

64. $Q(t) = 5t^3$

    a. $Q(2t)$ **b.** $2Q(t)$

    c. $Q(t^2)$ **d.** $[Q(t)]^2$

Evaluate each function for the given expressions and simplify.

65. $f(x) = x^3$

    a. $f(a^2)$ **b.** $a^3 \cdot f(a^3)$

    c. $f(ab)$ **d.** $f(a + b)$

66. $g(x) = x^4$

    a. $g(a^3)$ **b.** $a^4 \cdot g(a^4)$

    c. $g(ab)$ **d.** $g(a + b)$

67. $F(x) = 3x^5$

    a. $F(2a)$ **b.** $2F(a)$

    c. $F(a^2)$ **d.** $[F(a)]^2$

68. $G(x) = 4x^3$

    a. $G(3a)$ **b.** $3G(a)$

    c. $G(a^4)$ **d.** $[G(a)]^4$

For the functions in Problems 69–76, compute the following:
(a) $f(2) + f(3)$ (b) $f(2 + 3)$ (c) $f(a) + f(b)$ (d) $f(a + b)$

For which functions does $f(a + b) = f(a) + f(b)$ for all values of a and b?

69. $f(x) = 3x - 2$ **70.** $f(x) = 1 - 4x$ **71.** $f(x) = x^2 + 3$ **72.** $f(x) = x^2 - 1$

73. $f(x) = \sqrt{x + 1}$ **74.** $f(x) = \sqrt{6 - x}$ **75.** $f(x) = \dfrac{-2}{x}$ **76.** $f(x) = \dfrac{3}{x}$

77. Use a table of values to estimate a solution to
$f(x) = 800 + 6x - 0.2x^2 = 500$ as follows:
  a. Make a table starting at $x = 0$ and increasing by
$\Delta x = 10$, as shown in the accompanying tables. Find
two x-values a and b so that $f(a) > 500 > f(b)$.

x	0	10	20	30	40	50
$f(x)$						

x	60	70	80	90	100
$f(x)$					

  b. Make a new table starting at $x = a$ and increasing by
$\Delta x = 1$. Find two x-values, c and d, so that
$f(c) > 500 > f(d)$.
  c. Make a new table starting at $x = c$ and increasing by
$\Delta x = 0.1$. Find two x-values, p and q, so that
$f(p) > 500 > f(q)$.
  d. Take the average of p and q, that is, set $s = \dfrac{p + q}{2}$.
Then s is an approximate solution that is off by at
most 0.05.
  e. Evaluate $f(s)$ to check that the output is
approximately 500.

78. Use a table of values to estimate a solution to
$f(x) = x^3 - 4x^2 + 5x = 18,000$ as follows.
  a. Make a table starting at $x = 0$ and increasing by
$\Delta x = 10$, as shown in the accompanying tables. Find
two x-values a and b so that $f(a) < 18,000 < f(b)$.

x	0	10	20	30	40	50
$f(x)$						

x	60	70	80	90	100
$f(x)$					

  b. Make a new table starting at $x = a$ and increasing by
$\Delta x = 1$. Find two x-values, c and d, so that
$f(c) < 18,000 < f(d)$.
  c. Make a new table starting at $x = c$ and increasing by
$\Delta x = 0.1$. Find two x-values, p and q, so that
$f(p) < 18,000 < (q)$.
  d. Take the average of p and q, that is, set $s = \dfrac{p + q}{2}$.
Then s is an approximate solution that is off by at
most 0.05.
  e. Evaluate $f(s)$ to check that the output is
approximately 18,000.

79. Use tables of values to estimate the positive solution to
$f(x) = x^2 - \dfrac{1}{x} = 9000$, accurate to within 0.05.

80. Use tables of values to estimate the positive solution to
$f(x) = \dfrac{8}{x} + 500 - \dfrac{x^2}{9} = 300$, accurate to within 0.05.

Reading Function Values from a Graph

The graph in Figure 1.12 shows the Dow-Jones Industrial Average (the average value of the stock prices of 500 major companies) during the stock market correction of October 1987.

The Dow-Jones Industrial Average (DJIA) is given as a function of time during the 8 days from October 15 to October 22; that is, $f(t)$ is the DJIA recorded at noon on day t.

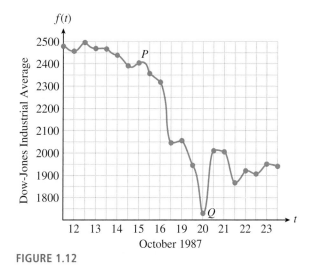

FIGURE 1.12

The values of the input variable, time, are displayed on the horizontal axis, and the values of the output variable, DJIA, are displayed on the vertical axis. There is no formula that gives the DJIA for a particular day; but it is still a function, defined by its graph. The value of $f(t)$ is specified by the vertical coordinate of the point with the given t-coordinate.

EXAMPLE 1

   a. The coordinates of point P in Figure 1.12 are $(15, 2412)$. What do the coordinates tell you about the function f?

   b. If the DJIA was 1726 at noon on October 20, what can you say about the graph of f?

Solutions   **a.** The coordinates of point P tell us that $f(15) = 2412$, so the DJIA was 2412 at noon on October 15.

   b. We can say that $f(20) = 1726$, so the point $(20, 1726)$ lies on the graph of f. This point is labeled Q in Figure 1.12.

Thus, the coordinates of each point on the graph of the function represent a pair of corresponding values of the two variables. In general, we can make the following statement.

Graph of a Function

The point (a, b) lies on the graph of the function f if and only if $f(a) = b$.

EXERCISE 1

The water level in Lake Huron alters unpredictably over time. The graph in Figure 1.13 gives the average water level, $L(t)$, in meters in the year t over a 20-year period. (Source: The Canadian Hydrographic Service)

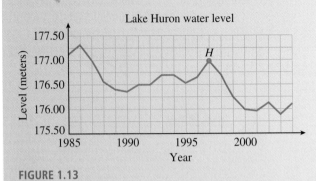

FIGURE 1.13

a. The coordinates of point H in Figure 1.13 are (1997, 176.98). What do the coordinates tell you about the function L?

b. The average water level in 2004 was 176.11 meters. Write this fact in function notation. What can you say about the graph of L?

Another way of describing how a graph depicts a function is as follows:

Functions and Coordinates

Each point on the graph of the function f has coordinates $(x, f(x))$ for some value of x.

EXAMPLE 2

Figure 1.14 shows the graph of a function g.

a. Find $g(-2)$ and $g(5)$.

b. For what value(s) of t is $g(t) = -2$?

c. What is the largest, or maximum, value of $g(t)$? For what value of t does the function take on its maximum value?

d. On what intervals is g increasing?

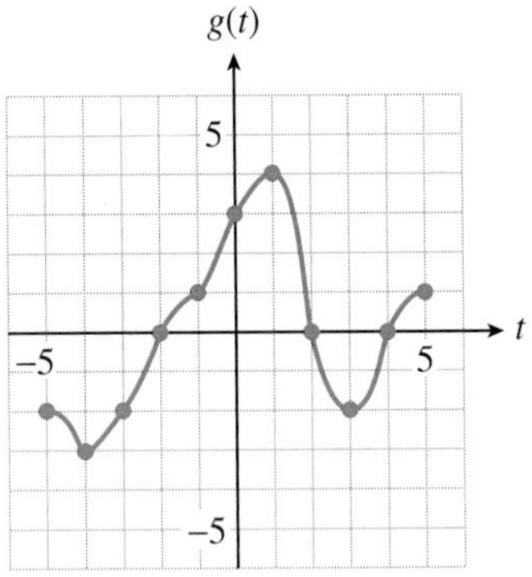

FIGURE 1.14

Solutions

a. To find $g(-2)$, we look for the point with x-coordinate -2. The point $(-2, 0)$ lies on the graph of g, so $g(-2) = 0$. Similarly, the point $(5, 1)$ lies on the graph, so $g(5) = 1$.

> See the Algebra Skills Refresher A.2 to review interval notation.

b. We look for points on the graph with y-coordinate -2. Because the points $(-5, -2)$, $(-3, -2)$, and $(3, -2)$ lie on the graph, we know that $g(-5) = -2$, $g(-3) = -2$, and $g(3) = -2$. Thus, the t-values we want are -5, -3 and 3.

c. The highest point on the graph is $(1, 4)$, so the largest y-value is 4. Thus, the maximum value of $g(t)$ is 4, and it occurs when $t = 1$.

d. A graph is increasing if the y-values get larger as we read from left to right. The graph of g is increasing for t-values between -4 and 1, and between 3 and 5. Thus, g is increasing on the intervals $(-4, 1)$ and $(3, 5)$.

EXERCISE 2

Refer to the graph of the function g shown in Figure 1.14 in Example 2.

a. Find $g(0)$.

b. For what value(s) of t is $g(t) = 0$?

c. What is the smallest, or minimum, value of $g(t)$? For what value of t does the function take on its minimum value?

d. On what intervals is g decreasing?

Finding Coordinates with a Graphing Calculator

We can use the ⎡TRACE⎤ feature of the calculator to find the coordinates of points on a graph. For example, graph the equation $y = -2.6x - 5.4$ in the window

$$\text{Xmin} = -5 \qquad \text{Xmax} = 4.4$$
$$\text{Ymin} = -20 \qquad \text{Ymax} = 15$$

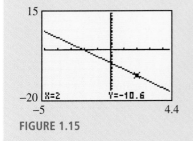

FIGURE 1.15

Press ⎡TRACE⎤, and a "bug" begins flashing on the display. The coordinates of the bug appear at the bottom of the display, as shown in Figure 1.15. Use the left and right arrows to move the bug along the graph. You can check that the coordinates of the point $(2, -10.6)$ do satisfy the equation $y = -2.6x - 5.4$.

The points identified by the **Trace** bug depend on the window settings and on the type of calculator. If we want to find the y-coordinate for a particular x-value, we enter the x-coordinate of the desired point and press ⎡ENTER⎤. Try entering 2 after **X =**, and you should find that $y = -10.6$, as before.

Constructing the Graph of a Function

Although some functions are defined by their graphs, we can also construct graphs for functions described by tables or equations. We make these graphs the same way we graph equations in two variables: by plotting points whose coordinates satisfy the equation.

EXAMPLE 3   Graph the function $f(x) = \sqrt{x+4}$.

Solution   Choose several convenient values for x and evaluate the function to find the corresponding $f(x)$-values. For this function we cannot choose x-values less than -4, because the square root of a negative number is not a real number. The results are shown in the table.

x	$f(x)$
-4	0
-3	1
0	2
2	$\sqrt{6}$
5	3

$$f(-4) = \sqrt{-4+4} = \sqrt{0} = 0$$
$$f(-3) = \sqrt{-3+4} = \sqrt{1} = 1$$
$$f(0) = \sqrt{0+4} = \sqrt{4} = 2$$
$$f(2) = \sqrt{2+4} = \sqrt{6} \approx 2.45$$
$$f(5) = \sqrt{5+4} = \sqrt{9} = 3$$

Points on the graph have coordinates $(x, f(x))$, so the vertical coordinate of each point is given by the value of $f(x)$. Plot the points and connect them with a smooth curve, as shown in Figure 1.16. Notice that no points on the graph have x-coordinates less than -4.

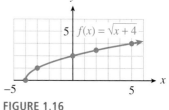

FIGURE 1.16

Using a Calculator to Graph a Function

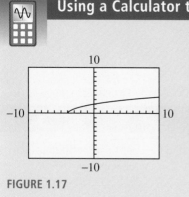

FIGURE 1.17

We can also use a graphing calculator to obtain a table and graph for the function in Example 3. We graph a function just as we graphed an equation. For this function, we enter

$$Y_1 = \sqrt{(X+4)}$$

and press $\boxed{\text{ZOOM}}$ $\boxed{6}$ for the standard window. (See Appendix B for details.) The calculator's graph is shown in Figure 1.17.

EXERCISE 3

$$f(x) = x^3 - 2$$

a. Complete the table of values and sketch a graph of the function.

x	-2	-1	$-\frac{1}{2}$	0	$\frac{1}{2}$	1	2
$f(x)$							

b. Use your calculator to make a table of values and graph the function.

The Vertical Line Test

In a function, two different outputs cannot be related to the same input. This restriction means that two different ordered pairs cannot have the same first coordinate. What does it mean for the graph of the function?

Consider the graph shown in Figure 1.18a. Every vertical line intersects the graph in at most one point, so there is only one point on the graph for each x-value. This graph represents a function. In Figure 1.18b, however, the line $x = 2$ intersects the graph at two points, (2, 1) and (2, 4). Two different y-values, 1 and 4, are related to the same x-value, 2. This graph cannot be the graph of a function.

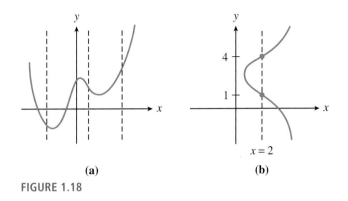

(a)   (b)

FIGURE 1.18

We summarize these observations as follows.

The Vertical Line Test

A graph represents a function if and only if every vertical line intersects the graph in at most one point.

EXAMPLE 4 Use the vertical line test to decide which of the graphs in Figure 1.19 represent functions.

Solution Graph (a) represents a function, because it passes the vertical line test. Graph (b) is not the graph of a function, because the vertical line at (for example) $x = 1$ intersects the graph at two points. For graph (c), notice the break in the curve at $x = 2$: The solid dot at (2, 1) is the only point on the graph with $x = 2$; the open circle at (2, 3) indicates that (2, 3) is not a point on the graph. Thus, graph (c) is a function, with $f(2) = 1$.

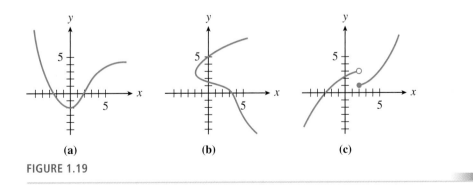

(a)   (b)   (c)

FIGURE 1.19

Use the vertical line test to determine which of the graphs in Figure 1.20 represent functions.

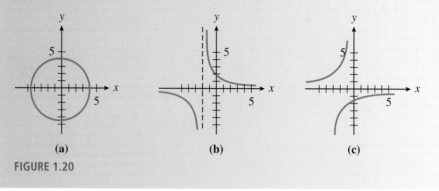

(a)   (b)   (c)

FIGURE 1.20

Graphical Solution of Equations and Inequalities

The graph of an equation in two variables is just a picture of its solutions. When we read the coordinates of a point on the graph, we are reading a pair of x- and y-values that make the equation true. For example, the point $(2, 7)$ lies on the graph of $y = 2x + 3$ shown in Figure 1.21, so we know that the ordered pair $(2, 7)$ is a solution of the equation $y = 2x + 3$. You can verify algebraically that $x = 2$ and $y = 7$ satisfy the equation:

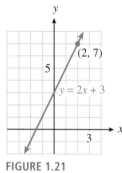

Does $\quad 7 = 2(2) + 3?\quad$ Yes

We can also say that $x = 2$ is a solution of the one-variable equation $2x + 3 = 7$. In fact, we can use the graph of $y = 2x + 3$ to solve the equation $2x + 3 = k$ for any value of k. Thus, we can use graphs to find solutions to equations in one variable.

FIGURE 1.21

EXAMPLE 5

Use the graph of $y = 285 - 15x$ to solve the equation $150 = 285 - 15x$.

Solution

Begin by locating the point P on the graph for which $y = 150$, as shown in Figure 1.22. Now find the x-coordinate of point P by drawing an imaginary line from P straight down to the x-axis. The x-coordinate of P is $x = 9$. Thus, P is the point $(9, 150)$, and $x = 9$ when $y = 150$. The solution of the equation

$$150 = 285 - 15x$$

is $x = 9$.

You can verify the solution algebraically by substituting $x = 9$ into the equation:

Does $\quad 150 = 285 - 15(9)?$

$285 - 15(9) = 285 - 135 = 150.\quad$ Yes

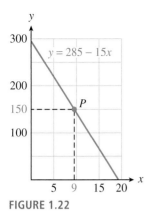

FIGURE 1.22

The relationship between an equation and its graph is an important one. For the previous example, make sure you understand that the following three statements are equivalent:

1. The point (9, 150) lies on the graph of $y = 285 - 15x$.
2. The ordered pair (9, 150) is a solution of the equation $y = 285 - 15x$.
3. $x = 9$ is a solution of the equation $150 = 285 - 15x$.

EXERCISE 5

a. Use the graph of $y = 30 - 8x$ shown in Figure 1.23 to solve the equation

$$30 - 8x = 50$$

b. Verify your solution algebraically.

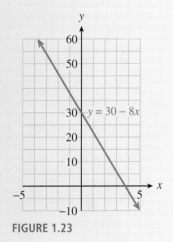

FIGURE 1.23

▶▶▶ See the Algebra Skills Refresher A.2 to review solving linear inequalities algebraically.

In a similar fashion, we can solve inequalities with a graph. Consider again the graph of $y = 2x + 3$, shown in Figure 1.24. We saw that $x = 2$ is the solution of the equation $2x + 3 = 7$. When we use $x = 2$ as the input for the function $f(x) = 2x + 3$, the output is $y = 7$. Which input values for x produce output values greater than 7? You can see in Figure 1.24 that x-values greater than 2 produce y-values greater than 7, because points on the graph with x-values greater than 2 have y-values greater than 7. Thus, the solutions of the inequality $2x + 3 > 7$ are $x > 2$. You can verify this result by solving the inequality algebraically.

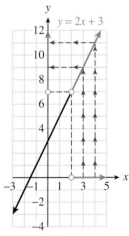

FIGURE 1.24

EXAMPLE 6   Use the graph of $y = 285 - 15x$ to solve the inequality

$$285 - 15x > 150$$

Solution We begin by locating the point P on the graph for which $y = 150$ and $x = 9$ (its x-coordinate). Now, because $y = 285 - 15x$ for points on the graph, the inequality $285 - 15x > 150$ is equivalent to $y > 150$. So we are looking for points on the graph with y-coordinate greater than 150. These points are shown in Figure 1.25. The x-coordinates of these points are the x-values that satisfy the inequality. From the graph, we see that the solutions are $x < 9$.

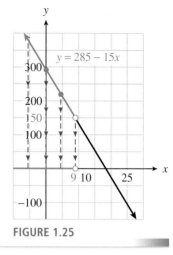

FIGURE 1.25

EXERCISE 6

a. Use the graph of $y = 30 - 8x$ in Figure 1.23 to solve the inequality

$$30 - 8x \le 50$$

b. Solve the inequality algebraically.

We can also use this graphical technique to solve nonlinear equations and inequalities.

EXAMPLE 7 Use a graph of $f(x) = -2x^3 + x^2 + 16x$ to solve the equation

$$-2x^3 + x^2 + 16x = 15$$

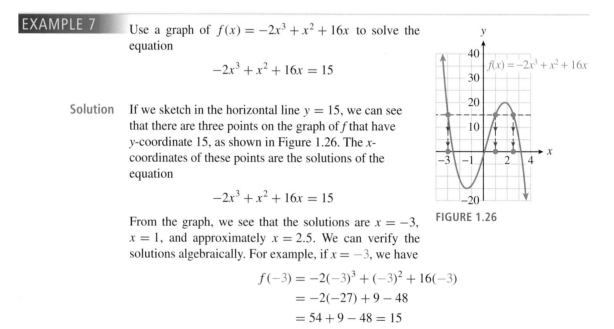

FIGURE 1.26

Solution If we sketch in the horizontal line $y = 15$, we can see that there are three points on the graph of f that have y-coordinate 15, as shown in Figure 1.26. The x-coordinates of these points are the solutions of the equation

$$-2x^3 + x^2 + 16x = 15$$

From the graph, we see that the solutions are $x = -3$, $x = 1$, and approximately $x = 2.5$. We can verify the solutions algebraically. For example, if $x = -3$, we have

$$
\begin{aligned}
f(-3) &= -2(-3)^3 + (-3)^2 + 16(-3) \\
&= -2(-27) + 9 - 48 \\
&= 54 + 9 - 48 = 15
\end{aligned}
$$

so -3 is a solution.

Use the graph of $y = \frac{1}{2}n^2 + 2n - 10$ shown in Figure 1.27 to solve

$$\frac{1}{2}n^2 + 2n - 10 = 6$$

and verify your solutions algebraically.

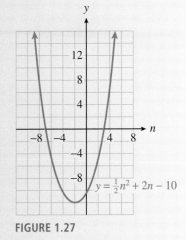

FIGURE 1.27

Using the Trace Feature

You can use the **Trace** feature on a graphing calculator to approximate solutions to equations. Graph the function $f(x)$ in Example 7 in the window

$$\text{Xmin} = -4 \quad \text{Xmax} = 4$$
$$\text{Ymin} = -20 \quad \text{Ymax} = 40$$

and trace along the curve to the point $(2.4680851, 15.512401)$. We are close to a solution, because the y-value is close to 15. Try entering x-values close to 2.4680851, for instance, $x = 2.4$ and $x = 2.5$, to find a better approximation for the solution.

> We can use the **intersect** feature on a graphing calculator to obtain more accurate estimates for the solutions of equations. See Appendix B for details.

EXAMPLE 8

Use the graph in Example 7 to solve the inequality

$$-2x^3 + x^2 + 16x \geq 15$$

Solution We first locate all points on the graph that have y-coordinates greater than or equal to 15. The x-coordinates of these points are the solutions of the inequality. Figure 1.28 shows the points, and their x-coordinates as intervals on the x-axis. The solutions are $x \leq -3$ and $1 \leq x \leq 2.5$, or in interval notation, $(-\infty, -3] \cup [1, 2.5]$.

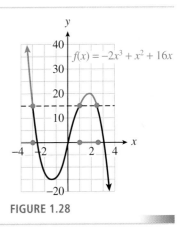

FIGURE 1.28

ANSWERS TO 1.3 EXERCISES

1. **a.** $L(1997) = 176.98$; the average water level was 176.98 meters in 1997.
   b. $L(2004) = 176.11$. The point $(2004, 176.11)$ lies on the graph of L.

2. **a.** 3   **b.** $-2, 2, 4$
   c. $-3; t = -4$ **d.** $(-5, -4)$ and $(1, 3)$

3.

x	-2	-1	$-\frac{1}{2}$	0	$\frac{1}{2}$	1	2
$f(x)$	-10	-3	$\frac{-17}{8}$	-2	$\frac{-15}{8}$	-1	6

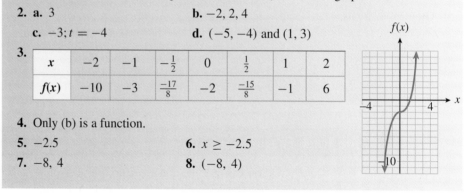

4. Only (b) is a function.

5. -2.5 6. $x \geq -2.5$

7. $-8, 4$ 8. $(-8, 4)$

SECTION 1.3 SUMMARY

VOCABULARY   *Look up the definitions of new terms in the Glossary.*

Coordinates   Algebraic solution   Vertical line test   Minimum
Interval   Maximum   Graphical solution   Inequality

CONCEPTS

1. The point (a, b) lies on the graph of the function f if and only if $f(a) = b$.
2. Each point on the graph of the function f has coordinates $(x, f(x))$ for some value of x.

3. The vertical line test tells us whether a graph represents a function.
4. We can use a graph to solve equations and inequalities in one variable.

STUDY QUESTIONS

1. How can you find the value of $f(3)$ from a graph of f?
2. If $f(8) = 2$, what point lies on the graph of f?
3. Explain how to construct the graph of a function from its equation.

4. Explain how to use the vertical line test.
5. How can you solve the equation $x + \sqrt{x} = 56$ using the graph of $y = x + \sqrt{x}$?

SKILLS   *Practice each skill in the Homework problems listed.*

1. Read function values from a graph: #1–8, 17–20, 33–36
2. Recognize the graph of a function: #9–10, 31 and 32
3. Construct a table of values and a graph of a function: #11–16

4. Solve equations and inequalities graphically: #21–30, 41–50

In Problems 1–8, use the graphs to answer the questions about the functions.

1. a. Find $h(-3)$, $h(1)$, and $h(3)$.
  b. For what value(s) of z is $h(z) = 3$?
  c. Find the intercepts of the graph. List the function values given by the intercepts.
  d. What is the maximum value of $h(z)$?
  e. For what value(s) of z does h take on its maximum value?
  f. On what intervals is the function increasing? Decreasing?

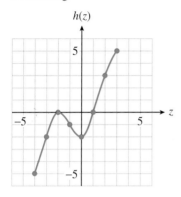

2. a. Find $G(-3)$, $G(-1)$, and $G(2)$.
  b. For what value(s) of s is $G(s) = 3$?
  c. Find the intercepts of the graph. List the function values given by the intercepts.
  d. What is the minimum value of $G(s)$?
  e. For what value(s) of s does G take on its minimum value?
  f. On what intervals is the function increasing? Decreasing?

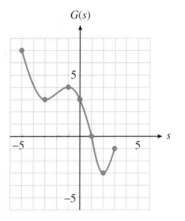

3. a. Find $R(1)$ and $R(3)$.
  b. For what value(s) of p is $R(p) = 2$?
  c. Find the intercepts of the graph. List the function values given by the intercepts.
  d. Find the maximum and minimum values of $R(p)$.
  e. For what value(s) of p does R take on its maximum and minimum values?
  f. On what intervals is the function increasing? Decreasing?

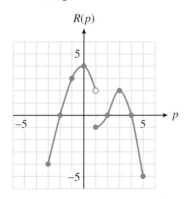

4. a. Find $f(-1)$ and $f(3)$.
  b. For what value(s) of t is $f(t) = 5$?
  c. Find the intercepts of the graph. List the function values given by the intercepts.
  d. Find the maximum and minimum values of $f(t)$.
  e. For what value(s) of t does f take on its maximum and minimum values?
  f. On what intervals is the function increasing? Decreasing?

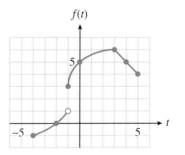

S(0), $S\left(\frac{1}{6}\right)$, and $S(-1)$.

ate the value of $S\left(\frac{1}{3}\right)$ from the graph.

hat value(s) of x is $S(x) = -\frac{1}{2}$?

d. Find the maximum and minimum values of $S(x)$.

e. For what value(s) of x does S take on its maximum and minimum values?

6. a. Find $C(0)$, $C\left(-\frac{1}{3}\right)$, and $C(1)$.

b. Estimate the value of $C\left(\frac{1}{6}\right)$ from the graph.

c. For what value(s) of x is $C(x) = \frac{1}{2}$?

d. Find the maximum and minimum values of $C(x)$.

e. For what value(s) of x does C take on its maximum and minimum values?

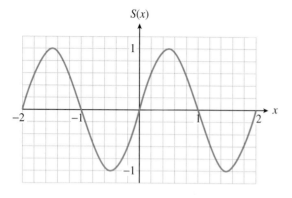

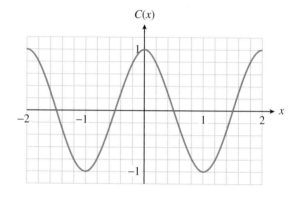

7. a. Find $F(-3)$, $F(-2)$, and $F(2)$.

b. For what value(s) of s is $F(s) = -1$?

c. Find the maximum and minimum values of $F(s)$.

d. For what value(s) of s does F take on its maximum and minimum values?

8. a. Find $P(-3)$, $P(-2)$, and $P(1)$.

b. For what value(s) of n is $P(n) = 0$?

c. Find the maximum and minimum values of $P(n)$.

d. For what value(s) of n does P take on its maximum and minimum values?

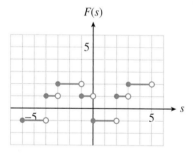

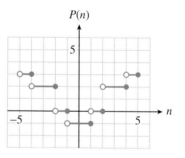

Which of the graphs in Problems 9 and 10 represent functions?

9.

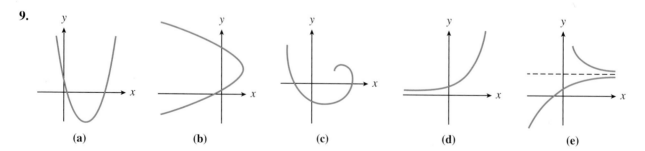

(a)   (b)   (c)   (d)   (e)

10.

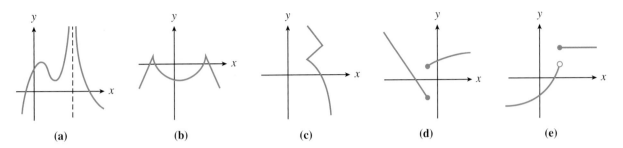

(a)   (b)   (c)   (d)   (e)

In Problems 11–16,
(a) Make a table of values and sketch a graph of the function by plotting points. (Use the suggested *x*-values.)
(b) Use your calculator to graph the function.
Compare the calculator's graph with your sketch.

11. $g(x) = x^3 + 4$; $x = -2, -1, \ldots, 2$

12. $h(x) = 2 + \sqrt{x}$; $x = 0, 1, \ldots, 9$

13. $G(x) = \sqrt{4 - x}$; $x = -5, -4, \ldots, 4$

14. $F(x) = \sqrt{x - 1}$; $x = 1, 2, \ldots, 10$

15. $v(x) = 1 + 6x - x^3$; $x = -3, -2, \ldots, 3$

16. $w(x) = x^3 - 8x$; $x = -4, -3, \ldots, 4$

17. The graph shows the speed of sound in the ocean as a function of depth, $S = f(d)$. The speed of sound is affected both by increasing water pressure and by dropping temperature. (Source: *Scientific American*)

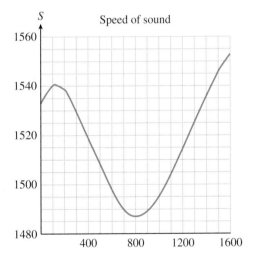

a. Evaluate $f(1000)$ and explain its meaning.
b. Solve $f(d) = 1500$ and explain its meaning.
c. At what depth is the speed of sound the slowest, and what is the speed? Write your answer with function notation.
d. Describe the behavior of $f(d)$ as d increases.

18. The graph shows the water level in Lake Superior as a function of time, $L = f(t)$. (Source: The Canadian Hydrographic Service)

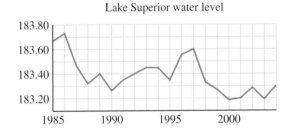

a. Evaluate $f(1997)$ and explain its meaning.
b. Solve $f(t) = 183.5$ and explain its meaning.
c. In which two years did Lake Superior reach its highest levels, and what were those levels? Write your answers with function notation.
d. Over which two-year period did the water level drop the most?

19. The graph shows the federal debt as a percentage of the gross domestic product (GDP), as a function of time, $D = f(t)$. (Source: Office of Management and Budget)

Federal debt as percent of GDP

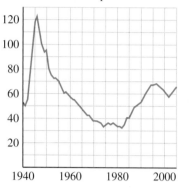

a. Evaluate $f(1985)$ and explain its meaning.
b. Solve $f(t) = 70$ and explain its meaning.
c. When did the federal debt reach its highest level since 1960, and what was that level? Write your answer with function notation.
d. What is the longest time interval over which the federal debt was decreasing?

20. The graph shows the elevation of the Los Angeles Marathon course as a function of the distance into the race, $a = f(t)$. (Source: *Los Angeles Times*, March 3, 2005)

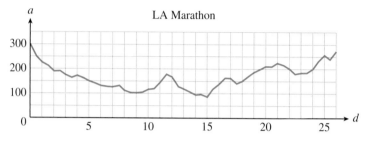

a. Evaluate $f(5)$ and explain its meaning.
b. Solve $f(d) = 200$ and explain its meaning.
c. When does the marathon course reach its lowest elevation, and what is that elevation? Write your answer with function notation.
d. Give three intervals over which the elevation is increasing.

21. The figure shows a graph of $y = -2x + 6$.
a. Use the graph to find all values of x for which
  i. $y = 12$
  ii. $y > 12$
  iii. $y < 12$
b. Use the graph to solve
  i. $-2x + 6 = 12$
  ii. $-2x + 6 > 12$
  iii. $-2x + 6 < 12$
c. Explain why your answers to parts (a) and (b) are the same.

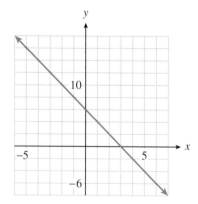

22. The figure shows a graph of $y = \dfrac{-x}{3} - 6$.

a. Use the graph to find all values of x for which the values of y are as follows:
  i. $y = -4$
  ii. $y > -4$
  iii. $y < -4$
b. Use the graph to solve the following:
  i. $\dfrac{-x}{3} - 6 = -4$
  ii. $\dfrac{-x}{3} - 6 > -4$
  iii. $\dfrac{-x}{3} - 6 < -4$
c. Explain why your answers to parts (a) and (b) are the same.

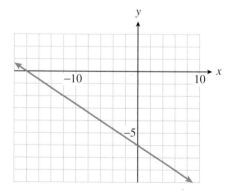

In Problems 23 and 24, use the graph to solve the equation or inequality, and then solve algebraically. (To review solving linear inequalities algebraically, see Algebra Skills Refresher A.2.)

23. The figure shows the graph of $y = 1.4x - 0.64$. Solve the following:
   a. $1.4x - 0.64 = 0.2$
   b. $-1.2 = 1.4x - 0.64$
   c. $1.4x - 0.64 > 0.2$
   d. $-1.2 > 1.4x - 0.64$

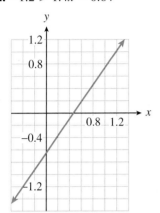

24. The figure shows the graph of $y = -2.4x + 2.32$. Solve the following:
   a. $1.6 = -2.4x + 2.32$
   b. $-2.4x + 2.32 = 0.4$
   c. $-2.4x + 2.32 \geq 1.6$
   d. $0.4 \geq -2.4x + 2.32$

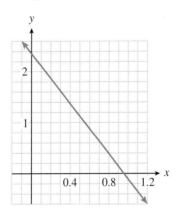

For Problems 25–30, use the graphs to estimate solutions to the equations and inequalities.

25. The figure shows the graph of $g(x) = \dfrac{12}{2 + x^2}$.

   a. Solve $\dfrac{12}{2 + x^2} = 4$.

   b. Solve $1 \leq \dfrac{12}{2 + x^2} \leq 2$.

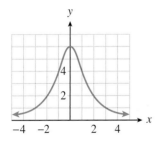

26. The figure shows the graph of $f(x) = \dfrac{30\sqrt{x}}{1 + x}$.

   a. Solve $\dfrac{30\sqrt{x}}{1 + x} = 15$.

   b. Solve $\dfrac{30\sqrt{x}}{1 + x} < 12$.

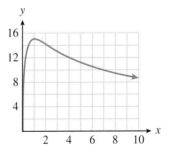

27. The figure shows a graph of
$$B = \frac{1}{3}p^3 - 3p + 2$$

a. Solve $\frac{1}{3}p^3 - 3p + 2 = 6$.

b. Solve $\frac{1}{3}p^3 - 3p + 2 = 5$.

c. Solve $\frac{1}{3}p^3 - 3p + 2 < 1$.

d. What range of values does B have for p between -2.5 and 0.5?

e. For what values of p is B increasing?

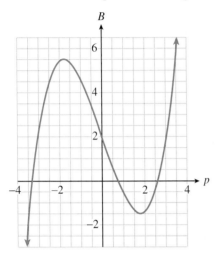

28. The figure shows a graph of
$$H = t^3 - 4t^2 - 4t + 12$$

a. Solve $t^3 - 4t^2 - 4t + 12 = -4$.

b. Solve $t^3 - 4t^2 - 4t + 12 = 16$.

c. Solve $t^3 - 4t^2 - 4t + 12 > 6$.

d. Estimate the horizontal and vertical intercepts of the graph.

e. For what values of t is H decreasing?

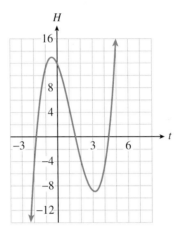

29. The figure shows a graph of $M = g(q)$.

(a) Find all values of q for which

   i. $g(q) = 0$

  ii. $g(q) = 16$

 iii. $g(q) < 6$

(b) For what values of q is $g(q)$ increasing?

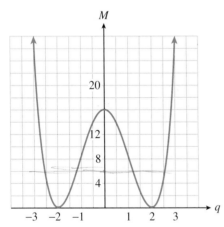

30. The figure shows a graph of $P = f(t)$.

(a) Find all values of t for which

   i. $f(t) = 3$

  ii. $f(t) > 4.5$

 iii. $2 \le f(t) \le 4$

(b) For what values of t is $f(t)$ decreasing?

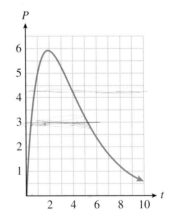

31. a. Delbert reads the following values from the graph of a function:

$$f(-3) = 5, \ f(-1) = 2, \ f(1) = 0,$$
$$f(-1) = -4, \ f(-3) = -2$$

Can his readings be correct? Explain why or why not.

b. Francine reads the following values from the graph of a function:

$$g(-2) = 6, \ g(0) = 0, \ g(2) = 6,$$
$$g(4) = 0, \ g(6) = 6$$

Can her readings be correct? Explain why or why not.

32. a. Sketch the graph of a function that has the following values:

$$F(-2) = 3, \ F(-1) = 3, \ F(0) = 3,$$
$$F(1) = 3, \ F(2) = 3$$

b. Sketch the graph of a function that has the following values:

$$G(-2) = 1, \ G(-1) = 0, \ G(0) = -1,$$
$$G(1) = 0, \ G(2) = 1$$

For Problems 33–36, graph each function in the friendly window

$$\text{Xmin} = -9.4, \quad \text{Xmax} = 9.4$$
$$\text{Ymin} = -10, \quad \text{Ymax} = 10$$

Then answer the questions about the graph. (See Appendix B for an explanation of friendly windows.)

33. $g(x) = \sqrt{36 - x^2}$
a. Complete the table. (Round values to tenths.)

x	-4	-2	3	5
$g(x)$				

b. Find all points on the graph for which $g(x) = 3.6$.

34. $f(x) = \sqrt{x^2 - 6}$
a. Complete the table. (Round values to tenths.)

x	-8	-2	3	6
$f(x)$				

b. Find all points on the graph for which $f(x) = -2$.

35. $F(x) = 0.5x^3 - 4x$
a. Estimate the coordinates of the turning points of the graph, where the graph changes from increasing to decreasing or vice versa.
b. Write an equation of the form $F(a) = b$ for each turning point.

36. $G(x) = 2 + 4x - x^3$
a. Estimate the coordinates of the turning points of the graph, that is, where the graph changes from increasing to decreasing or vice versa.
b. Write an equation of the form $G(a) = b$ for each turning point.

For Problems 37–40, graph each function
(a) First using the standard window.
(b) Then using the suggested window. Explain how the window alters the appearance of the graph in each case.

37. $h(x) = \dfrac{1}{x^2 + 10}$
$\text{Xmin} = -5, \quad \text{Xmax} = 5$
$\text{Ymin} = 0, \qquad \text{Ymax} = 0.5$

38. $H(x) = \sqrt{1 - x^2}$
$\text{Xmin} = -2, \quad \text{Xmax} = 2$
$\text{Ymin} = -2, \quad \text{Ymax} = 2$

39. $P(x) = (x - 8)(x + 6)(x - 15)$
$\text{Xmin} = -10, \quad \text{Xmax} = 20$
$\text{Ymin} = -250, \quad \text{Ymax} = 750$

40. $p(x) = 200x^3$
$\text{Xmin} = -5, \qquad \text{Xmax} = 5$
$\text{Ymin} = -10,000, \quad \text{Ymax} = 10,000$

Graph each equation with the ZInteger setting. (Press ZOOM 6 ,then ZOOM 8 ENTER .) Use the graph to answer each question. Use the equation to verify your answers.

41. Graph $y = 2x - 3$.
 a. For what value of x is $y = 5$?
 b. For what value of x is $y = -13$?
 c. For what values of x is $y > -1$?
 d. For what values of x is $y < 25$?

42. Graph $y = 4 - 2x$.
 a. For what value of x is $y = 6$?
 b. For what value of x is $y = -4$?
 c. For what values of x is $y > -12$?
 d. For what values of x is $y < 18$?

43. Graph $y = 6.5 - 1.8x$.
 a. For what value of x is $y = -13.3$?
 b. For what value of x is $y = 24.5$?
 c. For what values of x is $y \leq 15.5$?
 d. For what values of x is $y \geq -7.9$?

44. Graph $y = 0.2x + 1.4$.
 a. For what value of x is $y = -5.2$?
 b. For what value of x is $y = 2.8$?
 c. For what values of x is $y \leq -3.2$?
 d. For what values of x is $y \geq 4.4$?

In Problems 45–48, graph each equation with the ZInteger setting. Use the graph to solve each equation or inequality. Check your solutions algebraically.

45. Graph $y = -0.4x + 3.7$.
 a. Solve $-0.4x + 3.7 = 2.1$.
 b. Solve $-0.4x + 3.7 > -5.1$.

46. Graph $y = 0.4(x - 1.5)$.
 a. Solve $0.4(x - 1.5) = -8.6$.
 b. Solve $0.4(x - 1.5) < 8.6$.

47. Graph $y = \dfrac{2}{3}x - 24$.
 a. Solve $\dfrac{2}{3}x - 24 = -10\dfrac{2}{3}$.
 b. Solve $\dfrac{2}{3}x - 24 \leq -19\dfrac{1}{3}$.

48. Graph $y = \dfrac{80 - 3x}{5}$.
 a. Solve $\dfrac{80 - 3x}{5} = 22\dfrac{3}{5}$.
 b. Solve $\dfrac{80 - 3x}{5} \leq 9\dfrac{2}{5}$.

49. Graph $y = 0.01x^3 - 0.1x^2 - 2.75x + 15$.
 a. Use your graph to solve
  $0.01x^3 - 0.1x^2 - 2.75x + 15 = 0$.
 b. Press Y= and enter $Y_2 = 10$. Press GRAPH , and you should see the horizontal line $y = 10$ superimposed on your previous graph. How many solutions does the equation $0.01x^3 - 0.1x^2 - 2.75x + 15 = 10$ have? Estimate each solution to the nearest whole number.

50. Graph $y = 2.5x - 0.025x^2 - 0.005x^3$.
 a. Use your graph to solve
  $2.5x - 0.025x^2 - 0.005x^3 = 0$.
 b. Press Y= and enter $Y_2 = -5$. Press GRAPH , and you should see the horizontal line $y = -5$ superimposed on your previous graph. How many solutions does the equation $2.5x - 0.025x^2 - 0.005x^3 = -5$ have? Estimate each solution to the nearest whole number.

1.4 Slope and Rate of Change

Using Ratios for Comparison

Which is more expensive, a 64-ounce bottle of Velvolux dish soap that costs $3.52, or a 60-ounce bottle of Rainfresh dish soap that costs $3.36?

You are probably familiar with the notion of comparison shopping. To decide which dish soap is the better buy, we compute the unit price, or price per ounce, for each bottle. The unit price for Velvolux is

$$\frac{352 \text{ cents}}{64 \text{ ounces}} = 5.5 \text{ cents per ounce}$$

and the unit price for Rainfresh is

$$\frac{336 \text{ cents}}{60 \text{ ounces}} = 5.6 \text{ cents per ounce}$$

The Velvolux costs less per ounce, so it is the better buy. By computing the price of each brand for *the same amount of soap,* it is easy to compare them.

In many situations, a ratio, similar to a unit price, can provide a basis for comparison. Example 1 uses a ratio to measure a rate of growth.

EXAMPLE 1

Which grow faster, Hybrid A wheat seedlings, which grow 11.2 centimeters in 14 days, or Hybrid B seedlings, which grow 13.5 centimeters in 18 days?

Solution   We compute the growth rate for each strain of wheat. Growth rate is expressed as a ratio, $\frac{\text{centimeters}}{\text{days}}$, or centimeters per day. The growth rate for Hybrid A is

$$\frac{11.2 \text{ centimeters}}{14 \text{ days}} = 0.8 \text{ centimeters per day}$$ 3

and the growth rate for Hybrid B is

$$\frac{13.5 \text{ centimeters}}{18 \text{ days}} = 0.75 \text{ centimeters per day}$$

Because their rate of growth is larger, the Hybrid A seedlings grow faster.

By computing the growth of each strain of wheat seedling over *the same unit of time,* a single day, we have a basis for comparison. In this case, the ratio $\frac{\text{centimeters}}{\text{day}}$ measures the rate of growth of the wheat seedlings.

EXERCISE 1

Delbert traveled 258 miles on 12 gallons of gas, and Francine traveled 182 miles on 8 gallons of gas. Compute the ratio $\frac{\text{miles}}{\text{gallon}}$ for each car. Whose car gets the better gas mileage?

In Exercise 1, the ratio $\frac{\text{miles}}{\text{gallon}}$ measures the rate at which each car uses gasoline. By computing the mileage for each car for the same amount of gas, we have a basis for comparison. We can use this same idea, finding a common basis for comparison, to measure the steepness of an incline.

Measuring Steepness

Imagine you are an ant carrying a heavy burden along one of the two paths shown in Figure 1.29. Which path is more difficult? Most ants would agree that the steeper path is more difficult. But what exactly is steepness? It is not merely the gain in altitude, because even a gentle incline will reach a great height eventually. Steepness measures how sharply the altitude increases. An ant finds the second path more difficult, or steeper, because it rises 5 feet while the first path rises only 2 feet *over the same horizontal distance.*

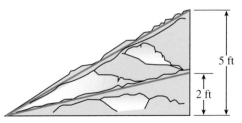

FIGURE 1.29

To compare the steepness of two inclined paths, we compute the ratio of change in altitude to change in horizontal distance for each path.

EXAMPLE 2

Which is steeper, Stony Point trail, which climbs 400 feet over a horizontal distance of 2500 feet, or Lone Pine trail, which climbs 360 feet over a horizontal distance of 1800 feet?

Solution   For each trail, we compute the ratio of vertical gain to horizontal distance. For Stony Point trail, the ratio is

$$\frac{400 \text{ feet}}{2500 \text{ feet}} = 0.16$$

and for Lone Pine trail, the ratio is

$$\frac{360 \text{ feet}}{1800 \text{ feet}} = 0.20$$

Lone Pine trail is steeper, because it has a vertical gain of 0.20 foot for every foot traveled horizontally. Or, in more practical units, Lone Pine trail rises 20 feet for every 100 feet of horizontal distance, whereas Stony Point trail rises only 16 feet over a horizontal distance of 100 feet.

EXERCISE 2

Which is steeper, a staircase that rises 10 feet over a horizontal distance of 4 feet, or the steps in the football stadium, which rise 20 yards over a horizontal distance of 12 yards?

Definition of Slope

To compare the steepness of the two trails in Example 2, it is not enough to know which trail has the greater gain in elevation overall. Instead, we compare their elevation gains over *the same horizontal distance*. Using the same horizontal distance provides a basis for comparison. The two trails are illustrated in Figure 1.30 as lines on a coordinate grid.

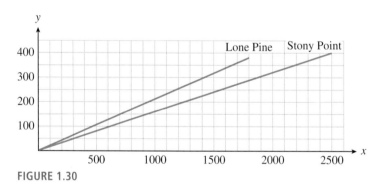

FIGURE 1.30

The ratio we computed in Example 2,

$$\frac{\text{change in elevation}}{\text{change in horizontal position}}$$

appears on the graphs in Figure 1.30 as

$$\frac{\text{change in } y\text{-coordinate}}{\text{change in } x\text{-coordinate}}$$

For example, as we travel along the line representing Stony Point trail, we move from the point $(0, 0)$ to the point $(2500, 400)$. The y-coordinate changes by 400 and the x-coordinate changes by 2500, giving the ratio 0.16 that we found in Example 2. We call this ratio the **slope** of the line.

> **Definition of Slope**
>
> The **slope** of a line is the ratio
>
> $$\frac{\textbf{change in } y\textbf{-coordinate}}{\textbf{change in } x\textbf{-coordinate}}$$
>
> as we move from one point to another on the line.

EXAMPLE 3   Compute the slope of the line that passes through points *A* and *B* in Figure 1.31.

Solution   As we move along the line from *A* to *B*, the *y*-coordinate changes by 3 units, and the *x*-coordinate changes by 4 units. The slope of the line is thus

$$\frac{\text{change in } y\text{-coordinate}}{\text{change in } x\text{-coordinate}} = \frac{3}{4}$$

FIGURE 1.31

> **EXERCISE 3**
>
> Compute the slope of the line through the indicated points in Figure 1.32. On both axes, one square represents one unit.
>
> $$\frac{\text{change in } y\text{-coordinate}}{\text{change in } x\text{-coordinate}} =$$

FIGURE 1.32

The slope of a line is a *number*. It tells us how much the *y*-coordinates of points on the line increase when we increase their *x*-coordinates by 1 unit. For instance, the slope $\frac{3}{4}$ in Example 3 means that the *y*-coordinate increases by $\frac{3}{4}$ unit when the *x*-coordinate increases by 1 unit. For increasing graphs, a larger slope indicates a greater increase in altitude, and hence a steeper line.

Notation for Slope

We use a shorthand notation for the ratio that defines slope,

$$\frac{\text{change in } y\text{-coordinate}}{\text{change in } x\text{-coordinate}}$$

The symbol Δ (the Greek letter *delta*) is used in mathematics to denote *change in*. In particular, Δy means *change in y-coordinate,* and Δx means *change in x-coordinate.* We also use the letter *m* to stand for slope. With these symbols, we can write the definition of slope as follows.

Notation for Slope

The **slope** of a line is given by

$$m = \frac{\Delta y}{\Delta x} = \frac{\text{change in } y\text{-coordinate}}{\text{change in } x\text{-coordinate}}, \quad \Delta x \neq 0$$

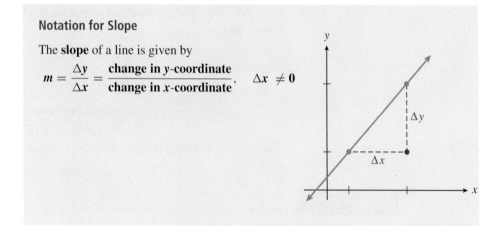

EXAMPLE 4

The Great Pyramid of Khufu in Egypt was built around 2550 B.C. It is 147 meters tall and has a square base 229 meters on each side. Calculate the slope of the sides of the pyramid, rounded to two decimal places.

Solution   From Figure 1.33, we see that Δx is only half the base of the Great Pyramid, so

$$\Delta x = 0.5(229) = 114.5$$

and the slope of the side is

$$m = \frac{\Delta y}{\Delta x} = \frac{147}{114.5} = 1.28$$

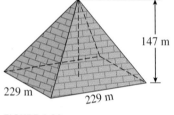

FIGURE 1.33

EXERCISE 4

The Kukulcan Pyramid at Chitin Itza in Mexico was built around 800 A.D. It is 24 meters high, with a temple built on its top platform, as shown in Figure 1.34. The square base is 55 meters on each side, and the top platform is 19.5 meters on each side. Calculate the slope of the sides of the pyramid. Which pyramid is steeper, Kukulcan or the Great Pyramid?

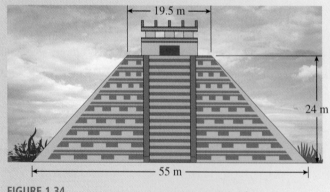

FIGURE 1.34

So far, we have only considered examples in which Δx and Δy are positive numbers, but they can also be negative.

$$\Delta x \text{ is } \begin{cases} \text{positive if } x \text{ increases (move to the right)} \\ \text{negative if } x \text{ decreases (move to the left)} \end{cases}$$

$$\Delta y \text{ is } \begin{cases} \text{positive if } y \text{ increases (move up)} \\ \text{negative if } y \text{ decreases (move down)} \end{cases}$$

EXAMPLE 5

Compute the slope of the line that passes through the points $P(-4, 2)$ and $Q(5, -1)$ shown in Figure 1.35. Illustrate Δy and Δx on the graph.

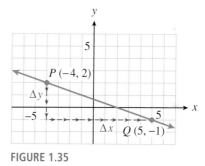

FIGURE 1.35

Solution As we move from the point $P(-4, 2)$ to the point $Q(5, -1)$, we move 3 units *down,* so $\Delta y = -3$. We then move 9 units to the right, so $\Delta x = 9$. Thus, the slope is

$$m = \frac{\Delta y}{\Delta x} = \frac{-3}{9} = \frac{-1}{3}$$

Δy and Δx are labeled on the graph.

We can move from point to point in either direction to compute the slope. The line graphed in Example 4 *decreases* as we move from left to right and hence has a negative slope. The slope is the same if we move from point Q to point P instead of from P to Q. (See Figure 1.36.) In that case, our computation looks like this:

$$m = \frac{\Delta y}{\Delta x} = \frac{3}{-9} = \frac{-1}{3}$$

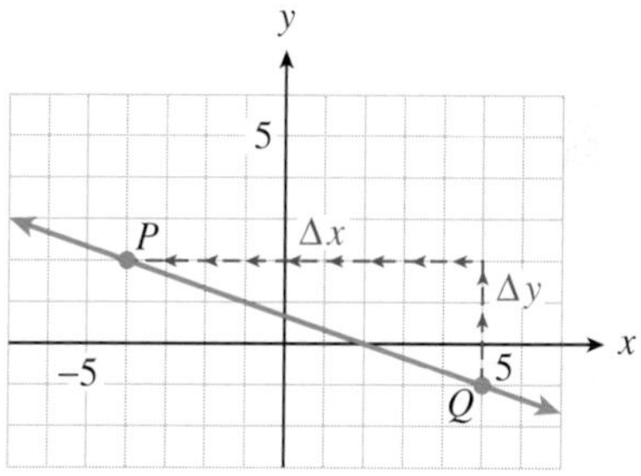

FIGURE 1.36

Lines Have Constant Slope

How do we know which two points to choose when we want to compute the slope of a line? It turns out that any two points on the line will do.

EXERCISE 5

a. Graph the line $4x - 2y = 8$ by finding the x- and y-intercepts.

b. Compute the slope of the line using the x-intercept and y-intercept.

c. Compute the slope of the line using the points $(4, 4)$ and $(1, -2)$.

Exercise 5 illustrates an important property of lines: They have constant slope. No matter which two points we use to calculate the slope, we will always get the same result. We will see later that lines are the only graphs that have this property.

We can think of the slope as a **scale factor** that tells us how many units y increases (or decreases) for each unit of increase in x. Compare the lines in Figure 1.37.

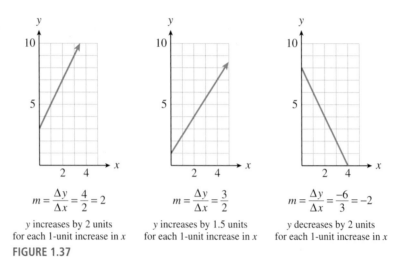

$$m = \frac{\Delta y}{\Delta x} = \frac{4}{2} = 2$$

y increases by 2 units
for each 1-unit increase in x

$$m = \frac{\Delta y}{\Delta x} = \frac{3}{2}$$

y increases by 1.5 units
for each 1-unit increase in x

$$m = \frac{\Delta y}{\Delta x} = \frac{-6}{3} = -2$$

y decreases by 2 units
for each 1-unit increase in x

FIGURE 1.37

Observe that a line with positive slope increases from left to right, and one with negative slope decreases. What sort of line has slope $m = 0$?

Meaning of Slope

In Example 1 of Section 1.1, we graphed the equation $C = 5 + 3t$ showing the cost of a bicycle rental in terms of the length of the rental. The graph is reproduced in Figure 1.38. We can choose any two points on the line to compute its slope. Using points P and Q as shown, we find that

$$m = \frac{\Delta C}{\Delta t} = \frac{9}{3} = 3$$

The slope of the line is 3.

What does this value mean for the cost of renting a bicycle? The expression

$$\frac{\Delta C}{\Delta t} = \frac{9}{3} \text{ stands for}$$

$$\frac{\text{change in cost}}{\text{change in time}} = \frac{9 \text{ dollars}}{3 \text{ hours}}$$

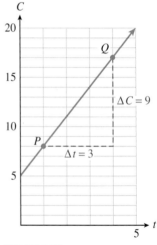

FIGURE 1.38

If we increase the length of the rental by 3 hours, the cost of the rental increases by 9 dollars. The slope gives the *rate of increase* in the rental fee, 3 dollars per hour.

In general, we can make the following statement.

Rate of Change

The slope of a line measures the **rate of change** of the output variable with respect to the input variable.

Depending on the variables involved, this rate might be interpreted as a rate of growth or a rate of speed. A negative slope might represent a rate of decrease or a rate of consumption. The slope of a graph can give us valuable information about the variables.

EXAMPLE 6

The graph in Figure 1.39 shows the distance in miles traveled by a big-rig truck driver after t hours on the road.

  a. Compute the slope of the graph.

  b. What does the slope tell us about the problem?

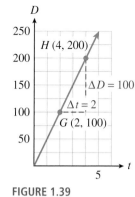

FIGURE 1.39

Solutions **a.** Choose any two points on the line, say $G(2, 100)$ and $H(4, 200)$, in Figure 1.39. As we move from G to H, we find

$$m = \frac{\Delta D}{\Delta t} = \frac{100}{2} = 50$$

The slope of the line is 50.

  b. The best way to understand the slope is to include units in the calculation. For our example,

$$\frac{\Delta D}{\Delta t} \quad \text{means} \quad \frac{\text{change in distance}}{\text{change in time}}$$

or

$$\frac{\Delta D}{\Delta t} = \frac{100 \text{ miles}}{2 \text{ hours}} = 50 \text{ miles per hour}$$

The slope represents the trucker's average speed or velocity.

EXERCISE 6

The graph in Figure 1.40 shows the altitude, a (in feet), of a skier t minutes after getting on a ski lift.

  a. Choose two points and compute the slope (including units).

  b. What does the slope tell us about the problem?

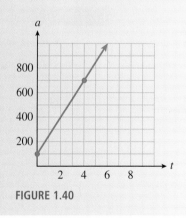

FIGURE 1.40

A Formula for Slope

We have defined the slope of a line to be the ratio $m = \frac{\Delta y}{\Delta x}$ as we move from one point to another on the line. So far, we have computed Δy and Δx by counting squares on the graph, but this method is not always practical. All we really need are the coordinates of two points on the graph.

We will use *subscripts* to distinguish the two points:

P_1 means *first point* and P_2 means *second point.*

We denote the coordinates of P_1 by (x_1, y_1) and the coordinates of P_2 by (x_2, y_2).

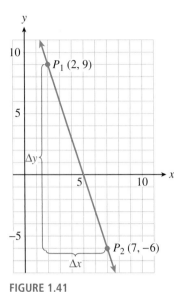

FIGURE 1.41

Now consider a specific example. The line through the two points $P_1\,(2, 9)$ and $P_2\,(7, -6)$ is shown in Figure 1.41. We can find Δx by subtracting the x-coordinates of the points:

$$\Delta x = 7 - 2 = 5$$

In general, we have

$$\Delta x = x_2 - x_1$$

and similarly

$$\Delta y = y_2 - y_1$$

These formulas work even if some of the coordinates are negative; in our example

$$\Delta y = y_2 - y_1 = -6 - 9 = -15$$

By counting squares *down* from P_1 to P_2, we see that Δy is indeed -15. The slope of the line in Figure 1.41 is

$$m = \frac{\Delta y}{\Delta x} = \frac{y_2 - y_1}{x_2 - x_1}$$

$$= \frac{-15}{5} = -3$$

We now have a formula for the slope of a line that works even if we do not have a graph.

Two-Point Slope Formula

The slope of the line passing through the points $P_1\,(x_1, y_1)$ and $P_2\,(x_2, y_2)$ is given by

$$\boldsymbol{m \;=\; \frac{\Delta y}{\Delta x} \;=\; \frac{y_2 - y_1}{x_2 - x_1}, \quad x_2 \neq x_1}$$

EXAMPLE 7 Compute the slope of the line in Figure 1.38 using the points $Q_1\,(6, -3)$ and $Q_2\,(4, 3)$.

Solution Substitute the coordinates of Q_1 and Q_2 into the slope formula to find

$$m = \frac{y_2 - y_1}{x_2 - x_1} = \frac{3 - (-3)}{4 - 6} = \frac{6}{-2} = -3$$

This value for the slope, -3, is the same value found above.

EXERCISE 7

a. Find the slope of the line passing through the points $(2, -3)$ and $(-2, -1)$.

b. Sketch a graph of the line by hand.

It will also be useful to write the slope formula with function notation. Recall that $f(x)$ is another symbol for y, and, in particular, that $y_1 = f(x_1)$ and $y_2 = f(x_2)$. Thus, if $x_2 \neq x_1$, we have

---

Slope Formula in Function Notation

$$\boldsymbol{m \;=\; \frac{y_2 - y_1}{x_2 - x_1} \;=\; \frac{f(x_2) - f(x_1)}{x_2 - x_1}, \quad x_2 \neq x_1}$$

EXAMPLE 8

Figure 1.42 shows a graph of

$$f(x) = x^2 - 6x$$

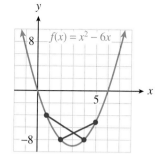

a. Compute the slope of the line segment joining the points at $x = 1$ and $x = 4$.

b. Compute the slope of the line segment joining the points at $x = 2$ and $x = 5$.

FIGURE 1.42

Solutions

a. We set $x_1 = 1$ and $x_2 = 4$ and find the function values at each point.

$$f(x_1) = f(1) = 1^2 - 6(1) = -5$$
$$f(x_2) = f(4) = 4^2 - 6(4) = -8$$

Then

$$m = \frac{f(x_2) - f(x_1)}{x_2 - x_1} = \frac{-8 - (-5)}{4 - 1} = \frac{-3}{3} = -1$$

b. We set $x_1 = 2$ and $x_2 = 5$ and find the function values at each point.

$$f(x_1) = f(2) = 2^2 - 6(2) = -8$$
$$f(x_2) = f(5) = 5^2 - 6(5) = -5$$

Then

$$m = \frac{f(x_2) - f(x_1)}{x_2 - x_1} = \frac{-5 - (-8)}{4 - 1} = \frac{3}{3} = 1$$

Note that the graph of f is not a straight line and that the slope is not constant.

EXERCISE 8

Figure 1.43 shows the graph of a function f.
a. Find $f(3)$ and $f(5)$.

b. Compute the slope of the line segment joining the points at $x = 3$ and $x = 5$.

c. Write an expression for the slope of the line segment joining the points at $x = a$ and $x = b$.

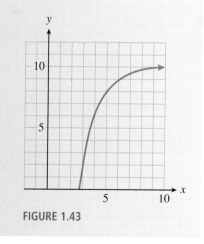

FIGURE 1.43

ANSWERS TO 1.4 EXERCISES

1. Delbert: 21.5 mpg, Francine: 22.75 mpg. Francine gets better mileage.

2. The staircase is steeper.   **3.** $\frac{1}{4}$

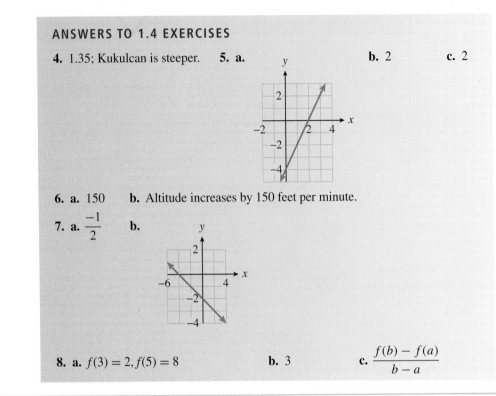
SECTION 1.4 SUMMARY

VOCABULARY   *Look up the definitions of new terms in the Glossary.*

Ratio   Scale factor   Rate of change   Slope

CONCEPTS

1. We can use ratios to compare quantities.

2. The slope ratio, $\dfrac{\text{change in } y\text{-coordinate}}{\text{change in } x\text{-coordinate}}$, measures the steepness of a line.

3. Notation for slope: $m = \dfrac{\Delta y}{\Delta x}, \quad \Delta x \neq 0$

4. Formula for slope: $m = \dfrac{y_2 - y_1}{x_2 - x_1}, \quad x_2 \neq x_1$

5. Function notation for slope:
$$m = \frac{f(x_2) - f(x_1)}{x_2 - x_1}, \quad x_2 \neq x_1$$

6. Lines have constant slope.

7. Slope is a scale factor that tells us how many units Δy increases for each unit increase in Δx as we move along the line.

8. The slope gives us the rate of change.

STUDY QUESTIONS

1. Explain how to compare prices with unit pricing.

2. Why is Δy the numerator of the slope ratio and Δx the denominator?

3. Which line is steeper, one with $m = -2$ or one with $m = -5$?

4. A classmate says that you must always use the intercepts to calculate the slope of a line. Do you agree? Explain.

5. In an application, what does the slope of the graph tell you about the situation?

SKILLS   *Practice each skill in the Homework problems listed.*

1. Use ratios for comparison: #1–4

2. Compute slope from a graph: #5–16, 23–26

3. Use slope to find Δy or Δx: #17–20, 27–30

4. Use slope to compare steepness: #21 and 22

5. Decide whether data points lie on a straight line: #41–46

6. Interpret slope as a rate of change: #31–40

7. Use function notation to discuss graphs and slope: #53–62

Compute ratios to answer the questions in Problems 1–4.

1. Carl runs 100 meters in 10 seconds. Anthony runs 200 meters in 19.6 seconds. Who has the faster average speed?

2. On his 512-mile round trip to Las Vegas and back, Corey needed 16 gallons of gasoline. He used 13 gallons of gasoline on a 429-mile trip to Los Angeles. On which trip did he get better fuel economy?

3. Grimy Gulch Pass rises 0.6 miles over a horizontal distance of 26 miles. Bob's driveway rises 12 feet over a horizontal distance of 150 feet. Which is steeper?

4. Which is steeper, the truck ramp for Acme Movers, which rises 4 feet over a horizontal distance of 9 feet, or a toy truck ramp, which rises 3 centimeters over a horizontal distance of 7 centimeters?

Compute the slope of the line through the indicated points. On both axes, one square represents one unit.

5.

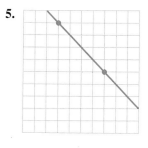

6.

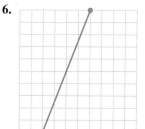

7.

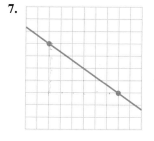

8.

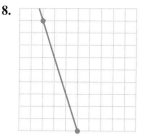

For Problems 9–14,
(a) Graph each line by the intercept method.
(b) Use the intercepts to compute the slope.

9. $3x - 4y = 12$ 10. $2y - 5x = 10$ 11. $2y + 6x = -18$ 12. $9x + 12y = 36$

13. $\dfrac{x}{5} - \dfrac{y}{8} = 1$ 14. $\dfrac{x}{7} - \dfrac{y}{4} = 1$

15. a. Use the points (0, 2) and (4, 8) to compute the slope of the line. Illustrate Δy and Δx on the graph.
  b. Use the points (−4, −4) and (4, 8) to compute the slope of the line. Illustrate Δy and Δx on the graph.
  c. Use the points (0, 2) and (−6, −7) to compute the slope of the line. Illustrate Δy and Δx on the graph.

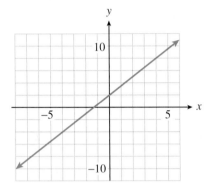

16. a. Use the points (0, −6) and (8, −12) to compute the slope of the line. Illustrate Δy and Δx on the graph.
  b. Use the points (−8, 0) and (4, −9) to compute the slope of the line. Illustrate Δy and Δx on the graph.
  c. Use the points (4, −9) and (0, −6) to compute the slope of the line. Illustrate Δy and Δx on the graph.

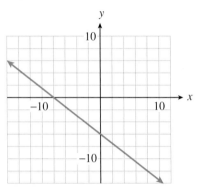

For Problems 17–20, use the formula $m = \dfrac{\Delta y}{\Delta x}$.

17. A line has slope $\frac{-3}{4}$.
  a. Find the vertical change associated with each horizontal change along the line.
  i. $\Delta x = 4$
  ii. $\Delta x = -8$
  iii. $\Delta x = 2$
  iv. $\Delta x = -6$
  b. Find the horizontal change associated with each vertical change along the line.
  i. $\Delta y = 3$
  ii. $\Delta y = -6$
  iii. $\Delta y = -2$
  iv. $\Delta y = 1$

18. A line has slope $\frac{5}{3}$.
  a. Find the vertical change associated with each horizontal change along the line.
  i. $\Delta x = 3$
  ii. $\Delta x = -6$
  iii. $\Delta x = 1$
  iv. $\Delta x = -24$
  b. Find the horizontal change associated with each vertical change along the line.
  i. $\Delta y = -5$
  ii. $\Delta y = 2.5$
  iii. $\Delta y = -1$
  iv. $\Delta y = 3$

19. Residential staircases are usually built with a slope of 70%, or $\frac{7}{10}$. If the vertical distance between stories is 10 feet, how much horizontal space does the staircase require?

20. A straight section of highway in the Midwest maintains a grade (slope) of 4%, or $\frac{1}{25}$, for 12 miles. How much does your elevation change as you travel the road?

21. Choose the line with the correct slope. The scales are the same on both axes.

   a. $m = 2$ **b.** $m = -\frac{1}{2}$

   c. $m = \frac{2}{3}$ **d.** $m = -\frac{5}{3}$

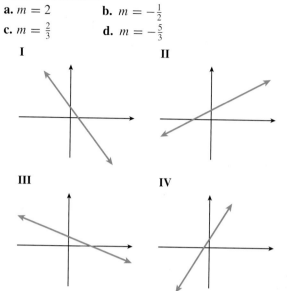

I   II

III   IV

22. Choose the line with the correct slope. The scales are the same on both axes.

   a. $0 < m < 1$ **b.** $m < -1$

   c. $m > 1$ **d.** $m = 0$

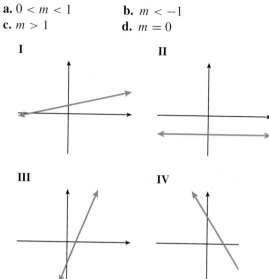

I   II

III   IV

Compute the slope of each line. Note the scales on the axes.

23.

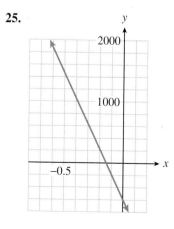

24.

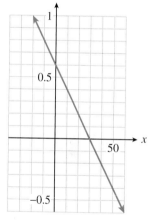

25.

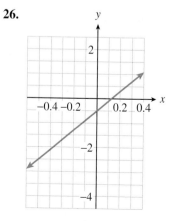

26.

Each table in Problems 27–30 gives the coordinates of points on a line.
(a) Find the slope of the line.
(b) Fill in the missing table entries.

27.

x	y
−4	−14
−2	−9
2	1
3	
	11

28.

x	y
−5	−3.8
−1	−0.6
2	1.8
	4.2
7	

29.

x	y
−3	36
−1	
	12
6	9
10	−3

30.

x	y
−10	800
−2	
5	440
	368
16	176

31. A temporary typist's paycheck (before deductions) is given, in dollars, by $S = 8t$, where t is the number of hours she worked.

   a. Make a table of values for the function.

t	4	8	20	40
S				

   b. Graph the function.
   c. Using two points on the graph, compute the slope $\frac{\Delta S}{\Delta t}$, including units.
   d. What does the slope tell us about the typist's paycheck?

32. The distance (in miles) covered by a cross-country competitor is given by $d = 6t$, where t is the number of hours she runs.

   a. Make a table of values for the function.

t	2	4	6	8
d				

   b. Graph the function.
   c. Using two points on the graph, compute the slope $\frac{\Delta d}{\Delta t}$, including units.
   d. What does the slope tell us about the cross-country runner?

In Problems 33–40,
(a) Choose two points and compute the slope of the graph (including units).
(b) Explain what the slope measures in the context of the problem.

33. The graph shows the number of barrels of oil, B, that has been pumped at a drill site t days after a new drill is installed.

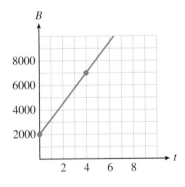

34. The graph shows the amount of garbage, G (in tons), that has been deposited at a dump site t years after new regulations go into effect.

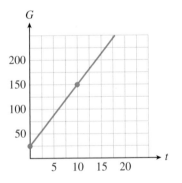

35. The graph shows the amount of emergency water, W (in liters), remaining in a southern California household t days after an earthquake.

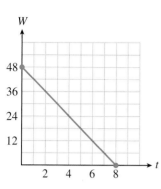

36. The graph shows the amount of money, M (in dollars), in Tammy's bank account w weeks after she loses all sources of income.

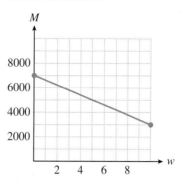

37. The graph shows the length in inches, i, corresponding to various lengths in feet f.

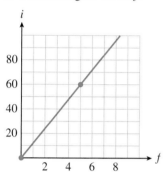

38. The graph shows the number of ounces, z, that correspond to various weights measured in pounds, p.

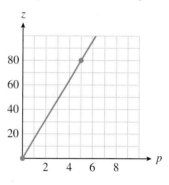

39. The graph shows the cost, C (in dollars), of coffee beans in terms of the amount of coffee, b (in kilograms).

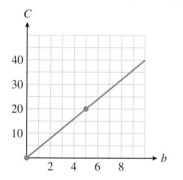

40. The graph shows Tracey's earnings, E (in dollars), in terms of the number of hours, h, that she babysits.

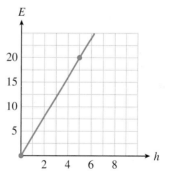

Which of the tables in Problems 41 and 42 represent variables that are related by a linear function? (*Hint*: Which relationships have constant slope?)

41. a.

x	y
2	12
3	17
4	22
5	27
6	32

b.

t	P
2	4
3	9
4	16
5	25
6	36

42. a.

h	w
−6	20
−3	18
0	16
3	14
6	12

b.

t	d
5	0
10	3
15	6
20	12
25	24

43. The table shows the amount of ammonium chloride salt, in grams, that can be dissolved in 100 grams of water at different temperatures.

Temperature, °C	10	12	15	21	25	40	52
Grams of salt	33	34	35.5	38.5	40.5	48	54

a. If you plot the data, will the points lie on a straight line? Why or why not?

b. Calculate the rate of change of salt dissolved with respect to temperature.

44. A spring is suspended from the ceiling. The table shows the length of the spring, in centimeters, as it is stretched by hanging various weights from it.

Weight, kg	3	4	8	10
Length, cm	25.76	25.88	26.36	26.6

Weight, kg	12	15	22
Length, cm	26.84	27.2	28.04

a. If you plot the data, will the points lie on a straight line? Why or why not?

b. Calculate the rate of change of length with respect to weight.

45. The table gives the radius and circumference of various circles, rounded to three decimal places.

r	C
4	25.133
6	37.699
10	62.832
15	94.248

a. If we plot the data, will the points lie on a straight line?

b. What familiar number does the slope turn out to be? (*Hint*: Recall a formula from geometry.)

46. The table gives the side and the diagonal of various squares, rounded to three decimal places.

s	d
3	4.243
6	8.485
8	11.314
10	14.142

a. If we plot the data, will the points lie on a straight line?

b. What familiar number does the slope turn out to be? (*Hint*: Draw a picture of one of the squares and use the Pythagorean theorem to compute its diagonal.)

47. Geologists can measure the depth of the ocean at different points using a technique called echo-sounding. Scientists on board a ship send a pulse of sound toward the ocean floor and measure the time interval until the echo returns to the ship. The speed of sound in seawater is about 1500 meters per second.

  a. Write the speed of sound as a ratio.

  b. If the echo returns in 4.5 seconds, what is the depth of the ocean at that point?

48. Niagara Falls was discovered by Father Louis Hennepin in 1682. In 1952, much of the water of the Niagara River was diverted for hydroelectric power, but until that time erosion caused the Falls to recede upstream by 3 feet per year.

  a. How far did the Falls recede from 1682 to 1952?

  b. The Falls were formed about 12,000 years ago during the end of the last ice age. How far downstream from their current position were they then? (Give your answer in miles.)

49. Geologists calculate the speed of seismic waves by plotting the travel times for waves to reach seismometers at known distances from the epicenter. The speed of the wave can help them determine the nature of the material it passes through. The graph shows a travel-time graph for P-waves from a shallow earthquake.

  a. Why do you think the graph is plotted with distance as the input variable?

  b. Use the graph to calculate the speed of the wave.

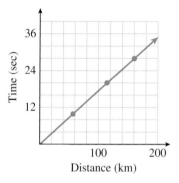

50. Energy (supplied by heat) is required to raise the temperature of a substance, and it is also needed to melt a solid substance to a liquid. The table shows data from heating a solid sample of stearic acid. Heat was applied at a constant rate throughout the experiment. (Source: J. A. Hunt and A. Sykes, 1984)

Time (minutes)	0	0.5	1.5	2	2.5
Temperature, °C	19	29	40	48	53

Time (minutes)	3	4	5	6	7
Temperature, °C	55	55	55	55	55

Time (minutes)	8	8.5	9	9.5	10
Temperature, °C	55	64	70	73	74

  a. Did the temperature rise at a constant rate? Describe the temperature as a function of time.

  b. Graph temperature as a function of time.

  c. What is the melting point of stearic acid? How long did it take the sample to melt?

51. The graph shows the temperature of the ocean as a function of depth.

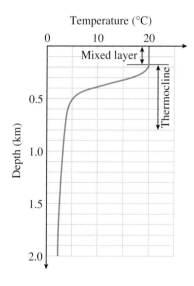

  a. What is the difference in temperature between the surface of the ocean and the deepest level shown?

  b. Over what depths does the temperature change most rapidly?

  c. What is the average rate of change of temperature with respect to depth in the region called the thermocline?

52. The graph shows the average air temperature as a function of altitude. (Figure (b) is an enlargement of the indicated region of Figure (a).) (Source: Ahrens, 1998)

a. Is temperature a decreasing function of altitude?

b. The *lapse rate* is the rate at which the temperature changes with altitude. In which regions of the atmosphere is the lapse rate positive?

c. The region where the lapse rate is zero is called the isothermal zone. Give an interval of altitudes that describes the isothermal zone.

d. What is the lapse rate in the mesosphere?

e. Describe the temperature for altitudes greater than 90 kilometers.

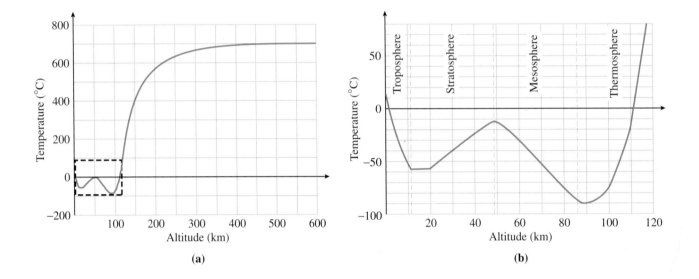

(a) (b)

In Problems 53–56, evaluate the function at $x = a$ and $x = b$, and then find the slope of the line segment joining the two corresponding points on the graph. Illustrate the line segment on a graph of the function.

53. $f(x) = x^2 - 2x - 8$

    a. $a = -2, b = 1$ **b.** $a = -1, b = 5$

54. $g(x) = \sqrt{x + 4}$

    a. $a = -2, b = 0$ **b.** $a = 0, b = 5$

55. $h(x) = \dfrac{4}{x + 2}$

    a. $a = 0, b = 6$ **b.** $a = -1, b = 2$

56. $q(x) = x^3 - 4x$

    a. $a = -1, b = 2$ **b.** $a = -1, b = 3$

In Problems 57–62, find the coordinates of the indicated points, and then write an algebraic expression using function notation for the indicated quantity.

57. The length of the vertical line segment on the *y*-axis

a.

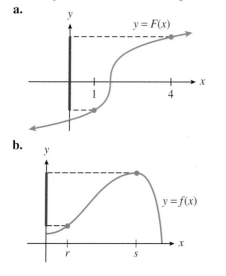

b.

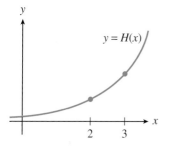

58. The length of the vertical line segment on the *y*-axis

a.

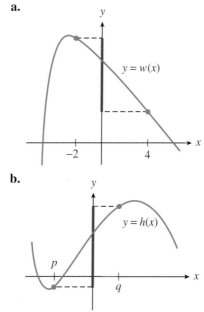

b.

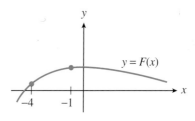

59. a. The increase in *y* as *x* increases from 2 to 3

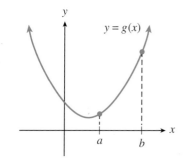

b. The increase in *y* as *x* increases from *a* to *b*

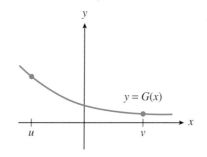

60. a. The increase in *y* as *x* increases from −4 to −1

b. The increase in *y* as *x* increases from *u* to *v*

61. The shaded area

a.

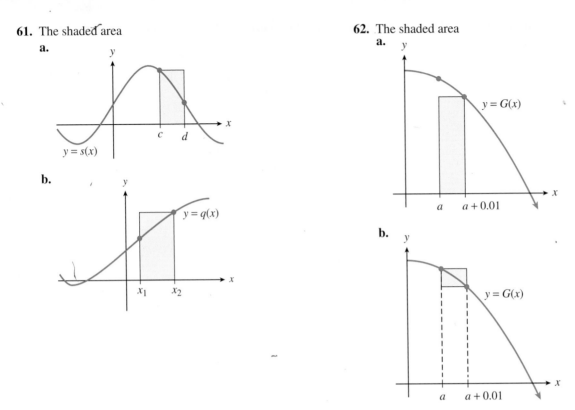

b.

62. The shaded area

a.

b.

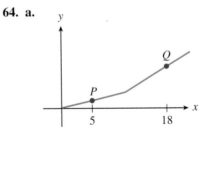

In Problems 63–66, find the coordinates of the indicated points on the graph of $y = f(x)$ and write an algebraic expression using function notation for the slope of the line segment joining points P and Q.

63. a.

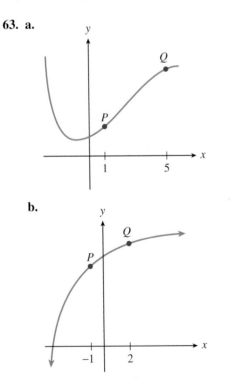

b.

64. a.

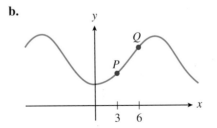

b.

65. a.

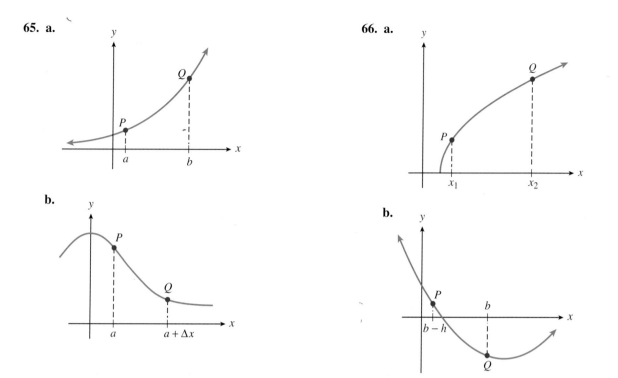

b.

66. a.

b.

Slope-Intercept Form

As we saw in Section 1.1, many linear models $y = f(x)$ have equations of the form

$$f(x) = \textbf{(starting value)} + \textbf{(rate of change)} \cdot x$$

The starting value, or the value of y at $x = 0$, is the y-intercept of the graph, and the rate of change is the slope of the graph. Thus, we can write the equation of a line as

$$f(x) = b + mx$$

where the constant term, b, is the y-intercept of the line, and m, the coefficient of x, is the slope of the line. This form for the equation of a line is called the **slope-intercept** form.

Slope-Intercept Form

If we write the equation of a linear function in the form,

$$f(x) = b + mx$$

then m is the **slope** of the line, and b is the y-intercept.

(You may have encountered the slope-intercept equation in the equivalent form $y = mx + b$.) For example, consider the two linear functions and their graphs shown in Figure 1.44.

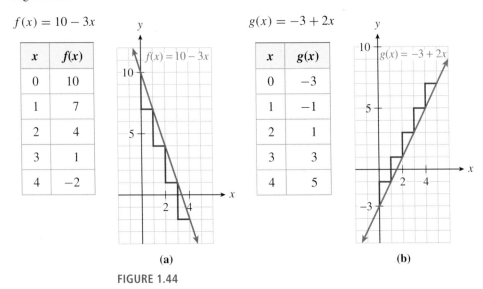

$f(x) = 10 - 3x$

x	$f(x)$
0	10
1	7
2	4
3	1
4	−2

$g(x) = -3 + 2x$

x	$g(x)$
0	−3
1	−1
2	1
3	3
4	5

(a)   (b)

FIGURE 1.44

We can see that the y-intercept of each line is given by the constant term, b. By examining the table of values, we can also see why the coefficient of x gives the slope of the line: For $f(x)$, each time x increases by 1 unit, y decreases by 3 units. For $g(x)$, each time x increases by 1 unit, y increases by 2 units. For each graph, the coefficient of x is a scale factor that tells us how many units y changes for 1 unit increase in x. But that is exactly what the slope tells us about a line.

EXAMPLE 1

Francine is choosing an Internet service provider. She paid $30 for a modem, and she is considering three companies for dialup service: Juno charges $14.95 per month, ISP.com charges $12.95 per month, and peoplepc charges $15.95 per month. Match the graphs in Figure 1.45 to Francine's Internet cost with each company.

Solution   Francine pays the same initial amount, $30 for the modem, under each plan. The monthly fee is the rate of change of her total cost, in dollars per month. We can write a formula for her cost under each plan.

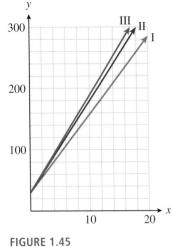

Juno: $f(x) = 30 + 14.95x$

ISP.com: $g(x) = 30 + 12.95x$

peoplepc: $h(x) = 30 + 15.95x$

The graphs of these three functions all have the same y-intercept, but their slopes are determined by the monthly fees. The steepest graph, III, is the one with the largest monthly fee, peoplepc, and ISP.com, which has the lowest monthly fee, has the least steep graph, I.

FIGURE 1.45

Delbert decides to use DSL for his Internet service. Earthlink charges a $99 activation fee and $39.95 per month, DigitalRain charges $50 for activation and $34.95 per month, and FreeAmerica charges $149 for activation and $34.95 per month.

a. Write a formula for Delbert's Internet costs under each plan.

b. Match Delbert's Internet cost under each company with its graph in Figure 1.46.

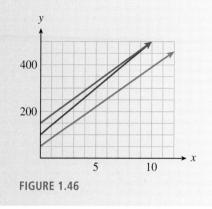

FIGURE 1.46

In the equation $f(x) = b + mx$, we call m and b **parameters**. Their values are fixed for any particular linear equation; for example, in the equation $y = 2x + 3$, $m = 2$ and $b = 3$, and the variables are x and y. By changing the values of m and b, we can write the equation for any line except a vertical line (see Figure 1.47). The collection of all linear functions $f(x) = b + mx$ is called a **two-parameter family** of functions.

These lines have the same
y-intercept but different slopes.

These lines have the same
slope but different y-intercepts.

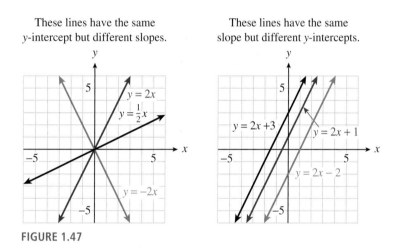

FIGURE 1.47

Slope-Intercept Method of Graphing

Look again at the lines in Figure 1.47: There is only one line that has a given slope and passes through a particular point. That is, the values of m and b determine the particular line. The value of b gives us a starting point, and the value of m tells us which direction to go to plot a second point. Thus, we can graph a line given in slope-intercept form without having to make a table of values.

EXAMPLE 2

a. Write the equation $4x - 3y = 6$ in slope-intercept form.

b. Graph the line by hand.

Solutions **a.** We solve the equation for y in terms of x.

$$-3y = 6 - 4x$$

$$y = \frac{6 - 4x}{-3} = \frac{6}{-3} + \frac{-4x}{-3}$$

$$y = -2 + \frac{4}{3}x$$

b. We see that the slope of the line is $m = \frac{4}{3}$ and its y-intercept is $b = -2$. We begin by plotting the y-intercept, $(0, -2)$. We then use the slope to find another point on the line. We have

$$m = \frac{\Delta y}{\Delta x} = \frac{4}{3}$$

so starting at $(0, -2)$, we move 4 units in the y-direction and 3 units in the x-direction, to arrive at the point $(3, 2)$. Finally, we draw the line through these two points. (See Figure 1.48.)

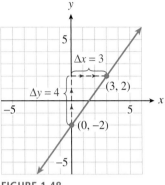

FIGURE 1.48

The slope of a line is a ratio and can be written in many equivalent ways. In Example 2, the slope is equal to $\frac{8}{6}$, $\frac{12}{9}$, and $\frac{-4}{-3}$. We can use any of these fractions to locate a third point on the line as a check. If we use $m = \frac{\Delta y}{\Delta x} = \frac{-4}{-3}$, we move *down* 4 units and *left* 3 units from the y-intercept to find the point $(-3, -6)$ on the line.

Slope-Intercept Method for Graphing a Line

1. Plot the y-intercept $(0, b)$.

2. Use the definition of slope to find a second point on the line: Starting at the y-intercept, move Δy units in the y-direction and Δx units in the x-direction. Plot a second point at this location.

3. Use an equivalent form of the slope to find a third point, and draw a line through the points.

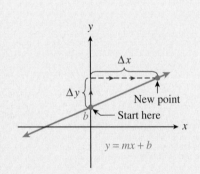

EXERCISE 2

a. Write the equation $2y + 3x + 4 = 0$ in slope-intercept form.

b. Use the slope-intercept method to graph the line.

Finding a Linear Equation from a Graph

We can also use the slope-intercept form to find the equation of a line from its graph. First, note the value of the y-intercept from the graph, and then calculate the slope using two convenient points.

EXAMPLE 3 Find an equation for the line shown in Figure 1.49.

Solution   The line crosses the y-axis at the point $(0, 3200)$, so the y-intercept is 3200. To calculate the slope of the line, locate another point, say $(20, 6000)$, and compute:

$$m = \frac{\Delta y}{\Delta x} = \frac{6000 - 3200}{20 - 0} = \frac{2800}{20} = 140$$

The slope-intercept form of the equation, with $m = 140$ and $b = 3200$, is $y = 3200 + 140x$.

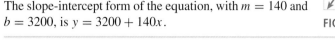

FIGURE 1.49

EXERCISE 3

Find an equation for the line shown in Figure 1.50.

$$b =$$

$$m =$$

$$y =$$

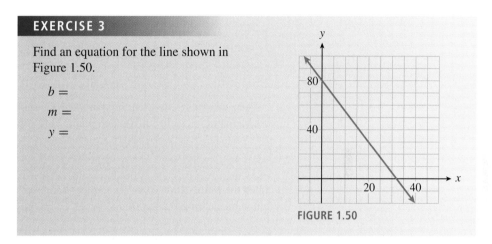

FIGURE 1.50

Point-Slope Form

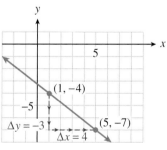

FIGURE 1.51

We can find the equation for a line if we know its slope and y-intercept. What if we do not know the y-intercept, but instead know some other point on the line? There is only one line that passes through a given point and has a given slope. For example, we can graph the line of slope $\frac{-3}{4}$ that passes through the point $(1, -4)$. We first plot the given point, $(1, -4)$, as shown in Figure 1.51. Then we use the slope to find another point on the line. The slope is

$$m = \frac{-3}{4} = \frac{\Delta y}{\Delta x}$$

so we move down 3 units and then 4 units to the right, starting from $(1, -4)$. This brings us to the point $(5, -7)$. We can then draw the line through these two points.

We can also find an equation for the line, as shown in Example 4.

EXAMPLE 4 Find an equation for the line that passes through $(1, -4)$ and has slope $\frac{-3}{4}$.

Solution   We will use the formula for slope,

$$m = \frac{y_2 - y_1}{x_2 - x_1}$$

We substitute $\frac{-3}{4}$ for the slope, m, and $(1, -4)$ for (x_1, y_1). For the second point, (x_2, y_2), we will use the variable point (x, y). Substituting these values into the slope formula gives us

$$\frac{-3}{4} = \frac{y - (-4)}{x - 1} = \frac{y + 4}{x - 1}$$

To solve for y we first multiply both sides by $x - 1$.

$$(x-1)\,\frac{-3}{4} = \frac{y+4}{x-1}\,(x-1)$$

$$\frac{-3}{4}(x - 1) = y + 4 \qquad\qquad \text{Apply the distributive law.}$$

$$\frac{-3}{4}x + \frac{3}{4} = y + 4 \qquad\qquad \text{Subtract 4 from both sides.}$$

$$\frac{-3}{4}x - \frac{13}{4} = y \qquad\qquad \tfrac{3}{4} - 4 = \tfrac{3}{4} - \tfrac{16}{4} = \tfrac{-13}{4}$$

When we use the slope formula in this way to find the equation of a line, we substitute a variable point (x, y) for the second point. This version of the formula,

$$m = \frac{y - y_1}{x - x_1}$$

is called the **point-slope form** for a linear equation. It is sometimes stated in another form obtained by clearing the fraction to get

$$(x - x_1)\,m = \frac{y - y_1}{x - x_1}\,(x - x_1) \qquad \text{Multiply both sides by } (x - x_1).$$

$$(x - x_1)\,m = y - y_1 \qquad\qquad \text{Clear fractions and solve for } y.$$

$$y = y_1 + m(x - x_1)$$

Point-Slope Form

The equation of the line that passes through the point (x_1, y_1) and has slope m is

$$y = y_1 + m(x - x_1)$$

EXERCISE 4

Use the point-slope form to find the equation of the line that passes through the point $(-3, 5)$ and has slope -1.4.

$$y = y_1 + m(x - x_1) \qquad\qquad \text{Substitute } -1.4 \text{ for } m \text{ and } (-3, 5) \text{ for } (x_1, y_1).$$
$$\qquad\qquad\qquad\qquad\qquad\qquad \text{Simplify: Apply the distributive law.}$$

The point-slope form is useful for modeling linear functions when we do not know the initial value but do know some other point on the line.

EXAMPLE 5

Under a proposed graduated income tax system, single taxpayers would owe $1500 plus 20% of the amount of their income over $13,000. (For example, if your income is $18,000, you would pay $1500 plus 20% of $5000.)

a. Complete the table of values for the tax, T, on various incomes, I.

I	15,000	20,000	22,000
T			

b. Write a linear equation in point-slope form for the tax, T, on an income I.

c. Write the equation in slope-intercept form.

Solutions   **a.** On an income of $15,000, the amount of income over $13,000 is $15,000 - $13,000 = $2000, so you would pay $1500 plus 20% of $2000, or

$$T = 1500 + 0.20(2000) = 1900$$

You can compute the other function values in the same way.

I	15,000	20,000	22,000
T	1900	2900	3300

b. On an income of I, the amount of income over $13,000 is $I - $13,000, so you would pay $1500 plus 20% of $I - 13,000$, or

$$T = 1500 + 0.20\,(I - 13,000)$$

c. Simplify the right side of the equation to get

$$T = 1500 + 0.20I - 2600$$
$$T = -1100 + 0.20I$$

EXERCISE 5

A healthy weight for a young woman of average height, 64 inches, is 120 pounds. To calculate a healthy weight for a woman taller than 64 inches, add 5 pounds for each inch of height over 64.

a. Write a linear equation in point-slope form for the healthy weight, W, for a woman of height, H, in inches.

b. Write the equation in slope-intercept form.

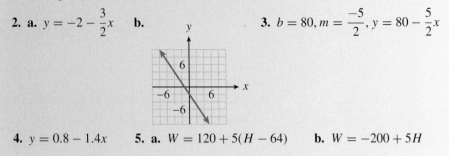
SECTION 1.5 SUMMARY

VOCABULARY   *Look up the definitions of new terms in the Glossary.*

Slope-intercept form   Point-slope form   Parameter

CONCEPTS

1. Linear functions form a two-parameter family,
$f(x) = b + mx$.

2. The initial value of the function and the y-intercept of its graph are given by b. The rate of change of the function and the slope of its graph are given by m.

3. The slope-intercept form, $\boldsymbol{y = b + mx}$, is useful when we know the initial value and the rate of change.

4. The point-slope form, $\boldsymbol{y = y_1 + m(x - x_1)}$, is useful when we know the rate of change and one point on the line.

STUDY QUESTIONS

1. How can you put a linear equation into slope-intercept form?

2. What do the coefficients in the slope-intercept form tell you about the line?

3. Explain how to graph a line using the slope-intercept method.

4. Explain how to find an equation for a line from its graph.

5. Explain how to use the point-slope form for a linear equation.

6. Francine says that the slope of the line $y = 4x - 6$ is $4x$. Is she correct? Explain your answer.

7. Delbert says that the slope of the line $3x - 4y = 8$ is 3. Is he correct? Explain your answer.

SKILLS   *Practice each skill in the Homework problems listed.*

1. Write a linear equation in slope-intercept form: #1–14
2. Identify the slope and y-intercept: #1–10
3. Graph a line by the slope-intercept method: #11–14
4. Find a linear equation from its graph: #21–26, 29–32, 53–56

5. Interpret the slope and y-intercept: #21–28, 63 and 64
6. Find a linear equation from one point and the slope: #33–50

ThomsonNOW™ Test yourself on key content at www.thomsonedu.com/login.

In Problems 1–10,
(a) Write each equation in slope-intercept form.
(b) State the slope and *y*-intercept of the line.

1. $3x + 2y = 1$ **2.** $5x - 4y = 0$ **3.** $\frac{1}{4}x + \frac{3}{2}y = \frac{1}{6}$ **4.** $\frac{7}{6}x - \frac{2}{9}y = 3$

5. $4.2x - 0.3y = 6.6$ **6.** $0.8x + 0.004y = 0.24$ **7.** $y + 29 = 0$ **8.** $y - 37 = 0$

9. $250x + 150y = 2450$ **10.** $80x - 360y = 6120$

In Problems 11–14,
(a) Sketch by hand the graph of the line with the given slope and *y*-intercept.
(b) Write an equation for the line.
(c) Find the *x*-intercept of the line.

11. $m = 3$ and $b = -2$ **12.** $m = -4$ and $b = 1$ **13.** $m = -\frac{5}{3}$ and $b = -6$ **14.** $m = \frac{3}{4}$ and $b = -2$

15. The point $(2, -1)$ lies on the graph of $f(x) = -3x + b$. Find b.

16. The point $(-3, -8)$ lies on the graph of $f(x) = \frac{2}{3}x + b$. Find b.

17. The point $(8, -5)$ lies on the graph of $f(x) = mx - 3$. Find m.

18. The point $(-5, -6)$ lies on the graph of $f(x) = mx + 2$. Find m.

19. Find the slope and intercepts of the line $Ax + By = C$.

20. Find the slope and intercepts of the line $\frac{x}{a} + \frac{y}{b} = 1$.

In Problems 21–26,
(a) Find a formula for the function whose graph is shown.
(b) Say what the slope and the vertical intercept tell us about the problem.

21. The graph shows the altitude, a (in feet), of a skier t minutes after getting on a ski lift.

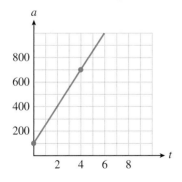

22. The graph shows the distance, d (in meters), traveled by a train t seconds after it passes an observer.

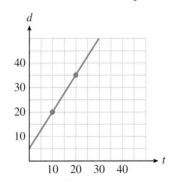

23. The graph shows the amount of garbage, G (in tons), that has been deposited at a dump site t years after new regulations go into effect.

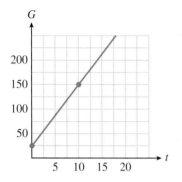

24. The graph shows the number of barrels of oil, B, that has been pumped at a drill site t days after a new drill is installed.

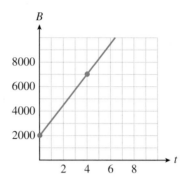

25. The graph shows the amount of money, M (in dollars), in Tammy's bank account w weeks after she loses all sources of income.

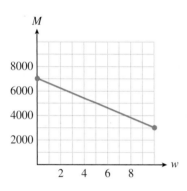

26. The graph shows the amount of emergency water, W (in liters), remaining in a southern California household t days after an earthquake.

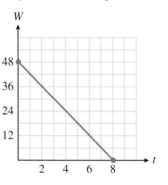

27. The formula $F = \frac{9}{5}C + 32$ converts the temperature in degrees Celsius to degrees Fahrenheit.
   a. What is the Fahrenheit temperature when it is $10°$ Celsius?
   b. What is the Celsius temperature when it is $-4°$ Fahrenheit?
   c. Choose appropriate $\boxed{\textbf{WINDOW}}$ settings and graph the equation $y = \frac{9}{5}x + 32$.
   d. Find the slope and explain its meaning for this problem.
   e. Find the intercepts and explain their meanings for this problem.

28. If the temperature on the ground is $70°$ Fahrenheit, the formula $T = 70 - \frac{3}{820}h$ gives the temperature at an altitude of h feet.
   a. What is the temperature at an altitude of 4100 feet?
   b. At what altitude is the temperature $34°$ Fahrenheit?
   c. Choose appropriate $\boxed{\textbf{WINDOW}}$ settings and graph the equation $y = 70 - \frac{3}{820}x$.
   d. Find the slope and explain its meaning for this problem.
   e. Find the intercepts and explain their meanings for this problem.

29. In England, oven cooking temperatures are often given as Gas Marks rather than degrees Fahrenheit. The table shows the equivalent oven temperatures for various Gas Marks.

Gas Mark	3	5	7	9
Degrees (F)	325	375	425	475

a. Plot the data and draw a line through the data points.
b. Calculate the slope of your line. Estimate the y-intercept from the graph.
c. Find an equation that gives the temperature in degrees Fahrenheit in terms of the Gas Mark.

30. European shoe sizes are scaled differently than American shoe sizes. The table shows the European equivalents for various American shoe sizes.

American shoe size	5.5	6.5	7.5	8.5
European shoe size	37	38	39	40

a. Plot the data and draw a line through the data points.
b. Calculate the slope of your line. Estimate the y-intercept from the graph.
c. Find an equation that gives the European shoe size in terms of American shoe size.

31. A spring is suspended from the ceiling. The table shows the length of the spring in centimeters as it is stretched by hanging various weights from it.

Weight, kg	3	4	8	10
Length, cm	25.76	25.88	26.36	26.6

Weight, kg	12	15	22
Length, cm	26.84	27.2	28.04

a. Plot the data on graph paper and draw a straight line through the points. Estimate the y-intercept of your graph.
b. Find an equation for the line.
c. If the spring is stretched to 27.56 cm, how heavy is the attached weight?

32. The table shows the amount of ammonium chloride salt, in grams, that can be dissolved in 100 grams of water at different temperatures.

Temperature, °C	10	12	15	21
Grams of salt	33	34	35.5	38.5

Temperature, °C	25	40	52
Grams of salt	40.5	48	54

a. Plot the data on graph paper and draw a straight line through the points. Estimate the y-intercept of your graph.
b. Find an equation for the line.
c. At what temperature will 46 grams of salt dissolve?

For Problems 33–36,
(a) Sketch by hand the graph of the line that passes through the given point and has the given slope.
(b) Write an equation for the line in point-slope form.
(c) Put your equation from part (b) into slope-intercept form.

33. $(2, -5); m = -3$
34. $(-6, -1); m = 4$
35. $(2, -1); m = \frac{5}{3}$
36. $(-1, 2); m = -\frac{3}{2}$

For Problems 37–40,
(a) Write an equation in point-slope form for the line that passes through the given point and has the given slope.
(b) Put your equation from part (a) into slope-intercept form.
(c) Use your graphing calculator to graph the line.

37. $(-6.4, -3.5), m = -0.25$
38. $(7.2, -5.6); m = 1.6$

39. $(80, -250), m = 2.4$
40. $(-150, 1800), m = -24$

For Problems 41 and 42,
(a) Find the slope of the line. (Note that not all the labeled points lie on the line.)
(b) Find an equation for the line.

41.

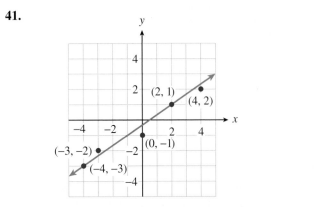

42.

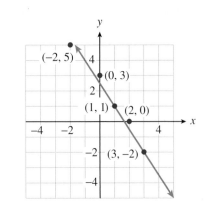

The equation of line l_1 is $y = q + px$, and the equation of line l_2 is $y = v + tx$.
(a) Decide whether the coordinates of each labeled point are
   i. a solution of $y = q + px$,
   ii. a solution of $y = v + tx$,
   iii. a solution of both equations, or
   iv. a solution of neither equation.
(b) Find p, q, t, and v.

43.

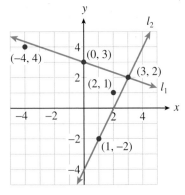

44.

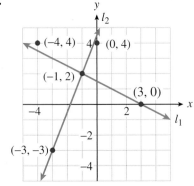

For Problems 45–50,
(a) Estimate the slope and vertical intercept of each line. (*Hint*: To calculate the slope, find two points on the graph that lie on the intersection of grid lines.)
(b) Using your estimates from (a), write an equation for the line.

45.

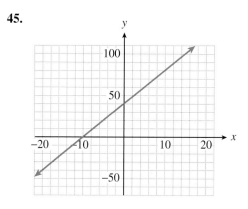

46.

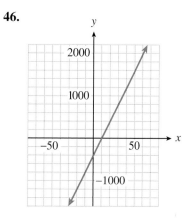

47.

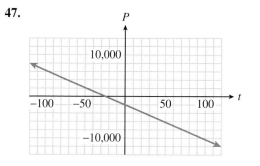

48.

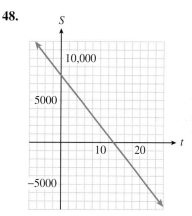

49.

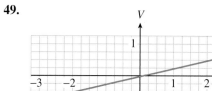

50.

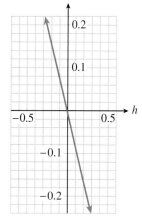

51. a. Write equations for three lines with slope $m = \frac{3}{4}$. (Many answers are possible.)
  b. Graph all three lines in the same window. What do you notice about the lines?

52. a. Write equations for three lines with slope $m = 0$. (Many answers are possible.)
  b. Graph all three lines in the same window. What do you notice about the lines?

In Problems 53–56, choose the correct graph for each equation. The scales on both axes are the same.

53. a. $y = \dfrac{3}{4}x + 2$ **b.** $y = \dfrac{-3}{4}x + 2$ **c.** $y = \dfrac{3}{4}x - 2$ **d.** $y = \dfrac{-3}{4}x - 2$

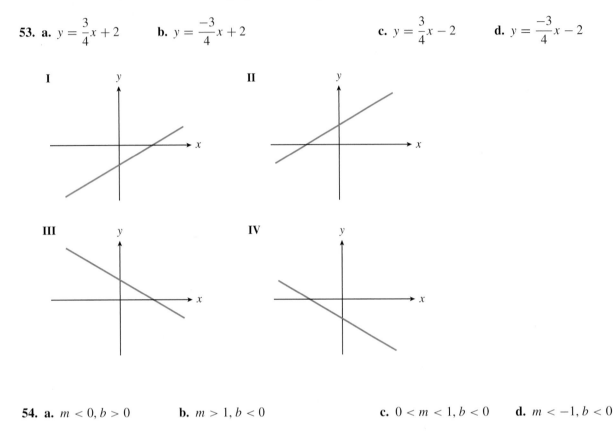

54. a. $m < 0, b > 0$ **b.** $m > 1, b < 0$ **c.** $0 < m < 1, b < 0$ **d.** $m < -1, b < 0$

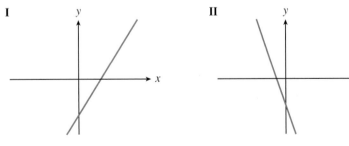

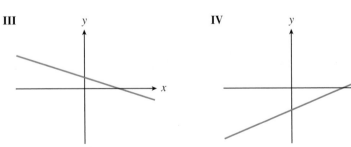

55. a. $y = 1 + 2(x + 3)$ **b.** $y = -1 + 2(x - 3)$ **c.** $y = -1 + 2(x + 3)$ **d.** $y = 1 + 2(x - 3)$

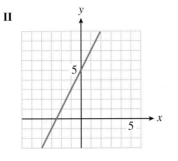

56. a. $y = 2 - \dfrac{2}{3}(x - 3)$ **b.** $y = 2 - \dfrac{3}{2}(x + 3)$ **c.** $y = 2 + \dfrac{3}{2}(x - 3)$ **d.** $y = 2 + \dfrac{2}{3}(x + 3)$

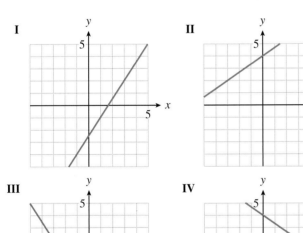

In Problems 57–60, find the slope of each line and the coordinates of one point on the line. (No calculation is necessary!)

57. $y + 1 = 2(x - 6)$ **58.** $2(y - 8) = 5(x + 2)$ **59.** $y = 3 - \dfrac{4}{3}(x + 5)$ **60.** $7x = -3y$

61. a. Draw a set of coordinate axes with a square grid (i.e., with units the same size in both directions). Sketch four lines through the point (0, 4) with the following slopes:

$$m = 3, \quad m = -3, \quad m = \frac{1}{3}, \quad m = \frac{-1}{3}$$

b. What do you notice about these lines? (*Hint:* Look for perpendicular lines.)

62. a. Draw a set of coordinate axes with a square grid (see Problem 61). Sketch four lines through the point (0, −3) with the following slopes.

$$m = \frac{2}{5}, \quad m = \frac{-2}{5}, \quad m = \frac{5}{2}, \quad m = \frac{-5}{2}$$

b. What do you notice about these lines? (*Hint:* Look for perpendicular lines.)

63. The boiling point of water changes with altitude and is approximated by the formula

$$B = f(h) = 212 - 0.0018h$$

where B is in degrees and h is in feet. State the slope and vertical intercept of the graph, including units, and explain their meaning in this context.

64. The height of a woman in centimeters is related to the length of her femur (in centimeters) by the formula $H = f(x) = 2.47x + 54.10$. State the slope and the vertical intercept of the graph, including units, and explain their meaning in this context.

1.6   Linear Regression

We have spent most of this chapter analyzing models described by graphs or equations. To create a model, however, we often start with a quantity of data. Choosing an appropriate function for a model is a complicated process. In this section, we consider only linear models and explore methods for fitting a linear function to a collection of data points. First, we fit a line through two data points.

Fitting a Line through Two Points

If we already know that two variables are related by a linear function, we can find a formula from just two data points. For example, variables that increase or decrease at a constant rate can be described by linear functions.

EXAMPLE 1   In 1993, Americans drank 188.6 million cases of wine. Wine consumption increased at a constant rate over the next decade, and we drank 258.3 million cases of wine in 2003.
(Source: *Los Angeles Times,* Adams Beverage Group)

a. Find a formula for wine consumption, W, in millions of cases, as a linear function of time, t, in years since 1990.

b. State the slope as a rate of change. What does the slope tell us about this problem?

Solutions   **a.** We have two data points of the form (t, W), namely (3, 188.6) and (13, 258.3). We use the point-slope formula to fit a line through these two points. First, we compute the slope.

$$\frac{\Delta W}{\Delta t} = \frac{258.3 - 188.6}{13 - 3} = 6.97$$

Next, we use the slope $m = 6.97$ and either of the two data points in the point-slope formula.

$$W = W_1 + m(t - t_1)$$
$$W = 188.6 + 6.97(t - 3)$$
$$W = 167.69 + 6.97t$$

Thus, $W = f(t) = 167.69 + 6.97t$.

b. The slope gives us the rate of change of the function, and the units of the variables can help us interpret the slope in context.

$$\frac{\Delta W}{\Delta t} = \frac{258.3 - 188.6}{13 - 3} \frac{\text{millions of cases}}{\text{years}} = 6.97 \text{ millions of cases/year}$$

Over the 10 years between 1993 and 2003, wine consumption in the United States increased at a rate of 6.97 million cases per year.

To Fit a Line through Two Points:

1. Compute the slope between the two points.

2. Substitute the slope and either point into the point-slope formula

$$y = y_1 + m(x - x_1)$$

EXERCISE 1

In 1991, there were 64.6 burglaries per 1000 households in the United States. The number of burglaries reported annually declined at a roughly constant rate over the next decade, and in 2001 there were 28.7 burglaries per 1000 households.
(Source: U.S. Department of Justice)

a. Find a function for the number of burglaries, B, as a function of time, t, in years, since 1990.

b. State the slope as a rate of change. What does the slope tell us about this problem?

Scatterplots

Empirical data points in a linear relation may not lie exactly on a line. There are many factors that can affect experimental data, including measurement error, the influence of environmental conditions, and the presence of related variable quantities.

EXAMPLE 2

A consumer group wants to test the gas mileage of a new model SUV. They test-drive six vehicles under similar conditions and record the distance each drove on various amounts of gasoline.

Gasoline used (gal)	9.6	11.3	8.8	5.2	10.3	6.7
Miles driven	155.8	183.6	139.6	80.4	167.1	99.7

a. Are the data linear?

b. Draw a line that fits the data.

c. What does the slope of the line tell us about the data?

Solutions **a.** No, the data are not strictly linear. If we compute the slopes between successive data points, the values are not constant. We can see from an accurate plot of the data, shown in Figure 1.52, that the points lie close to, but not precisely on, a straight line.

b. We would like to draw a line that comes as close as possible to all the data points, even though it may not pass precisely through any of them. In particular, we try to

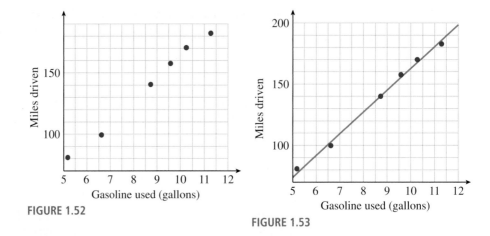

FIGURE 1.52

FIGURE 1.53

adjust the line so that we have the same number of data points above the line and below the line. One possible solution is shown in Figure 1.53.

c. To compute the slope of the line of best fit, we first choose two points on the line. Our line appears to pass through one of the data points, (8.8, 139.6). Look for a second point on the line whose coordinates are easy to read, perhaps (6.5,100). The slope is

$$m = \frac{139.6 - 100}{8.8 - 6.5} = 17.2 \text{ miles per gallon}$$

According to our data, the SUV gets about 17.2 miles to the gallon.

EXERCISE 2

a. Plot the data points. Do the points lie on a line?

b. Draw a line that fits the data.

x	1.49	3.68	4.95	5.49	7.88	8.41
y	2.69	3.7	4.6	5.2	7.2	7.3

The graph in Example 2 is called a **scatterplot.** The points on a scatterplot may or may not show some sort of pattern. Consider the three plots in Figure 1.54. In Figure 1.54a, the data points resemble a cloud of gnats; there is no apparent pattern to their locations. In Figure 1.54b, the data follow a generally decreasing trend, but certainly do not all lie on the same line. The points in Figure 1.54c are even more organized; they seem to lie very close to an imaginary line.

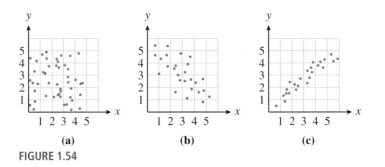

(a) **(b)** **(c)**

FIGURE 1.54

If the data in a scatterplot are roughly linear, we can estimate the location of an imaginary *line of best fit* that passes as close as possible to the data points. We can then use this line to make predictions about the data.

Linear Regression

One measure of a person's physical fitness is the *body mass index,* or BMI. Your BMI is the ratio of your weight in kilograms to the square of your height in centimeters. Thus, thinner people have lower BMI scores, and fatter people have higher scores. The Centers for Disease Control considers a BMI between 18.5 and 24.9 to be healthy. The points on the scatterplot in Figure 1.55 show the BMI of Miss America from 1918 to 1998. From the data in the scatterplot, can we see a trend in Americans' ideal of female beauty?

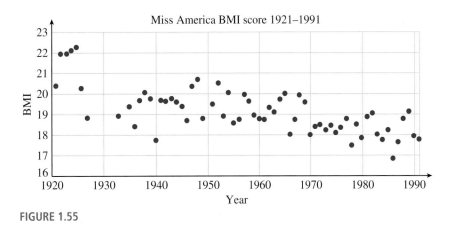

FIGURE 1.55

EXAMPLE 3

a. Estimate a line of best fit for the scatterplot in Figure 1.55. (Source: http://www.pbs.org)

b. Use your line to estimate the BMI of Miss America 1980.

Solutions

a. We draw a line that fits the data points as best we can, as shown in Figure 1.56. (Note that we have set $t = 0$ in 1920 on this graph.) We try to end up with roughly equal numbers of data points above and below our line.

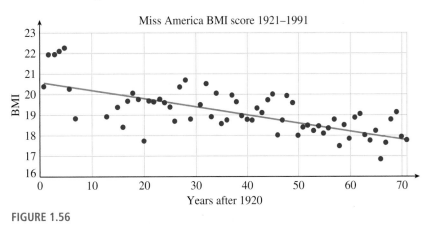

FIGURE 1.56

b. We see that when $t = 60$ on this line, the y-value is approximately 18.3. We therefore estimate that Miss America 1980 had a BMI of 18.3. (Her actual BMI was 17.85.)

Human brains consume a large amount of energy, about 16 times as much as muscle tissue per unit weight. In fact, brain metabolism accounts for about 25% of an adult human's energy needs, as compared to about 5% for other mammals. As hominid species evolved, their brains required larger and larger amounts of energy, as shown in Figure 1.57. (Source: *Scientific American*, December 2002)

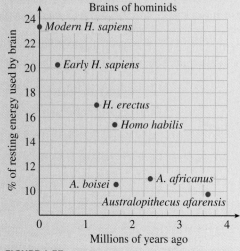

FIGURE 1.57

a. Draw a line of best fit through the data points.
b. Estimate the amount of energy used by the brain of a hominid species that lived three million years ago.

The process of predicting an output value based on a straight line that fits the data is called **linear regression,** and the line itself is called the **regression line**. The equation of the regression line is usually used (instead of a graph) to predict values.

EXAMPLE 4

a. Find the equation of the regression line in Example 3, Figure 1.56.
b. Use the regression equation to predict the BMI of Miss America 1980.

Solutions

a. We first calculate the slope by choosing two points on the regression line we drew in Figure 1.56. The points we choose are *not* necessarily any of the original data points; instead they should be points on the regression line itself. The line appears to pass through the points (17, 20) and (67, 18). The slope of the line is then

$$m = \frac{18 - 20}{67 - 17} \approx -0.04$$

Now we use the point-slope formula to find the equation of the line. (If you need to review the point-slope formula, see Section 1.5.) We substitute $m = -0.04$ and use either of the two points for (x_1, y_1); we will choose (17, 20). The equation of the regression line is

$$y = y_1 + m(x - x_1)$$
$$y = 20 - 0.04(x - 17) \quad \text{Simplify.}$$
$$y = 20.68 - 0.04t$$

b. We will use the regression equation to make our prediction. For Miss America 1980, $t = 60$ and

$$y = 20.68 - 0.04(60) = 18.28$$

This value agrees well with the estimate we made in Example 3.

EXERCISE 4

The number of manatees killed by watercraft in Florida waters has been increasing since 1975. Data are given at 5-year intervals in the table. (Source: Florida Fish and Wildlife Conservation Commission)

Year	Manatee deaths
1975	6
1980	16
1985	33
1990	47
1995	42
2000	78

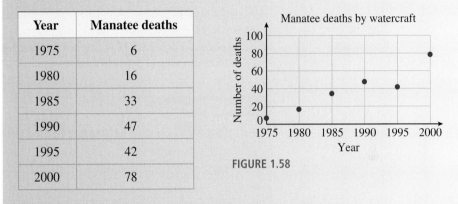

FIGURE 1.58

a. Draw a regression line through the data points shown in Figure 1.58.

b. Find an equation for the regression line, using $t = 0$ in 1975.

c. Use the regression equation to estimate the number of manatees killed by watercraft in 1998.

Linear Interpolation and Extrapolation

Using a regression line to estimate values between known data points is called **interpolation.** Making predictions beyond the range of known data is called **extrapolation.**

EXAMPLE 5

a. Use linear interpolation to estimate the BMI of Miss America 1960.

b. Use linear extrapolation to predict the BMI of Miss America 2001.

Solutions   **a.** For 1960, we substitute $t = 40$ into the regression equation we found in Example 4.

$$y = 20.68 - 0.04(40) = 19.08$$

We estimate that Miss America 1960 had a BMI of 19.08. (Her BMI was actually 18.79.)

b. For 2001, we substitute $t = 81$ into the regression equation.

$$y = 20.68 - 0.04(81) = 17.44$$

Our model predicts that Miss America 2001 had a BMI of 17.44. In fact, her BMI was 20.25. By the late 1990s, public concern over the self-image of young women had led to a reversal of the trend toward ever-thinner role models.

Example 5b illustrates an important fact about extrapolation: If we try to extrapolate too far, we may get unreasonable results. For example, if we use our model to predict the BMI of Miss America 2520 (when $t = 600$), we get

$$y = 20.68 - 0.04(600) = -3.32$$

Even if the Miss America pageant is still operating in 600 years, the winner cannot have a negative BMI. Our linear model provides a fair approximation for 1920–1990, but if we try to extrapolate too far beyond the known data, the model may no longer apply.

We can also use interpolation and extrapolation to make estimates for nonlinear functions. Sometimes a variable relationship is not linear, but a portion of its graph can be approximated by a line. The graph in Figure 1.59 shows a child's height each month. The graph is not linear because her rate of growth is not constant; her growth slows down as she approaches her adult height. However, over a short time interval the graph is close to a line, and that line can be used to approximate the coordinates of points on the curve.

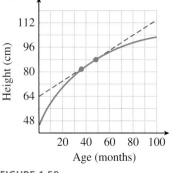

FIGURE 1.59

EXERCISE 5

Emily was 82 centimeters tall at age 36 months and 88 centimeters tall at age 48 months.

a. Find a linear equation that approximates Emily's height in terms of her age over the given time interval.

b. Use linear interpolation to estimate Emily's height when she was 38 months old, and extrapolate to predict her height at age 50 months.

c. Predict Emily's height at age 25 (300 months). Is your answer reasonable?

Estimating a line of best fit is a subjective process. Rather than base their estimates on such a line, statisticians often use the **least squares regression line.** This regression line minimizes the sum of the squares of all the vertical distances between the data points and the corresponding points on the line (see Figure 1.60). Many calculators are programmed to find the least squares regression line, using an algorithm that depends only on the data, not on the appearance of the graph.

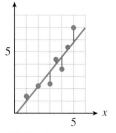

FIGURE 1.60

Using a Calculator for Linear Regression

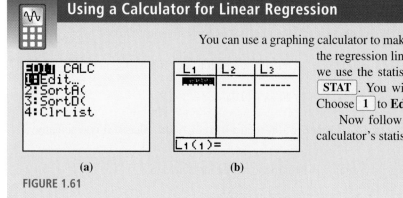

You can use a graphing calculator to make a scatterplot, find a regression line, and graph the regression line with the data points. On the TI-83 calculator, we use the statistics mode, which you can access by pressing STAT. You will see a display that looks like Figure 1.61a. Choose 1 to **Edit** (enter or alter) data.

Now follow the instructions in Example 6 for using your calculator's statistics features.

(a)   **(b)**

FIGURE 1.61

EXAMPLE 6

a. Find the equation of the least squares regression line for the following data:

$$(10, 12), \ (11, 14), \ (12, 14), \ (12, 16), \ (14, 20)$$

b. Plot the data points and the least squares regression line on the same axes.

Solutions **a.** We must first enter the data. Press [STAT] [ENTER] to select **Edit**. If there are data in column L_1 or L_2, clear them out: Use the [△] key to select L_1, press [CLEAR] [ENTER], then do the same for L_2. Enter the x-coordinates of the data points in the L_1 column and enter the y-coordinates in the L_2 column, as shown in Figure 1.62a.

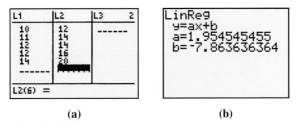

(a)   (b)

FIGURE 1.62

Now we are ready to find the regression equation for our data. Press [STAT] [▷] **4** to select linear regression, or **LinReg** $(ax + b)$, then press [ENTER]. The calculator will display the equation $y = ax + b$ and the values for a and b, as shown in Figure 1.62b. You should find that your regression line is approximately $y = 1.95x - 7.86$.

b. First, we first clear out any old definitions in the [Y=] list. Position the cursor after $Y_1 =$ and copy in the regression equation as follows: Press [VARS] 5 [▷] [▷] [ENTER]. To draw a scatterplot, press [2nd] [Y=] [1] and set the **Plot1** menu as shown in Figure 1.63a. Finally, press [ZOOM] **9** to see the scatterplot of the data and the regression line. The graph is shown in Figure 1.63b.

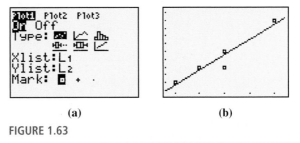

(a)   (b)

FIGURE 1.63

CAUTION

When you are through with the scatterplot, press [Y=] [△] [ENTER] to turn off the **Stat Plot**. If you neglect to do this, the calculator will continue to show the scatterplot even after you ask it to plot a new equation.

EXERCISE 6

a. Use your calculator's statistics features to find the least squares regression equation for the data in Exercise 2.

b. Plot the data and the graph of the regression equation.

For more information on using a graphing calculator for linear regression, see Appendix B.

ANSWERS TO 1.6 EXERCISES

1. a. $y = 68.19 - 3.59x$

b. −3.59 burglaries per 1000 households per year. From 1991 to 2001, the burglary rate declined by 3.59 burglaries per 1000 households every year.

2. a.

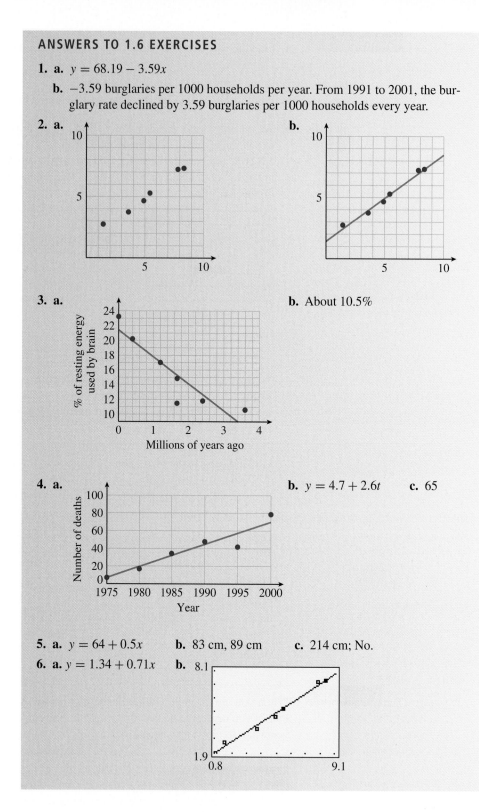

b.

3. a.

b. About 10.5%

4. a.

b. $y = 4.7 + 2.6t$ **c.** 65

5. a. $y = 64 + 0.5x$ **b.** 83 cm, 89 cm   **c.** 214 cm; No.

6. a. $y = 1.34 + 0.71x$ **b.**

VOCABULARY *Look up the definitions of new terms in the Glossary.*

Scatterplot   Least squares regression line   Extrapolate   Regression line
Interpolate   Linear regression

CONCEPTS

1. Data points may not lie exactly on the graph of an equation.
2. Points in a scatterplot may or may not exhibit a pattern.
3. We can approximate a linear pattern by a regression line.

4. We can use interpolation or extrapolation to make estimates and predictions.
5. If we extrapolate too far beyond the known data, we may get unreasonable results.

STUDY QUESTIONS

1. What is a regression line?
2. State two formulas you will need to calculate the equation of a line through two points.
3. Explain the difference between interpolation and extrapolation.

4. In general, should you have more confidence in figures obtained by interpolation or by extrapolation? Why?

SKILLS *Practice each skill in the Homework problems listed.*

1. Find the equation of a line through two points: #1–6, 29–36
2. Draw a line of best fit: #7–18

3. Find the equation of a regression line: #11–28, 37–40
4. Use interpolation and extrapolation to make predictions: #11–40

HOMEWORK 1.6

ThomsonNOW™ Test yourself on key content at **www.thomsonedu.com/login.**

In Problems 1–6, we find a linear model from two data points.
(a) Make a table showing the coordinates of two data points for the model.
   (Which variable should be plotted on the horizontal axis?)
(b) Find a linear equation relating the variables.
(c) State the slope of the line, including units, and explain its meaning in the context of the problem.

1. It cost a bicycle company $9000 to make 50 touring bikes in its first month of operation and $15,000 to make 125 bikes during its second month. Express the company's monthly production cost, C, in terms of the number, x, of bikes it makes.

2. Flying lessons cost $645 for an 8-hour course and $1425 for a 20-hour course. Both prices include a fixed insurance fee. Express the cost, C, of flying lessons in terms of the length, h, of the course in hours.

3. Under ideal conditions, Andrea's Porsche can travel 312 miles on a full tank (12 gallons of gasoline) and 130 miles on 5 gallons. Express the distance, d, Andrea can drive in terms of the amount of gasoline, g, she buys.

4. On an international flight, a passenger may check two bags each weighing 70 kilograms, or 154 pounds, and one carry-on bag weighing 50 kilograms, or 110 pounds. Express the weight, p, of a bag in pounds in terms of its weight, k, in kilograms.

5. A radio station in Detroit, Michigan, reports the high and low temperatures in the Detroit/Windsor area as 59°F and 23°F, respectively. A station in Windsor, Ontario, reports the same temperatures as 15°C and −5°C. Express the Fahrenheit temperature, F, in terms of the Celsius temperature, C.

6. Ms. Randolph bought a used car in 2000. In 2002, the car was worth $9000, and in 2005 it was valued at $4500. Express the value, V, of Ms. Randolph's car in terms of the number of years, t, she has owned it.

Each regression line can be improved by adjusting either *m* or *b*. Draw a line that fits the data points more closely.

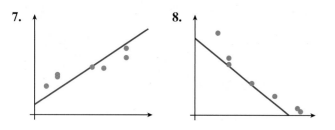

7.   8.   9.   10.

In Problems 11 and 12, use information from the graphs to answer the questions.

11. The scatterplot shows the ages of 10 army drill sergeants and the time it took each to run 100 meters, in seconds.

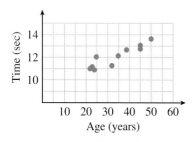

    a. What was the hundred-meter time for the 25-year-old drill sergeant?

    b. How old was the drill sergeant whose hundred-meter time was 12.6 seconds?

    c. Use a straightedge to draw a line of best fit through the data points.

    d. Use your line of best fit to predict the hundred-meter time of a 28-year-old drill sergeant.

    e. Choose two points on your regression line and find its equation.

    f. Use the equation to predict the hundred-meter time of a 40-year-old drill sergeant and a 12 year-old drill sergeant. Are these predictions reasonable?

12. The scatterplot shows the outside temperature and the number of cups of cocoa sold at an outdoor skating rink snack bar on 13 consecutive nights.

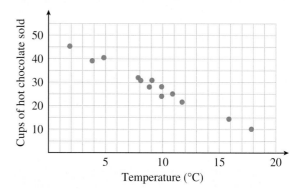

    a. How many cups of cocoa were sold when the temperature was 2°C?

    b. What was the temperature on the night when 25 cups of cocoa were sold?

    c. Use a straightedge to draw a line of best fit through the data points.

    d. Use your line of best fit to predict the number of cups of cocoa that will be sold at the snack bar if the temperature is 7°C.

    e. Choose two points on your regression line and find its equation.

    f. Use the equation to predict the number of cups of cocoa that will be sold when the temperature is 10°C and when the temperature is 24°C. Are these predictions reasonable?

13. With Americans' increased use of faxes, pagers, and cell phones, new area codes are being created at a steady rate. The table shows the number of area codes in the United States each year. (Source: *USA Today,* NeuStar, Inc.)

Year	1997	1998	1999	2000
Number of area codes	151	186	204	226

Year	2001	2002	2003
Number of area codes	239	262	274

a. Let t represent the number of years after 1995 and plot the data. Draw a line of best fit for the data points.

b. Find an equation for your regression line.

c. How many area codes do you predict for 2010?

14. The number of mobile homes in the United States has been increasing since 1960. The data in the table are given in millions of mobile homes. (Source: *USA Today,* U.S. Census Bureau)

Year	1960	1970	1980	1990	2000
Number of mobile homes	0.8	2.1	4.7	7.4	8.8

a. Let t represent the number of years after 1960 and plot the data. Draw a line of best fit for the data points.

b. Find an equation for your regression line.

c. How many mobile homes do you predict for 2010?

15. Teenage birth rates in the United States declined from 1991 to 2000. The table shows the number of births per 1000 women in selected years. (Source: U.S. National Health Statistics)

Year	1991	1993	1995	1996	1997	1998
Births	62.1	59.6	56.8	54.4	52.3	51.1

a. Let t represent the number of years after 1990 and plot the data. Draw a line of best fit for the data points.

b. Find an equation for your regression line.

c. Estimate the teen birth rate in 1994.

d. Predict the teen birth rate in 2010.

16. The table shows the minimum wage in the United States at five-year intervals. (Source: Economic Policy Institute)

Year	1960	1965	1970	1975	1980
Minimum wage	1.00	1.25	1.60	2.10	3.10

Year	1985	1990	1995	2000
Minimum wage	3.35	3.80	4.25	5.15

a. Let t represent the number of years after 1960 and plot the data. Draw a line of best fit for the data points.

b. Find an equation for your regression line.

c. Estimate the minimum wage in 1972.

d. Predict the minimum wage in 2010.

17. Life expectancy in the United States has been rising since the nineteenth century. The table shows the U.S. life expectancy in selected years. (Source: http://www. infoplease.com)

Year	1950	1960	1970	1980	1990	2000
Life expectancy at birth	68.2	69.7	70.8	73.7	75.4	77

a. Let t represent the number of years after 1950, and plot the data. Draw a line of best fit for the data points.
b. Find an equation for your regression line.
c. Estimate the life expectancy of someone born in 1987.
d. Predict the life expectancy of someone born in 2010.

18. The table shows the per capita cigarette consumption in the United States at five-year intervals. (Source: http://www. infoplease.com)

Year	1980	1985	1990	1995	2000
Per capita cigarette consumption	3,851	3,461	2,827	2,515	2,092

a. Let t represent the number of years after 1980, and plot the data. Draw a line of best fit for the data points.
b. Find an equation for your regression line.
c. Estimate the per capita cigarette consumption in 1998.
d. Predict the per capita cigarette consumption in 2010.

19. "The earnings gap between high-school and college graduates continues to widen, the Census Bureau says. On average, college graduates now earn just over $51,000 a year, almost twice as much as high-school graduates. And those with no high-school diploma have actually seen their earnings drop in recent years." The table shows the unemployment rate and the median weekly earnings for employees with different levels of education. (Source: Morning Edition, National Public Radio, March 28, 2005)

a. Plot years of education on the horizontal axis and weekly earnings on the vertical axis.
b. Find an equation for the regression line.
c. State the slope of the regression line, including units, and explain what it means in the context of the data.
d. Do you think this model is useful for extrapolation or interpolation? For example, what weekly earnings does the model predict for someone with 15 years of education? For 25 years? Do you think these predictions are valid? Why or why not?

	Years of education	Unemployment rate	Weekly earnings ($)
Some high school, no diploma	10	8.8	396
High-school graduate	12	5.5	554
Some college, no degree	13	5.2	622
Associate's degree	14	4.0	672
Bachelor's degree	16	3.3	900
Master's degree	18	2.9	1064
Professional degree	20	1.7	1307

20. The table shows the birth rate (in births per woman) and the female literacy rate (as a percent of the adult female population) in a number of nations. (Source: UNESCO, *The World Fact Book,* EarthTrends)

 a. Plot the data with literacy rate on the horizontal axis. Draw a line of best fit for the data points.
 b. Find an equation for the regression line.
 c. What values for the input variable make sense for the model? What are the largest and smallest values predicted by the model for the output variable?
 d. State the slope of the regression line, including units, and explain what it means in the context of the data.

Country	Literacy rate	Birth rate
Brazil	88.6	1.93
Egypt	43.6	2.88
Germany	99	1.39
Iraq	53	4.28
Japan	99	1.39
Saudi Arabia	69.3	4.05
Niger	9.4	6.75
Pakistan	35.2	4.14
Peru	82.1	2.56
Philippines	92.7	3.16
Portugal	91	1.47
Russian Federation	99.2	1.27
United States	97	2.08

21. The table shows the amount of carbon released into the atmosphere annually from burning fossil fuels, in billions of tons, at 5-year intervals from 1950 to 1995. (Source: www.worldwatch.org)

Year	50	55	60	65	70
Carbon emissions	1.6	2.0	2.5	3.1	4.0

Year	75	80	85	90	95
Carbon emissions	4.5	5.2	5.3	5.9	6.2

 a. Let t represent the number of years after 1950 and plot the data. Draw a line of best fit for the data points.
 b. Find an equation for your regression line.
 c. Estimate the amount of carbon released in 1992.

22. High-frequency radiation is harmful to living things because it can cause changes in their genetic material. The data below, collected by C. P. Oliver in 1930, show the frequency of genetic transmutations induced in fruit flies by doses of X-rays, measured in roentgens. (Source: C. P. Oliver, 1930)

Dosage (roentgens)	285	570	1640	3280	6560
Percent of mutated genes	1.18	2.99	4.56	9.63	15.85

 a. Plot the data and draw a line of best fit through the data points.
 b. Find the equation of your regression line.
 c. Use the regression equation to predict the percent of mutations that might result from exposure to 5000 roentgens of radiation.

23. Bracken, a type of fern, is one of the most successful plants in the world, growing on every continent except Antarctica. New plants, which are genetic clones of the original, spring from a network of underground stems, or rhizomes, to form a large circular colony. The graph shows the diameters of various colonies plotted against their age. (Source: Chapman et al.,1992)

a. Calculate the rate of growth of the diameter of a bracken colony, in meters per year.

b. Find an equation for the line of best fit. (What should the vertical intercept of the line be?)

c. In Finland, bracken colonies over 450 meters in diameter have been found. How old are these colonies?

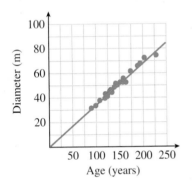

24. The European sedge warbler can sing several different songs consisting of trills, whistles, and buzzes. Male warblers who sing the largest number of songs are the first to acquire mates in the spring. The data below show the number of different songs sung by several male warblers and the day on which they acquired mates, where day 1 is April 20. (Source: Krebs and Davies, 1993)

Number of songs	41	38	34	32	30
Pairing day	20	24	25	21	24

Number of songs	25	24	24	23	14
Pairing day	27	31	35	40	42

a. Plot the data points, with number of songs on the horizontal axis. A regression line for the data is $y = -0.85x + 53$. Graph this line on the same axes with the data.

b. What does the slope of the regression line represent?

c. When can a sedge warbler that knows 10 songs expect to find a mate?

d. What do the intercepts of the regression line represent? Do these values make sense in context?

25. One of the factors that determines the strength of a muscle is its cross-sectional area. The data below show the cross-sectional area of the arm flexor muscle for several men and women, and their strength, measured by the maximum force they exerted against a resistance.

(Source: Davis, Kimmet, Autry, 1986)

Women	Area (sq cm)	11.5	10.8	11.7	12.0	12.5
	Strength (kg)	11.3	13.2	13.2	14.5	15.6
Men	Area (sq cm)	13.5	13.8	15.4	15.4	17.7
	Strength (kg)	15.0	17.3	19.0	19.8	20.6

Women	Area (sq cm)	12.7	14.4	14.4	15.7
	Strength (kg)	14.8	15.6	16.1	18.4
Men	Area (sq cm)	18.6	20.8	–	–
	Strength (kg)	20.8	26.3	–	–

a. Plot the data for both men and women on the same graph using different symbols for the data points for men and the data points for women.

b. Are the data for both men and women described reasonably well by the same regression line? Draw a line of best fit through the data.

c. Find the equation of your line of best fit, or use a calculator to find the regression line for the data.

d. What does the slope mean in this context?

26. Astronomers use a numerical scale called magnitude to measure the brightness of a star, with brighter stars assigned smaller magnitudes. When we view a star from Earth, dust in the air absorbs some of the light, making the star appear fainter than it really is. Thus, the observed magnitude of a star, m, depends on the distance its light rays must travel through the Earth's atmosphere. The observed magnitude is given by

$$m = m_0 + kx$$

where m_0 is the actual magnitude of the star outside the atmosphere, x is the air mass (a measure of the distance through the atmosphere), and k is a constant called the *extinction coefficient*. To calculate m_0, astronomers observe the same object several times during the night at different positions in the sky, and hence for different values of x. Here are data from such observations.

(Source: Karttunen et al., 1987)

Altitude	Air mass, x	Magnitude, m
50°	1.31	0.90
35°	1.74	0.98
25°	2.37	1.07
20°	2.92	1.17

a. Plot observed magnitude against air mass, and draw a line of best fit through the data.

b. Find the equation of your line of best fit, or use a calculator to find the regression line for the data.

c. What is the value of the extinction coefficient? What is the apparent magnitude of the star outside Earth's atmosphere?

27. Six students are trying to identify an unknown chemical compound by heating the substance and measuring the density of the gas that evaporates. (*Density = mass/volume.*) The students record the mass lost by the solid substance and the volume of the gas that evaporated from it. They know that the mass lost by the solid must be the same as the mass of the gas that evaporated. (Source: Hunt and Sykes, 1984)

Student	A	B	C	D	E	F
Volume of gas (cm^3)	48	60	24	81	76	54
Loss in mass (mg)	64	81	32	107	88	72

   a. Plot the data with volume on the horizontal axis. Which student made an error in the experiment?
   b. Ignoring the incorrect data point, draw a line of best fit through the other points.
   c. Find an equation of the form $y = kx$ for the data. Why should you expect the regression line to pass through the origin?
   d. Use your equation to calculate the mass of 1000 cm^3 (one liter) of the gas.
   e. Here are the densities of some gases at room temperature:

Hydrogen	8	mg/liter
Nitrogen	1160	mg/liter
Oxygen	1330	mg/liter
Carbon dioxide	1830	mg/liter

   Which of these might have been the gas that evaporated from the unknown substance? (*Hint:* Use your answer to part d to calculate the density of the gas. 1 cm^3 = 1 millimeter.)

28. The formulas for many chemical compounds involve ratios of small integers. For example, the formula for water, H_2O, means that two atoms of hydrogen combine with one atom of oxygen to make one water molecule. Similarly, magnesium and oxygen combine to produce magnesium oxide. In this problem, we will discover the chemical formula for magnesium oxide. (Source: Hunt and Sykes, 1984)

   a. Twenty-four grams of magnesium contain the same number of atoms as sixteen grams of oxygen. Complete the table showing the amount of oxygen needed if the formula for magnesium oxide is MgO, Mg_2O, or MgO_2.

Grams of Mg	Grams of O (if MgO)	Grams of O (if Mg_2O)	Grams of O (if MgO_2)
24	16		
48			
12			
6			

   b. Graph three lines on the same axes to represent the three possibilities, with grams of magnesium on the horizontal axis and grams of oxygen on the vertical axis.
   c. Here are the results of some experiments synthesizing magnesium oxide.

Experiment	Grams of magnesium	Grams of oxygen
1	15	10
2	22	14
3	30	20
4	28	18
5	10	6

   Plot the data on your graph from part (b). Which is the correct formula for magnesium oxide?

For Problems 29–32,
(a) Use linear interpolation to give approximate answers.
(b) What is the meaning of the slope in the context of the problem?

29. The temperature in Encino dropped from 81°C at 1 a.m. to 73°C at 5 a.m. Estimate the temperature at 4 a.m.

30. Newborn blue whales are about 24 feet long and weigh 3 tons. The young whale nurses for 7 months, at which time it is 53 feet long. Estimate the length of a 1-year-old blue whale.

31. A car starts from a standstill and accelerates to a speed of 60 miles per hour in 6 seconds. Estimate the car's speed 2 seconds after it began to accelerate.

32. A truck on a slippery road is moving at 24 feet per second when the driver steps on the brakes. The truck needs 3 seconds to come to a stop. Estimate the truck's speed 2 seconds after the brakes were applied.

In Problems 33–36, use linear interpolation or extrapolation to answer the questions.

33. The temperature of an automobile engine is 9° Celsius when the engine is started and 51°C 7 minutes later. Use a linear model to predict the engine temperature for both 2 minutes and 2 hours after it started. Are your predictions reasonable?

34. The temperature in Death Valley is 95° Fahrenheit at 5 a.m. and rises to 110° Fahrenheit by noon. Use a linear model to predict the temperature at 2 p.m. and at midnight. Are your predictions reasonable?

35. Ben weighed 8 pounds at birth and 20 pounds at age 1 year. How much will he weigh at age 10 if his weight increases at a constant rate?

36. The elephant at the City Zoo becomes ill and loses weight. She weighed 10,012 pounds when healthy and only 9641 pounds a week later. Predict her weight after 10 days of illness.

37. Birds' nests are always in danger from predators. If there are other nests close by, the chances of predators finding the nest increase. The table shows the probability of a nest being found by predators and the distance to the nearest neighboring nest. (Source: Perrins, 1979)

Distance to nearest neighbor (meters)	20	40	60	80	100
Probability of predators (%)	47	34	32	17	1.5

   a. Plot the data and the least squares regression line.
   b. Use the regression line to estimate the probability of predators finding a nest if its nearest neighbor is 50 meters away.
   c. If the probability of predators finding a nest is 10%, how far away is its nearest neighbor?
   d. What is the probability of predators finding a nest if its nearest neighbor is 120 meters away? Is your answer reasonable?

38. A trained cyclist pedals faster as he increases his cycling speed, even with a multiple-gear bicycle. The table shows the pedal frequency, p (in revolutions per minute), and the cycling speed, c (in kilometers per hour), of one cyclist. (Source: Pugh, 1974)

Speed (km/hr)	8.8	12.5	16.2	24.4	31.9	35.0
Pedal frequency (rpm)	44.5	50.7	60.6	77.9	81.9	95.3

   a. Plot the data and the least squares regression line.
   b. Estimate the cyclist's pedal frequency at a speed of 20 kilometers per hour.
   c. Estimate the cyclist's speed when he is pedaling at 70 revolutions per minute.
   d. Does your regression line give a reasonable prediction for the pedaling frequency when the cyclist is not moving? Explain.

39. In this problem we will calculate the efficiency of swimming as a means of locomotion. A swimmer generates power to maintain a constant speed in the water. If she must swim against an opposing force, the power increases. The following table shows the power expended by a swimmer while working against different amounts of force. (A positive force opposes the swimmer, and a negative force helps her.) (Source: diPrampero et al., 1974, and Alexander, 1992)

Force (newtons)	−3.5	0	0	6	8	10	17	17
Metabolic power (watts)	100	190	230	320	380	450	560	600

  a. Plot the data on the grid, or use the **StatPlot** feature on your calculator. Use your calculator to find the least squares regression line. Graph the regression line on top of the data.
  b. Use your regression line to estimate the power needed for the swimmer to overcome an opposing force of 15 newtons.
  c. Use your regression line to estimate the power generated by the swimmer when there is no force either hindering or helping her.
  d. Estimate the force needed to tow the swimmer at 0.4 meters per second while she rests. (If she is resting, she is not generating any power).
  e. The swimmer's **mechanical** power (or rate of work) is computed by multiplying her speed times the force needed to tow her at rest. Use your answer to part (d) to calculate the mechanical power she generates by swimming at 0.4 meters per second.
  f. The ratio of mechanical power to metabolic power is a measure of the swimmer's efficiency. Compute the efficiency of the swimmer when there is no external force opposing or helping her.

40. In this problem, we calculate the amount of energy generated by a cyclist. An athlete uses oxygen slowly when resting but more quickly during physical exertion. In an experiment, several trained cyclists took turns pedaling on a bicycle ergometer, which measures their work rate. The table shows the work rate of the cyclists, in watts, measured against their oxygen intake, in liters per minute. (Source: Pugh, 1974)

Oxygen consumption (liters/min)	1	1.7	2	3.3	3.9	3.6	4.3	5
Work rate (watts)	40	100	180	220	280	300	320	410

  a. Plot the data on the grid, or use the **StatPlot** feature on your calculator. Use your calculator to find the least squares regression line. Graph the regression line on top of the data.
  b. Find the horizontal intercept of the regression line. What does the horizontal intercept tell you about this situation?
  c. Estimate the power produced by a cyclist consuming oxygen at 5.9 liters per minute.
  d. What is the slope of the regression line? The slope represents the amount of power, in watts, generated by a cyclist for each liter of oxygen consumed per minute. How many watts of power does a cyclist generate from each liter of oxygen?
  e. One watt of power represents an energy output of one joule per second. How many joules of energy does the cyclist generate in one minute?
  f. How many joules of energy can be extracted from each cubic centimeter of oxygen used? (One liter is equal to 1000 cubic centimeters.)

KEY CONCEPTS

1. We can describe a relationship between variables with a table of values, a graph, or an equation.
2. Linear models have equations of the following form:

$$y = \text{(starting value)} + \text{(rate of change)} \cdot x$$

3. The **general form** for a linear equation is $Ax + By = C$.
4. We can use the **intercepts** to graph a line. The intercepts are also useful for interpreting a model.
5. A **function** is a rule that assigns to each value of the input variable a unique value of the output variable.
6. Function notation: $y = f(x)$, where x is the input and y is the output.
7. The point (a, b) lies on the graph of the function f if and only if $f(a) = b$.
8. Each point on the graph of the function f has coordinates $(x, f(x))$ for some value of x.
9. The **vertical line test** tells us whether a graph represents a function.
10. Lines have constant slope.
11. The slope of a line gives us the **rate of change** of one variable with respect to another.

12.

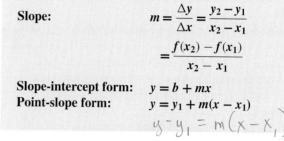

Formulas for Linear Functions

Slope:	$m = \dfrac{\Delta y}{\Delta x} = \dfrac{y_2 - y_1}{x_2 - x_1}$
	$= \dfrac{f(x_2) - f(x_1)}{x_2 - x_1}$
Slope-intercept form:	$y = b + mx$
Point-slope form:	$y = y_1 + m(x - x_1)$

$$y - y_1 = m(x - x_1)$$

13. The **slope-intercept form** is useful when we know the initial value and the rate of change.
14. The **point-slope form** is useful when we know the rate of change and one point on the line.
15. Linear functions form a **two-parameter family**, $f(x) = b + mx$.
16. We can approximate a linear pattern by a **regression line**.
17. We can use **interpolation** or **extrapolation** to make estimates and predictions.
18. If we extrapolate too far beyond the known data, we may get unreasonable results.

REVIEW PROBLEMS

Write and graph a linear equation for each situation. Then answer the questions.

1. Last year, Pinwheel Industries introduced a new model calculator. It cost $2000 to develop the calculator and $20 to manufacture each one.
   a. Complete the table of values showing the total cost, C, of producing n calculators.

n	100	500	800	1200	1500
C					

   b. Write an equation that expresses C in terms of n.
   c. Graph the equation by hand.
   d. What is the cost of producing 1000 calculators? Illustrate this as a point on your graph.
   e. How many calculators can be produced for $10,000? Illustrate this as a point on your graph.

2. Megan weighed 5 pounds at birth and gained 18 ounces per month during her first year.
   a. Complete the table of values for Megan's weight, w, in terms of her age, m, in months.

m	2	4	6	9	12
w					

   b. Write an equation that expresses w in terms of m.
   c. Graph the equation by hand.
   d. How much did Megan weigh at 9 months? Illustrate this as a point on your graph.
   e. When did Megan weigh 9 pounds? Illustrate this as a point on your graph.

3. The total amount of oil remaining in 2005 is estimated at 2.1 trillion barrels, and total annual consumption is about 28 billion barrels.

  a. Assuming that oil consumption continues at the same level, write an equation for the remaining oil, R, as a function of time, t (in years since 2005).

  b. Find the intercepts and graph the equation by hand.

  c. What is the significance of the intercepts to the world's oil supply?

4. The world's copper reserves were 950 million tons in 2004; total annual consumption was 16.8 million tons.

  a. Assuming that copper consumption continues at the same level, write an equation for the remaining copper reserves, R, as a function of time, t (in years since 2004).

  b. Find the intercepts and graph the function by hand.

  c. What is the significance of the intercepts to the world's copper supply?

5. The owner of a movie theater needs to bring in $1000 at each screening in order to stay in business. He sells adult tickets at $5 apiece and children's tickets at $2 each.

  a. Write an equation that relates the number of adult tickets, A, he must sell and the number of children's tickets, C.

  b. Find the intercepts and graph the equation by hand.

  c. If the owner sells 120 adult tickets, how many children's tickets must he sell?

  d. What is the significance of the intercepts to the sale of tickets?

6. Alida plans to spend part of her vacation in Atlantic City and part in Saint-Tropez. She estimates that after airfare her vacation will cost $60 per day in Atlantic City and $100 per day in Saint-Tropez. She has $1200 to spend after airfare.

  a. Write an equation that relates the number of days, C, Alida can spend in Atlantic City and the number of days, T, in Saint-Tropez.

  b. Find the intercepts and graph the equation by hand.

  c. If Alida spends 10 days in Atlantic City, how long can she spend in Saint-Tropez?

  d. What is the significance of the intercepts to Alida's vacation?

Graph each equation on graph paper. Use the most convenient method for each problem.

7. $4x - 3y = 12$

8. $\dfrac{x}{6} - \dfrac{y}{12} = 1$

9. $50x = 40y - 20{,}000$

10. $1.4x + 2.1y = 8.4$

11. $3x - 4y = 0$

12. $x = -4y$

13. $4x = -12$

14. $2y - x = 0$

Which of the following tables describe functions? Explain.

15.

x	−2	−1	0	1	2	3
y	6	0	1	2	6	8

16.

p	3	−3	2	−2	−2	0
q	2	−1	4	−4	3	0

17.

Student	Score on IQ test	Score on SAT test
(A)	118	649
(B)	98	450
(C)	110	590
(D)	105	520
(E)	98	490
(F)	122	680

18.

Student	Correct answers on math quiz	Quiz grade
(A)	13	85
(B)	15	89
(C)	10	79
(D)	12	82
(E)	16	91
(F)	18	95

19. The total number of barrels of oil pumped by the AQ oil company is given by the formula $N(t) = 2000 + 500t$, where N is the number of barrels of oil t days after a new well is opened. Evaluate $N(10)$ and explain what it means.

20. The number of hours required for a boat to travel upstream between two cities is given by the formula $H(v) = \dfrac{24}{v - 8}$, where v represents the boat's top speed in miles per hour. Evaluate $H(16)$ and explain what it means.

Which of the following graphs represent functions?

21.
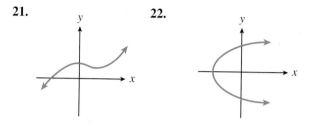

22.

23.
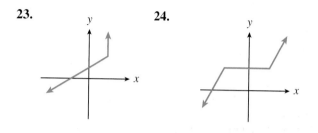

24.

Evaluate each function for the given values.

25. $F(t) = \sqrt{1 + 4t^2}$, $F(0)$ and $F(-3)$

26. $G(x) = \sqrt[3]{x - 8}$, $G(0)$ and $G(20)$

27. $h(v) = 6 - |4 - 2v|$, $h(8)$ and $h(-8)$

28. $m(p) = \dfrac{120}{p + 15}$, $m(5)$ and $m(-40)$

Refer to the graphs shown for Problems 29 and 30.

29. a. Find $f(-2)$ and $f(2)$.
  b. For what value(s) of t is $f(t) = 4$?
  c. Find the t- and $f(t)$-intercepts of the graph.
  d. What is the maximum value of f? For what value(s) of t does f take on its maximum value?

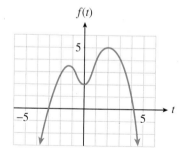

30. a. Find $P(-3)$ and $P(3)$.
  b. For what value(s) of z is $P(z) = 2$?
  c. Find the z- and $P(z)$-intercepts of the graph.
  d. What is the minimum value of P? For what value(s) of z does P take on its minimum value?

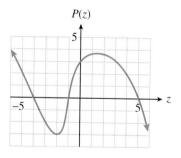

Graph the given function on a graphing calculator. Then use the graph to solve the equations and inequalities. Round your answers to one decimal place if necessary.

31. $y = \sqrt[3]{x}$
   a. Solve $\sqrt[3]{x} = 0.8$ **b.** Solve $\sqrt[3]{x} = 1.5$
   c. Solve $\sqrt[3]{x} > 1.7$ **d.** Solve $\sqrt[3]{x} \leq 1.26$

32. $y = \dfrac{1}{x}$
   a. Solve $\dfrac{1}{x} = 2.5$.   **b.** Solve $\dfrac{1}{x} = 0.3125$.
   c. Solve $\dfrac{1}{x} \geq 0.\overline{2}$.   **d.** Solve $\dfrac{1}{x} < 5$.

33. $f(x) = \dfrac{1}{x^2}$
   a. Solve $\dfrac{1}{x^2} = 0.03$.   **b.** Solve $\dfrac{1}{x^2} = 6.25$.
   c. Solve $\dfrac{1}{x^2} > 0.16$.   **d.** Solve $\dfrac{1}{x^2} \leq 4$.

34. $g(x) = \sqrt{x}$
   a. Solve $\sqrt{x} = 0.707$.   **b.** Solve $\sqrt{x} = 1.7$.
   c. Solve $\sqrt{x} < 1.5$.   **d.** Solve $\sqrt{x} \geq 1.3$.

Evaluate each function.

35. $H(t) = t^2 + 2t$, $H(2a)$ and $H(a + 1)$

36. $F(x) = 2 - 3x$, $F(2) + F(3)$ and $F(2 + 3)$

37. $f(x) = 2x^2 - 4$, $f(a) + f(b)$ and $f(a + b)$

38. $G(t) = 1 - t^2$, $G(3w)$ and $G(s + 1)$

39. A spiked volleyball travels 6 feet in 0.04 seconds. A pitched baseball travels 66 feet in 0.48 seconds. Which ball travels faster?

40. Kendra needs $4\frac{1}{2}$ gallons of Luke's Brand primer to cover 1710 square feet of wall. She uses $5\frac{1}{3}$ gallons of Slattery's Brand primer for 2040 square feet of wall. Which brand covered more wall per gallon?

41. Which is steeper, Stone Canyon Drive, which rises 840 feet over a horizontal distance of 1500 feet, or Highway 33, which rises 1150 feet over a horizontal distance of 2000 feet?

42. The top of Romeo's ladder is on Juliet's window sill that is 11 feet above the ground, and the bottom of the ladder is 5 feet from the base of the wall. Is the incline of this ladder as steep as a firefighter's ladder that rises a height of 35 feet over a horizontal distance of 16 feet?

43. The table shows the amount of oil, B (in thousands of barrels), left in a tanker t minutes after it hits an iceberg and springs a leak.

t	0	10	20	30
B	800	750	700	650

   a. Write a linear function for B in terms of t.
   b. Choose appropriate window settings on your calculator and graph your function.
   c. Give the slope of the graph, including units, and explain the meaning of the slope in terms of the oil leak.

44. A traditional first experiment for chemistry students is to make 98 observations about a burning candle. Delbert records the height, h, of the candle in inches at various times t minutes after he lit it.

t	0	10	30	45
h	12	11.5	10.5	9.75

   a. Write a linear function for h in terms of t.
   b. Choose appropriate window settings on your calculator and graph your function.
   c. Give the slope of the graph, including units, and explain the meaning of the slope in terms of the candle.

45. An interior decorator bases her fee on the cost of a remodeling job. The accompanying table shows her fee, F, for jobs of various costs, C, both given in dollars.

C	5000	10,000	20,000	50,000
F	1000	1500	2500	5500

  a. Write a linear function for F in terms of C.
  b. Choose appropriate window settings on your calculator and graph your function.
  c. Give the slope of the graph, and explain the meaning of the slope in terms of the decorator's fee.

46. Auto registration fees in Connie's home state depend on the value of the automobile. The table below shows the registration fee, R, for a car whose value is V, both given in dollars.

V	5000	10,000	15,000	20,000
R	135	235	335	435

  a. Write a linear function for R in terms of V.
  b. Choose appropriate window settings on your calculator and graph your function.
  c. Give the slope of the graph, and explain the meaning of the slope in terms of the registration fee.

Find the slope of the line segment joining each pair of points.

47. $(-1, 4), (3, -2)$ **48.** $(5, 0), (2, -6)$ **49.** $(6.2, 1.4), (-2.1, 4.8)$ **50.** $(0, -6.4), (-5.6, 3.2)$

51. The planners at AquaWorld want the small water slide to have a slope of 25%. If the slide is 20 feet tall, how far should the end of the slide be from the base of the ladder?

52. In areas with heavy snowfall, the pitch (or slope) of the roof of an A-frame house should be at least 1.2. If a small ski chalet is 40 feet wide at its base, how tall is the center of the roof?

Find the coordinates of the indicated points, and then write an algebraic expression using function notation for the indicated quantities.

53. **a.** Δy as x increases from x_1 to x_2
     b. The slope of the line segment joining P and Q

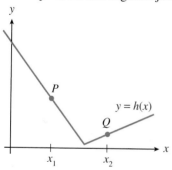

54. **a.** Δy as x increases from 2 to $2 + h$
     b. The slope of the line segment joining P and Q

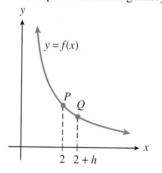

Which of the following tables represent linear functions?

55. **a.**

r	E
1	5
2	$\frac{5}{2}$
3	$\frac{5}{3}$
4	$\frac{5}{4}$
5	1

b.

s	t
10	6.2
20	9.7
30	12.6
40	15.8
50	19.0

56. **a.**

w	A
2	-13
4	-23
6	-33
8	-43
10	-53

b.

x	C
0	0
2	5
4	10
8	20
16	40

Each table gives values for a linear function. Fill in the missing values.

57.

d	V
−5	−4.8
−2	−3
	−1.2
6	1.8
10	

58.

q	S
−8	−8
−4	56
3	
	200
9	264

Find the slope and *y*-intercept of each line.

59. $2x - 4y = 5$

60. $\dfrac{1}{2}x + \dfrac{2}{3}y = \dfrac{5}{6}$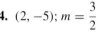

61. $8.4x + 2.1y = 6.3$

62. $y - 3 = 0$

For Problems 63 and 64,
(a) Graph by hand the line that passes through the given point with the given slope.
(b) Find an equation for the line.

63. $(-4, 6);\ m = -\dfrac{2}{3}$

64. $(2, -5);\ m = \dfrac{3}{2}$

For Problems 65 and 66,
(a) Find the slope and *y*-intercept of each line.
(b) Write an equation for the line.

65.

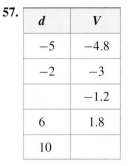

66.

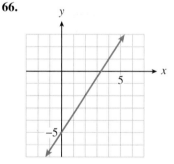

67. What is the slope of the line whose intercepts are $(-5, 0)$ and $(0, 3)$?

68. a. Find the *x*- and *y*-intercepts of the line $\dfrac{x}{4} - \dfrac{y}{6} = 1$.
   b. What is the slope of the line in part (a)?

69. a. What is the slope of the line $y = 2 + \dfrac{3}{2}(x - 4)$?
   b. Find the point on the line whose *x*-coordinate is 4. Can there be more than one such point?
   c. Use your answers from parts (a) and (b) to find another point on the line.

70. A line passes through the point $(-5, 3)$ and has slope $\dfrac{2}{3}$. Find the coordinates of two more points on the line.

71. A line passes through the point $(-2, -6)$ and has slope $-\frac{8}{5}$. Find the coordinates of two more points on the line.

72. Find an equation in point-slope form for the line of slope $\frac{6}{5}$ that passes through $(-3, -4)$.

73. The rate at which air temperature decreases with altitude is called the lapse rate. In the troposphere, the layer of atmosphere that extends from the Earth's surface to a height of about 7 miles, the lapse rate is about $3.6°F$ for every 1000 feet. (Source: Ahrens, 1998)

    a. If the temperature on the ground is $62°F$, write an equation for the temperature, T, at an altitude of h feet.

    b. What is the temperature outside an aircraft flying at an altitude of 30,000 feet? How much colder is that than the ground temperature?

    c. What is the temperature at the top of the troposphere?

74. In his television program *Notes from a Small Island*, aired in February 1999, Bill Bryson discussed the future of the British aristocracy. Because not all families produce an heir, 4 or 5 noble lines die out each year. At this rate, Mr. Bryson says, if no more peers are created, there will be no titled families left by the year 2175.

    a. Assuming that on average 4.5 titled families die out each year, write an equation for the number, N, of noble houses left in year t, where $t = 0$ in the year 1999.

    b. Graph your equation.

    c. According to your graph, how many noble families existed in 1999? Which point on the graph corresponds to this information?

Find an equation for the line passing through the two given points.

75. $(3, -5), (-2, 4)$

76. $(0, 8), (4, -2)$

For Problems 77 and 78,
(a) Make a table of values showing two data points.
(b) Find a linear equation relating the variables.
(c) State the slope of the line, including units, and explain its meaning in the context of the problem.

77. The population of Maple Rapids was 4800 in 1990 and had grown to 6780 by 2005. Assume that the population increases at a constant rate. Express the population, P, of Maple Rapids in terms of the number of years, t, since 1990.

78. Cicely's odometer read 112 miles when she filled up her 14-gallon gas tank and 308 when the gas gauge read half full. Express her odometer reading, m, in terms of the amount of gas, g, she used.

79. In 1986, the space shuttle *Challenger* exploded because of O-ring failure on a morning when the temperature was about $30°F$. Previously, there had been one incident of O-ring failure when the temperature was $70°F$ and three incidents when the temperature was $54°F$. Use linear extrapolation to estimate the number of incidents of O-ring failure you would expect when the temperature is $30°F$.

80. Thelma typed a 19-page technical report in 40 minutes. She required only 18 minutes for an 8-page technical report. Use linear interpolation to estimate how long Thelma would require to type a 12-page technical report.

81. The scatterplot shows weights (in pounds) and heights (in inches) for a team of distance runners.

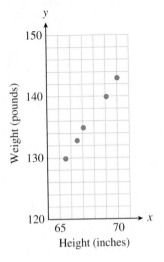

Height (inches)

a. Use a straightedge to draw a line that fits the data.
b. Use your line to predict the weight of a 65-inch-tall runner and the weight of a 71-inch-tall runner.
c. Use your answers from part (b) to approximate the equation of a regression line.
d. Use your answer to part (c) to predict the weight of a runner who is 68 inches tall.
e. The points on the scatterplot are (65.5, 130), (66.5, 133), (67, 135), (69, 140), and (70, 143). Use your calculator to find the least squares regression line.
f. Use the regression line to predict the weight of a runner who is 68 inches tall.

82. The scatterplot shows best times for various women running 400 meters and 100 meters.

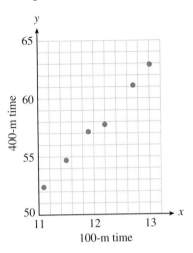

100-m time

a. Use a straightedge to draw a line that fits the data.
b. Use your line to predict the 400-meter time of a woman who runs the 100-meter dash in 11.2 seconds and the 400-meter time of a woman who runs the 100-meter dash in 13.2 seconds.
c. Use your answers from part (b) to approximate the equation of a regression line.
d. Use your answer to part (c) to predict the 400-meter time of a woman who runs the 100-meter dash in 12.1 seconds.
e. The points on the scatterplot are (11.1, 52.4), (11.5, 54.7), (11.9, 57.4), (12.2, 57.9), (12.7, 61.3), and (13.0, 63.0). Use your calculator to find the least squares regression line.
f. Use the regression line to predict the 400-meter time of a woman who runs the 100-meter dash in 12.1 seconds.

83. *Archaeopteryx* is an extinct creature with characteristics of both birds and reptiles. Only six fossil specimens are known, and only five of those include both a femur (leg bone) and a humerus (forearm bone) The scatterplot shows the lengths of femur and humerus for the five *Archaeopteryx* specimens.

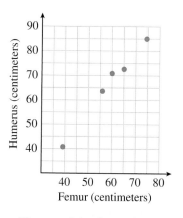

a. Use a straightedge to draw a line that fits the data.

b. Predict the humerus length of an *Archaeopteryx* whose femur is 40 centimeters.

c. Predict the humerus length of an *Archaeopteryx* whose femur is 75 centimeters.

d. Use your answers from parts (a) and (b) to find the equation of a regression line.

e. Use your answer to part (d) to predict the humerus length of an *Archaeopteryx* whose femur is 60 centimeters.

f. Use your calculator and the given points on the scatterplot to find the least squares regression line. Compare the score this equation gives for part (d) with what you predicted earlier. The ordered pairs defining the data are (38, 41), (56, 63), (59, 70), (64, 72), (74, 84).

84. The scatterplot shows the boiling temperature of various substances on the horizontal axis and their heats of vaporization on the vertical axis. (The heat of vaporization is the energy needed to change the substance from liquid to gas at its boiling point.)

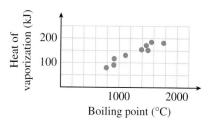

a. Use a straightedge to estimate a line of best fit for the scatterplot.

b. Use your line to predict the heat of vaporization of silver, whose boiling temperature is 2160°C.

c. Find the equation of the regression line.

d. Use the regression line to predict the heat of vaporization of potassium bromide, whose boiling temperature is 1435°C.

PROJECTS FOR CHAPTER 1

1. The number of eggs (clutch size) that a bird lays varies greatly. Is there an optimal clutch size for birds of a given species, or does it depend on the individual bird? In 1980, biologists in Sweden conducted an experiment on magpies as follows: They reduced or enlarged the natural clutch size by adding or removing eggs from the nests. They then computed the average number of fledglings successfully raised by the parent birds in each case. The graph shows the results for magpies that initially laid 5, 6, 7, or 8 eggs. (Source: Högstedt, 1980, via Krebs as developed in Davies, 1993)

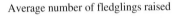

Average number of fledglings raised

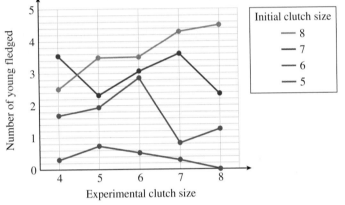

Experimental clutch size

a. Use the graph to fill in the table of values for the number of fledglings raised in each situation.

Initial clutch size laid	Experimental clutch size				
	4	5	6	7	8
5					
6					
7					
8					

b. For each initial clutch size, which experimental clutch size produced the most fledglings? Record your answers in the table.

Initial clutch size	5	6	7	8
Optimum clutch size				

c. What conclusions can you draw in response to the question in the problem?

2. The Big Island of Hawaii is the last island in a chain of islands and submarine mountain peaks that stretch almost 6000 kilometers across the Pacific Ocean. All are extinct volcanoes except for the Big Island itself, which is still active. The ages of the extinct peaks are roughly proportional to their distance from the Big Island. Geologists believe that the volcanic islands were formed as the tectonic plate drifted across a hot spot in the Earth's mantle. The figure shows a map of the islands, scaled in kilometers. (Source: Open University, 1998)

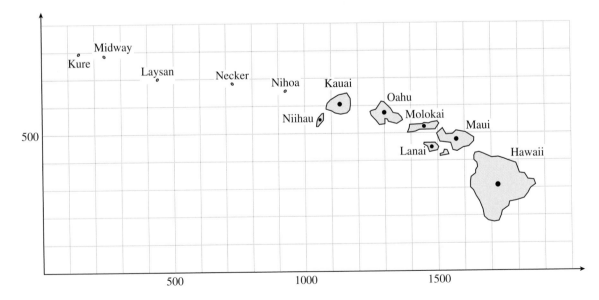

a. The table gives the ages of the islands, in millions of years. Estimate the distance from each island to the Big Island, along a straight-line path through their centers. Fill in the third row of the table.

b. Make a scatterplot showing the age of each island along the horizontal axis and its distance from Hawaii on the vertical axis.

c. Draw a line of best fit through the data.

d. Calculate the slope of the line of best fit, including units.

e. Explain why the slope provides an estimate for the speed of the Pacific plate.

Island	Hawaii	Maui	Lanai	Molokai	Oahu	Kauai	Niihau	Nihoa	Necker	Laysan	Midway
Age	0.5	0.8	1.3	1.8	3.8	5.1	4.9	7.5	10	20	27
Distance											

3. The graph shows a cross section of Earth's surface along an east–west line from the coast of Africa through the Atlantic Ocean to South America. Both axes are scaled in kilometers. Use the figure to estimate the distances in this problem. (Source: Open University, 1998)

a. What is the highest land elevation shown in the figure? What is the lowest ocean depth shown? Give the horizontal coordinates of these two points, in kilometers west of the 75° W longitude line.

b. How deep is the Atlantic Ocean directly above the crest of the Mid-Atlantic Ridge? How deep is the ocean above the abyssal plain on either side of the ridge?

c. What is the height of the Mid-Atlantic Ridge above the abyssal plain? What is the width of the Mid-Atlantic Ridge?

d. Using your answers to part (c), calculate the slope from the abyssal plain to the crest of the Mid-Atlantic Ridge, rounded to five decimal places.

e. Estimate the slopes of the continental shelf, the continental slope, and the continental rise. Use the coordinates of the points indicated on the figure.

f. Why do these slopes look much steeper in the accompanying figure than their numerical values suggest?

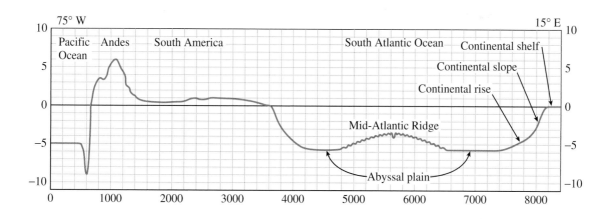

4. The Mid-Atlantic Ridge is a mountain range on the sea floor beneath the Atlantic Ocean. It was discovered in the late nineteenth century during the laying of transatlantic telephone cables. The ridge is volcanic, and the ocean floor is moving away from the ridge on either side. Geologists have estimated the speed of this sea-floor spreading by recording the age of the rocks on the sea floor and their distance from the ridge. (The age of the rocks is calculated by measuring their magnetic polarity. At known intervals over the last four million years, the Earth reversed its polarity, and this information is encoded in the rocks.) (Source: Open University, 1998)

a. According to the table, rocks that are 0.78 million years old have moved 17 kilometers from the ridge. What was the speed of spreading over the past 0.78 million years? (This is the rate of spreading closest to the ridge.)

b. Plot the data in the table, with age on the horizontal axis and separation distance on the vertical axis. Draw a line of best fit through the data.

c. Calculate the slope of the regression line. What are the units of the slope?

d. The slope you calculated in part (c) represents the average spreading rate over the past 3.58 million years. Is the average rate greater or smaller than the rate of spreading closest to the ridge?

e. Convert the average spreading rate to millimeters per year.

Age (millions of years)	0.78	0.99	1.07	1.79	1.95	2.60	3.04	3.11	3.22	3.33	3.58
Distance (km)	17	18	21	32	39	48	58	59	62	65	66

5. Naismith's rule is used by runners and walkers to estimate journey times in hilly terrain. In 1892, Naismith wrote in the *Scottish Mountaineering Club Journal* that a person "in fair condition should allow for easy expeditions an hour for every three miles on the map, with an additional hour for every 2000 feet of ascent." (Source: Scarf, 1998)

a. According to Naismith, one unit of ascent requires the same time as how many units of horizontal travel? (Convert miles to feet.) This is called Naismith's number. Round your answer to one decimal place.

b. A walk in the Brecon Beacons in Wales covers 3.75 kilometers horizontally and climbs 582 meters. What is the equivalent flat distance?

c. If you can walk at a pace of 15 minutes per kilometer over flat ground, how long will the walk in the Brecon Beacons take?

6. Empirical investigations have improved Naismith's number (see Problem 5) to 8.0 for men and 9.5 for women. Part of the Karrimor International Mountain Marathon in the Arrochar Alps in Scotland has a choice of two routes. Route A is 1.75 kilometers long with a 240-meter climb, and route B is 3.25 kilometers long with a 90-meter climb. (Source: Scarf, 1998)

a. Which route is faster for women?

b. Which route is faster for men?

c. At a pace of 6 minutes per flat kilometer, how much faster is the preferred route for women?

d. At a pace of 6 minutes per flat kilometer, how much faster is the preferred route for men?

Christian Michaels/Getty Images

Modeling with Functions

World3 is a computer model developed by a team of researchers at MIT. The model tracks population growth, use of resources, land development, industrial investment, pollution, and many other variables that describe human impact on the planet. The figure below is taken from the researchers' book, *Limits to Growth: The 30-Year Update*. The graphs represent four possible answers to World3's core question: How may the expanding global population and material economy interact with and adapt to Earth's limited carrying capacity (the maximum it can sustain) over the coming decades?

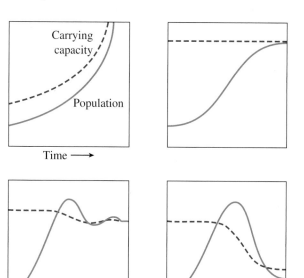

Source: Meadows, Randers, and Meadows, 2004

ThomsonNOW™

Throughout this chapter, the ThomsonNOW logo indicates an opportunity for online self-study, linking you to interactive tutorials and videos based on your level of understanding.

In this chapter, we examine the properties and features of some basic nonlinear functions and how they may be used as mathematical models.

Investigation 2

Epidemics

A contagious disease whose spread is unchecked can devastate a confined population. For example, in the early sixteenth century Spanish troops introduced smallpox into the Aztec population in Central America, and the resulting epidemic contributed significantly to the fall of Montezuma's empire.

Suppose that an outbreak of cholera follows severe flooding in an isolated town of 5000 people. Initially (on Day 0), 40 people are infected. Every day after that, 25% of those still healthy fall ill.

1. At the beginning of the first day (Day 1), how many people are still healthy? _____

   How many will fall ill during the first day? _____

   What is the total number of people infected after the first day?

   _____

2. Check your results against the first two rows of the table. Subtract the total number of infected residents from 5000 to find the number of healthy residents at the beginning of the second day. Then fill in the rest of the table for 10 days. (Round off decimal results to the nearest whole number.)

3. Use the last column of the table to plot the total number of infected residents, I, against time, t. Connect your data points with a smooth curve.

4. Do the values of I approach some largest value? Draw a dotted horizontal line at that value of I. Will the values of I ever exceed that value?

5. What is the first day on which at least 95% of the population is infected?

6. Look back at the table. What is happening to the number of new patients each day as time goes on? How is this phenomenon reflected in the graph? How would your graph look if the number of new patients every day were a constant?

7. Summarize your work: In your own words, describe how the number of residents infected with cholera changes with time. Include a description of your graph.

Day	Number healthy	New patients	Total infected
0	5000	40	40
1	4960	1240	1280
2			
3			
4			
5			
6			
7			
8			
9			
10			

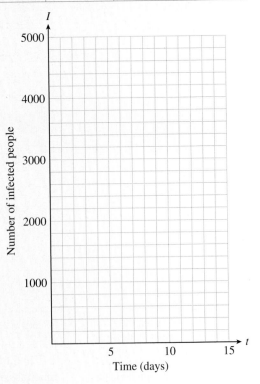

Perimeter and Area

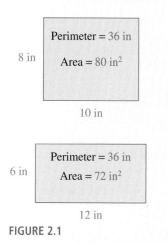

FIGURE 2.1

Do all rectangles with the same perimeter, say 36 inches, have the same area? Two different rectangles with perimeter 36 inches are shown in Figure 2.1. The first rectangle has base 10 inches and height 8 inches, and its area is 80 square inches. The second rectangle has base 12 inches and height 6 inches. Its area is 72 square inches.

1. The table shows the bases of various rectangles, in inches. Each rectangle has a perimeter of 36 inches. Fill in the height and the area of each rectangle. (To find the height of the rectangle, reason as follows: The base plus the height makes up half of the rectangle's perimeter.)

2. What happens to the area of the rectangle when we change its base but still keep the perimeter at 36 inches? Plot the points with coordinates *(Base, Area)*. (For this graph, we will not use the heights of the rectangles.) The first two points, (10, 80) and (12, 72), are shown. Connect your data points with a smooth curve.

3. What are the coordinates of the highest point on your graph?

4. Each point on your graph represents a particular rectangle with perimeter 36 inches. The first coordinate of the point gives the base of the rectangle, and the second coordinate gives the area of the rectangle. What is the largest area you found among rectangles with perimeter 36 inches? What is the base for that rectangle? What is its height?

5. Give the dimensions of the rectangle that corresponds to the point (13, 65).

Base	Height	Area
10	8	80
12	6	72
3		
14		
5		
17		
9		
2		
11		
4		
16		
15		
1		
6		
8		
13		
7		

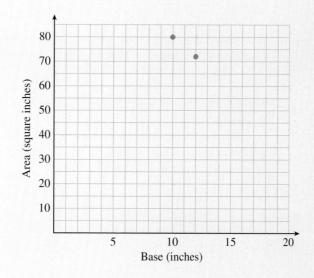

6. Find two points on your graph with vertical coordinate 80.

7. If the rectangle has area 80 square inches, what is its base? Why are there two different answers? Describe the rectangle corresponding to each answer.

8. Now we will write an algebraic expression for the area of the rectangle in terms of its base. Let x represent the base of the rectangle. First, express the height of the rectangle in terms of x. (*Hint:* If the perimeter of the rectangle is 36 inches, what is the sum of the base and the height?) Now write an expression for the area of the rectangle in terms of x.

9. Use your formula from part (8) to compute the area of the rectangle when the base is 5 inches. Does your answer agree with the values in your table and the point on your graph?

10. Use your formula to compute the area of the rectangle when $x = 0$ and when $x = 18$. Describe the rectangles that correspond to these data points.

11. Continue your graph to include the points corresponding to $x = 0$ and $x = 18$.

<table>
<tr><td></td><td></td></tr>
</table>

2.1   Nonlinear Models

In Chapter 1, we considered models described by linear functions. In this chapter, we begin our study of nonlinear models.

Solving Nonlinear Equations

When studying nonlinear models, we will need to solve nonlinear equations. For example, in Investigation 3 we used a graph to solve the quadratic equation

$$18x - x^2 = 80$$

Here is another example. Figure 2.2 shows a table and graph for the function $y = 2x^2 - 5$.

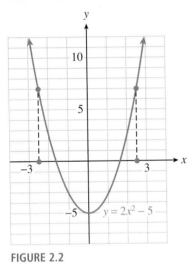

FIGURE 2.2

x	-3	-2	-1	0	1	2	3
y	13	3	-3	-5	-3	3	13

You can see that there are two points on the graph for each y-value greater than -5. For example, the two points with y-coordinate 7 are shown. To solve the equation

$$2x^2 - 5 = 7$$

we need only find the x-coordinates of these points. From the graph, the solutions appear to be about 2.5 and -2.5.

How can we solve this equation algebraically? The opposite operation for squaring a number is taking a square root. So we can undo the operation of squaring by extracting square roots. We first solve for x^2 to get

$$2x^2 = 12$$
$$x^2 = 6$$

and then take square roots to find

$$x = \pm\sqrt{6}$$

Don't forget that every positive number has *two* square roots. The symbol ± (read "*plus or minus*") is a shorthand notation used to indicate both square roots of 6. The *exact* solutions are thus $\sqrt{6}$ and $-\sqrt{6}$. We can also find decimal *approximations* for the solutions using a calculator. Rounded to two decimal places, the approximate solutions are 2.45 and −2.45.

In general, we can solve equations of the form

$$ax^2 + c = 0$$

by isolating x^2 on one side of the equation and then taking the square root of each side. This method for solving equations is called **extraction of roots.**

Extraction of Roots

To solve the equation

$$ax^2 + c = 0$$

1. Isolate x^2.
2. Take square roots of both sides. There are two solutions.

EXAMPLE 1

If a cat falls off a tree branch 20 feet above the ground, its height t seconds later is given by $h = 20 - 16t^2$.

a. What is the height of the cat 0.5 second later?

b. How long does the cat have to get in position to land on its feet before it reaches the ground?

Solutions **a.** In this question, we are given the value of t and asked to find the corresponding value of h. To do this, we evaluate the formula for $t = 0.5$. We substitute **0.5** for t into the formula and simplify.

$$\begin{aligned} h &= 20 - 16(\mathbf{0.5})^2 \quad &\text{Compute the power.} \\ &= 20 - 16\,(0.25) \quad &\text{Multiply; then subtract.} \\ &= 20 - 4 = 16 \end{aligned}$$

The cat is 16 feet above the ground after 0.5 second.

b. We would like to find the value of t when the height, h, is known. We substitute $h = 0$ into the equation to obtain

$$0 = 20 - 16t^2$$

To solve this equation, we use extraction of roots. First isolate t^2 on one side of the equation.

$$16t^2 = 20 \quad \text{Divide by 16.}$$

$$t^2 = \frac{20}{16} = 1.25$$

Now take the square root of both sides of the equation to find

$$t = \pm\sqrt{1.25} \approx \pm 1.118$$

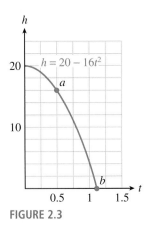

$h = 20 - 16t^2$

FIGURE 2.3

Only the positive solution makes sense here, so the cat has approximately 1.12 seconds to get into position for landing. A graph of the cat's height after t seconds is shown in Figure 2.3. The points corresponding to parts (a) and (b) are labeled.

Note that in Example 1a we *evaluated* the expression $20 - 16t^2$ to find a value for h, and in part (b) we *solved* the equation $0 = 20 - 16t^2$ to find a value for t.

EXERCISE 1

a. Solve by extracting roots $\dfrac{3x^2 - 8}{5} = 10$.   First, isolate x^2.
Take the square root of both sides.

b. Give exact answers; then give approximations rounded to two decimal places.

Solving Formulas

We can use extraction of roots to solve many formulas involving the square of the variable.

EXAMPLE 2   The formula $V = \frac{1}{3}\pi r^2 h$ gives the volume of a cone in terms of its height and radius. Solve the formula for r in terms of V and h.

Solution   Because the variable we want is squared, we use extraction of roots. First, multiply both sides by 3 to clear the fraction.

$$3V = 3\left(\frac{1}{3}\pi r^2 h\right)$$

$$3V = \pi r^2 h \qquad \text{Divide both sides by } \pi h.$$

$$\frac{3V}{\pi h} = r^2 \qquad \text{Take square roots.}$$

$$\pm\sqrt{\frac{3V}{\pi h}} = r$$

Because the radius of a cone must be a positive number, we use only the positive square root: $r = \sqrt{\frac{3V}{\pi h}}$.

EXERCISE 2

Find a formula for the radius of a circle in terms of its area.

Start with the formula for the area of a circle: $A = $ _____
Solve for r in terms of A.

More Extraction of Roots

Equations of the form

$$a(px + q)^2 + r = 0$$

can also be solved by extraction of roots after isolating the squared expression, $(px + q)^2$.

EXAMPLE 3 Solve the equation $3(x-2)^2 = 48$.

Solution First, isolate the perfect square, $(x-2)^2$.

$$3(x-2)^2 = 48 \qquad \text{Divide both sides by 3.}$$
$$(x-2)^2 = 16 \qquad \text{Take the square root of each side.}$$
$$x - 2 = \pm\sqrt{16} = \pm 4$$

This gives us two equations for x,

$$x - 2 = 4 \quad \text{or} \quad x - 2 = -4 \quad \text{Solve each equation.}$$
$$x = 6 \quad \text{or} \qquad x = -2$$

The solutions are 6 and -2.

Here is a general strategy for solving equations by extraction of roots.

Extraction of Roots

To solve the equation

$$a(px + q)^2 + r = 0$$

1. Isolate the squared expression, $(px + q)^2$.
2. Take the square root of each side of the equation. Remember that a positive number has two square roots.
3. Solve each equation. There are two solutions.

EXERCISE 3

Solve $2(5x + 3)^2 = 38$ by extracting roots.

a. Give your answers as exact values.

b. Find approximations for the solutions to two decimal places.

Compound Interest and Inflation

Many savings institutions offer accounts on which the interest is *compounded annually*. At the end of each year, the interest earned is added to the principal, and the interest for the next year is computed on this larger sum of money.

Compound Interest

If interest is compounded annually for n years, the amount, A, of money in an account is given by

$$A = P(1 + r)^n$$

where P is the principal and r is the interest rate, expressed as a decimal fraction.

EXAMPLE 4 Carmella invests $3000 in an account that pays an interest rate, r, compounded annually.

a. Write an expression for the amount of money in Carmella's account after two years.

b. What interest rate would be necessary for Carmella's account to grow to $3500 in two years?

Solutions   **a.** Use the formula above with $P = 3000$ and $n = 2$. Carmella's account balance will be

$$A = 3000(1 + r)^2$$

b. Substitute 3500 for A in the equation.

$$3500 = 3000(1 + r)^2$$

We can solve this equation in r by extraction of roots. First, isolate the perfect square.

$$3500 = 3000(1 + r)^2 \qquad \text{Divide both sides by 3000.}$$
$$1.1\overline{6} = (1 + r)^2 \qquad \text{Take the square root of both sides.}$$
$$\pm 1.0801 \approx 1 + r \qquad \text{Subtract 1 from both sides.}$$
$$r \approx 0.0801 \quad \text{or} \quad r \approx -2.0801$$

Because the interest rate must be a positive number, we discard the negative solution. Carmella needs an account with interest rate $r \approx 0.0801$, or just over 8%, to achieve an account balance of $3500 in two years.

The formula for compound interest also applies to the effects of inflation. For instance, if there is a steady inflation rate of 4% per year, in two years an item that now costs $100 will cost

$$A = P(1 + r)^2$$
$$= 100(1 + 0.04)^2 = \$108.16$$

EXERCISE 4

Two years ago, the average cost of dinner and a movie was $24. This year the average cost is $25.44. What was the rate of inflation over the past two years?

Other Nonlinear Equations

Because squaring and taking square roots are opposite operations, we can solve the equation

$$\sqrt{x} = 8.2$$

by squaring both sides to get

$$\left(\sqrt{x}\right)^2 = 8.2^2$$
$$x = 67.24$$

Similarly, we can solve

$$x^3 = 258$$

by taking the cube root of both sides, because cubing and taking cube roots are opposite operations. Rounding to three places, we find

$$\sqrt[3]{x^3} = \sqrt[3]{258}$$
$$x = 6.366$$

See the Algebra Skills Refresher A.1 to review cube roots.

The notion of undoing operations can help us solve a variety of simple nonlinear equations. The operation of taking a reciprocal is its own opposite, so we solve the equation

$$\frac{1}{x} = 50$$

by taking the reciprocal of both sides to get

$$x = \frac{1}{50} = 0.02$$

EXAMPLE 5

Solve $\dfrac{3}{x-2} = 4$.

Solution We begin by taking the reciprocal of both sides of the equation to get

$$\frac{x-2}{3} = \frac{1}{4}$$

We continue to undo the operations in reverse order. Multiply both sides by 3.

$$x - 2 = \frac{3}{4} \qquad \text{Add 2 to both sides.}$$
$$x = 2 + \frac{3}{4} = \frac{11}{4} \qquad \frac{2}{1} = \frac{8}{4}, \text{ so } \frac{2}{1} + \frac{3}{4} = \frac{8}{4} + \frac{3}{4} = \frac{11}{4}$$

The solution is $\frac{11}{4}$, or 2.75.

EXERCISE 5

Solve $2\sqrt{x+4} = 6$.

Using the Intersect Feature

We can use the **intersect** feature on a graphing calculator to solve equations.

EXAMPLE 6

Use a graphing calculator to solve $\frac{3}{x-2} = 4$.

Solution We would like to find the points on the graph of $y = \frac{3}{x-2}$ that have y-coordinate equal to 4. Graph the two functions

$$Y_1 = 3/(X - 2)$$
$$Y_2 = 4$$

in the window

$$\text{Xmin} = -9.4 \qquad \text{Xmax} = 9.4$$
$$\text{Ymin} = -10 \qquad \text{Ymax} = 10$$

The point where the two graphs intersect locates the solution of the equation. If we trace along the graph of Y_1, the closest we can get to the intersection point is $(2.8, 3.75)$, as shown in Figure 2.4a. We get a better approximation using the **intersect** feature.

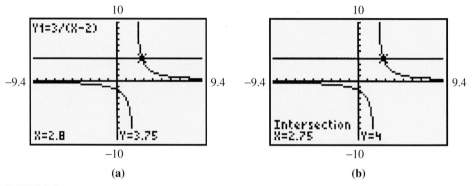

(a)   **(b)**

FIGURE 2.4

Use the arrow keys to position the **Trace** bug as close to the intersection point as you can. Then press $\boxed{\text{2nd}}$ $\boxed{\text{CALC}}$ to see the **Calculate** menu. Press $\boxed{5}$ for **intersect**; then respond to each of the calculator's questions, **First curve?, Second curve?,** and **Guess?** by pressing $\boxed{\text{ENTER}}$. The calculator will then display the intersection point, $x = 2.75$, $y = 4$, as shown in Figure 2.4b. The solution of the original equation is $x = 2.75$.

EXERCISE 6

Use the intersect feature to solve the equation $2x^2 - 5 = 7$. Round your answers to three decimal places.

ANSWERS TO 2.1 EXERCISES

1. $x = \pm\sqrt{\dfrac{58}{3}} \approx \pm 4.40$ **2.** $r = \sqrt{A/\pi}$ **3a.** $x = \dfrac{-3 \pm \sqrt{19}}{5}$

3b. $x \approx -1.47$ or $x \approx 0.27$ **4.** $r \approx 2.96\%$ **5.** $x = 5$

6. $x = \pm 2.449$

SECTION 2.1 SUMMARY

VOCABULARY   *Look up the definitions of new terms in the Glossary.*

Quadratic	Compound interest	Exact solution	Base
Perfect square	Extraction of roots	Inflation	Approximation
Area	Cube root	Isolate	
Height	Perimeter	Reciprocal	

CONCEPTS

3. We can give exact answers to a simple nonlinear equation, or we can give decimal approximations.

2.

Compound Interest
If interest is compounded annually for n years, the amount, A, of money in an account is given by

$$A = P(1 + r)^n$$

where P is the principal and r is the interest rate, expressed as a decimal fraction.

4. Simple nonlinear equations can be solved by undoing the operations on the variable.

STUDY QUESTIONS

1. How many square roots does a positive number have?
2. What is the first step in solving the equation $a(px + q)^2 = r$ by extraction of roots?
3. Give the exact solutions of the equation $x^2 = 10$, and then give decimal approximations rounded to hundredths.
4. State a formula for the amount in an account on which 5% interest is compounded annually.

5. Give an example of two rectangles with the same perimeter but different areas.
6. The perimeter of a rectangle is 50 meters. Write an expression for the length of the rectangle in terms of its width.
7. What is the opposite operation for taking a reciprocal?
8. What is the reciprocal of $\frac{1}{\sqrt{x}}$?

SKILLS   *Practice each skill in the Homework problems listed.*

1. Solve equations by extraction of roots: #1–12, 31–42
2. Solve formulas: #13–16, 63–68
3. Use the Pythagorean theorem: #19–24

4. Solve equations graphically: #25–30
5. Solve simple nonlinear equations: #43–54
6. Solve problems: #55–62

HOMEWORK 2.1

ThomsonNOW™ Test yourself on key content at **www.thomsonedu.com/login.**

Solve by extracting roots. Give exact values for your answers.

1. $9x^2 = 25$ **2.** $4x^2 = 9$ **3.** $4x^2 - 24 = 0$ **4.** $3x^2 - 9 = 0$

5. $\dfrac{2x^2}{3} = 4$ **6.** $\dfrac{3x^2}{5} = 6$

Solve by extracting roots. Round your answers to two decimal places.

7. $2x^2 = 14$ **8.** $3x^2 = 15$

9. $1.5x^2 = 0.7x^2 + 26.2$ **10.** $0.4x^2 = 2x^2 - 8.6$

11. $5x^2 - 97 = 3.2x^2 - 38$ **12.** $17 - \dfrac{x^2}{4} = 43 - x^2$

Solve the formulas for the specified variable.

13. $F = \dfrac{mv^2}{r}$, for v

14. $A = \dfrac{\sqrt{3}}{4} s^2$, for s

15. $s = \frac{1}{2}gt^2$, for t

16. $S = 4\pi r^2$, for r

For Problems 17 and 18, refer to the geometric formulas at the front of the book.

17. A conical coffee filter is 8.4 centimeters tall.
   a. Write a formula for the filter's volume in terms of its widest radius (at the top of the filter).
   b. Complete the table of values for the volume equation. If you double the radius of the filter, by what factor does the volume increase?

r	1	2	3	4	5	6	7	8
V								

   c. If the volume of the filter is 302.4 cubic centimeters, what is its radius?
   d. Use your calculator to graph the volume equation. Locate the point on the graph that corresponds to the filter in part (c).

18. A large bottle of shampoo is 20 centimeters tall and cylindrical in shape.
   a. Write a formula for the volume of the bottle in terms of its radius.
   b. Complete the table of values for the volume equation. If you halve the radius of the bottle, by what factor does the volume decrease?

r	1	2	3	4	5	6	7	8
V								

   c. What radius should the bottle have if it must hold 240 milliliters of shampoo? (One milliliter is equal to 1 cubic centimeter.)
   d. Use your calculator to graph the volume equation. Locate the point on the graph that corresponds to the bottle in part (c).

For Problems 19–24,
(a) Make a sketch of the situation described, and label a right triangle.
(b) Use the Pythagorean theorem to solve each problem. (See Algebra Skills Refresher A.11 to review the Pythagorean theorem.)

19. The size of a TV screen is the length of its diagonal. If the width of a 35-inch TV screen is 28 inches, what is its height?

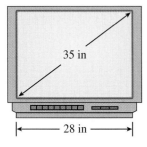

20. How high on a building will a 25-foot ladder reach if its foot is 15 feet away from the base of the wall?

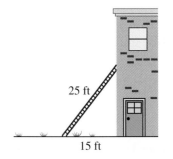

21. If a 30-meter pine tree casts a shadow of 30 meters, how far is the tip of the shadow from the top of the tree?

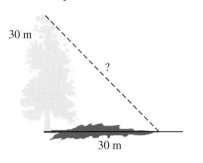

30 m

?

30 m

22. A baseball diamond is a square whose sides are 90 feet in length. Find the straight-line distance from home plate to second base.

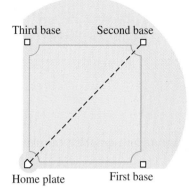

Third base   Second base

Home plate   First base

23. What size square can be inscribed in a circle of radius 8 inches?

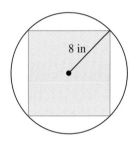

8 in

24. What size rectangle can be inscribed in a circle of radius 30 feet if the length of the rectangle must be 3 times its width?

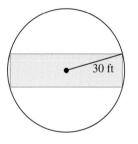

30 ft

For Problems 25–30,
(a) Use a calculator or computer to graph the function in the suggested window.
(b) Use your graph to find two solutions for the given equation. (See Section 1.3 to review graphical solution of equations.)
(c) Check your solutions algebraically, using mental arithmetic.

25. a. $y = \dfrac{1}{4}x^2$

Xmin $= -15$ Xmax $= 15$
Ymin $= -10$ Ymax $= 40$

b. $\dfrac{1}{4}x^2 = 36$

26. a. $y = 8x^2$

Xmin $= -15$ Xmax $= 15$
Ymin $= -50$ Ymax $= 450$

b. $8x^2 = 392$

27. a. $y = (x - 5)^2$

Xmin $= -5$ Xmax $= 15$
Ymin $= -5$ Ymax $= 25$

b. $(x - 5)^2 = 16$

28. a. $y = (x + 2)^2$

Xmin $= -10$ Xmax $= 10$
Ymin $= -2$ Ymax $= 12$

b. $(x + 2)^2 = 9$

29. a. $y = 3(x - 4)^2$

Xmin $= -5$ Xmax $= 15$
Ymin $= -20$ Ymax $= 130$

b. $3(x - 4)^2 = 108$

30. a. $y = \dfrac{1}{2}(x + 3)^2$

Xmin $= -15$ Xmax $= 5$
Ymin $= -5$ Ymax $= 15$

b. $\dfrac{1}{2}(x + 3)^2 = 8$

Solve by extraction of roots.

31. $(x - 2)^2 = 9$

32. $(x + 3)^2 = 4$

33. $(2x - 1)^2 = 16$

34. $(3x + 1)^2 = 25$

35. $4(x + 2)^2 = 12$

36. $6(x - 5)^2 = 42$

37. $\left(x - \dfrac{1}{2}\right)^2 = \dfrac{3}{4}$

38. $\left(x - \dfrac{2}{3}\right)^2 = \dfrac{5}{9}$

39. $81\left(x + \dfrac{1}{3}\right)^2 = 1$

40. $16\left(x + \dfrac{1}{2}\right)^2 = 1$

41. $3(8x - 7)^2 = 24$

42. $2(5x - 12)^2 = 48$

For Problems 43–54,
(a) Solve algebraically.
(b) Use the *intersect* feature on a graphing calculator to solve.

43. $4x^3 - 12 = 852$

44. $\dfrac{8x^3 + 6}{3} = 74$

45. $5\sqrt{x} - 9 = 31$

46. $25 - 2\sqrt{x} = 1$

47. $\dfrac{1}{2x - 3} = \dfrac{3}{4}$

48. $\dfrac{15}{x + 16} = 3$

49. $8 - 6\sqrt[3]{x} = -4$

50. $\dfrac{4\sqrt[3]{x}}{5} + 3 = 7$

51. $\sqrt{3x - 2} + 3 = 8$

52. $6\sqrt{1 - 2x} = 30$

53. $\dfrac{2}{\sqrt{4x - 2}} = 8$

54. $\dfrac{1}{\sqrt{x + 2}} = \dfrac{3}{4}$

55. Cyril plans to invest $5000 in a money market account that pays interest compounded annually.
 a. Write a formula for the balance, B, in Cyril's account after two years as a function of the interest rate, r.
 b. If Cyril would like to have $6250 in two years, what interest rate must the account pay?
 c. Use your calculator to graph the formula for Cyril's account balance. Locate the point on the graph that corresponds to the amount in part (b).

56. You plan to deposit your savings of $1600 in an account that compounds interest annually.
 a. Write a formula for the amount in your savings account after two years as a function of the interest rate, r.
 b. To the nearest tenth of a percent, what interest rate will you require if you want your $1600 to grow to $2000 in two years?
 c. Use your calculator to graph the formula for the account balance. Locate the point on the graph that corresponds to the amount in part (b).

57. Carol's living expenses two years ago were $1200 per month. This year, the same items cost Carol $1400 per month. What was the annual inflation rate for the past two years?

58. Two years ago, the average price of a house in the suburbs was $188,600. This year, the average price is $203,700. What was the annual percent increase in the cost of a house?

59. A machinist wants to make a metal section of pipe that is 80 millimeters long and has an interior volume of 9000 cubic millimeters. If the pipe is 2 millimeters thick, its interior volume is given by the formula

$$V = \pi(r - 2)^2 h$$

 where h is the length of the pipe and r is its radius. What should the radius of the pipe be?

60. A storage box for sweaters is constructed from a square sheet of corrugated cardboard measuring x inches on a side. The volume of the box, in cubic inches, is

$$V = 10(x - 20)^2$$

 If the box should have a volume of 1960 cubic inches, what size cardboard square is needed?

61. The area of an equilateral triangle is given by the formula $A = \frac{\sqrt{3}}{4}s^2$, where s is the length of the side.
  a. Find the areas of equilateral triangles with sides of length 2 centimeters, 4 centimeters, and 10 centimeters. First give exact values, then approximations to hundredths.
  b. Graph the area equation in the window

$$\text{Xmin} = 0 \qquad \text{Xmax} = 14.1$$
$$\text{Ymin} = 0 \qquad \text{Ymax} = 60$$

  Use the $\boxed{\textbf{TRACE}}$ or **value** feature to verify your answers to part (a).
  c. Trace along the curve to the point (5.1, 11.26266). What do the coordinates of this point represent?
  d. Use your graph to estimate the side of an equilateral triangle whose area is 20 square centimeters.
  e. Write and solve an equation to answer part (d).
  f. If the area of an equilateral triangle is $100\sqrt{3}$ square centimeters, what is the length of its side?

62. The area of the ring in the figure is given by the formula $A = \pi R^2 - \pi r^2$, where R is the radius of the outer circle and r is the radius of the inner circle.

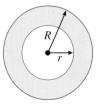

  a. Suppose the inner radius of the ring is kept fixed at $r = 4$ centimeters, but the radius of the outer circle, R, is allowed to vary. Find the area of the ring when the outer radius is 6 centimeters, 8 centimeters, and 12 centimeters. First give exact values, then approximations to hundredths.
  b. Graph the area equation, with $r = 4$, in the window

$$\text{Xmin} = 0 \qquad \text{Xmax} = 14.1$$
$$\text{Ymin} = 0 \qquad \text{Ymax} = 400$$

  Use the $\boxed{\textbf{TRACE}}$ feature to verify your answers to part (a).
  c. Trace along the curve to the point (9.75, 248.38217). What do the coordinates of this point represent?
  d. Use your graph to estimate the outer radius of the ring when its area is 100 square centimeters.
  e. Write and solve an equation to answer part (d).
  f. If the area of the ring is 9π square centimeters, what is the radius of the outer circle?

For Problems 63–68, solve for x in terms of a, b, and c.

63. $\dfrac{ax^2}{b} = c$

64. $\dfrac{bx^2}{c} - a = 0$

65. $(x - a)^2 = 16$

66. $(x + a)^2 = 36$

67. $(ax + b)^2 = 9$

68. $(ax - b)^2 = 25$

69. You have 36 feet of rope and you want to enclose a rectangular display area against one wall of an exhibit hall. The area enclosed depends on the dimensions of the rectangle you make. Because the wall makes one side of the rectangle, the length of the rope accounts for only three sides. Thus

$$\text{Base} + 2\,(\text{Height}) = 36$$

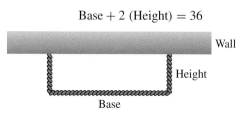

a. Complete the table showing the base and the area of the rectangle for the given heights.

Height	Base	Area	Height	Base	Area
1	34	34	10		
2	32	64	11		
3			12		
4			13		
5			14		
6			15		
7			16		
8			17		
9			18		

b. Make a graph with *Height* on the horizontal axis and *Area* on the vertical axis. Draw a smooth curve through your data points.

c. What is the area of the largest rectangle you can enclose in this way? What are its dimensions? On your graph, label the point that corresponds to this rectangle with the letter *M*.

d. Let *x* stand for the height of a rectangle and write algebraic expressions for the base and the area of the rectangle.

e. Enter your algebraic expression for the area in your calculator, then use the **Table** feature to verify the entries in your table in part (a).

f. Graph your formula for area on your graphing calculator. Use your table of values and your hand-drawn graph to help you choose appropriate **WINDOW** settings.

g. Use the **intersect** command to find the height of the rectangle whose area is 149.5 square feet.

70. We are going to make an open box from a square piece of cardboard by cutting 3-inch squares from each corner and then turning up the edges as shown in the figure.

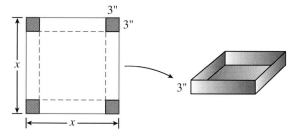

a. Complete the table showing the side of the original sheet of cardboard, the dimensions of the box created from it, and the volume of the box.

Side	Length of box	Width of box	Height of box	Volume of box
7	1	1	3	3
8	2	2	3	12
9				
10				
11				
12				
13				
14				
15				

Explain why the side of the cardboard square cannot be smaller than 6 inches. What happens if the cardboard is exactly 6 inches on a side?

b. Make a graph with *Side* on the horizontal axis and *Volume* on the vertical axis. Draw a smooth curve through your data points. (Use your table to help you decide on appropriate scales for the axes.)

c. Let *x* represent the side of the original sheet of cardboard. Write algebraic expressions for the dimensions of the box and for its volume.

d. Enter your expression for the volume of the box in your calculator; then use the **Table** feature to verify the values in your table in part (a).

e. Graph your formula for volume on your graphing calculator. Use your table of values and your hand-drawn graph to help you choose appropriate **WINDOW** settings.

f. Use the **intersect** command to find out how large a square of cardboard you need to make a box with volume 126.75 cubic inches.

g. Does your graph have a highest point? What happens to the volume of the box as you increase *x*?

71. The jump height, J, in meters, achieved by a pole vaulter is given approximately by $J = v^2/2g$, where v is the vaulter's speed in meters per second at the end of his run, and $g = 9.8$ is the gravitational acceleration. (Source: Alexander, 1992)

a. Fill in the table of values for jump heights achieved with values of v from 0 to 11 meters per second.

v	0	1	2	3	4	5	6	7	8	9	10	11
J												

b. Graph the jump height versus final speed. (Use the table values to help you choose a window for the graph.)

c. The jump height should be added to the height of the vaulter's center of gravity (at about hip level) to give the maximum height, H, he can clear. For a typical pole vaulter, his center of gravity at the end of the run is 0.9 meters from the ground. Complete the table of values for maximum heights, H, and graph H on your graph of J.

v	0	1	2	3	4	5	6	7	8	9	10	11
H												

d. A good pole vaulter can reach a final speed of 9.5 meters per second. What height will he clear?

e. In 2005, the world record in pole vaulting, established by Sergey Bubka in 1994, was 6.14 meters. What was the vaulter's speed at the end of his run?

72. To be launched into space, a satellite must travel fast enough to escape Earth's gravity. This escape velocity, v, satisfies the equation

$$\frac{1}{2}mv^2 = \frac{GMm}{R}$$

where m is the mass of the satellite, M is the mass of the Earth, R is the radius of the Earth, and G is the universal gravitational constant.

a. Solve the equation for v in terms of the other variables.

b. The equation

$$mg = \frac{GMm}{R^2}$$

gives the force of gravity at the Earth's surface. We can use this equation to simplify the expression for v: First, multiply both sides of the equation by $\frac{R}{M}$. You now have an expression for $\frac{GM}{R}$. Substitute this new expression into your formula for v.

c. The radius of the Earth is about 6400 km, and $g = 0.0098$. Calculate the escape velocity from Earth in kilometers per second. Convert your answer to miles per hour. (One kilometer is 0.621 miles.)

d. The radius of the moon is 1740 km, and the value of g at the moon's surface is 0.0016. Calculate the escape velocity from the moon in kilometers per second and convert to miles per hour.

In this section, we study the graphs of some important basic functions. Many functions fall into families or classes of similar functions, and recognizing the appropriate family for a given situation is an important part of modeling.

We begin by reviewing the absolute value.

Absolute Value

The absolute value is used to discuss problems involving distance. For example, consider the number line in Figure 2.5. Starting at the origin, we travel in opposite *directions* to reach the two numbers 6 and -6, but the *distance* we travel in each case is the same.

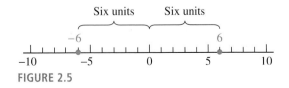

FIGURE 2.5

The distance from a number c to the origin is called the **absolute value** of c, denoted by $|c|$. Because distance is never negative, the absolute value of a number is always positive (or zero). Thus, $|6| = 6$ and $|-6| = 6$. In general, we define the absolute value of a number x as follows.

Absolute Value

The **absolute value** of x is defined by

$$|x| = \begin{cases} x & \text{if } x \geq 0 \\ -x & \text{if } x < 0 \end{cases}$$

This definition says that the absolute value of a positive number (or zero) is the same as the number. To find the absolute value of a negative number, we take the opposite of the number, which results in a positive number. For instance,

$$|-6| = -(-6) = 6$$

Absolute value bars act like grouping devices in the order of operations: You should complete any operations that appear inside absolute value bars before you compute the absolute value.

EXAMPLE 1   Simplify each expression.

a. $|3 - 8|$

b. $|3| - |8|$

Solutions   **a.** Simplify the expression inside the absolute value bars first.

$$|3 - 8| = |-5| = 5$$

b. Simplify each absolute value; then subtract.

$$|3| - |8| = 3 - 8 = -5$$

EXERCISE 1

Simplify each expression.

a. $12 - 3|-6|$

b. $-7 - 3|2 - 9|$

Examples of Models

Many situations can be modeled by a handful of simple functions. The following examples represent applications of eight useful functions.

The contractor for a new hotel is estimating the cost of the marble tile for a circular lobby. The cost is a function of the *square* of the diameter of the lobby.

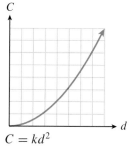

$$C = kd^2$$

The number of board-feet that can be cut from a Ponderosa pine is a function of the *cube* of the circumference of the tree at a standard height.

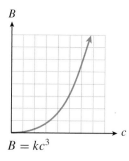

$$B = kc^3$$

The manager of an appliance store must decide how many coffee-makers to order every quarter. The optimal order size is a function of the *square root* of the annual demand for coffee-makers.

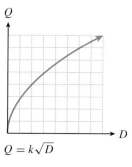

$$Q = k\sqrt{D}$$

Investors are deciding whether to support a windmill farm. The wind speed needed to generate a given amount of power is a function of the *cube root* of the power.

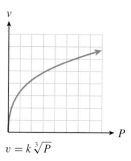

$$v = k\sqrt[3]{P}$$

The frequency of the note produced by a violin string is a function of the *reciprocal* of the length of the string.

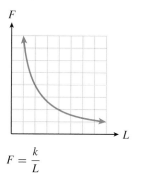

$$F = \frac{k}{L}$$

The loudness, or intensity, of the music at a concert is a function of the *reciprocal of the square* of your distance from the speakers.

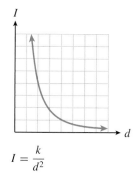

$$I = \frac{k}{d^2}$$

The annual return on an investment is a function of the interest rate.

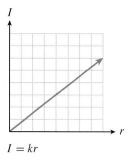

$$I = kr$$

You are flying from Los Angeles to New York. Your distance from the Mississippi River is an *absolute value* function of time.

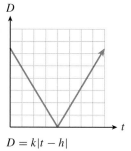

$$D = k|t - h|$$

We will consider each of these functions and their applications in more detail in later sections. For now, you should become familiar with the properties of each graph and be able to sketch them easily from memory.

Investigation 4

Eight Basic Functions

Part I Some Powers and Roots

1. Complete Table 2.1, a table of values for the squaring function, $f(x) = x^2$, and the cubing function, $g(x) = x^3$. Then sketch each function on graph paper, using the table values to help you scale the axes.

2. Verify both graphs with your graphing calculator.

3. State the intervals on which each graph is increasing.

4. Write a few sentences comparing the two graphs. The graph of $y = x^2$ is called a **parabola,** and the graph of $y = x^3$ is called a **cubic.**

5. Complete Table 2.2 and Table 2.3 for the square root function, $f(x) = \sqrt{x}$, and the cube root function, $g(x) = \sqrt[3]{x}$. (Round your answers to two decimal places.) Then sketch each function on graph paper, using the table values to help you scale the axes.

x	$f(x) = x^2$	$g(x) = x^3$
-3		
-2		
-1		
$-\frac{1}{2}$		
0		
$\frac{1}{2}$		
1		
2		
3		

TABLE 2.1

x	$f(x) = \sqrt{x}$
0	
$\frac{1}{2}$	
1	
2	
3	
4	
5	
7	
9	

TABLE 2.2

x	$g(x) = \sqrt[3]{x}$
-8	
-4	
-1	
$-\frac{1}{2}$	
0	
$\frac{1}{2}$	
1	
4	
8	

TABLE 2.3

x	$f(x) = \frac{1}{x}$	$g(x) = \frac{1}{x^2}$
-4		
-3		
-2		
-1		
$-\frac{1}{2}$		
0		
$\frac{1}{2}$		
1		
2		
3		
4		

TABLE 2.4

6. Verify both graphs with your graphing calculator.

7. State the intervals on which each graph is increasing.

8. Write a few sentences comparing the two graphs.

Part II Asymptotes

1. Complete Table 2.4 for the functions

$$f(x) = \frac{1}{x} \quad \text{and} \quad g(x) = \frac{1}{x^2}$$

What is true about $f(0)$ and $g(0)$?

2. Prepare a grid on graph paper, scaling both axes from -5 to 5. Plot the points from Table 2.4 and connect them with smooth curves.

3. As x increases through larger and larger values, what happens to the values of $f(x)$? Extend your graph to reflect your answer.

   What happens to $f(x)$ as x decreases through larger and larger negative values (that is, for $x = -5, -6, -7, \ldots$)? Extend your graph for these x-values.

   As the values of x get larger in absolute value, the graph approaches the x-axis. However, because $\frac{1}{x}$ never *equals* zero for any x-value, the graph never actually touches the x-axis. We say that the x-axis is a **horizontal asymptote** for the graph.

4. Repeat step (3) for the graph of $g(x)$.

5. Next we will examine the graphs of f and g near $x = 0$. Use your calculator to evaluate f for several x-values close to zero and record the results in Table 2.5.

x	$f(x) = \frac{1}{x}$	$g(x) = \frac{1}{x^2}$
-2		
-1		
-0.1		
-0.01		
-0.001		

(a)

x	$f(x) = \frac{1}{x}$	$g(x) = \frac{1}{x^2}$
2		
1		
0.1		
0.01		
0.001		

(b)

TABLE 2.5

What happens to the values of $f(x)$ as x approaches zero? Extend your graph of f to reflect your answer.

As x approaches zero from the left (through negative values), the function values decrease toward $-\infty$. As x approaches zero from the right (through positive values), the function values increase toward ∞. The graph approaches but never touches the vertical line $x = 0$ (the y-axis.) We say that the graph of f has a **vertical asymptote** at $x = 0$.

6. Repeat step (5) for the graph of $g(x)$.

7. The functions $f(x) = \frac{1}{x}$ and $g(x) = \frac{1}{x^2}$ are examples of **rational functions,** so called because they are fractions, or ratios. Verify both graphs with your graphing calculator. Use the window

$$\text{Xmin} = -4 \quad \text{Xmax} = 4$$
$$\text{Ymin} = -4 \quad \text{Ymax} = 4$$

x	$f(x) = x$	$g(x) = \lvert x \rvert$
-4		
-2		
-1		
$-\frac{1}{2}$		
0		
$\frac{1}{2}$		
1		
2		
4		

TABLE 2.6

8. State the intervals on which each graph is increasing.

9. Write a few sentences comparing the two graphs.

Part III Absolute Value

1. Complete Table 2.6 for the two functions $f(x) = x$ and $g(x) = \lvert x \rvert$. Then sketch each function on graph paper, using the table values to help you scale the axes.

2. Verify both graphs with your graphing calculator. Your calculator uses the notation abs (x) instead of $\lvert x \rvert$ for the absolute value of x. First, position the cursor after $Y_1 =$ in the graphing window. Now access the absolute value function by pressing **2nd** **0** for **CATALOG**; then **ENTER** for **abs (.** Don't forget to press X if you want to graph $y = \lvert x \rvert$.

3. State the intervals on which each graph is increasing.

4. Write a few sentences comparing the two graphs.

Graphs of Eight Basic Functions

The graphs of the eight basic functions considered in Investigation 4 are shown in Figure 2.6 on page 145. Once you know the shape of each graph, you can sketch an accurate picture by plotting a few guidepoints and drawing the curve through those points. Usually, points (or vertical asymptotes!) at $x = -1, 0,$ and 1 make good guidepoints.

Properties of the Basic Functions

In Section 1.2, we saw that for most functions, $f(a + b)$ is *not* equal to $f(a) + f(b)$. We may be able to find *some* values of a and b for which $f(a + b) = f(a) + f(b)$ is true, but not for *all* values of a and b. If there is even one value of a or b for which $f(a + b)$ is not equal to $f(a) + f(b)$, we cannot claim that $f(a + b) = f(a) + f(b)$ for that function. For example, for the function $f(x) = x^2$, if we choose $a = 3$ and $b = 4$, then

$$f(3 + 4) = f(7) = 7^2 = 49$$
$$\text{but} \qquad f(3) + f(4) = 3^2 + 4^2 = 9 + 16 = 25$$

so we have proved that $f(a + b) \neq f(a) + f(b)$ for the squaring function. (In fact, we already knew this because $(a + b)^2 \neq a^2 + b^2$ as long as neither a nor b is 0.)

What about multiplication? Which of the basic functions have the property that $f(ab) = f(a)f(b)$ for all a and b? You will consider this question in the homework problems, but in particular you will need to recall the following properties of absolute value.

Properties of Absolute Value	
$\lvert a + b \rvert \leq \lvert a \rvert + \lvert b \rvert$	Triangle inequality
$\lvert ab \rvert = \lvert a \rvert \lvert b \rvert$	Multiplicative property

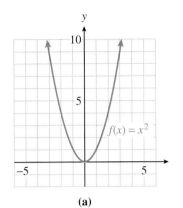

(a)

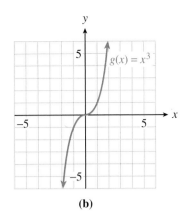

(b)

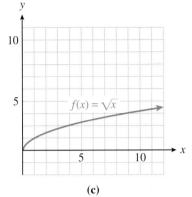

(c)

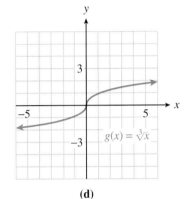

(d)

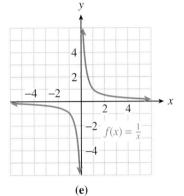

(e)

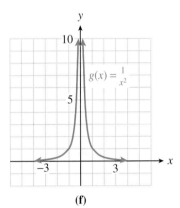

(f)

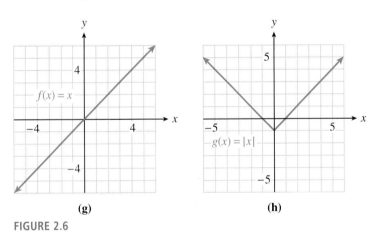

(g)

(h)

FIGURE 2.6

EXAMPLE 2   Verify the triangle inequality for three cases: a and b are both positive, a and b are both negative, and a and b have opposite signs.

Solution   We choose positive values for a and b, say $a = 3$ and $b = 5$. Then

$$|3 + 5| = |8| = 8 \quad \text{and} \quad |3| + |5| = 3 + 5 = 8$$

so $|3 + 5| = |3| + |5|$. For the second case, we choose $a = -3$ and $b = -5$. Then

$$|-3 + (-5)| = |-8| = 8 \quad \text{and} \quad |-3| + |-5| = 3 + 5 = 8$$

so $|-3 + (-5)| = |-3| + |-5|$. For the third case, we choose $a = 3$ and $b = -5$. Then

$$|3 + (-5)| = |-2| = 2 \quad \text{and} \quad |3| + |-5| = 3 + 5 = 8$$

so $|3 + (-5)| < |3| + |-5|$. In each case, $|a + b| \leq |a| + |b|$.

Note that *verifying* a statement for one or two values of the variables does not *prove* the statement is true for *all* values of the variables. However, working with examples can help us understand the meaning and significance of mathematical properties.

EXERCISE 2

Verify the multiplicative property of absolute value for the three cases in Example 2.

Functions Defined Piecewise

A function may be defined by different formulas on different portions of the *x*-axis. Such a function is said to be defined *piecewise*. To graph a function defined piecewise, we consider each piece of the *x*-axis separately.

EXAMPLE 3 Graph the function defined by

$$f(x) = \begin{cases} x + 1 & \text{if } x \leq 1 \\ 3 & \text{if } x > 1 \end{cases}$$

Solution Think of the plane as divided into two regions by the vertical line $x = 1$, as shown in Figure 2.7. In the left-hand region ($x \leq 1$), we graph the line $y = x + 1$. (The fastest way to graph the line is to plot its intercepts, $(-1, 0)$ and $(0, 1)$.) Notice that the value $x = 1$ is included in the first region, so $f(1) = 1 + 1 = 2$, and the point $(1, 2)$ is included on the graph. We indicate this with a solid dot at the point $(1, 2)$. In the right-hand region ($x > 1$), we graph the horizontal line $y = 3$. The value $x = 1$ is *not* included in the second region, so the point $(1, 3)$ is *not* part of the graph. We indicate this with an open circle at the point $(1, 3)$.

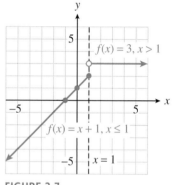

FIGURE 2.7

EXERCISE 3

Graph the piecewise defined function

$$g(x) = \begin{cases} -1 - x & \text{if } x \leq -1 \\ x^3 & \text{if } x > -1 \end{cases}$$

The absolute value function $f(x) = |x|$ is an example of a function that is defined piecewise.

$$f(x) = |x| = \begin{cases} x & \text{if } x \geq 0 \\ -x & \text{if } x < 0 \end{cases}$$

To sketch the absolute value function, we graph the line $y = x$ in the first quadrant and the line $y = -x$ in the second quadrant.

EXAMPLE 4

a. Write a piecewise definition for $g(x) = |x - 3|$.

b. Sketch a graph of $g(x) = |x - 3|$.

Solutions   a. In the definition for $|x|$, we replace x by $x - 3$ to get

$$g(x) = |x - 3| = \begin{cases} x - 3 & \text{if } x - 3 \geq 0 \\ -(x - 3) & \text{if } x - 3 < 0 \end{cases}$$

We can simplify this expression to

$$g(x) = |x - 3| = \begin{cases} x - 3 & \text{if } x \geq 3 \\ -x + 3 & \text{if } x < 3 \end{cases}$$

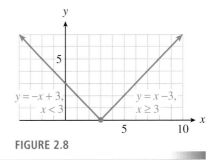

b. In the first region, $x \geq 3$, we graph the line $y = x - 3$. Because $x = 3$ is included in this region, the endpoint of this portion of the graph, $(3, 0)$, is included, too. In the second region, $x < 3$, we graph the line $y = -x + 3$. Note that the two pieces of the graph meet at the point $(0, 3)$, as shown in Figure 2.8.

FIGURE 2.8

EXERCISE 4

a. Use your calculator to graph $g(x) = |x - 3|$ and $h(x) = |x| + |-3|$. Are the graphs the same?

b. Explain why the functions $f(x) = |x + k|$ and $g(x) = |x| + |k|$ are not the same if $k \neq 0$.

ANSWERS TO 2.2 EXERCISES

1. a. -6 **b.** -28

2. $|3||5| = 15 = |3 \cdot 5|$, $|-3||-5| = 15 = |(-3)(-5)|$,
$|3||-5| = 15 = |3(-5)|$

3.

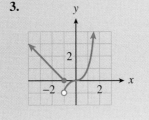

4. a. No

b. Because $|x + k| \neq |x| + |k|$ when x and k have opposite signs

---

SECTION 2.2 SUMMARY

VOCABULARY   *Look up the definitions of new terms in the Glossary.*

Absolute value	Verify	Triangle inequality	Vertical asymptote
Rational function	Parabola	Piecewise defined function	Multiplicative property
Guidepoints	Horizontal asymptote	Cubic	

CONCEPTS

1. The **absolute value** of x is *defined* by

$$|x| = \begin{cases} x & \text{if } x \geq 0 \\ -x & \text{if } x < 0 \end{cases}$$

2. The absolute value has the following properties:

$$|a + b| \leq |a| + |b| \qquad \text{Triangle inequality}$$
$$|ab| = |a||b| \qquad \text{Multiplicative property}$$

3. Many useful functions fall into families or classes of variations on basic functions.
4. We can make sketches of the eight basic functions using guidepoints.
5. Functions can be defined piecewise, with different formulas on different intervals.

STUDY QUESTIONS

1. Is it true that $-x$ must be a negative number? Why or why not?
2. Are there any numbers for which $x = -x$?
3. If $0 < x < 1$, which is larger, x^2 or x^3?
4. If $0 < x < 1$, which is larger, $\sqrt{x}$ or $\sqrt[3]{x}$?
5. List the eight basic functions considered in this section.
6. Which of the eight basic functions have a horizontal asymptote? A vertical asymptote?
7. What does an open circle on a graph mean?
8. For what value(s) of x does $|x + 6| = 0$?

SKILLS   *Practice each skill in the Homework problems listed.*

1. Simplify expressions containing absolute values: #1–10
2. Sketch graphs of the basic functions by hand: #15–18
3. Identify the graph of a basic function: #19–26
4. Solve equations and inequalities graphically: #11–14, 27–34
5. Graph functions defined piecewise: #41–58

HOMEWORK 2.2

Simplify each expression according to the order of operations.

1. **a.** $-|-9|$
   b. $-(-9)$

2. **a.** $2 - (-6)$
   b. $2 - |-6|$

3. **a.** $|-8| - |12|$
   b. $|-8 - 12|$

4. **a.** $|-3| + |-5|$
   b. $|-3 + (-5)|$

5. $4 - 9|2 - 8|$

6. $2 - 5|-6 - 3|$

7. $|-4 - 5||1 - 3(-5)|$

8. $|-3 + 7||-2(6 - 10)|$

9. $||-5| - |-6||$

10. $||4| - |-6||$

In Problems 11–14, show how to use the graphs to find the values. Estimate your answers to one decimal point. Compare your estimates to values obtained with a calculator.

11. Refer to the graph of $f(x) = x^3$.

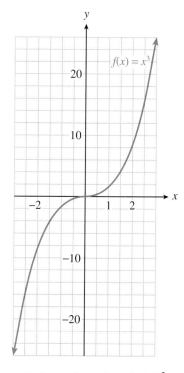

a. Estimate the value of $(1.4)^3$.
b. Find all numbers whose cube is -20.
c. Find all solutions of the equation $x^3 = 6$.
d. Estimate the value of $\sqrt[3]{24}$.

12. Refer to the graph of $f(x) = x^2$.

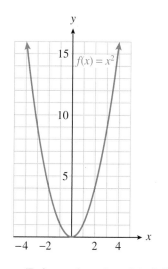

a. Estimate the value of $(-2.5)^2$.
b. Find all numbers whose square is 12.
c. Find all solutions of the equation $x^2 = 15$.
d. Estimate the value of $\sqrt{10.5}$.

13. Refer to the graph of $y = \dfrac{1}{x}$.

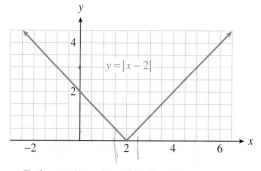

a. Estimate the value of $\dfrac{1}{3.4}$.

b. Find all numbers whose reciprocal is -2.5.
c. Find all solutions of the equation $\dfrac{1}{x} = 4.8$.

14. Refer to the graph of $y = |x-2|$.

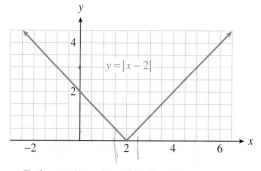

a. Estimate the value of $|1.6 - 2|$.
b. Find all values of x for which $|x - 2| = 3$.
c. Find all solutions of the equation $|x - 2| = 0.4$.

For Problems 15–18,

(a) Sketch both functions on the same grid, paying attention to the shape of the graph. Plot at least three guidepoints for each graph to ensure accuracy.

(b) Use the graph to find all solutions of the equation $f(x) = g(x)$.

(c) On what intervals is $f(x) > g(x)$?

15. $f(x) = x^2$, $g(x) = x^3$

16. $f(x) = \sqrt{x}$, $g(x) = \sqrt[3]{x}$

17. $f(x) = \dfrac{1}{x}$, $g(x) = \dfrac{1}{x^2}$

18. $f(x) = x$, $g(x) = |x|$

Graph each set of functions together in the *ZDecimal* window. Describe how graphs (b) and (c) are different from the basic graph.

19. a. $f(x) = x^3$
  b. $g(x) = x^3 - 2$
  c. $h(x) = x^3 + 1$

20. a. $f(x) = |x|$
  b. $g(x) = |x - 2|$
  c. $h(x) = |x + 1|$

21. a. $f(x) = \dfrac{1}{x}$
  b. $g(x) = \dfrac{1}{x + 1.5}$
  c. $h(x) = \dfrac{1}{x - 1}$

22. a. $f(x) = \dfrac{1}{x^2}$
  b. $g(x) = \dfrac{1}{x^2} + 2$
  c. $h(x) = \dfrac{1}{x^2} - 1$

23. a. $f(x) = \sqrt{x}$
  b. $g(x) = -\sqrt{x}$
  c. $h(x) = \sqrt{-x}$

24. a. $f(x) = \sqrt[3]{x}$
  b. $g(x) = -\sqrt[3]{x}$
  c. $h(x) = \sqrt[3]{-x}$

Each graph is a variation of one of the eight basic graphs of Investigation 4. Identify the basic graph for each problem.

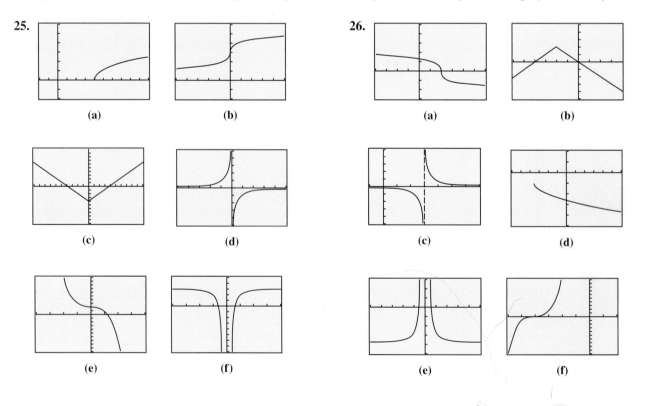

25.
(a)   (b)
(c)   (d)
(e)   (f)

26.
(a)   (b)
(c)   (d)
(e)   (f)

In Problems 27–30, use the graph to estimate the solution to the equation or inequality. Show the solution or solutions on the graph. Then check your answers algebraically.

27. The figure shows a graph of $f(x) = \sqrt{x} - 2$, for $x > 0$. Solve the following:

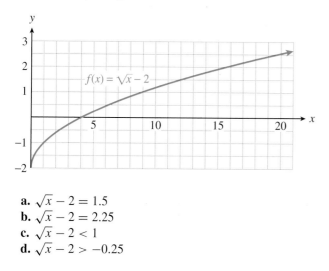

a. $\sqrt{x} - 2 = 1.5$
b. $\sqrt{x} - 2 = 2.25$
c. $\sqrt{x} - 2 < 1$
d. $\sqrt{x} - 2 > -0.25$

28. The figure shows a graph of $g(x) = \dfrac{4}{x+2}$, for $x > -2$. Solve the following:

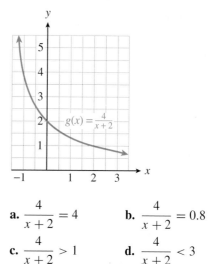

a. $\dfrac{4}{x+2} = 4$ **b.** $\dfrac{4}{x+2} = 0.8$

c. $\dfrac{4}{x+2} > 1$ **d.** $\dfrac{4}{x+2} < 3$

29. The figure shows a graph of $w(t) = -10(t+1)^3 + 10$. Solve the following:

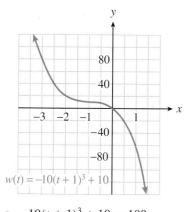

a. $-10(t+1)^3 + 10 = 100$
b. $-10(t+1)^3 + 10 = -140$
c. $-10(t+1)^3 + 10 > -50$
d. $-20 < -10(t+1)^3 + 10 < 40$

30. The figure shows a graph of $H(z) = 4\sqrt[3]{z-4} + 6$. Solve the following:

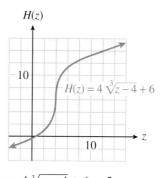

a. $4\sqrt[3]{z-4} + 6 = 2$
b. $4\sqrt[3]{z-4} + 6 = 12$
c. $4\sqrt[3]{z-4} + 6 > 14$
d. $4\sqrt[3]{z-4} + 6 < 6$

Graph each function with the *ZInteger* setting. Use the graph to solve each equation or inequality. Check your solutions algebraically.

31. Graph $F(x) = 4\sqrt{x - 25}$.

    a. Solve $4\sqrt{x - 25} = 16$.

    b. Solve $8 < 4\sqrt{x - 25} \leq 24$.

32. Graph $G(x) = 15 - 0.01(x - 2)^3$.

    a. Solve $15 - 0.01(x - 2)^3 = -18.75$.

    b. Solve $15 - 0.01(x - 2)^3 \leq 25$.

33. Graph $H(x) = 24 - 0.25(x - 6)^2$.

    a. Solve $24 - 0.25(x - 6)^2 = -6.25$.

    b. Solve $24 - 0.25(x - 6)^2 > 11.75$.

34. Graph $R(x) = 0.1(x + 12)^2 - 18$.

    a. Solve $0.1(x + 12)^2 - 18 = 14.4$.

    b. Solve $0.1(x + 12)^2 - 18 < 4.5$.

For Problems 35–40,
(a) Graph the equation by completing the table and plotting points.
(b) Does the equation define *y* as a function of *x*? Why or why not?

35. $x = y^2$

x							
y	-2	-1	$-\frac{1}{2}$	0	$\frac{1}{2}$	1	2

36. $x = y^3$

x							
y	-2	-1	$-\frac{1}{2}$	0	$\frac{1}{2}$	1	2

37. $x = |y|$

x							
y	-2	-1	$-\frac{1}{2}$	0	$\frac{1}{2}$	1	2

38. $|x| = |y|$

x							
y	-2	-1	$-\frac{1}{2}$	0	$\frac{1}{2}$	1	2

39. $x = \dfrac{1}{y}$

x							
y	-2	-1	$-\frac{1}{2}$	0	$\frac{1}{2}$	1	2

40. $x = \dfrac{1}{y^2}$

x							
y	-2	-1	$-\frac{1}{2}$	0	$\frac{1}{2}$	1	2

For Problems 41–51, graph the following piecewise defined functions. Indicate whether the endpoints of each piece are included on the graph.

41. $f(x) = \begin{cases} -2 & \text{if } x \leq 1 \\ x - 3 & \text{if } x > 1 \end{cases}$

42. $h(x) = \begin{cases} -x + 2 & \text{if } x \leq -1 \\ 3 & \text{if } x > -1 \end{cases}$

43. $G(t) = \begin{cases} 3t + 9 & \text{if } t < -2 \\ -3 - \frac{1}{2}t & \text{if } t \geq -2 \end{cases}$

44. $F(s) = \begin{cases} \frac{1}{3}s + 3 & \text{if } s < 3 \\ 2s - 3 & \text{if } s \geq 3 \end{cases}$

45. $H(t) = \begin{cases} t^2 & \text{if } t \leq 1 \\ \frac{1}{2}t + \frac{1}{2} & \text{if } t > 1 \end{cases}$

46. $g(t) = \begin{cases} \frac{3}{2}t + 7 & \text{if } t \leq -2 \\ t^2 & \text{if } t > -2 \end{cases}$

47. $k(x) = \begin{cases} |x| & \text{if } x \leq 2 \\ \sqrt{x} & \text{if } x > 2 \end{cases}$

48. $S(x) = \begin{cases} \frac{1}{x} & \text{if } x < 1 \\ |x| & \text{if } x \geq 1 \end{cases}$

49. $D(x) = \begin{cases} |x| & \text{if } x < -1 \\ x^3 & \text{if } x \geq -1 \end{cases}$

50. $m(x) = \begin{cases} x^2 & \text{if } x \leq \frac{1}{2} \\ |x| & \text{if } x > \frac{1}{2} \end{cases}$

51. $P(t) = \begin{cases} t^3 & \text{if } t \leq 1 \\ \frac{1}{t^2} & \text{if } t > 1 \end{cases}$

52. $Q(t) = \begin{cases} t^2 & \text{if } t \leq -1 \\ \sqrt[3]{t} & \text{if } t > -1 \end{cases}$

Write a piecewise definition for the function and sketch its graph.

53. $f(x) = |2x - 8|$

54. $g(x) = |3x + 6|$

55. $g(t) = \left| 1 + \dfrac{t}{3} \right|$

56. $f(t) = \left| \dfrac{1}{2}t - 3 \right|$

57. $F(x) = |x^3|$

58. $G(x) = \left| \dfrac{1}{x} \right|$

In Problems 59–64, decide whether each statement is true for all values of *a* and *b*. If the statement is true, give an algebraic justification. If it is false, find values of *a* and *b* to disprove it.
(a) $f(a + b) = f(a) + f(b)$
(b) $f(ab) = f(a)f(b)$

59. $f(x) = x^2$

60. $f(x) = x^3$

61. $f(x) = \dfrac{1}{x}$

62. $f(x) = \sqrt{x}$

63. $f(x) = mx + b$

64. $f(x) = kx$

65. Verify that $|a - b|$ gives the distance between *a* and *b* on a number line.
    a. $a = 3, b = 8$ **b.** $a = -2, b = -6$
    c. $a = 4, b = -3$ **d.** $a = -2, b = 5$

66. Which of the following statements is true for all values of *a* and *b*?
    (1) $|a - b| = |a| - |b|$
    (2) $|a - b| \leq |a| - |b|$
    (3) $|a - b| \geq |a| - |b|$

67. Explain how the distributive law, $a(b + c) = ab + ac$, is different from the equation $f(a + b) = f(a) + f(b)$.

68. For each function, decide whether $f(kx) = kf(x)$ for all $x \neq 0$, where $k \neq 0$ is a constant.
    a. $f(x) = x^2$ **b.** $f(x) = \dfrac{1}{x}$
    c. $f(x) = \sqrt{x}$ **d.** $f(x) = |x|$

Models for real situations are often variations of the basic functions introduced in Section 2.2. In this section, we explore how certain changes in the formula for a function affect its graph. In particular, we will compare the graph of $y = f(x)$ with the graphs of $y = f(x) + k$, $y = f(x + h)$, and $y = af(x)$ for different values of the constants k, h, and a. Such variations are called **transformations** of the graph.

Vertical Translations

Figure 2.9 shows the graphs of $f(x) = x^2 + 4$, $g(x) = x^2 - 4$, and the basic parabola, $y = x^2$. By comparing tables of values, we can see exactly how the graphs of f and g are related to the basic parabola.

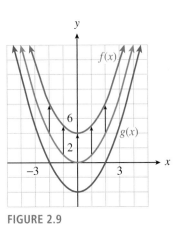

FIGURE 2.9

x	-2	-1	0	1	2
$y = x^2$	4	1	0	1	4
$f(x) = x^2 + 4$	8	5	4	5	8

x	-2	-1	0	1	2
$y = x^2$	4	1	0	1	4
$g(x) = x^2 - 4$	0	-3	-4	-3	0

Each y-value in the table for $f(x)$ is four units greater than the corresponding y-value for the basic parabola. Consequently, each point on the graph of $f(x)$ is four units higher than the corresponding point on the basic parabola, as shown by the arrows. Similarly, each point on the graph of $g(x)$ is four units lower than the corresponding point on the basic parabola.

The graphs of $y = f(x)$ and $y = g(x)$ are said to be **translations** of the graph of $y = x^2$. They are shifted to a different location in the plane but retain the same size and shape as the original graph. In general, we have the following principles.

> **Vertical Translations**
>
> Compared with the graph of $y = f(x)$,
>
> **1.** The graph of $y = f(x) + k$ $(k > 0)$ is shifted *upward* k units.
> **2.** The graph of $y = f(x) - k$ $(k > 0)$ is shifted *downward* k units.

EXAMPLE 1 Graph the following functions.

a. $g(x) = |x| + 3$

b. $h(x) = \dfrac{1}{x} - 2$

Solutions

a. The table shows that the *y*-values for $g(x)$ are each three units greater than the corresponding *y*-values for the absolute value function. The graph of $g(x) = |x| + 3$ is a translation of the basic graph of $y = |x|$, shifted upward three units, as shown in Figure 2.10.

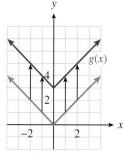

x	-2	-1	0	1	2
$y = |x|$	2	1	0	1	2
$g(x) = |x| + 3$	5	4	3	4	5

FIGURE 2.10

b. The table shows that the *y*-values for $h(x)$ are each two units smaller than the corresponding *y*-values for $y = \frac{1}{x}$. The graph of $h(x) = \frac{1}{x} - 2$ is a translation of the basic graph of $y = \frac{1}{x}$, shifted downward two units, as shown in Figure 2.11.

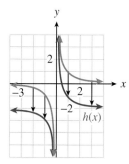

x	-2	-1	$\frac{1}{2}$	1	2
$y = \frac{1}{x}$	$\frac{-1}{2}$	-1	2	1	$\frac{1}{2}$
$h(x) = \frac{1}{x} - 2$	$\frac{-5}{2}$	-3	0	-1	$\frac{-3}{2}$

FIGURE 2.11

EXERCISE 1

a. Graph the function $f(x) = |x| + 1$.

b. How is the graph of f different from the graph of $y = |x|$?

EXAMPLE 2

The function $E = f(h)$ graphed in Figure 2.12 gives the amount of electrical power, in megawatts, drawn by a community from its local power plant as a function of time during a 24-hour period in 2002. Sketch a graph of $y = f(h) + 300$ and interpret its meaning.

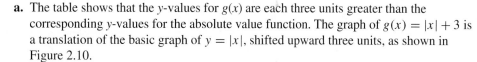

FIGURE 2.12

Solution The graph of $y = f(h) + 300$ is a vertical translation of the graph of f, as shown in Figure 2.13.

At each hour of the day, or for each value of h, the *y*-coordinate is 300 greater than on the graph of f. So at each hour, the community is drawing 300 megawatts more power than in 2002.

FIGURE 2.13

An evaporative cooler, or swamp cooler, is an energy-efficient type of air conditioner used in dry climates. A typical swamp cooler can reduce the temperature inside a house by 15 degrees. Figure 2.14a shows the graph of $T = f(t)$, the temperature inside Kate's house t hours after she turns on the swamp cooler. Write a formula in terms of f for the function g shown in Figure 2.14b and give a possible explanation of its meaning.

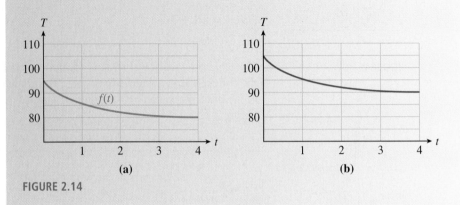

(a)   (b)

FIGURE 2.14

Horizontal Translations

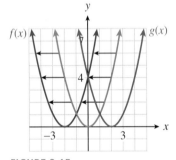

FIGURE 2.15

Now consider the graphs of $f(x) = (x + 2)^2$ and $g(x) = (x - 2)^2$ shown in Figure 2.15. Compared with the graph of the basic function $y = x^2$, the graph of $f(x) = (x + 2)^2$ is shifted two units to the *left*, as shown by the arrows. You can see why this happens by studying the function values in the table. Locate a particular y-value for $y = x^2$, say, $y = 1$. You must move two units to the left in the table to find the same y-value for $f(x)$, as shown by the arrow. In fact, each y-value for $f(x)$ occurs two units to the left when compared to the same y-value for $y = x^2$.

x	-3	-2	-1	0	1	2	3
$y = x^2$	9	4	1	0	1	4	9
$f(x) = (x + 2)^2$	1	0	1	4	9	16	25

x	-3	-2	-1	0	1	2	3
$y = x^2$	9	4	1	0	1	4	9
$g(x) = (x - 2)^2$	25	16	9	4	1	0	1

Similarly, the graph of $g(x) = (x - 2)^2$ is shifted two units to the *right* compared to the graph of $y = x^2$. In the table for g, each y-value for $g(x)$ occurs two units to the right of the same y-value for $y = x^2$. In general, we have the following principle.

Horizontal Translations

Compared with the graph of $y = f(x)$,

1. The graph of $y = f(x + h)$ $(h > 0)$ is shifted h units to the *left*.
2. The graph of $y = f(x - h)$ $(h > 0)$ is shifted h units to the *right*.

EXAMPLE 3   Graph the following functions.

a. $g(x) = \sqrt{x + 1}$

b. $h(x) = \dfrac{1}{(x - 3)^2}$

Solutions   **a.** The table shows that each y-value for $g(x)$ occurs one unit to the left of the same y-value for the graph of $y = \sqrt{x}$. Consequently, each point on the graph of $y = g(x)$ is shifted one unit to the left of $y = \sqrt{x}$, as shown in Figure 2.16.

x	-1	0	1	2	3
$y = \sqrt{x}$	undefined	0	1	1.414	1.732
$g(x) = \sqrt{x + 1}$	0	1	1.414	1.732	2

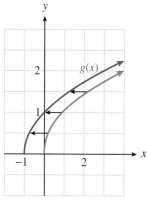

FIGURE 2.16

b. The table shows that each y-value for $h(x)$ occurs three units to the right of the same y-value for the graph of $y = \frac{1}{x^2}$. Consequently, each point on the graph of $y = h(x)$ is shifted three units to the right of $y = \frac{1}{x^2}$, as shown in Figure 2.17.

x	-1	0	1	2	3	4
$y = \frac{1}{x^2}$	1	undefined	1	$\frac{1}{4}$	$\frac{1}{9}$	$\frac{1}{16}$
$h(x) = \dfrac{1}{(x - 3)^2}$	$\frac{1}{16}$	$\frac{1}{9}$	$\frac{1}{4}$	1	undefined	1

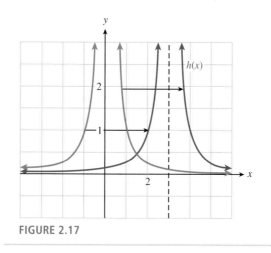

FIGURE 2.17

EXERCISE 3

a. Graph the function $f(x) = |x + 1|$.

b. How is the graph of f different from the graph of $y = |x|$?

EXAMPLE 4

The function $N = f(p)$ graphed in Figure 2.18 gives the number of people who have a given eye pressure level p from a sample of 100 people with healthy eyes, and the function g gives the number of people with pressure level p in a sample of 100 glaucoma patients.

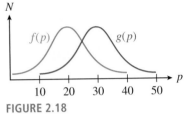

FIGURE 2.18

a. Write a formula for g as a transformation of f.

b. For what pressure readings could a doctor be fairly certain that a patient has glaucoma?

Solutions

a. The graph of g is translated 10 units to the right of f, so $g(p) = f(p - 10)$.

b. Pressure readings above 40 are a strong indication of glaucoma. Readings between 10 and 40 cannot conclusively distinguish healthy eyes from those with glaucoma.

EXERCISE 4

The function $C = f(t)$ in Figure 2.19 gives the caffeine level in Delbert's bloodstream at time t hours after he drinks a cup of coffee, and $g(t)$ gives the caffeine level in Francine's bloodstream. Write a formula for g in terms of f, and explain what it tells you about Delbert and Francine.

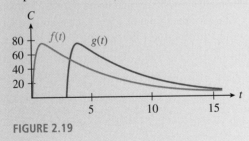

FIGURE 2.19

The graphs of some functions involve both horizontal and vertical translations.

EXAMPLE 5

Graph $f(x) = (x + 4)^3 + 2$.

Solution

We identify the basic graph from the structure of the formula for $f(x)$. In this case, the basic graph is $y = x^3$, so we begin by locating a few points on that graph, as shown in Figure 2.20. We will perform the translations separately, following the order of operations. First, we sketch a graph of $y = (x + 4)^3$ by shifting each point on the basic graph four units to the left. We then move each point up two units to obtain the graph of $f(x) = (x + 4)^3 + 2$. All three graphs are shown in Figure 2.20.

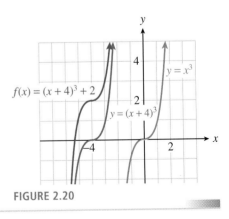

FIGURE 2.20

▶▶▶
See Appendix B for more about graphing
transformations with a calculator.

EXERCISE 5

a. Graph the function $f(x) = |x - 2| - 1$.

b. How is the graph of f different from the graph of $y = |x|$?

Scale Factors

We have seen that *adding* a constant to the expression defining a function results in a translation of its graph. What happens if we *multiply* the expression by a constant? Consider the graphs of the functions

$$f(x) = 2x^2, \quad g(x) = \frac{1}{2}x^2, \quad \text{and} \quad h(x) = -x^2$$

shown in Figures 2.21a, b, and c, and compare each to the graph of $y = x^2$.

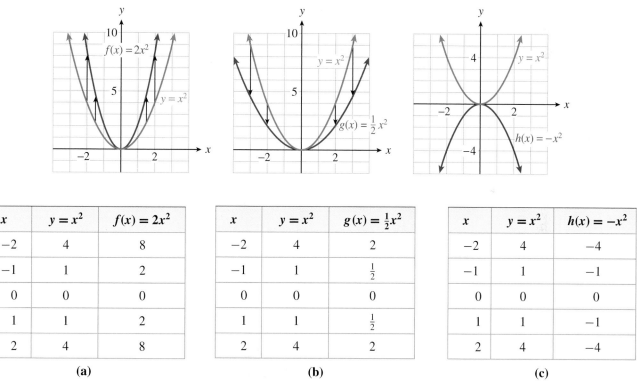

x	$y = x^2$	$f(x) = 2x^2$
-2	4	8
-1	1	2
0	0	0
1	1	2
2	4	8

(a)

x	$y = x^2$	$g(x) = \frac{1}{2}x^2$
-2	4	2
-1	1	$\frac{1}{2}$
0	0	0
1	1	$\frac{1}{2}$
2	4	2

(b)

x	$y = x^2$	$h(x) = -x^2$
-2	4	-4
-1	1	-1
0	0	0
1	1	-1
2	4	-4

(c)

FIGURE 2.21

Compared to the graph of $y = x^2$, the graph of $f(x) = 2x^2$ is expanded, or stretched, vertically by a factor of 2. The y-coordinate of each point on the graph has been doubled, as you can see in the table of values, so each point on the graph of f is twice as far from the x-axis as its counterpart on the basic graph $y = x^2$. The graph of $g(x) = \frac{1}{2}x^2$ is compressed vertically by a factor of $\frac{1}{2}$; each point is half as far from the x-axis as its counterpart on the graph of $y = x^2$. The graph of $h(x) = -x^2$ is flipped, or reflected, about the x-axis; the y-coordinate of each point on the graph of $y = x^2$ is replaced by its opposite.

In general, we have the following principles.

Scale Factors and Reflections

Compared with the graph of $y = f(x)$, the graph of $y = af(x)$, where $a \neq 0$, is

1. stretched vertically by a factor of $|a|$ if $|a| > 1$,
2. compressed vertically by a factor of $|a|$ if $0 < |a| < 1$, and
3. reflected about the x-axis if $a < 0$.

The constant a is called the **scale factor** for the graph.

EXAMPLE 6   Graph the following functions.

a. $g(x) = 3\sqrt[3]{x}$

b. $h(x) = \dfrac{-1}{2}|x|$

Solutions   **a.** The graph of $g(x) = 3\sqrt[3]{x}$ is a vertical expansion of the basic graph $y = \sqrt[3]{x}$ by a factor of 3, as shown in Figure 2.22. Each point on the basic graph has its y-coordinate tripled.

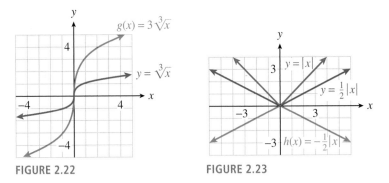

FIGURE 2.22   **FIGURE 2.23**

b. The graph of $h(x) = \dfrac{-1}{2}|x|$ is a vertical compression of the basic graph $y = |x|$ by a factor of $\frac{1}{2}$, combined with a reflection about the x-axis. You may find it helpful to graph the function in two steps, as shown in Figure 2.23.

EXERCISE 6

a. Graph the function $f(x) = 2|x|$.

b. How is the graph of f different from the graph of $y = |x|$?

EXAMPLE 7   The function $A = f(t)$ graphed in Figure 2.24 gives a person's blood alcohol level t hours after drinking a martini. Sketch a graph of $g(t) = 2f(t)$ and explain what it tells you.

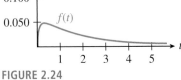

FIGURE 2.24

Solution   To sketch a graph of *g*, we stretch the graph of *f* vertically by a factor of 2, as shown in Figure 2.25. At each time *t*, the person's blood alcohol level is twice the value given by *f*. The function *g* could represent a person's blood alcohol level *t* hours after drinking two martinis.

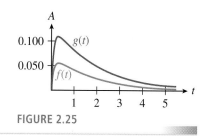

FIGURE 2.25

EXERCISE 7

If the Earth were not tilted on its axis, there would be 12 daylight hours every day all over the planet. But in fact, the length of a day in a particular location depends on the latitude and the time of year. The graph in Figure 2.26 shows $H = f(t)$, the length of a day in Helsinki, Finland, *t* days after January 1, and $R = g(t)$, the length of a day in Rome. Each is expressed as the number of hours greater or less than 12. Write a formula for *f* in terms of *g*. What does this formula tell you?

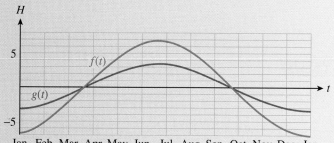

FIGURE 2.26

ANSWERS TO 2.3 EXERCISES

1. a.

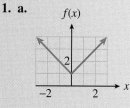

b. Translate $y = |x|$ one unit up.

2. $g(t) = f(t) + 10$. The outside temperature was 10° hotter.

3. a.

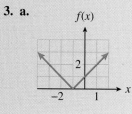

b. Translate $y = |x|$ one unit left.

4. $g(t) = f(t - 3)$. Francine drank her coffee 3 hours after Delbert drank his.

5. a. $f(x)$

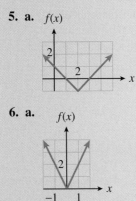

b. Translate $y = |x|$ one unit down and two units right.

6. a. $f(x)$

b. Stretch $y = |x|$ vertically by a factor of 2.

7. $f(t) \approx 2g(t)$. On any given day, the number of daylight hours varies from 12 hours about twice as much in Helsinki as it does in Rome.

SECTION 2.3 SUMMARY

VOCABULARY   *Look up the definitions of new terms in the Glossary.*

Transformation	Scale factor	Vertical stretch	Vertical compression
Reflection	Vertical translation	Horizontal translation	

CONCEPTS

1.

Vertical Translations

Compared with the graph of $y = f(x)$,

1. The graph of $y = f(x) + k$ $(k > 0)$ is shifted *upward* k units.
2. The graph of $y = f(x) - k$ $(k > 0)$ is shifted *downward* k units.

2.

Horizontal Translations

Compared with the graph of $y = f(x)$,

1. The graph of $y = f(x + h)$ $(h > 0)$ is shifted h units to the *left*.
2. The graph of $y = f(x - h)$ $(h > 0)$ is shifted h units to the *right*.

3.

Scale Factors and Reflections

Compared with the graph of $y = f(x)$, the graph of $y = af(x)$, where $a \neq 0$, is

1. stretched vertically by a factor of $|a|$ if $|a| > 1$,
2. compressed vertically by a factor of $|a|$ if $0 < |a| < 1$, and
3. reflected about the x-axis if $a < 0$.

1. How does a vertical translation affect the formula for a function? Give an example.
2. How does a horizontal translation affect the formula for a function? Give an example.

3. How does a scale factor affect the formula for a function? Give an example.
4. How is the graph of $y = -f(x)$ different from the graph of $y = f(x)$?

SKILLS   *Practice each skill in the Homework problems listed.*

1. Write formulas for transformations of functions: #1–6, 19–22, 35–38
2. Recognize and sketch translations of the basic graphs: #7–18
3. Recognize and sketch expansions, compression, and reflections of the basic graphs: #23–34, 43–50

4. Identify transformations from tables of values: #39–42
5. Sketch graphs obtained by two or more transformations of a basic graph: #51–62
6. Write a formula for a transformation of a graph: #63–76
7. Interpret transformations of graphs in context: #71–76

HOMEWORK 2.3

ThomsonNOW™ Test yourself on key content at www.thomsonedu.com/login.

Identify each graph as a translation of a basic function, and write a formula for the graph.

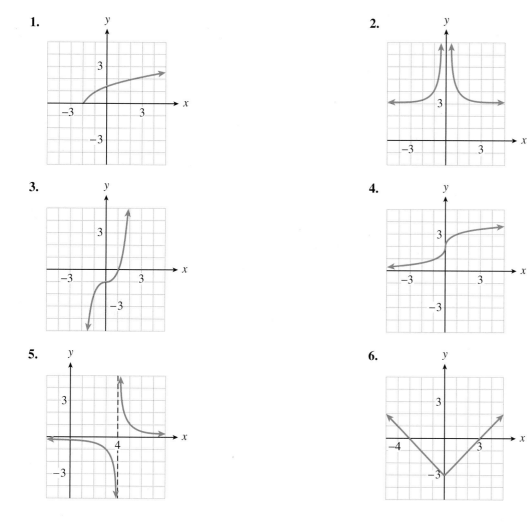

1.

2.

3.

4.

5.

6.

For Problems 7–18,
(a) Describe how to transform one of the basic graphs to obtain the graph of the given function.
(b) Using guidepoints, sketch the basic graph and the graph of the given function on the same axes. Label the
  coordinates of three points on the graph of the given function.

7. $f(x) = |x| - 2$

8. $g(x) = (x + 1)^3$

9. $g(s) = \sqrt[3]{s - 4}$

10. $f(s) = s^2 + 3$

11. $F(t) = \dfrac{1}{t^2} + 1$

12. $G(t) = \sqrt{t - 2}$

13. $G(r) = (r + 2)^3$

14. $F(r) = \dfrac{1}{r - 4}$

15. $H(d) = \sqrt{d} - 3$

16. $h(d) = \sqrt[3]{d} + 5$

17. $h(v) = \dfrac{1}{v + 6}$

18. $H(v) = \dfrac{1}{v^2} - 2$

Identify each graph as a stretch, compression, or reflection of a basic function, and write a formula for the graph.

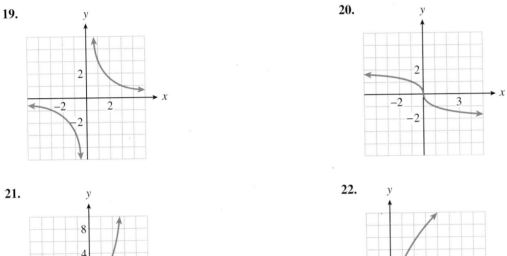

19.

20.

21.

22.

For Problems 23–32,
(a) Identify the scale factor for each function and describe how it affects the graph of the corresponding basic function.
(b) Using guidepoints, sketch the basic graph and the graph of the given function on the same axes. Label the
  coordinates of three points on the graph of the given function.

23. $f(x) = \dfrac{1}{3}|x|$

24. $H(x) = -3|x|$

25. $h(z) = \dfrac{-2}{z^2}$

26. $g(z) = \dfrac{2}{z}$

27. $G(v) = -3\sqrt{v}$

28. $F(v) = -4\sqrt[3]{v}$

29. $g(s) = \dfrac{-1}{2}s^3$

30. $f(s) = \dfrac{1}{8}s^3$

31. $H(x) = \dfrac{1}{3x}$

32. $h(x) = \dfrac{-1}{4x^2}$

In Problems 33 and 34, match each graph with its equation.

33.

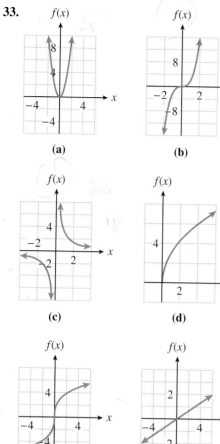

(a)

(b)

(c)

(d)

(e)

(f)

i. $f(x) = 3\sqrt{x}$

ii. $f(x) = 2x^3$

iii. $f(x) = \dfrac{x}{3}$

iv. $f(x) = \dfrac{3}{x}$

v. $f(x) = 2\sqrt[3]{x}$

vi. $f(x) = 3x^2$

34.

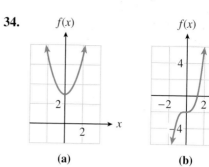

(a)

(b)

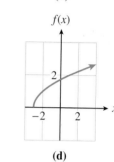

(c)

(d)

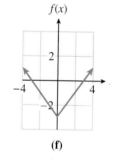

(e)

(f)

i. $f(x) = x^3 - 2$

ii. $f(x) = \sqrt[3]{x} + 2$

iii. $f(x) = \dfrac{1}{(x-3)^2}$

iv. $f(x) = |x| - 3$

v. $f(x) = x^2 + 3$

vi. $f(x) = \sqrt{x-3}$

In Problems 35–38, the graph of a function is shown. Describe each transformation of the graph; then give a formula for each in terms of the original function.

35.

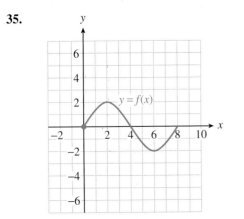

(a)

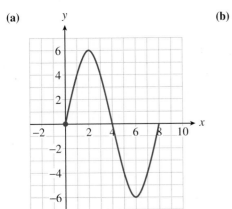

(b)

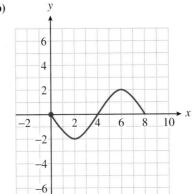

(c)

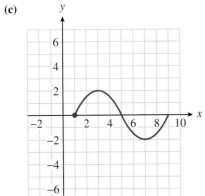

(d)

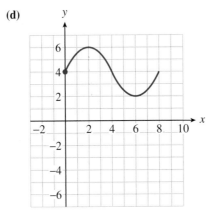

36.

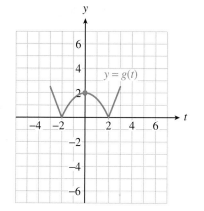

(a)

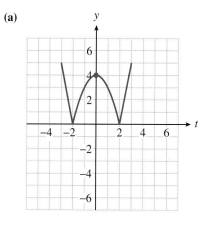

(b)

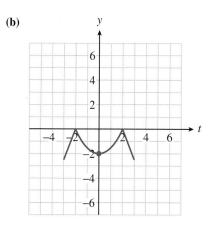

(c)

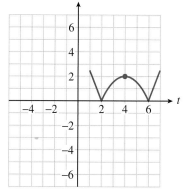

(d)

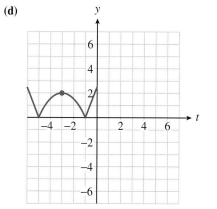

37.

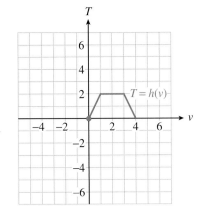

(a)

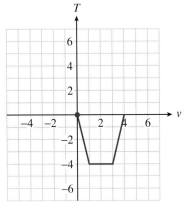

(b)

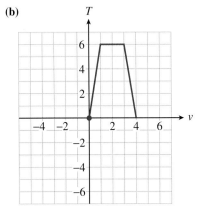

(c)

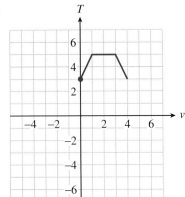

(d)

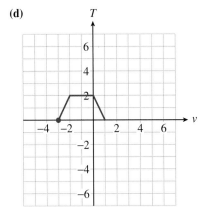

38.

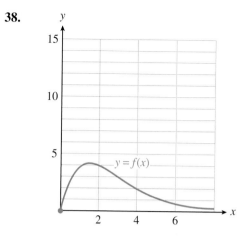

(a)

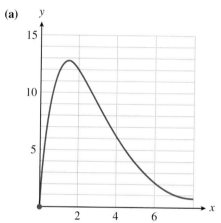

(b)

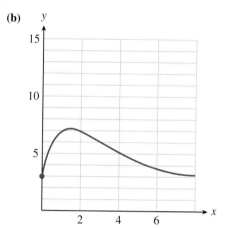

(c)

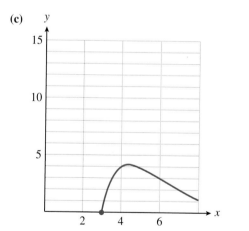

(d)

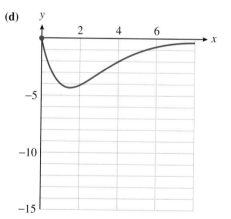

In Problems 39–42, each table in parts (a)–(d) describes a transformation of $f(x)$. Identify the transformation and write a formula for the new function in terms of f.

39.

x	1	2	3	4	5	6
$f(x)$	8	6	4	2	0	2

a.

x	1	2	3	4	5	6
y	10	8	6	4	2	4

b.

x	1	2	3	4	5	6
y	4	2	0	−2	−4	−2

c.

x	1	2	3	4	5	6
y	4	3	2	1	0	1

d.

x	1	2	3	4	5	6
y	10	8	6	4	2	0

40.

x	−3	−2	−1	0	1	2
$f(x)$	13	3	−3	−5	−3	3

a.

x	−3	−2	−1	0	1	2
y	−26	−6	6	10	6	−6

b.

x	−3	−2	−1	0	1	2
y	18	8	2	0	2	8

c.

x	−3	−2	−1	0	1	2
y	−3	−5	−3	3	13	27

d.

x	−3	−2	−1	0	1	2
y	2.6	0.6	−0.6	−1	−.6	0.6

41.

x	−2	−1	0	1	2	3
$f(x)$	−9	−8	−7	−6	1	20

a.

x	−2	−1	0	1	2	3
y	−34	−9	−8	−7	−6	1

b.

x	−2	−1	0	1	2	3
y	−4	21	22	23	24	31

c.

x	−2	−1	0	1	2	3
y	18	16	14	12	−2	−40

d.

x	−2	−1	0	1	2	3
y	8	6	4	2	−12	−50

42.

x	1	2	3	4	5	6
$f(x)$	60	30	20	15	12	10

a.

x	1	2	3	4	5	6
y	30	15	10	7.5	6	5

b.

x	1	2	3	4	5	6
y	35	20	15	12.5	11	10

c.

x	1	2	3	4	5	6
y	−12	−6	−4	−3	−2.4	−2

d.

x	1	2	3	4	5	6
y	−10	−4	−2	−1	1.4	0

Write each function in the form $y = kf(x)$, where $f(x)$ is one of the basic functions. Describe how the graph differs from that of the basic function.

43. $y = \dfrac{1}{2x^2}$

44. $y = \sqrt{9x}$

45. $y = \sqrt[3]{8x}$

46. $y = \dfrac{1}{4x}$

47. $y = |3x|$

48. $y = \left(\dfrac{x}{2}\right)^2$

49. $y = \left(\dfrac{x}{2}\right)^3$

50. $y = \left|\dfrac{x}{5}\right|$

For Problems 51–62,
(a) The graph of each function can be obtained from one of the basic graphs by two or more transformations. Describe the transformations.
(b) Sketch the basic graph and the graph of the given function by hand on the same axes. Label the coordinates of three points on the graph of the given function.

51. $f(x) = 2 + (x - 3)^2$

52. $f(x) = (x + 4)^2 + 1$

53. $g(z) = \dfrac{1}{z + 2} - 3$

54. $g(z) = \dfrac{1}{z - 1} + 1$

55. $F(u) = -3\sqrt{u + 4} + 4$

56. $F(u) = 4\sqrt{u - 3} - 5$

57. $G(t) = 2|t - 5| - 1$

58. $G(t) = 2 - |t + 4|$

59. $H(w) = 6 - \dfrac{2}{(w - 1)^2}$

60. $H(w) = \dfrac{3}{(w + 2)^2} - 1$

61. $f(t) = \sqrt[3]{t - 8} - 1$

62. $f(t) = \sqrt[3]{t + 1} + 8$

In Problems 63 and 64, each graph can be obtained by two transformations of the given graph. Describe the transformations and write a formula for the new graph in terms of f.

63.

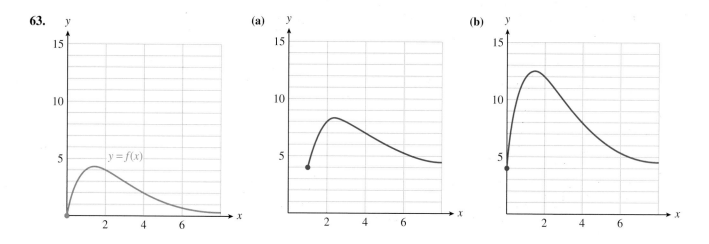

64.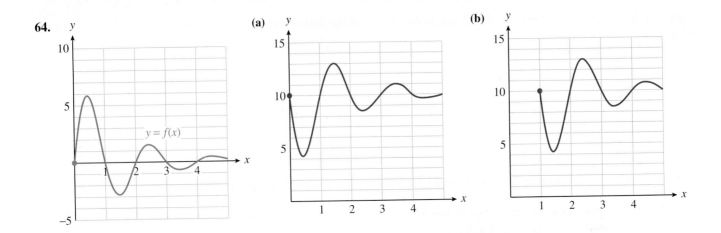

(a)

(b)

For Problems 65–70,
(a) Describe the graph as a transformation of a basic function.
(b) Give an equation for the function shown.

65.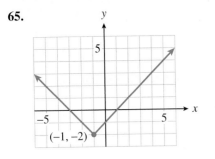

(−1, −2)

66.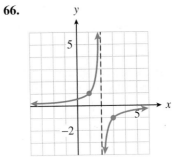

67.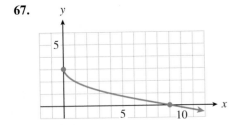

68.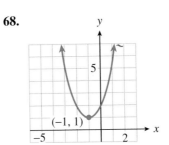

(−1, 1)

69.

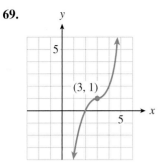

70.

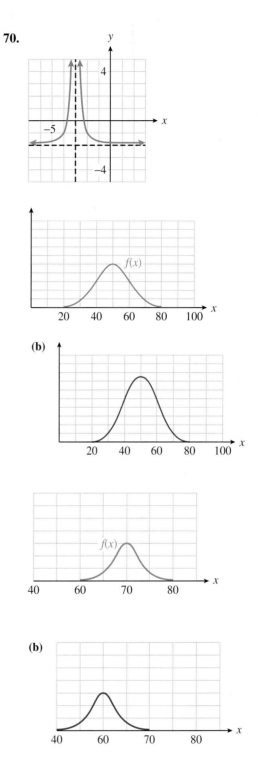

71. The graph of $f(x)$ shows the number of students in Professor Hilbert's class who scored x points on a quiz. Write a formula for each transformation of f ((a) and (b) of the figure below); then explain how the quiz results in that class compare to the results in Professor Hilbert's class.

(a)

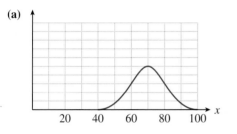

(b)

72. The graph of $f(x)$ shows the number of men at Tyler College who are x inches tall. Write a formula for each transformation of f; then explain how the heights in that population compare to the Tyler College men.

(a)

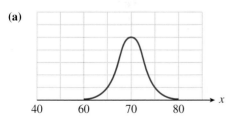

(b)

73. The graph of $f(x)$ shows the California state income tax rate, in percent, for a single taxpayer whose annual taxable income is x dollars. Write a formula for each transformation of f; then explain what it tells you about the income tax scheme in that state.

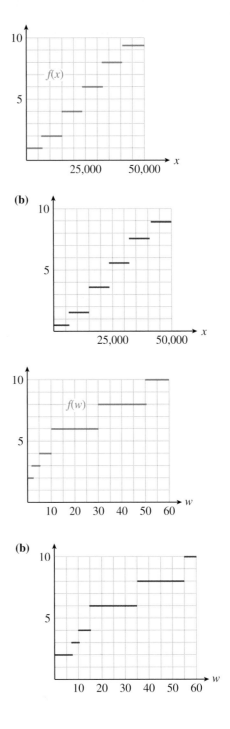

(a)

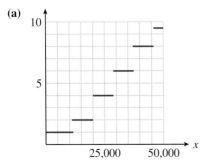

(b)

74. The graph of $f(w)$ shows the shipping rate at SendIt for a package that weighs w pounds. Write a formula for each transformation of f and explain how the shipping rates compare to the rates at SendIt.

(a)

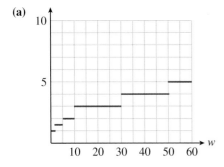

(b)

75. The graph of $g(t)$ shows the population of marmots in a national park t months after January 1. Write a formula for each transformation of f and explain how the population of that species compares to the population of marmots.

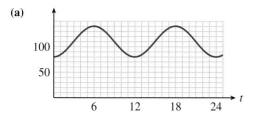

(a)

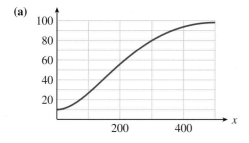

(b)

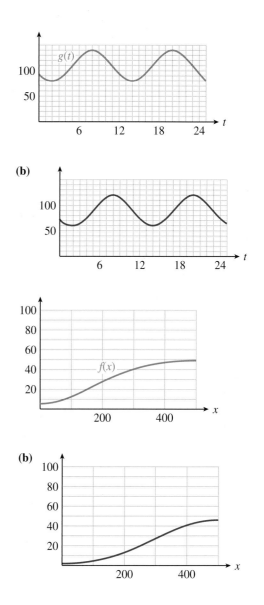

76. The graph of $f(x)$ is a *dose-response curve*. It shows the intensity of the response to a drug as a function of the dosage x milligrams administered. The intensity is given as a percentage of the maximum response. Write a formula for each transformation of f and explain what it tells you about the response to that drug.

(a)

(b)

2.4 ■ Functions as Mathematical Models

The Shape of the Graph

Creating a good model for a situation often begins with deciding what kind of function to use. An appropriate model can depend on very qualitative considerations, such as the general shape of the graph. What sort of function has the right shape to describe the process we want to model? Should it be increasing or decreasing, or some combination of both? Is the slope constant or is it changing? In Examples 1 and 2, we investigate how the shape of a graph illustrates the nature of the process it models.

EXAMPLE 1

Forrest leaves his house to go to school. For each of the following situations, sketch a possible graph of Forrest's distance from home as a function of time.

a. Forrest walks at a constant speed until he reaches the bus stop.

b. Forrest walks at a constant speed until he reaches the bus stop; then he waits there until the bus arrives.

c. Forrest walks at a constant speed until he reaches the bus stop, waits there until the bus arrives, and then the bus drives him to school at a constant speed.

Solutions   a. The graph is a straight-line segment, as shown in Figure 2.27a. It begins at the origin because at the instant that Forrest leaves the house, his distance from home is 0. (In other words, when $t = 0$, $y = 0$.) The graph is a straight line because Forrest has a constant speed. The slope of the line is equal to Forrest's walking speed.

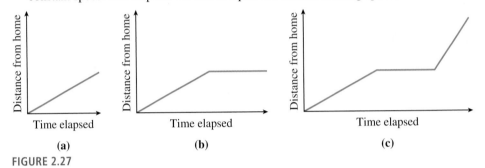

FIGURE 2.27

b. The graph begins just as the graph in part (a) does. But while Forrest waits for the bus, his distance from home remains constant, so the graph at that time is a horizontal line, as shown in Figure 2.27b. The line has slope 0 because while Forrest is waiting for the bus, his speed is 0.

c. The graph begins just as the graph in part (b) does. The last section of the graph represents the bus ride. It has a constant slope because the bus is moving at a constant speed. Because the bus (probably) moves faster than Forrest walks, the slope of this segment is greater than the slope for the walking section. The graph is shown in Figure 2.27c.

EXERCISE 1

Erin walks from her home to a convenience store, where she buys some cat food, and then walks back home. Sketch a possible graph of her distance from home as a function of time.

The graphs in Example 1 are piecewise linear, because Forrest traveled at a constant rate in each segment. In addition to choosing a graph that is increasing, decreasing, or constant to model a process, we can consider graphs that bend upward or downward. The bend is called the **concavity** of the graph.

EXAMPLE 2

The two functions described in this example are both increasing functions, but they increase in different ways. Match each function to its graph in Figure 2.28 and to the appropriate table of values.

a. The number of flu cases reported at an urban medical center during an epidemic is an increasing function of time, and it is growing at a faster and faster rate.

b. The temperature of a potato placed in a hot oven increases rapidly at first, then more slowly as it approaches the temperature of the oven.

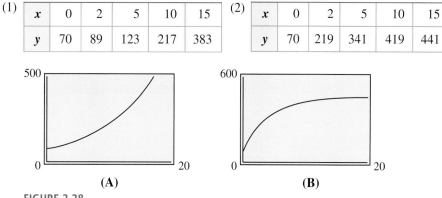

(1)

x	0	2	5	10	15
y	70	89	123	217	383

(2)

x	0	2	5	10	15
y	70	219	341	419	441

(A)   **(B)**

FIGURE 2.28

Solutions   **a.** The number of flu cases is described by graph (A) and table (1). The function values in table (1) increase at an increasing rate. We can see this by computing the rate of change over successive time intervals.

$x = 0$ to $x = 5$:

$$m = \frac{\Delta y}{\Delta x} = \frac{123 - 70}{5 - 0} = 10.6$$

$x = 5$ to $x = 10$:

$$m = \frac{\Delta y}{\Delta x} = \frac{217 - 123}{10 - 5} = 18.8$$

$x = 10$ to $x = 15$:

$$m = \frac{\Delta y}{\Delta x} = \frac{383 - 217}{15 - 10} = 33.2$$

The increasing rates can be seen in Figure 2.29; the graph bends upward as the slopes increase.

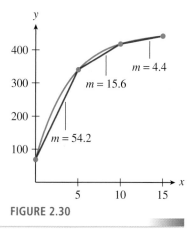

FIGURE 2.29

b. The temperature of the potato is described by graph (B) and table (2). The function values in table (2) increase, but at a decreasing rate.

$x = 0$ to $x = 5$:

$$m = \frac{\Delta y}{\Delta x} = \frac{341 - 70}{5 - 0} = 54.2$$

$x = 5$ to $x = 10$:

$$m = \frac{\Delta y}{\Delta x} = \frac{419 - 341}{10 - 5} = 15.6$$

$x = 10$ to $x = 15$:

$$m = \frac{\Delta y}{\Delta x} = \frac{441 - 419}{15 - 10} = 4.4$$

The decreasing slopes can be seen in Figure 2.30. The graph is increasing but bends downward.

FIGURE 2.30

A graph that bends upward is called **concave up,** and one that bends downward is **concave down.**

EXERCISE 2

Francine bought a cup of cocoa at the cafeteria. The cocoa cooled off rapidly at first, and then gradually approached room temperature. Which graph in Figure 2.31 more accurately reflects the temperature of the cocoa as a function of time? Explain why. Is the graph you chose concave up or concave down?

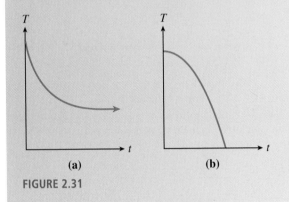

(a)   **(b)**

FIGURE 2.31

Using the Basic Functions as Models

We have considered some situations that can be modeled by linear functions. In this section, we will look at a few of the other basic functions. Example 3 illustrates an application of the function $f(x) = \sqrt{x}$.

EXAMPLE 3   The speed of sound is a function of the temperature of the air in kelvins. (The temperature, T, in kelvins is given by $T = C + 273$, where C is the temperature in degrees Celsius.) The table shows the speed of sound, s, in meters per second, at various temperatures, T.

T (°K)	0	20	50	100	200	400
s (m/sec)	0	89.7	141.8	200.6	283.7	401.2

  a. Plot the data to obtain a graph. Which of the basic functions does your graph most resemble?

  b. Find a value of k so that $s = kf(T)$ fits the data.

  c. On a summer night when the temperature is 20° Celsius, you see a flash of lightning, and 6 seconds later you hear the thunderclap. Use your function to estimate your distance from the thunderstorm.

Solutions   **a.** The graph of the data is shown in Figure 2.32. The shape of the graph reminds us of the square root function, $y = \sqrt{x}$.

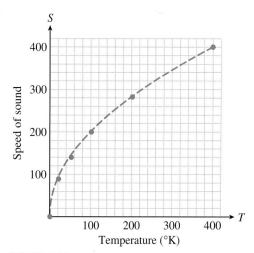

FIGURE 2.32

b. We are looking for a value of k so that the function $f(T) = k\sqrt{T}$ fits the data. We substitute one of the data points into the formula and solve for k. If we choose the point (100, 200.6), we obtain

$$200.6 = k\sqrt{100}$$

and solving for k yields $k = 20.06$. We can check that the formula $s = 20.06\sqrt{T}$ is a good fit for the rest of the data points as well. Thus, we suggest the function

$$f(T) = 20.06\sqrt{T}$$

as a model for the speed of sound.

c. First, use the model to calculate the speed of sound at a temperature of 20° Celsius. The Kelvin temperature is

$$T = 20 + 273 = 293$$

so we evaluate $s = f(T)$ for $T = 293$.

$$f(293) = 20.06\sqrt{293} \approx 343.4$$

Thus, s is approximately 343.4 meters per second.

The lightning and the thunderclap occur simultaneously, and the speed of light is so fast (about 30,000,000 meters per second) that we see the lightning flash as it occurs. So if the sound of the thunderclap takes 6 seconds after the flash to reach us, we can use our calculated speed of sound to find our distance from the storm.

$$\begin{aligned} \text{distance} &= \text{speed} \times \text{time} \\ &= (343.4 \text{ m/sec})\,(6 \text{ sec}) = 2060.4 \text{ meters} \end{aligned}$$

The thunderstorm is 2060 meters, or about 1.3 miles, away.

The ultraviolet index (UVI) is issued by the National Weather Service as a forecast of the amount of ultraviolet radiation expected to reach Earth around noon on a given day. The data show how much exposure to the sun people can take before risking sunburn.

UVI	2	3	4	5	6	8	10	12
Minutes to burn (more sensitive)	30	20	15	12	10	7.5	6	5
Minutes to burn (less sensitive)	150	100	75	60	50	37.5	30	25

a. Plot m, the minutes to burn, against u, the UVI, to obtain two graphs, one for people who are more sensitive to sunburn, and another for people less sensitive to sunburn. Which of the basic functions do your graphs most resemble?

b. For each graph, find a value of k so that $m = kf(u)$ fits the data.

At this point, a word of caution is in order. There is more to choosing a model than finding a curve that fits the data. A model based purely on the data is called an **empirical model.** However, many functions have similar shapes over small intervals of their input variables, and there may be several candidates that model the data. Such a model simply describes the general shape of the data set; the parameters of the model do not necessarily correspond to any actual process.

In contrast, **mechanistic models** provide insight into the biological, chemical, or physical process that is thought to govern the phenomenon under study. Parameters derived from mechanistic models are quantitative estimates of real system properties. Here is what GraphPad Software has to say about modeling:

> Choosing a model is a scientific decision. You should base your choice on your understanding of chemistry or physiology (or genetics, etc.). The choice should not be based solely on the shape of the graph.
>
> Some programs ... automatically fit data to hundreds or thousands of equations and then present you with the equation(s) that fit the data best. Using such a program is appealing because it frees you from the need to choose an equation. The problem is that the program has no understanding of the scientific context of your experiment. The equations that fit the data best are unlikely to correspond to scientifically meaningful models. You will not be able to interpret the best-fit values of the variables, and the results are unlikely to be useful for data analysis.
>
> (Source: *Fitting Models to Biological Data Using Linear and Nonlinear Regression,* Motulsky & Christopoulos, GraphPad Software, 2003)

Modeling with Piecewise Functions

Recall that a piecewise function is defined by different formulas on different portions of the *x*-axis.

EXAMPLE 4

In 2005, the income tax for a single taxpayer with a taxable income x under \$150,000 was given by the following table.

If taxpayer's income is . . .		Then the estimated tax is . . .		
Over	But not over	Base tax	+ Rate	Of the amount over
\$0	\$7300	\$0	10%	\$0
\$7300	\$29,700	\$730	15%	\$7300
\$29,700	\$71,950	\$4090	25%	\$29,700
\$71,950	\$150,150	\$14,652.50	28%	\$71,950

(Source: www.savewealth.com/taxes/rates)

a. Calculate the tax on incomes of \$500, \$29,700, and \$40,000.

b. Write a piecewise function for $T(x)$.

c. Graph the function $T(x)$.

Solutions   **a.** An income of $x = 500$ is in the first tax bracket, so the tax is

$$T(500) = 0 + 0.10(500) = 50$$

The income $x = 29,700$ is just on the upper edge of the second tax bracket. The amount over \$7300 is \$29,700 − \$7300, so

$$T(29,700) = 730 + 0.15(29,700 - 7300) = 4090$$

The income $x = 40,000$ is in the third bracket, so the tax is

$$T(40,000) = 4090 + 0.25(40,000 - 29,700) = 6665$$

> ▶▶▶
> See Section 1.5 to review the point-slope formula.

b. The first two columns of the table give the tax brackets, or the x-intervals on which each piece of the function is defined. In each bracket, the tax $T(x)$ is given by

Base tax + Rate · (Amount over bracket base)

For example, the tax in the second bracket is

$$T(x) = 730 + 0.15(x - 7300)$$

Writing the formulas for each of the four tax brackets gives us

$$T(x) = \begin{cases} 0.10x & 0 \le x \le 7300 \\ 730 + 0.15(x - 7300) & 7300 < x \le 29,700 \\ 4090 + 0.25(x - 29,700) & 29,700 < x \le 71,950 \\ 14,652.50 + 0.28(x - 71,950) & 71,950 < x \le 150,150 \end{cases}$$

c. The graph of T is piecewise linear. The first piece starts at the origin and has slope 0.10. The second piece is in point-slope form, $y = y_1 + m(x - x_1)$, so it has slope 0.15 and passes through the point $(7300, 730)$. Similarly, the third piece has slope 0.25 and passes through $(29,700, 40,490)$, and the fourth piece has slope 0.28 and passes through $(71,950, 14,652.5)$. You can check that for this function, all four pieces are connected at their endpoints, as shown in Figure 2.33.

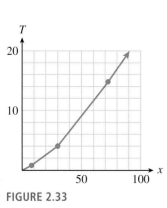

FIGURE 2.33

ANSWERS TO 2.4 EXERCISES

1.

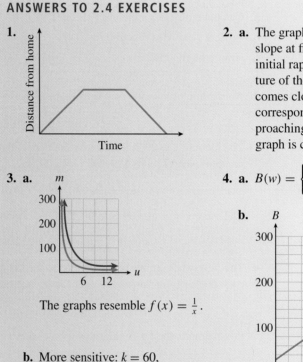

2. a. The graph has a steep negative slope at first, corresponding to an initial rapid drop in the temperature of the cocoa. The graph becomes closer to a horizontal line, corresponding to the cocoa approaching room temperature. The graph is concave up.

3. a.

The graphs resemble $f(x) = \frac{1}{x}$.

b. More sensitive: $k = 60$,
Less sensitive: $k = 300$

4. a. $B(w) = \begin{cases} 30 + 2w & 0 \le w \le 50 \\ 50 + 3w & w > 50 \end{cases}$

b.

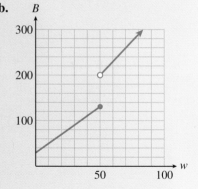

SECTION 2.4 SUMMARY

VOCABULARY   *Look up the definitions of new terms in the Glossary.*

Increasing	Decreasing	Concave up	Concave down
Empirical model	Mechanistic model		

1. The shape of a graph describes how the output variable changes.
2. A nonlinear graph may be concave up or concave down. If a graph is concave up, its slope is increasing. If it is concave down, its slope is decreasing.

3. The basic functions can be used to model physical situations.
4. Some situations can be modeled by piecewise functions.
5. Fitting a curve to the data is not enough to produce a useful model; appropriate scientific principles should also be considered.

STUDY QUESTIONS

1. Sketch the graph of a function whose slope is positive and increasing.
2. Sketch the graph of a function whose slope is positive and decreasing.
3. Which basic function is increasing but bending downward?

4. Which basic function is decreasing but bending upward?
5. Why is it bad practice to choose a model purely on the shape of the data plot?

SKILLS   *Practice each skill in the Homework problems listed.*

1. Sketch a graph whose shape models a situation: #1–18
2. Choose one of the basic graphs to fit a situation or a set of data: #19–24, 35–44
3. Decide whether the graph of a function is increasing or decreasing, concave up or concave down from a table of values: #25–28

4. Write and sketch a piecewise define function to model a situation: #45–48

HOMEWORK 2.4

ThomsonNOW™ Test yourself on key content at **www.thomsonedu.com/login**.

In Problems 1–4, which graph best illustrates each of the following situations?

1. Your pulse rate during an aerobics class

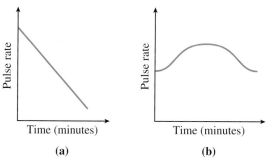

(a)   (b)

2. The stopping distances for cars traveling at various speeds

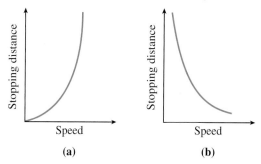

(a)   (b)

3. Your income in terms of the number of hours you worked

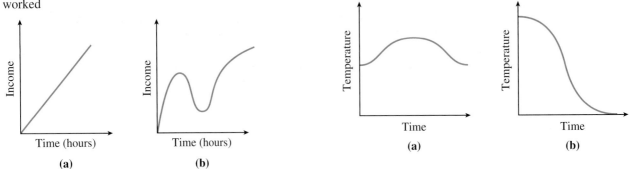

(a)

(b)

4. Your temperature during an illness

(a)

(b)

In Problems 5–8, sketch graphs to illustrate the following situations.

5. Halfway from your English class to your math class, you realize that you left your math book in the classroom. You retrieve the book, then walk to your math class. Graph the distance between you and your English classroom as a function of time, from the moment you originally leave the English classroom until you reach the math classroom.

6. After you leave your math class, you start off toward your music class. Halfway there you meet an old friend, so you stop and chat for a while. Then you continue to the music class. Graph the distance between you and your math classroom as a function of time, from the moment you leave the math classroom until you reach the music classroom.

7. Toni drives from home to meet her friend at the gym, which is halfway between their homes. They work out together at the gym; then they both go to the friend's home for a snack. Finally Toni drives home. Graph the distance between Toni and her home as a function of time, from the moment she leaves home until she returns.

8. While bicycling from home to school, Greg gets a flat tire. He repairs the tire in just a few minutes but decides to backtrack a few miles to a service station, where he cleans up. Finally, he bicycles the rest of the way to school. Graph the distance between Greg and his home as a function of time, from the moment he leaves home until he arrives at school.

Choose the graph that depicts the function described in Problems 9 and 10.

9. Inflation is still rising, but by less each month.

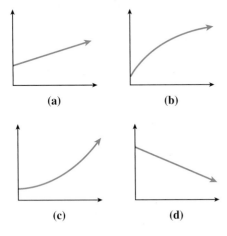

(a)   **(b)**

(c)   **(d)**

10. The price of wheat was rising more rapidly in 1996 than at any time during the previous decade.

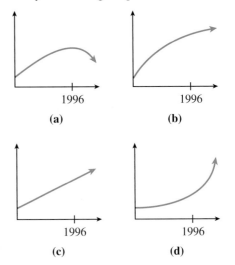

(a)   **(b)**

(c)   **(d)**

In Problems 11 and 12, match each graph with the function it illustrates.

11. **a.** The volume of a cylindrical container of constant height as a function of its radius
    b. The time it takes to travel a fixed distance as a function of average speed
    c. The simple interest earned at a given interest rate as a function of the investment
    d. The number of Senators present versus the number absent in the U.S. Senate

12. **a.** Unemployment was falling but is now steady.
    b. Inflation, which rose slowly until last month, is now rising rapidly.
    c. The birthrate rose steadily until 1990 but is now beginning to fall.
    d. The price of gasoline has fallen steadily over the past few months.

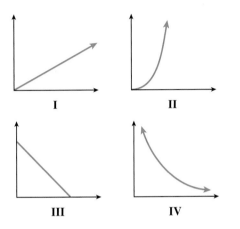

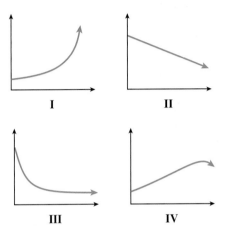

Sketch possible graphs to illustrate the situations described in Problems 13–18.

13. The height of a man as a function of his age, from birth to adulthood

14. The number of people willing to buy a new high-definition television, as a function of its price

15. The height of your head above the ground during a ride on a Ferris wheel

16. The height above the ground of a rubber ball dropped from the top of a 10-foot ladder

17. The average age at which women first marry decreased from 1940 to 1960, but it has been increasing since then.

18. When you learn a foreign language, the number of vocabulary words you know increases slowly at first, then increases more rapidly, and finally starts to level off.

Each situation in Problems 19–24 can be modeled by a transformation of a basic function. Name the basic function and sketch a possible graph.

19. The volume of a hot air balloon, as a function of its radius

20. The length of a rectangle as a function of its width, if its area is 24 square feet

21. The time it takes you to travel 600 miles, as a function of your average speed

22. The sales tax on a purchase, as a function of its price

23. The width of a square skylight, as a function of its area

24. The number of calories in a candy bar, as a function of its weight

In Problems 25–28, use the table of values to answer the questions.
(a) Based on the given values, is the function increasing or decreasing?
(b) Could the function be concave up, concave down, or linear?

25.

x	0	1	2	3	4
$f(x)$	1	1.5	2.25	3.375	5.0625

26.

x	0	1	2	3	4
$g(x)$	1	0.8	0.64	0.512	0.4096

27.

x	0	1	2	3	4
$s(x)$	0	0.174	0.342	0.5	0.643

28.

x	0	1	2	3	4
$c(x)$	1	0.985	0.940	0.866	0.766

In Problems 29–34,
(a) Is the graph increasing or decreasing, concave up or concave down?
(b) Match the graph of the function with the graph of its rate of change, shown in Figures A–F.

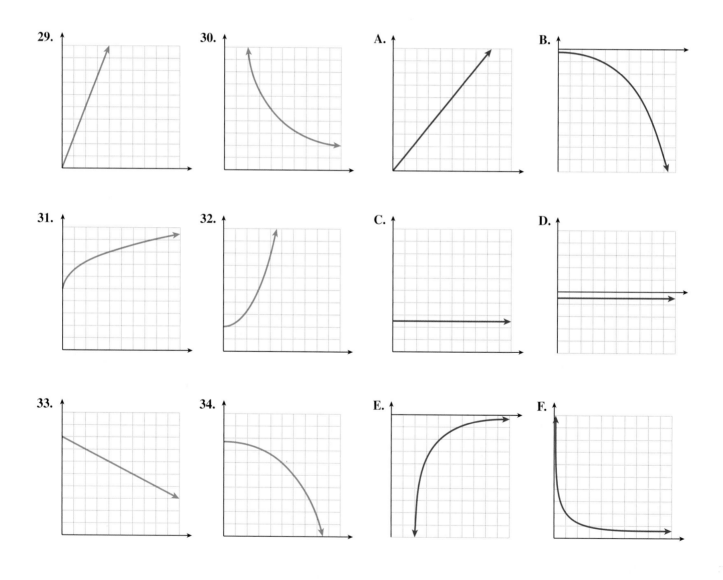

For Problems 35–40, plot the data; then decide which of the basic functions could describe the data.

35.

x	0	0.5	1	2	4
y	0	3.17	4	5.04	6.35

36.

x	0	0.5	1	2	4
y	0	5.66	8	11.31	16

37.

x	0.5	1	2	3	4
y	12	3	0.75	0.33	0.1875

38.

x	0.5	1	2	3	4
y	12	6	3	2	1.5

39.

x	0	0.5	1	2	3
y	0	0.125	0.5	2	4.5

40.

x	0	0.5	1	2	3
y	0	0.0125	0.1	0.8	2.7

41. Four different functions are described below. Match each description with the appropriate table of values and with its graph.

a. As a chemical pollutant pours into a lake, its concentration is a function of time. The concentration of the pollutant initially increases quite rapidly, but due to the natural mixing and self-cleansing action of the lake, the concentration levels off and stabilizes at some saturation level.

b. An overnight express train travels at a constant speed across the Great Plains. The train's distance from its point of origin is a function of time.

c. The population of a small suburb of a Florida city is a function of time. The population began increasing rather slowly, but it has continued to grow at a faster and faster rate.

d. The level of production at a manufacturing plant is a function of capital outlay, that is, the amount of money invested in the plant. At first, small increases in capital outlay result in large increases in production, but eventually the investors begin to experience diminishing returns on their money, so that although production continues to increase, it is at a disappointingly slow rate.

(1)

x	1	2	3	4	5	6	7	8
y	60	72	86	104	124	149	179	215

(2)

x	1	2	3	4	5	6	7	8
y	60	85	103	120	134	147	159	169

(3)

x	1	2	3	4	5	6	7	8
y	60	120	180	240	300	360	420	480

(4)

x	1	2	3	4	5	6	7	8
y	60	96	118	131	138	143	146	147

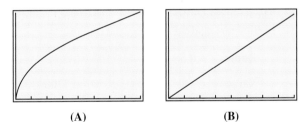

(A)   **(B)**

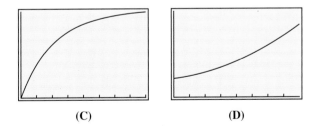

(C)   **(D)**

42. Four different functions are described below. Match each description with the appropriate table of values and with its graph.

a. Fresh water flowing through Crystal Lake has gradually reduced the phosphate concentration to its natural level, and it is now stable.

b. The number of bacteria in a person during the course of an illness is a function of time. It increases rapidly at first, then decreases slowly as the patient recovers.

c. A squirrel drops a pine cone from the top of a California redwood. The height of the pine cone is a function of time, decreasing ever more rapidly as gravity accelerates its descent.

d. Enrollment in Ginny's Weight Reduction program is a function of time. It began declining last fall. After the holidays, enrollment stabilized for a while but soon began to fall off again.

(1)

x	0	1	2	3	4
y	160	144	96	16	0

(2)

x	0	1	2	3	4
y	20	560	230	90	30

(3)

x	0	1	2	3	4
y	480	340	240	160	120

(4)

x	0	1	2	3	4
y	250	180	170	150	80

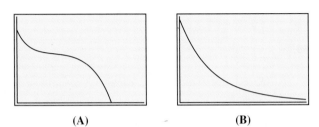

(A)   (B)

(C)   (D)

43. The table shows the radii, r, of several gold coins, in centimeters, and their value, v, in dollars.

Radius	0.5	1	1.5	2	2.5
Value	200	800	1800	3200	5000

a. Which graph represents the data?

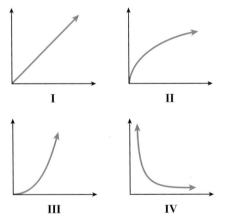

I   II

III   IV

b. Which equation describes the function?

I. $v = k\sqrt{r}$ **II.** $v = kr$

III. $v = kr^2$ **IV.** $v = \dfrac{k}{r}$

44. The table shows how the amount of water, A, flowing past a point on a river is related to the width, W, of the river at that point.

Width (feet)	11	23	34	46
Amount of water (ft³/sec)	23	34	41	47

a. Which graph represents the data?

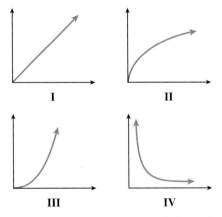

I   II

III   IV

b. Which equation describes the information?

I. $A = k\sqrt{W}$ **II.** $A = kW$

III. $A = kW^2$ **IV.** $A = \dfrac{k}{W}$

45. If you order from Coldwater Creek, the shipping charges are given by the following table.

Purchase amount	Shipping charge
Up to $25	$5.95
$25.01 to $50	$7.95
$50.01 to $75	$9.95
$75.01 to $100	$10.95

  a. Write a piecewise formula for $S(x)$, the shipping charge as a function of the purchase amount, x.
  b. Graph $S(x)$.

46. The Bopp-Busch Tool and Die Company markets its products to individuals, to contractors, and to wholesale distributors. The company offers three different price structures for its toggle bolts. If you order 20 or fewer boxes, the price is $2.50 each. If you order more than 20 but no more than 50 boxes, the price is $2.25 each. If you order more than 50 boxes, the price is $2.10 each.
  a. Write a piecewise formula for $C(x)$, the cost of ordering x boxes of toggle bolts.
  b. Graph $C(x)$.

47. Bob goes skydiving on his birthday. The function $h(t)$ approximates Bob's altitude t seconds into the trip.

$$h(t) = \begin{cases} 25t & 0 \le t < 400 \\ 10{,}000 & 400 \le t < 500 \\ 10{,}000 - 16(t-500)^2 & 500 \le t < 520 \\ 3600 - 120(t-520) & 520 \le t \le 550 \end{cases}$$

  a. Graph $h(t)$. Describe what you think is happening during each piece of the graph.
  b. Find two times when Bob is at an altitude of 6000 feet.

48. Jenni lives in the San Fernando Valley, where it is hot during summer days but cools down at night. Jenni uses the air conditioner as little as possible. The function $T(h)$ approximates the temperature in Jenni's house h hours after midnight.

$$T(h) = \begin{cases} 65 & 0 \le h < 8 \\ 65 + 5h & 8 \le h < 14 \\ 95 - \dfrac{40}{h} & 14 \le h < 16 \\ 75 & 16 \le h < 20 \\ 75 - 2h & 20 \le h < 24 \end{cases}$$

  a. Graph $T(h)$. Describe what you think is happening during each piece of the graph.
  b. Find two times when the temperature inside the house is 85° Fahrenheit.

49. Lead nitrate and potassium iodide react in solution to produce lead iodide, which settles out, or precipitates, as a yellow solid at the bottom of the container. As you add more lead nitrate to the solution, more lead iodide is produced until all the potassium iodide is used up. The table shows the height of the precipitate in the container as a function of the amount of lead nitrate added.

(Source: Hunt and Sykes, 1984)

Lead nitrate solution (cc)	0.5	1.0	1.5	2.0
Height of precipitate (mm)	2.8	4.8	6.2	7.4

Lead nitrate solution (cc)	2.5	3.0	3.5	4.0
Height of precipitate (mm)	9.5	9.6	9.6	9.6

a. Plot the data. Sketch a piecewise linear function with two parts to fit the data points.
b. Calculate the slope of the increasing part of the graph, including units. What is the significance of the slope?
c. Write a formula for your piecewise function.
d. Interpret your graph in the context of the problem.

50. The graph shows the temperature of 1 gram of water as a function of the amount of heat applied, in calories. Recall that water freezes at 0°C and boils at 100°C.

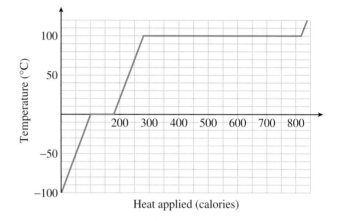

a. How much heat is required to raise the temperature of 1 gram of water by 1 degree?
b. How much heat is required to convert 1 gram of ice to water?
c. How much heat is required to convert 1 gram of water to steam?
d. Write a piecewise function to describe the graph.

51. As the global population increases, many scientists believe it is approaching, or has already exceeded, the maximum number the Earth can sustain. This maximum number, or *carrying capacity,* depends on the finite natural resources of the planet—water, land, air, and materials—but also on how people use and preserve the resources. The graphs show four different ways that a growing population can approach its carrying capacity over time. (Source: Meadows, Randers, and Meadows, 2004)

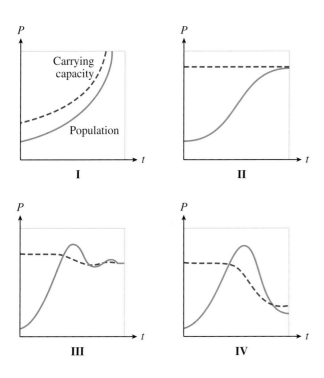

Match each graph to one of the scenarios described in (a)–(d) and explain your choice.

a. Sigmoid growth: The population levels off smoothly below the carrying capacity.

b. Overshoot and collapse: The population exceeds the carrying capacity with severe damage to the resource base and is forced to decline rapidly to achieve a new balance with a reduced carrying capacity.

c. Continued growth: The carrying capacity is far away, or growing faster than the population.

d. Overshoot and oscillation: The population exceeds the carrying capacity without inflicting permanent damage, then oscillates around the limit before leveling off.

52. The introduction of a new species into an environment can affect the growth of an existing species in various ways. The graphs show four hypothetical scenarios after Species A is introduced into an environment where Species B is established.

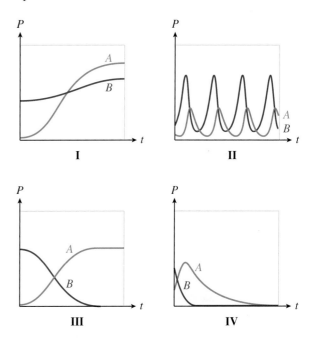

Match each graph to one of the scenarios described in (a)–(d) and explain your choice.

a. Predator-prey (sustained): Species A becomes a predator population that grows when its prey, Species B, is abundant, but declines when the prey population is small. The prey population grows when predators are scarce but shrinks when predators are abundant.

b. Predator-prey (extinction): Species A becomes a predator population that annihilates Species B, but then Species A itself declines toward extinction.

c. Competition: Species A and B have a common food source, and the Species A replaces Species B in the environment.

d. Symbiosis: Species A and B help each other to grow.

The absolute value function is used to model problems involving distance. Recall that the absolute value of a number gives the distance from the origin to that number on the number line.

> **Distance and Absolute Value**
>
> The **distance** between two points x and a is given by $|x - a|$.

For example, the equation $|x - 2| = 6$ means "the distance between x and 2 is 6 units." The number x could be to the left or the right of 2 on the number line. Thus, the equation has two solutions, 8 and -4, as shown in Figure 2.34.

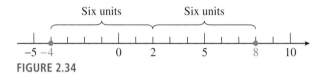

FIGURE 2.34

EXAMPLE 1   Write each statement using absolute value notation. Illustrate the solutions on a number line.

a. x is three units from the origin.

b. p is two units from 5.

c. a is within four units of -2.

Solutions   First, restate each statement in terms of distance.

a. The distance between x and the origin is three units, or $|x| = 3$. Thus, x can be 3 or -3. See Figure 2.35a.

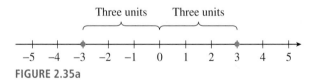

FIGURE 2.35a

b. The distance between p and 5 is two units, or $|p - 5| = 2$. If we count two units on either side of 5, we see that p can be 3 or 7. See Figure 2.35b.

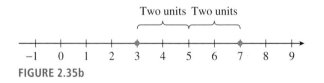

FIGURE 2.35b

c. The distance between a and -2 is less than four units, or $|a - (-2)| < 4$, or $|a + 2| < 4$. Count four units on either side of -2, to find -6 and 2. Then a is between -6 and 2, or $-6 < a < 2$. See Figure 2.35c on page 193.

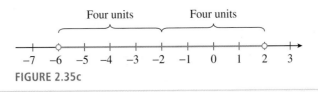

FIGURE 2.35c

EXERCISE 1

Write each statement using absolute value notation; then illustrate the solutions on a number line.

a. x is five units away from -3.

b. x is at least six units away from 4.

Absolute Value Equations

We can use number lines to solve equations such as

$$|3x - 6| = 9$$

First, factor out the coefficient of x, to get $|3(x - 2)| = 9$. Because of the multiplicative property of the absolute value, namely that $|ab| = |a||b|$, we can write the left side as

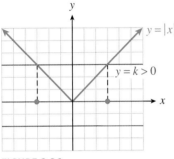

FIGURE 2.36

$$|3||x - 2| = 9$$
$$3|x - 2| = 9 \quad \text{Divide both sides by 3.}$$
$$|x - 2| = 3$$

which tells us that the distance between x and 2 is 3 units, so $x = -1$ or $x = 5$.

Graphs can be helpful for working with absolute values. Consider the simple equation $|x| = 5$, which has two solutions, $x = 5$ and $x = -5$. In fact, we can see from the graph in Figure 2.36 that the equation $|x| = k$ has two solutions if $k > 0$, one solution if $k = 0$, and no solution if $k < 0$.

EXAMPLE 2

a. Use a graph of $y = |3x - 6|$ to solve the equation $|3x - 6| = 9$.

b. Use a graph of $y = |3x - 6|$ to solve the equation $|3x - 6| = -2$.

Solutions

a. Figure 2.37 shows the graphs of $y = |3x - 6|$ and $y = 9$. We see that there are two points on the graph of $y = |3x - 6|$ that have $y = 9$, and those points have x-coordinates $x = -1$ and $x = 5$. We can verify algebraically that the solutions are -1 and 5.

$$x = -1: \quad |3(-1) - 6| = |-9| = 9$$
$$x = 5: \quad |3(5) - 6| = |9| = 9$$

b. There are no points on the graph of $y = |3x - 6|$ with $x = -2$, so the equation $|3x - 6| = -2$ has no solutions.

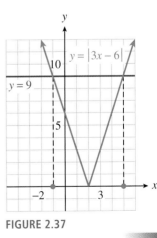

FIGURE 2.37

Solving Absolute Value Equations

We can use a graphing calculator to solve the equations in Example 2. Figure 2.38 shows the graphs of $Y_1 = \mathbf{abs}(3X - 6)$ and $Y_2 = 9$ in the window

$$\text{Xmin} = -2.7 \quad \text{Xmax} = 6.7$$
$$\text{Ymin} = -2 \quad \text{Ymax} = 12$$

We use the **Trace** or the **intersect** feature to locate the intersection points at $(-1, 9)$ and $(5, 9)$.

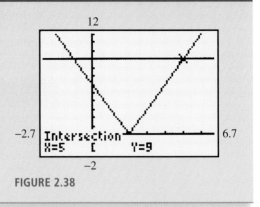

FIGURE 2.38

EXERCISE 2

a. Graph $y = |2x + 7|$ for $-12 \leq x \leq 8$.

b. Use your graph to solve the equation $|2x + 7| = 11$.

To solve an absolute value equation algebraically, we use the definition of absolute value.

EXAMPLE 3 Solve the equation $|3x - 6| = 9$ algebraically.

Solution We write the piecewise definition of $|3x - 6|$.

$$|3x - 6| = \begin{cases} 3x - 6 & \text{if } 3x - 6 \geq 0, \text{ or } x \geq 2 \\ -(3x - 6) & \text{if } 3x - 6 < 0, \text{ or } x < 2 \end{cases}$$

Thus, the absolute value equation $|3x - 6| = 9$ is equivalent to two regular equations:

$$3x - 6 = 9 \quad \text{or} \quad -(3x - 6) = 9$$

or, by simplifying the second equation,

$$3x - 6 = 9 \quad \text{or} \quad 3x - 6 = -9$$

Solving these two equations gives us the same solutions we found in Example 5, $x = 5$ and -1.

In general, we have the following strategy for solving absolute value equations.

Absolute Value Equations

The equation

$$|ax + b| = c \quad (c > 0)$$

is equivalent to

$$ax + b = c \quad \text{or} \quad ax + b = -c$$

Solve $|2x + 7| = 11$ algebraically.

Absolute Value Inequalities

We can also use graphs to solve absolute value inequalities. Look again at the graph of $y = |3x - 6|$ in Figure 2.39a. Because of the V-shape of the graph, all points with y-values less than 9 lie *between* the two solutions of $|3x - 6| = 9$, that is, between -1 and 5. Thus, the solutions of the inequality $|3x - 6| < 9$ are $-1 < x < 5$. (In the Homework problems, you will be asked to show this algebraically.)

On the other hand, to solve the inequality $|3x - 6| > 9$, we look for points on the graph with y-values greater than 9. In Figure 2.39b, we see that these points have x-values *outside* the interval between -1 and 5. In other words, the solutions of the inequality $|3x - 6| > 9$ are $x < -1$ or $x > 5$.

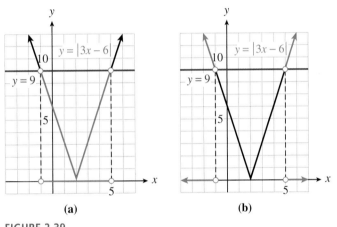

(a) (b)

FIGURE 2.39

Thus, we can solve an absolute value inequality by first solving the related equation.

Absolute Value Inequalities

Suppose the solutions of the equation $|ax + b| = c$ are r and s, with $r < s$. Then

1. The solutions of $|ax + b| < c$ are

$$r < x < s$$

2. The solutions of $|ax + b| > c$ are

$$x < r \quad \text{or} \quad x > s$$

EXAMPLE 4   Solve $|4x - 15| < 0.01$.

Solution   First, we solve the equation $|4x - 15| = 0.01$. There are two cases:

$$4x - 15 = 0.01 \quad \text{or} \quad 4x - 15 = -0.01$$
$$4x = 15.01 \qquad\qquad 4x = 14.99$$
$$x = 3.7525 \qquad\qquad x = 3.7475$$

Because the inequality symbol is $<$, the solutions of the inequality are between these two values: $3.7475 < x < 3.7525$. In interval notation, the solutions are $(3.7475, 3.7525)$.

EXERCISE 4

a. Solve the inequality $|2x + 7| < 11$.

b. Solve the inequality $|2x + 7| > 11$.

Using the Absolute Value in Modeling

In Example 5, we use the absolute value to model a problem about distances.

EXAMPLE 5   Marlene is driving to a new outlet mall on Highway 17. There is a gas station at Marlene's on-ramp, where she buys gas and resets her odometer to zero before getting on the highway. The mall is only 15 miles from Marlene's on-ramp, but she mistakenly drives past the mall and continues down the highway. Marlene's distance from the mall is a function of how far she has driven on Highway 17. See Figure 2.40.

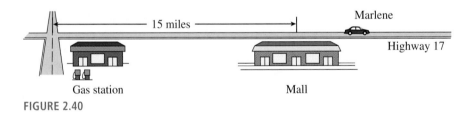

FIGURE 2.40

a. Make a table of values showing how far Marlene has driven on Highway 17 and how far she is from the mall.

b. Make a graph of Marlene's distance from the mall versus the number of miles she has driven on the highway. Which of the basic graphs from Section 2.2 does your graph most resemble?

c. Find a piecewise defined formula that describes Marlene's distance from the mall as a function of the distance she has driven on the highway.

Solutions   **a.** Marlene gets closer to the mall for each mile that she has driven on the highway until she has driven 15 miles, and after that she gets farther from the mall.

Miles on highway	0	5	10	15	20	25	30
Miles from mall	15	10	5	0	5	10	15

b. Plot the points in the table to obtain the graph shown in Figure 2.41. This graph looks like the absolute value function defined in Section 2.2, except that the vertex is the point (15, 0) instead of the origin.

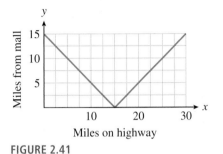

FIGURE 2.41

c. Let x represent the number of miles on the highway and $f(x)$ the number of miles from the mall. For x-values less than 15, the graph is a straight line with slope -1 and y-intercept at (0, 15), so its equation is $y = -x + 15$. Thus,

$$f(x) = -x + 15 \quad \text{when} \quad 0 \le x < 15$$

On the other hand, when $x \ge 15$, the graph of f is a straight line with slope 1 that passes through the point (15, 0). The point-slope form of this line is

$$y = 0 + 1(x - 15)$$

so $y = x - 15$. Thus,

$$f(x) = x - 15 \quad \text{when } x \ge 15$$

Combining the two pieces, we obtain

$$f(x) = \begin{cases} -x + 15, & \text{when } 0 \le x < 15 \\ x - 15, & \text{when } x \ge 15 \end{cases}$$

The graph of $f(x)$ is a part of the graph of $y = |x - 15|$. If we think of the highway as a portion of the real number line, with Marlene's on-ramp located at the origin, then the outlet mall is located at 15. Marlene's coordinate as she drives along the highway is x, and the distance from Marlene to the mall is given by $f(x) = |x - 15|$.

EXERCISE 5

a. Use the graph in Example 5 to determine how far Marlene has driven when she is within 5 miles of the mall. Write and solve an absolute value inequality to verify your answer.

b. Write and solve an absolute value inequality to determine how far Marlene has driven when she is at least 10 miles from the mall.

Measurement Error

If you weigh a sample in chemistry lab, the scale's digital readout might show 6.0 grams. But it is unlikely that the sample weighs *exactly* 6 grams; there is always some error in measured values. Because the scale shows the weight as 6.0 grams, we know that the true weight of the sample must be between 5.95 grams and 6.05 grams: If the weight were less than 5.95 grams, the scale would round down to 5.9 grams, and if the weight were more than 6.05 grams, the scale would round up to 6.1 grams. We should report the mass of the sample as 6 ± 0.05 grams, which tells the reader that the error in the measurement is no more than 0.05 grams.

We can also describe this measurement error, or **error tolerance,** using an absolute value inequality. Because the measured mass m can be no more than 0.05 from 6, we write

$$|m - 6| \le 0.05$$

Note that the solution of this inequality is $5.95 \le m \le 6.05$.

EXAMPLE 6

a. The specifications for a computer chip state that its thickness in millimeters must satisfy $|t - 0.023| < 0.001$. What are the acceptable values for the thickness of the chip?

b. The safe dosage of a new drug is between 250 and 450 milligrams, inclusive. Write the safe dosage as an error tolerance involving absolute values.

Solutions

a. The error tolerance can also be stated as $t = 0.023 \pm 0.001$ millimeters, so the acceptable values are between 0.022 and 0.024 millimeters.

b. The safe dosage d satisfies $250 \le d \le 450$, as shown in Figure 2.42.

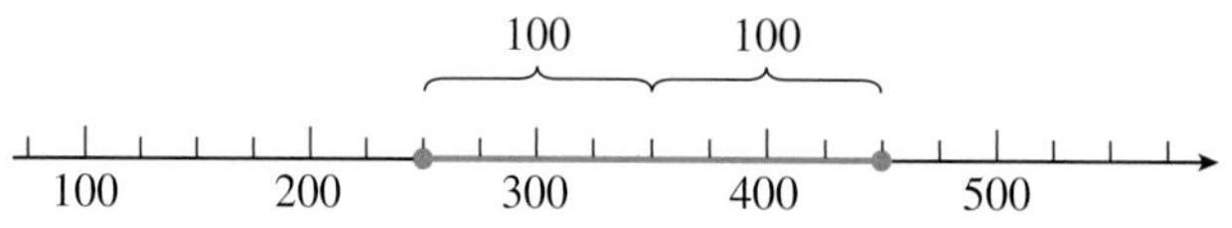

FIGURE 2.42

The center of this interval is 350, and the endpoints are each 100 units from the center. Thus, the safe values are within 100 units of 350, or

$$|d - 350| \le 100$$

EXERCISE 6

The temperature, T, in a laboratory must remain between 9°C and 12°C.

a. Write the error tolerance as an absolute value inequality.

b. For a special experiment, the temperature in degrees celsius must satisfy $|T - 6.7| \le 0.03$. Give the interval of possible temperatures.

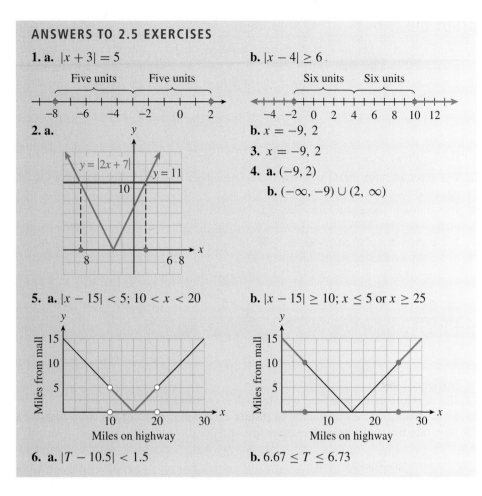

ANSWERS TO 2.5 EXERCISES

1. a. $|x + 3| = 5$

Five units   Five units

$$\begin{array}{c} -8 \quad -6 \quad -4 \quad -2 \quad 0 \quad 2 \end{array}$$

b. $|x - 4| \geq 6$

Six units   Six units

$$\begin{array}{c} -4 \quad -2 \quad 0 \quad 2 \quad 4 \quad 6 \quad 8 \quad 10 \quad 12 \end{array}$$

2. a.

$y = |2x + 7|$, $y = 11$

b. $x = -9, 2$

3. $x = -9, 2$

4. a. $(-9, 2)$

    b. $(-\infty, -9) \cup (2, \infty)$

5. a. $|x - 15| < 5; \ 10 < x < 20$

b. $|x - 15| \geq 10; \ x \leq 5 \text{ or } x \geq 25$

Miles from mall / Miles on highway

6. a. $|T - 10.5| < 1.5$

b. $6.67 \leq T \leq 6.73$

SECTION 2.5 SUMMARY

VOCABULARY   *Look up the definitions of new terms in the Glossary.*

Absolute value equation   Absolute value inequality   Error tolerance

CONCEPTS

1. The absolute value is used to model distance: The **distance** between two points x and a is given by $|x - a|$.

2.

Absolute Value Equations

The equation

$$|ax + b| = c \quad (c > 0)$$

is equivalent to

$$ax + b = c \quad \text{or} \quad ax + b = -c$$

3.

Absolute Value Inequalities

Suppose the solutions of the equation $|ax + b| = c$ are r and s, with $r < s$. Then
1. The solutions of $|ax + b| < c$ are

$$r < x < s$$

2. The solutions of $|ax + b| > c$ are

$$x < r \quad \text{or} \quad x > s$$

4. The **error tolerance** e in a measurement M can be expressed as $|x - M| < e$, or as $x = M \pm e$.

Both indicate that $M - e < x < M + e$.

STUDY QUESTIONS

1. Write a function that models the distance between x and a fixed point k on the number line.
2. For what values of c does the equation $|ax + b| = c$ have one solution? No solution?
3. If you know that the solutions of $|ax + b| < c$ are $-3 < x < 6$, what are the solutions of $|ax + b| > c$?
4. What is the center of the interval $[220, 238]$?
5. What is the center of the interval $[a, b]$?

SKILLS   *Practice each skill in the Homework problems listed.*

1. Use absolute value notation to write statements about distance: #1–8
2. Use graphs to solve absolute value equations and inequalities: #9–12
3. Solve absolute value equations: #13–24
4. Solve absolute value inequalities: #25–40
5. Express error tolerances using absolute value notation: #41–48
6. Analyze absolute value functions: #49–56
7. Model problems about distance using the absolute value function: #57–60

HOMEWORK 2.5

ThomsonNOW™ Test yourself on key content at www.thomsonedu.com/login.

In Problems 1–8,
(a) Use absolute value notation to write each expression as an equation or an inequality. (It may be helpful to restate each sentence using the word *distance*.)
(b) Illustrate the solutions on a number line.

1. x is six units from the origin.

2. a is seven units from the origin.

3. The distance from p to -3 is five units.

4. The distance from q to -7 is two units.

5. t is within three units of 6.

6. w is no more than one unit from -5.

7. b is at least 0.5 unit from -1.

8. m is more than 0.1 unit from 8.

9. Graph $y = |x + 3|$. Use your graph to solve the following equations and inequalities.
   a. $|x + 3| = 2$ **b.** $|x + 3| \leq 4$
   c. $|x + 3| > 5$

10. Graph $y = |x - 2|$. Use your graph to solve the following equations and inequalities.
    a. $|x - 2| = 5$ **b.** $|x - 2| < 8$
    c. $|x - 2| \geq 4$

11. Graph $y = |2x - 8|$. Use your graph to solve the following equations and inequalities.
    a. $|2x - 8| = 0$ **b.** $|2x - 8| = -2$
    c. $|2x - 8| < -6$

12. Graph $y = |4x + 8|$. Use your graph to solve the following equations and inequalities.
    a. $|4x + 8| = 0$ **b.** $|4x + 8| < 0$
    c. $|4x + 8| > -3$

Solve.

13. $|2x - 1| = 4$ **14.** $|3x - 1| = 5$ **15.** $0 = |7 + 3q|$ **16.** $|-11 - 5t| = 0$

17. $4 = \dfrac{|b + 2|}{3}$ **18.** $6|n + 2| = 9$ **19.** $|2(w - 7)| = 1$ **20.** $2 = \left|\dfrac{a - 4}{5}\right|$

21. $|c - 2| + 3 = 1$ **22.** $5 = 4 - |h + 3|$ **23.** $-7 = |2m + 3|$ **24.** $|5r - 3| = -2$

Solve.

25. $|2x + 6| < 3$ **26.** $|5 - 3x| \le 1$ **27.** $7 \le |3 - 2d|$ **28.** $10 < |3r + 2|$

29. $|6s + 15| > -3$ **30.** $|8b - 12| < -4$ **31.** $|t - 1.5| < 0.1$ **32.** $|z - 2.6| \le 0.1$

33. $|T - 3.25| \ge 0.05$ **34.** $|P - 0.6| > 0.01$ **35.** $-1 \ge \left|\dfrac{n - 3}{2}\right|$ **36.** $-0.1 \le |9(p + 2)|$

In Problems 37–40, give an interval of possible values for the measurement.

37. The length, l, of a rod is given by $|l - 4.3| < 0.001$, in centimeters.

38. The mass, m, of the device shall be $|m - 450| < 4$, in grams.

39. The candle will burn for t minutes, where $|t - 300| \le 50$.

40. The ramp will have angle of inclination α, and $|\alpha - 10°| \le 0.5°$.

In Problems 41–44, write the error tolerance using absolute values.

41. The chemical compound must be maintained at a temperature, T, between $4.7°$ and $5.3°$C.

42. The diameter, d, of the hole shall be in the range of 24.98 to 25.02 centimeters.

43. The subject will receive a dosage D from 95 to 105 milligrams of the drug.

44. The pendulum swings out and back in a time period t between 0.9995 and 1.0005 seconds.

45. An electrical component of a high-tech sensor requires 0.25 ounces of gold. Assume that the actual amount of gold used, g, is not in error by more than 0.001 ounce. Write an absolute value inequality for the possible error and show the possible values of g on a number line.

46. In a pasteurization process, milk is to be irradiated for 10 seconds. The actual period t of irradiation cannot be off by more than 0.8 second. Write an absolute value inequality for the possible error and show the possible values of t on a number line.

47. In a lab assignment, a student reports that a chemical reaction required 200 minutes to complete. Let t represent the actual time of the reaction.

   a. Write an absolute value inequality for t, assuming that the student rounded his answer to the nearest 100 minutes. Give the smallest and largest possible value for t. (*Hint:* What is the shortest time that would round to 200 minutes? The greatest time?)

   b. Write an absolute value inequality for t, assuming that the student rounded his answer to the nearest minute. Give the smallest and largest possible value for t.

   c. Write an absolute value inequality for t, assuming that the student rounded his answer to the nearest 0.1 minute. Give the smallest and largest possible value for t.

48. An espresso machine has a square metal plate. The side of the plate is 2 ± 0.01 cm.

   a. Write an absolute value inequality for the length of the side, s. Give the smallest and largest possible value for s.

   b. Compute the smallest and largest possible area of the plate, including units.

   c. Write an absolute value inequality for the area, A.

49. **a.** Write the piecewise definition for $|3x - 6|$.

   b. Use your answer to part (a) to write two inequalities that together are equivalent to $|3x - 6| < 9$.

   c. Solve the inequalities in part (b) and check that the solutions agree with the solutions of $|3x - 6| < 9$.

   d. Show that $|3x - 6| < 9$ is equivalent to the compound inequality $-9 < 3x - 6 < 9$.

50. **a.** Write the piecewise definition for $|3x - 6|$.

   b. Use your answer to part (a) to write two inequalities that together are equivalent to $|3x - 6| > 9$.

   c. Solve the inequalities in part (b) and check that the solutions agree with the solutions of $|3x - 6| > 9$.

   d. Show that $|3x - 6| > 9$ is equivalent to the compound inequality $3x - 6 < -9$ or $3x - 6 > 9$.

51. **a.** Write the piecewise definition for $|2x + 5|$.

   b. Use your answer to part (a) to write two inequalities that together are equivalent to $|2x + 5| > 7$.

   c. Solve the inequalities in part (b) and check that the solutions agree with the solutions of $|2x + 5| > 7$.

   d. Show that $|2x + 5| > 7$ is equivalent to the compound inequality $2x + 5 < -7$ or $2x + 5 > 7$.

52. **a.** Write the piecewise definition for $|2x + 5|$.

   b. Use your answer to part (a) to write two inequalities that together are equivalent to $|2x + 5| < 7$.

   c. Solve the inequalities in part (b) and check that the solutions agree with the solutions of $|2x + 5| < 7$.

   d. Show that $|2x + 5| < 9$ is equivalent to the compound inequality $-7 < 2x + 5 < 7$.

For Problems 53–56, graph the function and answer the questions.

53. $f(x) = |x + 4| + |x - 4|$

   a. Using your graph, write a piecewise formula for $f(x)$.

   b. Experiment by graphing $g(x) = |x + p| + |x - q|$ for different positive values of p and q. Make a conjecture about how the graph depends on p and q.

   c. Write a piecewise formula for $g(x) = |x + p| + |x - q|$.

54. $f(x) = |x + 4| - |x - 4|$

   a. Using your graph, write a piecewise formula for $f(x)$.

   b. Experiment by graphing $g(x) = |x + p| - |x - q|$ for different positive values of p and q. Make a conjecture about how the graph depends on p and q.

   c. Write a piecewise formula for $g(x) = |x + p| - |x - q|$.

55. $f(x) = |x + 4| + |x| + |x - 4|$

   a. Using your graph, write a piecewise formula for $f(x)$.

   b. What is the minimum value of $f(x)$?

   c. If $p, q \geq 0$, what is the minimum value of $g(x) = |x + p| + |x| + |x - q|$?

56. $f(x) = |x + 4| - |x| + |x - 4|$

   a. Using your graph, write a piecewise formula for $f(x)$.

   b. What is the minimum value of $f(x)$?

   c. If $p, q \geq 0$, what is the minimum value of $g(x) = |x + p| - |x| + |x - q|$?

Problems 57–60 use the absolute value function to model distance. Use the strategy outlined in Problems 57 and 58 to solve Problems 59 and 60.

57. A small pottery is setting up a workshop to produce mugs. Three machines are located on a long table, as shown in the figure. The potter must use each machine once in the course of producing a mug. Let x represent the coordinate of the potter's station.

a. Write expressions for the distance from the potter's station to each of the machines.
b. Write a function that gives the sum of the distances from the potter's station to the three machines.
c. Graph your function for $-20 \le x \le 30$. Where should the potter stand in order to minimize the distance she must walk to the machines?

59. Richard and Marian are moving to Parkville to take jobs after they graduate. The main road through Parkville runs east and west, crossing a river in the center of town. Richard's job is located 10 miles east of the river on the main road, and Marian's job is 6 miles west of the river. There is a health club they both like located 2 miles east of the river. If they plan to visit the health club every workday, where should Richard and Marian look for an apartment to minimize their total daily driving distance?

58. Suppose the pottery in Problem 57 adds a fourth machine to the procedure for producing a mug, located at $x = 16$ in the figure.
a. Write and graph a new function for the sum of the potter's distances to the four machines.
b. Where should the potter stand now to minimize the distance she has to walk while producing a mug?

60. Romina's Bakery has just signed contracts to provide baked goods for three new restaurants located on Route 28 outside of town. The Coffee Stop is 2 miles north of town center, Sneaky Pete's is 8 miles north, and the Sea Shell is 12 miles south. Romina wants to open a branch bakery on Route 28 to handle the new business. Where should she locate the bakery in order to minimize the distance she must drive for deliveries?

2.6 ■ Domain and Range

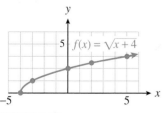

FIGURE 2.43

Definitions of Domain and Range

In Example 3 of Section 1.3, we graphed the function $f(x) = \sqrt{x + 4}$ and observed that $f(x)$ is undefined for x-values less than -4. For this function, we must choose x-values in the interval $[-4, \infty)$. All the points on the graph have x-coordinates greater than or equal to -4, as shown in Figure 2.43. The set of all permissible values of the input variable is called the **domain** of the function f.

We also see that there are no points with negative $f(x)$-values on the graph of f: All the points have $f(x)$-values greater than or equal to zero. The set of all outputs or function values corresponding to the domain is called the **range** of the function. Thus, the domain of the function $f(x) = \sqrt{x + 4}$ is the interval $[-4, \infty)$, and its range is the interval $[0, \infty)$. In general, we make the following definitions.

Domain and Range

The **domain** of a function is the set of permissible values for the input variable. The **range** is the set of function values (that is, values of the output variable) that correspond to the domain values.

Using the notions of domain and range, we restate the definition of a function as follows.

Definition of Function

A relationship between two variables is a **function** if each element of the domain is paired with exactly one element of the range.

Finding Domain and Range from a Graph

We can identify the domain and range of a function from its graph. The domain is the set of x-values of all points on the graph, and the range is the set of y-values.

EXAMPLE 1

a. Determine the domain and range of the function h graphed in Figure 2.44.

b. For the indicated points, show the domain values and their corresponding range values in the form of ordered pairs.

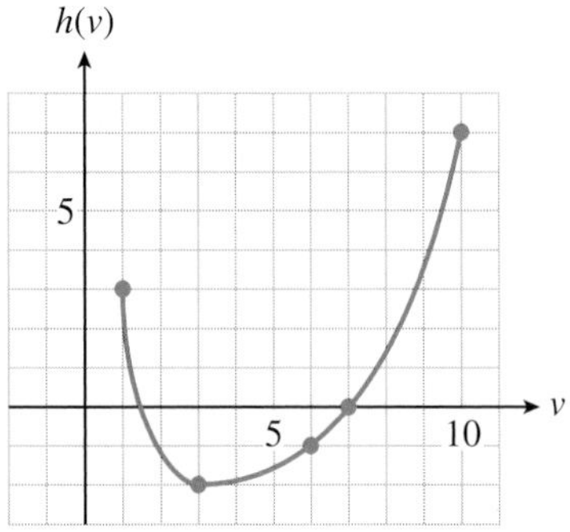

Solutions   **a.** All the points on the graph have v-coordinates between 1 and 10, inclusive, so the domain of the function h is the interval $[1, 10]$. The $h(v)$-coordinates have values between -2 and 7, inclusive, so the range of the function is the interval $[-2, 7]$.

b. Recall that the points on the graph of a function have coordinates $(v, h(v))$. In other words, the coordinates of each point are made up of a domain value and its corresponding range value. Read the coordinates of the indicated points to obtain the ordered pairs $(1, 3)$, $(3, -2)$, $(6, -1)$, $(7, 0)$, and $(10, 7)$.

FIGURE 2.44

Figure 2.45 shows the graph of the function h in Example 1 with the domain values marked on the horizontal axis and the range values marked on the vertical axis. Imagine a rectangle whose length and width are determined by those segments, as shown in Figure 2.45. All the points $(v, h(v))$ on the graph of the function lie within this rectangle. This rectangle is a convenient window in the plane for viewing the function. Of course, if the domain or range of the function is an infinite interval, we can never include the whole graph within a viewing rectangle and must be satisfied with studying only the important parts of the graph.

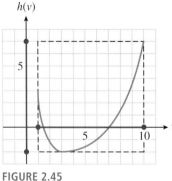

FIGURE 2.45

a. Draw the smallest viewing window possible around the graph shown in Figure 2.46.

b. Find the domain and range of the function.

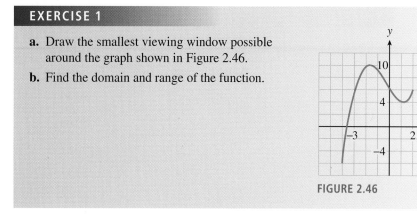

FIGURE 2.46

Sometimes the domain is given as part of the definition of a function.

EXAMPLE 2   Graph the function $f(x) = x^2 - 6$ on the domain $0 \leq x \leq 4$ and give its range.

Solution   The graph is part of a parabola that opens upward. Obtain several points on the graph by evaluating the function at convenient x-values in the domain.

x	$f(x)$
0	−6
1	−5
2	−2
3	3
4	10

since $f(0) = 0^2 - 6 = -6$

since $f(1) = 1^2 - 6 = -5$

since $f(2) = 2^2 - 6 = -2$

since $f(3) = 3^2 - 6 = 3$

since $f(4) = 4^2 - 6 = 10$

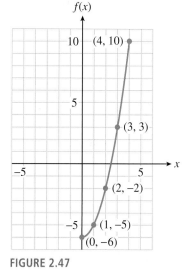

FIGURE 2.47

The range of the function is the set of all $f(x)$-values that appear on the graph. We can see in Figure 2.47 that the lowest point on the graph is $(0, -6)$, so the smallest $f(x)$-value is −6. The highest point on the graph is $(4, 10)$, so the largest $f(x)$-value is 10. Thus, the range of the function f is the interval $[-6, 10]$.

Graph the function $g(x) = x^3 - 4$ on the domain $[-2, 3]$ and give its range.

Not all functions have domains and ranges that are intervals.

EXAMPLE 3

a. The table gives the postage for sending printed material by first-class mail in 2006. Graph the postage function $p = g(w)$.

Weight in ounces (w)	Postage (p)
$0 < w \leq 1$	$0.39
$1 < w \leq 2$	$0.63
$2 < w \leq 3$	$0.87
$3 < w \leq 4$	$1.11
$4 < w \leq 5$	$1.35
$5 < w \leq 6$	$1.59
$6 < w \leq 7$	$1.83

b. Determine the domain and range of the function.

Solutions

a. From the table, we see that articles of any weight up to 1 ounce require $0.39 postage. This means that for all w-values greater than 0 but less than or equal to 1, the p-value is 0.39. Thus, the graph of $p = g(w)$ between $w = 0$ and $w = 1$ looks like a small piece of the horizontal line $p = 0.39$. Similarly, for all w-values greater than 1 but less than or equal to 2, the p-value is 0.63, so the graph on this interval looks like a small piece of the line $p = 0.63$. Continue in this way to obtain the graph shown in Figure 2.48.

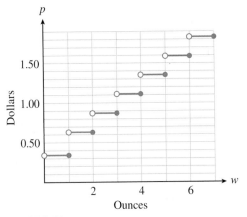

FIGURE 2.48

The open circles at the left endpoint of each horizontal segment indicate that that point is not included in the graph; the closed circles are points on the graph. For instance, if $w = 3$, the postage, p, is $0.87, not $1.11. Consequently, the point $(3, 0.87)$ is part of the graph of g, but the point $(3, 1.11)$ is not.

b. Postage rates are given for all weights greater than 0 ounces up to and including 7 ounces, so the domain of the function is the half-open interval $(0, 7]$. (The domain is an interval because there is a point on the graph for *every* w-value from 0 to 7.) The range of the function is not an interval, however, because the possible values for p do not include *all* the real numbers between 0.3 and 1.75. The range is the set of discrete values 0.39, 0.63, 0.87, 1.11, 1.35, 1.59, and 1.83.

Finding the Domain from a Formula

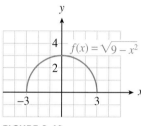

FIGURE 2.49

See Algebra Skills Refresher A.11 to review equations for circles.

If the domain of a function is not given as part of its definition, we assume that the domain is as large as possible. We include in the domain all x-values that make sense when substituted into the function's formula. For example, the domain of $f(x) = \sqrt{9 - x^2}$ is the interval $[-3, 3]$, because x-values less than -3 or greater than 3 result in square roots of negative numbers. You may recognize the graph of f as the upper half of the circle $x^2 + y^2 = 9$, as shown in Figure 2.49.

EXAMPLE 4

Find the domain of the function $g(x) = \frac{1}{x-3}$.

Solution

We must omit any x-values that do not make sense in the function's formula. Because division by zero is undefined, we cannot allow the denominator of $\frac{1}{x-3}$ to be zero. Since $x - 3 = 0$ when $x = 3$, we exclude $x = 3$ from the domain of g. Thus, the domain of g is the set of all real numbers except 3.

For the functions we have studied so far, there are only two operations we must avoid when finding the domain: division by zero and taking the square root of a negative number. Many common functions have as their domain the entire set of real numbers. In particular, a linear function $f(x) = b + mx$ can be evaluated at any real number value of x, so its domain is the set of all real numbers. This set is represented in interval notation as $(-\infty, \infty)$.

The range of the linear function $f(x) = b + mx$ (if $m \ne 0$) is also the set of all real numbers, because the graph continues infinitely at both ends. (See Figure 2.50a.) If $m = 0$, then $f(x) = b$, and the graph of f is a horizontal line. In this case, the range consists of a single number, b.

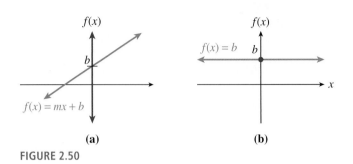

(a)   **(b)**

FIGURE 2.50

Restricting the Domain

In many applications, we may restrict the domain of a function to suit the situation at hand.

EXAMPLE 5

The function $h = f(t) = 1454 - 16t^2$ gives the height of an algebra book dropped from the top of the Sears Tower as a function of time. Give a suitable domain for this application, and the corresponding range.

Solution   You can use the window

$$\text{Xmin} = -10 \quad \text{Xmax} = 10$$
$$\text{Ymin} = -100 \quad \text{Ymax} = 1500$$

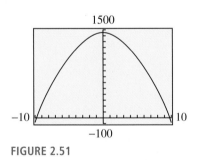

FIGURE 2.51

to obtain the graph shown in Figure 2.51. Because t represents the time in seconds after the book was dropped, only positive t-values make sense for the problem. The book stops falling when it hits the ground, at $h = 0$. You can verify that this happens at approximately $t = 9.5$ seconds. Thus, only t-values between 0 and 9.5 are realistic for this application, so we restrict the domain of the function f to the interval [0, 9.5]. During that time period, the height, h, of the book decreases from 1454 feet to 0 feet. The range of the function on the domain [0, 9.5] is [0, 1454]. The graph is shown in Figure 2.52.

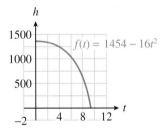

FIGURE 2.52

EXERCISE 5

The children in Francine's art class are going to make cardboard boxes. Each child is given a sheet of cardboard that measures 18 inches by 24 inches. To make a box, the child will cut out a square from each corner and turn up the edges, as shown in Figure 2.53.

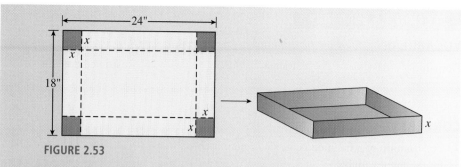

FIGURE 2.53

a. Write a formula $V = f(x)$ for the volume of the box in terms of x, the side of the cut-out square. (See the geometric formulas inside the front cover for the formula for the volume of a box.)

b. What is the domain of the function? (What are the largest and smallest possible values of x?)

c. Graph the function and estimate its range.

ANSWERS TO 2.6 EXERCISES

1. domain: $[-4, 2]$; range: $[-6, 10.1]$ **2.** range: $[-12, 23]$

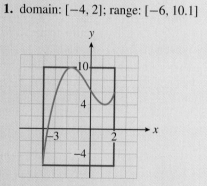

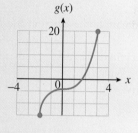

3. domain: $[0, 120]$; range: $[30, 130] \cup (200, 410]$

4. a. $x \neq 4$ **b.** range: $y > 0$

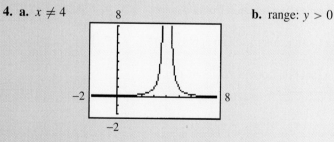

5. a. $V = f(x) = x(24 - 2x)(18 - 2x)$ **b.** $(0, 9)$

c. $(0, 655)$

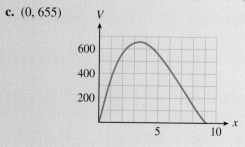

VOCABULARY   *Look up the definitions of new terms in the Glossary.*

Domain   Range   Restricted domain

CONCEPTS

1. The **domain** of a function is the set of permissible values for the input variable.
2. The **range** is the set of function values (that is, values of the output variable) that correspond to the domain values.
3. A relationship between two variables is a **function** if each element of the domain is paired with only one element of the range.
4. We can identify the domain and range of a function from its graph. The domain is the set of x-values of all points on the graph, and the range is the set of y-values.
5. If the domain of a function is not given as part of its definition, we assume that the domain is as large as possible.
6. In applications, we may restrict the domain and range of a function to suit the situation at hand.

STUDY QUESTIONS

1. Explain how to find the domain and range of a function from its graph.
2. What is the domain of the function $f(x) = 4$? What is its range?
3. Which of the eight basic functions are increasing on their entire domain? Which are decreasing on their entire domain?
4. Which of the eight basic functions are concave up on their entire domain? Which are concave down on their entire domain?
5. Which of the eight basic functions can be evaluated at any real number? Which can take on any real number as a function value?
6. Which of the eight basic functions can be graphed in one piece, without lifting the pencil from the paper?

SKILLS   *Practice each skill in the Homework problems listed.*

1. Find the domain and range of a function from its graph: #1–16
2. Restrict the domain of a function to suit an application: #17–24
3. Find the domain of a function from its algebraic formula: #25–30
4. Find the corresponding domain value for a given range value: #31–38
5. Find the range of a function on a given domain: #39–50

Find the domain and range of each function from its graph.

1.

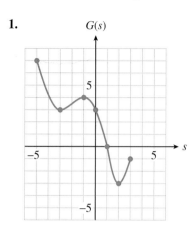

2.

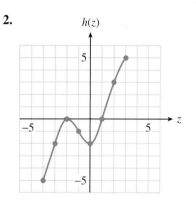

3.

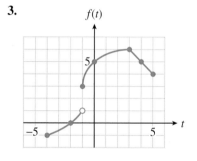

4.

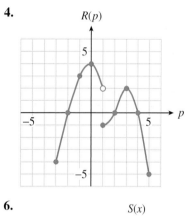

5.

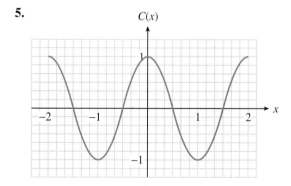

6.

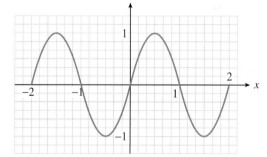

7.

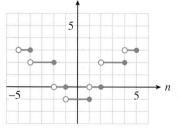

8.

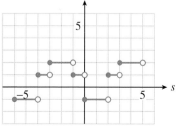

State the domain and range of each basic function.

9. a. $f(x) = x^3$
b. $g(x) = x^2$

10. a. $F(x) = |x|$
b. $G(x) = x$

11. a. $H(x) = \dfrac{1}{x^2}$
b. $M(x) = \dfrac{1}{x}$

12. a. $p(x) = \sqrt[3]{x}$
b. $q(x) = \sqrt{x}$

13. The graph shows the elevation of the Los Angeles Marathon course as a function of the distance into the race, $a = f(t)$. Estimate the domain and range of the function. (Source: *Los Angeles Times*)

14. The graph shows the federal debt as a percentage of the gross domestic product, as a function of time, $D = f(t)$. Estimate the domain and range of the function.
(Source: Office of Management and Budget)

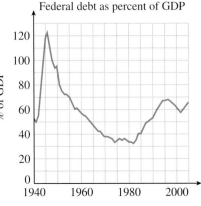

15. The graph shows the average air temperature as a function of altitude, $T = f(h)$. Estimate the domain and range of the function. (Source: Ahrens, 1998)

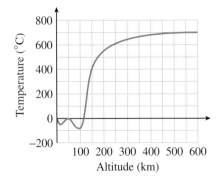

16. The graph shows the speed of sound in the ocean as a function of depth, $S = f(d)$. Estimate the domain and range of the function. (Source: *Scientific American*)

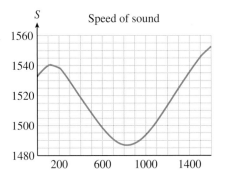

17. Clinton purchases $6000 of photographic equipment to set up his studio. He estimates a salvage value of $500 for the equipment in 10 years, and for tax purposes he uses straight-line depreciation.
a. Write a formula for the value of the equipment, $V(t)$, after t years.
b. State the domain and range of the function $V(t)$.

18. Leslie plans to invest some money in two CD accounts. The first account pays 3.6% interest per year, and the second account pays 2.8% interest per year. Leslie would like to earn $500 per year on her investment.
a. Write a linear equation in general form that relates x, the amount Leslie invests at 3.4%, and y, the amount she invests at 2.8%.
b. Use your equation from part (a) to write y as a function of x, $y = f(x)$.
c. Find the domain and range of f.

19. The height of a golfball, in feet, t seconds after being hit is given by the function $h = -16t^2 + 64t$.
  a. Graph the function.
  b. State the domain and range of the function and explain what they tell us about the golfball.

20. Gameworld is marketing a new boardgame called Synaps. If Gameworld charges p dollars for the game, its revenue, R, is given by the function $R = 1000p - 50p^2$.
  a. Graph the function.
  b. State the domain and range of the function, and explain what they tell us about the revenue.

21. In New York City, taxi cabs charge $2.50 for distances up to $\frac{1}{3}$ mile, plus $0.40 for each additional $\frac{1}{5}$ mile or portion thereof. (Source: www.visitnyc.com)
  a. Sketch a graph of $F(d)$, which gives taxi fare as a function of distance traveled, on the domain $0 < d \leq 1$.
  b. State the range of $F(d)$ on that domain.
  c. How much will it cost Renee to travel by taxi from Columbia University to Rockefeller Center, a distance of 5.7 miles?

22. If you order from Coldwater Creek, the shipping charges are given by the following table.

Purchase amount	Shipping charge
Up to $25	$5.95
$25.01 to $50	$7.95
$50.01 to $75	$9.95
$75.01 to $100	$10.95

State the domain and range of $S(x)$, the shipping charge as a function of the purchase amount, x.

23. The Bopp-Busch Tool and Die Company markets its products to individuals, to contractors, and to wholesale distributors. The company offers three different price structures for its toggle bolts. If you order 20 or fewer boxes, the price is $2.50 each. If you order more than 20 but no more than 50 boxes, the price is $2.25 each. If you order more than 50 boxes, the price is $2.10 each. State the domain and range of $C(x)$, the cost of ordering x boxes of toggle bolts.

24. The Java Stop uses paper cups at a rate of 300 per day. At opening on Tuesday morning Java Stop has on hand 1200 paper cups. On Friday mornings Java Stop takes delivery of a week's worth of cups.
  a. Write a piecewise function for the number of cups Java Stop has on hand for one week, starting Tuesday morning.
  b. Graph the function.
  c. State the domain and range of the function.

Find the domain of each function algebraically. Then graph the function, and use the graph to help you find the range.

25. a. $f(x) = \dfrac{1}{(x-4)^2}$
  b. $h(x) = \dfrac{1}{x^2} - 4$

26. a. $g(t) = \dfrac{1}{t} + 2$
  b. $F(t) = \dfrac{1}{t+2}$

27. a. $G(v) = v^3 + 5$
  b. $H(v) = (v+5)^3$

28. a. $h(n) = 3 + (n-1)^2$
  b. $g(n) = 3 - (n+1)^2$

29. a. $T(z) = \sqrt{z-2}$
  b. $S(z) = \sqrt{z} - 2$

30. a. $Q(x) = 4 - |x|$
  b. $P(x) = |4 - x|$

Decide whether the given value is in the range of the function. If so, find the domain value(s) that produce each range value.

31. $f(x) = 6 - |2x + 4|$
  a. $f(x) = 8$
  b. $f(x) = -2$

32. $g(x) = (x-5)^3 + 1$
  a. $g(x) = 0$
  b. $g(x) = -7$

33. $h(t) = 4 + 2\sqrt[3]{t}$
 a. $h(t) = -4$
 b. $h(t) = 0$

34. $F(t) = 12 + 0.5(t-2)^2$
 a. $F(t) = 10$
 b. $F(t) = 20$

35. $G(w) = 3 + \dfrac{2}{w-1}$
 a. $G(w) = -1$
 b. $G(w) = 3$

36. $H(n) = \dfrac{4}{(n+2)^2} - 5$
 a. $H(n) = -6$
 b. $H(n) = -1$

37. $Q(h) = 2 + \sqrt{h+5}$
 a. $Q(h) = 1$
 b. $Q(h) = 5$

38. $P(q) = 8 - \sqrt{4-q}$
 a. $P(q) = 4$
 b. $P(q) = 12$

For Problems 39–50,
(a) Use a graphing calculator to graph each function on the given domain. Using the $\boxed{\text{TRACE}}$ key, adjust *Ymin* and *Ymax* until you can estimate the range of the function.
(b) Verify your answer algebraically by evaluating the function. State the domain and range in interval notation.

39. $f(x) = x^2 - 4x; \quad -2 \le x \le 5$

40. $g(x) = 6x - x^2; \quad -1 \le x \le 5$

41. $g(t) = -t^2 - 2t; \quad -5 \le t \le 3$

42. $f(t) = -t^2 - 4t; \quad -6 \le t \le 2$

43. $h(x) = x^3 - 1; \quad -2 \le x \le 2$

44. $q(x) = x^3 + 4; \quad -3 \le x \le 2$

45. $F(t) = \sqrt{8-t}; \quad -1 \le t \le 8$

46. $G(t) = \sqrt{t+6}; \quad -6 \le t \le 3$

47. $G(x) = \dfrac{1}{3-x}; \quad -1.25 \le x \le 2.75$

48. $H(x) = \dfrac{1}{x-1}; \quad 3.25 \le x \le -1.25$

49. $G(x) = \dfrac{1}{3-x}; \quad 3 < x \le 6$

50. $H(x) = \dfrac{1}{x-1}; \quad 1 < x \le 4$

51. a. Show that the graph of $y = \sqrt{16 - x^2}$ is a semicircle. (*Hint*: Write the equation in the form $x^2 + y^2 = r^2$. See Algebra Skills Refresher A.11 to review circles.)
 b. State the domain and range of the function.
 c. Graph the function in the window

$$\text{Xmin} = -6 \quad \text{Xmax} = 6$$
$$\text{Ymin} = 0 \quad \text{Ymax} = 8$$

In what way is the calculator's graph misleading?

52. a. For what values of x is the function $y = \dfrac{2x-8}{x-2}$ undefined?
 b. Graph the function in the standard window. In what way is the calculator's graph misleading?
 c. Graph the function again in the window

$$\text{Xmin} = -9.4 \quad \text{Xmax} = 9.4$$
$$\text{Ymin} = -10 \quad \text{Ymax} = 10$$

State the domain and range of the function.

In Problems 53–60, find the domain and range of each transformation of the given function.

53. $f(x) = \dfrac{1}{x^2}$
 a. $y = f(x-2)$
 b. $y = f(x) - 2$
 c. $y = f(x-3) - 5$

54. $f(x) = \sqrt{x}$
 a. $y = -f(x)$
 b. $y = 4 + f(x)$
 c. $y = 4 - f(x)$

55. $f(x) = x^2$
  a. $y = -2f(x)$
  b. $y = 6 - 2f(x)$
  c. $y = 6 - 2f(x + 3)$

56. $f(x) = \dfrac{1}{x}$
  a. $y = 3f(x)$
  b. $y = 3 + f(x - 1)$
  c. $y = 3 - f(x - 1)$

57. The domain of f is $[0, 10]$ and the range is $[-2, 2]$.
  a. $y = f(x - 3)$
  b. $y = 3f(x)$
  c. $y = 2f(x - 5)$

58. The domain of f is $[-4, 4]$ and the range is $[3, 10]$.
  a. $y = f(x) + 10$
  b. $y = f(x + 10)$
  c. $y = f(x - 1) + 4$

59. The domain of f is $(0, +\infty)$ and the range is $(0, 1)$.
  a. $y = 5f(x)$
  b. $y = 3f(x + 2)$
  c. $y = 2f(x - 3) + 2$

60. The domain of f is $(-1, 1)$ and the range is $(-\infty, 0)$.
  a. $y = f(x + 1)$
  b. $y = 3 - f(x + 1)$
  c. $y = 4 + 2f(x - 1)$

In Problems 61–64, use a graphing calculator to explore some properties of the basic functions.

61. a. Graph $f(x) = x^2$ and $g(x) = x^3$ on the domain $[0, 1]$ and state the range of each function. On the interval $(0, 1)$, which is greater, $f(x)$ or $g(x)$?
  b. Graph $f(x) = x^2$ and $g(x) = x^3$ on the domain $[1, 10]$ and state the range of each function. On the interval $(1, 100)$, which is greater, $f(x)$ or $g(x)$?

62. a. Graph $f(x) = \sqrt{x}$ and $g(x) = \sqrt[3]{x}$ on the domain $[0, 1]$ and state the range of each function. On the interval $(0, 1)$, which is greater, $f(x)$ or $g(x)$?
  b. Graph $f(x) = \sqrt{x}$ and $g(x) = \sqrt[3]{x}$ on the domain $[1, 100]$ and state the range of each function. On the interval $(1, 100)$, which is greater, $f(x)$ or $g(x)$?

63. a. Graph $f(x) = \dfrac{1}{x}$ and $g(x) = \dfrac{1}{x^2}$ on the domain $[0.01, 1]$ and state the range of each function. On the interval $(0, 1)$, which is greater, $f(x)$ or $g(x)$?
  b. Graph $f(x) = \dfrac{1}{x}$ and $g(x) = \dfrac{1}{x^2}$ on the domain $[1, 10]$ and state the range of each function. On the interval $(1, \infty)$, which is greater, $f(x)$ or $g(x)$?

64. a. Graph $F(x) = |x^3|$ in the **ZDecimal** window. How does the graph compare to the graph of $y = x^3$?
  b. Graph $G(x) = \left|\dfrac{1}{x}\right|$ in the **ZDecimal** window. How does the graph compare to the graph of $y = \dfrac{1}{x}$?

65. The number of hours of daylight on the summer solstice is a function of latitude in the northern hemisphere. Give the domain and range of the function.

66. A semicircular window has a radius of 2 feet. The area of a sector of the window (a pie-shaped wedge) is a function of the angle at the center of the circle. Give the domain and range of this function.

CHAPTER 2 Summary and Review

KEY CONCEPTS

1. We can solve equations of the form $a(px + q)^2 + r = 0$ by extraction of roots.

2. The formula for compound interest is $A = P(1 + r)^n$.

3. Simple nonlinear equations can be solved by undoing the operations on the variable.

4.

The **absolute value** of x is defined by

$$|x| = \begin{cases} x & \text{if } x \geq 0 \\ -x & \text{if } x < 0 \end{cases}$$

5.

The absolute value has the following properties:

$$|a + b| \le |a| + |b| \quad \text{Triangle inequality}$$

$$|ab| = |a|\,|b| \quad \text{Multiplicative property}$$

6. Many situations can be modeled by one of eight basic functions:

$$y = x \quad y = |x| \quad y = x^2 \quad y = x^3$$

$$y = \frac{1}{x} \quad y = \frac{1}{x^2} \quad y = \sqrt{x} \quad y = \sqrt[3]{x}$$

7. Functions can be defined piecewise, with different formulas on different intervals.

8.

Transformations of Functions

- The graph of $y = f(x) + k$ is **shifted vertically** compared to the graph of $y = f(x)$.
- The graph of $y = f(x + h)$ is **shifted horizontally** compared to the graph of $y = f(x)$.
- The graph of $y = af(x)$ is **stretched or compressed vertically** compared to the graph of $y = f(x)$.
- The graph of $y = -f(x)$ is **reflected about the x-axis** compared to the graph of $y = f(x)$.

9. A nonlinear graph may be **concave up** or **concave down.** If a graph is concave up, its slope is increasing. If it is concave down, its slope is decreasing.

10. The absolute value is used to model distance: The **distance** between two points x and a is given by $|x - a|$.

11.

Absolute Value Equations and Inequalities

- The equation $|ax + b| = c \;(c > 0)$ is equivalent to

$$ax + b = c \quad \text{or} \quad ax + b = -c$$

- If the solutions of the equation $|ax + b| = c$ are r and s, with $r < s$, then the solutions of $|ax + b| < c$ are $r < x < s$.
- If the solutions of the equation $|ax + b| = c$ are r and s, with $r < s$, then the solutions of $|ax + b| > c$ are $x < r$ or $x > s$.

12. We can use absolute value notation to express error tolerances in measurements.

13. The **domain** of a function is the set of permissible values for the input variable. The **range** is the set of function values (that is, values of the output variable) that correspond to the domain values.

14. A relationship between two variables is a **function** if each element of the domain is paired with only one element of the range.

15. We can identify the domain and range of a function from its graph. The domain is the set of input values of all points on the graph, and the range is the set of output values.

16. If the domain of a function is not given as part of its definition, we assume that the domain is as large as possible. In many applications, however, we may restrict the domain and range of a function to suit the situation at hand.

REVIEW PROBLEMS

Solve by extraction of roots.

1. $(2x - 5)^2 = 9$

2. $(7x - 1)^2 = 15$

3. $6\left(\dfrac{w - 1}{3}\right)^2 - 4 = 2$

4. $\left(\dfrac{2p}{5}\right)^2 = -3$

Solve the formulas for the specified variable.

5. $A = P(1 + r)^2$, for r

6. $V = \dfrac{4}{3}\pi r^3$, for r

7. Lewis invested $2000 in an account that compounds interest annually. He made no deposits or withdrawals after that. Two years later, he closed the account, withdrawing $2464.20. What interest rate did Lewis earn?

8. Earl borrowed $5500 from his uncle for two years with interest compounded annually. At the end of two years, he owed his uncle $6474.74. What was the interest rate on the loan?

Solve.

9. $\sqrt[3]{P-1} = 0.1$

10. $\dfrac{1}{1-t} = \dfrac{2}{3}$

11. $\dfrac{3}{\sqrt{m+7}} = \dfrac{1}{2}$

12. $15 = 3\sqrt{w+1}$

13. $4r^3 - 8 = 100$

14. $5s^2 + 6 = 3s^2 + 31$

Use the Pythagorean theorem to write and solve an equation.

15. A widescreen television measures 96 cm by 54 cm. How long is the diagonal?

16. A 15-foot ladder leans to the top of a 12-foot fence. How far is the foot of the ladder from the base of the fence?

Simplify.

17. $|-18| - |20|$

18. $|-2 \cdot (3 - 18)|$

19. $|-2 \cdot 3 - 18|$

20. $-2 \cdot |3 - 18|$

Use the graph to solve the equation or inequality.

21. Refer to the graph of $y = \left|\dfrac{x}{2} - 1\right|$.

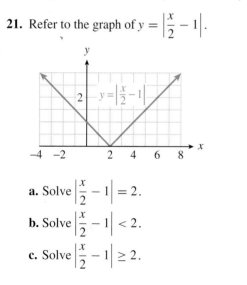

a. Solve $\left|\dfrac{x}{2} - 1\right| = 2$.

b. Solve $\left|\dfrac{x}{2} - 1\right| < 2$.

c. Solve $\left|\dfrac{x}{2} - 1\right| \geq 2$.

22. Refer to the graph of $y = \dfrac{-x^2}{2} + x + 1$.

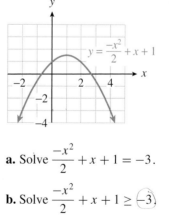

a. Solve $\dfrac{-x^2}{2} + x + 1 = -3$.

b. Solve $\dfrac{-x^2}{2} + x + 1 \geq -3$.

c. Solve $\dfrac{-x^2}{2} + x + 1 \leq -3$.

23. Refer to the graph of $y = \dfrac{6}{x^2 - 3x + 3}$.

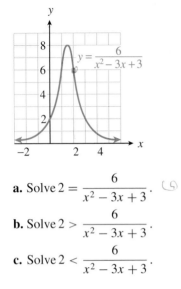

a. Solve $2 = \dfrac{6}{x^2 - 3x + 3}$.

b. Solve $2 > \dfrac{6}{x^2 - 3x + 3}$.

c. Solve $2 < \dfrac{6}{x^2 - 3x + 3}$.

24. Refer to the graph of $y = \left| x^3 + 3x^2 + 3x + 1 \right|$.

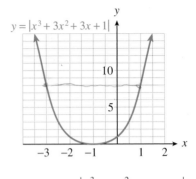

a. Solve $8 = \left| x^3 + 3x^2 + 3x + 1 \right|$.

b. Solve $8 > \left| x^3 + 3x^2 + 3x + 1 \right|$.

c. Solve $8 < \left| x^3 + 3x^2 + 3x + 1 \right|$.

Graph each piecewise defined function.

25. $f(x) = \begin{cases} x + 1 & \text{if } x \le 0 \\ x^2 & \text{if } x > 0 \end{cases}$

26. $g(x) = \begin{cases} x - 1 & \text{if } x \le 1 \\ x^3 & \text{if } x > 1 \end{cases}$

27. $H(x) = \begin{cases} x^2 & \text{if } x \le 0 \\ \sqrt{x} & \text{if } x > 0 \end{cases}$

28. $F(x) = \begin{cases} |x| & \text{if } x \le 0 \\ \frac{1}{x} & \text{if } x > 0 \end{cases}$

29. $S(x) = \begin{cases} x^3 & \text{if } x \le 1 \\ |x| & \text{if } x > 1 \end{cases}$

30. $T(x) = \begin{cases} \frac{1}{x^2} & \text{if } x < 0 \\ \sqrt{x} & \text{if } x \ge 0 \end{cases}$

For Problems 31–36,
(a) Describe each function as transformation of a basic function.
(b) Sketch a graph of the basic function and the given function on the same axes.

31. $g(x) = |x| + 2$

32. $F(t) = \dfrac{1}{t} - 2$

33. $f(s) = \sqrt{s} + 3$

34. $g(u) = \sqrt{u + 2} - 3$

35. $G(t) = |t + 2| - 3$

36. $H(t) = \dfrac{1}{(t - 2)^2} + 3$

37. $h(s) = -2\sqrt{s}$

38. $g(s) = \dfrac{1}{2}|s|$

In Problems 39–42, write a formula for each transformation of the given function.

39.

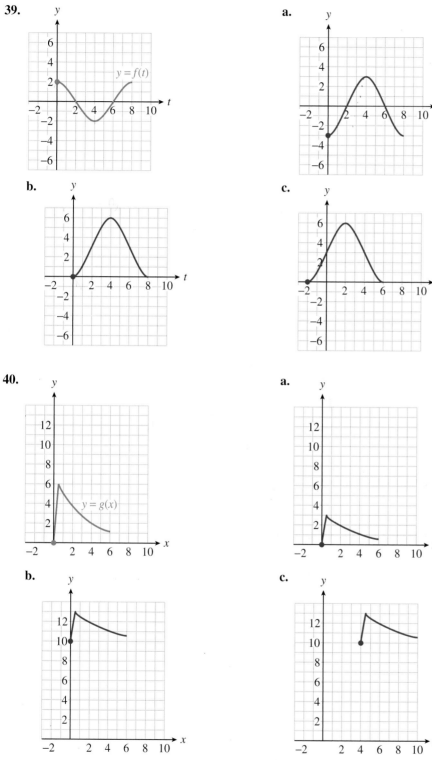

a.

b.

c.

40.

a.

b.

c.

41.

t	0	1	2	3	4	5
$f(t)$	243	81	27	9	3	1

a.

t	1	2	3	4	5	6
y	243	81	27	9	3	1

b.

t	1	2	3	4	5	6
y	−243	−81	−27	−9	−3	−1

c.

t	1	2	3	4	5	6
y	57	219	273	291	297	299

42.

x	1	2	3	4	5	6
$f(x)$	25	24	21	16	9	0

a.

x	−1	0	1	2	3	4
y	25	24	21	16	9	0

b.

x	−1	0	1	2	3	4
y	50	48	42	32	18	0

c.

x	−1	0	1	2	3	4
y	70	68	62	52	38	20

Give an equation for the function graphed.

43.

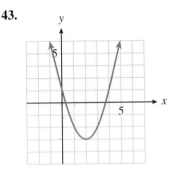

44.

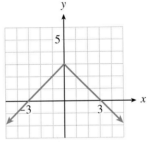

Sketch graphs to illustrate the situations in Problems 45 and 46.

45. Inga runs hot water into the bathtub until it is about half full. Because the water is too hot, she lets it sit for a while before getting into the tub. After several minutes of bathing, she gets out and drains the tub. Graph the water level in the bathtub as a function of time, from the moment Inga starts filling the tub until it is drained.

46. David turns on the oven and it heats up steadily until the proper baking temperature is reached. The oven maintains that temperature during the time David bakes a pot roast. When he turns the oven off, David leaves the oven door open for a few minutes, and the temperature drops fairly rapidly during that time. After David closes the door, the temperature continues to drop, but at a much slower rate. Graph the temperature of the oven as a function of time, from the moment David first turns on the oven until shortly after David closes the door when the oven is cooling.

Match each table with its graph.

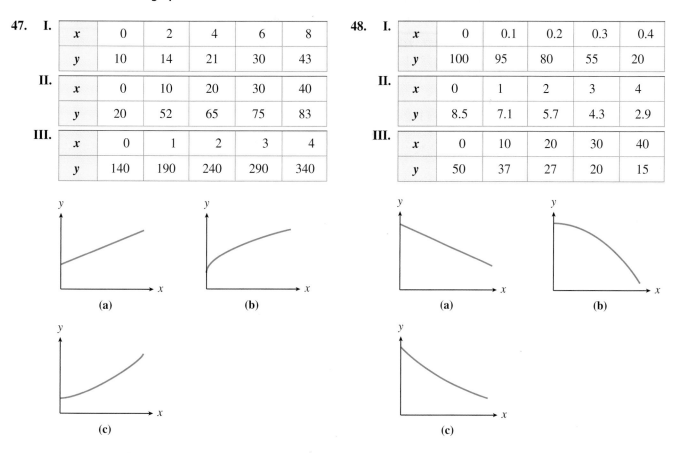

47.

I.

x	0	2	4	6	8
y	10	14	21	30	43

II.

x	0	10	20	30	40
y	20	52	65	75	83

III.

x	0	1	2	3	4
y	140	190	240	290	340

48.

I.

x	0	0.1	0.2	0.3	0.4
y	100	95	80	55	20

II.

x	0	1	2	3	4
y	8.5	7.1	5.7	4.3	2.9

III.

x	0	10	20	30	40
y	50	37	27	20	15

(a)   (b)   (c)

Write and graph a piecewise function for Problems 49 and 50.

49. The fluid level in a tank is a function of the number of days since the year began. The level was initially at 60 inches and rose an inch a day for 10 days, remained constant for the next 20 days, then dropped a half-inch each day for 30 days.

50. The temperature at different locations in a large room is a function of distance from the window. Within 2 feet of the window, the temperature is 66° Fahrenheit, but the temperature rises by 0.5° for each of the next 10 feet, then maintains the temperature at 12 feet for the rest of the room.

Use absolute value notation to write each expression as an equation or inequality.

51. x is four units from the origin.

52. The distance from y to -5 is three units.

53. p is within four units of 7.

54. q is at least $\frac{3}{10}$ of a unit from -4.

Solve.

55. $|9 - 5t| = 3$

56. $1 = |4q - 7|$

57. $-29 = |2w + 3|$

58. $\left|\dfrac{8n + 3}{5}\right| = -11$

59. $1 = \left| \dfrac{7 - 2p}{5} \right|$

60. $|6(r - 10)| = 30$

61. $|3x - 2| < 4$

62. $|2x + 0.3| \le 0.5$

63. $|3y + 1.2| \ge 1.5$

64. $\left| 3z + \dfrac{1}{2} \right| > \dfrac{1}{3}$

Express the error tolerance using absolute value.

65. The height, H, of a female trainee must be between 56 inches and 75 inches.

66. The time, t, in freefall must be at least 3.5 seconds but no more than 8.1 seconds.

Give an interval of possible values for the measurement.

67. The mass, M, of the sample must satisfy $|M - 2.1| \le 0.05$.

68. The temperature, T, of the refrigerator is specified by $|T - 4.0| < 0.5$.

In Problems 69 and 70,
(a) Plot the points and sketch a smooth curve through them.
(b) Use your graph to help you discover the equation that describes the function.

69.

x	$g(x)$
2	12
3	8
4	6
6	4
8	3
12	2

70.

x	$F(x)$
-2	8
-1	1
0	0
1	-1
2	-8
3	-27

In Problems 71–76,
(a) Use the graph to complete the table of values.
(b) By finding a pattern in the table of values, write an equation for the graph.

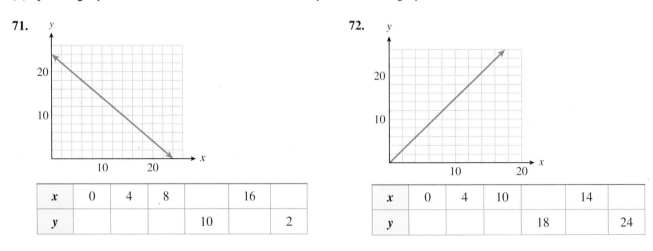

71.

x	0	4	8		16	
y				10		2

72.

x	0	4	10		14	
y				18		24

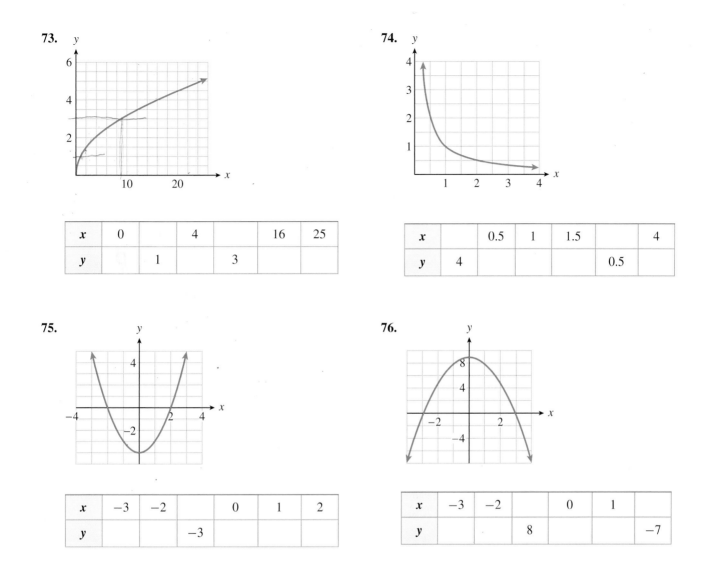

73.

x	0		4		16	25
y		1		3		

74.

x		0.5	1	1.5		4
y	4				0.5	

75.

x	−3	−2		0	1	2
y			−3			

76.

x	−3	−2		0	1	
y			8			−7

Use a graphing calculator to graph each function on the given domain. Adjust *Ymin* and *Ymax* until you can determine the range of the function using the ☐ TRACE ☐ key. Then verify your answer algebraically by evaluating the function. State the domain and corresponding range in interval notation.

77. $f(t) = -t^2 + 3t; \quad -2 \le t \le 4$

78. $g(s) = \sqrt{s - 2}; \quad 2 \le s \le 6$

79. $F(x) = \dfrac{1}{x + 2}; \quad -2 < x \le 4$

80. $H(x) = \dfrac{1}{2 - x}; \quad -4 \le x < 2$

PROJECTS FOR CHAPTER 2: PERIODIC FUNCTIONS

Part I

A **periodic function** is one whose values repeat at evenly spaced intervals, or **periods,** of the input variable. Periodic functions are used to model phenomena that exhibit cyclical behavior, such as growth patterns in plants and animals, radio waves, and planetary motion. In this project, we consider some applications of periodic functions.

EXAMPLE 1   Which of the functions in Figure 2.54 are periodic? If the function is periodic, give its period.

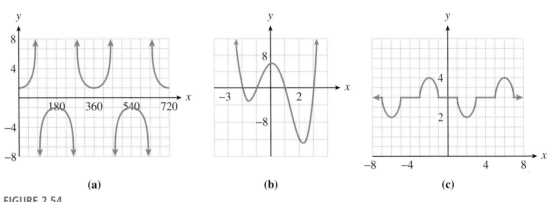

(a)   (b)   (c)

FIGURE 2.54

Solutions   **a.** This graph is periodic with period 360.

          b. This graph is not periodic.

          c. This graph is periodic with period 8.

EXERCISE SET 1

1. Which of the functions are periodic? If the function is periodic, give its period.

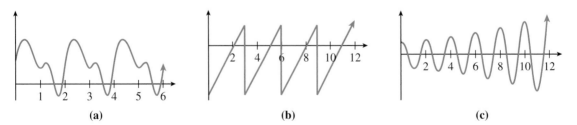

(a)   (b)   (c)

2. Which of the following graphs are periodic? If the graph is periodic, give its period.

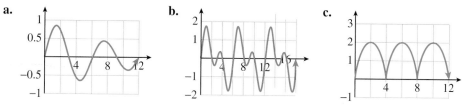

a. b. c. d.

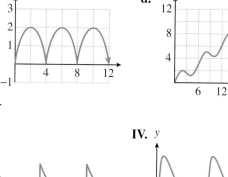

3. Match each of the following situations with the appropriate graph.

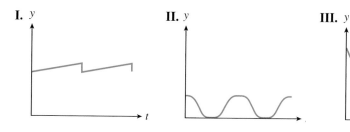

I. *y*   II. *y*   III. *y*   IV. *y*

a. When the heart contracts, blood pressure in the arteries rises rapidly to a peak (systolic blood pressure) and then falls off quickly to a minimum (diastolic blood pressure). Blood pressure is a periodic function of time.

b. After an injection is given to a patient, the amount of the drug present in his bloodstream decreases exponentially. The patient receives injections at regular intervals to restore the drug level to the prescribed level. The amount of the drug present is a periodic function of time.

c. The monorail shuttle train between the north and south terminals at Gatwick Airport departs from the south terminal every 12 minutes. The distance from the train to the south terminal is a periodic function of time.

d. Delbert gets a haircut every two weeks. The length of his hair is a periodic function of time.

4. A patient receives regular doses of medication to maintain a certain level of the drug in his body. After each dose, the patient's body eliminates a certain percent of the medication before the next dose is administered. The graph shows the amount of the drug, in milliliters, in the patient's body as a function of time in hours.

a. How much of the medication is administered with each dose?

b. How often is the medication administered?

c. What percent of the drug is eliminated from the body between doses?

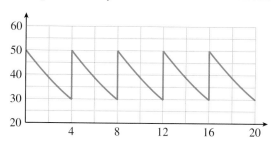

5. You are sitting on your front porch late one evening, and you see a light coming down the road tracing out the path shown below, with distances in inches. You realize that you are seeing a bicycle light, fixed to the front wheel of the bike.

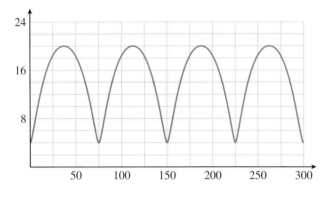

   a. Approximately what is the period of the graph?
   b. How far above the ground is the light?
   c. What is the diameter of the bicycle wheel?

6. The graph shows arterial blood pressure, measured in millimeters of mercury (mmHg), as a function of time.

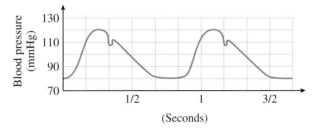

   a. What are the maximum (systolic) and minimum (diastolic) pressures? The *pulse pressure* is the difference of systolic and diastolic pressures. What is the pulse pressure?
   b. The *mean arterial pressure* is the diastolic pressure plus one-third of the pulse pressure. Calculate the mean arterial pressure and draw a horizontal line on the graph at that pressure.
   c. The blood pressure graph repeats its cycle with each heartbeat. What is the heart rate, in beats per minute, of the person whose blood pressure is shown in the graph?

For Problems 7–10, sketch a periodic function that models the situation.

7. At a ski slope, the lift chairs take 5 minutes to travel from the bottom, at an elevation of 3000 feet, to the top, at elevation 4000 feet. The cable supporting the ski lift chairs is a loop turning on pulleys at a constant speed. At the top and bottom, the chairs are at a constant elevation for a few seconds to allow skiers to get on and off.
   a. Sketch a graph of $h(t)$, the height of one chair at time t. Show at least two complete up-and-down trips.
   b. What is the period of $h(t)$?

8. The heater in Paul's house doesn't have a thermostat; it runs on a timer. It uses 300 watts when it is running. Paul sets the heater to run from 6 a.m. to noon, and again from 4 p.m. to 10 p.m.
   a. Sketch a graph of $P(t)$, the power drawn by the heater as a function of time. Show at least two days of heater use.
   b. What is the period of $P(t)$?

9. Francine adds water to her fish pond once a week to keep the depth at 30 centimeters. During the week, the water evaporates at a constant rate of 0.5 centimeter per day.
   a. Sketch a graph of $D(t)$, the depth of the water, as a function of time. Show at least two weeks.
   b. What is the period of $D(t)$?

10. Erin's fox terrier, Casey, is very energetic and bounces excitedly at dinner time. Casey can jump 30 inches high, and each jump takes him 0.8 second.
   a. Sketch a graph of Casey's height, $h(t)$, as a function of time. Show at least two jumps.
   b. What is the period of $h(t)$?

Part II

Many periodic functions have a characteristic wave shape like the graph shown in Figure 2.55. These graphs are called **sinusoidal,** after the trigonometric functions sine and cosine. They are often described by three parameters: the **period, midline,** and **amplitude.**

   The *period* of the graph is the smallest interval of input values on which the graph repeats. The *midline* is the horizontal line at the average of the maximum and minimum values of the output variable. The *amplitude* is the vertical distance between the maximum output value and the midline.

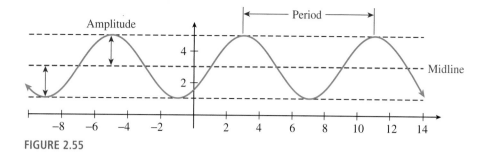

FIGURE 2.55

EXAMPLE 2

The table shows the number of hours of daylight in Glasgow, Scotland, on the first of each month.

Month	Jan	Feb	Mar	Apr	May	Jun
Daylight hours	7.1	8.7	10.7	13.1	15.3	17.2

Month	Jul	Aug	Sep	Oct	Nov	Dec
Daylight hours	17.5	16.7	13.8	11.5	9.2	7.5

a. Sketch a sinusoidal graph of daylight hours as a function of time, with $t = 1$ in January.

b. Estimate the period, amplitude, and midline of the graph.

Solutions **a.** Plot the data points and fit a sinusoidal curve by eye, as shown in Figure 2.56.

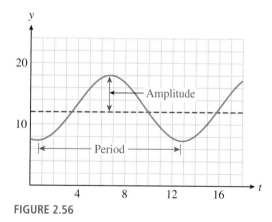

FIGURE 2.56

b. The period of the graph is 12 months. The midline is approximately $y = 12.25$, and the amplitude is approximately 5.25.

EXERCISE SET 2

1. The graph shows the number of daylight hours in Jacksonville, in Anchorage, at the Arctic Circle, and at the Equator.

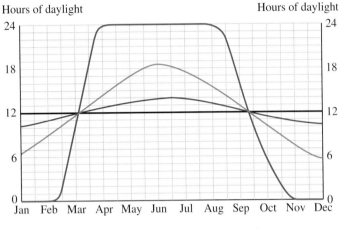

Hours of daylight

a. Which graph corresponds to each location?
b. What are the maximum and minimum number of daylight hours in Jacksonville?
c. For how long are there 24 hours of daylight per day at the Arctic Circle?

2. Match each of the following situations with the appropriate graph.

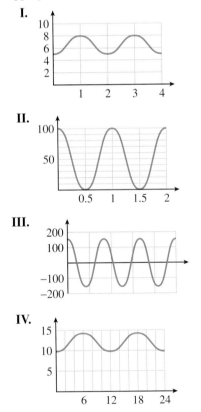

I.

II.

III.

IV.

a. The number of hours of daylight in Salt Lake City varies from a minimum of 9.6 hours on the winter solstice to a maximum of 14.4 hours on the summer solstice.
b. A weight is 6.5 feet above the floor, suspended from the ceiling by a spring. The weight is pulled down to 5 feet above the floor and released, rising past 6.5 feet in 0.5 second before attaining its maximum height of 8 feet. Neglecting the effects of friction, the height of the weight will continue to oscillate between its minimum and maximum height.
c. The voltage used in U.S. electrical current changes from 155 V to -155 V and back 60 times each second.
d. Although the moon is spherical, what we can see from Earth looks like a (sometimes only partly visible) disk. The percentage of the moon's disk that is visible varies between 0 (at new moon) to 100 (at full moon).

3. As the moon revolves around the Earth, the percent of the disk that we see varies sinusoidally with a period of approximately 30 days. There are eight phases, starting with the new moon, when the moon's disk is dark, followed by waxing crescent, first quarter, waxing gibbous, full moon (when the disk is 100% visible), waning gibbous, last quarter, and waning crescent. Which graph best represents the phases of the moon?

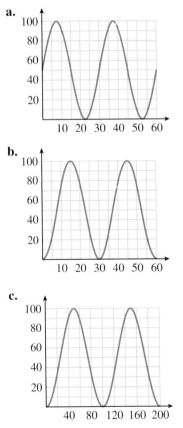

a.

b.

c.

4. The table shows sunrise and sunset times in Los Angeles on the fifteenth of each month.

Month	Oct	Nov	Dec	Jan	Feb	Mar
Sunrise	5:58	6:26	6:51	6:59	6:39	6:04
Sunset	17:20	16:50	16:45	17:07	17:37	18:01

Month	Apr	May	Jun	Jul	Aug	Sep
Sunrise	5:22	4:52	4:42	4:43	5:15	5:37
Sunset	18:25	18:48	19:07	19:05	18:40	18:00

a. Use the grid (a) to plot the sunrise times and sketch a sinusoidal graph through the points.

b. Use the grid (b) to plot the sunset times and sketch a sinusoidal graph through the points.

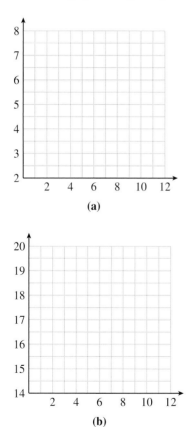

(a)

(b)

5. a. Use the data from Problem 4 to complete the table with the hours of sunlight in Los Angeles on the fifteenth of each month.

Month	Oct	Nov	Dec	Jan	Feb	Mar
Hours of daylight						

Month	Apr	May	Jun	Jul	Aug	Sep
Hours of daylight						

b. Plot the daylight hours and sketch a sinusoidal graph through the points.

6. Many people who believe in astrology also believe in biorhythms. The graph shows an individual's three biorhythms—physical, emotional, and intellectual—for 36 days, from $t = 0$ on September 30 to November 5.

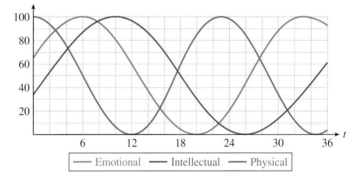

a. Find the dates of highest and lowest activity for each biorhythm during the month of October.

b. Find the period of each biorhythm in days.

c. On the day of your birth, all three biorhythms are at their maximum. How old will you be before all three are again at the maximum level?

7. a. Is the function shown periodic? If so, what is its period? If not, explain why not.

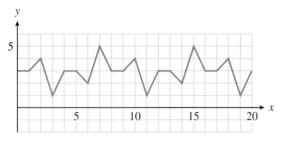

b. Compute the difference between the maximum and minimum function values. Sketch in the midline of the graph.

c. Find the smallest positive value of k for which $f(x) = f(x + k)$ for all x.

d. Find the smallest positive values of a and b for which $f(b) - f(a)$ is a maximum.

8. a. Find the period, the maximum and minimum values, and the midline of the graph of $y = f(x)$.

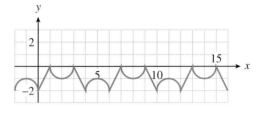

b. Sketch a graph of $y = 2f(x)$.

c. Sketch a graph of $y = 2 + f(x)$.

d. Modify the graph of $f(x)$ so that the period is twice its current value.

9. The apparent magnitude of a star is a measure of its brightness as seen from Earth. Smaller values for the apparent magnitude correspond to brighter stars. The graph below, called a light curve, shows the apparent magnitude of the star Algol as a function of time. Algol is an eclipsing binary star, which means that it is actually a system of two stars, a bright principal star and its dimmer companion, in orbit around each other. As each star passes in front of the other, it eclipses some of the light that reaches Earth from the system. (Source: Gamow, 1965, Brandt & Maran, 1972)

a. The light curve is periodic. What is its period?

b. What is the range of apparent magnitudes of the Algol system?

c. Explain the large and small dips in the light curve. What is happening to cause the dips?

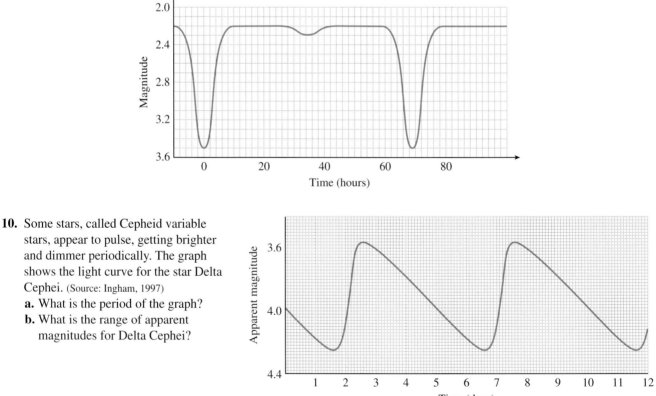

10. Some stars, called Cepheid variable stars, appear to pulse, getting brighter and dimmer periodically. The graph shows the light curve for the star Delta Cephei. (Source: Ingham, 1997)

a. What is the period of the graph?

b. What is the range of apparent magnitudes for Delta Cephei?

11. The figure is a tide chart for Los Angeles for the week of December 17–23, 2000. The horizontal axis shows time in hours, with $t = 12$ corresponding to noon on December 17. The vertical axis shows the height of the tide in feet above mean sea level.

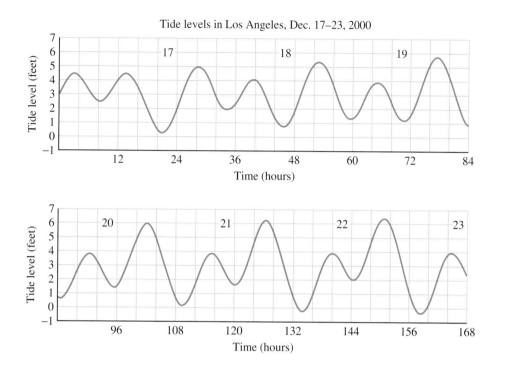

Tide levels in Los Angeles, Dec. 17–23, 2000

a. High tides occurred at 3:07 a.m. and 2:08 p.m. on December 17, and low tides at 8:41 a.m. and 9:02 p.m. Estimate the heights of the high and low tides on that day.

b. Is tide height a periodic function of time? Use the information from part (a) to justify your answer.

c. Make a table showing approximate times and heights for the high tides throughout the week. Make a similar table for the low tides.

d. Describe the trend in the heights of the high tides over the week. Describe the trend in the heights of the low tides.

e. What is the largest height difference between consecutive high and low tides during the week shown? When does it occur?

CHAPTER **3**

Power Functions

W e next turn our attention to a large and useful family of functions called **power functions.** This family includes transformations of several of the basic functions, such as

$$F(d) = \frac{k}{d^2} \quad \text{and} \quad S(T) = 20.06\sqrt{T}$$

The first function gives the gravitational force, F, exerted by the sun on an object at a distance, d. The second function gives the speed of sound, S, in terms of the air temperature, T. By extending our definition of exponent to include negative numbers and fractions, we will be able to express such functions in the form $f(x) = kx^n$. Here is an example of a power function with a fractional exponent.

In 1932, Max Kleiber published a remarkable equation for the metabolic rate of an animal as a function of its mass. The table on page 234 shows the mass of various animals in kilograms and their metabolic rates, in kilocalories per day. A plot of the data, resulting in the famous "mouse-to-elephant" curve, is shown in the figure.

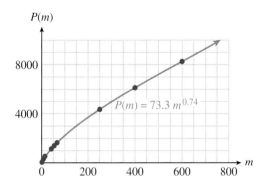

$P(m) = 73.3 \, m^{0.74}$

ThomsonNOW

Throughout this chapter, the ThomsonNOW logo indicates an opportunity for online self-study, linking you to interactive tutorials and videos based on your level of understanding.

Animal	Mass (kg)	Metabolic rate (kcal/day)
Baboon	6.2	300
Cat	3.0	150
Chimpanzee	38	1110
Cow	400	6080
Dog	15.5	520
Elephant	3670	48,800
Guinea pig	0.8	48
Human	65	1660
Mouse	0.02	3.4
Pig	250	4350
Polar bear	600	8340
Rabbit	3.5	165
Rat	0.2	28
Sheep	50	1300

Kleiber modeled his data by the power function

$$P(m) = 73.3\, m^{0.74}$$

where P is the metabolic rate and m is the mass of the animal. Kleiber's rule initiated the use of **allometric equations,** or power functions of mass, in physiology.

Investigation 5 Inflating a Balloon

If you blow air into a balloon, what do you think will happen to the air pressure inside the balloon as it expands? Here is what two physics books have to say:

> The greater the pressure inside, the greater the balloon's volume.
>
> Boleman, Jay, *Physics, a Window on Our World*

> Contrary to the process of blowing up a toy balloon, the pressure required to force air into a bubble decreases with bubble size.
>
> Sears, Francis, *Mechanics, Heat, and Sound*

1. What do the books say will happen to the air pressure as the volume of the ballon increases?

2. Based on these two quotes and your own intuition, sketch a graph showing how pressure changes as a function of the diameter of the balloon. Describe your graph: Is it increasing or decreasing? Is it concave up (bending upward) or concave down (bending downward)?

3. In 1998, two high school students, April Leonardo and Tolu Noah, decided to see for themselves how the pressure inside a balloon changes as the balloon expands. Using a column of water to measure pressure, they collected the following data while blowing up a balloon. Graph their data on the grid.

Diameter (cm)	Pressure (cm H_2O)
5.7	60.6
7.3	57.2
8.2	47.9
10.7	38.1
12	37.1
14.6	31.9
17.5	28.1
20.5	26.4
23.5	28
25.2	31.4
26.1	34
27.5	37.2
28.4	37.9
29	40.7
30	43.3
30.6	46.6
31.3	50
32.2	61.9

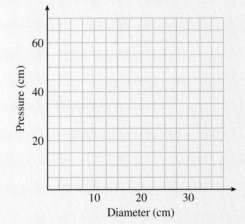

4. Describe the graph of April and Tolu's data. How does it compare to your graph in part (1)? Do their data confirm the predictions of the physics books? (We will return to April and Tolu's experiment in Section 3.4.)

3.1   Variation

Two types of functions are widely used in modeling and are known by special names: **direct variation** and **inverse variation.**

Direct Variation

Two variables are **directly proportional** (or just **proportional**) if the ratios of their corresponding values are always equal. Consider the functions described in Tables 3.1 and 3.2. The first table shows the price of gasoline as a function of the number of gallons purchased.

Gallons of gasoline	Total price	Price/ Gallons
4	$9.60	$\dfrac{9.60}{4} = 2.40$
6	$14.40	$\dfrac{14.40}{6} = 2.40$
8	$19.20	$\dfrac{19.20}{8} = 2.40$
12	$28.80	$\dfrac{28.80}{12} = 2.40$
15	$36.00	$\dfrac{36.00}{15} = 2.40$

TABLE 3.1

Years	Population	People/Years
10	432	$\dfrac{432}{10} \approx 43$
20	932	$\dfrac{932}{20} \approx 47$
30	2013	$\dfrac{2013}{30} \approx 67$
40	4345	$\dfrac{4345}{40} \approx 109$
50	9380	$\dfrac{9380}{50} \approx 188$
60	20,251	$\dfrac{20,251}{60} \approx 338$

TABLE 3.2

The ratio $\frac{\text{total price}}{\text{number of gallons}}$, or price per gallon, is the same for each pair of values in Table 3.1. This agrees with everyday experience: The price per gallon of gasoline is the same no matter how many gallons you buy. Thus, the total price of a gasoline purchase is directly proportional to the number of gallons purchased.

The second table, Table 3.2, shows the population of a small town as a function of the town's age. The ratio $\frac{\text{number of people}}{\text{number of years}}$ gives the average rate of growth of the population in people per year. You can see that this ratio is *not* constant; in fact, it increases as time goes on. Thus, the population of the town is *not* proportional to its age.

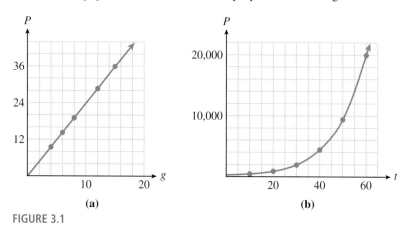

FIGURE 3.1

The graphs of these two functions are shown in Figure 3.1. We see that the price, P, of a fill-up is a linear function of the number of gallons, g, purchased. This should not be surprising if we write an equation relating the variables g and P. Because the ratio of their values is constant, we can write

$$\frac{P}{g} = k$$

where k is a constant. In this example, the constant k is 2.40, the price of gasoline per gallon. Solving for P in terms of g, we have

$$P = kg = 2.40g$$

which we recognize as the equation of a line through the origin.

In general, we make the following definition.

> **Direct Variation**
>
> y **varies directly** with x if
>
> $$y = kx$$
>
> where k is a positive constant called the **constant of variation.**

From the preceding discussion, we see that *vary directly* means exactly the same thing as *are directly proportional.* The two phrases are interchangeable.

EXAMPLE 1

a. The circumference, C, of a circle varies directly with its radius, r, because

$$C = 2\pi r$$

The constant of variation is 2π, or about 6.28.

b. The amount of interest, I, earned in one year on an account paying 7% simple interest, varies directly with the principal, P, invested, because

$$I = 0.07P$$

Direct variation defines a linear function of the form

$$y = f(x) = kx$$

The positive constant k in the equation $y = kx$ is just the slope of the graph, so it tells us how rapidly the graph increases. Compared to the standard form for a linear function, $y = b + mx$, the constant term, b, is zero, so the graph of a direct variation passes through the origin.

> **EXERCISE 1**
>
> Which of the graphs in Figure 3.2 could represent direct variation? Explain why.
>
>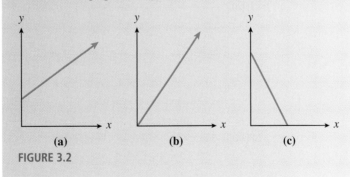
>
> (a)   (b)   (c)
>
> **FIGURE 3.2**

The Scaling Property of Direct Variation

The fact that the constant term is zero in a direct variation is significant: If we double the value of x, then the value of y will double also. In fact, increasing x by any factor causes y to increase by the same factor. For example, in Table 3.1 doubling the number of gallons of gas purchased, say, from 4 gallons to 8 gallons or from 6 gallons to 12 gallons, causes the total price to double also. Or, consider investing $800 for one year at 7% simple interest, as in Example 1b. The interest earned is

$$I = 0.07(800) = \$56$$

If we increase the investment by a factor of 1.6 to 1.6 (800), or $1280, the interest will be

$$I = 0.07(1280) = \$89.60$$

You can check that multiplying the original interest of $56 by a factor of 1.6 does give the same figure for the new interest, $89.60.

EXAMPLE 2

a. Tuition at Woodrow University is $400 plus $30 per unit. Is the tuition proportional to the number of units you take?

b. Imogen makes a 15% commission on her sales of environmentally friendly products marketed by her co-op. Do her earnings vary directly with her sales?

Solutions

a. Let u represent the number of units you take, and let $T(u)$ represent your tuition. Then

$$T(u) = 400 + 30u$$

Thus, $T(u)$ is a linear function of u, but the T-intercept is 400, not 0. Your tuition is *not* proportional to the number of units you take, so this is not an example of direct variation. You can check that doubling the number of units does not double the tuition. For example,

$$T(6) = 400 + 30(6) = 580$$

and

$$T(12) = 400 + 30(12) = 760$$

Tuition for 12 units is not double the tuition for 6 units. The graph of $T(u)$ in Figure 3.3a does not pass through the origin.

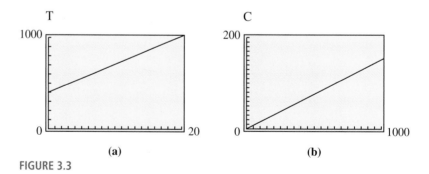

(a)   **(b)**

FIGURE 3.3

b. Let S represent Imogen's sales, and let $C(S)$ represent her commission. Then

$$C(S) = 0.15S$$

Thus, $C(S)$ is a linear function of S with a C-intercept of zero, so Imogen's earnings do vary directly with her sales. This is an example of direct variation. (See Figure 3.3b.)

Which table could represent direct variation? Explain why. (*Hint:* What happens to *y* if you multiply *x* by a constant?)

a.

x	1	2	3	6	8	9
y	2.5	5	7.5	15	20	22.5

b.

x	2	3	4	6	8	9
y	2	3.5	5	7	8.5	10

Finding a Formula for Direct Variation

If we know any one pair of values for the variables in a direct variation, we can find the constant of variation. We can then use the constant to write a formula for one of the variables as a function of the other.

EXAMPLE 3

If an object is dropped from a great height, say, off the rim of the Grand Canyon, its speed, *v*, varies directly with the time, *t*, the object has been falling. A rock dropped off the edge of the Canyon is falling at a speed of 39.2 meters per second when it passes a lizard on a ledge 4 seconds later.

a. Express *v* as a function of *t*.

b. What is the speed of the rock after it has fallen for 6 seconds?

c. Sketch a graph of $v(t)$ versus *t*.

Solutions

a. Because *v* varies directly with *t*, there is a positive constant *k* for which

$$v = kt$$

Substitute $v = 39.2$ when $t = 4$ and solve for *k* to find

$$39.2 = k(4) \quad \text{Divide both sides by 4.}$$
$$k = 9.8$$

Thus, $v(t) = 9.8\,t$.

b. Evaluate the function you found in part (a) for $t = 6$.

$$v(6) = 9.8(6) = 58.8$$

At $t = 6$ seconds, the rock is falling at a speed of 58.8 meters per second.

c. Use your calculator to graph the function $v(t) = 9.8t$. The graph is shown in Figure 3.4.

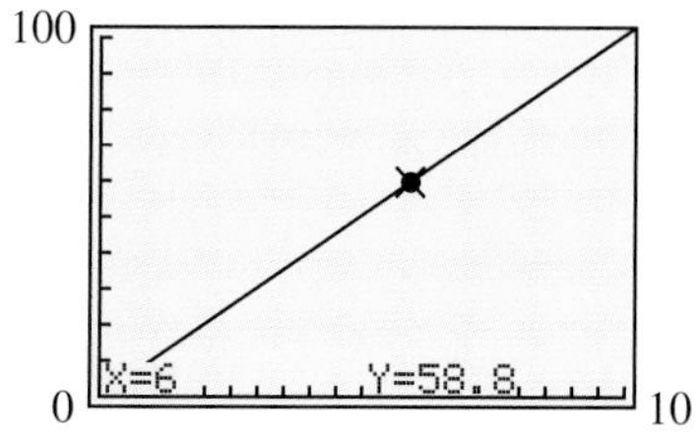

FIGURE 3.4

The volume of a bag of rice, in cups, is directly proportional to the weight of the bag. A 2-pound bag contains 3.5 cups of rice.

a. Express the volume, *V*, of a bag of rice as a function of its weight, *w*.

b. How many cups of rice are in a 15-pound bag?

Direct Variation with a Power of x

We can generalize the notion of direct variation to include situations in which y is proportional to a power of x, instead of x itself.

> **Direct Variation with a Power**
>
> y **varies directly** with a power of x if
>
> $$y = kx^n$$
>
> where k and n are positive constants.

EXAMPLE 4

The surface area of a sphere varies directly with the *square* of its radius. A balloon of radius 5 centimeters has surface area 100π square centimeters, or about 314 square centimeters. Find a formula for the surface area of a sphere as a function of its radius.

Solution If S stands for the surface area of a sphere of radius r, then

$$S = f(r) = kr^2$$

To find the constant of variation, k, we substitute the values of S and r.

$$100\pi = k(5)^2$$
$$4\pi = k$$

Thus, $S = f(r) = 4\pi r^2$.

EXERCISE 4

The volume of a sphere varies directly with the *cube* of its radius. A balloon of radius 5 centimeters has volume $\frac{500\pi}{3}$ cubic centimeters, or about 524 cubic centimeters. Find a formula for the volume of a sphere as a function of its radius.

In any example of direct variation, as the input variable increases through positive values, the output variable increases also. Thus, a direct variation is an increasing function, as we can see when we consider the graphs of some typical direct variations in Figure 3.5.

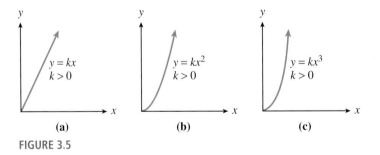

(a) $\quad\quad$ **(b)** $\quad\quad$ **(c)**

FIGURE 3.5

CAUTION The graph of a direct variation always passes through the origin, so when the input is zero, the output is also zero. Thus, the functions $y = 3x + 2$ and $y = 0.4x^2 - 2.3$, for example, are *not* direct variation, even though they are increasing functions for positive x.

Even without an equation, we can check whether a table of data describes direct variation or merely an increasing function. If y varies directly with x^n, then $y = kx^n$, or, equivalently, $\frac{y}{x^n} = k$.

> **Test for Direct Variation**
>
> If the ratio $\frac{y}{x^n}$ is constant, then y varies directly with x^n.

EXAMPLE 5

Delbert collects the following data and would like to know if y varies directly with the square of x. What should he calculate?

x	2	5	8	10	12
y	6	16.5	36	54	76

Solution

If y varies directly with x^2, then $y = kx^2$, or $\frac{y}{x^2} = k$. Delbert should calculate the ratio $\frac{y}{x^2}$ for each data point.

x	2	5	8	10	12
y	6	16.5	36	54	76
$\frac{y}{x^2}$	1.5	0.66	0.56	0.54	0.53

Because the ratio $\frac{y}{x^2}$ is not constant, y does not vary directly with x^2.

> **EXERCISE 5**
>
> Does B vary directly with the cube of r? Explain your decision.
>
> | r | 0.1 | 0.3 | 0.5 | 0.8 | 1.2 |
> |---|---|---|---|---|---|
> | B | 0.072 | 1.944 | 9.0 | 36.864 | 124.416 |

Scaling

Recall that if y varies directly with x, then doubling x causes y to double also. But is the area of a 16-inch circular pizza double the area of an 8-inch pizza? If you double the dimensions of a model of a skyscraper, will its weight double also? You probably know that the answer to both of these questions is *No.* The area of a circle is proportional to the *square* of its radius, and the volume (and hence the weight) of an object is proportional to the *cube* of its linear dimension. Variation with a power of x produces a different scaling effect.

EXAMPLE 6

The Taipei 101 building is 1671 feet tall, and in 2006 it was the tallest skyscraper in the world. Show that doubling the dimensions of a model of the Taipei 101 building produces a model that weighs 8 times as much.

Solution The Taipei 101 skyscraper is approximately box shaped, so its volume is given by the product of its linear dimensions, $V = lwh$. The weight of an object is proportional to its volume, so the weight, W, of the model is

$$W = klwh$$

where the constant k depends on the material of the model. If we double the length, width, and height of the model, then

$$W_{\text{new}} = k(2l)(2w)(2h)$$
$$= 2^3(klwh) = 8W_{\text{old}}$$

The weight of the new model is $2^3 = 8$ times the weight of the original model.

EXERCISE 6

Use the formula for the area of a circle to show that doubling the diameter of a pizza quadruples its area.

In general, if y varies directly with a power of x, that is, if $y = kx^n$, then doubling the value of x causes y to increase by a factor of 2^n. In fact, if we multiply x by any positive number c, then

$$y_{\text{new}} = k(cx)^n$$
$$= c^n(kx^n) = c^n(y_{\text{old}})$$

so the value of y is multiplied by c^n. We will call n the **scaling exponent,** and you will often see variation described in terms of *scaling*. For example, we might say that "the area of a circle scales as the square of its radius." (In many applications, the power n is called the *scale factor,* even though it is not a factor but an exponent.)

Inverse Variation

How long does it take to travel a distance of 600 miles? The answer depends on your average speed. If you are on a bicycle trip, your average speed might be 15 miles per hour. In that case, your traveling time will be

$$T = \frac{D}{R} = \frac{600}{15} = 40 \text{ hours}$$

(Of course, you will have to add time for rest stops; the 40 hours are just your travel time.) If you are driving your car, you might average 50 miles per hour. Your travel time is then

$$T = \frac{D}{R} = \frac{600}{50} = 12 \text{ hours}$$

If you take a commercial air flight, the plane's speed might be 400 miles per hour, and the flight time would be

$$T = \frac{D}{R} = \frac{600}{400} = 1.5 \text{ hours}$$

You can see that for higher average speeds, the travel time is shorter. In other words, the time needed for a 600-mile journey is a decreasing function of average speed. In fact, a formula for the function is

$$T = f(R) = \frac{600}{R}$$

This function is an example of **inverse variation.** A table of values and a graph of the function are shown in Figure 3.6.

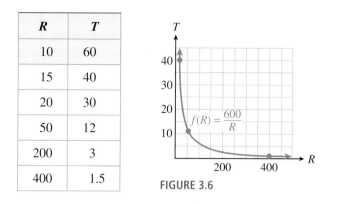

R	T
10	60
15	40
20	30
50	12
200	3
400	1.5

$$f(R) = \frac{600}{R}$$

FIGURE 3.6

Inverse Variation

y **varies inversely** with *x* if

$$y = \frac{k}{x}, \qquad x \neq 0$$

where *k* is a positive constant.

Inverse variation describes a decreasing function, but not every decreasing function represents inverse variation. People sometimes mistakenly use the phrase *varies inversely* to describe any decreasing function, but if *y* varies inversely with *x*, the variables must satisfy an equation of the form $y = \frac{k}{x}$, or $xy = k$. To decide whether two variables truly vary inversely, we can check whether their product is constant. For instance, in the preceding travel-time example, we see from the table that $RT = 600$.

R	10	15	20	50	200	400
T	60	40	30	12	3	1.5
RT	600	600	600	600	600	600

We can also define inverse variation with a power of the variable.

Inverse Variation with a Power

y **varies inversely** with x^n if

$$y = \frac{k}{x^n}, \qquad x \neq 0$$

where *k* and *n* are positive constants.

We may also say that *y* is **inversely proportional** to x^n.

EXAMPLE 7

The weight, w, of an object varies inversely with the square of its distance, d, from the center of the Earth. Thus,

$$w = \frac{k}{d^2}$$

If you double your distance from the center of the Earth, what happens to your weight? What if you triple the distance?

Solution

Suppose you weigh W pounds at distance D from the center of the Earth. Then $W = \frac{k}{D^2}$. At distance $2D$, your weight will be

$$w = \frac{k}{(2D)^2} = \frac{k}{4D^2} = \frac{1}{4} \cdot \frac{k}{D^2} = \frac{1}{4}W$$

Your new weight will be $\frac{1}{4}$ of your old weight. By a similar calculation, you can check that by tripling the distance, your weight will be reduced to $\frac{1}{9}$ of its original value.

EXERCISE 7

The amount of force, F, (in pounds) needed to loosen a rusty bolt with a wrench is inversely proportional to the length, l, of the wrench. Thus,

$$F = \frac{k}{l}$$

If you increase the length of the wrench by 50% so that the new length is $1.5l$, what happens to the amount of force required to loosen the bolt?

In Example 7 and Exercise 7, as the independent variable increases through positive values, the dependent variable decreases. An inverse variation is an example of a decreasing function. The graphs of some typical inverse variations are shown in Figure 3.7.

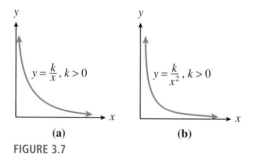

(a)   (b)

FIGURE 3.7

Finding a Formula for Inverse Variation

If we know that two variables vary inversely and we can find one pair of corresponding values for the variables, we can determine k, the constant of variation.

EXAMPLE 8

The intensity of electromagnetic radiation, such as light or radio waves, varies inversely with the square of the distance from its source. Radio station KPCC broadcasts a signal that is measured at 0.016 watt per square meter by a receiver 1 kilometer away.

a. Write a formula that gives signal strength as a function of distance.

b. If you live 5 kilometers from the station, what is the strength of the signal you will receive?

Solutions **a.** Let I stand for the intensity of the signal in watts per square meter, and d for the distance from the station in kilometers. Then $I = \frac{k}{d^2}$. To find the constant k, we substitute 0.016 for I and 1 for d. Solving for k gives us

$$0.016 = \frac{k}{1^2}$$

$$k = 0.016\,(1^2) = 0.016$$

Thus, $I = \dfrac{0.016}{d^2}$.

b. Now we can substitute 5 for d and solve for I.

$$I = \frac{0.016}{5^2} = 0.00064$$

At a distance of 5 kilometers from the station, the signal strength is 0.00064 watt per square meter.

EXERCISE 8

Delbert's officemates want to buy a $120 gold watch for a colleague who is retiring. The cost per person is inversely proportional to the number of people who contribute.

a. Express the cost per person, C, as a function of the number of people, p, who contribute.

b. Sketch the function on the domain $0 \le p \le 20$.

ANSWERS TO 3.1 EXERCISES

1. (b): The graph is a straight line through the origin.

2. (a): If we multiply x by c, y is also multiplied by c.

3. a. $V = 1.75w$ **b.** 26.25   **4.** $V = \frac{4}{3}\pi r^3$ **5.** Yes, $\frac{B}{r^3}$ is constant.

6. $A = \pi r^2$, so $A_{\text{new}} = \pi(2r)^2 = 4\pi r^2 = 4A_{\text{old}}$ **7.** $F_{\text{new}} = \frac{2}{3}F_{\text{old}}$

8. a. $C = \dfrac{120}{p}$ **b.**

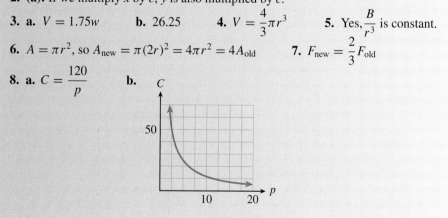

VOCABULARY   *Look up the definitions of new terms in the Glossary.*

Direct variation
Inverse variation

Directly proportional
Inversely proportional

Constant of variation

Scaling exponent

CONCEPTS

1.

> **Direct and Inverse Variation**
>
> - *y* **varies directly with** *x* if the ratio $\frac{y}{x}$ is constant, that is, if $y = kx$.
> - *y* **varies directly with a power** *x* if the ratio $\frac{y}{x^n}$ is constant, that is, if $y = kx^n$.
> - *y* **varies inversely with** *x* if the product xy is constant, that is, if $y = \frac{k}{x}$.
> - *y* **varies inversely with a power of** *x* if the product $x^n y$ is constant, that is, if $y = \frac{k}{x^n}$.

2. The graph of a direct variation passes through the origin. The graph of an inverse variation has a vertical asymptote at the origin.

3. If $y = kx^n$, we say that *y* **scales** as x^n.

STUDY QUESTIONS

1. Describe the graph of $y = f(x)$ if *y* varies directly with *x*.

2. What is true about the ratio of two variables if they are directly proportional?

3. If *y* is inversely proportional to *x*, then the graph of *y* versus *x* is a transformation of which basic graph?

4. If *y* varies directly with a power of *x*, write a formula for *y* as a function of *x*.

5. If *y* varies inversely with a power of *x*, write a formula for *y* as a function of *x*.

6. If $y = kx^4$, what happens to *y* if you double *x*?

7. State a test to determine whether *y* varies inversely with x^n.

8. If $y = \frac{k}{x^2}$, and we double the value of *x*, what happens to the value of *y*?

SKILLS   *Practice each skill in the Homework problems listed.*

1. Find the constant of variation: #1–4, 13–26

2. Write a formula for direct or inverse variation: #1–4, 13–26, 35–46

3. Recognize direct and inverse variation from a table of values: #27–34, 39–42

4. Recognize direct or inverse variation from a graph: #9–12, 35–38

5. Use scaling in direct and inverse variation: #13–20, 43–46

ThomsonNOW™ Test yourself on key content at www.thomsonedu.com/login.

1. Delbert's credit card statement lists three purchases he made while on a business trip in the Midwest. His company's accountant would like to know the sales tax rate on the purchases.

Price of item	18	28	12
Tax	1.17	1.82	0.78
Tax/Price			

   a. Compute the ratio of the tax to the price of each item. Is the tax proportional to the price? What is the tax rate?
   b. Express the tax, T, as a function of the price, p, of the item.
   c. Sketch a graph of the function by hand, and label the scales on the axes.

2. At constant acceleration from rest, the distance traveled by a race car is proportional to the square of the time elapsed. The highest recorded road-tested acceleration is 0 to 60 miles per hour in 3.07 seconds, which produces the following data.

Time (seconds)	2	2.5	3
Distance (feet)	57.32	89.563	128.97
Distance/Time^2			

   a. Compute the ratios of the distance traveled to the square of the time elapsed. What was the acceleration, in feet per second squared?
   b. Express the distance traveled, d, as a function of time in seconds, t.
   c. Sketch a graph of the function by hand and label the scales on the axes.

3. The marketing department for a paper company is testing wrapping paper rolls in various dimensions to see which shape consumers prefer. All the rolls contain the same amount of wrapping paper.

Width (feet)	2	2.5	3
Length (feet)	12	9.6	8
Length × width			

   a. Compute the product of the length and width for each roll of wrapping paper. What is the constant of inverse proportionality?
   b. Express the length, L, of the paper as a function of the width, w, of the roll.
   c. Sketch a graph of the function by hand and label the scales on the axes.

4. The force of gravity on a 1-kilogram mass is inversely proportional to the square of the object's distance from the center of the Earth. The table shows the force on the object, in newtons, at distances that are multiples of the Earth's radius.

Distance (Earth radii)	1	2	4
Force (newtons)	9.8	2.45	0.6125
Force × distance^2			

   a. Compute the products of the force and the square of the distance. What is the constant of inverse proportionality
   b. Express the gravitational force, F, on a 1-kilogram mass as a function of its distance, r, from the Earth's center, measured in Earth radii.
   c. Sketch a graph of the function by hand, and label the scales on the axes.

5. a. How can you tell from a table of values whether y varies directly with x?
   b. How can you tell from a table of values whether y varies inversely with x?

6. a. How can you tell from a table of values whether y varies directly with a power of x?
   b. How can you tell from a table of values whether y varies inversely with a power of x?

7. The length of a rectangle is 10 inches, and its width is 8 inches. Suppose we increase the length of the rectangle while holding the width constant.

a. Fill in the table.

Length	Width	Perimeter	Area
10	8		
12	8		
15	8		
20	8		

b. Does the perimeter vary directly with the length?

c. Write a formula for the perimeter of the rectangle in terms of its length.

d. Does the area vary directly with the length?

e. Write a formula for the area of the rectangle in terms of its length.

8. The base of an isosceles triangle is 12 centimeters, and the equal sides have length 15 centimeters. Suppose we increase the base of the triangle while holding the sides constant.

a. Fill in the table. (*Hint:* Use the Pythagorean theorem to find the height of the triangle.)

Base	Sides	Height	Perimeter	Area
12	15			
15	15			
18	15			
20	15			

b. Does the perimeter vary directly with the base?

c. Write a formula for the perimeter of the triangle in terms of its base.

d. Write a formula for the area of the triangle in terms of its base.

e. Does the area vary directly with the base?

9. Which of the graphs could describe direct variation? Explain your answer.

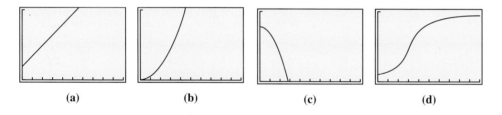

    (a)   (b)   (c)   (d)

10. Which of the graphs could describe direct variation? Explain your answer.

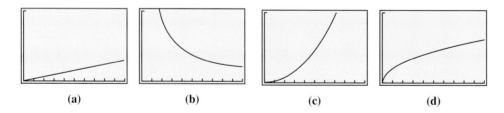

    (a)   (b)   (c)   (d)

11. Which of the graphs could describe inverse variation? Explain your answer.

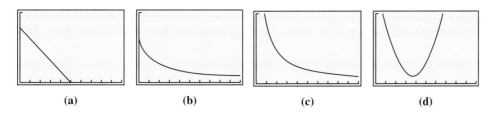

| (a) | (b) | (c) | (d) |

12. Which of the graphs could describe inverse variation? Explain your answer.

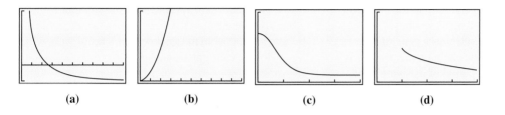

| (a) | (b) | (c) | (d) |

13. The weight of an object on the Moon varies directly with its weight on Earth. A person who weighs 150 pounds on Earth would weigh only 24.75 pounds on the Moon.

  a. Find a function that gives the weight m of an object on the Moon in terms of its weight w on Earth. Complete the table and graph your function in a suitable window.

w	50	100	200	400
m				

  b. How much would a person weigh on the Moon if she weighs 120 pounds on Earth?

  c. A piece of rock weighs 50 pounds on the Moon. How much will it weigh back on Earth?

  d. If you double the weight of an object on Earth, what will happen to its weight on the Moon?

14. Hubble's law says that distant galaxies are receding from us at a rate that varies directly with their distance. (The speeds of the galaxies are measured using a phenomenon called redshifting.) A galaxy in the constellation Ursa Major is 980 million light-years away and is receding at a speed of 15,000 kilometers per second.

  a. Find a function that gives the speed, v, of a galaxy in terms of its distance, d, from Earth. Complete the table and graph your function in a suitable window. (Distances are given in millions of light-years.)

d	500	1000	2000	4000
v				

  b. How far away is a galaxy in the constellation Hydra that is receding at 61,000 kilometers per second?

  c. A galaxy in Leo is 1240 million light-years away. How fast is it receding from us?

  d. If one constellation is twice as distant as another, how do their speeds compare?

15. The length, L, of a pendulum varies directly with the square of its period, T, the time required for the pendulum to make one complete swing back and forth. The pendulum on a grandfather clock is 3.25 feet long and has a period of 2 seconds.

    a. Express L as a function of T. Complete the table and graph your function in a suitable window.

T	1	5	10	20
L				

    b. How long is the Foucault pendulum in the Pantheon in Paris, which has a period of 17 seconds?

    c. A hypnotist uses a gold pendant as a pendulum to mesmerize his clients. If the chain on the pendant is 9 inches long, what is the period of its swing?

    d. In order to double the period of a pendulum, how must you vary its length?

16. The load, L, that a beam can support varies directly with the square of its vertical thickness, h. A beam that is 4 inches thick can support a load of 2000 pounds.

    a. Express L as a function of h. Complete the table and graph your function in a suitable window.

h	1	2	4	8
L				

    b. What size load can be supported by a beam that is 6 inches thick?

    c. How thick a beam is needed to support a load of 100 pounds?

    d. If you double the thickness of a beam, how will the load it can support change?

17. Computer monitors produce a magnetic field. The effect of the field, B, on the user varies inversely with his or her distance, d, from the screen. The field from a certain color monitor was measured at 22 milligauss 4 inches from the screen.

    a. Express the field strength as a function of distance from the screen. Complete the table and graph your function in a suitable window.

d	1	2	12	24
B				

    b. What is the field strength 10 inches from the screen?

    c. An elevated risk of cancer can result from exposure to field strengths of 4.3 milligauss. How far from the screen should the computer user sit to keep the field level below 4.3 milligauss?

    d. If you double your distance from the screen, how does the field strength change?

18. The amount of current, I, that flows through a circuit varies inversely with the resistance, R, on the circuit. An iron with a resistance of 12 ohms draws 10 amps of current.

    a. Express the current as a function of the resistance. Complete the table and graph your function in a suitable window.

R	1	2	10	20
I				

    b. How much current is drawn by a compact fluorescent light bulb with a resistance of 533.3 ohms?

    c. What is the resistance of a toaster that draws 12.5 amps of current?

    d. If the resistance of one appliance is double the resistance of a second appliance, how does the current they draw compare?

19. The amount of power, P, generated by a windmill varies directly with the cube of the wind speed, w. A windmill on Oahu, Hawaii, produces 7300 kilowatts of power when the wind speed is 32 miles per hour.

   a. Express the power as a function of wind speed. Complete the table and graph your function in a suitable window.

w	10	20	40	80
p				

   b. How much power would the windmill produce in a light breeze of 15 miles per hour?

   c. What wind speed is needed to produce 10,000 kilowatts of power?

   d. If the wind speed doubles, what happens to the amount of power generated?

20. A crystal form of pyrite (a compound of iron and sulfur) has the shape of a regular solid with 12 faces. Each face is a regular pentagon. This compound is called pyritohedron, and its mass, M, varies directly with the cube of the length, L, of 1 edge. If each edge is 1.1 centimeters, then the mass is 51 grams.

   a. Express the mass of pyritohedron as a function of the length of 1 edge. Complete the table and graph your function in a suitable window.

L	0.5	1	2	4
M				

   b. What is the weight of a chunk of pyritohedron if each edge is 2.2 centimeters?

   c. How long would each edge be for a 1643-gram piece of pyritohedron?

   d. If one chunk has double the length of a second chunk, how do their weights compare?

For Problems 21–26,
(a) Use the values in the table to find the constant of variation, k, and write y as a function of x.
(b) Fill in the rest of the table with the correct values.
(c) What happens to y when you double the value of x?

21. y varies directly with x.

x	y
2	
5	1.5
	2.4
12	
	4.5

22. y varies directly with x.

x	y
0.8	
1.5	54
	108
	126
6	

23. y varies directly with the square of x.

x	y
3	
6	24
	54
12	
	150

24. y varies directly with the cube of x.

x	y
2	120
3	
	1875
6	
	15,000

25. y varies inversely with x.

x	y
4	
	15
20	6
30	
	3

26. y varies inversely with the square of x.

x	y
0.2	
	80
2	
4	1.25
	0.8

For each table, decide whether (a) *y* varies directly with *x*, (b) *y* varies directly with *x*², or (c) *y* does not vary directly with a power of *x*. Explain why your choice is correct. If your choice is (a) or (b), find the constant of variation.

27.

x	y
2	2.0
3	4.5
5	12.5
8	32.0

28.

x	y
2	12
4	28
6	44
9	68

29.

x	y
1.5	3.0
2.4	5.3
5.5	33
8.2	73.8

30.

x	y
1.2	7.20
2.5	31.25
6.4	204.80
12	720.00

For each table, decide whether (a) *y* varies inversely with *x*, (b) *y* varies inversely with *x*², or (c) *y* does not vary inversely with a power of *x*. Explain why your choice is correct. If your choice is (a) or (b), find the constant of variation.

31.

x	y
0.5	288
2.0	18
3.0	8
6.0	2

32.

x	y
0.5	100.0
2.0	25.0
4.0	12.5
5.0	10.0

33.

x	y
1.0	4.0
1.3	3.7
3.0	2.0
4.0	1.0

34.

x	y
0.5	180.00
2.0	11.25
3.0	5.00
5.0	1.80

The functions described by a table of data or by a graph in Problems 35–42 are examples of direct or inverse variation.
(a) Find an algebraic formula for the function, including the constant of variation, *k*.
(b) Answer the question in the problem.

35. The faster a car moves, the more difficult it is to stop. The graph shows the distance, *d*, required to stop a car as a function of its velocity, *v*, before the brakes were applied. What distance is needed to stop a car moving at 100 kilometers per hour?

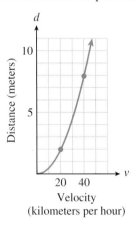

36. A wide pipe can handle a greater water flow than a narrow pipe. The graph shows the water flow through a pipe, *w*, as a function of its radius, *r*. How great is the water flow through a pipe of radius of 10 inches?

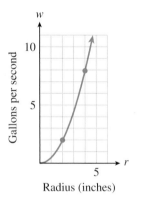

37. If the price of mushrooms goes up, the amount consumers are willing to buy goes down. The graph shows the number of tons of shiitake mushrooms, m, sold in California each week as a function of their price, p. If the price of shiitake mushrooms rises to $10 per pound, how many tons will be sold?

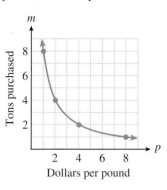

38. When an adult plays with a small child on a seesaw, the adult must sit closer to the pivot point to balance the seesaw. The graph shows this distance, d, as a function of the adult's weight, w. How far from the pivot must Kareem sit if he weighs 280 pounds?

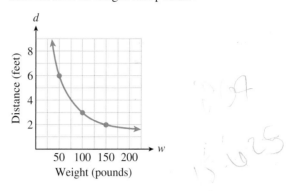

39. Ocean temperatures are generally colder at the greater depths. The table shows the temperature of the water as a function of depth. What is the ocean temperature at a depth of 6 kilometers?

Depth (km)	Temperature (°C)
0.5	12
1	6
2	3
3	2

40. The shorter the length of a vibrating guitar string, the higher the frequency of the vibrations. The fifth string is 65 centimeters long and is tuned to A (with a frequency of 220 vibrations per second). The placement of the fret relative to the bridge changes the effective length of the guitar string. The table shows frequency as a function of effective length. How far from the bridge should the fret be placed for the note C (256 vibrations per second)?

Length (cm)	Frequency
55	260
57.2	250
65	220
71.5	200

41. The strength of a cylindrical rod depends on its diameter. The greater the diameter of the rod, the more weight it can support before collapsing. The table shows the maximum weight supported by a rod as a function of its diameter. How much weight can a 1.2-centimeter rod support before collapsing?

Diameter (cm)	Weight (newtons)
0.5	150
1.0	600
1.5	1350
2.0	2400

42. The maximum height attained by a cannonball depends on the speed at which it was shot. The table shows maximum height as a function of initial speed. What height is attained by a cannonball whose initial upward speed was 100 feet per second?

Speed (ft/sec)	Height (ft)
40	200
50	312.5
60	450
70	612.5

43. The wind resistance, W, experienced by a vehicle on the freeway varies directly with the square of its speed, v.
  a. If you double your speed, what happens to the wind resistance?
  b. If you drive one-third as fast, what happens to the wind resistance?
  c. If you decrease your speed by 10%, what happens to the wind resistance?

44. The weight, w, of a bronze statue varies directly with the cube of its height, h.
  a. If you double the height of the statue, what happens to its weight?
  b. If you make the statue one-fourth as tall, what happens to its weight?
  c. If you increase the height of the statue by 50%, what happens to its weight?

45. The intensity of illumination, I, from a lamp varies inversely with the square of your distance, d, from the lamp.
  a. If you double your distance from a reading lamp, what happens to the illumination?
  b. If you triple the distance, what happens to the illumination?
  c. If you increase the distance by 25%, what happens to the illumination?

46. The resistance, R, of an electrical wire varies inversely with the square of its diameter, d.
  a. If you replace an old wire with a new one whose diameter is half that of the old one, what happens to the resistance?
  b. If you replace an old wire with a new one whose diameter is two-thirds of the old one, what happens to the resistance?
  c. If you decrease the diameter of the wire by 30%, what happens to the resistance?

The quoted material in Problems 47–50 (in italics) is taken from the article "Quantum Black Holes," by Bernard J. Carr and Steven B. Giddings, in the May 2005 issue of *Scientific American*. (See Algebra Skills Refresher A.1 to review scientific notation.)

47. *The density to which matter must be squeezed [to create a black hole] scales as the inverse square of the mass. For a hole with the mass of the Sun, the density is about 10^9 kilograms per cubic meter, higher than that of an atomic nucleus.*
  a. Recall that the density of an object is its mass per unit volume. Given that the mass of the sun is about 2×10^{30} kilograms, write a formula for the density, D, of a black hole as a function of its mass, m.
  b. *The known laws of physics allow for a matter density up to the so-called Planck value of 10^{97} kilograms per cubic meter.* If a black hole with this density could be created, it would be the smallest possible black hole. What would its mass be?
  c. Assuming that a black hole is spherical, what would be the radius of the smallest possible black hole?

48. *A black hole radiates thermally, like a hot coal, with a temperature inversely proportional to its mass. For a solar-mass black hole, the temperature is around a millionth of a kelvin.*
  a. The solar mass is given in Problem 47. Write a formula for the temperature, T, of a black hole as a function of its mass, m.
  b. What is the temperature of a black hole of mass 10^{12} kilograms, about the mass of a mountain?

49. *The total time for a black hole to evaporate away is proportional to the cube of its initial mass. For a solar-mass hole, the lifetime is an unobservably long 10^{64} years.*
  a. The solar mass is given in Problem 47. Write a formula for the lifetime, L, of a black hole as a function of its mass, m.
  b. The present age of the universe is about 10^{10} years. What would be the mass of a black hole as old as the universe?

50. *String theory . . . predicts that space has dimensions beyond the usual three. In three dimensions, the force of gravity quadruples as you halve the distance between two objects. But in nine dimensions, gravity would get 256 times as strong. In three dimensions, the force of gravity varies inversely with the square of distance.* Write a formula for the force of gravity in nine dimensions.

Use algebra to support your answers to Problems 51–56. Begin with a formula for direct or inverse variation.

51. Suppose y varies directly with x. Show that if you multiply x by any constant c, then y will be multiplied by the same constant.

52. Suppose y varies inversely with x. Show that if you multiply x by any constant c, then y will be divided by the same constant.

53. Explain why the ratio $\frac{y}{x^2}$ is a constant when y varies directly with x^2.

54. Explain why the product yx^2 is a constant when y varies inversely with x^2.

55. If x varies directly with y and y varies directly with z, does x vary directly with z?

56. If x varies inversely with y and y varies inversely with z, does x vary inversely with z?

3.2 Integer Exponents

Before studying this section, you may want to review Algebra Skills Refresher A.6, the Laws of Exponents.

Recall that a positive integer exponent tells us how many times its base occurs as a factor in an expression. For example,

$$4a^3b^2 \qquad \text{means} \qquad 4aaabb$$

Negative Exponents

Study the list of powers of 2 shown in Table 3.3a, and observe the pattern as we move up the list from bottom to top. Each time the exponent increases by 1 we multiply by another factor of 2. We can continue up the list as far as we like.

If we move back down the list, we divide by 2 at each step, until we get to the bottom of the list, $2^1 = 2$. What if we continue the list in the same way, dividing by 2 each time we decrease the exponent? The results are shown in Table 3.3b.

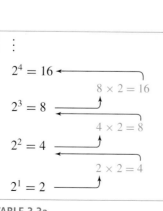

TABLE 3.3a

TABLE 3.3b

As we continue to divide by 2, we generate fractions whose denominators are powers of 2. In particular,

$$2^{-1} = \frac{1}{2} = \frac{1}{2^1} \qquad \text{and} \qquad 2^{-2} = \frac{1}{4} = \frac{1}{2^2}$$

Based on these observations, we make the following definitions.

Definition of Negative and Zero Exponents

$$a^{-n} = \frac{1}{a^n} \qquad\qquad (a \neq 0)$$

$$a^0 = 1 \qquad\qquad (a \neq 0)$$

These definitions tell us that if the base a is not zero, then any number raised to the zero power is 1, and that a negative exponent denotes a reciprocal.

EXAMPLE 1

a. $2^{-3} = \frac{1}{2^3} = \frac{1}{8}$ **b.** $9x^{-2} = 9 \cdot \frac{1}{x^2} = \frac{9}{x^2}$

CAUTION

1. A negative exponent does *not* mean that the power is negative! For example,

$$2^{-3} \neq -2^3$$

2. In Example 1b, note that

$$9x^{-2} \neq \frac{1}{9x^2}$$

The exponent, -2, applies *only* to the base x, not to 9.

EXERCISE 1

Write each expression without using negative exponents.

a. 5^{-4} **b.** $5x^{-4}$

In the next example, we see how to evaluate expressions that contain negative exponents and how to solve equations involving negative exponents.

EXAMPLE 2

The body mass index, or BMI, is one measure of a person's physical fitness. Your body mass index is defined by

$$\text{BMI} = wh^{-2}$$

where w is your weight in kilograms and h is your height in meters. The World Health Organization classifies a person as obese if his or her BMI is 25 or higher.

   a. Calculate the BMI for a woman who is 1.625 meters (64 inches) tall and weighs 54 kilograms (120 pounds).

   b. For a fixed weight, how does BMI vary with height?

   c. The world's heaviest athlete is the amateur sumo wrestler Emanuel Yarbrough, who weighs 319 kg (704 pounds). What height would Yarbrough have to be to have a BMI under 25?

Solutions   **a.** $\text{BMI} = 54\left(1.625^{-2}\right) = 54\left(\dfrac{1}{1.625^2}\right) = 20.45$

   b. $\text{BMI} = wh^{-2} = \dfrac{w}{h^2}$, so BMI varies inversely with the square of height. That is, for a fixed weight, BMI decreases as height increases.

   c. To find the height that gives a BMI of 25, we solve the equation $25 = 319h^{-2}$. Note that the variable h appears in the denominator of a fraction, so we begin by clearing the denominator—in this case we multiply both sides of the equation by h^2.

$$
\begin{aligned}
25 &= \frac{319}{h^2} &&\text{Multiply both sides by } h^2. \\
25h^2 &= 319 &&\text{Divide both sides by 25.} \\
h^2 &= 12.76 &&\text{Extract square roots.} \\
h &\approx 3.57
\end{aligned}
$$

To have a BMI under 25, Yarbrough would have to be over 3.57 meters, or 11 feet 8 inches tall. (In fact, he is 6 feet 8 inches tall.)

EXERCISE 2

Solve the equation $0.2x^{-3} = 1.5$.
Rewrite without a negative exponent.
Clear the fraction.
Isolate the variable.

Power Functions

The functions that describe direct and inverse variation are part of a larger family of functions called **power functions.**

Power Functions

A function of the form

$$f(x) = kx^p$$

where k and p are nonzero constants, is called a **power function.**

Examples of power functions are

$$V(r) = \frac{4}{3}\pi r^3 \quad \text{and} \quad L(T) = 0.8125T^2$$

In addition, the basic functions

$$f(x) = \frac{1}{x} \quad \text{and} \quad g(x) = \frac{1}{x^2}$$

which we studied in Chapter 2, can be written as

$$f(x) = x^{-1} \quad \text{and} \quad g(x) = x^{-2}$$

Their graphs are shown in Figure 3.8. Note that the domains of power functions with negative exponents do not include zero.

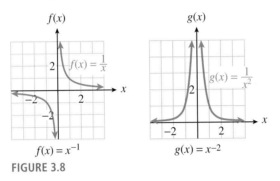

$f(x) = x^{-1}$ $\qquad$ $g(x) = x^{-2}$

FIGURE 3.8

EXAMPLE 3

Which of the following are power functions?

a. $f(x) = \frac{1}{3}x^4 + 2$ $\qquad$ **b.** $g(x) = \frac{1}{3x^4}$ $\qquad$ **c.** $h(x) = \frac{x+6}{x^3}$

Solutions

a. This is not a power function, because of the addition of the constant term.

b. We can write $g(x) = \frac{1}{3}x^{-4}$, so g is a power function.

c. This is not a power function, but it can be treated as the sum of two power functions, because $h(x) = x^{-2} + 6x^{-3}$.

EXERCISE 3

Write each function as a power function in the form $y = kx^p$.

a. $f(x) = \frac{12}{x^2}$ $\qquad$ **b.** $g(x) = \frac{1}{4x}$ $\qquad$ **c.** $h(x) = \frac{2}{5x^6}$

Most applications are concerned with positive variables only, so many models use only the portion of the graph in the first quadrant.

EXAMPLE 4

In the Middle Ages in Europe, castles were built as defensive strongholds. An attacking force would build a huge catapult called a trebuchet to hurl rocks and scrap metal inside the castle walls. The engineers could adjust its range by varying the mass of the projectiles. The mass, m, of the projectile should be inversely proportional to the square of the distance, d, to the target.

a. Use a negative exponent to write m as a function of d, $m = f(d)$.

b. The engineers test the trebuchet with a 20-kilogram projectile, which lands 250 meters away. Find the constant of proportionality; then rewrite your formula for m.

c. Graph $m = f(d)$.

d. The trebuchet is 180 meters from the courtyard within the castle. What size projectile will hit the target?

e. The attacking force would like to hurl a 100-kilogram projectile at the castle. How close must the attackers bring their trebuchet?

Solutions

a. If we use k for the constant of proportionality, then $m = \frac{k}{d^2}$. Rewriting this equation with a negative exponent gives $m = kd^{-2}$.

b. Substitute $m = 20$ and $d = 250$ to obtain

$$20 = k(250)^{-2} \qquad \text{Multiply both sides by } 250^2.$$
$$1{,}250{,}000 = k$$

Thus, $m = 1{,}250{,}000\,d^{-2}$.

c. Evaluate the function for several values of m, or use your calculator to obtain the graph in Figure 3.9.

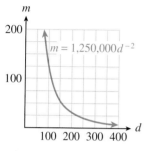

FIGURE 3.9

d. Substitute $d = 180$ into the formula:

$$m = 1{,}250{,}000\,(180)^{-2}$$
$$= \frac{1{,}250{,}000}{32{,}400} \approx 38.58$$

The attackers should use a mass of approximately 38.6 kilograms.

e. Substitute $m = 100$ into the formula and solve for d.

$$100 = 1{,}250{,}000\,d^{-2} \qquad \text{Multiply by } d^2.$$
$$100d^2 = 1{,}250{,}000 \qquad \text{Divide by 100.}$$
$$d^2 = 12{,}500 \qquad \text{Take square roots.}$$
$$d = \pm\sqrt{12{,}500}$$

They must locate the trebuchet $\sqrt{12{,}500} \approx 111.8$ meters from the castle.

The function $m = \frac{k}{d^2}$ is an example of an **inverse square law,** because m varies inversely with the square of d. Such laws are fairly common in physics and its applications, because gravitational and other forces behave in this way. Here is a more modern example of an inverse square law.

EXERCISE 4

Cell phone towers typically transmit signals at 10 watts of power. The signal strength varies inversely with the square of distance from the tower, and 1 kilometer away the signal strength is 0.8 picowatt. (A picowatt is 10^{-12} watt.) Cell phones can receive a signal as small as 0.01 picowatt. How far can you be from the nearest tower and still hope to have cell phone reception?

Working with Negative Exponents

A negative exponent denotes the *reciprocal* of a power. Thus, to simplify a fraction with a negative exponent, we compute the positive power of its reciprocal.

EXAMPLE 5

a. $\left(\dfrac{3}{5}\right)^{-2} = \dfrac{1}{\left(\frac{3}{5}\right)^2} = \left(\dfrac{5}{3}\right)^2 = \dfrac{25}{9}$

> ▶▶▶
> To review operations on fractions, see Algebra Skills Refresher A.9.

b. $\left(\dfrac{x^3}{4}\right)^{-3} = \left(\dfrac{4}{x^3}\right)^3 = \dfrac{(4)^3}{(x^3)^3} = \dfrac{64}{x^9}$

EXERCISE 5

Simplify $\left(\dfrac{2}{x^2}\right)^{-4}$.

Dividing by a power with a negative exponent is equivalent to multiplying by a power with a positive exponent.

EXAMPLE 6

a.
$$\dfrac{1}{5^{-3}} = 1 \div 5^{-3}$$
$$= 1 \div \dfrac{1}{5^3}$$
$$= 1 \times 5^3 = 125$$

b.
$$\dfrac{k^2}{m^{-4}} = k^2 \div m^{-4}$$
$$= k^2 \div \dfrac{1}{m^4}$$
$$= k^2 m^4$$

EXERCISE 6

Write each expression without using negative exponents.

a. $\left(\dfrac{3}{b^4}\right)^{-2}$ **b.** $\dfrac{12}{x^{-6}}$

Laws of Exponents

The laws of exponents, reviewed in Algebra Skills Refresher A.6, apply to all integer exponents, positive, negative, and zero. When we allow negative exponents, we can simplify the rule for computing quotients of powers.

II. $\dfrac{a^m}{a^n} = a^{m-n} \quad (a \neq 0)$

For example, by applying this new version of the law for quotients, we find

$$\dfrac{x^2}{x^5} = x^{2-5} = x^{-3}$$

which is consistent with our previous version of the rule,

$$\dfrac{x^2}{x^5} = \dfrac{1}{x^{5-2}} = \dfrac{1}{x^3}$$

For reference, we restate the laws of exponents below. The laws are valid for all integer exponents m and n, and for $a, b \neq 0$.

Laws of Exponents

I. $a^m \cdot a^n = a^{m+n}$

II. $\left(\dfrac{a^m}{a^n}\right) = a^{m-n}$

III. $(a^m)^n = a^{mn}$

IV. $(ab)^n = a^n b^n$

V. $\left(\dfrac{a}{b}\right)^n = \dfrac{a^n}{b^n}$

EXAMPLE 7

a. $x^3 \cdot x^{-5} = x^{3-5} = x^{-2}$ Apply the first law: Add exponents.

b. $\dfrac{8x^{-2}}{4x^{-6}} = \dfrac{8}{4}x^{-2-(-6)} = 2x^4$ Apply the second law: Subtract exponents.

c. $(5x^{-3})^{-2} = 5^{-2}(x^{-3})^{-2} = \dfrac{x^6}{25}$ Apply laws IV and III.

You can check that each of the calculations in Example 7 is shorter when we use negative exponents instead of converting the expressions into algebraic fractions.

EXERCISE 7

Simplify by applying the laws of exponents.

a. $(2a^{-4})(-4a^2)$ **b.** $\dfrac{(r^2)^{-3}}{3r^{-4}}$

CAUTION

The laws of exponents do not apply to sums or differences of powers. We can add or subtract *like terms,* that is, powers with the same exponent. For example,

$$6x^{-2} + 3x^{-2} = 9x^{-2}$$

but **we cannot add or subtract terms with different exponents.** Thus, for example,

$$4x^2 - 3x^{-2} \qquad \text{cannot be simplified}$$
$$x^{-1} + x^{-3} \qquad \text{cannot be simplified}$$

In Table 3.3b, we saw that $2^0 = 1$, and in fact $a^0 = 1$ as long as $a \neq 0$. Now we can see that this definition is consistent with the laws of exponents. The quotient of any (nonzero) number divided by itself is 1. But by applying the second law of exponents, we also have

$$1 = \dfrac{a^m}{a^m} = a^{m-m} = a^0$$

Thus,

$$a^0 = 1, \quad \text{if } a \neq 0$$

For example,

$$3^0 = 1, \quad (-528)^0 = 1, \quad \text{and} \quad (0.024)^0 = 1$$

SECTION 3.2 SUMMARY

VOCABULARY   *Look up the definitions of new terms in the Glossary.*

Power function   Inverse square law

CONCEPTS

1. A negative exponent denotes a reciprocal: $a^{-n} = \dfrac{1}{a^n}$, if $a \neq 0$.
2. Any number (except zero) raised to the zero power is 1: $a^0 = 1$, if $a \neq 0$.
3. A function of the form $f(x) = kx^p$, where k and p are constants, is called a **power function.**
4. The laws of exponents are valid for all integer exponents m and n, and for $a, b \neq 0$.

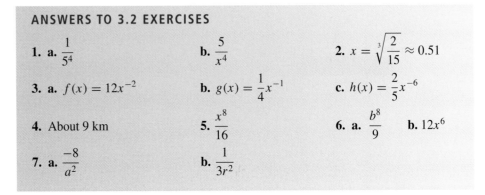

Laws of Exponents

I. $a^m \cdot a^n = a^{m+n}$ **II.** $\left(\dfrac{a^m}{a^n}\right) = a^{m-n}$

III. $(a^m)^n = a^{mn}$ **IV.** $(ab)^n = a^n b^n$

V. $\left(\dfrac{a}{b}\right)^n = \dfrac{a^n}{b^n}$

STUDY QUESTIONS

1. Explain the difference between each pair of expressions.
   a. -2^3 and 2^{-3} **b.** $-x^4$ and x^{-4} **c.** -2^n and 2^{-n}
2. Write a power function for "y varies inversely with the cube of x."
3. Explain why it makes sense to define $10^0 = 1$.
4. Why is zero excluded from the domain of $f(x) = 3x^{-2}$?
5. Choose a value for x to show that the following statement is false:

$$2x^{-2} + 4x^{-1} = 6x^{-3} \quad \text{False!}$$

1. Simplify expressions with negative exponents: #1–12
2. Solve equations involving negative exponents: #19–24
3. Write formulas for power functions: #17 and 18, 25–34

4. Evaluate and analyze power functions: #13–16, 25–34
5. Apply the laws of exponents to simplify expressions: #35–62

HOMEWORK 3.2

Thomson**NOW**™ Test yourself on key content at **www.thomsonedu.com/login.**

1. Make a table showing powers of 3 from 3^{-5} to 3^5. Illustrate why defining $3^0 = 1$ makes sense.

2. Make a table showing powers of 5 from 5^{-4} to 5^4. Illustrate why defining $5^0 = 1$ makes sense.

Compute each power.

3. a. 2^3 **b.** $(-2)^3$ **c.** 2^{-3} **d.** $(-2)^{-3}$

4. a. 4^2 **b.** $(-4)^2$ **c.** 4^{-2} **d.** $(-4)^{-2}$

5. a. $\left(\dfrac{1}{2}\right)^3$ **b.** $\left(-\dfrac{1}{2}\right)^3$ **c.** $\left(\dfrac{1}{2}\right)^{-3}$ **d.** $\left(-\dfrac{1}{2}\right)^{-3}$

6. a. $\left(\dfrac{1}{4}\right)^2$ **b.** $\left(-\dfrac{1}{4}\right)^2$ **c.** $\left(\dfrac{1}{4}\right)^{-2}$ **d.** $\left(-\dfrac{1}{4}\right)^{-2}$

Write without negative exponents and simplify.

7. a. 2^{-1} **b.** $(-5)^{-2}$
   c. $\left(\dfrac{1}{3}\right)^{-3}$ **d.** $\dfrac{1}{(-2)^{-4}}$

8. a. 3^{-2} **b.** $(-2)^{-3}$
   c. $\left(\dfrac{3}{5}\right)^{-2}$ **d.** $\dfrac{1}{(-3)^{-3}}$

9. a. $\dfrac{5}{4^{-3}}$ **b.** $(2q)^{-5}$
   c. $-4x^{-2}$ **d.** $\dfrac{8}{b^{-3}}$

10. a. $\dfrac{3}{2^{-6}}$ **b.** $(4k)^{-3}$
   c. $-7x^{-4}$ **d.** $\dfrac{5}{a^{-5}}$

11. a. $(m-n)^{-2}$ **b.** $y^{-2} + y^{-3}$
   c. $2pq^{-4}$ **d.** $\dfrac{-5y^{-2}}{x^{-5}}$

12. a. $(p+q)^{-3}$ **b.** $z^{-1} - z^{-2}$
   c. $8m^{-2}n^2$ **d.** $\dfrac{-6y^{-3}}{x^{-3}}$

Use your calculator to fill in the tables in Problems 13 and 14. Round your answers to two decimal places.

13. $f(x) = x^{-2}$

a.

x	1	2	4	8	16
$f(x)$					

b. What happens to the values of $f(x)$ as the values of x increase? Explain why.

c.

x	1	0.5	0.25	0.125	0.0625
$f(x)$					

d. What happens to the values of $f(x)$ as the values of x decrease toward 0? Explain why.

14. $g(x) = x^{-3}$

a.

x	1	2	4.5	6.2	9.3
$g(x)$					

b. What happens to the values of $g(x)$ as the values of x increase? Explain why.

c.

x	1.5	0.6	0.1	0.03	0.002
$g(x)$					

d. What happens to the values of $g(x)$ as the values of x decrease toward 0? Explain why.

15. a. Use your calculator to graph each of the following functions on the window

$$\text{Xmin} = -5, \quad \text{Xmax} = 5$$
$$\text{Ymin} = -2, \quad \text{Ymax} = 10$$

i. $f(x) = x^2$ **ii.** $f(x) = x^{-2}$

iii. $f(x) = \dfrac{1}{x^2}$ **iv.** $f(x) = \left(\dfrac{1}{x}\right)^2$

b. Which functions have the same graph? Explain your results.

16. a. Use your calculator to graph each of the following functions on the window

$$\text{Xmin} = -3, \quad \text{Xmax} = 3$$
$$\text{Ymin} = -5, \quad \text{Ymax} = 5$$

i. $f(x) = x^3$ **ii.** $f(x) = x^{-3}$

iii. $f(x) = \dfrac{1}{x^3}$ **iv.** $f(x) = \left(\dfrac{1}{x}\right)^3$

b. Which functions have the same graph? Explain your results.

Write each expression as a power function using negative exponents.

17. a. $F(r) = \dfrac{3}{r^4}$ **b.** $G(w) = \dfrac{2}{5w^3}$

    c. $H(z) = \dfrac{1}{(3z)^2}$

18. a. $h(s) = \dfrac{9}{s^3}$ **b.** $f(v) = \dfrac{3}{8v^6}$

    c. $g(t) = \dfrac{1}{(5t)^4}$

Solve

19. $6x^{-2} = 3.84$

20. $0.8w^{-2} = 1.25$

21. $12 + 0.04t^{-3} = 175.84$

22. $854 - 48z^{-3} = 104$

23. $100 - 0.15v^{-4} = 6.25$

24. $8100p^{-4} - 250 = 3656.25$

25. When an automobile accelerates, the power, P, needed to overcome air resistance varies directly with a power of the speed, v.

a. Use the data and the graph to find the scaling exponent and the constant of variation. Then write a formula for P as a power function of v.

v (mph)	10	20	30	40
P (watts)	355	2840	9585	22,720

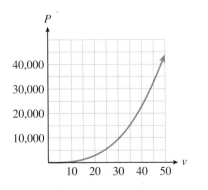

b. Find the speed that requires 50,000 watts of power.
c. If you increase your speed by 50%, by what factor does the power requirement increase?

26. The power, P, generated by a windmill varies directly with a power of wind velocity, v.

a. Use the data and the graph to find the scaling exponent and the constant of variation. Then write a formula for P as a power function of v.

v (mph)	10	20	30	40
P (watts)	15	120	405	960

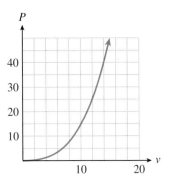

b. Find the wind velocity needed to generate 500 watts of power.
c. If the wind speed drops by half, what happens to the power generated?

27. The "Rule of 70" is used to estimate how long it takes an investment to double in value when interest is compounded annually. The doubling time, D, is inversely proportional to the interest rate, i. (Note that i is expressed as a percent, not as a decimal fraction. For example, if the interest rate is 8%, then $i = 8$.)

a. Use the data and the graph to find the constant of proportionality and write D as a power function of i.

i	4	6	8	10
D	17.5	11.67	8.75	7

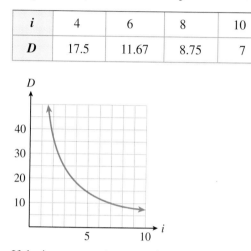

b. If the interest rate increases from 5% to 6%, how will the doubling time change?

28. The f-stop setting on a camera regulates the size of the aperture and thus the amount of light entering the camera. The f/stop is inversely proportional to the diameter, d, of the aperture.

a. Use the data and the graph to find the constant of proportionality and write d as a power function of f. Values of d have been rounded to one decimal place.

f	2.8	4	5.6	8	11
d	17.9	12.5	8.9	6.3	4.5

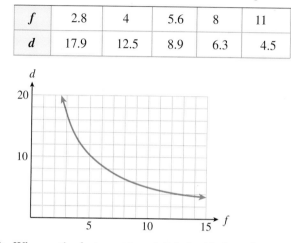

b. Why are the f-stop settings labeled with the values given in the table? (*Hint:* As you stop down the aperture from one f-value to the next, by what factor does d increase?)

29. The Stefan-Boltzmann law relates the total amount of radiation emitted by a star to its temperature, T, in kelvins, by the following formula:

$$sT^4 = \frac{L}{4\pi R^2}$$

where R is the radius of the star, L is its luminosity, and $s = 5.7 \times 10^{-8}$ watt/m^2 is a constant governing radiation. (See Algebra Skills Refresher A.1 to review scientific notation.)

a. Write a formula for luminosity as a power function of temperature for a fixed radius.

b. The radius of the Sun is $R = 9.96 \times 10^8$ meters, and its luminosity is $L = 3.9 \times 10^{26}$ watts. Calculate the temperature of the Sun.

30. Poiseuille's law for the flow of liquid through a tube can be used to describe blood flow through an artery. The rate of flow, F, in liters per minute is proportional to the fourth power of the radius, r, divided by the length, L, of the artery.

a. Write a formula for the rate of flow as a power function of radius.

b. If the radius and length of the artery are measured in centimeters, then the constant of variation, $k = 7.8 \times 10^5$, is determined by blood pressure and viscosity. If a certain artery is 20 centimeters long, what should its radius be in order to allow a blood flow of 5 liters per minute?

31. Airplanes use radar to detect the distances to other objects. A radar unit transmits a pulse of energy, which bounces off a distant object, and the echo of the pulse returns to the sender. The power, P, of the returning echo is inversely proportional to the fourth power of the distance, d, to the object. A radar operator receives an echo of 5×10^{-10} watts from an aircraft 2 nautical miles away.
   a. Express the power of the echo received in picowatts. (1 picowatt $= 10^{-12}$ watts.)
   b. Write a function that expresses P in terms of d using negative exponents. Use picowatts for the units of power.
   c. Complete the table of values for the power of the echo received from objects at various distances.

d (nautical miles)	4	5	7	10
P (picowatts)				

   d. Radar unit scan typically detect signals as low as 10^{-13} watts. How far away is an aircraft whose echo is 10^{-13} watts? (*Hint:* Convert 10^{-13} watts to picowatts.)
   e. Sketch a graph of P as a function of d. Use units of picowatts on the vertical axis.

32. The lifetime of a star is roughly inversely proportional to the cube of its mass. Our Sun, which has a mass of one solar mass, will last for approximately 10 billion years.
   a. Write a power function for the lifetime, L, of a star in terms of its mass, m.
   b. Sketch a graph of the function using units of solar mass on the horizontal axis.
   c. How long will a star that is 10 times more massive than the Sun last?
   d. One solar mass is about 2×10^{30} kilograms. Rewrite your formula for L with the units of mass in kilograms.
   e. How long will a star that is half as massive as the Sun last?

33. The amount of force or thrust generated by the propeller of a ship is a function of two variables: the diameter of the propeller and its speed, in rotations per minute. The thrust, T, in pounds, is proportional to the square of the speed, r, and the fourth power of the diameter, d, in feet.
   a. Write a formula for the thrust in terms of the speed if the diameter of the propeller is 2 feet.
   b. A propeller of diameter 2 feet generates a thrust of 1000 pounds at 100 rotations per minute. Find the constant of variation in the formula for thrust.
   c. Sketch a graph of the thrust as a function of the propeller speed for a propellor of diameter 2 feet. If the speed of the propeller is doubled, by what factor does the thrust increase?

34. Refer to Problem 33.
   a. Write a formula for the thrust, T, in terms of the diameter of the propeller if its speed is 100 rotations per minute.
   b. A propeller of diameter 4 feet generates a thrust of 32,000 pounds at 100 rotations per minute. Find the constant of variation in the formula for thrust.
   c. Sketch a graph of the thrust as a function of the diameter of the propeller at a speed of 100 rotations per minute. If the diameter of the propeller is doubled, by what factor does the thrust increase?

Use the laws of exponents to simplify and write without negative exponents.

35. a. $a^{-3} \cdot a^8$ b. $5^{-4} \cdot 5^{-3}$
    c. $\dfrac{p^{-7}}{p^{-4}}$ d. $(7^{-2})^5$

36. a. $b^2 \cdot b^{-6}$ b. $4^{-2} \cdot 4^{-6}$
    c. $\dfrac{w^{-9}}{w^2}$ d. $(9^{-4})^3$

37. a. $(4x^{-5})(5x^2)$ b. $\dfrac{3u^{-3}}{9u^9}$ c. $\dfrac{5^6 t^0}{5^{-2} t^{-1}}$

38. a. $(3y^{-8})(2y^4)$ b. $\dfrac{4c^{-4}}{8c^{-8}}$ c. $\dfrac{3^{10} s^{-1}}{3^{-5} s^0}$

39. a. $(3x^{-2}y^3)^{-2}$ b. $\left(\dfrac{6a^{-3}}{b^2}\right)^{-2}$
    c. $\dfrac{5h^{-3}(h^4)^{-2}}{6h^{-5}}$

40. a. $(2x^3 y^{-4})^{-3}$ b. $\left(\dfrac{a^4}{4b^{-5}}\right)^{-3}$
    c. $\dfrac{4v^{-5}(v^{-2})^{-4}}{3v^{-8}}$

Write each expression as a sum of terms of the form kx^p.

41. a. $\dfrac{x}{3} + \dfrac{3}{x}$

    b. $\dfrac{x - 6x^2}{4x^3}$

42. a. $\dfrac{2}{x^2} - \dfrac{x^2}{2}$

    b. $\dfrac{5x + 1}{(3x)^2}$

43. a. $\dfrac{2}{x^4}\left(\dfrac{x^2}{4} + \dfrac{x}{2} - \dfrac{1}{4}\right)$

    b. $\dfrac{x^2}{3}\left(\dfrac{2}{x^4} - \dfrac{1}{3x^2} + \dfrac{1}{2}\right)$

44. a. $\dfrac{9}{x^3}\left(\dfrac{x^3}{3} - 1 - \dfrac{1}{x^3}\right)$

    b. $\dfrac{x^2}{2}\left(\dfrac{3}{x} - \dfrac{5}{x^3} + \dfrac{7}{x^5}\right)$

Use the distributive law to write each product as a sum of power functions.

45. $x^{-1}(x^2 - 3x + 2)$

46. $3x^{-2}(2x^4 + x^2 - 4)$

47. $-3t^{-2}(t^2 - 2 - 4t^{-2})$

48. $-t^{-3}(3t^2 - 1 - t^{-2})$

49. $2u^{-3}(-2u^3 - u^2 + 3u)$

50. $2u^{-1}(-1 - u - 2u^2)$

Factor as indicated, writing the second factor with positive exponents only.

51. $4x^2 + 16x^{-2} = 4x^{-2}(\,?\,)$

52. $20y - 15y^{-1} = 5y^{-1}(\,?\,)$

53. $3a^{-3} - 3a + a^3 = a^{-3}(\,?\,)$

54. $2 - 4q^{-2} - 8q^{-4} = 2q^{-4}(\,?\,)$

55. a. Is it true that $(x + y)^{-2} = x^{-2} + y^{-2}$? Explain why or why not.

    b. Give a numerical example to support your answer.

56. a. Is it true that $(a - b)^{-1} = a^{-1} - b^{-1}$? Explain why or why not.

    b. Give a numerical example to support your answer.

57. a. Show that $x + x^{-1} = \dfrac{x^2 + 1}{x}$.

    b. Show that $x^3 + x^{-3} = \dfrac{x^6 + 1}{x^3}$.

    c. Write $x^n + x^{-n}$ as an algebraic fraction. Justify your answer.

58. a. Show that $x^{-m} + x^{-n} = \dfrac{x^n + x^m}{x^{n+m}}$.

    b. If $m < n$, show that $x^{-m} + x^{-n} = \dfrac{x^{n-m} + 1}{x^n}$.

By rewriting expressions as fractions, verify that the laws of exponents hold for negative exponents. Show where you apply the corresponding law for positive exponents. Here is the fourth law as an example:

$$(ab)^{-3} = \frac{1}{(ab)^3} = \frac{1}{a^3 b^3} \qquad \text{By the fourth law of exponents.}$$

$$= \frac{1}{a^3} \cdot \frac{1}{b^3} = a^{-3} b^{-3}$$

59. $a^{-2} a^{-3} = a^{-5}$

60. $\dfrac{a^{-6}}{a^{-2}} = a^{-4}$

61. $\dfrac{a^{-2}}{a^{-6}} = a^4$

62. $\left(a^{-2}\right)^{-3} = a^6$

In Section 3.2, we saw that inverse variation can be expressed as a power function by using negative exponents. We can also use exponents to denote square roots and other radicals.

*n*th Roots

Recall that s is a square root of b if $s^2 = b$, and s is a cube root of b if $s^3 = b$. In a similar way, we can define the fourth, fifth, or sixth root of a number. For instance, the fourth root of b is a number s whose fourth power is b. In general, we make the following definition.

> **nth Roots**
>
> s is called an **nth root of b** if $s^n = b$.

We use the symbol $\sqrt[n]{b}$ to denote the nth root of b. An expression of the form $\sqrt[n]{b}$ is called a **radical,** b is called the **radicand,** and n is called the index of the radical.

EXAMPLE 1

a. $\sqrt[4]{81} = 3$ because $3^4 = 81$
b. $\sqrt[5]{32} = 2$ because $2^5 = 32$
c. $\sqrt[6]{64} = 2$ because $2^6 = 64$
d. $\sqrt[4]{1} = 1$ because $1^4 = 1$
e. $\sqrt[5]{100,000} = 10$ because $10^5 = 100,000$

> **EXERCISE 1**
>
> Evaluate each radical.
>
> a. $\sqrt[4]{16}$
> b. $\sqrt[5]{243}$

Exponential Notation for Radicals

A convenient notation for radicals uses fractional exponents. Consider the expression $9^{1/2}$. What meaning can we attach to an exponent that is a fraction? The third law of exponents says that when we raise a power to a power, we multiply the exponents together:

$$(x^a)^b = x^{ab}$$

Therefore, if we square the number $9^{1/2}$, we get

$$(9^{1/2})^2 = 9^{(1/2)(2)} = 9^1 = 9$$

Thus, $9^{1/2}$ is a number whose square is 9. But this means that $9^{1/2}$ is a square root of 9, or

$$9^{1/2} = \sqrt{9} = 3$$

In general, any nonnegative number raised to the 1/2 power is equal to the positive square root of the number, or

$$a^{1/2} = \sqrt{a}$$

EXAMPLE 2

a. $25^{1/2} = 5$
b. $-25^{1/2} = -5$
c. $(-25)^{1/2}$ is not a real number.
d. $0^{1/2} = 0$

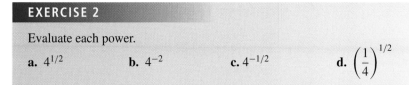

The same reasoning works for roots with any index. For instance, $8^{1/3}$ is the cube root of 8, because

$$(8^{1/3})^3 = 8^{(1/3)(3)} = 8^1 = 8$$

In general, we make the following definition for fractional exponents.

Exponential Notation for Radicals

For any integer $n \geq 2$ and for $a \geq 0$,

$$a^{1/n} = \sqrt[n]{a}$$

EXAMPLE 3

a. $81^{1/4} = \sqrt[4]{81} = 3$ **b.** $125^{1/3} = \sqrt[3]{125} = 5$

CAUTION

Note that

$$25^{1/2} \neq \frac{1}{2}(25) \quad \text{and} \quad 125^{1/3} \neq \frac{1}{3}(125)$$

An exponent of $\frac{1}{2}$ denotes the square root of its base, and an exponent of $\frac{1}{3}$ denotes the cube root of its base.

Of course, we can use decimal fractions for exponents as well. For example,

$$\sqrt{a} = a^{1/2} = a^{0.5} \quad \text{and} \quad \sqrt[4]{a} = a^{1/4} = a^{0.25}$$

EXAMPLE 4

a. $100^{0.5} = \sqrt{100} = 10$ **b.** $16^{0.25} = \sqrt[4]{16} = 2$

Irrational Numbers

You can read more about irrational numbers in Algebra Skills Refresher, A.12.

What about nth roots such as $\sqrt{23}$ and $5^{1/3}$ that cannot be evaluated easily? These are examples of *irrational numbers*. We can use a calculator to obtain decimal approximations to irrational numbers. For example, you can verify that

$$\sqrt{23} \approx 4.796 \quad \text{and} \quad 5^{1/3} \approx 1.710$$

It is not possible to write down an exact decimal equivalent for an irrational number, but we can find an approximation to as many decimal places as we like.

CAUTION

The following keying sequence for evaluating the irrational number $7^{1/5}$ is **incorrect:**

$$7 \boxed{\wedge} 1 \boxed{\div} 5 \boxed{\text{ENTER}}$$

You can check that this sequence calculates $\frac{7^1}{5}$, instead of $7^{1/5}$. Recall that according to the order of operations, powers are computed before multiplications or divisions. We must enclose the exponent $\frac{1}{5}$ in parentheses and enter

$$7 \boxed{\wedge} \boxed{(} 1 \boxed{\div} 5 \boxed{)} \boxed{\text{ENTER}}$$

Or, because $\frac{1}{5} = 0.2$, we can enter

$$7 \boxed{\wedge} 0.2 \boxed{\text{ENTER}}$$

Working with Fractional Exponents

Fractional exponents simplify many calculations involving radicals. You should learn to convert easily between exponential and radical notation. Remember that a negative exponent denotes a reciprocal.

EXAMPLE 5

Convert each radical to exponential notation.

a. $\sqrt[3]{12} = 12^{1/3}$ **b.** $\sqrt[4]{2y} = (2y)^{1/4}$ or $(2y)^{0.25}$

In Example 5b, note that the parentheses around $2y$ must not be omitted.

> **EXERCISE 5**
>
> Convert each radical to exponential notation.
>
> **a.** $\dfrac{1}{\sqrt[5]{ab}}$ **b.** $3\sqrt[6]{w}$

EXAMPLE 6

Convert each power to radical notation.

a. $5^{1/2} = \sqrt{5}$ **b.** $x^{0.2} = \sqrt[5]{x}$

c. $2x^{1/3} = 2\sqrt[3]{x}$ **d.** $8a^{-1/4} = \dfrac{8}{\sqrt[4]{a}}$

In Example 6d, note that the exponent $-1/4$ applies only to a, not to $8a$.

a. Convert $\dfrac{3}{\sqrt[4]{2x}}$ to exponential notation.

b. Convert $-5b^{0.125}$ to radical notation.

Using Fractional Exponents to Solve Equations

In Chapter 2, we learned that raising to powers and taking roots are *inverse operations,* that is, each operation undoes the effects of the other. This relationship is especially easy to see when the root is denoted by a fractional exponent. For example, to solve the equation

$$x^4 = 250$$

we would take the fourth root of each side. But instead of using radical notation, we can raise both sides of the equation to the power $\frac{1}{4}$:

$$(x^4)^{1/4} = 250^{1/4}$$
$$x \approx 3.98$$

The third law of exponents tells us that $(x^a)^b = x^{ab}$, so $(x^4)^{1/4} = x^{(1/4)(4)} = x^1$. In general, to solve an equation involving a power function x^n, we first isolate the power, then raise both sides to the exponent $\frac{1}{n}$.

EXAMPLE 7

For astronomers, the mass of a star is its most important property, but it is also the most difficult to measure directly. For many stars, their luminosity, or brightness, varies roughly as the fourth power of the mass.

a. Our Sun has luminosity 4×10^{26} watts and mass 2×10^{30} kilograms. Because the numbers involved are so large, astronomers often use these solar constants as units of measure: The luminosity of the Sun is 1 *solar luminosity,* and its mass is 1 *solar mass.* Write a power function for the luminosity, L, of a star in terms of its mass, M, using units of solar mass and solar luminosity.

b. The star Sirius is 23 times brighter than the Sun, so its luminosity is 23 solar luminosities. Estimate the mass of Sirius in units of solar mass.

Solutions **a.** Because L varies as the fourth power of M, we have

$$L = kM^4$$

Substituting the values of L and M for the Sun (namely, $L = 1$ and $M = 1$), we find

$$1 = k(1)^4$$

so $k = 1$ and $L = M^4$.

b. We substitute the luminosity of Sirius, $L = 23$, to get

$$23 = M^4$$

To solve the equation for M, we raise both sides to the $\frac{1}{4}$ power.

$$(23)^{1/4} = (M^4)^{1/4}$$
$$2.1899 = M$$

The mass of Sirius is about 2.2 solar masses, or about 2.2 times the mass of the Sun.

See the list of geometric formulas inside the front cover to find the volume of a sphere.

EXERCISE 7

A spherical fish tank in the lobby of the Atlantis Hotel holds about 905 cubic feet of water. What is the radius of the fish tank?

Power Functions

The basic functions $y = \sqrt{x}$ and $y = \sqrt[3]{x}$ are power functions of the form $f(x) = x^{1/n}$, and the graphs of all such functions have shapes similar to those two, depending on whether the index of the root is even or odd. Figure 3.10a shows the graphs of $y = x^{1/2}$, $y = x^{1/4}$, and $y = x^{1/6}$. Figure 3.10b shows the graphs of $y = x^{1/3}$, $y = x^{1/5}$, and $y = x^{1/7}$.

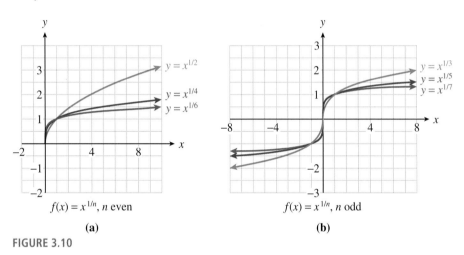

$f(x) = x^{1/n}$, n even $f(x) = x^{1/n}$, n odd

(a)   (b)

FIGURE 3.10

We cannot take an even root of a negative number. (See "A Note on Roots of Negative Numbers" at the end of this section.) Hence, if n is even, the domain of $f(x) = x^{1/n}$ is restricted to nonnegative real numbers, but if n is odd, the domain of $f(x) = x^{1/n}$ is the set of all real numbers.

We will also encounter power functions with negative exponents. For example, an animal's heart rate is related to its size or mass, with smaller animals generally having faster heart rates. The heart rates of mammals are given approximately by the power function

$$H(m) = km^{-1/4}$$

where m is the animal's mass and k is a constant.

EXAMPLE 8

A typical human male weighs about 70 kilograms and has a resting heart rate of 70 beats per minute.

a. Find the constant of proportionality, k, and write a formula for $H(m)$.

b. Fill in the table with the heart rates of the mammals whose masses are given.

Animal	Shrew	Rabbit	Cat	Wolf	Horse	Polar bear	Elephant	Whale
Mass (kg)	0.004	2	4	80	300	600	5400	70,000
Heart rate								

c. Sketch a graph of H for masses up to 6000 kilograms.

Solutions **a.** Substitute $H = 70$ and $m = 70$ into the
equation; then solve for k.

$$70 = k \cdot 70^{-1/4}$$

$$k = \frac{70}{70^{-1/4}} = 70^{5/4} \approx 202.5$$

Thus, $H(m) = 202.5 m^{-1/4}$.

b. Evaluate the function H for each of the masses
given in the table.

c. Plot the points in the table to obtain the graph
shown in Figure 3.11.

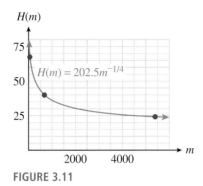

FIGURE 3.11

Animal	Shrew	Rabbit	Cat	Wolf	Horse	Polar bear	Elephant	Whale
Mass (kg)	0.004	2	4	80	300	600	5400	70,000
Heart rate	805	170	143	68	49	41	24	12

Many properties relating to the growth of plants and animals can be described by
power functions of their mass. The study of the relationship between the growth rates of
different parts of an organism, or of organisms of similar type, is called *allometry*. An equation of the form

$$\textbf{variable} = \textbf{\textit{k}}(\textbf{mass})^{\textbf{\textit{p}}}$$

used to describe such a relationship is called an *allometric equation.*

Of course, power functions can be expressed using any of the notations we have discussed. For example, the function in Example 8 can be written as

$$H(m) = 202.5 m^{-1/4} \quad \text{or} \quad H(m) = 202.5 m^{-0.25} \quad \text{or} \quad H(m) = \frac{202.5}{\sqrt[4]{m}}$$

EXERCISE 8

a. Complete the table of values for the power function $f(x) = x^{-1/2}$.

x	0.1	0.25	0.5	1	2	4	8	10	20	100
$f(x)$										

b. Sketch the graph of $y = f(x)$.

c. Write the formula for $f(x)$ with a decimal exponent, and with radical notation.

Solving Radical Equations

A **radical equation** is one in which the variable appears under a square root or other radical. The radical may be denoted by a fractional exponent. For example, the equation

$$5x^{1/3} = 32$$

is a radical equation because $x^{1/3} = \sqrt[3]{x}$. To solve the equation, we first isolate the power
to get

$$x^{1/3} = 6.4$$

Then we raise both sides of the equation to the reciprocal of $\frac{1}{3}$, or 3.

$$(x^{1/3})^3 = 6.4^3$$
$$x = 262.144$$

EXAMPLE 9

When a car brakes suddenly, its speed can be estimated from the length of the skid marks it leaves on the pavement. A formula for the car's speed, in miles per hour, is $v = f(d) = (24d)^{1/2}$, where the length of the skid marks, d, is given in feet.

a. If a car leaves skid marks 80 feet long, how fast was the car traveling when the driver applied the brakes?

b. How far will a car skid if its driver applies the brakes while traveling 80 miles per hour?

Solutions   **a.** To find the velocity of the car, we evaluate the function for $d = 80$.

$$v = (24 \cdot 80)^{1/2} \qquad \text{Substitute 80 for } d.$$
$$= (1920)^{1/2} \qquad \text{Multiply inside the radical.}$$
$$\approx 43.8178046 \qquad \text{Take the square root.}$$

The car was traveling at approximately 44 miles per hour.

b. We would like to find the value of d when the value of v is known. We substitute $v = 80$ into the formula and solve the equation

$$80 = (24d)^{1/2} \qquad \text{Solve for } d.$$

Because d appears to the power $\frac{1}{2}$, we first square both sides of the equation to get

$$80^2 = ((24d)^{1/2})^2 \qquad \text{Square both sides.}$$
$$6400 = 24d \qquad \text{Divide by 24.}$$
$$266.\overline{6} = d$$

You can check that this value for d works in the original equation. Thus, the car will skid approximately 267 feet. A graph of the function $v = (24d)^{1/2}$ is shown in Figure 3.12, along with the points corresponding to the values in parts (a) and (b).

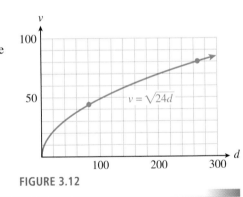

FIGURE 3.12

See the Algebra Skills Refresher A.10 to review extraneous solutions.

Thus, we can solve an equation where one side is an nth root of x by raising both sides of the equation to the nth power. We must be careful when raising both sides of an equation to an even power, since extraneous solutions may be introduced. However, because most applications of power functions deal with positive domains only, they do not usually involve extraneous solutions.

EXERCISE 9

In Example 8, we found the heart-rate function, $H(m) = 202.5m^{-1/4}$. What would be the mass of an animal whose heart rate is 120 beats per minute?

A Note on Roots of Negative Numbers

You already know that $\sqrt{-9}$ is not a real number, because there is no real number whose square is -9. Similarly, $\sqrt[4]{-16}$ is not a real number, because there is no real number r for which $r^4 = -16$. (Both of these radicals are *complex numbers*. Complex numbers are discussed in Chapter 7.) In general, we cannot find an even root (square root, fourth root, and so on) of a negative number.

On the other hand, every positive number has *two* even roots that are real numbers. For example, both 3 and -3 are square roots of 9. The symbol $\sqrt{9}$ refers only to the positive, or **principal root,** of 9. If we want to refer to the negative square root of 9, we must write $-\sqrt{9} = -3$. Similarly, both 2 and -2 are fourth roots of 16, because $2^4 = 16$ and $(-2)^4 = 16$. However, the symbol $\sqrt[4]{16}$ refers to the principal, or positive, fourth root only. Thus,

$$\sqrt[4]{16} = 2 \qquad \text{and} \qquad -\sqrt[4]{16} = -2$$

Things are simpler for odd roots (cube roots, fifth roots, and so on). Every real number, whether positive, negative, or zero, has exactly one real-valued odd root. For example,

$$\sqrt[5]{32} = 2 \qquad \text{and} \qquad \sqrt[5]{-32} = -2$$

Here is a summary of our discussion.

Roots of Real Numbers

1. Every positive number has two real-valued roots, one positive and one negative, if the index is even.

2. A negative number has no real-valued root if the index is even.

3. Every real number, positive, negative, or zero, has exactly one real-valued root if the index is odd.

EXAMPLE 10

a. $\sqrt[4]{-625}$ is not a real number.

b. $-\sqrt[4]{625} = -5$

c. $\sqrt[5]{-1} = -1$

d. $\sqrt[4]{-1}$ is not a real number.

The same principles apply to powers with fractional exponents. Thus

$$(-32)^{1/5} = -2$$

but $(-64)^{1/6}$ is not a real number. On the other hand,

$$-64^{1/6} = -2$$

because the exponent $1/6$ applies only to 64, and the negative sign is applied after the root is computed.

EXERCISE 10

Evaluate each power, if possible.

a. $-81^{1/4}$

b. $(-81)^{1/4}$

c. $-64^{1/3}$

d. $(-64)^{1/3}$

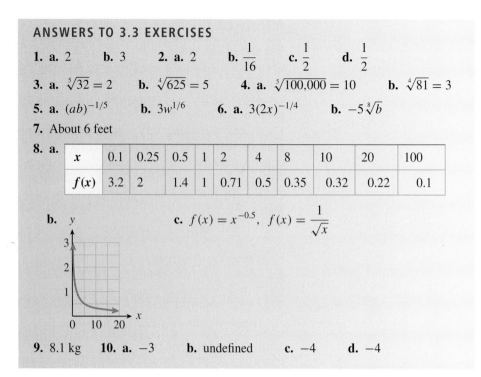

ANSWERS TO 3.3 EXERCISES

1. a. 2   **b.** 3   **2. a.** 2   **b.** $\dfrac{1}{16}$ **c.** $\dfrac{1}{2}$ **d.** $\dfrac{1}{2}$

3. a. $\sqrt[5]{32} = 2$ **b.** $\sqrt[4]{625} = 5$ **4. a.** $\sqrt[5]{100,000} = 10$ **b.** $\sqrt[4]{81} = 3$

5. a. $(ab)^{-1/5}$ **b.** $3w^{1/6}$ **6. a.** $3(2x)^{-1/4}$ **b.** $-5\sqrt[8]{b}$

7. About 6 feet

8. a.

x	0.1	0.25	0.5	1	2	4	8	10	20	100
$f(x)$	3.2	2	1.4	1	0.71	0.5	0.35	0.32	0.22	0.1

b.   **c.** $f(x) = x^{-0.5}$, $f(x) = \dfrac{1}{\sqrt{x}}$

9. 8.1 kg   **10. a.** -3 **b.** undefined   **c.** -4 **d.** -4

SECTION 3.3 SUMMARY

VOCABULARY   *Look up the definitions of new terms in the Glossary.*

*n*th root   Radical   Radical equation   Radical notation
Index   Exponential notation Radicand   Allometric equation
Irrational number

CONCEPTS

1. *n*th roots: s is called an *n*th **root of *b*** if $s^n = b$.

2. Exponential notation: For any integer $n \geq 2$ and for $a \geq 0$, $a^{1/n} = \sqrt[n]{a}$.

3. We cannot write down an exact decimal equivalent for an **irrational number,** but we can approximate an irrational number to as many decimal places as we like.

4. We can solve the equation $x^n = b$ by raising both sides to the $\frac{1}{n}$ power.

5. An **allometric equation** is a power function of the form **variable = *k* (mass)^p**.

6. We can solve the equation $x^{1/n} = b$ by raising both sides to the *n*th power.

7.

> **Roots of Real Numbers**
>
> - Every positive number has two real-valued roots, one positive and one negative, if the index is even.
> - A negative number has no real-valued root if the index is even.
> - Every real number, positive, negative, or zero, has exactly one real-valued root if the index is odd.

STUDY QUESTIONS

1. Use an example to illustrate the terms *radical*, *radicand*, *index*, and *principal root*.
2. Explain why $x^{1/4}$ is a reasonable notation for $\sqrt[4]{x}$.
3. What does the notation $x^{0.2}$ mean?

4. Express each of the following algebraic notations in words; then evaluate each for $x = 16$:

$$4x, \quad x^4, \quad \frac{x}{4}, \quad \frac{1}{4}x, \quad x^{1/4}, \quad x-4, \quad x^{-4}, \quad x^{-1/4}$$

5. How is the third law of exponents, $(x^a)^b = x^{ab}$, useful in solving equations?

SKILLS *Practice each skill in the Homework problems listed.*

1. Evaluate powers and roots: #1–8, 17–20
2. Convert between radical and exponential notation: #9–16, 21 and 22

3. Solve radical equations: #23–38, 59 and 60
4. Graph and analyze power functions: #39–58
5. Work with fractional exponents: #61–68

HOMEWORK 3.3

ThomsonNOW™ Test yourself on key content at www.thomsonedu.com/login.

Find the indicated root without using a calculator; then check your answers.

1. a. $\sqrt{121}$ **b.** $\sqrt[3]{27}$ **c.** $\sqrt[4]{625}$ **2. a.** $\sqrt{169}$ **b.** $\sqrt[3]{64}$ **c.** $\sqrt[4]{81}$

3. a. $\sqrt[5]{32}$ **b.** $\sqrt[4]{16}$ **c.** $\sqrt[3]{729}$ **4. a.** $\sqrt[5]{100{,}000}$ **b.** $\sqrt[4]{1296}$ **c.** $\sqrt[3]{343}$

Find the indicated power without using a calculator; then check your answers.

5. a. $9^{1/2}$ **b.** $81^{1/4}$ **c.** $64^{1/6}$ **6. a.** $25^{1/2}$ **b.** $16^{1/4}$ **c.** $27^{1/3}$

7. a. $32^{0.2}$ **b.** $8^{-1/3}$ **c.** $64^{-0.5}$ **8. a.** $625^{0.25}$ **b.** $243^{-1/5}$ **c.** $49^{-0.5}$

Write each expression in radical form.

9. a. $3^{1/2}$ **b.** $4x^{1/3}$ **c.** $(4x)^{0.2}$ **10. a.** $7^{1/2}$ **b.** $3x^{1/4}$ **c.** $(3x)^{0.25}$

11. a. $6^{-1/3}$ **b.** $3(xy)^{-0.125}$ **c.** $(x-2)^{1/4}$ **12. a.** $8^{-1/4}$ **b.** $y(5x)^{-0.5}$ **c.** $(y+2)^{1/3}$

Write each expression in exponential form.

13. a. $\sqrt{7}$ **b.** $\sqrt[3]{2x}$ **c.** $2\sqrt[5]{z}$ **14. a.** $\sqrt{5}$ **b.** $\sqrt[3]{4y}$ **c.** $5\sqrt[3]{x}$

15. a. $\dfrac{-3}{\sqrt[4]{6}}$ **b.** $\sqrt[4]{x-3y}$ **c.** $\dfrac{-1}{\sqrt[5]{1+3b}}$ **16. a.** $\dfrac{2}{\sqrt[5]{3}}$ **b.** $\sqrt[3]{y+2x}$ **c.** $\dfrac{-1}{\sqrt[4]{3a-2b}}$

Simplify.

17. a. $\left(\sqrt[3]{125}\right)^3$ **b.** $\left(\sqrt[4]{2}\right)^4$

   c. $\left(3\sqrt{7}\right)^2$ **d.** $\left(-x^2\sqrt[3]{2x}\right)^3$

18. a. $\left(\sqrt[4]{16}\right)^4$ **b.** $\left(\sqrt[3]{6}\right)^3$

   c. $\left(2\sqrt[3]{12}\right)^3$ **d.** $\left(-a^3\sqrt[4]{a^2}\right)^4$

Use a calculator to approximate each irrational number to the nearest thousandth.

19. a. $2^{1/2}$ **b.** $\sqrt[3]{75}$ **c.** $\sqrt[4]{1.6}$ **20. a.** $3^{1/2}$ **b.** $\sqrt[4]{60}$ **c.** $\sqrt[3]{1.4}$

    d. $365^{-1/3}$ **e.** $0.006^{-0.2}$ **d.** $1058^{-1/4}$ **e.** $1.05^{-0.1}$

Write each expression as a power function.

21. a. $g(x) = 3.7\sqrt[3]{x}$ **b.** $H(x) = \sqrt[4]{85x}$ **22. a.** $h(v) = 12.7\sqrt{v}$ **b.** $F(p) = \sqrt[3]{2.9p}$

    c. $F(t) = \dfrac{25}{\sqrt[5]{t}}$ **c.** $G(w) = \dfrac{5}{8\sqrt[8]{w}}$

Solve.

23. $6.5x^{1/3} + 3.8 = 33.05$ **24.** $9.8 - 76x^{1/4} + 15 = 9.6$ **25.** $4(x + 2)^{1/5} = 12$

26. $-9(x - 3)^{1/5} = 18$ **27.** $(2x - 3)^{-1/4} = \dfrac{1}{2}$ **28.** $(5x + 2)^{-1/3} = \dfrac{1}{4}$

29. $\sqrt[3]{x^2 - 3} = 3$ **30.** $\sqrt[4]{x^3 - 7} = 2$

Solve each formula for the indicated variable.

31. $T = 2\pi\sqrt{\dfrac{L}{g}}$ for L **32.** $T = 2\pi\sqrt{\dfrac{m}{k}}$ for m **33.** $r = \sqrt{t^2 - s^2}$ for s **34.** $c = \sqrt{a^2 - b^2}$ for b

35. $r = \sqrt[3]{\dfrac{3V}{4\pi}}$ for V **36.** $d = \sqrt[3]{\dfrac{16Mr^2}{m}}$ for M **37.** $R = \sqrt[4]{\dfrac{8Lvf}{\pi p}}$ for p **38.** $T = \sqrt[4]{\dfrac{E}{SA}}$ for A

39. The period of a pendulum is the time it takes for the pendulum to complete one entire swing, from left to right and back again. The greater the length, L, of the pendulum, the longer its period, T. In fact, if L is measured in feet, then the period is given in seconds by

$$T = 2\pi\sqrt{\dfrac{L}{32}}$$

   a. Write the formula for T as a power function in the form $f(x) = kx^p$.
   b. Suppose you are standing in the Convention Center in Portland, Oregon, and you time the period of its Foucault pendulum (the longest in the world). Its period is approximately 10.54 seconds. How long is the pendulum?
   c. Choose a reasonable domain for the function $T = f(L)$ and graph the function.

40. If you are flying in an airplane at an altitude of h miles, on a clear day you can see a distance of d miles to the horizon, where $d = \sqrt{7920h}$.
   a. Write the formula for d as a power function in the form $f(x) = kx^p$.
   b. Choose a reasonable domain for the function $d = f(h)$ and graph the function.
   c. At what altitude will you be able to see for a distance of 100 miles? How high is that in feet?

41. If you walk in the normal way, your maximum speed, v, in meters per second, is limited by the length of your legs, r, according to the formula

$$v = \sqrt{gr}$$

where the constant g is approximately 10 meters per second squared. (Source: Alexander, 1992)

   a. A typical adult man has legs about 0.9 meter long. How fast can he walk?

   b. A typical four-year-old has legs 0.5 meter long. How fast can she walk?

   c. Graph maximum walking speed as a function of leg length.

   d. Race-walkers can walk as fast as 4.4 meters per second by rotating their hips so that the effective length of their legs is increased. What is that effective length?

   e. On the Moon the value of g is 1.6 meters per second squared. How fast can a typical adult man walk on the Moon?

42. When a ship moves through the water, it creates waves that impede its own progress. Because of this resistance, there is an upper limit to the speed at which a ship can travel, given, in knots, by

$$v_{max} = 1.3\sqrt{L}$$

where L is the length of the vessel, in feet.
(Source: Gilner, 1972)

   a. Graph maximum speed as a function of vessel length.

   b. The world's largest ship, the oil tanker *Jahre Viking*, is 1054 feet long. What is its top speed?

   c. As a ship approaches its maximum speed, the power required increases sharply. Therefore, most merchant ships are designed to cruise at speeds no higher than $v_c = 0.8\sqrt{L}$. Graph v_c on the same axes with v_{max}.

   d. What is the cruising speed of the *Jahre Viking?* What percent of its maximum speed is that?

43. A rough estimate for the radius of the nucleus of an atom is provided by the formula

$$r = kA^{1/3}$$

where A is the mass number of the nucleus and $k \approx 1.3 \times 10^{-13}$ centimeter.

   a. Estimate the radius of the nucleus of an atom of iodine-127, which has mass number 127. If the nucleus is roughly spherical, what is its volume?

   b. The nuclear mass of iodine-127 is 2.1×10^{-22} gram. What is the density of the nucleus? (Density is mass per unit volume.)

   c. Complete the table of values for the radii of various radioisotopes.

Element	Carbon	Potassium	Cobalt	Technetium	Radium
Mass number, A	14	40	60	99	226
Radius, r					

   d. Sketch a graph of r as a function of A. (Use units of 10^{-13} centimeter on the vertical axis.)

44. In the sport of crew racing, the best times vary closely with the number of men in the crew, according to the formula

$$t = kn^{-1/9}$$

where n is the number of men in the crew and t is the winning time, in minutes, for a 2000-meter race.

   a. If the winning time for the 8-man crew was 5.73 minutes, estimate the value of k.

   b. Complete the table of values of predicted winning times for the other racing classes.

Size of crew, n	1	2	4	8
Winning time, t				

   c. Sketch a graph of t as a function of n.

In Problems 45–48, one quantity varies directly with the square root of the other, that is, $y = k\sqrt{x}$.
(a) Find the value of k and write a power function relating the variables.
(b) Use your function to answer the question.
(c) Graph your function and verify your answer to part (b) on the graph.

45. The stream speed necessary to move a granite particle is a function of the diameter of the particle; faster river currents can move larger particles. The table shows the stream speed necessary to move particles of different sizes. What speed is needed to carry a particle with diameter 0.36 centimeter?

Diameter, d (cm)	Speed, s (cm/sec)
0.01	5
0.04	10
0.09	15
0.16	20

46. The speed at which water comes out of the spigot at the bottom of a water jug is a function of the water level in the jug; it slows down as the water level drops. The table shows different water levels and the resulting flow speeds. What is the flow speed when the water level is at 16 inches?

Level, L (in)	Speed, s (gal/min)
9	1.5
6.25	1.25
4	1
2.25	0.75

47. The rate, r, in feet per second, at which water flows from a fire hose is a function of the water pressure, P, in psi (pounds per square inch). What is the rate of water flow at a typical water pressure of 60 psi?

P (psi)	10	20	30	40
r (ft/sec)	38.3	54.1	66.3	76.5

48. When a layer of ice forms on a pond, the thickness of the ice, d, in centimeters, is a function of time, t, in minutes. How thick is the ice after 3 hours?

t (min)	10	30	40	60
d (cm)	0.50	0.87	1.01	1.24

49. Membership in the County Museum has been increasing since it was built in 1980. The number of members is given by the function

$$M(t) = 72 + 100\,t^{1/3}$$

where t is the number of years since 1980.
  a. How many members were there in 1990? In 2000?
  b. In what year will the museum have 400 members? If the membership continues to grow according to the given function, when will the museum have 500 members?
  c. Graph the function $M(t)$. How would you describe the growth of the membership over time?

50. Due to improvements in technology, the annual electricity cost of running most major appliances has decreased steadily since 1970. The average annual cost of running a refrigerator is given, in dollars, by the function

$$C(t) = 148 - 28t^{1/3}$$

where t is the number of years since 1970.
  a. How much did it cost to run a refrigerator in 1980? In 1990?
  b. When was the cost of running a refrigerator half of the cost in 1970? If the cost continues to decline according to the given function, when will it cost $50 per year to run a refrigerator?
  c. Graph the function $C(t)$. Do you think that the cost will continue to decline indefinitely according to the given function? Why or why not?

51. Match each function with the description of its graph in the first quadrant.
  I. $f(x) = x^2$
  II. $f(x) = x^{-2}$
  III. $f(x) = x^{1/2}$
  a. Increasing and concave up
  b. Increasing and concave down
  c. Decreasing and concave up
  d. Decreasing and concave down

52. In each pair, match the functions with their graphs.

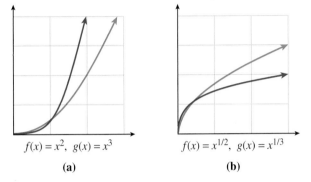

$f(x) = x^2, \ g(x) = x^3$

(a)

$f(x) = x^{1/2}, \ g(x) = x^{1/3}$

(b)

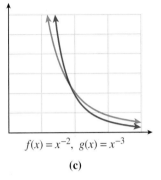

$f(x) = x^{-2}, \ g(x) = x^{-3}$

(c)

53. a. Graph the functions

$$y_1 = x^{1/2}, \quad y_2 = x^{1/3}, \quad y_3 = x^{1/4}, \quad y_4 = x^{1/5}$$

in the window

$$\text{Xmin} = 0, \quad \text{Xmax} = 100$$
$$\text{Ymin} = 0, \quad \text{Ymax} = 10$$

What do you observe?
  b. Use your graphs to evaluate $100^{1/2}$, $100^{1/3}$, $100^{1/4}$, and $100^{1/5}$.
  c. Use your calculator to evaluate $100^{1/n}$ for $n = 10$, $n = 100$, and $n = 1000$. What happens when n gets large?

54. a. Graph the functions

$$y_1 = x^{1/2}, \quad y_2 = x^{1/3}, \quad y_3 = x^{1/4}, \quad y_4 = x^{1/5}$$

in the window

$$\text{Xmin} = 0, \quad \text{Xmax} = 1$$
$$\text{Ymin} = 0, \quad \text{Ymax} = 1$$

What do you observe?
  b. Use your graphs to evaluate $0.5^{1/2}$, $0.5^{1/3}$, $0.5^{1/4}$, and $0.5^{1/5}$.
  c. Use your calculator to evaluate $0.5^{1/n}$ for $n = 10, n = 100$, and $n = 1000$. What happens when n gets large?

For Problems 55–58, graph each set of functions in the given window. What do you observe?

55. $y_1 = \sqrt{x}, \quad y_2 = x^2, \quad y_3 = x$
$$\text{Xmin} = 0, \text{Xmax} = 4$$
$$\text{Ymin} = 0, \text{Ymax} = 4$$

56. $y_1 = \sqrt[3]{x}, \quad y_2 = x^3, \quad y_3 = x$
$$\text{Xmin} = -4, \text{Xmax} = 4$$
$$\text{Ymin} = -4, \text{Ymax} = 4$$

57. $y_1 = \sqrt[5]{x}, \quad y_2 = x^5, \quad y_3 = x$
$$\text{Xmin} = -2, \text{Xmax} = 2,$$
$$\text{Ymin} = -2, \text{Ymax} = 2$$

58. $y_1 = \sqrt[4]{x}, \quad y_2 = x^4, \quad y_3 = x$
$$\text{Xmin} = 0, \quad \text{Xmax} = 2$$
$$\text{Ymin} = 0, \quad \text{Ymax} = 2$$

59. **a.** Graph the functions $f(x) = 4\sqrt[3]{x} - 9$ and $g(x) = 12$ in the window

$$\text{Xmin} = 0 \qquad \text{Xmax} = 47$$
$$\text{Ymin} = -8 \qquad \text{Ymax} = 16$$

b. Use the graph to solve the equation $4\sqrt[3]{x} - 9 = 12$.

c. Solve the equation algebraically.

60. **a.** Graph the functions $f(x) = 6 + 2\sqrt[4]{12 - x}$ and $g(x) = 10$ in the window

$$\text{Xmin} = -27 \quad \text{Xmax} = 20$$
$$\text{Ymin} = 4 \qquad \text{Ymax} = 12$$

b. Use the graph to solve the equation $6 + 2\sqrt[4]{12 - x} = 10$.

c. Solve the equation algebraically.

61. **a.** Write $\sqrt{x}$ with a fractional exponent.

b. Write $\sqrt{\sqrt{x}}$ with fractional exponents.

c. Use the laws of exponents to show that $\sqrt{\sqrt{x}} = \sqrt[4]{x}$.

62. **a.** Write $\sqrt[3]{x}$ with a fractional exponent.

b. Write $\sqrt{\sqrt[3]{x}}$ with fractional exponents.

c. Use the laws of exponents to show that $\sqrt{\sqrt[3]{x}} = \sqrt[6]{x}$.

Write each expression as a sum of terms of the form kx^p.

63. $\dfrac{\sqrt{x}}{4} - \dfrac{2}{\sqrt{x}} + \dfrac{x}{\sqrt{2}}$

64. $\dfrac{\sqrt{3}}{x} + \dfrac{3}{\sqrt{x}} - \dfrac{\sqrt{x}}{3}$

65. $\dfrac{6 - \sqrt[3]{x}}{2\sqrt[3]{x}}$

66. $\dfrac{\sqrt[4]{x} + 2}{2\sqrt[4]{x}}$

67. $x^{-0.5}(x + x^{0.25} - x^{0.5})$

68. $x^{0.5}(x^{-1} + x^{-0.5} + x^{-0.25})$

3.4 Rational Exponents

Powers of the Form $a^{m/n}$

In the last section, we considered powers of the form $a^{1/n}$, such as $x^{1/3}$ and $x^{-1/4}$, and saw that $a^{1/n}$ is equivalent to the root $\sqrt[n]{a}$. What about other fractional exponents? What meaning can we attach to a power of the form $a^{m/n}$?

Consider the power $x^{3/2}$. Notice that the exponent $\frac{3}{2} = 3\left(\frac{1}{2}\right)$, and thus by the third law of exponents, we can write

$$(x^{1/2})^3 = x^{(1/2)3} = x^{3/2}$$

In other words, we can compute $x^{3/2}$ by first taking the square root of x and then cubing the result. For example,

$$100^{3/2} = (100^{1/2})^3 \qquad \text{Take the square root of 100.}$$
$$= 10^3 = 1000 \qquad \text{Cube the result.}$$

We will define fractional powers only when the base is a positive number.

Rational Exponents

$$a^{m/n} = (a^{1/n})^m = (a^m)^{1/n}, \quad a > 0, n \neq 0$$

To compute $a^{m/n}$, we can compute the nth root first, or the mth power, whichever is easier. For example,

$$8^{2/3} = (8^{1/3})^2 = 2^2 = 4$$

or

$$8^{2/3} = (8^2)^{1/3} = 64^{1/3} = 4$$

EXAMPLE 1

a. $81^{3/4} = (81^{1/4})^3$
$= 3^3 = 27$

b. $-27^{5/3} = -(27^{1/3})^5$
$= -3^5 = -243$

c. $27^{-2/3} = \dfrac{1}{(27^{1/3})^2}$
$= \dfrac{1}{3^2} = \dfrac{1}{9}$

d. $5^{3/2} = (5^{1/2})^3$
$\approx (2.236)^3 \approx 11.180$

You can verify all the calculations in Example 1 on your calculator. For example, to evaluate $81^{3/4}$, key in

$$81 \;\boxed{\wedge}\; \boxed{(}\; 3 \;\boxed{\div}\; 4 \;\boxed{)}\; \boxed{\textbf{ENTER}}$$

or simply

$$81 \;\boxed{\wedge}\; 0.75 \;\boxed{\textbf{ENTER}}$$

EXERCISE 1

Evaluate each power.

a. $32^{-3/5}$

b. $-81^{1.25}$

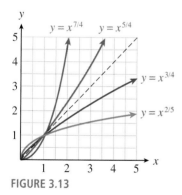

FIGURE 3.13

Power Functions

The graphs of power functions $y = x^{m/n}$, where m/n is positive, are all increasing for $x \geq 0$. If $m/n > 1$, the graph is concave up. If $0 < m/n < 1$, the graph is concave down. Some examples are shown in Figure 3.13.

Perhaps the single most useful piece of information a scientist can have about an animal is its metabolic rate. The metabolic rate is the amount of energy the animal uses per unit of time for its usual activities, including locomotion, thermoregulation, growth, and reproduction. The basal metabolic rate, or BMR, sometimes called the resting metabolic rate, is the minimum amount of energy the animal can expend in order to survive.

EXAMPLE 2

A revised form of Kleiber's rule states that the basal metabolic rate for many groups of animals is given by

$$B(m) = 70m^{0.75}$$

where m is the mass of the animal in kilograms and the BMR is measured in kilocalories per day.

a. Calculate the BMR for various animals whose masses are given in the table.

Animal	Bat	Squirrel	Raccoon	Lynx	Human	Moose	Rhinoceros
Weight (kg)	0.1	0.6	8	30	70	360	3500
BMR (kcal/day)							

b. Sketch a graph of Kleiber's rule for $0 < m \le 400$.

c. Do larger species eat more or less, relative to their body mass, than smaller ones?

Solutions **a.** Evaluate the function for the values of m given. For example, to calculate the BMR of a bat, we compute

$$B(0.1) = 70(0.1)^{0.75} = 12.1$$

A bat expends, and hence must consume, at least 12 kilocalories per day. Evaluate the function to complete the rest of the table.

Animal	Bat	Squirrel	Raccoon	Lynx	Human	Moose	Rhinoceros
Weight (kg)	0.1	0.6	8	30	70	360	3500
BMR (kcal/day)	12	48	333	897	1694	5785	31,853

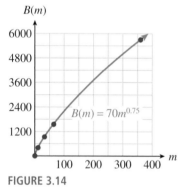

FIGURE 3.14

c. Plot the data from the table to obtain the graph in Figure 3.14.

d. If energy consumption were proportional to body weight, the graph would be a straight line. But because the exponent in Kleiber's rule, $\frac{3}{4}$, is less than 1, the graph is concave down, or bends downward. Therefore, larger species eat less than smaller ones, relative to their body weight. For example, a moose weighs 600 times as much as a squirrel, but its energy requirement is only 121 times the squirrel's.

EXERCISE 2

a. Complete the table of values for the function $f(x) = x^{-3/4}$.

x	0.1	0.2	0.5	1	2	5	8	10
$f(x)$								

b. Sketch the graph of the function.

More about Scaling

In Example 2, we saw that large animals eat less than smaller ones, relative to their body weight. This is because the scaling exponent in Kleiber's rule is less than 1. For example, let s represent the mass of a squirrel. The mass of a moose is then $600s$, and its metabolic rate is

$$B(600s) = 70(600s)^{0.75}$$
$$= 600^{0.75} \cdot 70s^{0.75} = 121B(s)$$

or 121 times the metabolic rate of the squirrel. Metabolic rate scales as $k^{0.75}$, compared to the mass of the animal.

In a famous experiment in the 1960s, an elephant was given LSD. The dose was determined from a previous experiment in which a 2.6-kg cat was given 0.26 gram of LSD. Because the elephant weighed 2970 kg, the experimenters used a direct proportion to calculate the dose for the elephant:

$$\frac{0.26 \text{ g}}{2.6 \text{ kg}} = \frac{x \text{ g}}{2970 \text{ kg}}$$

and arrived at the figure 297 g of LSD. Unfortunately, the elephant did not survive the experiment.

EXAMPLE 3

Use Kleiber's rule and the dosage for a cat to estimate the corresponding dose for an elephant.

Solution If the experimenters had taken into account the scaling exponent of 0.75 in metabolic rate, they would have used a smaller dose. Because the elephant weighs $\frac{2970}{2.6}$, or about 1142 times as much as the cat, the dose would be $1142^{0.75} = 196$ times the dosage for a cat, or about 51 grams.

EXERCISE 3

A human being weighs about 70 kg, and 0.2 mg of LSD is enough to induce severe psychotic symptoms. Use these data and Kleiber's rule to predict what dosage would produce a similar effect in an elephant.

Radical Notation

Because $a^{1/n} = \sqrt[n]{a}$, we can write any power with a fractional exponent in radical form as follows.

Rational Exponents and Radicals

$$a^{m/n} = \sqrt[n]{a^m} = (\sqrt[n]{a})^m$$

EXAMPLE 4

a. $125^{4/3} = \sqrt[3]{125^4}$ or $\left(\sqrt[3]{125}\right)^4$

b. $x^{0.4} = x^{2/5} = \sqrt[5]{x^2}$ **c.** $6w^{-3/4} = \dfrac{6}{\sqrt[4]{w^3}}$

Usually, we will want to convert from radical notation to fractional exponents, since exponential notation is easier to use.

EXAMPLE 5

a. $\sqrt{x^5} = x^{5/2}$ **b.** $5\sqrt[4]{p^3} = 5p^{3/4}$

c. $\dfrac{3}{\sqrt[5]{t^2}} = 3t^{-2/5}$ **d.** $\sqrt[3]{2y^2} = (2y^2)^{1/3} = 2^{1/3}y^{2/3}$

Operations with Rational Exponents

Powers with rational exponents—positive, negative, or zero—obey the laws of exponents, which we discussed in Section 3.1. You may want to review those laws before studying the following examples.

EXAMPLE 6

a. $\dfrac{7^{0.75}}{7^{0.5}} = 7^{0.75-0.5} = 7^{0.25}$ Apply the second law of exponents.

b. $v \cdot v^{-2/3} = v^{1+(-2/3)} = v^{1/3}$ Apply the first law of exponents.

c. $(x^8)^{0.5} = x^{8(0.5)} = x^4$ Apply the third law of exponents.

d. $\dfrac{(5^{1/2}y^2)^2}{(5^{2/3}y)^3} = \dfrac{5y^4}{5^2y^3}$ Apply the fourth law of exponents.

       $= \dfrac{y^{4-3}}{5^{2-1}} = \dfrac{y}{5}$ Apply the second law of exponents.

Solving Equations

According to the third law of exponents, when we raise a power to another power, we multiply the exponents together. In particular, if the two exponents are reciprocals, then their product is 1. For example,

$$(x^{2/3})^{3/2} = x^{(2/3)\,(3/2)} = x^1 = x$$

This observation can help us to solve equations involving fractional exponents. For instance, to solve the equation

$$x^{2/3} = 4$$

we raise both sides of the equation to the reciprocal power, $3/2$. This gives us

$$(x^{2/3})^{3/2} = 4^{3/2}$$
$$x = 8$$

The solution is 8.

EXAMPLE 7   Solve $(2x + 1)^{3/4} = 27$.

Solution   Raise both sides of the equation to the reciprocal power, $\frac{4}{3}$.

$$[(2x + 1)^{3/4}]^{4/3} = 27^{4/3} \qquad \text{Apply the third law of exponents.}$$
$$2x + 1 = 81 \qquad \text{Solve as usual.}$$
$$x = 40$$

EXERCISE 7

Solve the equation $3.2z^{0.6} - 9.7 = 8.7$. Round your answer to two decimal places.

Isolate the power.
Raise both sides to the reciprocal power.

Investigation 6   ## Vampire Bats

Small animals such as bats cannot **survive** for long without eating. The graph in Figure 3.15 shows how the weight, W, of a typical vampire bat decreases over time until its next meal, until the bat reaches the point of starvation. The curve is the graph of the function

$$W(h) = 130.25h^{-0.126}$$

where h is the number of hours since the bat's most recent meal. (Source: Wilkinson, 1984)

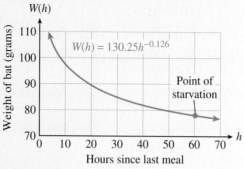

FIGURE 3.15

1. Use the graph to estimate answers to the following questions: How long can the bat survive after eating until its next meal? What is the bat's weight at the point of starvation?

2. Use the formula for $W(h)$ to verify your answers.

3. Write and solve an equation to answer the question: When the bat's weight has dropped to 90 grams, how long can it survive before eating again?

4. Complete the table showing the number of hours since the bat last ate when its weight has dropped to the given values.

Weight (grams)	97.5	92.5	85	80
Hours since eating				
Point on graph	A	B	C	D

5. Compute the slope of the line segments from point A to point B, and from point C to point D. (See Figure 3.16.) Include units in your answers.

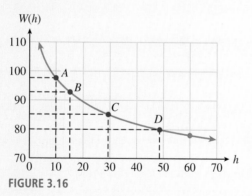

FIGURE 3.16

6. What happens to the slope of the curve as h increases? What does this tell you about the concavity of the curve?

7. Suppose a bat that weighs 80 grams consumes 5 grams of blood. How many hours of life does it gain? Suppose a bat that weighs 97.5 grams gives up a meal of 5 grams of blood. How many hours of life does it forfeit?

8. Vampire bats sometimes donate blood (through regurgitation) to other bats that are close to starvation. Suppose a bat at point A on the curve donates 5 grams of blood to a bat at point D. Explain why this strategy is effective for the survival of the bat community.

ANSWERS TO 3.4 EXERCISES

1. a. $\dfrac{1}{8}$ **b.** -243

2. a.

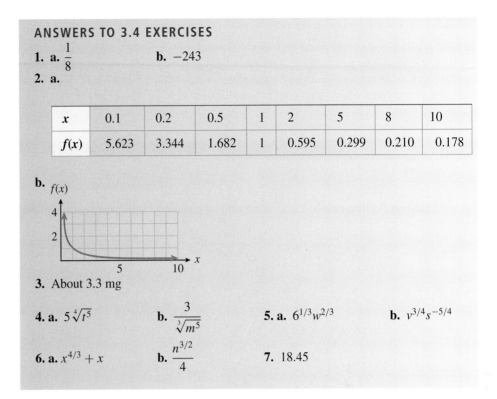

x	0.1	0.2	0.5	1	2	5	8	10
$f(x)$	5.623	3.344	1.682	1	0.595	0.299	0.210	0.178

b.

3. About 3.3 mg

4. a. $5\sqrt[4]{t^5}$ **b.** $\dfrac{3}{\sqrt[3]{m^5}}$ **5. a.** $6^{1/3}w^{2/3}$ **b.** $v^{3/4}s^{-5/4}$

6. a. $x^{4/3} + x$ **b.** $\dfrac{n^{3/2}}{4}$ **7.** 18.45

SECTION 3.4 SUMMARY

VOCABULARY *Look up the definition of new term in the Glossary.*

Rational exponent

CONCEPTS

1. Rational exponents:
$a^{m/n} = (a^{1/n})^m = (a^m)^{1/n}, \quad a > 0, \ n \neq 0.$

2. To compute $a^{m/n}$, we can compute the nth root first, or the mth power, whichever is easier.

3. The graphs of power functions $y = x^{m/n}$, where m/n is positive, are all increasing for $x \geq 0$. If $m/n > 1$, the graph is concave up. If $0 < m/n < 1$, the graph is concave down.

4. Radical notation: $a^{m/n} = \sqrt[n]{a^m} = (\sqrt[n]{a})^m$.

5. Powers with rational exponents—positive, negative, or zero—obey the laws of exponents.

6. To solve the equation $x^{m/n} = k$, we raise both sides to the power n/m.

STUDY QUESTIONS

1. What does the notation $a^{0.98}$ mean?

2. Explain how to evaluate the function $f(x) = x^{-3/4}$ for $x = 625$, without using a calculator.

3. Explain why $x\sqrt{x} = x^{1.5}$.

4. What is the first step in solving the equation $(x - 2)^{-5/2} = 1.8$?

5. If the graph of $f(x) = x^{a/b}$ is concave down, and $a/b > 0$, what else can you say about a/b?

1. Simplify and evaluate powers with rational exponents: #1–4, 13–18
2. Convert between exponential and radical notation: #5–12
3. Graph power functions with rational exponents: #19–22
4. Analyze power functions with rational exponents: #23–36
5. Simplify expressions using the laws of exponents: #37–44, 57–70
6. Solve equations involving rational exponents: #45–56

HOMEWORK 3.4

ThomsonNOW™ Test yourself on key content at www.thomsonedu.com/login.

For the problems in Homework 3.4, assume that all variables represent positive numbers. Evaluate each power in Problems 1–4.

1. a. $81^{3/4}$ **b.** $125^{2/3}$ **c.** $625^{0.75}$ **2. a.** $-8^{2/3}$ **b.** $-64^{2/3}$ **c.** $243^{0.4}$

3. a. $16^{-3/2}$ **b.** $8^{-4/3}$ **c.** $32^{-1.6}$ **4. a.** $-125^{-4/3}$ **b.** $-32^{-3/5}$ **c.** $100^{-2.5}$

Write each power in radical form.

5. a. $x^{4/5}$ **b.** $b^{-5/6}$ **c.** $(pq)^{-2/3}$ **6. a.** $y^{3/4}$ **b.** $a^{-2/7}$ **c.** $(st)^{-3/5}$

7. a. $3x^{0.4}$ **b.** $4z^{-4/3}$ **c.** $-2x^{0.25}y^{0.75}$ **8. a.** $5y^{2/3}$ **b.** $6w^{-1.5}$ **c.** $-3x^{0.4}y^{0.6}$

Write each expression with fractional exponents.

9. a. $\sqrt[3]{x^2}$ **b.** $2\sqrt[5]{ab^3}$ **c.** $\dfrac{-4m}{\sqrt[6]{p^7}}$ **10. a.** $\sqrt{y^3}$ **b.** $6\sqrt[5]{(ab)^3}$ **c.** $\dfrac{-2n}{\sqrt[8]{q^{11}}}$

11. a. $\sqrt[3]{(ab)^2}$ **b.** $\dfrac{8}{\sqrt[4]{x^3}}$ **c.** $\dfrac{R}{3\sqrt{TK^5}}$ **12. a.** $\sqrt[3]{ab^2}$ **b.** $\dfrac{5}{\sqrt[3]{y^2}}$ **c.** $\dfrac{S}{4\sqrt{VH^3}}$

Evaluate each root without using a calculator.

13. a. $\sqrt[5]{32^3}$ **b.** $-\sqrt[3]{27^4}$ **c.** $\sqrt[4]{16y^{12}}$ **14. a.** $\sqrt[4]{16^5}$ **b.** $-\sqrt[3]{125^2}$ **c.** $\sqrt[5]{243x^{10}}$

15. a. $-\sqrt{a^8b^{16}}$ **b.** $\sqrt[3]{8x^9y^{27}}$ **c.** $-\sqrt[4]{81a^8b^{12}}$ **16. a.** $-\sqrt{a^{10}b^{36}}$ **b.** $\sqrt[3]{64x^6y^{18}}$ **c.** $-\sqrt[5]{32x^{25}y^5}$

Use a calculator to approximate each power or root to the nearest thousandth.

17. a. $12^{5/6}$ **b.** $\sqrt[3]{6^4}$
    c. $37^{-2/3}$ **d.** $4.7^{2.3}$

18. a. $20^{5/4}$ **b.** $\sqrt[5]{8^3}$
    c. $128^{-3/4}$ **d.** $16.1^{0.29}$

19. During a flu epidemic in a small town, health officials estimate that the number of people infected t days after the first case was discovered is given by

$$I(t) = 50\,t^{3/5}$$

a. Make a table of values for $I(t)$ on the domain $0 \le t \le 20$. What is the range of the function on that domain?

t	5	10	15	20
$I(t)$				

b. How long will it be before 300 people are ill?

c. Graph the function $I(t)$ and verify your answer to part (b) on your graph.

20. The research division of an advertising firm estimates that the number of people who have seen their ads t days after the campaign begins is given by the function

$$N(t) = 2000\,t^{5/4}$$

a. Make a table of values for $N(t)$ on the domain $0 \le t \le 20$. What is the range of the function on that domain?

t	6	10	14	20
$N(t)$				

b. How long will it be before 75,000 people have seen the ads?

c. Graph the function $N(t)$ and verify your answer to part (b) on your graph.

In Problems 21 and 22, graph each set of power functions in the suggested window and compare the graphs.

21. $y_1 = x$, $y_2 = x^{5/4}$, $y_3 = x^{3/2}$, $y_4 = x^2$, $y_5 = x^{5/2}$

Xmin $= 0$, Xmax $= 6$, Ymin $= 0$, Ymax $= 10$

22. $y_1 = x^{2/5}$, $y_2 = x^{1/2}$, $y_3 = x^{2/3}$, $y_4 = x^{3/4}$, $y_5 = x$

Xmin $= 0$, Xmax $= 6$, Ymin $= 0$, Ymax $= 4$

23. The *surface to volume ratio* is important in studying how organisms grow and why animals of different sizes have different characteristics.

a. Write formulas for the volume, V, and the surface area, A, of a cube in terms of its length, L.

b. Express the length of the cube as a function of its volume. Express the length of the cube as a function of its surface area.

c. Express the surface area of the cube as a function of its volume.

d. Express the surface to volume ratio of a cube in terms of its length. What happens to the surface to volume ratio as L increases?

24. Repeat Problem 23 for the volume and surface area of a sphere in terms of its radius, R.

a. Write formulas for the volume, V, and the surface area, A, of a sphere in terms of its radius, R.

b. Express the radius of the sphere as a function of its volume. Express the radius of the sphere as a function of its surface area.

c. Express the surface area of the sphere as a function of its volume.

d. Express the surface to volume ratio of a sphere in terms of its radius. What happens to the surface to volume ratio as R increases?

25. A brewery wants to replace its old vats with larger ones. To estimate the cost of the new equipment, the accountant uses the 0.6 rule for industrial costs, which states that the cost of a new container is approximately $N = Cr^{0.6}$, where C is the cost of the old container and r is the ratio of the capacity of the new container to the old one.

a. If an old vat cost $5000, graph N as a function of r.

b. How much should the accountant budget for a new vat that holds 1.8 times as much as the old one?

26. If a quantity of air expands without changing temperature, its pressure, in pounds per square inch, is given by $P = kV^{-1.4}$, where V is the volume of the air in cubic inches and $k = 2.79 \times 10^4$.

a. Graph P as a function of V.

b. Find the air pressure of an air sample when its volume is 50 cubic inches.

27. In the 1970s, Jared Diamond studied the number of bird species on small islands near New Guinea. He found that larger islands support a larger number of different species, according to the formula

$$S = 15.1A^{0.22}$$

where S is the number of species on an island of area A square kilometers. (Source: Chapman and Reiss, 1992)

a. Fill in the table.

A	10	100	1000	5000	10,000
S					

b. Graph the function on the domain $0 < A \leq 10,000$.

c. How many species of birds would you expect to find on Manus Island, with an area of 2100 square kilometers?
On Lavongai, which bird's area is 1140 square kilometers?

d. How large must an island be in order to support 200 different species of bird?

28. The drainage basin of a river channel is the area of land that contributes water to the river. The table gives the lengths in miles of some of the world's largest rivers and the areas of their drainage basins in square miles.
(Source: Leopold, Wolman, and Miller 1992)

a. Plot the data, using units of 100,000 on the horizontal axis and units of 500 on the vertical axis.

b. The length, L, of the channel is related to the area, A, of its drainage basin according to the formula

$$L = 1.05A^{0.58}$$

Graph this function on top of the data points.

c. The drainage basin for the Congo covers about 1,600,000 square miles. Estimate the length of the Congo River.

d. The Rio Grande is 1700 miles long. What is the area of its drainage basin?

River	Area of drainage basin	Length
Amazon	2,700,000	4200
Nile	1,400,000	4200
Mississippi	1,300,000	4100
Yangtze	580,000	2900
Volga	480,000	2300
St. Lawrence	460,000	1900
Ganges	440,000	1400
Orinoco	380,000	1400
Indus	360,000	2000
Danube	350,000	1800
Colorado	250,000	1700
Platte	72,000	800
Rhine	63,000	900
Seine	48,000	500
Delaware	12,000	200

29. The table below shows the exponent, b, in the allometric equation

$$\text{variable} = k \, (\text{body mass})^p$$

for some variables related to mammals.

(Source: Chapman and Reiss, 1992)

Variable	Exponent, p
Home range size	1.26
Lung volume	1.02
Brain mass	0.70
Respiration rate	−0.26

a. Match each equation to one of the graphs shown in the figure.

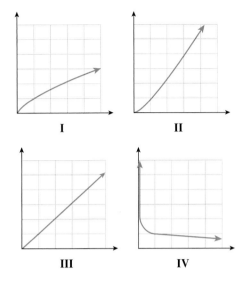

I   II

III   IV

b. Explain how the value of p in the allometric equation determines the shape of the graph. Consider the cases $p > 1$, $0 < p < 1$, and $p < 0$.

30. The average body mass of a dolphin is about 140 kilograms, twice the body mass of an average human male.

a. Using the allometric equations in Problem 29, calculate the ratio of the brain mass of a dolphin to that of a human.

b. A good-sized brown bear weighs about 280 kilograms, twice the weight of a dolphin. Calculate the ratio of the brain mass of a brown bear to that of a dolphin.

c. Use a ratio to compare the heartbeat frequencies of a dolphin and a human, and those of a brown bear and a dolphin. (See Example 8 of Section 3.3.)

31. The gourd species *Tricosanthes* grows according to the formula $L = ad^{2.2}$, where L is its length and d is its width. The species *Lagenaria* has the growth law $L = ad^{0.81}$. (Source: Burton, 1998)

a. By comparing the exponents, predict which gourd grows into a long, thin shape, and which is relatively fatter. Which species is called the snake gourd, and which is the bottle gourd?

b. The snake gourd reaches a length of 2 meters (200 cm), with a diameter of only 4 cm. Find the value of a in its growth law.

c. The bottle gourd is 10 cm long and 7 cm in diameter at maturity. Find the value of a in its growth law.

d. The giant bottle gourd grows to a length of 23 cm with a diameter of 20 cm. Does it grow according to the same law as standard bottle gourds?

32. As a fiddler crab grows, one claw (called the chela) grows much faster than the rest of the body. The table shows the mass of the chela, C, versus the mass of the rest of the body, b, for a number of fiddler crabs.

(Source: Burton, 1998)

b	65	110	170	205	300	360	615
C	6	15	30	40	68	110	240

a. Plot the data.

b. On the same axes, graph the function $C = 0.007b^{1.63}$. How well does the function fit the data?

c. Using the function in part (b), predict the chela mass of a fiddler crab if the rest of its body weighs 400 mg.

d. The chela from a fiddler crab weighs 250 mg. How much does the rest of its body weigh?

e. As the body mass of a fiddler crab doubles from 100 mg to 200 mg, by what factor does the mass of its chela increase? As the body mass doubles from 200 mg to 400 mg?

33. The climate of a region has a great influence on the types of animals that can survive there. Extreme temperatures create difficult living conditions, so the diversity of wildlife decreases as the annual temperature range increases. Along the west coast of North America, the number of species of mammals, M, is approximately related to the temperature range, R, (in degrees Celsius) by the function $M = f(R) = 433.8R^{-0.742}$.

(Source: Chapman and Reiss, 1992)

   a. Graph the function for temperature ranges up to 30°C.
   b. How many species would you expect to find in a region where the temperature range is 10°C? Label the corresponding point on your graph.
   c. If 50 different species are found in a certain region, what temperature range would you expect the region to experience? Label the corresponding point on your graph.
   d. Evaluate the function to find $f(9), f(10), f(19)$, and $f(20)$. What do these values represent? Calculate the change in the number of species as the temperature range increases from 9°C to 10°C and from 19°C to 20°C. Which 1° increase results in a greater decrease in diversity? Explain your answer in terms of slopes on your graph.

34. A bicycle ergometer is used to measure the amount of power generated by a cyclist. The scatterplot shows how long an athlete was able to sustain various levels of power output. The curve is the graph of $y = 500x^{-0.29}$, which approximately models the data. (Source: Alexander, 1992)

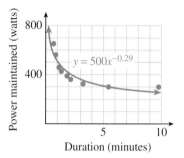

   a. In this graph, which variable is independent and which is dependent?
   b. The athlete maintained 650 watts of power for 40 seconds. What power output does the equation predict for 40 seconds?
   c. The athlete maintained 300 watts of power for 10 minutes. How long does the equation predict that power output can be maintained?
   d. In 1979, a remarkable pedal-powered aircraft called the Gossamer Albatross was successfully flown across the English Channel. The flight took 3 hours. According to the equation, what level of power can be maintained for 3 hours?
   e. The Gossamer Albatross needed 250 watts of power to keep it airborne. For how long can 250 watts be maintained according to the given equation?

35. Investigation 5 at the start of this chapter gives data for the pressure inside April and Tolu's balloon as a function of its diameter. As the diameter of the balloon increases from 5 cm to 20 cm, the pressure inside decreases. Can we find a function that describes this portion of the graph?

   a. Pressure is the force per unit area exerted by the balloon on the air inside, or $P = \frac{F}{A}$. Because the balloon is spherical, its surface area, A, is given by $A = \pi d^2$. Because the force increases as the balloon expands, we will try a power function $F = kd^p$, where k and p are constants, and see if this fits the data. Combine the three equations, $P = \frac{F}{A}$, $A = \pi d^2$, and $F = kd^p$, to express P as a power function of d.
   b. Use your calculator's power regression feature to find a power function that fits the data. Graph the function $P = 211d^{-0.7}$ on top of the data. Do the data support the hypothesis that P is a power function of d?
   c. What is the value of the exponent p in $F = kd^p$?

36. The table shows the total number of frequent flyer miles redeemed by customers through the given year.
(Source: www.hotelnewsresource.com)

  a. Plot the data, with $t = 0$ in 1980. What type of function might model the data?
  b. Graph the function $f(t) = 3.13t^{2.33}$ on top of the data.
  c. Evaluate $f(5)$ and $f(25)$. What do those values mean in this context?
  d. Use the regression equation to predict when the total number of miles redeemed will reach 10 trillion. (*Hint*: How many billions make a trillion?)

Year	Cumulative miles redeemed (billions)
1982	14.8
1984	85.3
1986	215.4
1988	387.5
1990	641.3
1992	975.2
1994	1455.9
1996	1996
1998	2670.8
2000	3379.1
2002	4123.6

Simplify by applying the laws of exponents. Write your answers with positive exponents only.

37. a. $4a^{6/5}a^{4/5}$ **b.** $9b^{4/3}b^{1/3}$

38. a. $(-2m^{2/3})^4$ **b.** $(-5n^{3/4})^3$

39. a. $\dfrac{8w^{9/4}}{2w^{3/4}}$ **b.** $\dfrac{12z^{11/3}}{4z^{5/3}}$

40. a. $(-3u^{5/3})(5u^{-2/3})$ **b.** $(-2v^{7/8})(-3v^{-3/8})$

41. a. $\dfrac{k^{3/4}}{2k}$ **b.** $\dfrac{4h^{2/3}}{3h}$

42. a. $c^{-2/3}\left(\dfrac{2}{3}c^2\right)$ **b.** $\dfrac{r^3}{4}(r^{-5/2})$

43. The incubation time for a bird's egg is a function of the mass, m, of the egg, and has been experimentally determined as

$$I(m) = 12.0m^{0.217}$$

where m is measured in grams and I is in days. (Source: Burton, 1998)

  a. Calculate the incubation time (to the nearest day) for the wren, whose eggs weigh about 2.5 grams, and the greylag goose, whose eggs weigh 46 grams.
  b. During incubation, birds' eggs lose water vapor through their porous shells. The rate of water loss from the egg is also a function of its mass, and it appears to follow the rule

$$W(m) = 0.015m^{0.742}$$

  in grams per day. Combine the functions $I(m)$ and $W(m)$ to calculate the fraction of the initial egg mass that is lost during the entire incubation period.
  c. Explain why your result shows that most eggs lose about 18% of their mass during incubation.

44. The incubation time for birds' eggs is given by

$$I(m) = 12.0m^{0.217}$$

where m is the weight of the egg in grams, and I is in days. (See Problem 43.) Before hatching, the eggs take in oxygen at the rate of

$$O(m) = 22.2m^{0.77}$$

in milliliters per day. (Source: Burton, 1998)
  a. Combine the functions $I(m)$ and $O(m)$ to calculate the total amount of oxygen taken in by the egg during its incubation.
  b. Use your result from part (a) to explain why total oxygen consumption per unit mass is approximately inversely proportional to incubation time.
  c. Predict the oxygen consumption per gram of a herring gull's eggs, given that their incubation time is 26 days. (The actual value is 11 milliliters per day.)

Solve. Round your answers to the nearest thousandth if necessary.

45. $x^{2/3} - 1 = 15$ **46.** $x^{3/4} + 3 = 11$ **47.** $x^{-2/5} = 9$ **48.** $x^{-3/2} = 8$

49. $2(5.2 - x^{5/3}) = 1.4$ **50.** $3(8.6 - x^{5/2}) = 6.5$

51. Kepler's law gives a relation between the period, p, of a planet's revolution, in years, and its average distance, a, from the sun:

$$p^2 = Ka^3$$

where $K = 1.243 \times 10^{-24}$, a is measured in miles, and p is in years.
  a. Solve Kepler's law for p as a function of a.
  b. Find the period of Mars if its average distance from the sun is 1.417×10^8 miles.

52. Refer to Kepler's law, $p^2 = Ka^3$, in Problem 51.
  a. Solve Kepler's law for a as a function of p.
  b. Find the distance from Venus to the sun if its period is 0.615 years.

53. If $f(x) = (3x - 4)^{3/2}$, find x so that $f(x) = 27$.

54. If $g(x) = (6x - 2)^{5/3}$, find x so that $g(x) = 32$.

55. If $S(x) = 12x^{-5/4}$, find x so that $S(x) = 20$.

56. If $T(x) = 9x^{-6/5}$, find x so that $T(x) = 15$.

Use the distributive law to find the product.

57. $2x^{1/2}(x - x^{1/2})$ **58.** $x^{1/3}(2x^{2/3} - x^{1/3})$ **59.** $\dfrac{1}{2}y^{-1/3}(y^{2/3} + 3y^{-5/6})$

60. $3y^{-3/8}\left(\dfrac{1}{4}y^{-1/4} + y^{3/4}\right)$ **61.** $(2x^{1/4} + 1)(x^{1/4} - 1)$ **62.** $(2x^{1/3} - 1)(x^{1/3} + 1)$

63. $(a^{3/4} - 2)^2$ **64.** $(a^{2/3} + 3)^2$

Factor out the smallest power from each expression. Write your answers with positive exponents only.

65. $x^{3/2} + x = x(?)$

66. $y - y^{2/3} = y^{2/3}(?)$

67. $y^{3/4} - y^{-1/4} = y^{-1/4}(?)$

68. $x^{-3/2} + x^{-1/2} = x^{-1/2}(?)$

69. $a^{1/3} + 3 - a^{-1/3} = a^{-1/3}(?)$

70. $3b - b^{3/4} + 4b^{-3/4} = b^{-3/4}(?)$

3.5   Joint Variation

Functions of Two or More Variables

So far, we have studied functions that relate values of an output variable to values of a single input variable. But it is not uncommon for an output variable to depend on two or more inputs. Many familiar formulas describe functions of several variables. For example, the perimeter of a rectangle depends on its length and width. The volume of a cylinder depends on its radius and height. The distance you travel depends on your speed and the time you spent traveling. Each of these formulas can be written with function notation.

$$P = f(l, w) = 2l + 2w \qquad \text{Perimeter is a function of length and width.}$$
$$V = f(r, h) = \pi r^2 h \qquad \text{Volume is a function of radius and height.}$$
$$d = f(r, t) = rt \qquad \text{Distance is a function of rate and time.}$$

EXAMPLE 1   The cost, C, of driving a rental car is given by the function

$$C = f(t, g) = 29.95t + 2.80g$$

where t is the number of days you rent the car and g is the number of gallons of gas you buy.

a. Evaluate $f(3,10)$ and explain what it means.

b. You have $100 to rent a car for 2 days. How much gas can you buy?

Solutions   **a.** We substitute 3 for t and 10 for g to find

$$f(3, 10) = 29.95(3) + 2.80(10) = 117.85$$

It will cost $117.85 to rent a car for 3 days and buy 10 gallons of gas.

b. We would like to find the value of g when $t = 2$ and $C = 100$. That is, we want to solve the equation

$$100 = f(2, g) = 29.95(2) + 2.80g$$
$$100 = 59.90 + 2.80g$$
$$g = 14.32$$

You can buy 14.32 gallons of gas.

The maximum height that the water stream from a fire hose can reach depends on the water pressure and the diameter of the nozzle, and is given by the function

$$H = f(P, n) = 26 + \frac{5}{8}P + 5n$$

where P is the nozzle pressure in psi, and n is measured in $\frac{1}{8}$-inch increments over the standard nozzle diameter of $\frac{3}{4}$ inch.

a. Evaluate $f(40, 2)$ and explain what it means.

b. What nozzle pressure is needed to reach a height of 91 feet with a $1\frac{1}{8}$-inch nozzle?

Tables of Values

Just as for functions of a single variable, we can use tables to describe functions of two variables, $z = f(x, y)$. The row and column headings show the values of the two input variables, and the table entries show the values of the output variable.

EXAMPLE 2

Windchill is a function of two variables, temperature and wind speed, or $W = f(s, t)$. The table shows the windchill factor for various combinations of temperature and wind speed.

Windchill Factors								
	Temperature (°F)							
Wind speed (mph)	35	30	25	20	15	10	5	0
5	33	27	21	16	12	7	0	−5
10	22	16	10	3	−3	−9	−15	−22
15	16	9	2	−5	−11	−18	−25	−31
20	12	4	−3	−10	−17	−24	−31	−39
25	8	1	−7	−15	−22	−29	−36	−44

a. What is the windchill factor when the temperature is 15°F and the wind is blowing at 20 mph? Write this fact with function notation.

b. Find a value for t so that $f(10, t) = -15$. What does this equation tell you about the windchill factor?

c. Solve the equation $f(s, 30) = 1$. What does this tell you about the windchill factor?

a. We look in the row for 20 mph and the column for 15°. The associated windchill factor is -17, so $f(20, 15) = -17$.

b. We look in the row for $s = 10$ until we find the windchill factor of $W = -15$. The column heading for that entry is 5, so $t = 5$. When the wind speed is 10 mph and the windchill factor is -15, the temperature is 5°F.

c. In the $t = 30$°F column, we find the windchill factor of 1 in the 25-mph row, so $s = 25$. The wind speed is 25 mph when the temperature is 30°F and the windchill factor is 1.

EXERCISE 2

A retirement plan requires employees to put aside a fixed amount of money each year until retirement. The amount accumulated, A, includes 8% annual interest on employee's annual contribution, c. A is a function of c and the number of years, t, that the employee makes contributions, so $A = f(c, t)$.

Retirement Fund Balance

Annual contribution	Number of years of contributions				
	10	20	30	40	50
500	7243	22,881	56,642	129,528	286,885
1000	14,487	45,762	113,283	259,057	573,770
1500	21,730	68,643	169,925	388,585	860,655
2000	28,973	91,524	226,566	518,113	1,147,540
2500	43,460	137,286	339,850	777,170	1,721,310
3000	50,703	160,167	396,491	906,698	2,008,196

a. How much will an employee accumulate if she contributes $500 a year for 40 years? Write your answer with function notation.

b. How much must she contribute each year in order to accumulate $573,770 after 50 years? Write your answer with function notation.

c. Find a value of t that solves the equation $137{,}286 = f(2500, t)$. What does this equation tell you about the retirement fund?

Joint Variation

Sometimes we can find patterns relating the entries in a table.

EXAMPLE 3   Rectangular beams of a given length can support a load, L, that depends on both the width and the depth of the beam, so that $L = f(w, d)$. The table shows some of the values.

Maximum Load (kilograms)						
	Depth (cm)					
Width (cm)	0	1	2	3	4	5
0	0	0	0	0	0	0
1	0	10	40	90	160	250
2	0	20	80	180	320	500
3	0	30	120	270	480	750
4	0	40	160	360	640	1000
5	0	50	200	450	800	1250

a. Evaluate the function at $f(2, 5)$. Interpret your answer for the problem situation.

b. Is it true that $f(2, 5) = f(5, 2)$?

c. Consider the row corresponding to a width of 3 cm. How does the load depend on the depth?

Solutions

Depth	Load
0	0
1	30
2	120
3	270
4	480
5	750

a. In the row for 2 cm and the column for 5 cm, we find that $f(2, 5) = 500$. A beam of width 2 cm and depth 5 cm can support a maximum load of 500 kilograms.

b. In the row for 5 cm and the column for 2 cm, we find that $f(5, 2) = 200$, so
$f(2, 5) \neq f(5, 2)$.

c. Using the row for width 3 cm, we make a new table showing the relationship between load and depth. The increase in load for each increase of 1 cm in depth is not a constant, so the graph in Figure 3.17 is not a straight line. The curve does pass through the origin, so perhaps the data describe direct variation with a power of depth. If we try the equation $L = kd^2$ and use the point $(1, 30)$, we find that $30 = k \cdot 1^2$, so $k = 30$. You can check that the equation $L = 30d^2$ does fit the rest of the data points.

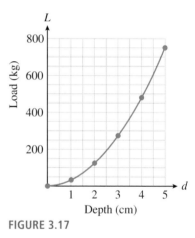

FIGURE 3.17

EXERCISE 3

a. For the table in Example 3, consider the column corresponding to a beam depth of 3 cm. Graph L as a function of w when the depth is constant at $d = 3$.

b. Find a formula for L as a function of w for $d = 3$.

In Exercise 3, you should find that the load varies directly with width when the depth is 3 centimeters. In fact, the load varies directly with width for any fixed depth. In Example 3, we saw that the load varies with the square of depth when the width is 3 centimeters, and this relationship also holds for any value of w. Consequently, we can find a constant k such that

$$\text{load} = k \cdot \text{width} \cdot \text{depth}^2$$

This relationship between variables is an example of **joint variation.**

Joint Variation

If

$$z = kxy, \quad k > 0$$

we say that z **varies jointly** with x and y. If

$$z = k\frac{x}{y}, \quad k > 0, y \neq 0$$

we say that z varies directly with x and inversely with y.

EXAMPLE 4   Find a formula for load as a function of width and depth for the data in Example 3.

Solution   The function we want has the form

$$L = f(w, d) = kwd^2$$

for some value of k. We use the fact that $L = 10$ when $w = 1$ and $d = 1$. Then

$$10 = k(1)(1^2)$$

so $k = 10$. The formula for load as a function of width and depth is

$$L = 10wd^2$$

You can check that this formula works for all the values in the table.

EXERCISE 4

The cost, C, of tiling a rectangular floor depends on the dimensions (length and width) of the floor, so $C = f(w, l)$. The table shows the costs in dollars for some dimensions.

Cost of Tiling a Floor

Width (ft)	\multicolumn{6}{c}{Length (ft)}
	5	6	7	8	9	10
5	400	480	560	640	720	800
6	480	576	672	768	864	960
7	560	672	784	896	1008	1120
8	640	768	896	1024	1152	1280
9	720	864	1008	1152	1296	1440
10	800	960	1120	1280	1440	1600

a. Consider the row corresponding to 6 feet in width. Does cost vary directly with length?

b. Consider the column corresponding to a length of 10 feet. Does the cost vary directly with width?

c. Given that the cost varies jointly with the length and width of the floor, find a formula for $C = f(w, l)$.

Graphs

It is possible to make graphs in three dimensions for functions of two variables, but we will not do that here. Instead, we will represent such functions graphically by holding one of the two variables constant.

EXAMPLE 5

In Example 4, we found a formula for the load a beam can support,

$$L = 10wd^2$$

a. Graph L as a function of w for $d = 1, 2, 3,$ and 4.

b. Graph L as a function of d for $w = 1, 2, 3,$ and 4.

Solutions

a. We make four graphs on the same grid, one for each value of d:

when	$d = 1,$	$L = 10w$
when	$d = 2,$	$L = 40w$
when	$d = 3,$	$L = 90w$
when	$d = 4,$	$L = 160w$

The graph is shown in Figure 3.18. We can see that L varies directly with the width of the beam for any fixed value of its depth.

b. We make one graph for each value of w:

when	$w = 1,$	$L = 10d^2$
when	$w = 2,$	$L = 20d^2$
when	$w = 3,$	$L = 30d^2$
when	$w = 4,$	$L = 40d^2$

The graph is shown in Figure 3.19. For any fixed value of its width, L varies directly with the square of depth.

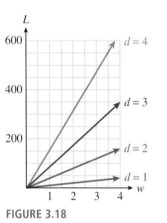

FIGURE 3.18

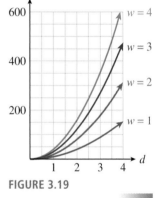

FIGURE 3.19

EXERCISE 5

The period of a satellite orbiting the Earth varies directly with the radius of the orbit and inversely with the speed of the satellite.

a. Write a formula for the period, T, as a function of orbital radius, r, and velocity, v.

b. GPS satellites orbit at an altitude of 20,200 kilometers and a speed of 233 kilometers per second. The period of a GPS satellite is 11 hours and 58 minutes. Find the constant of variation in your formula for T. (The radius of the Earth is 6360 km.)

c. Satellites in polar orbits are used to measure ozone concentrations in the atmosphere. They orbit at an altitude of 300 km and have a period of 90 minutes. What is the speed of a satellite in polar orbit?

d. Graph T as a function of v for $r = 5000$, $r = 10,000$, $r = 20,000$, and $r = 30,000$.

1. a. $f(40, 2) = 61$. With nozzle diameter 1 inch and nozzle pressure 40 psi, the water will reach 61 feet.

b. 80 psi **2. a.** \$129,528, $f(500, 40) = 129,528$ **b.** \$1000, $f(c, 50) = 573,770$

c. 20 years. If you contribute \$2500 per year for 20 years, you will accumulate \$137,286.

3. a.   **b.** $L = 90w$

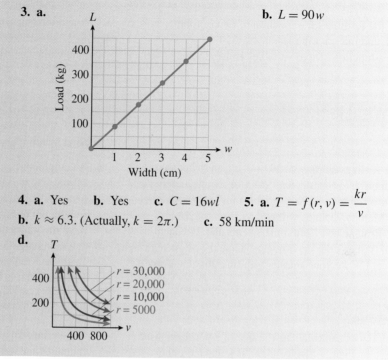

4. a. Yes   **b.** Yes   **c.** $C = 16wl$ **5. a.** $T = f(r, v) = \dfrac{kr}{v}$

b. $k \approx 6.3$. (Actually, $k = 2\pi$.)   **c.** 58 km/min

d.

SECTION 3.5 SUMMARY

VOCABULARY   *Look up the definitions of new terms in the Glossary.*

Function of two variables   Joint variation

CONCEPTS

1. The notation $z = f(x, y)$ indicates that z is a function of two variables, x and y.

2. We can use a table with rows and columns to display the output values for a function of two variables.

3.

> **Joint Variation:**
>
> • We say that z **varies jointly** with x and y if
> $$z = kxy, \quad k \neq 0$$
>
> • We say that z varies directly with x and inversely with y if
> $$z = k\frac{x}{y}, \quad k \neq 0, y \neq 0$$

4. We can represent a function of two variables graphically by showing a set of graphs for several fixed values of one of the variables.

1. Explain the difference between the symbols $f(ab)$ and $f(a, b)$.
2. Why is it true that $f(ab) = f(ba)$, but not usually true that $f(a, b) = f(b, a)$?

3. If z varies jointly with x and y, and we hold one of the input variables constant, what will the graph look like?
4. What is wrong with the statement "z varies jointly with x and y, so $z = f(xy)$"?

SKILLS   *Practice each skill in the Homework problems listed.*

1. Evaluate the formula for a function of two or more variables, and interpret the result: #1–6
2. Evaluate a function of two variables from a table: #7–10

3. Write a formula for joint variation: #11–18
4. Graph a function of two variables by fixing values of one of the variables: #11, 12, 15, 16, 19, 20

HOMEWORK 3.5

ThomsonNOW™ Test yourself on key content at www.thomsonedu.com/login.

1. Melody Airlines charges $129 for a coach ticket from San Francisco to Seattle, and $240 for a first-class ticket.
   a. Write a function of two variables for the revenue, R, that Melody Airlines will collect from the flight.
   b. The airplane has 12 first-class seats and 24 coach seats. What is the maximum revenue the airline can collect? Write your answer with function notation.

2. A manufacturing firm calculates its profit (or loss) by subtracting the cost of production from its revenue. The firm can produce 100 items per week, with fixed costs (overhead) of $500.
   a. Write a function for the firm's weekly profit, P, if they charge a price of p dollars per item and it costs them c dollars to produce each item.
   b. If each item costs $80 to produce, what price should the firm charge in order to make a profit? Write your answer with function notation.

3. Archaeologists can calculate the size of a pot from just a fragment, or sherd, of the original. If L and h are the dimensions of an arc of a circle, as shown in the figure, then the radius of the entire circle is given by the function

$$r = f(L, h) = \frac{L^2}{8h} + \frac{h}{2}$$

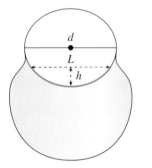

   a. A pottery sherd has dimensions $L = 4$ inches and $h = \frac{3}{2}$ inch. What was the radius of the whole pot? Write your answer with function notation.
   b. Does r vary directly with L^2? Why or why not?
   c. Show that the formula gives the correct value for r when the sherd is actually a semicircle. (*Hint:* What are the values of L and h in this case?)

4. The surface area of a cylinder is a function of its diameter and height,

$$S = f(d, h) = \pi dh + \frac{\pi}{2}d^2$$

   a. What is the surface area of a cylindrical oatmeal container with diameter 4 inches and height 7 inches? Write your answer with function notation.
   b. Write a formula in terms of d for the surface area of a cylinder whose height is equal to its diameter.
   c. How does the surface area of the cylinder in part (b) compare with the surface area of a sphere of the same diameter? Sketch both surfaces with the same center.

5. The Dubois formula is used to estimate the surface area, S, of a person in terms of his or her weight and height. A good estimate of surface area is critical to some forms of cancer treatment. In square centimeters, S is given by

$$S = f(w, h) = 71.84 w^{0.425} h^{0.725}$$

where w is in kilograms and h is in centimeters.

a. Use the Dubois formula to estimate the surface area of a person who weighs 60 kg and is 160 cm tall.

b. Does surface area increase more rapidly with weight or with height?

c. What percent increase in surface area does a 10% increase in weight produce?

6. In the 1970s, McNeill Alexander proposed a relationship between an animal's running speed, v, its hip height, h, and its stride length, s. If stride and hip height are measured in meters, the running speed is given in meters per second by

$$v = f(h, s) = 0.78 s^{1.67} h^{-1.17}$$

a. What is the speed of a racehorse whose hip height is 1.6 meters and stride length is 7 meters?

b. A cheetah can run at 33 meters per second. If its hip height is 0.8 meters, what is its stride length?

c. A pronghorn antelope has the same stride length as a cheetah and its hip height is 12% greater than the cheetah's. How does its running speed compare to a cheetah's?

7. If you walk for exercise, the number of calories, C, you burn per mile depends on your walking speed, s, and your weight, w.

a. Write this fact in function notation.

b. Use the table to evaluate $f(4.5, 160)$ and explain its meaning.

c. Solve the inequality $f(s, 160) > 110$ and explain its meaning.

d. If you weigh 140 pounds, how fast should you walk to burn the most calories per mile?

e. How can you use the table to calculate how many calories you burn per hour while walking?

Calories Burned per Mile							
	Weight (pounds)						
Speed (mph)	100	120	140	160	180	200	220
2.0	65	80	93	105	120	133	145
2.5	62	74	88	100	112	124	138
3.0	60	72	83	95	108	120	132
3.5	59	71	83	93	107	119	130
4.0	59	70	81	94	105	118	129
4.5	69	82	97	110	122	138	151
5.0	77	92	108	123	138	154	169
6.0	86	99	114	130	147	167	190
7.0	96	111	128	146	165	187	212

8. The BMI (body mass index) is used to determine whether a person is a healthy weight, overweight, or obese. *B* is a function of height, *h*, and weight, *w*.
   a. Write this fact in function notation.
   b. Use the table to evaluate $f(68,150)$, and explain its meaning.
   c. A person is deemed overweight if his or her BMI is at least 25 but less than 30. Write this fact in function notation. (A person is obese if the BMI is over 30.)
   d. Solve the inequality $f(66, w) < 25$ and explain its meaning.
   e. Solve the inequality $f(h, 200) > 30$ and explain its meaning.

Body Mass Index											
	Weight (pounds)										
Height (inches)	120	130	140	150	160	170	180	190	200	210	220
58	25	27	29	31	34	36	38	40	42	44	46
60	23	25	27	29	31	33	35	37	39	41	43
62	22	24	26	27	29	31	33	35	37	38	40
64	21	22	24	26	28	29	31	33	34	36	38
66	19	21	23	24	26	27	29	31	32	34	36
68	18	20	21	23	24	26	27	29	30	32	34
70	17	19	20	22	23	24	26	27	29	30	32
72	16	18	19	20	22	23	24	26	27	28	30
74	15	17	18	19	21	22	23	24	26	27	28
76	15	16	17	18	20	21	22	23	24	26	27

9. An *amortization table* shows the monthly payments for a loan or mortgage. The table below gives monthly payments for a loan of $100,000. The monthly payment, *P*, is a function of the annual interest rate, *r*, and the length of the loan, *t*, in years.

Monthly Payment						
	Length of loan (years)					
Interest rate	5	10	15	20	25	30
0.05	1879	1056	788	657	582	535
0.06	1924	1105	840	713	641	597
0.07	1969	1154	894	771	703	661
0.08	2014	1205	949	831	767	729
0.09	2060	1257	1007	893	833	799
0.10	2107	1311	1066	957	901	870

a. You would like to borrow $100,000 for 20 years. What interest rate, to the nearest percent, can you accept if your monthly payments must be no more than $800? What interest rate can you accept if the loan is for 30 years? If $P = f(r, t)$, use function notation to write both of these questions as inequalities.
b. Suppose you borrow $100,000 for 10 years at 10% interest. Which would cause a greater reduction in your monthly payment: Reducing the interest rate by 5% or increasing the length of the loan by 5 years?
c. At a fixed interest rate of 8%, is the monthly payment a linear function of the length of the loan?
d. For a fixed loan period of 25 years, is the monthly payment a linear function of interest rate?
e. Does a 1% increase in the interest rate have a greater affect on the monthly payment for a 15-year loan or a 30-year loan?

10. Warmer air can hold more moisture than cooler air. *Relative humidity* is the amount of moisture in the air, as a fraction of the saturation level at the current temperature. A common measure of humidity is the *dewpoint:* the temperature at which the current humidity would saturate the air, so that dew forms. The table gives dewpoints, D, in °F, as a function of temperature, T, and relative humidity, H, $D = f(T, H)$.

   a. Estimate the relative humidity if the temperature is 70°F and the dewpoint is 40°F. Write your answer in function notation.

   b. Does the dewpoint rise or fall with temperature? (*Hint:* Consider any column in the table, and notice how dewpoint changes with increasing temperature.)

   c. Does the dewpoint rise or fall with humidity? (*Hint:* Consider any row in the table, and notice how dewpoint changes with increasing humidity.)

   d. Suppose that the temperature is 70°F and the relative humidity is 70%. Which would cause a larger change in dewpoint: a rise in temperature to 80°F or an increase in humidity to 80%?

   e. Does dewpoint change more rapidly with temperature when the humidity is low or when the humidity is high?

Dewpoint

	Relative humidity (%)										
Temperature (°F)	1	10	20	30	40	50	60	70	80	90	100
30	−60	−20	−6	2	9	14	18	21	25	27	30
40	−53	−12	2	11	18	23	28	31	34	37	40
50	−47	−4	11	20	27	32	37	40	44	47	50
60	−41	3	19	29	36	41	46	50	54	57	60
70	−35	11	27	37	45	50	55	60	64	67	70
80	−29	19	35	46	54	60	65	69	73	77	80
90	−23	26	43	54	62	69	74	79	83	87	90

11. The table shows automobile fuel efficiency, E, as a function of gasoline used, g, and miles driven, m. Values of E are rounded to tenths where necessary.

   a. Choose one row of the table and decide if E varies directly with m or inversely with m. Explain your method.

   b. Choose one column of the table and decide if E varies directly with g or inversely with g. Explain your method.

   c. Find the constant of variation and write E as a function of g and m. What are the units of E?

   d. Sketch a graph of E as a function of m for $g = 8$, 10, 12, and 14.

   e. Sketch a graph of E as a function of g for $m = 100$, 200, 300, and 400.

Fuel Efficiency

	Distance (miles)							
Gas (gallons)	100	150	200	250	300	350	400	450
8	12.5	18.75	25	31.25	37.5	43.75	50	56.25
10	10	15	20	25	30	35	40	45
12	8.3	12.5	16.7	20.8	25	29.2	33.3	37.5
14	7.1	10.7	14.3	17.9	21.4	25	28.6	32.1
16	6.3	9.4	12.5	15.6	18.8	21.9	25	28.1
18	5.6	8.3	11.1	13.9	16.7	19.4	22.2	25

12. The table shows the productivity, P, of a manufacturing plant as a function of the number of items produced, I, and the hours of labor used, w. Values of P are rounded to tenths where necessary.
    a. Choose one row of the table and decide if P varies directly with w or inversely with w. Explain your method.
    b. Choose one column of the table and decide if P varies directly with I or inversely with I. Explain your method.
    c. Find the constant of variation and write P as a function of I and w. What are the units of P?
    d. Sketch a graph of P as a function of w for $I = 500, 600, 800,$ and 1000.
    e. Sketch a graph of P as a function of I for $w = 100, 200, 300,$ and 400.

Productivity								
Items produced	**Labor (hours)**							
	100	150	200	250	300	350	400	450
500	5	3.3	2.5	2	1.7	1.4	1.25	1.1
600	6	4	3	2.4	2	1.7	1.5	1.3
700	7	4.7	3.5	2.8	2.3	2	1.75	1.6
800	8	5.3	4	3.2	2.7	2.3	2	1.8
900	9	6	4.5	3.6	3	2.6	2.25	2.1
1000	10	6.7	5	4	3.3	2.94	2.5	2.2

13. The water gushing out of a fire hose exerts a backward force that the firefighter must control. This force, called the nozzle reaction, R, is a function of the diameter, d, of the nozzle and the water pressure, P, at the nozzle.
    a. Show that R varies directly with P.
    b. Show that R varies directly with a power of d. What is the power?
    c. Find the constant of variation and write a formula for R as a function of d and P.
    d. A typical fire hose has nozzle diameter $2\frac{1}{2}$ inches and nozzle pressure 60 psi. Use your formula to calculate the nozzle reaction.

Nozzle Reaction (lb)						
Nozzle diameter (in)	**Water pressure (psi)**					
	30	40	50	60	70	80
1.00	47.1	62.8	78.5	94.2	109.9	125.6
1.25	73.6	98.1	122.7	147.2	171.7	196.3
1.50	106	141.3	176.6	212	247.3	282.6
1.75	144.2	192.3	240.4	288.5	336.6	384.7
2.00	188.4	251.2	314	376.8	439.6	502.4

14. The resistance, R, of a wire depends on its length, L, and diameter, d. The table shows the resistance of copper wires of gauges from 10 to 20. The diameters of the wires are given in mils, where 1 mil $= 0.001$ inch.
    a. Show that R varies directly with L.
    b. Show that R varies inversely with a power of d. What is the power?
    c. Find the constant of variation and write a formula for R as a function of L and d.
    d. Household current uses 12-gauge wire, with a diameter of 0.081 inches. Use your formula to calculate the resistance of a 100-foot length of 12-gauge wire, and verify with the table.

Resistance (ohms)						
Length (ft)	**Diameter (mils)**					
	102	81	64	51	40	32
50	0.0500	0.0793	0.1270	0.1999	0.325	0.5078
100	0.1000	0.1585	0.2539	0.3999	0.65	1.0156
150	0.1499	0.2378	0.3809	0.5998	0.975	1.5234
200	0.1999	0.3170	0.5078	0.7997	1.3	2.0313

15. In 2005, *Popular Mechanics* tested the acceleration from rest for eight sports cars.

Car	Time (sec)	Distance (ft)	Acceleration (ft/sec²)
Mercedes-Benz E55 AMG	5.06	222.64	17.39
Lamborghini Gallardo	4.68	205.92	18.80
Chevrolet Corvette Z06	4.58	201.52	19.21
Mercedes-Benz SL600	4.62	203.28	
Porsche 911 GT2	4.46	196.24	
Dodge Viper SRT-10	4.30	189.20	
Saleen S7 Competition	3.73	164.12	
Ford Gran Torino	3.43	150.92	

   a. Acceleration, a, varies directly with distance, d, and inversely with a power of time, t. Use the information in the table to write a as a function of d and t.

   b. Use your formula to compute the acceleration of the other five cars.

   c. Use your formula to plot a against d for $t = 1, 2, 3,$ and 4 seconds.

   d. Use your formula to plot a against t for $d = 100, 200, 300,$ and 400 feet.

16. The rate of water flow, F, through a fire hose is a function of the diameter, d, of the nozzle and the nozzle pressure, p.

Water Flow (gal/min)						
	Water pressure (psi)					
Nozzle diameter (in)	30	40	50	60	70	80
1.00	164	190	212	232	251	268
1.25	257	296	331	363	392	419
1.50	370	427	477	523	565	604
1.75	503	581	650	712	769	822
2.00	657	759	849	930	1004	1073

   a. Plot F as a function of d for $p = 30, 40, 50,$ and 60. Which basic function do your graphs resemble?

   b. Plot F as a function of p for $d = 1, 1.25, 1.5,$ and 1.75. Which basic function do your graphs resemble?

   c. Write a formula for F as a function of d and p, and find the constant of variation. Check your formula against the table.

17. Railroad engineers use a transition curve between straight sections of track and bends, which are designed as arcs of circles. The length, L, of the transition curve varies directly with the cube of the train's speed, v, and inversely with the radius, R, of the circular arc.

   a. On a section of track where the speed limit is 50 mph, a circular bend has a radius of 2000 feet, and the transition curve is 200 feet long. Find the constant of variation and write a formula for L as a function of v and R.

   b. If the speed limit is increased by 20%, how is the length of the transition curve affected?

   c. If the radius of the bend is increased by 20%, how is the length of the transition curve affected?

18. The density, D, of a planet varies directly with its mass, M, and inversely with the cube of its radius, r.

   a. Use the data for Earth to find the constant of variation, then write a formula for D as a function of M and r.

   b. Calculate the densities of the other planets.

   c. The planets are composed of three broad categories of materials: rocky materials, icy materials (including water), and the materials that dominate the sun, namely hydrogen and helium. The density of rock varies from 3000 to 8000 kg/m³. Which of the planets could be composed mainly of rock?

Planet	Radius (km)	Mass (10^{20} kg)	Density (kg/m³)
Mercury	2440	3302	
Venus	6052	48,690	
Earth	6378	59,740	5497
Mars	3397	6419	
Jupiter	71,490	18,990,000	
Saturn	60,270	5,685,000	
Uranus	25,560	866,200	
Neptune	24,765	1,028,000	
Pluto	1150	150	

19. Ammonia has many uses in industry and agriculture, including the production of fertilizers. It is produced in the laboratory from nitrogen and hydrogen, but the process requires high pressure and temperature for significant yield. The graph illustrates the relationship. (Source: Hunt and Sykes, 1984)

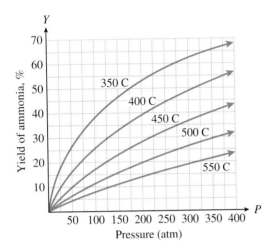

a. Complete the table showing the yield of ammonia, as a percent of the gas mixture leaving the reactor, at various pressures and temperatures.

Percent Ammonia

Temperature (°C)	Pressure (atmospheres)							
	50	100	150	200	250	300	350	400
350								
400								
450								
500								
550								

b. What happens to the yield of ammonia if the pressure is held constant but the temperature is increased beyond 350°C?

c. Sketch a graph of the yield of ammonia as a function of temperature when the pressure is 300 atmospheres.

20. The graph shows the *heat index,* which combines air temperature and relative humidity to determine an apparent temperature, or what the temperature actually feels like. (Source: Ahrens, 1998)

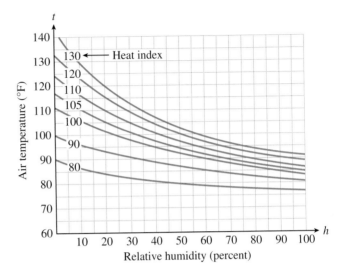

a. Complete the table showing the heat index for various combinations of air temperature and relative humidity.

Heat Index

Air temperature (°F)	Relative humidity (%)					
	0	20	40	60	80	100
80						
90						
100						
110						
120						

b. Complete the table showing the relative humidity at which the heat index is equal to the actual air temperature.

Air temperature (°F)	80	90	100	110	120
Relative humidity (%)					

c. Sketch a graph of the heat index as a function of air temperature if the relative humidity is 70%.

KEY CONCEPTS

1.

> ### Direct and Inverse Variation
>
> - **y varies directly with x** if the ratio $\frac{y}{x}$ is constant, that is, if $y = kx$.
> - **y varies directly with a power x** if the ratio $\frac{y}{x^n}$ is constant, that is, if $y = kx^n$.
> - **y varies inversely with x** if the product xy is constant, that is, if $y = \frac{k}{x}$.
> - **y varies inversely with a power of x** if the product $x^n y$ is constant, that is, if $y = \frac{k}{x^n}$.

2. The graph of a direct variation passes through the origin. The graph of an inverse variation has a vertical asymptote at the origin.

3. If $y = kx^n$, we say that y **scales** as x^n.

4. nth roots: s is called an ***n*th root of b** if $s^n = b$.

5.

> ### Exponential Notation
>
> $a^{-n} = \dfrac{1}{a^n}, \qquad a \neq 0$
>
> $a^0 = 1, \qquad a \neq 0.$
>
> $a^{1/n} = \sqrt[n]{a}, \qquad n$ an integer, $n > 2$
>
> $a^{m/n} = (a^{1/n})^m = (a^m)^{1/n}, \quad a > 0, n \neq 0$

6. In particular, a negative exponent denotes a reciprocal, and a fractional exponent denotes a root.

7. $a^{m/n} = \sqrt[n]{a^m} = (\sqrt[n]{a})^m$

8. To compute $a^{m/n}$, we can compute the nth root first, or the mth power, whichever is easier.

9. We cannot write down an exact decimal equivalent for an **irrational number,** but we can approximate an irrational number to as many decimal places as we like.

10. The laws of exponents are valid for all exponents m and n, and for $b \neq 0$.

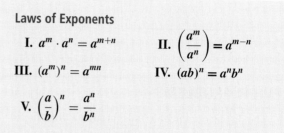

> ### Laws of Exponents
>
> **I.** $a^m \cdot a^n = a^{m+n}$ **II.** $\left(\dfrac{a^m}{a^n}\right) = a^{m-n}$
>
> **III.** $(a^m)^n = a^{mn}$ **IV.** $(ab)^n = a^n b^n$
>
> **V.** $\left(\dfrac{a}{b}\right)^n = \dfrac{a^n}{b^n}$

11. A function of the form $f(x) = kx^p$, where k and P are constants, is called a **power function.**

12. An **allometric equation** is a power function of the form **variable** $= k(\mathbf{mass})^p$.

13. We can solve the equation $x^n = b$ by raising both sides to the $\frac{1}{n}$ power.

14. We can solve the equation $x^{1/n} = b$ by raising both sides to the nth power.

15. To solve the equation $x^{m/n} = k$, we raise both sides to the power n/m.

16. The graphs of power functions $y = x^{m/n}$, where m/n is positive are all increasing for $x \geq 0$. If $m/n > 1$, the graph is concave up. If $0 < m/n < 1$, the graph is concave down.

17. The notation $z = f(x, y)$ indicates that z is a function of two variables, x and y.

18. We can use a table with rows and columns to display the output values for a function of two variables.

19.

> ### Joint Variation
>
> - We say that z **varies jointly** with x and y if
>
> $$z = kxy, \quad k \neq 0$$
>
> - We say that z varies directly with x and inversely with y if
>
> $$z = k\frac{x}{y}, \quad k \neq 0, \ y \neq 0$$

20. We can represent a function of two variables graphically by showing a set of graphs for several fixed values of one of the variables.

21.

> **Roots of Real Numbers**
>
> - Every positive number has two real-valued roots, one positive and one negative, if the index is even.
> - A negative number has no real-valued root if the index is even.
> - Every real number, positive, negative, or zero, has exactly one real-valued root if the index is odd.

REVIEW PROBLEMS

1. The distance s a pebble falls through a thick liquid varies directly with the square of the length of time t it falls.
  a. If the pebble falls 28 centimeters in 4 seconds, express the distance it will fall as a function of time.
  b. Find the distance the pebble will fall in 6 seconds.

2. The volume, V, of a gas varies directly with the temperature, T, and inversely with the pressure, P, of the gas.
  a. If $V = 40$ when $T = 300$ and $P = 30$, express the volume of the gas as a function of the temperature and pressure of the gas.
  b. Find the volume when $T = 320$ and $P = 40$.

3. The demand for bottled water is inversely proportional to the price per bottle. If Droplets can sell 600 bottles at \$8 each, how many bottles can the company sell at \$10 each?

4. The intensity of illumination from a light source varies inversely with the square of the distance from the source. If a reading lamp has an intensity of 100 lumens at a distance of 3 feet, what is its intensity 8 feet away?

5. A person's weight, w, varies inversely with the square of his or her distance, r, from the center of the Earth.
  a. Express w as a function of r. Let k stand for the constant of variation.
  b. Make a rough graph of your function.
  c. How far from the center of the Earth must Neil be in order to weigh one-third of his weight on the surface? The radius of the Earth is about 3960 miles.

6. The period, T, of a pendulum varies directly with the square root of its length, L.
  a. Express T as a function of L. Let k stand for the constant of variation.
  b. Make a rough graph of your function.
  c. If a certain pendulum is replaced by a new one four-fifths as long as the old one, what happens to the period?

For each table, y varies directly or inversely with a power of x. Find the power of x and the constant of variation, k. Write a formula for each function of the form $y = kx^n$ or $y = \dfrac{k}{x^n}$.

7.

x	y
2	4.8
5	30.0
8	76.8
11	145.2

8.

x	y
1.4	75.6
2.3	124.2
5.9	318.6
8.3	448.2

9.

x	y
0.5	40.0
2.0	10.0
4.0	5.0
8.0	2.5

10.

x	y
1.5	320.0
2.5	115.2
4.0	45.0
6.0	20.0

Write without negative exponents and simplify.

11. a. $(-3)^{-4}$ **b.** 4^{-3} **12. a.** $\left(\dfrac{1}{3}\right)^{-2}$ **b.** $\dfrac{3}{5^{-2}}$

13. a. $(3m)^{-5}$ **b.** $-7y^{-8}$ **14. a.** $a^{-1}+a^{-2}$ **b.** $\dfrac{3q^{-9}}{r^{-2}}$

15. a. $6c^{-7}\cdot 3^{-1}c^4$ **b.** $\dfrac{11z^{-7}}{3^{-2}z^{-5}}$ **16. a.** $(2d^{-2}k^3)^{-4}$ **b.** $\dfrac{2w^3(w^{-2})^{-3}}{5w^{-5}}$

Write each power in radical form.

17. a. $25m^{1/2}$ **b.** $8n^{-1/3}$ **18. a.** $(13d)^{2/3}$ **b.** $6x^{2/5}y^{3/5}$

19. a. $(3q)^{-3/4}$ **b.** $7(uv)^{3/2}$ **20. a.** $(a^2+b^2)^{0.5}$ **b.** $(16-x^2)^{0.25}$

Write each radical as a power with a fractional exponent.

21. a. $2\sqrt[3]{x^2}$ **b.** $\dfrac{1}{4}\sqrt[4]{x}$ **22. a.** $z^2\sqrt{z}$ **b.** $z\sqrt[3]{z}$

23. a. $\dfrac{6}{\sqrt[4]{b^3}}$ **b.** $\dfrac{-1}{3\sqrt[3]{b}}$ **24. a.** $\dfrac{-4}{(\sqrt[4]{a})^2}$ **b.** $\dfrac{2}{(\sqrt{a})^3}$

Sketch graphs by hand for each function on the domain $(0, \infty)$.

25. y varies directly with x^2. The constant of variation is $k = 0.25$.

26. y varies directly with x. The constant of variation is $k = 1.5$.

27. y varies inversely with x. The constant of variation is $k = 2$.

28. y varies inversely with x^2. The constant of variation is $k = 4$.

Write each function in the form $y = kx^p$.

29. $f(x) = \dfrac{2}{3x^4}$

30. $g(x) = \dfrac{8x^7}{29}$

For Problems 31–34,
(a) Evaluate each function for the given values.
(b) Graph the function.

31. $Q(x) = 4x^{5/2}$

x	16	$\frac{1}{4}$	3	100
$Q(x)$				

32. $T(w) = -3w^{2/3}$

w	27	$\frac{1}{8}$	20	1000
$T(w)$				

33. $f(x) = x^{0.3}$

x	0	1	5	10	20	50	70	100
$f(x)$								

34. $g(x) = x^{-0.7}$

x	0.1	0.2	0.5	1	2	5	8	10
$g(x)$								

35. According to the theory of relativity, the mass of an object traveling at velocity v is given by the function

$$m = \frac{M}{\sqrt{1 - \frac{v^2}{c^2}}}$$

where M is the mass of the object at rest and c is the speed of light. Find the mass of a man traveling at a velocity of $0.7c$ if his rest mass is 80 kilograms.

36. The cylinder of smallest surface area for a given volume has a radius and height both equal to $\sqrt[3]{\frac{V}{\pi}}$. Find the dimensions of the tin can of smallest surface area with volume 60 cubic inches.

37. Membership in the Wildlife Society has grown according to the function

$$M(t) = 30t^{3/4}$$

where t is the number of years since its founding in 1970.
a. Sketch a graph of the function $M(t)$.
b. What was the society's membership in 1990?
c. In what year will the membership be 810 people?

38. The heron population in Saltmarsh Refuge is estimated by conservationists at

$$P(t) = 360t^{-2/3}$$

where t is the number of years since the refuge was established in 1990.
a. Sketch a graph of the function $P(t)$.
b. How many heron were there in 1995?
c. In what year will there be only 40 herons left?

39. Manufacturers of ships (and other complex products) find that the average cost of producing a ship decreases as more of those ships are produced. This relationship is called the *experience curve,* given by the equation

$$C = ax^{-b}$$

where C is the average cost per ship in millions of dollars and x is the number of ships produced. The value of the constant b depends on the complexity of the ship. (Source: Storch, Hammon, and Bunch, 1988)

  a. What is the significance of the constant of proportionality a? (*Hint:* What is the value of C if only one ship is built?)
  b. For one kind of ship, $b = \frac{1}{8}$, and the cost of producing the first ship is $12 million. Write the equation for C as a function of x using radical notation.
  c. Compute the cost per ship when 2 ships have been built. By what percent does the cost per ship decrease? By what percent does the cost per ship decrease from building 2 ships to building 4 ships?
  d. By what percent does the average cost decrease from building n ships to building $2n$ ships? (In the shipbuilding industry, the average cost per ship usually decreases by 5 to 10% each time the number of ships doubles.)

40. A population is in a period of *supergrowth* if its rate of growth, R, at any time is proportional to P^k, where P is the population at that time and k is a constant greater than 1. Suppose R is given by

$$R = 0.015\,P^{1.2}$$

where P is measured in thousands and R is measured in thousands per year.
  a. Find R when $P = 20$, when $P = 40$, and when $P = 60$.
  b. What will the population be when its rate of growth is 5000 per year?
  c. Graph R and use your graph to verify your answers to parts (a) and (b).

Solve.

41. $6t^{-3} = \dfrac{3}{500}$

42. $3.5 - 2.4p^{-2} = -6.1$

43. $\sqrt[3]{x+1} = 2$

44. $x^{2/3} + 2 = 6$

45. $(x-1)^{-3/2} = \dfrac{1}{8}$

46. $(2x+1)^{-1/2} = \dfrac{1}{3}$

47. $8\sqrt[4]{x+6} = 24$

48. $9.8 = 7\sqrt[3]{z-4}$

49. $\dfrac{2}{3}(2y+1)^{0.2} = 6$

50. $1.3w^{0.3} + 4.7 = 5.2$

Solve each formula for the indicated variable.

51. $t = \sqrt{\dfrac{2v}{g}}$, for g

52. $q - 1 = 2\sqrt{\dfrac{r^2-1}{3}}$, for r

53. $R = \dfrac{1+\sqrt{p^2+1}}{2}$, for p

54. $q = \sqrt[3]{\dfrac{1+r^2}{2}}$, for r

Simplify by applying the laws of exponents.

55. $(7t)^3 (7t)^{-1}$

56. $\dfrac{36r^{-2}s}{9r^{-3}s^4}$

57. $\dfrac{(2k^{-1})^{-4}}{4k^{-3}}$

58. $(2w^{-3})^{-2} (2w^{-3})^5 (-5w^2)$

59. $\dfrac{8a^{-3/4}}{a^{-11/4}}$

60. $b^{2/3}(4b^{-2/3} - b^{1/3})$

61. When the Concorde lands at Heathrow Airport in London, the width, w, of the sonic boom felt on the ground is given in kilometers by the following formula:

$$w = 4 \left(\frac{Th}{m}\right)^{1/2}$$

where T stands for the temperature on the ground in kelvins, h is the altitude of the Concorde when it breaks the sound barrier, and m is the drop in temperature for each gain in altitude of one kilometer.
   a. Find the width of the sonic boom if the ground temperature is 293 K, the altitude of the Concorde is 15 kilometers, and the temperature drop is 4 K per kilometer of altitude.
   b. Graph w as a function of h if $T = 293$ and $m = 4$.

62. The manager of an office supply store must decide how many of each item in stock she should order. The Wilson lot size formula gives the most cost-efficient quantity, Q, as a function of the cost, C, of placing an order, the number of items, N, sold per week, and the weekly inventory cost, I, per item (cost of storage, maintenance, and so on).

$$Q = \left(\frac{2CN}{I}\right)^{1/2}$$

   a. How many reams of computer paper should she order if she sells on average 80 reams per week, the weekly inventory cost for a ream is $0.20, and the cost of ordering, including delivery charges, is $25?
   b. Graph Q as a function of N if $C = 25$ and $I = 0.2$.

63. Two businesswomen start a small company to produce saddle bags for bicycles. The number of saddle bags, q, they can produce depends on the amount of money, m, they invest and the number of hours of labor, w, they employ, according to the Cobb-Douglas formula

$$q = 0.6\, m^{1/4} w^{3/4}$$

where m is measured in thousands of dollars.
   a. If the businesswomen invest $100,000 and employ 1600 hours of labor in their first month of production, how many saddle bags can they expect to produce?
   b. With the same initial investment, how many hours of labor would they need in order to produce 200 saddle bags?

64. A child who weighs w pounds and is h inches tall has a surface area (in square inches) given approximately by

$$S = 8.5\, h^{0.35} w^{0.55}$$

   a. What is the surface area of a child who weighs 60 pounds and is 40 inches tall?
   b. What is the weight of a child who is 50 inches tall and whose surface area is 397 square inches?

65. The cost, C, of insulating the ceiling in a building depends on the thickness of the insulation and the area of the ceiling. The table shows values of $C = f(t, A)$, where t is the thickness of the insulation and A is the area of the ceiling.

	Cost of Insulation (dollars)					
	Area (sq m)					
Thickness (cm)	100	200	300	400	500	600
4	72	144	216	288	360	432
5	90	180	270	360	450	540
6	108	216	324	432	540	648
7	126	252	378	504	630	756
8	144	288	432	576	720	864
9	162	324	486	648	810	972

a. What does it cost to insulate a ceiling with an area of 500 square meters with 5 cm of insulation? Write your answer in function notation.

b. Solve the equation $864 = f(t, 600)$ and interpret your answer.

c. Consider the row corresponding to a thickness of 4 cm. How does the cost of insulating the ceiling depend on the area of the ceiling?

d. Consider the column corresponding to an area of 100 square meters. How does the cost depend on the thickness of the insulation?

e. Given that the cost varies jointly with the thickness of the insulation and the area of the ceiling, write an equation for cost as a function of area and thickness of insulation. $C = At$

f. Use your formula from part (e) to determine the cost of insulating a building with 10 centimeters of insulation if the area of the ceiling is 800 square meters.

66. The volume, V, of a quantity of helium depends on both the temperature and the pressure of the gas. The table shows values of $V = f(P, T)$ for temperature in kelvins and pressure in atmospheres.

	Volume (cubic meters)					
	Temperature (K)					
Pressure (atmospheres)	100	150	200	250	300	350
1	18	27	36	45	54	63
2	9	13.5	18	22.5	27	31.5
3	6	9	12	15	18	21
4	4.5	6.75	9	11.25	13.5	15.75

a. What is the volume of helium when the pressure is 4 atmospheres and the temperature is 350 K? Write your answer in function notation.

b. Solve the equation $15 = f(3, T)$ and interpret your answer.

c. Consider the row corresponding to 2 atmospheres. How is the volume related to the absolute temperature?

d. Consider the column corresponding to 300 K. How is the volume related to the pressure?

e. Given that the volume of the gas varies directly with temperature and inversely with pressure, write an equation for volume as a function of temperature and pressure.

f. Use your formula from part (e) to determine the volume of the helium at 50 K and pressure of 0.4 atmospheres.

67. In his hiking guidebook, *Afoot and Afield in Los Angeles County*, Jerry Schad notes that the number of people on a wilderness trail is inversely proportional to "the square of the distance and the cube of the elevation gain from the nearest road."

  a. Choose variables and write a formula for this relationship.

  b. On a sunny Saturday afternoon, you count 42 people enjoying the Rock Pool at Malibu Creek State Park. The Rock Pool is 1.5 miles from the main parking lot, and the trail includes an elevation gain of 250 feet. Calculate the constant of variation in your formula from part (a). (*Hint:* Convert the elevation gain to miles.)

  c. Lookout Trail leads 1.9 miles from the parking lot and involves an elevation gain of 500 feet. How many people would you expect to encounter at the end of the trail?

68. A company's monthly production, P, depends on the capital, C, the company has invested and the amount of labor, L, available each month. The Cobb-Douglas model for production assumes that P varies jointly with C^a and L^b, where a and b are positive constants less than 1. The Aztech Chip Company invested 625 units of capital and hired 256 workers and produces 8000 computer chips each month.

  a. Suppose that $a = 0.25$, $b = 0.75$. Find the constant of variation and a formula giving P in terms of C and L.

  b. If Aztech increases its labor force to 300 workers, what production level can they expect?

  c. If Aztech maintains its labor force at 256 workers, what amount of capital outlay would be required for monthly production to reach 16,000 computer chips?

PROJECTS FOR CHAPTER 3

1. A hot object such as a light bulb or a star radiates energy over a range of wavelengths, but the wavelength with maximum energy is inversely proportional to the temperature of the object. If temperature is measured in kelvins, and wavelength in micrometers, the constant of proportionality is 2898. (One micrometer is one thousandth of a millimeter, or $1\ \mu m = 10^{-6}$ meter.)

  a. Write a formula for the wavelength of maximum energy, λ_{max}, as a function of temperature, T. This formula, called Wien's law, was discovered in 1894.

  b. Our sun's temperature is about 5765 K. At what wavelength is most of its energy radiated?

  c. The color of light depends on its wavelength, as shown in the table. Can you explain why the sun does not appear to be green? Use Wien's law to describe how the color of a star depends on its temperature.

Color	Wavelength (μm)
Red	0.64–0.74
Orange	0.59–0.64
Yellow	0.56–0.59
Green	0.50–0.56
Blue	0.44–0.50
Violet	0.39–0.44

  d. Astronomers cannot measure the temperature of a star directly, but they can determine the color or wavelength of its light. Write a formula for T as a function of λ_{max}.

  e. Estimate the temperatures of the following stars, given the approximate value of λ_{max} for each.

Star	λ_{max}	Temperature
R Cygni	1.115	
Betelgeuse	0.966	
Arcturus	0.725	
Polaris	0.414	
Sirius	0.322	
Rigel	0.223	

  f. Sketch a graph of T as a function of λ_{max} and locate each star on the graph.

2. Halley's comet which orbits the sun every 76 years, was first observed in 240 B.C. Its orbit is highly elliptical, so that its closest approach to the Sun (*perihelion*) is only 0.587 AU, while at its greatest distance (*aphelion*) the comet is 34.39 AU from the Sun. (An AU, or astronomical unit, is the distance from the Earth to the Sun, 1.5×10^8 kilometers.)

  a. Calculate the distances in meters from the Sun to Halley's comet at perihelion and aphelion.

  b. Halley's comet has a volume of 700 cubic kilometers, and its density is about 0.1 gram per cubic centimeter. Calculate the mass of the comet in kilograms.

c. The gravitational force (in newtons) exerted by the Sun on its satellites is inversely proportional to the square of the distance to the satellite in meters. The constant of variation is Gm_1m_2, where $m_1 = 1.99 \times 10^{30}$ kilograms is the mass of the Sun, m_2 is the mass of the satellite, and $G = 6.67 \times 10^{-11}$ is the gravitational constant. Write a formula for the force, F, exerted by the sun on Halley's comet at a distance of d meters.

d. Calculate the force exerted by the sun on Halley's comet at perihelion and at aphelion.

3. Are world record times for track events proportional to the length of the race? The table gives the men's and women's world records in 2005 for races from 1 kilometer to 100 kilometers in length.

Distance (km)	Men's record (min)	Women's record (min)
1	2.199	2.483
1.5	3.433	3.841
2	4.747	5.423
3	7.345	8.102
5	12.656	14.468
10	26.379	29.530
20	56.927	65.443
25	73.93	87.098
30	89.313	105.833

Source: International Association of Athletics Federations

a. On separate graphs, plot the men's and women's times against distance. Does time appear to be proportional to distance?

b. Use slopes to decide whether the graphs of time versus distance are in fact linear.

c. Both sets of data can be modeled by power functions of the form $t = kx^b$, where b is called the *fatigue index*. Graph the function $M(x) = 2.21x^{1.086}$ over the men's data points, and $W(x) = 2.46x^{1.099}$ over the women's data. Describe how the graphs of the two functions differ. Explain why b is called the fatigue index.

4. Fell running is a popular sport in the hills, or *fells,* of the British Isles. Fell running records depend on the altitude gain over the course of the race as well as its length. The equivalent horizontal distance for a race of length x kilometers with an ascent of y kilometers is given by $x + Ny$, where N is Naismith's number (see Project 5 for Chapter 1). The record times for women's races are approximated in minutes by $t = 2.43(x + 9.5y)^{1.15}$, and men's times by $t = 2.18(x + 8.0y)^{1.14}$.
(Source: Scarf, 1998)

a. Whose times show a greater fatigue index, men or women? (See Problem 3.)

b. Whose times are more strongly affected by ascents?

c. Predict the winning times for both men and women in a 56-kilometer race with an ascent of 2750 meters.

5. Elasticity is the property of an object that causes it to regain its original shape after being compressed or deformed. One measure of elasticity considers how high the object bounces when dropped onto a hard surface,

$$e = \sqrt{\frac{\text{height bounced}}{\text{height dropped}}}$$

(Source: Davis, Kimmet, and Autry, 1986)

a. The table gives the value of e for various types of balls. Calculate the bounce height for each ball when it is dropped from a height of 6 feet onto a wooden floor.

Type of ball	Bounce height	e
Baseball		0.50
Basketball		0.75
Golfball		0.60
Handball		0.80
Softball		0.55
Superball		0.90
Tennisball		0.74
Volleyball		0.75

b. Write a formula for e in terms of H, the bounce height, for the data in part (a).

c. Graph the function from part (b).

d. If Ball A has twice the elasticity of Ball B, how much higher will Ball A bounce than Ball B?

6. The tone produced by a vibrating string depends on the frequency of the vibration. The frequency in turn depends on the length of the string, its weight, and its tension. In 1636, Marin Mersenne quantified these relationships as follows. The frequency, f, of the vibration is

   i. inversely proportional to the string's length, L,

  ii. directly proportional to the square root of the string's tension, T, and

 iii. inversely proportional to the square root of the string's weight per unit length, w. (Source: Berg and Stork, 1982)

  a. Write a formula for f that summarizes Mersenne's laws.

  b. Sketch a graph of f as a function of L, assuming that T and w are constant. (You do not have enough information to put scales on the axes, but you can show the shape of the graph.)

  c. On a piano, the frequency of the highest note is about 4200 hertz. This frequency is 150 times the frequency of the lowest note, at about 28 hertz. Ideally, only the lengths of the strings should change, so that all the notes have the same tonal quality. If the string for the highest note is 5 centimeters long, how long should the string for the lowest note be?

  d. Sketch a graph of f as a function of T, assuming that L and w are constant.

  e. Sketch a graph of f as a function of w, assuming that L and T are constant.

  f. The tension of all the strings in a piano should be about the same to avoid warping the frame. Suggest another way to produce a lower note. (*Hint:* Look at a piano's strings.)

  g. The longest string on the piano in part (c) is 133.5 cm long. How much heavier (per unit length) is the longest string than the shortest string?

7. In 1981, John Damuth collected data on the average body mass, m, and the average population density, D, for 307 species of herbivores. He found that, very roughly,

$$D = km^{-0.75}$$

(Source: Burton, 1998)

  a. Explain why you might expect an animal's rate of food consumption to be proportional to its metabolic rate. (See Example 2 in Section 3.4 for an explanation of metabolic rate.)

  b. Explain why you might expect the population density of a species to be inversely proportional to the rate of food consumption of an individual animal.

  c. Use Kleiber's rule and your answers to parts (a) and (b) to explain why Damuth's proposed formula for population density is reasonable.

  d. Sketch a graph of the function D. You do not have enough information to put scales on the axes, but you can show the shape of the graph. (*Hint:* Graph the function for $k = 1$.)

8. Studies on pine plantations in the 1930s showed that as the trees grow and compete for space, some of the die, so that the density of trees per unit area decreases. The average mass of an individual tree is a power function of the density, d, of the trees per unit area, given by

$$M(d) = kd^{-1.5}$$

This formula is known as the $-\frac{3}{2}$ *self-thinning law.*
(Source: Chapman and Reiss, 1992)

  a. To simplify the calculations, suppose that a pine tree is shaped like a tall circular cone and that as it grows, its height is always a constant multiple of its base radius, r. Explain why the base radius of the tree is proportional to the square root of the area the tree covers. Write r as a power function of d.

  b. Write a formula for the volume of the tree in terms of its base radius, r. Use part (b) to write the volume as a power function of d.

  c. The mass (or weight) of a pine tree is roughly proportional to its volume, and the area taken up by a single tree is inversely proportional to the plant density, d. Use these facts to justify the self-thinning law.

  d. Sketch a graph of the function M. You do not have enough information to put scales on the axes, but you can show the shape of the graph. (*Hint:* Graph the function for $K = 1$.)

CHAPTER **4**

Exponential Functions

We next consider another important family of functions, called exponential functions. These functions describe growth by a constant factor in equal time periods. Exponential functions model many familiar processes, including the growth of populations, compound interest, and radioactive decay.

In 1965, Gordon Moore, the cofounder of Intel, observed that the number of transistors on a computer chip had doubled every year since the integrated circuit was invented. Moore predicted that the pace would slow down a bit, but the number of transistors would continue to double every 2 years. More recently, data density has doubled approximately every 18 months, and this is the current definition of Moore's law. Most experts, including Moore himself, expect Moore's law to hold for at least another two decades.

Year	Name of circuit	Transistors
1971	4004	2300
1972	8008	3300
1974	8080	6000
1978	8086	29,000
1979	8088	30,000
1982	80286	134,000
1985	80386	275,000
1989	90486	1,200,000
1993	Pentium	3,000,000
1995	Pentium Pro	5,500,000
1997	Pentium II	7,500,000
1998	Pentium II Xeon	7,500,000
1999	Pentium III	9,500,000

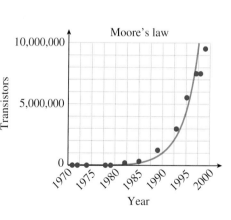

ThomsonNOW

Throughout this chapter, the ThomsonNOW logo indicates an opportunity for online self-study, linking you to interactive tutorials and videos based on your level of understanding.

The data shown are modeled by the exponential function $N(t) = 2200(1.356)^t$, where t is the number of years since 1970.

Population Growth

A. In a laboratory experiment, researchers establish a colony of 100 bacteria and monitor its growth. The colony triples in population every day.

1. Fill in the table showing the population $P(t)$ of bacteria t days later.

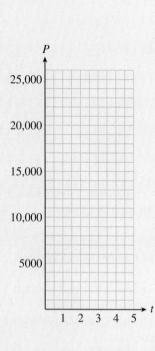

t	$P(t)$
0	100
1	
2	
3	
4	
5	

$P(0) = 100$
$P(1) = 100 \cdot 3 =$
$P(2) = [100 \cdot 3] \cdot 3 =$
$P(3) =$
$P(4) =$
$P(5) =$

2. Plot the data points from the table and connect them with a smooth curve.

3. Write a function that gives the population of the colony at any time t, in days. (*Hint:* Express the values you calculated in part (1) using powers of 3. Do you see a connection between the value of t and the exponent on 3?)

4. Graph your function from part (3) using a calculator. (Use the table to choose an appropriate domain and range.) The graph should resemble your hand-drawn graph from part (2).

5. Evaluate your function to find the number of bacteria present after 8 days. How many bacteria are present after 36 hours?

B. Under ideal conditions, the number of rabbits in a certain area can double every 3 months. A rancher estimates that 60 rabbits live on his land.

1. Fill in the table showing the population $P(t)$ of rabbits t months later.

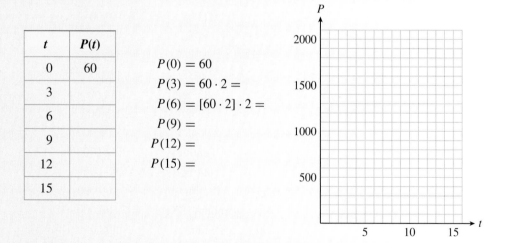

t	$P(t)$
0	60
3	
6	
9	
12	
15	

$P(0) = 60$
$P(3) = 60 \cdot 2 =$
$P(6) = [60 \cdot 2] \cdot 2 =$
$P(9) =$
$P(12) =$
$P(15) =$

2. Plot the data points and connect them with a smooth curve.

3. Write a function that gives the population of rabbits at any time t, in months. (*Hint:* Express the values you calculated in part (a) using powers of 2. Note that the population of rabbits is multiplied by 2 every 3 months. If you know the value of t, how do you find the corresponding exponent in $P(t)$?)

4. Graph your function from part (3) using a calculator. (Use the table to choose an appropriate domain and range.) The graph should resemble your hand-drawn graph from part (2).

5. Evaluate your function to find the number of rabbits present after 2 years. How many rabbits are present after 8 months?

4.1 Exponential Growth and Decay

Exponential Growth

The functions in Investigation 7 describe **exponential growth.** During each time interval of a fixed length, the population is *multiplied* by a certain constant amount. In Part A, the bacteria population grows by a factor of 3 every day.

t	0	1	2	3	4
$P(t)$	100	300	900	2700	8100

For this reason, we say that 3 is the **growth factor** for the function. Functions that describe exponential growth can be expressed in a standard form.

Exponential Growth

The function

$$P(t) = P_0 b^t$$

models exponential growth.

$P_0 = P(0)$ is the **initial value** of P;

b is the **growth factor.**

For the bacteria population, we have

$$P(t) = 100 \cdot 3^t$$

so $P_0 = 100$ and $b = 3$.

EXAMPLE 1 A colony of bacteria starts with 300 organisms and doubles every week.

    a. Write a formula for the population of the bacteria colony after t weeks.

    b. How many bacteria will there be after 8 weeks? After 5 days?

Solutions   **a.** The initial value of the population was $P_0 = 300$, and its weekly growth factor is $b = 2$. Thus, a formula for the population after t weeks is

$$P(t) = 300 \cdot 2^t$$

    b. After 8 weeks, the population will be

$$P(8) = 300 \cdot 2^8 = 76{,}800 \text{ bacteria}$$

Because 5 days is $\frac{5}{7}$ of a week, after 5 days the population will be

$$P\left(\frac{5}{7}\right) = 300 \cdot 2^{5/7} = 492.2$$

We cannot have a fraction of a bacterium, so we round to the nearest whole number, 492.

CAUTION   In Example 1a, note that

$$300 \cdot 2^8 \neq 600^8$$

According to the order of operations, we compute the power 2^8 first, then multiply by 300.

EXERCISE 1

A population of 24 fruit flies triples every month.

    a. Write a formula for the population of the bacteria colony after t weeks.

    b. How many fruit flies will there be after 6 months? After 3 weeks? (Assume that a month equals 4 weeks.)

Growth Factors

In Part B of Investigation 7, the rabbit population grew by a factor of 2 every 3 months.

t	0	3	6	9	12
$P(t)$	60	120	240	480	960

       ×2   ×2   ×2   ×2

To write the growth formula for this population, we divide the value of t by 3 to find the number of doubling periods.

$$P(t) = 60 \cdot 2^{t/3}$$

To see the growth factor for the function, we use the third law of exponents to write $2^{t/3}$ in another form. Recall that to raise a power to a power, we multiply exponents, so

$$(2^{1/3})^t = 2^{t(1/3)} = 2^{t/3}$$

The growth law for the rabbit population is thus

$$P(t) = 60 \cdot (2^{1/3})^t$$

The initial value of the function is $P_0 = 60$, and the growth factor is $b = 2^{1/3}$, or approximately 1.26. The rabbit population grows by a factor of about 1.26 every month.

If the units are the same, a population with a larger growth factor grows faster than one with a smaller growth factor.

EXAMPLE 2

A lab technician compares the growth of 2 species of bacteria. She starts 2 colonies of 50 bacteria each. Species A doubles in population every 2 days, and species B triples every 3 days. Find the growth factor for each species.

Solution   A function describing the growth of species A is

$$P(t) = 50 \cdot 2^{t/2} = 50 \cdot (2^{1/2})^t$$

so the growth factor for species A is $2^{1/2}$, or approximately 1.41. For species B,

$$P(t) = 50 \cdot 3^{t/3} = 50 \cdot (3^{1/3})^t$$

so the growth factor for species B is $3^{1/3}$, or approximately 1.44. Species B grows faster than species A.

EXERCISE 2

In 1999, analysts expected the number of Internet service providers to double in five years.

a. What was the annual growth factor for the number of Internet service providers?

b. If there were 5078 Internet service providers in April 1999, estimate the number of providers in April 2000 and in April 2001.

c. Write a formula for $I(t)$, the number of Internet service providers t years after 1999.

Source: *LA Times*, Sept. 6, 1999

Percent Increase

Exponential growth occurs in other circumstances, too. For example, if the interest on a savings account is compounded annually, the amount of money in the account grows exponentially.

Consider a principal of $100 invested at 5% interest compounded annually. At the end of 1 year, the amount is

$$\text{Amount} = \text{Principal} + \text{Interest}$$
$$A = P + Pr$$
$$= 100 + 100(0.05) = 105$$

It will be more useful to write the formula for the amount after 1 year in factored form.

$$A = P + Pr \qquad \text{Factor out } P.$$
$$= P(1 + r)$$

With this version of the formula, the calculation for the amount at the end of 1 year looks like this:

$$A = P(1+r)$$
$$= 100(1 + 0.05)$$
$$= 100(1.05) = 105$$

The amount, $105, becomes the new principal for the second year. To find the amount at the end of the second year, we apply the formula again, with $P = 105$.

$$A = P(1+r)$$
$$= 105(1 + 0.05)$$
$$= 105(1.05) = 110.25$$

Observe that to find the amount at the end of each year, we multiply the principal by a factor of $1 + r = 1.05$. Thus, we can express the amount at the end of the second year as

$$A = [100(1.05)](1.05)$$
$$= 100(1.05)^2$$

and at the end of the third year as

$$A = [100(1.05)^2](1.05)$$
$$= 100(1.05)^3$$

At the end of each year, we multiply the old balance by another factor of 1.05 to get the new amount. We organize our results into Table 4.1, where $A(t)$ represents the amount of money in the account after t years. For this example, a formula for the amount after t years is

$$A(t) = 100(1.05)^t$$

t	$P(1+r)^t$	$A(t)$
0	100	100
1	100(1.05)	105
2	$100(1.05)^2$	110.25
3	$100(1.05)^3$	115.76

TABLE 4.1

In general, for an initial investment of P dollars at an interest rate r compounded annually, we have the following formula for the amount accumulated after t years.

Compound Interest

The **amount $A(t)$** accumulated (principal plus interest) in an account bearing interest compounded annually is

$$A(t) = P(1+r)^t$$

where	P	is the principal invested,
	r	is the interest rate,
	t	is the time period, in years.

This function describes exponential growth with an initial value of P and a growth factor of $b = 1 + r$. The interest rate r, which indicates the *percent increase* in the account each year, corresponds to a *growth factor* of $1 + r$. The notion of percent increase is often used to describe the growth factor for quantities that grow exponentially.

EXAMPLE 3

During a period of rapid inflation, prices rose by 12% over 6 months. At the beginning of the inflationary period, a pound of butter cost $2.

a. Make a table of values showing the rise in the cost of butter over the next 2 years.

b. Write a function that gives the price of a pound of butter t years after inflation began.

c. How much did a pound of butter cost after 3 years? After 15 months?

d. Graph the function you found in part (b).

Solutions

a. The *percent increase* in the price of butter is 12% every 6 months. Therefore, the *growth factor* for the price of butter is $1 + 0.12 = 1.12$ every half-year. If $P(t)$ represents the price of butter after t years, then $P(0) = 2$, and every half-year we multiply the price by 1.12, as shown in Table 4.2.

t	$P(t)$		
0	2	$\Big)\times 1.12$	$P(0) = 2.00$
$\frac{1}{2}$	$2(1.12)$	$\Big)\times 1.12$	$P\!\left(\frac{1}{2}\right) = 2.24$
1	$2(1.12)^2$	$\Big)\times 1.12$	$P(1) = 2.51$
$\frac{3}{2}$	$2(1.12)^3$	$\Big)\times 1.12$	$P\!\left(\frac{3}{2}\right) = 2.81$
2	$2(1.12)^4$	$\Big)\times 1.12$	$P(2) = 3.15$

TABLE 4.2

b. Look closely at the second column of Table 4.2. After t years of inflation, the original price of $2 has been multiplied by $2t$ factors of 1.12. Thus,

$$P = 2(1.12)^{2t}$$

c. To find the price of butter at any time after inflation began, we evaluate the function at the appropriate value of t.

$$P(3) = 2(1.12)^{2(3)}$$
$$= 2(1.12)^6 \approx 3.95$$

After 3 years, the price was $3.95. Fifteen months is 1.25 years, so we evaluate $P(1.25)$.

$$P(1.25) = 2(1.12)^{2(1.25)}$$
$$= 2(1.12)^{2.5} \approx 2.66$$

After 15 months, the price of butter was $2.66.

d. Evaluate the function

$$P(t) = 2(1.12)^{2t}$$

for several values, as shown in Table 4.3. Plot the points and connect them with a smooth curve to obtain the graph shown in Figure 4.1.

t	$P(t)$
0	2.00
1	2.51
2	3.15
3	3.95
4	4.95

TABLE 4.3

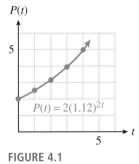

FIGURE 4.1

In Example 3, we can rewrite the formula for $P(t)$ as follows:

$$P(t) = 2(1.12)^{2t}$$
$$= 2[(1.12)^2]^t = 2(1.2544)^t$$

Thus, the annual growth factor for the price of butter is 1.2544, and the annual percent growth rate is 25.44%.

In 1998, the average annual cost of attending a public college was $10,069, and costs were climbing by 6% per year.

a. Write a formula for $C(t)$, the cost of one year of college t years after 1998.

b. Complete the table and sketch a graph of $C(t)$.

t	0	5	10	15	20	25
$C(t)$						

c. If the percent growth rate remained steady, how much did a year of college cost in 2005?

d. If the percent growth rate continues to remain steady, how much will a year of college cost in 2020?

Exponential Decay

In the preceding examples, exponential growth was modeled by increasing functions of the form

$$P(t) = P_0 b^t$$

where $b > 1$. The function $P(t) = P_0 b^t$ is a *decreasing* function if $0 < b < 1$. In this case, we say that the function describes exponential decay, and the constant b is called the **decay factor**. In Investigation 8, we consider two examples of exponential decay.

Investigation 8

Exponential Decay

A. A small coal-mining town has been losing population since 1940, when 5000 people lived there. At each census thereafter (taken at 10-year intervals), the population declined to approximately 0.90 of its earlier figure.

1. Fill in the table showing the population $P(t)$ of the town t years after 1940.

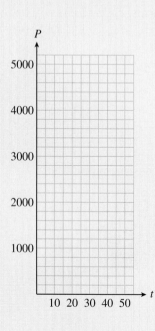

t	$P(t)$
0	5000
10	
20	
30	
40	
50	

$P(0) = 5000$

$P(10) = 5000 \cdot 0.90 =$

$P(20) = [5000 \cdot 0.90] \cdot 0.90 =$

$P(30) =$

$P(40) =$

$P(50) =$

2. Plot the data points and connect them with a smooth curve.

3. Write a function that gives the population of the town at any time t in years after 1940. (*Hint:* Express the values you calculated in part (1) using powers of 0.90. Do you see a connection between the value of t and the exponent on 0.90?)

4. Graph your function from part (3) using a calculator. (Use the table to choose an appropriate domain and range.) The graph should resemble your hand-drawn graph from part (2).

5. Evaluate your function to find the population of the town in 1995. What was the population in 2000?

B. A plastic window coating 1 millimeter thick decreases the light coming through a window by 25%. This means that 75% of the original amount of light comes through 1 millimeter of the coating. Each additional millimeter of coating reduces the light by another 25%.

1. Fill in the table showing the percent of the light, $P(x)$, that shines through x millimeters of the window coating.

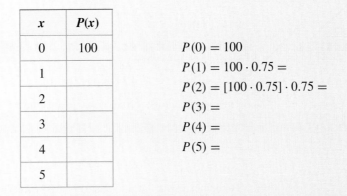

x	$P(x)$
0	100
1	
2	
3	
4	
5	

$P(0) = 100$
$P(1) = 100 \cdot 0.75 =$
$P(2) = [100 \cdot 0.75] \cdot 0.75 =$
$P(3) =$
$P(4) =$
$P(5) =$

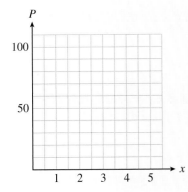

2. Plot the data points and connect them with a smooth curve.

3. Write a function that gives the percent of the light that shines through x millimeters of the coating. (*Hint:* Express the values you calculated in part (1) using powers of 0.75. Do you see a connection between the value of x and the exponent on 0.75?)

4. Graph your function from part (3) using a calculator. (Use your table of values to choose an appropriate domain and range.) The graph should resemble your hand-drawn graph from part (2).

5. Evaluate your function to find the percent of the light that comes through 6 millimeters of plastic coating. What percent comes through $\frac{1}{2}$ millimeter?

Decay Factors

Before Example 3, we noted that a percent increase of r (in decimal form) corresponds to a growth factor of $b = 1 + r$. A percent decrease of r corresponds to a decay factor of $b = 1 - r$. In Part B of Investigation 8, each millimeter of plastic reduced the amount of light by 25%, so $r = 0.25$, and the decay factor for the function $P(x)$ is

$$b = 1 - r$$
$$= 1 - 0.25 = 0.75$$

EXAMPLE 4 David Reed writes in *Context* magazine: "Computing prices have been falling exponentially—50% every 18 months—for the past 30 years and will probably stay on that curve for another couple of decades." An accounting firm invests $50,000 in new computer equipment.

a. Write a formula for the value of the equipment t years from now.

b. By what percent does the equipment depreciate each year?

c. What will the equipment be worth in 5 years?

Solutions   **a.** The initial value of the equipment is $V_0 = 50,000$. Every 18 months, the value of the equipment is multiplied by

$$b = 1 - r = 1 - 0.50 = 0.50$$

However, because 18 months is 1.5 years, we must divide t by 1.5 in our formula, giving us

$$V(t) = 50,000(0.50)^{t/1.5}$$

b. After 1 year, we have

$$V(1) = 50,000(0.50)^{1/1.5} = 50,000(0.63)$$

The equipment is worth 63% of its original value, so it has depreciated by $1 - 0.63$, or 37%.

c. After 5 years,

$$V(5) = 50,000(0.50)^{5/1.5} = 4960.628$$

To the nearest dollar, the equipment is worth \$4961.

EXERCISE 4

The number of butterflies visiting a nature station is declining by 18% per year. In 1998, 3600 butterflies visited the nature station.

a. What is the decay factor in the annual butterfly count?

b. Write a formula for $B(t)$, the number of butterflies t years after 1998.

c. Complete the table and sketch a graph of $B(t)$.

t	0	2	4	6	8	10
$B(t)$						

We summarize our observations about exponential growth and decay functions as follows.

Exponential Growth and Decay

The function

$$P(t) = P_0 b^t$$

models exponential growth and decay.

$P_0 = P(0)$ is the **initial value** of P;

b is the **growth** or **decay factor.**

1. If $b > 1$, then $P(t)$ is increasing, and $b = 1 + r$, where r represents percent increase.

2. If $0 < b < 1$, then $P(t)$ is decreasing, and $b = 1 - r$, where r represents percent decrease.

Comparing Linear Growth and Exponential Growth

It may be helpful to compare linear growth and exponential growth. Consider the two functions

$$L(t) = 5 + 2t \quad \text{and} \quad E(t) = 5 \cdot 2^t \quad (t \geq 0)$$

whose graphs are shown in Figure 4.2.

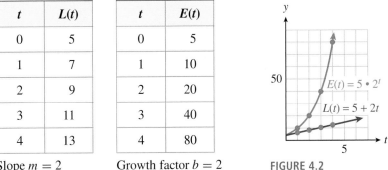

t	$L(t)$
0	5
1	7
2	9
3	11
4	13

Slope $m = 2$
TABLE 4.4

t	$E(t)$
0	5
1	10
2	20
3	40
4	80

Growth factor $b = 2$

FIGURE 4.2

L is a linear function with initial value 5 and slope 2; E is an exponential function with initial value 5 and growth factor 2. In a way, the growth factor of an exponential function is analogous to the slope of a linear function: Each measures how quickly the function is increasing (or decreasing). However, for each unit increase in t, 2 units are *added* to the value of $L(t)$, whereas the value of $E(t)$ is *multiplied* by 2. An exponential function with growth factor 2 eventually grows much more rapidly than a linear function with slope 2, as you can see by comparing the graphs in Figure 4.2 or the function values in Table 4.4.

EXAMPLE 5

A solar energy company sold $80,000 worth of solar collectors last year, its first year of operation. This year its sales rose to $88,000, an increase of 10%. The marketing department must estimate its projected sales for the next 3 years.

  a. If the marketing department predicts that sales will grow linearly, what should it expect the sales total to be next year? Graph the projected sales figures over the next 3 years, assuming that sales will grow linearly.

  b. If the marketing department predicts that sales will grow exponentially, what should it expect the sales total to be next year? Graph the projected sales figures over the next 3 years, assuming that sales will grow exponentially.

Solutions **a.** Let $L(t)$ represent the company's total sales t years after starting business, where $t = 0$ is the first year of operation. If sales grow linearly, then $L(t)$ has the form $L(t) = mt + b$. Now $L(0) = 80,000$, so the intercept b is 80,000. The slope m of the graph is

$$\frac{\Delta S}{\Delta t} = \frac{8000 \text{ dollars}}{1 \text{ year}} = 8000$$

where $\Delta S = 8000$ is the increase in sales during the first year. Thus, $L(t) = 8000t + 80{,}000$, and sales grow by *adding* $8000 each year. The expected sales total for the next year is

$$L(2) = 8000(2) + 80{,}000 = 96{,}000$$

b. Let $E(t)$ represent the company's sales assuming that sales will grow exponentially. Then $E(t)$ has the form $E(t) = E_0 b^t$. The percent increase in sales over the first year was $r = 0.10$, so the growth rate is

$$b = 1 + r = 1.10$$

The initial value, E_0, is 80,000. Thus, $E(t) = 80{,}000(1.10)^t$, and sales grow by being *multiplied* each year by 1.10. The expected sales total for the next year is

$$E(2) = 80{,}000(1.10)^2 = 96{,}800$$

Evaluate each function at several points (as shown in Table 4.5) to obtain the graphs shown in Figure 4.3.

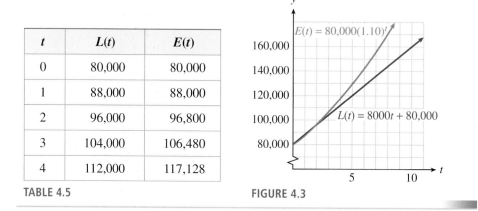

t	$L(t)$	$E(t)$
0	80,000	80,000
1	88,000	88,000
2	96,000	96,800
3	104,000	106,480
4	112,000	117,128

TABLE 4.5   FIGURE 4.3

EXERCISE 5

A new car begins to depreciate in value as soon as you drive it off the lot. Some models depreciate linearly, and others depreciate exponentially. Suppose you buy a new car for $20,000, and 1 year later its value has decreased to $17,000.

a. If the value decreased linearly, what was its annual rate of decrease?

b. If the value decreased exponentially, what was its annual decay factor? What was its annual percent depreciation?

c. Calculate the value of your car when it is 5 years old under each assumption, linear or exponential depreciation.

ANSWERS TO 4.1 EXERCISES

1. a. $P(t) = 24 \cdot 3^t$ **b.** 17,496; 55

2. a. $2^{1/5}$ **b.** 5833 and 6700   **c.** $I(t) = 5078 \cdot 2^{t/5}$

3. a. $C(t) = 10{,}069 \cdot 1.06^t$

b.

t	0	5	10	15	20	25
$C(t)$	10,069	13,475	18,032	24,131	32,293	43,215

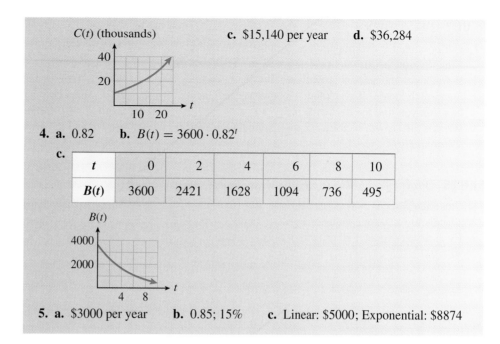

4. a. 0.82   **b.** $B(t) = 3600 \cdot 0.82^t$

c.

t	0	2	4	6	8	10
$B(t)$	3600	2421	1628	1094	736	495

5. a. $3000 per year   **b.** 0.85; 15%   **c.** Linear: $5000; Exponential: $8874

SECTION 4.1 SUMMARY

VOCABULARY *Look up the definitions of new terms in the Glossary.*

Exponential growth   Initial value   Exponential decay   Principal
Percent increase   Compound interest   Growth factor   Decay factor
Amount

CONCEPTS

1. If a quantity is multiplied by a constant factor, b, in each time period, we say that it undergoes **exponential growth** or **decay.** The constant b is called the **growth factor** if $b > 1$ and the **decay factor** if $0 < b < 1$.

2. Quantities that increase or decrease by a constant percent in each time period grow or decay exponentially.

3.

Exponential Growth and Decay

The function

$$P(t) = P_0 \, b^t$$

models exponential growth and decay.

$P_0 = P(0)$ is the **initial value** of P;

b is the **growth** or **decay factor.**

1. If $b > 1$, then $P(t)$ is increasing, and $b = 1 + r$, where r represents percent increase.

2. If $0 < b < 1$, then $P(t)$ is decreasing, and $b = 1 - r$, where r represents percent decrease.

4.

Compound Interest

The **amount** $A(t)$ accumulated (principal plus interest) in an account bearing interest compounded annually is

$$A(t) = P(1 + r)^t$$

where P is the principal invested,

r is the interest rate,

t is the time period, in years.

5. In linear growth, a constant amount is *added* to the output for each unit increase in the input. In exponential growth, the output is *multiplied* by a constant factor for each unit increase in the input.

STUDY QUESTIONS

1. Is it possible for two populations with the same initial value to grow at different percent rates?
2. If you know the percent growth rate, how can you find the growth factor? If you know the percent decay rate, how can you find the decay factor?
3. What is the growth factor for a population that grows 4% annually?
4. What is the decay factor for a population that declines by 4% annually?
5. What is the growth factor for a population that grows by 100% annually?
6. Explain the difference between the slope in linear growth and the growth factor in exponential growth.

SKILLS *Practice each skill in the Homework problems listed.*

1. Calculate percent increase or decrease: #1–10
2. Write a formula for exponential growth or decay: #11–22
3. Evaluate an exponential growth or decay function: #11–22
4. Simplify exponential expressions: #23–32
5. Solve power equations: #33–40
6. Find the growth factor or initial value: #41–58
7. Solve for percent increase or decrease: #63–66

HOMEWORK 4.1

ThomsonNOW™ Test yourself on key content at www.thomsonedu.com/login.

1. **a.** A parking permit at Huron College cost $25 last year, but this year the price increased by 12%. What is the price this year?
   b. If the price of a parking permit increases by 12% again next year, what will the price be then?

2. **a.** The computer you want cost $1200 when it first came on the market, but after 3 months the price was reduced by 15%. What was the price then?
   b. If the price falls by another 15% next month, what will the price be then?

3. The value of your stock portfolio fell 10% last year, but this year it increased by 10%. How does the current value of your portfolio compare to what it was two years ago?

4. You got a 5% raise in January, but then in March everyone took a pay cut of 5%. How does your new salary compare to what it was last December?

5. The population of Summerville is currently 12 hundred people.
   a. Write a formula for the population if it grows at a constant rate of 1.5 hundred people per year. What is the population after 3 years?
   b. Write a formula for the population if it has a constant growth factor of 1.5 per year. What is the population after 3 years?

6. Delbert's sports car was worth $45,000 when he bought it.
   a. Write a formula for the value of the car if it depreciates at a constant rate of $7000 per year. What is the value of the car after 4 years?
   b. Write a formula for the value of the car if it has a constant depreciation factor of 0.70 per year. What is the value of the car after 4 years?

7. Francine's truck was worth $18,000 when she bought it.
   a. Write a formula for the value of the truck if it depreciates by $2000 per year. What is the value of the truck after 5 years?
   b. Write a formula for the value of the truck if it depreciates by 20% per year. What is the value of the truck after 5 years?

8. The population of Lakeview is currently 150,000 people.
   a. Write a formula for the population if it grows by 6000 people per year. What is the population after 2 years?
   b. Write a formula for the population if grows by 6% per year. What is the population after 2 years?

9. The table shows the growth factor for a number of different populations. For each population, find the percent growth rate.

Population	A	B	C	D	E
Growth factor	1.2	1.02	1.075	2.0	2.15
Percent growth rate					

10. The table shows the decay factor for a number of different populations. For each population, find the percent decay rate.

Population	A	B	C	D	E
Decay factor	0.6	0.06	0.96	0.996	0.096
Percent decay rate					

For Problems 11–16,
(a) Write a function that describes exponential growth.
(b) Use your calculator to graph the function.
(c) Evaluate the function at the given values.

11. A typical beehive contains 20,000 insects. The population can increase in size by a factor of 2.5 every 6 weeks. How many bees could there be after 4 weeks? After 20 weeks?

12. A rancher who started with 800 head of cattle finds that his herd increases by a factor of 1.8 every 3 years. How many head of cattle will he have after 1 year? After 10 years?

13. A sum of $4000 is invested in an account that pays 8% interest compounded annually. How much is in the account after 2 years? After 10 years?

14. Otto invests $600 in an account that pays 7.3% interest compounded annually. How much is in Otto's account after 3 years? After 6 years?

15. Paul bought a house for $200,000 in 1983. Since 1983, housing prices have risen an average of 5% per year. How much was the house worth in 1995? How much will it be worth in 2010?

16. Sales of Windsurfers have increased 12% per year since 1990. If Sunsails sold 1500 Windsurfers in 1990, how many did it sell in 1995? How many should it have expected to sell in 2002?

For Problems 17–22,
(a) Write a function that describes exponential decay.
(b) Graph the function.
(c) Evaluate the function at the given values.

17. During a vigorous spraying program, the mosquito population was reduced to $\frac{3}{4}$ of its previous size every 2 weeks. If the mosquito population was originally estimated at 250,000, how many mosquitoes remained after 3 weeks of spraying? After 8 weeks?

18. The number of perch in Hidden Lake has declined to half of its previous value every 5 years since 1970, when the perch population was estimated at 8000. How many perch were there in 1980? In 1998?

19. Scuba divers find that the water in Emerald Lake filters out 15% of the sunlight for each 4 feet that they descend. How much sunlight penetrates to a depth of 20 feet? To a depth of 45 feet?

20. Arch's motorboat cost $15,000 in 1990 and has depreciated by 10% every 3 years. How much was the boat worth in 1999? In 2000?

21. Plutonium-238 is a radioactive element that decays over time into a less harmful element at a rate of 0.8% per year. A power plant has 50 pounds of plutonium-238 to dispose of. How much plutonium-238 will be left after 10 years? After 100 years?

22. Iodine-131 is a radioactive element that decays at a rate of 8.3% per day. How much of a 12-gram sample will be left after 1 week? After 15 days?

In Problems 23–26, use the laws of exponents to simplify.

23. a. $3^x 3^4$ **b.** $(3^x)^4$ **c.** $3^x 4^x$

24. a. $8^x 8^x$ **b.** $8^{x+2} 8^{x-1}$ **c.** $\dfrac{8^{2x}}{8^x}$

25. a. $b^{-4t} b^{2t}$ **b.** $(b^t)^{1/2}$ **c.** $b^{t-1} b^{1-t}$

26. a. $b^{t/2} b^{t/2}$ **b.** $\dfrac{b^{2t}}{b}$ **c.** $b^{1/t} b^t$

27. Let $P(t) = 12(3)^t$. Show that $P(t+1) = 3P(t)$.

28. Let $N(t) = 8(5)^t$. Show that $\dfrac{N(t+k)}{N(t)} = 5^k$.

29. Let $P(x) = P_0 a^x$. Show that $P(x+k) = a^k P(x)$.

30. Let $N(x) = N_0 b^x$. Show that $\dfrac{N(x+1)}{N(x)} = b$.

31. a. Explain why $P(t) = 2 \cdot 3^t$ and $Q(t) = 6^t$ are not the same function.
    b. Complete the table of values for P and Q, showing that their values are not the same.

t	0	1	2
$P(t)$			
$Q(t)$			

32. a. Explain why $P(t) = 4 \cdot \left(\frac{1}{2}\right)^t$ and $Q(t) = 2^t$ are not the same function.
    b. Complete the table of values for P and Q, showing that their values are not the same.

t	0	1	2
$P(t)$			
$Q(t)$			

Solve each equation. (See Section 3.3 to review solving equations involving powers of the variable.) Round your answer to two places if necessary.

33. $768 = 12b^3$

34. $1875 = 3b^4$

35. $14{,}929.92 = 5000b^6$

36. $151{,}875 = 20{,}000b^5$

37. $1253 = 260(1+r)^{12}$

38. $116{,}473 = 48{,}600(1+r)^{15}$

39. $56.27 = 78(1-r)^8$

40. $10.56 = 12.4(1-r)^{20}$

41. a. Riverside County is the fastest growing county in California. In 2000, the population was 1,545,387. Write a formula for the population of Riverside County. (You do not know the value of the growth factor, b, yet.)
    b. In 2004, the population had grown to 1,871,950. Find the growth factor and the percent rate of growth.
    c. Predict the population of Riverside County in 2010.

42. a. In 2006, a new Ford Focus cost $15,574. The value of a Focus decreases exponentially over time. Write a formula for the value of a Focus. (You do not know the value of the decay factor, b, yet.)
    b. A 2-year old Focus cost $11,788. Find the decay factor and the percent rate of depreciation.
    c. How much would a 4-year old Focus cost?

43. In the 1940s, David Lack undertook a study of the European robin. He tagged 130 one-year-old robins and found that on average 35.6% of the birds survived each year. (Source: Burton, 1998)

   a. According to the data, how many robins would have originally hatched to produce 130 one-year-olds?

   b. Write a formula for the number of the original robins still alive after t years.

   c. Graph your function.

   d. One of the original robins actually survived for 9 years. How many robins does the model predict will survive for 9 years?

44. Many insects grow by discrete amounts each time they shed their exoskeletons. Dyar's rule says that the size of the insect increases by a constant ratio at each stage. (Source: Burton, 1998)

   a. Dyar measured the width of the head of a caterpillar of a swallowtail butterfly at each stage. The caterpillar's head was initially approximately 42 millimeters wide, and 63.84 millimeters wide after its first stage. Find the growth ratio.

   b. Write a formula for the width of the caterpillar's head at the nth stage.

   c. Graph your function.

   d. What head width does the model predict after 5 stages?

For Problems 45–54,
(a) Each table describes exponential growth or decay. Find the growth or decay factor.
(b) Complete the table. Round values to two decimal places if necessary.

45.

t	0	1	2	3	4
P	8	12	18		

46.

t	0	1	2	3	4
P	4	5	6.25		

47.

x	0	1	2	3	4
Q	20	24			

48.

x	0	1	2	3	4
Q	100	105			

49.

w	0	1	2	3	4
N	120	96			

50.

w	0	1	2	3	4
N	640	480			

51.

t	0	1	2	3	4
C	10		6.4		

52.

t	0	1	2	3	4
C	20			2.5	

53.

n	0	1	2	3	4
B	200			266.2	

54.

n	0	1	2	3	4
B	40		62.5		

Each graph in Problems 55–58 represents exponential growth or decay.
(a) Find the initial value and the growth or decay factor.
(b) Write a formula for the function.

55.

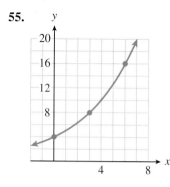

56.

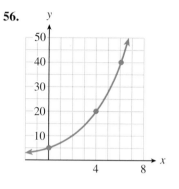

57.

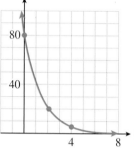

58.

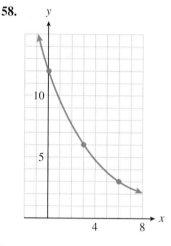

59. If 8% of the air leaks out of Brian's bicycle tire every day, what percent of the air will be left after 2 days? After a week?

60. If housing prices are increasing by 15% per year, by what percent will they increase in 2 years? In 3 years?

61. Francine says that if a population grew by 48% in 6 years, then it grew by 8% per year. Is she correct? Either justify or correct her calculation.

62. Delbert says that if a population decreased by 60% in 5 years, then it decreased by 12% per year. Is he correct? Either justify or correct his calculation.

In Problems 63–66, assume that each population grows exponentially with constant annual percent increase, _r_.

63. a. The population of the state of Texas was 16,986,335 in 1990. Write a formula in terms of r for the population of Texas t years later.
b. In 2000, the population was 20,851,820. Write an equation and solve for r. What was the annual percent increase to the nearest hundredth of a percent?

64. a. The population of the state of Florida was 12,937,926 in 1990. Write a formula in terms of r for the population of Florida t years later.
b. In 2000, the population was 15,982,378. Write an equation and solve for r. What was the annual percent increase to the nearest hundredth of a percent?

65. **a.** The population of Rainville was 10,000 in 1980 and doubled in 20 years. What was the annual percent increase to the nearest hundredth percent?
   b. The population of Elmira was 350,000 in 1980 and doubled in 20 years. What was the annual percent increase to the nearest hundredth of a percent?
   c. If a population doubles in 20 years, does the percent increase depend on the size of the original population?
   d. The population of Grayling doubled in 20 years. What was the annual percent increase to the nearest hundredth of a percent?

66. **a.** The population of Boomtown was 300 in 1908 and tripled in 7 years. What was the annual percent increase to the nearest hundredth of a percent?
   b. The population of Fairview was 15,000 in 1962 and tripled in 7 years. What was the annual percent increase to the nearest hundredth of a percent?
   c. If a population triples in 7 years, does the percent increase depend on the size of the original population?
   d. The population of Pleasant Lake tripled in 7 years. What was the annual percent increase to the nearest hundredth of a percent?

67. A researcher starts 2 populations of fruit flies of different species, each with 30 flies. Species A increases by 30% in 6 days and species B increases by 20% in 4 days.
   a. What was the population of species A after 6 days? Find the daily growth factor for species A.
   b. What was the population of species B after 4 days? Find the daily growth factor for species B.
   c. Which species multiplies more rapidly?

68. A biologist isolates two strains of a particular virus and monitors the growth of each, starting with samples of 0.01 gram. Strain A increases by 10% in 8 hours and strain B increases by 12% in 9 hours.
   a. How much did the sample of strain A weigh after 8 hours? What was its hourly growth factor?
   b. How much did the sample of strain B weigh after 9 hours? What was its hourly growth factor?
   c. Which strain of virus grows more rapidly?

In Problems 69–72, we compare linear and exponential growth.

69. At a large university 3 students start a rumor that final exams have been canceled. After 2 hours, 6 students (including the first 3) have heard the rumor.
   a. Assuming that the rumor grows linearly, complete the table below for $L(t)$, the number of students who have heard the rumor after t hours. Then write a formula for the function $L(t)$. Graph the function.

t	0	2	4	6	8
$L(t)$					

   b. Complete the table below, assuming that the rumor grows exponentially. Write a formula for the function $E(t)$ and graph it on the same set of axes with $L(t)$.

t	0	2	4	6	8
$E(t)$					

70. Over the weekend, the Midland Infirmary identifies 4 cases of Asian flu. It has treated a total of 10 cases 3 days later.
   a. Assuming that the number of flu cases grows linearly, complete the table below for $L(t)$, the number of people infected after t days. Then write a formula for the function $L(t)$. Graph the function.

t	0	3	6	9	12
$L(t)$					

   b. Complete the table below assuming that the flu grows exponentially. Write a formula for the function $E(t)$ and graph it on the same set of axes with $L(t)$.

t	0	3	6	9	12
$E(t)$					

71. The world's population of tigers declined from 10,400 in 1980 to 6000 in 1998.
  a. If the population declined linearly, what was its annual rate of decrease?
  b. If the population declined exponentially, what was its annual decay factor? What was its annual percent decrease?
  c. Predict the number of tigers in 2010 under each assumption, linear or exponential decline.

72. In 2003, the Center for Biological Diversity filed a lawsuit against the federal government for failing to protect Alaskan sea otters. The population of sea otters, which numbered between 150,000 and 300,000 before hunting began in 1741, declined from about 20,000 in 1992 to 6000 in 2000. (Source: Center for Biological Diversity)
  a. If the population declined linearly after 1992, what was its annual rate of decrease?
  b. If the population declined exponentially after 1992, what was its annual decay factor? What was its annual percent decrease?
  c. Predict the number of sea otters in 2010 under each assumption, linear or exponential decline.

4.2 Exponential Functions

In Section 4.1, we studied functions that describe exponential growth or decay. More formally, we define an **exponential function** as follows,

> **Exponential Function**
>
> $f(x) = ab^x$, where $b > 0$ and $b \neq 1$, $a \neq 0$

Some examples of exponential functions are

$$f(x) = 5^x, \quad P(t) = 250(1.7)^t, \quad \text{and} \quad g(n) = 2.4(0.3)^n$$

The constant a is the y-intercept of the graph because

$$f(0) = a \cdot b^0 = a \cdot 1 = a$$

For the examples above, we find that the y-intercepts are

$$f(0) = 5^0 = 1,$$
$$P(0) = 250(1.7)^0 = 250, \quad \text{and}$$
$$g(0) = 2.4(0.3)^0 = 2.4$$

The positive constant b is called the base of the exponential function. We do not allow b to be negative, because if $b < 0$, then b^x is not a real number for some values of x. For example, if $b = -4$ and $f(x) = (-4)^x$, then $f(1/2) = (-4)^{1/2}$ is an imaginary number. We also exclude $b = 1$ as a base because $1^x = 1$ for all values of x; hence the function $f(x) = 1^x$ is actually the constant function $f(x) = 1$.

Graphs of Exponential Functions

The graphs of exponential functions have two characteristic shapes, depending on whether the base, b, is greater than 1 or less than 1. As typical examples, consider the graphs of $f(x) = 2^x$ and $g(x) = \left(\frac{1}{2}\right)^x$ shown in Figure 4.4. Some values for f and g are recorded in Table 4.6.

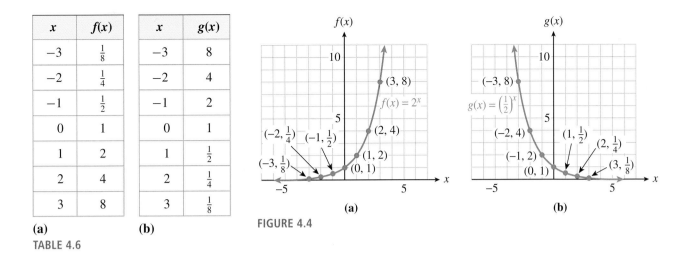

x	f(x)
−3	$\frac{1}{8}$
−2	$\frac{1}{4}$
−1	$\frac{1}{2}$
0	1
1	2
2	4
3	8

(a)

x	g(x)
−3	8
−2	4
−1	2
0	1
1	$\frac{1}{2}$
2	$\frac{1}{4}$
3	$\frac{1}{8}$

(b)

TABLE 4.6

FIGURE 4.4

Notice that $f(x) = 2^x$ is an increasing function and $g(x) = \left(\frac{1}{2}\right)^x$ is a decreasing function. Both are concave up. In general, exponential functions have the following properties.

> **Properties of Exponential Functions, $f(x) = ab^x$, $a > 0$**
>
> 1. Domain: all real numbers.
> 2. Range: all positive numbers.
> 3. If $b > 1$, the function is increasing and concave up;
>   if $0 < b < 1$, the function is decreasing and concave up.
> 4. The y-intercept is $(0, a)$. There is no x-intercept.

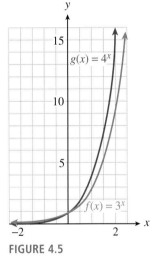

FIGURE 4.5

In Table 4.6a, you can see that as the x-values decrease toward negative infinity, the corresponding y-values decrease toward zero. As a result, the graph of f decreases toward the x-axis as we move to the left. Thus, the negative x-axis is a horizontal asymptote for exponential functions with $b > 1$, as shown in Figure 4.4a. For exponential functions with $0 < b < 1$, the positive x-axis is an asymptote, as illustrated in Figure 4.4b. (See Section 2.2 to review asymptotes.)

In Example 1, we compare two increasing exponential functions. The larger the value of the base, b, the faster the function grows. In this example, both functions have $a = 1$.

EXAMPLE 1 Compare the graphs of $f(x) = 3^x$ and $g(x) = 4^x$.

Solution Evaluate each function for several convenient values, as shown in Table 4.7.

Plot the points for each function and connect them with smooth curves. For positive x-values, $g(x)$ is always larger than $f(x)$, and is increasing more rapidly. In Figure 4.5, $g(x) = 4^x$ climbs more rapidly than $f(x) = 3^x$. Both graphs cross the y-axis at $(0, 1)$.

x	f(x)	g(x)
−2	$\frac{1}{9}$	$\frac{1}{16}$
−1	$\frac{1}{3}$	$\frac{1}{4}$
0	1	1
1	3	4
2	9	16

TABLE 4.7

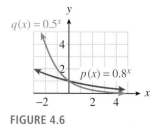

FIGURE 4.6

For decreasing exponential functions, those with bases between 0 and 1, the smaller the base, the more steeply the graph decreases. For example, compare the graphs of $p(x) = 0.8^x$ and $q(x) = 0.5^x$ shown in Figure 4.6.

EXERCISE 1

a. State the ranges of the functions f and g in Figure 4.5 from Example 1 on the domain $[-2, 2]$.

b. State the ranges of the functions p and q shown in Figure 4.6 on the domain $[-2, 2]$. Round your answers to two decimal places.

Transformations of Exponential Functions

See Section 2.3 to review transformations of functions.

In Chapter 2, we considered transformations of the basic graphs. For instance, the graphs of the functions $y = x^2 - 4$ and $y = (x - 4)^2$ are shifts of the basic parabola, $y = x^2$. In a similar way, we can shift or stretch the graph of an exponential function while the basic shape is preserved.

EXAMPLE 2

Use your calculator to graph the following functions. Describe how these graphs compare with the graph of $h(x) = 2^x$.

a. $f(x) = 2^x + 3$ **b.** $g(x) = 2^{x+3}$

Solutions   Enter the formulas for the three functions as shown below. Note the parentheses around the exponent in the keying sequence for $Y_3 = g(x)$

$$Y_1 = 2\,\boxed{\wedge}\,X$$
$$Y_2 = 2\,\boxed{\wedge}\,X\,\boxed{+}\,3$$
$$Y_3 = 2\,\boxed{\wedge}\,\boxed{(}\,X\,\boxed{+}\,3\,\boxed{)}$$

The graphs of $h(x) = 2^x$, $f(x) = 2^x + 3$, and $g(x) = 2^{x+3}$ are shown using the standard window in Figure 4.7.

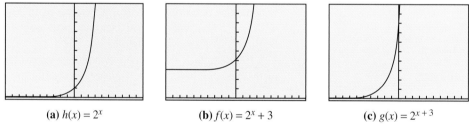

(a) $h(x) = 2^x$ (b) $f(x) = 2^x + 3$ (c) $g(x) = 2^{x+3}$

FIGURE 4.7

a. The graph of $f(x) = 2^x + 3$, shown in Figure 4.7b, has the same basic shape as that of $h(x) = 2^x$, but it has a horizontal asymptote at $y = 3$ instead of at $y = 0$ (the x-axis). In fact, $f(x) = h(x) + 3$, so the graph of f is a vertical translation of the graph of h by 3 units. If every point on the graph of $h(x) = 2^x$ is moved 3 units upward, the result is the graph of $f(x) = 2^x + 3$.

b. First note that $g(x) = 2^{x+3} = h(x + 3)$. In fact, the graph of $g(x) = 2^{x+3}$ shown in Figure 4.7c has the same basic shape as $h(x) = 2^x$ but has been translated 3 units to the left.

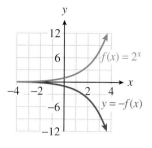

(a)

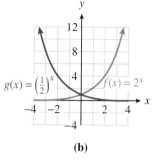

(b)

FIGURE 4.8

Recall that the graph of $y = -f(x)$ is the reflection about the x-axis of the graph of $y = f(x)$. The graphs of $y = 2^x$ and $y = -2^x$ are shown in Figure 4.8a. You may have also noticed a relationship between the graphs of $f(x) = 2^x$ and $g(x) = \left(\frac{1}{2}\right)^x$, which are shown in Figure 4.8b.

The graph of g is the reflection of the graph of f about the y-axis. We can see why this is true by writing the formula for $g(x)$ in another way:

$$g(x) = \left(\frac{1}{2}\right)^x = (2^{-1})^x = 2^{-x}$$

We see that $g(x)$ is the same function as $f(-x)$. Replacing x by $-x$ in the formula for a function switches every point (p, q) on the graph with the point $(-p, q)$ and thus reflects the graph about the y-axis.

Reflections of Graphs

1. The graph of $y = -f(x)$ is the reflection of the graph of $y = f(x)$ about the x-axis.
2. The graph of $y = f(-x)$ is the reflection of the graph of $y = f(x)$ about the y-axis.

EXERCISE 2

Which of the functions below have the same graph? Explain why.

a. $f(x) = \left(\frac{1}{4}\right)^x$ **b.** $g(x) = -4^x$ **c.** $h(x) = 4^{-x}$

Comparing Exponential and Power Functions

Exponential functions are not the same as the power functions we studied in Chapter 3. Although both involve expressions with exponents, it is the location of the variable that makes the difference.

	Power Functions	**Exponential Functions**
General formula	$h(x) = kx^p$	$f(x) = ab^x$
Description	variable base and constant exponent	constant base and variable exponent
Example	$h(x) = 2x^3$	$f(x) = 2(3^x)$

These two families of functions have very different properties, as well.

EXAMPLE 3 Compare the power function $h(x) = 2x^3$ and the exponential function $f(x) = 2(3^x)$.

Solution

First, compare the values for these two functions in Table 4.8. The scaling exponent for $h(x)$ is 3, so that when x doubles, say, from 1 to 2, the output is multiplied by 2^3, or 8. On the other hand, we can tell that f is exponential because its values increase by a factor of 3 for each unit increase in x. (To see this, divide any function value by the previous one. For example, $54 \div 18 = 3$, and $18 \div 6 = 3$.)

As you would expect, the graphs of the two functions are also quite different. For starters, note that the power function goes through the origin, while the exponential function has y-intercept $(0, 2)$. (See Figure 4.9a.) From the table, we see that $h(3) = f(3) = 54$, so the two graphs intersect at $x = 3$. (They also intersect at approximately $x = 2.48$.) However, if you compare the values of $h(x) = 2x^3$ and $f(x) = 2(3^x)$ for larger values of x, you will see that eventually the exponential function overtakes the power function, as shown in Figure 4.9b.

x	$h(x) = 2x^3$	$f(x) = 2(3^x)$
-3	-54	$\frac{2}{27}$
-2	-16	$\frac{1}{4}$
-1	-2	$\frac{2}{3}$
0	0	2
1	2	6
2	16	18
3	54	54

TABLE 4.8

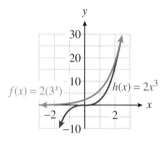

(a)

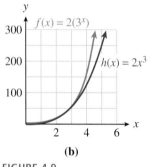

(b)

FIGURE 4.9

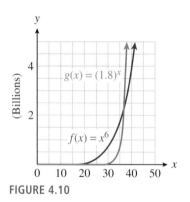

FIGURE 4.10

The relationship in Example 3 holds true for all increasing power and exponential functions: For large enough values of x, the exponential function will always be greater than the power function, regardless of the parameters in the functions. Figure 4.10 shows the graphs of $f(x) = x^6$ and $g(x) = 1.8^x$. At first, $f(x) > g(x)$, but at around $x = 37$, $g(x)$ overtakes $f(x)$, and $g(x) > f(x)$ for all $x > 37$.

EXERCISE 3

Which of the following functions are exponential functions, and which are power functions?

a. $F(x) = 1.5^x$ **b.** $G(x) = 3x^{1.5}$

c. $H(x) = 3^{1.5x}$ **d.** $K(x) = (3x)^{1.5}$

Exponential Equations

An **exponential equation** is one in which the variable is part of an exponent. For example, the equation

$$3^x = 81$$

is exponential. Many exponential equations can be solved by writing both sides of the equation as powers with the same base. To solve the above equation, we write

$$3^x = 3^4$$

which is true if and only if $x = 4$. In general, if two equivalent powers have the same base, then their exponents must be equal also, as long as the base is not 0 or ± 1.

Sometimes the laws of exponents can be used to express both sides of an equation as single powers of a common base.

EXAMPLE 4   Solve the following equations.

a. $3^{x-2} = 9^3$

b. $27 \cdot 3^{-2x} = 9^{x+1}$

Solutions   **a.** Using the fact that $9 = 3^2$, write each side of the equation as a power of 3:

$$3^{x-2} = (3^2)^3$$
$$3^{x-2} = 3^6$$

Now equate the exponents to obtain

$$x - 2 = 6$$
$$x = 8$$

b. Write each factor as a power of 3.

$$3^3 \cdot 3^{-2x} = (3^2)^{x+1}$$

Use the laws of exponents to simplify each side:

$$3^{3-2x} = 3^{2x+2}$$

Now equate the exponents to obtain

$$3 - 2x = 2x + 2$$
$$-4x = -1$$

The solution is $x = \frac{1}{4}$.

EXERCISE 4

Solve the equation $2^{x+2} = 128$.

Write each side as a power of 2.
Equate exponents.

Exponential equations arise frequently in the study of exponential growth.

EXAMPLE 5   During the summer a population of fleas doubles in number every 5 days. If a population starts with 10 fleas, how long will it be before there are 10,240 fleas?

Solution   Let P represent the number of fleas present after t days. The original population of 10 is multiplied by a factor of 2 every 5 days, or

$$P(t) = 10 \cdot 2^{t/5}$$

Set $P = 10{,}240$ and solve for t:

$$10{,}240 = 10 \cdot 2^{t/5} \qquad \text{Divide both sides by 10.}$$
$$1024 = 2^{t/5} \qquad \text{Write 1024 as a power of 2.}$$
$$2^{10} = 2^{t/5}$$

Equate the exponents to get $10 = \frac{t}{5}$, or $t = 50$. The population will grow to 10,240 fleas in 50 days.

During an advertising campaign in a large city, the makers of Chip-O's corn chips estimate that the number of people who have heard of Chip-O's increases by a factor of 8 every 4 days.

a. If 100 people are given trial bags of Chip-O's to start the campaign, write a function, $N(t)$, for the number of people who have heard of Chip-O's after t days of advertising.

b. Use your calculator to graph the function $N(t)$ on the domain $0 \leq t \leq 15$.

c. How many days should the makers run the campaign in order for Chip-O's to be familiar to 51,200 people? Use algebraic methods to find your answer and verify on your graph.

Graphical Solution of Exponential Equations

It is not always so easy to express both sides of the equation as powers of the same base. In the following sections, we will develop more general methods for finding exact solutions to exponential equations. But we can use a graphing calculator to obtain approximate solutions.

EXAMPLE 6   Use the graph of $y = 2^x$ to find an approximate solution to the equation $2^x = 5$ accurate to the nearest hundredth.

Solution   Enter $Y_1 = 2 \wedge X$ and use the standard graphing window (**ZOOM** **6**) to obtain the graph shown in Figure 4.11a. We are looking for a point on this graph with y-coordinate 5. Using the **TRACE** feature, we see that the y-coordinates are too small when $x < 2.1$

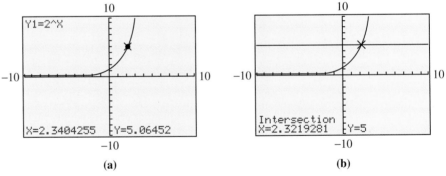

(a)   **(b)**

FIGURE 4.11

and too large when $x > 2.4$. The solution we want lies somewhere between $x = 2.1$ and $x = 2.4$, but this approximation is not accurate enough. To improve our approximation, we will use the **intersect** feature. Set $Y_2 = 5$ and press **GRAPH**. The x-coordinate of the intersection point of the two graphs is the solution of the equation $2^x = 5$ Activating the **intersect** command results in Figure 4.11b, and we see that, to the nearest hundredth, the solution is 2.32.

We can verify that our estimate is reasonable by substituting into the equation:

$$2^{2.32} \stackrel{?}{=} 5$$

We enter $2 \wedge 2.32$ **ENTER** to get 4.993322196. This number is not equal to 5, but it is close, so we believe that $x = 2.32$ is a reasonable approximation to the solution of the equation $2^x = 5$.

See Section 2.1 to review the intersect feature.

Use the graph of $y = 5^x$ to find an approximate solution to $5^x = 285$, accurate to two decimal places.

ANSWERS TO 4.2 EXERCISES

1. a. f: $\left[\frac{1}{9}, 9\right]$; g: $\left[\frac{1}{16}, 16\right]$ **b.** p: $[0.64, 1.56]$; q: $[0.25, 4]$

2. (a) and (c)   **3.** exponential: (a), (c); power: (b), (d)

4. $x = 5$

5. a. $N(t) = 100 \cdot 8^{t/4}$ **b.** 250,000

   c. 12 days

6. $x \approx 3.51$

15

SECTION 4.2 SUMMARY

VOCABULARY   *Look up the definitions of new terms in the Glossary.*

Exponential function   Base   Exponential equation

CONCEPTS

1. An **exponential function** has the form

$$f(x) = ab^x, \quad \text{where} \quad b > 0 \quad \text{and} \quad b \neq 1, a \neq 0$$

2.

> **Properties of Exponential Functions,**
> **$f(x) = ab^x$, $a > 0$**
>
> **1.** Domain: all real numbers.
> **2.** Range: all positive numbers.
> **3.** If $b > 1$, the function is increasing and concave up; if $0 < b < 1$, the function is decreasing and concave up.
> **4.** The y-intercept is $(0, a)$. There is no x-intercept.

3. The graphs of exponential functions can be transformed by shifts, stretches, and reflections.

4.

> **Reflections of Graphs**
>
> **1.** The graph of $y = -f(x)$ is the reflection of the graph of $y = f(x)$ about the x-axis.
> **2.** The graph of $y = f(-x)$ is the reflection of the graph of $y = f(x)$ about the y-axis.

5. Exponential functions $f(x) = ab^x$ have different properties than power functions $f(x) = kx^p$.

6. We can solve some **exponential equations** by writing both sides with the same **base** and equating the exponents.

7. We can use graphs to find approximate solutions to exponential equations.

STUDY QUESTIONS

1. Give the general form for an exponential function. What restrictions do we place on the base of the function?

2. Explain why the output of an exponential function $f(x) = b^x$ is always positive, even if x is negative.

3. How are the graphs of the functions $f(x) = b^x$ and $g(x) = \left(\frac{1}{b}\right)^x$ related?

4. How is an exponential function different from a power function?

5. Delbert says that $8\left(\frac{1}{2}\right)^x$ is equivalent to 4^x. Convince him that he is mistaken.

6. Explain the algebraic technique for solving exponential equations described in this section.

SKILLS *Practice each skill in the Homework problems listed.*

1. Describe the graph of an exponential function: #1–14
2. Graph transformations of exponential functions: #15–18, 53–60
3. Evaluate exponential functions: #19–22
4. Find the equation of an exponential function from its graph: #23–26

5. Solve exponential equations: #27–44
6. Distinguish between power and exponential functions. #45–52, 65 and 66

HOMEWORK 4.2

ThomsonNOW™ Test yourself on key content at www.thomsonedu.com/login.

Find the *y*-intercept of each exponential function and decide whether the graph is increasing or decreasing.

1. **a.** $f(x) = 26(1.4)^x$
   b. $g(x) = 1.2(0.84)^x$

   c. $h(x) = 75\left(\frac{4}{5}\right)^x$
   d. $k(x) = \frac{2}{3}\left(\frac{9}{8}\right)^x$

2. **a.** $M(x) = 1.5(0.05)^x$
   b. $N(x) = 0.05(1.05)^x$

   c. $P(x) = \left(\frac{5}{8}\right)^x$
   d. $Q(x) = \left(\frac{4}{3}\right)^x$

Make a table of values and graph each pair of functions by hand on the domain $[-3, 3]$. Describe the similarities and differences between the two graphs.

3. **a.** $f(x) = 3^x$
   b. $g(x) = \left(\frac{1}{3}\right)^x$

4. **a.** $F(x) = \left(\frac{1}{10}\right)^x$
   b. $G(x) = 10^x$

5. **a.** $h(t) = 4^{-t}$
   b. $q(t) = -4^t$

6. **a.** $P(t) = -5^t$
   b. $R(t) = 5^{-t}$

Match each function with its graph.

7.

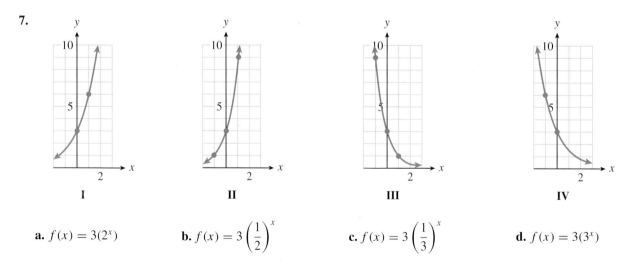

I   II   III   IV

a. $f(x) = 3(2^x)$
b. $f(x) = 3\left(\frac{1}{2}\right)^x$
c. $f(x) = 3\left(\frac{1}{3}\right)^x$
d. $f(x) = 3(3^x)$

8.

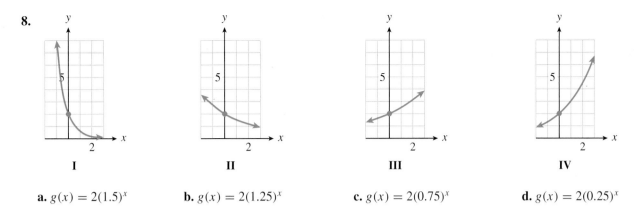

I	**II**	**III**	**IV**

a. $g(x) = 2(1.5)^x$ **b.** $g(x) = 2(1.25)^x$ **c.** $g(x) = 2(0.75)^x$ **d.** $g(x) = 2(0.25)^x$

For Problems 9–12,
(a) Use a graphing calculator to graph the functions on the domain [−5, 5].
(b) Give the range of the function on that domain, accurate to hundredths.

9. $g(t) = 4(1.3^t)$ **10.** $h(t) = 3(2.4^t)$ **11.** $N(x) = 50(0.8^x)$ **12.** $P(x) = 80(0.7^x)$

In each group of functions, which have identical graphs? Explain why.

13. a. $h(x) = 6^x$ **b.** $k(x) = \left(\dfrac{1}{6}\right)^x$ **14. a.** $Q(t) = 5^t$ **b.** $R(t) = \left(\dfrac{1}{5}\right)^t$

   c. $m(x) = 6^{-x}$ **d.** $n(x) = \dfrac{1}{6^x}$ **c.** $F(t) = \left(\dfrac{1}{5}\right)^{-t}$ **d.** $G(t) = \dfrac{1}{5^{-t}}$

For Problems 15–18,
(a) Use the order of operations to explain why the two functions are different.
(b) Complete the table of values and graph both functions in the same window.
(c) Describe each as a transformation of $y = 2^x$ or $y = 3^x$.

15. $f(x) = 2^{x-1}$, $g(x) = 2^x - 1$

x	$y = 2^x$	$f(x)$	$g(x)$
−2			
−1			
0			
1			
2			

16. $f(x) = 3^x + 2$, $g(x) = 3^{x+2}$

x	$y = 3^x$	$f(x)$	$g(x)$
−2			
−1			
0			
1			
2			

17. $f(x) = -3^x$, $g(x) = 3^{-x}$

x	$y = 3^x$	$f(x)$	$g(x)$
-2			
-1			
0			
1			
2			

18. $f(x) = 2^{-x}$, $g(x) = -2^x$

x	$y = 2^x$	$f(x)$	$g(x)$
-2			
-1			
0			
1			
2			

For the given function, evaluate each pair of expressions. Are they equivalent?

19. $f(x) = 3(5^x)$
  a. $f(a + 2)$ and $9f(a)$
  b. $f(2a)$ and $2f(a)$

20. $g(x) = 1.8^x$
  a. $g(h + 3)$ and $g(h)g(3)$
  b. $g(2h)$ and $[g(h)]^2$

21. $P(t) = 8^t$
  a. $P(w) - P(z)$ and $P(w - z)$
  b. $P(-x)$ and $\dfrac{1}{P(x)}$

22. $Q(t) = 5(0.2)^t$
  a. $Q(b - 1)$ and $5Q(b)$
  b. $Q(a)Q(b)$ and $5Q(a + b)$

23. The graph of $f(x) = P_0 b^x$ is shown in the figure.

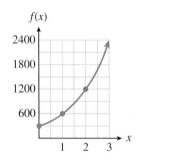

  a. Read the value of P_0 from the graph.
  b. Make a short table of values for the function by reading values from the graph. Does your table confirm that the function is exponential?
  c. Use your table to calculate the growth factor, b.
  d. Using your answers to parts (a) and (c), write a formula for $f(x)$.

24. The graph of $g(x) = P_0 b^x$ is shown in the figure.

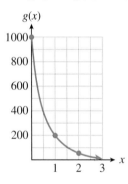

  a. Read the value of P_0 from the graph.
  b. Make a short table of values for the function by reading values from the graph. Does your table confirm that the function is exponential?
  c. Use your table to calculate the decay factor, b.
  d. Using your answers to parts (a) and (c), write a formula for $g(x)$.

25. For several days after the Northridge earthquake on January 17, 1994, the area received a number of significant aftershocks. The red graph shows that the number of aftershocks decreased exponentially over time. The graph of the function $S(d) = S_0 b^d$, shown in black, approximates the data. (Source: Los Angeles Times, June 27, 1995)

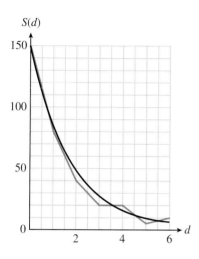

a. Read the value of S_0 from the graph.
b. Find an approximation for the decay factor, b, by comparing two points on the graph. (Some of the points on the graph of $S(d)$ are approximately (1, 82), (2, 45), (3, 25), and (4, 14).)
c. Using your answers to (a) and (b), write a formula for $S(d)$.

26. The frequency of a musical note depends on its pitch. The graph shows that the frequency increases exponentially. The function $F(p) = F_0 b^p$ gives the frequency as a function of the number of half-tones, p, above the starting point on the scale.

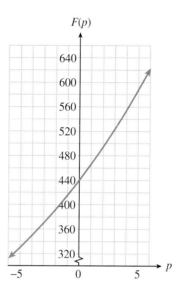

a. Read the frequency F_0 from the graph. (This is the frequency of the note A above middle C.)
b. Find an approximation for the growth factor, b, by comparing two points on the graph. (Some of the points on the graph of $F(p)$ are approximately (1, 466), (2, 494), (3, 523), and (4, 554).)
c. Using your answers to (a) and (b), write a formula for $F(p)$.
d. The frequency doubles when you raise a note by one octave, which is equivalent to 12 half-tones. Use this information to find an *exact* value for b.

Solve each equation algebraically.

27. $5^{x+2} = 25^{4/3}$

28. $3^{x-1} = 27^{1/2}$

29. $3^{2x-1} = \dfrac{\sqrt{3}}{9}$

30. $2^{3x-1} = \dfrac{\sqrt{2}}{16}$

31. $4 \cdot 2^{x-3} = 8^{-2x}$

32. $9 \cdot 3^{x+2} = 81^{-x}$

33. $27^{4x+2} = 81^{x-1}$

34. $16^{2-3x} = 64^{x+5}$

35. $10^{x^2-1} = 1000$

36. $5^{x^2-x-4} = 25$

37. Before the advent of antibiotics, an outbreak of cholera might spread through a city so that the number of cases doubled every 6 days.
a. Twenty-six cases were discovered on July 5. Write a function for the number of cases of cholera t days later.
b. Use your calculator to graph your function on the interval $0 \le t \le 90$.
c. When should hospitals expect to be treating 106,496 cases? Use algebraic methods to find your answer, and verify it on your graph.

38. An outbreak of ungulate fever can sweep through the livestock in a region so that the number of animals affected triples every 4 days.
  a. A rancher discovers 4 cases of ungulate fever among his herd. Write a function for the number of cases of ungulate fever t days later.
  b. Use your calculator to graph your function on the interval $0 \le t \le 20$.
  c. If the rancher does not act quickly, how long will it be until 324 head are affected? Use algebraic methods to find your answer, and verify it on your graph.

39. A color television set loses 30% of its value every 2 years.
  a. Write a function for the value of a television set t years after it was purchased if it cost $700 originally.
  b. Use your calculator to graph your function on the interval $0 \le t \le 20$.
  c. How long will it be before a $700 television set depreciates to $343? Use algebraic methods to find your answer, and verify it on your graph.

40. A mobile home loses 20% of its value every 3 years.
  a. A certain mobile home costs $20,000. Write a function for its value after t years.
  b. Use your calculator to graph your function on the interval $0 \le t \le 30$.
  c. How long will it be before a $20,000 mobile home depreciates to $12,800? Use algebraic methods to find your answer, and verify it on your graph.

Use a graph to find an approximate solution accurate to the nearest hundredth.

41. $3^{x-1} = 4$

42. $2^{x+3} = 5$

43. $4^{-x} = 7$

44. $6^{-x} = 3$

Decide whether each function is an exponential function, a power function, or neither.

45. a. $g(t) = 3t^{0.4}$ **b.** $h(t) = 4(0.3)^t$
  c. $D(x) = 6x^{1/2}$ **d.** $E(x) = 4x + x^4$

46. a. $R(w) = 5(5)^{w-1}$ **b.** $Q(w) = 2^w - w^2$
  c. $M(z) = 0.2z^{1.3}$ **d.** $N(z) = z^{-3}$

Decide whether the table could describe a linear function, a power function, on exponential function, or none of these. Find a formula for each linear, power, or exponential function.

47. a.

x	y
0	3
1	6
2	12
3	24
4	48

b.

t	P
0	0
1	0.5
2	2
3	4.5
4	8

48. a.

x	N
0	0
1	2
2	16
3	54
4	128

b.

p	R
0	405
1	135
2	45
3	15
4	5

49. a.

t	y
1	100
2	50
3	$33\frac{1}{3}$
4	25
5	20

b.

x	P
1	$\frac{1}{2}$
2	1
3	2
4	4
5	8

50. a.

h	a
0	70
1	7
2	0.7
3	0.07
4	0.007

b.

t	Q
0	0
1	$\frac{1}{4}$
2	1
3	$\frac{9}{4}$
4	4

Fill in the tables. Graph each pair of functions in the same window. Then answer the questions below.

51.

x	$f(x) = x^2$	$g(x) = 2^x$
-2		
-1		
0		
1		
2		
3		
4		
5		

a. Give the range of f and the range of g.
b. For how many values of x does $f(x) = g(x)$?
c. Estimate the value(s) of x for which $f(x) = g(x)$.
d. For what values of x is $f(x) < g(x)$?
e. Which function grows more rapidly for large values of x?

52.

x	$f(x) = x^3$	$g(x) = 3^x$
-2		
-1		
0		
1		
2		
3		
4		
5		

a. Give the range of f and the range of g.
b. For how many values of x does $f(x) = g(x)$?
c. Estimate the value(s) of x for which $f(x) = g(x)$.
d. For what values of x is $f(x) < g(x)$?
e. Which function grows more rapidly for large values of x?

For Problems 53–60, write a formula for each transformation of the given function and sketch its graph. State the domain and range of each transformation, its intercept(s), and any asymptotes.

53. $f(x) = 3^x$
　a. $y = f(x) - 4$
　b. $y = f(x - 4)$
　c. $y = -4f(x)$

54. $g(x) = 4^x$
　a. $y = g(x) + 2$
　b. $y = g(x + 2)$
　c. $y = 2g(x)$

55. $h(t) = 6^t$
　a. $y = -h(t)$
　b. $y = h(-t)$
　c. $y = -h(-t)$

56. $j(t) = \left(\dfrac{1}{3}\right)^t$
　a. $y = j(-t)$
　b. $y = -j(t)$
　c. $y = -j(-t)$

57. $g(x) = 2^x$
　a. $y = g(x - 3)$
　b. $y = g(x - 3) + 4$

58. $f(x) = 10^x$
　a. $y = f(x + 5)$
　b. $y = f(x + 5) - 20$

59. $N(t) = \left(\dfrac{1}{2}\right)^t$
　a. $y = -N(t)$
　b. $y = 6 - N(t)$

60. $P(t) = 0.4^t$
　a. $y = -P(t)$
　b. $y = 8 - P(t)$

For Problems 61–64,
(a) Describe the graph as a transformation of $y = 2^x$.
(b) Give an equation for the function graphed.

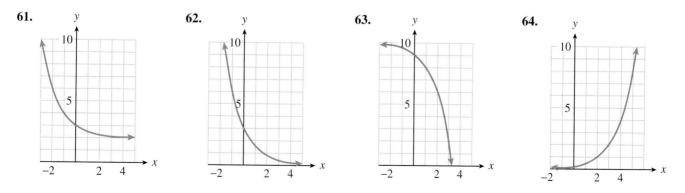

61.　**62.**　**63.**　**64.**

Match the graph of each function to its formula. In each formula, $a > 0$ and $b > 1$.

65. a. $y = ab^x$ **b.** $y = ab^{-x}$ **c.** $y = ax^b$

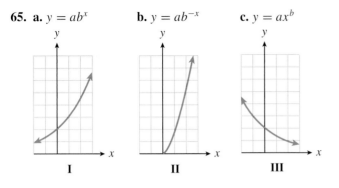

I   II   III

66. a. $y = ax^{-b}$ **b.** $y = -ab^x$ **c.** $y = ax^{1/b}$

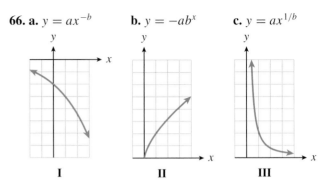

I   II   III

67. The function $f(t)$ describes a volunteer's heart rate during a treadmill test.

$$f(t) = \begin{cases} 100 & 0 \le t < 3 \\ 56t - 68 & 3 \le t < 4 \\ 186 - 500(0.5)^t & 4 \le t < 9 \\ 100 + 6.6(0.6)^{t-14} & 9 \le t < 20 \end{cases}$$

The heart rate is given in beats per minute and t is in minutes. (See Section 2.2 to review functions defined piecewise.) (Source: Davis, Kimmet, and Autry, 1986)
a. Evaluate the function to complete the table.

t	3.5	4	8	10	15
$f(t)$					

b. Sketch the graph of the function.
c. The treadmill test began with walking at 5.5 kilometers per hour, then jogging, starting at 12 kilometers per hour and increasing to 14 kilometers per hour, and finished with a cool-down walking period. Identify each of these activities on the graph and describe the volunteer's heart rate during each phase.

68. Carbon dioxide (CO_2) is called a greenhouse gas because it traps part of the Earth's outgoing energy. Animals release CO_2 into the atmosphere, and plants remove CO_2 through photosynthesis. In modern times, deforestation and the burning of fossil fuels both contribute to CO_2 levels. The figure shows atmospheric concentrations of CO_2, in parts per million, measured at the Mauna Loa Observatory in Hawaii.
a. The red curve shows annual oscillations in CO_2 levels. Can you explain why CO_2 levels vary throughout the year? (*Hint*: Why would photosynthesis vary throughout the year?)
b. The blue curve shows the average annual CO_2 readings. By approximately how much does the CO_2 level vary from its average value during the year?
c. In 1960, the average CO_2 level was 316.75 parts per million, and the average level has been rising by 0.4% per year. If the level continues to rise at this rate, what CO_2 readings can we expect in the year 2100?

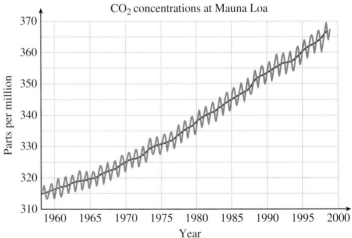

In this section, we introduce a new mathematical tool called a *logarithm,* which will help us solve exponential equations.

Suppose that a colony of bacteria doubles in size every day. If the colony starts with 50 bacteria, how long will it be before there are 800 bacteria? We answered questions of this type in Section 4.2 by writing and solving an exponential equation. The function

$$P(t) = 50 \cdot 2^t$$

gives the number of bacteria present on day t, so we must solve the equation

$$800 = 50 \cdot 2^t$$

Dividing both sides by 50 yields

$$16 = 2^t$$

The solution of this equation is the answer to the following question: To what power must we raise 2 in order to get 16?

The value of t that solves the equation is called the base 2 **logarithm** of 16. Since $2^4 = 16$, the base 2 logarithm of 16 is 4. We write this as

$$\log_2 16 = 4$$

In other words, we solve an exponential equation by computing a logarithm. You can check that $t = 4$ solves the problem stated above:

$$P(4) = 50 \cdot 2^4 = 800$$

Thus, the unknown exponent is called a logarithm. In general, for positive values of b and x, we make the following definition.

Definition of Logarithm

The **base b logarithm of x,** written **$\log_b x$,** is the exponent to which b must be raised in order to yield x.

It will help to keep in mind that *a logarithm is just an exponent.* Some logarithms, like some square roots, are easy to evaluate, while others require a calculator. We will start with the easy ones.

EXAMPLE 1

Compute the logarithms.

a. $\log_3 9$ **b.** $\log_5 125$ **c.** $\log_4 \dfrac{1}{16}$ **d.** $\log_5 \sqrt{5}$

Solutions   **a.** To evaluate $\log_3 9$, we ask what exponent on base 3 will produce 9. In symbols, we want to fill in the blank in the equation $3^{\underline{?}} = 9$. The exponent we need is 2, so

$$\log_3 9 = 2 \quad \text{because} \quad 3^2 = 9$$

We use similar reasoning to compute the other logarithms.

b. $\log_5 125 = 3$ because $5^3 = 125$

c. $\log_4 \dfrac{1}{16} = -2$ because $4^{-2} = \dfrac{1}{16}$

d. $\log_5 \sqrt{5} = \dfrac{1}{2}$ because $5^{1/2} = \sqrt{5}$

EXERCISE 1

Find each logarithm.

a. $\log_3 81$

b. $\log_{10} \dfrac{1}{1000}$

From the definition of a logarithm and the examples above, we see that the following two statements are equivalent.

Logarithms and Exponents: Conversion Equations

If $b > 0$ and $x > 0$,

$$y = \log_b x \quad \text{if and only if} \quad x = b^y$$

In other words, the logarithm y, is the same as the *exponent* in $x = b^y$. We see again that *a logarithm is an exponent;* it is the exponent to which b must be raised to yield x.

These equations allow us to convert from logarithmic to exponential form, or vice versa. You should memorize the conversion equations, because we will use them frequently.

As special cases of the equivalence in (1), we can compute the following useful logarithms. For any base $b > 0$,

Some Useful Logarithms

$$\log_b b = 1 \quad \text{because} \quad b^1 = b$$
$$\log_b 1 = 0 \quad \text{because} \quad b^0 = 1$$
$$\log_b b^x = x \quad \text{because} \quad b^x = b^x$$

EXAMPLE 2

a. $\log_2 2 = 1$ **b.** $\log_5 1 = 0$ **c.** $\log_3 3^4 = 4$

EXERCISE 2

Find each logarithm.

a. $\log_n 1$

b. $\log_n n^3$

Using the Conversion Equations

We use logarithms to solve exponential equations, just as we use square roots to solve quadratic equations. Consider the two equations

$$x^2 = 25 \qquad \text{and} \qquad 2^x = 8$$

We solve the first equation by taking a square root, and we solve the second equation by computing a logarithm:

$$x = \pm\sqrt{25} = \pm 5 \qquad \text{and} \qquad x = \log_2 8 = 3$$

The operation of taking a base b logarithm is the *inverse* operation for raising the base b to a power, just as extracting square roots is the inverse of squaring a number.

Every exponential equation can be rewritten in logarithmic form by using the conversion equations. Thus,

$$3 = \log_2 8 \qquad \text{and} \qquad 8 = 2^3$$

are equivalent statements, just as

$$5 = \sqrt{25} \qquad \text{and} \qquad 25 = 5^2$$

are equivalent statements. Rewriting an equation in logarithmic form is a basic strategy for finding its solution.

EXAMPLE 3   Rewrite each equation in logarithmic form.

    a. $2^{-1} = \dfrac{1}{2}$ **b.** $a^{1/5} = 2.8$

    c. $6^{1.5} = T$ **d.** $M^v = 3K$

Solutions   First identify the base b, and then the exponent or logarithm y. Use the conversion equations to rewrite $b^y = x$ in the form $\log_b x = y$.

    a. The base is *2* and the exponent is -1. Thus, $\log_2 \dfrac{1}{2} = -1$.

    b. The base is *a* and the exponent is $\dfrac{1}{5}$. Thus, $\log_a 2.8 = \dfrac{1}{5}$.

    c. The base is *6* and the exponent is 1.5. Thus, $\log_6 T = 1.5$.

    d. The base is *M* and the exponent is *v*. Thus, $\log_M 3K = v$.

EXERCISE 3

Rewrite each equation in logarithmic form.

    a. $8^{-1/3} = \dfrac{1}{2}$ **b.** $5^x = 46$

Approximating Logarithms

Suppose we would like to solve the equation

$$2^x = 26$$

The solution of this equation is $x = \log_2 26$, but can we find a decimal approximation for this value? There is no integer power of 2 that equals 26, because

$$2^4 = 16$$

$$\text{and} \quad 2^5 = 32$$

Thus, $\log_2 26$ must be between 4 and 5. We can use trial and error to find the value of $\log_2 26$ to the nearest tenth. Use your calculator to make a table of values for $y = 2^x$, starting with $x = 4$ and using increments of 0.1.

x	2^x		x	2^x
4	$2^4 = 16$		4.5	$2^{4.5} = 22.627$
4.1	$2^{4.1} = 17.148$		4.6	$2^{4.6} = 24.251$
4.2	$2^{4.2} = 18.379$		4.7	$2^{4.7} = 25.992$
4.3	$2^{4.3} = 19.698$		4.8	$2^{4.8} = 27.858$
4.4	$2^{4.4} = 21.112$		4.9	$2^{4.9} = 29.857$

TABLE 4.9

From Table 4.9, we see that 26 is between $2^{4.7}$ and $2^{4.8}$, and is closer to $2^{4.7}$. To the nearest tenth, $\log_2 26 \approx 4.7$.

Trial and error can be a time-consuming process. In Example 4, we illustrate a graphical method for estimating the value of a logarithm.

EXAMPLE 4   Approximate $\log_3 7$ to the nearest hundredth.

Solution   If $\log_3 7 = x$, then $3^x = 7$. We will use the graph of $y = 3^x$ to approximate a solution to $3^x = 7$. Graph $Y_1 = 3 \; \boxed{\wedge} \; X$ and $Y_2 = 7$ in the standard window ($\boxed{\textbf{ZOOM}}$ **6**) to obtain the graph shown in Figure 4.12. Activate the **intersect** feature to find that the two graphs intersect at the point (1.7712437, 7). Because this point lies on the graph of $y = 3^x$, we know that

$$3^{1.7712437} \approx 7, \quad \text{or} \quad \log_3 7 \approx 1.7712437$$

To the nearest hundredth, $\log_3 7 \approx 1.77$.

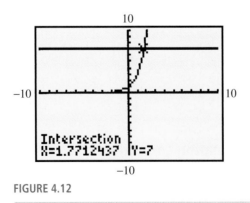

FIGURE 4.12

a. Rewrite the equation $3^x = 90$ in logarithmic form.

b. Use a graph to approximate the solution to the equation in part (a). Round your answer to three decimal places.

Base 10 Logarithms

Some logarithms are used so frequently in applications that their values are programmed into scientific and graphing calculators. These are the base 10 logarithms, such as

$$\log_{10} 1000 = 3 \qquad \text{and} \qquad \log_{10} 0.01 = -2$$

Base 10 logarithms are called **common logarithms,** and the subscript 10 is often omitted, so that $\log x$ is understood to mean $\log_{10} x$.

To evaluate a base 10 logarithm, we use the $\boxed{\text{LOG}}$ key on a calculator. Many logarithms are irrational numbers, and the calculator gives as many digits as its display allows. We can then round off to the desired accuracy.

EXAMPLE 5

Approximate the following logarithms to 2 decimal places.

a. $\log 6.5$ **b.** $\log 256$

Solutions **a.** The keying sequence $\boxed{\text{LOG}}$ 6.5 $\boxed{)}$ $\boxed{\text{ENTER}}$ produces the display

log (6.5)
.812913566

so $\log 6.5 \approx 0.81$.

b. The keying sequence $\boxed{\text{LOG}}$ 256 $\boxed{)}$ $\boxed{\text{ENTER}}$ yields 2.408239965, so $\log 256 \approx 2.41$.

We can check the approximations found in Example 5 with our conversion equations. Remember that a logarithm is an exponent, and in this example the base is 10. We find that

$$10^{0.81} \approx 6.45654229$$

$$\text{and} \qquad 10^{2.41} \approx 257.0395783$$

so our approximations are reasonable, although you can see that rounding a logarithm to 2 decimal places does lose some accuracy. For this reason, rounding logarithms to 4 decimal places is customary.

a. Evaluate $\log 250$, and round your answer to two decimal places. Check your answer using the conversion equations.

b. Evaluate $\log 250$, and round your answer to four decimal places. Check your answer using the conversion equations.

Solving Exponential Equations

We can now solve any exponential equation with base 10. For instance, to solve the equation

$$16 \cdot 10^t = 360$$

we first divide both sides by 16 to obtain

$$10^t = 22.5$$

Then we convert the equation to logarithmic form and evaluate:

$$t = \log_{10} 22.5 \approx 1.352182518$$

To 4 decimal places, the solution is 1.3522.

To solve exponential equations involving powers of 10, we can use the following steps.

Steps for Solving Base 10 Exponential Equations

1. Isolate the power on one side of the equation.
2. Rewrite the equation in logarithmic form.
3. Use a calculator, if necessary, to evaluate the logarithm.
4. Solve for the variable.

EXAMPLE 6   Solve the equation $38 = 95 - 15 \cdot 10^{0.4x}$.

Solution   First, isolate the power of 10: Subtract 95 from both sides of the equation and divide by -15 to obtain

$$-57 = -15 \cdot 10^{0.4x} \qquad \text{Divide by } -15.$$

$$3.8 = 10^{0.4x}$$

Convert the equation to logarithmic form as

$$\log_{10} 3.8 = 0.4x$$

Solving for x yields

$$\frac{\log_{10} 3.8}{0.4} = x$$

We can evaluate this expression on the calculator by entering

$$\boxed{\text{LOG}}\ 3.8\ \boxed{)}\ \boxed{\div}\ 0.4\ \boxed{\text{ENTER}}$$

which yields 1.449458992. Thus, to four decimal places, $x \approx 1.4495$.

CAUTION   Be careful when using a calculator to evaluate expressions involving logs. We can evaluate a single logarithm like log 3.8 by entering $\boxed{\text{LOG}}$ 3.8 without an ending parenthesis, so that the calculator shows

log 3.8
.5795835966

But if we want to evaluate $\frac{\log 3.8}{0.4}$, we *must* enclose 3.8 in parentheses, as shown in Example 6. If we omit the parenthesis after 3.8, the calculator will interpret the expression as $\log\left(\frac{3.8}{0.4}\right)$, which is not the expression we wanted.

Solve $12 - 30(10^{-0.2x}) = 11.25$.

Application to Exponential Models

We have seen that exponential functions are used to describe some applications of growth and decay, $P(t) = P_0 b^t$. There are two common questions that arise in connection with exponential models:

1. Given a value of t, what is the corresponding value of $P(t)$?
2. Given a value of $P(t)$, what is find the corresponding value of t?

To answer the first question, we evaluate the function $P(t)$ at the appropriate value. To answer the second question, we must solve an exponential equation, and this usually involves logarithms.

EXAMPLE 7

The value of a large tractor originally worth 30,000 depreciates exponentially according to the formula $V(t) = 30,000(10)^{-0.04t}$, where t is in years. When will the tractor be worth half its original value?

Solution We want to find the value of t for which $V(t) = 15,000$. That is, we want to solve the equation

$$15,000 = 30,000(10)^{-0.04t}$$

Divide both sides by 30,000 to obtain

$$0.5 = 10^{-0.04t}$$

Convert the equation to logarithmic form as

$$\log_{10} 0.5 = -0.04t$$

and divide by -0.04 to obtain

$$\frac{\log_{10} 0.5}{-0.04} = t$$

To evaluate this expression, key in

$$\boxed{\text{LOG}} \;\; \boxed{0.5} \;\boxed{)}\; \boxed{\div}\; \boxed{(-)}\; 0.04 \;\boxed{\text{ENTER}}$$

to find $t \approx 7.525749892$. The tractor will be worth \$15,000 in approximately $7\frac{1}{2}$ years.

EXERCISE 7

The percentage of American homes with computers grew exponentially from 1994 to 1999. For $t = 0$ in 1994, the growth law was $P(t) = 25.85(10)^{0.052t}$.

a. What percent of American homes had computers in 1994?

b. If the percentage of homes with computers continued to grow at the same rate, when did 90% of American homes have a computer?

c. Do you think that the function $P(t)$ will continue to model the percentage of American homes with computers? Why or why not?

Source: *Los Angeles Times,* August 20, 1999

At this stage, it seems we will only be able to solve exponential equations in which the base is 10. However, we will see in Section 4.4 how the properties of logarithms enable us to solve exponential equations with any base.

ANSWERS TO 4.3 EXERCISES

1. a. 4 **b.** -3 **2. a.** 0 **b.** 3

3. a. $\log_8\left(\dfrac{1}{2}\right) = \dfrac{-1}{3}$ **b.** $\log_5(46) = x$ **4. a.** $\log_3(90) = x$ **b.** $x \approx 4.096$

5. a. 2.40 **b.** 2.3979 **6.** 8.01

7. a. 25.85% **b.** $t \approx 10.4$ (year 2004)

   c. No, the percent of homes with computers cannot exceed 100%.

SECTION 4.3 SUMMARY

VOCABULARY *Look up the definitions of new terms in the Glossary.*

Logarithm   Common logarithm

CONCEPTS

1. We use logarithms to help us solve exponential equations.

2. The **base b logarithm of x**, written **$\log_b x$**, is the exponent to which b must be raised in order to yield x.

3. If $b > 0$ and $x > 0$,

$$y = \log_b x \quad \text{if and only if} \quad x = b^y$$

4. The operation of taking a base b logarithm is the *inverse* operation for raising the base b to a power.

5. Base 10 logarithms are called **common logarithms**, and $\log x$ means $\log_{10} x$.

6.

> **Steps for Solving Base 10 Exponential Equations**
>
> **1.** Isolate the power on one side of the equation.
>
> **2.** Rewrite the equation in logarithmic form.
>
> **3.** Use a calculator, if necessary, to evaluate the logarithm.
>
> **4.** Solve for the variable.

STUDY QUESTIONS

1. To find $\log_6 27$ means to find an exponent x that satisfies the equation _____.

2. Can a logarithm be a negative number?

3. Evaluate the following logarithms:
  a. $\log_8 8^{15}$ **b.** $\log_5 5^{\sqrt{13}}$
  c. $\log_b b^{2.63}$

4. Guess the solution of $10^x = 750$. Now find an approximation correct to four decimal places. Was your guess too big or too small?

SKILLS   *Practice each skill in the Homework problems listed.*

1. Compute logs base b using the definition: #1–10, 59–66

2. Convert from exponential to logarithmic form: #11–22

3. Approximate logarithms: #23–34

4. Solve exponential equations base 10: #35–48

5. Solve application problems: #49–58

HOMEWORK 4.3

ThomsonNOW Test yourself on key content at www.thomsonedu.com/login.

Find each logarithm without using a calculator.

1. a. $\log_7 49$ **b.** $\log_2 32$ **2. a.** $\log_4 64$ **b.** $\log_3 27$

3. a. $\log_3 \sqrt{3}$ **b.** $\log_3 \dfrac{1}{3}$ **4. a.** $\log_5 \dfrac{1}{5}$ **b.** $\log_5 \sqrt{5}$

5. a. $\log_4 4$ **b.** $\log_6 1$ **6. a.** $\log_{10} 1$ **b.** $\log_{10} 10$

7. a. $\log_8 8^5$ **b.** $\log_7 7^6$ **8. a.** $\log_{10} 10^{-4}$ **b.** $\log_{10} 10^{-6}$

9. a. $\log_{10} 0.1$ **b.** $\log_{10} 0.001$ **10. a.** $\log_{10} 10{,}000$ **b.** $\log_{10} 1000$

Rewrite each equation in logarithmic form.

11. $2^{10} = 1024$ **12.** $11^4 = 14{,}641$ **13.** $10^{0.699} \approx 5$ **14.** $10^{-0.602} \approx 0.25$

15. $t^{3/2} = 16$ **16.** $v^{5/3} = 12$ **17.** $0.8^{1.2} = M$ **18.** $3.7^{2.5} = Q$

19. $x^{5t} = W - 3$ **20.** $z^{-3t} = 2P + 5$ **21.** $3^{-0.2t} = 2N_0$ **22.** $10^{1.3t} = 3M_0$

For Problems 23–26,
(a) Solve each equation, writing your answer as a logarithm.
(b) Use trial and error to approximate the logarithm to one decimal place.

23. $4^x = 2.5$ **24.** $2^x = 0.2$ **25.** $10^x = 0.003$ **26.** $10^x = 4500$

For Problems 27–30,
(a) By computing successive powers of the base, trap each log between two integers.
(b) Use a graph to approximate each logarithm to the nearest hundredth. (*Hint:* Use the conversion equations to rewrite
    $x = \log_b y$ **as an appropriate exponential function.)**

27. $\log_{10} 7$ **28.** $\log_{10} 50$ **29.** $\log_3 67.9$ **30.** $\log_5 86.3$

Use a calculator to approximate each logarithm to four decimal places. Make a conjecture about logarithms based on the results of each problem.

31. a. $\log_{10} 5.43$
    b. $\log_{10} 54.3$
    c. $\log_{10} 543$
    d. $\log_{10} 5430$

32. a. $\log_{10} 0.625$
    b. $\log_{10} 0.0625$
    c. $\log_{10} 0.00625$
    d. $\log_{10} 0.000625$

33. a. $\log 2$
    b. $\log 4$
    c. $\log 8$
    d. $\log 16$

34. a. $\log 4$
    b. $\log 0.25$
    c. $\log 5$
    d. $\log 0.2$

Solve for *x*. Round your answers to hundredths.

35. $10^{-3x} = 5$ **36.** $10^{-5x} = 76$

37. $25 \cdot 10^{0.2x} = 80$ **38.** $8 \cdot 10^{1.6x} = 312$

39. $12.2 = 2(10^{1.4x}) - 11.6$

40. $163 = 3(10^{0.7x}) - 49.3$

41. $3(10^{-1.5x}) - 14.7 = 17.1$

42. $4(10^{-0.6x}) + 16.1 = 28.2$

43. $80(1 - 10^{-0.2x}) = 65$

44. $250(1 - 10^{-0.3x}) = 100$

In Problems 45–48, each calculation contains an error. Find and correct it.

45. $2 \cdot 5^x = 848$
$$10^x = 848$$
$$x = \log 848 \quad \text{(Incorrect!)}$$

46. $15 \cdot 10^x = 20$
$$10^x = 5$$
$$x = \log 5 \quad \text{(Incorrect!)}$$

47. $10^{4x} = 20$
$$10^x = 5$$
$$x = \log 5 \quad \text{(Incorrect!)}$$

48. $12 + 6^x = 42$
$$6^x = 30$$
$$x = 5 \quad \text{(Incorrect!)}$$

49. The population of the state of California increased during the years 1990 to 2000 according to the formula $P(t) = 29{,}760{,}021\,(10)^{0.0056t}$, where t is measured in years since 1990.
  a. What was the population in 2000?
  b. Assuming the same rate of growth, estimate the population of California in the years 2010, 2015, and 2020.
  c. When did the population of California reach 35,000,000?
  d. When should the population reach 40,000,000?
  e. Graph the function P with a suitable domain and range, then verify your answers to parts (a) through (d).

50. The population of the state of New York increased during the years 1990 to 2000 according to the formula $P(t) = 17{,}990{,}455\,(10)^{0.0023t}$, where t is measured in years since 1990.
  a. What was the population in 2000?
  b. Assuming the same rate of growth, estimate the population of New York in the years 2010, 2015, and 2020.
  c. When should the population of New York reach 20,000,000?
  d. When should the population reach 30,000,000?
  e. Graph the function P with a suitable domain and range, then verify your answers to parts (a) through (d).

51. The absolute magnitude, M, of a star is a measurement of its brightness. For example, our Sun, not a particularly bright star, has magnitude $M = 4.83$. The magnitude in turn is a measure of the luminosity, L, or amount of light energy emitted by the star, where
$$L = L_0 10^{-0.4M}$$
  a. The luminosity of a star is measured in solar units, so that our Sun has luminosity $L = 1$. Use the values of L and M for the Sun to calculate a value of L_0 in the equation above.
  b. Is luminosity an increasing or decreasing function of magnitude? Graph the function on the domain $[-3, 3]$. What is its range on that domain?
  c. The luminosity of Sirius is 22.5 times that of the Sun, or $L = 22.5$. Calculate the magnitude of Sirius.
  d. If two stars differ in magnitude by 5, what is the ratio of their luminosities?
  e. A decrease in magnitude by 1 corresponds to an increase in luminosity by what factor? Give an exact value and an approximation to four decimal places.
  f. Normal stars have magnitudes between -10 and 19. What range of luminosities do stars exhibit?

52. The loudness of a sound is a consequence of its intensity, I, or the amount of energy it generates, in watts per square meter. The intensity is related to the decibel level, D, which is another measure of loudness, by
$$I = 10^{-12+D/10}$$
  a. Is intensity an increasing or decreasing function of decibel level? The faintest sound a healthy human can hear is 0 decibels. What is the intensity of a 0 decibel sound?
  b. A whisper produces an energy intensity of 10^{-9} watts per square meter. What is the decibel level of a whisper?
  c. If two sounds differ in loudness by 10 decibels, what is the ratio of their intensities?
  d. An increase in loudness of 1 decibel produces a *just noticeable difference* to the human ear. By what factor does the intensity increase?
  e. Sounds of 130 decibels are at the threshold of pain for people. What is the range of the intensity function on the domain $[0, 130]$?

The atmospheric pressure decreases with altitude above the surface of the Earth. For Problems 53–58, use the function

$$P(h) = 30\,(10)^{-0.09h}$$

where altitude, *h*, is given in miles and atmospheric pressure, *P*, in inches of mercury. Graph this function in the window

$$\text{Xmin} = 0, \qquad \text{Xmax} = 9.4$$
$$\text{Ymin} = 0, \qquad \text{Ymax} = 30$$

Solve the problems below algebraically, and verify with your graph.

53. The elevation of Mount Everest, the highest mountain in the world, is 29,028 feet. What is the atmospheric pressure at the top? (*Hint*: 1 mile = 5280 feet)

54. The elevation of Mount McKinley, the highest mountain in the United States, is 20,320 feet. What is the atmospheric pressure at the top?

55. How high above sea level is the atmospheric pressure 20.2 inches of mercury?

56. How high above sea level is the atmospheric pressure 16.1 inches of mercury?

57. Find the height above sea level at which the atmospheric pressure is equal to one-half the pressure at sea level. (*Hint*: What is the altitude at sea level?)

58. Find the height above sea level at which the atmospheric pressure is equal to one-fourth the pressure at sea level. (*Hint*: What is the altitude at sea level?)

Simplify each expression.

59. $\log_2(\log_4 16)$

60. $\log_5(\log_5 5)$

61. $\log_{10}[\log_3(\log_5 125)]$

62. $\log_{10}[\log_2(\log_3 9)]$

63. $\log_2[\log_2(\log_3 81)]$

64. $\log_4[\log_2(\log_3 81)]$

65. $\log_b(\log_b b)$

66. $\log_b(\log_a a^b)$

4.4   Properties of Logarithms

See the Algebra Skills Refresher A.6 to review the laws of exponents.

Because logarithms are actually exponents, they have several properties that can be derived from the laws of exponents. Here are the laws we will need at present.

1. To multiply two powers with the same base, add the exponents and leave the base unchanged.

$$a^m \cdot a^n = a^{m+n}$$

2. To divide two powers with the same base, subtract the exponents and leave the base unchanged.

$$\frac{a^m}{a^n} = a^{m-n}$$

3. To raise a power to a power, keep the same base and multiply the exponents.

$$(a^m)^n = a^{mn}$$

Each of these laws corresponds to one of three properties of logarithms.

4.4 ■ Properties of Logarithms   **365**

Properties of Logarithms

If x, y, and $b > 0$, then

1. $\log_b(xy) = \log_b x + \log_b y$
2. $\log_b \dfrac{x}{y} = \log_b x - \log_b y$
3. $\log_b x^k = k \log_b x$

We will consider proofs of the three properties of logarithms in the Homework problems. For now, study the examples below, keeping in mind that a logarithm is an exponent.

1. Property (1):

$$\log_2 32 = \log_2 (4 \cdot 8) = \log_2 4 + \log_2 8 \qquad \text{because} \qquad 2^5 = 2^2 \cdot 2^3$$
$$5 \qquad\qquad\qquad = 2 + 3 \qquad\qquad\qquad\qquad 32 = 4 \cdot 8$$

2. Property (2):

$$\log_2 8 = \log_2 \frac{16}{2} = \log_2 16 - \log_2 2 \qquad \text{because} \qquad 2^3 = \frac{2^4}{2^1}$$
$$3 \qquad\qquad = 4 - 1 \qquad\qquad\qquad\qquad 8 = \frac{16}{2}$$

3. Property (3):

$$\log_2 64 = \log_2 (4)^3 = 3 \log_2 4 \qquad \text{because} \qquad (2^2)^3 = 2^6$$
$$6 \qquad\qquad = 3 \cdot 2 \qquad\qquad\qquad\qquad (4)^3 = 64$$

Using the Properties of Logarithms

Of course, these properties are useful not so much for computing logs but rather for simplifying expressions that contain variables. We will use them to solve exponential equations. But first, we will practice applying the properties. In Example 1, we rewrite one log in terms of simpler logs.

EXAMPLE 1

Simplify $\log_b \sqrt{xy}$.

Solution   First, we write $\sqrt{xy}$ using a fractional exponent:

$$\log_b \sqrt{xy} = \log_b (xy)^{1/2}$$

Then we apply Property (3) to rewrite the exponent as a coefficient:

$$\log_b (xy)^{1/2} = \frac{1}{2} \log_b (xy)$$

Finally, by Property (1) we write the log of a product as a sum of logs:

$$\frac{1}{2}(\log_b xy) = \frac{1}{2}(\log_b x + \log_b y)$$

Thus, $\log_b \sqrt{xy} = \frac{1}{2}(\log_b x + \log_b y)$.

Simplify $\log_b \dfrac{x}{y^2}$.

CAUTION

Be careful when using the properties of logarithms! Compare the statements below:

1. $\log_b(2x) = \log_b 2 + \log_b x$ by Property 1,

but

$\log_b(2 + x) \neq \log_b 2 + \log_b x$

2. $\log_b\left(\dfrac{x}{5}\right) = \log_b x - \log_b 5$ by Property 2,

but

$\log_b\left(\dfrac{x}{5}\right) \neq \dfrac{\log_b x}{\log_b 5}$

We can also use the properties of logarithms to combine sums and differences of logarithms into one logarithm.

EXAMPLE 2   Express $3(\log_b x - \log_b y)$ as a single logarithm with a coefficient of 1.

Solution   We begin by applying Property (2) to combine the logs.

$$3(\log_b x - \log_b y) = 3\log_b\left(\dfrac{x}{y}\right)$$

Then, using Property (3), we replace the coefficient 3 by an exponent 3.

$$3\log_b\left(\dfrac{x}{y}\right) = \log_b\left(\dfrac{x}{y}\right)^3$$

EXERCISE 2

Express $2\log_b x + 4\log_b(x + 3)$ as a single logarithm with a coefficient of 1.

Solving Exponential Equations

By using Property (3), we can now solve exponential equations in which the base is not 10. For example, to solve the equation

$$5^x = 7 \tag{1}$$

we could rewrite the equation in logarithmic form to obtain the exact solution

$$x = \log_5 7$$

However, we cannot evaluate $\log_5 7$; there is no log base 5 button on the calculator. If we want a decimal approximation for the solution, we begin by taking the base 10 logarithm of both sides, even though the base of the power is not 10. This gives us

$$\log_{10}(5^x) = \log_{10} 7$$

Then we use Property (3) to rewrite the left side as

$$x \log_{10} 5 = \log_{10} 7$$

Note how using Property (3) allows us to solve the equation: The variable, x, is no longer in the exponent, and it is multiplied by a constant, $\log_{10} 5$. To finish the solution, we divide both sides by $\log_{10} 5$ to get

$$x = \frac{\log_{10} 7}{\log_{10} 5}$$

On your calculator, enter the sequence

$$\boxed{\text{LOG}}\ 7\ \boxed{)}\ \boxed{\div}\ \boxed{\text{LOG}}\ 5\ \boxed{)}\ \boxed{\text{ENTER}}$$

to find that $x \approx 1.2091$.

CAUTION   Do not confuse the expression $\frac{\log_{10} 7}{\log_{10} 5}$ with $\log_{10}\left(\frac{7}{5}\right)$; they are not the same! Property (2) allows us to simplify $\log\left(\frac{x}{y}\right)$, but not $\frac{\log x}{\log y}$. We cannot rewrite $\frac{\log_{10} 7}{\log_{10} 5}$, so we must evaluate it as $(\log 7)/(\log 5)$. You can check on your calculator that

$$\frac{\log_{10} 7}{\log_{10} 5} \neq \log_{10}\left(\frac{7}{5}\right) = \log_{10} 1.4.$$

EXAMPLE 3   Solve $1640 = 80 \cdot 6^{0.03x}$.

Solution   Divide both sides by 80 to obtain

$$20.5 = 6^{0.03x}$$

Take the base 10 logarithm of both sides of the equation and use Property (3) of logarithms to get

$$\log_{10} 20.5 = \log_{10} 6^{0.03x}$$
$$= 0.03x \log_{10} 6$$

On the right side of the equation, x is multiplied by two constants, 0.03 and $\log_{10} 6$. So, to solve for x we must divide both sides of the equation by $0.03 \log_{10} 6$. We use a calculator to evaluate the answer:

$$x = \frac{\log_{10} 20.5}{0.03 \log_{10} 6} \approx 56.19$$

(On your calculator, remember to enclose the denominator, $0.03 \log_{10} 6$, in parentheses.)

CAUTION   In Example 2, do not try to simplify

$$80 \cdot 6^{0.03x} \rightarrow 480^{0.03x} \qquad \text{Incorrect!}$$

Remember that the order of operations tells us to compute the power $6^{0.03x}$ *before* multiplying by 80.

We summarize our method for solving exponential equations as follows.

Steps for Solving Exponential Equations

1. Isolate the power on one side of the equation.
2. Take the log base 10 of both sides.
3. Simplify by applying Log Property (3).
4. Solve for the variable.

EXERCISE 3

Solve $5(1.2)^{2.5x} = 77$.
Divide both sides by 5.
Take the log of both sides.
Apply Property (3) to simplify the left side.
Solve for x.

Applications

By using the properties of logarithms, we can now solve equations that arise in exponential growth and decay models, no matter what base the exponential function uses.

EXAMPLE 4   The population of Silicon City was 6500 in 1990 and has been tripling every 12 years. When will the population reach 75,000?

Solution   The population of Silicon City grows according to the formula

$$P(t) = 6500 \cdot 3^{t/12}$$

See Section 4.1 to review exponential growth.

where t is the number of years after 1990. We want to find the value of t for which $P(t) = 75,000$; that is, we want to solve the equation

$$6500 \cdot 3^{t/12} = 75,000 \quad \text{Divide both sides by 6500.}$$

$$3^{t/12} = \frac{150}{13}$$

Take the base 10 logarithm of both sides and solve for t.

$$\log_{10}(3^{t/12}) = \log_{10}\left(\frac{150}{13}\right) \quad \text{Apply Property (3).}$$

$$\frac{t}{12}\log_{10} 3 = \log_{10}\left(\frac{150}{13}\right) \quad \text{Divide by } \log_{10} 3; \text{ multiply by 12.}$$

$$t = \frac{12\left(\log_{10}\frac{150}{13}\right)}{\log_{10} 3} \approx 26.71$$

The population of Silicon City will reach 75,000 about 27 years after 1990, or in 2017.

Traffic on U.S. highways is growing by 2.7% per year. (Source: *Time,* Jan. 25, 1999)

a. Write a formula for the volume, V, of traffic as a function of time, using V_0 for the current volume.

b. How long will it take the volume of traffic to double? (*Hint:* Find the value of t that gives $V = 2V_0$.)

Compound Interest

The amount of money in an account that earns interest compounded annually grows exponentially according to the formula

$$A(t) = P(1 + r)^t$$

(See Section 4.1 to review compound interest.) Many accounts compound interest more frequently than once a year. If the interest is compounded n times per year, then in t years there will be nt compounding periods, and in each period the account earns interest at a rate of $\frac{r}{n}$. The amount accumulated is given by a generalization of our earlier formula.

> **Compound Interest**
>
> The **amount $A(t)$** accumulated (principal plus interest) in an account bearing interest compounded n times annually is
>
> $$A(t) = P\left(1 + \frac{r}{n}\right)^{nt}$$
>
> where P is the principal invested,
>
> r is the interest rate,
>
> t is the time period, in years.

EXAMPLE 5

Rashad deposited $1000 in an account that pays 4% interest. Calculate the amount in his account after 5 years if the interest is compounded

a. semiannually   **b.** quarterly   **c.** monthly

Solutions

a. *Semiannually* means "twice a year," so we use the formula for compound interest with $P = 1000, r = 0.04, n = 2,$ and $t = 5$.

$$A(5) = 1000\left(1 + \frac{0.04}{2}\right)^{2(5)}$$

$$= 1000(1.02)^{10} = 1218.99$$

If interest is compounded semiannually, the balance in the account after 5 years is $1218.99.

b. *Quarterly* means "4 times a year," so we use the formula for compound interest with $P = 1000, r = 0.04, n = 4,$ and $t = 5$.

$$A(5) = 1000\left(1 + \frac{0.04}{4}\right)^{4(5)}$$

$$= 1000(1.01)^{20} = 1220.19$$

If interest is compounded quarterly, the balance in the account after 5 years is $1220.19.

c. There are 12 months in a year, so we use the formula for compound interest with $P = 1000$, $r = 0.04$, $n = 12$, and $t = 5$.

$$A(5) = 1000\left(1 + \frac{0.04}{12}\right)^{12(5)}$$

$$= 1000(1.00\overline{3})^{60} = 1221.00$$

If interest is compounded monthly, the balance in the account after 5 years is $1221.

In Example 5, you can see that the larger the value of n, the greater the value of A, keeping the other parameters fixed. More frequent compounding periods result in a higher account balance.

EXERCISE 5

Calculate the amount in Rashad's account after 5 years if the interest is compounded daily. (See Example 5. There are 365 days in a year.)

Solving Formulas

The techniques for solving exponential equations can also be used to solve formulas involving exponential expressions for one variable in terms of the others.

EXAMPLE 6

Solve $2C = Cb^{kt}$ for t. (Assume that C and $k \neq 0$.)

Solution   First, we divide both sides by C to isolate the power.

$$b^{kt} = 2$$

Next, we take the log base 10 of both sides.

$$\log b^{kt} = \log 2 \quad \text{Apply Log Property (3).}$$
$$kt \log b = \log 2$$

Finally, we divide both sides by $k \log b$ to solve for t.

$$t = \frac{\log 2}{k \log b}$$

EXERCISE 6

Solve $A = P(1 + r)^t$ for t.

ANSWERS TO 4.4 EXERCISES

1. $\log_b x - 2\log_b y$ **2.** $\log_b x^2(x + 3)^4$ **3.** $x = \dfrac{\log 15.4}{2.5 \log 1.2} \approx 5.999$

4. a. $V(t) = V_0(1.027)^t$ **b.** about 26 years   **5.** $1221.39

6. $t = \dfrac{\log(A/P)}{\log(1 + r)}$

VOCABULARY   *Look up the definition of the new term in the Glossary.*

Compounding period

CONCEPTS

1.

Properties of Logarithms

If x, y, and $b > 0$, then

1. $\log_b (xy) = \log_b x + \log_b y$

2. $\log_b \dfrac{x}{y} = \log_b x - \log_b y$

3. $\log_b x^k = k \log_b x$

2. We can use the properties of logarithms to solve exponential equations with any base.

Steps for Solving Exponential Equations

1. Isolate the power on one side of the equation.

2. Take the log base 10 of both sides.

3. Simplify by applying Log Property (3).

4. Solve for the variable.

3. The amount in an account earning interest compounded n times per year is an exponential function of time.

Compound Interest

The **amount** A (t) accumulated (principal plus interest) in an account bearing interest compounded n times annually is

$$A(t) = P \left(1 + \frac{r}{n}\right)^{nt}$$

where P is the principal invested,

r is the interest rate,

t is the time period, in years.

STUDY QUESTIONS

1. The properties of logs are really another form of which familiar laws?

2. Which log property allows us to solve an exponential equation whose base is not 10?

3. Explain why $12 \cdot 10^{3x}$ is not the same as 120^{3x}.

4. Which of the following expressions are equivalent?

$$\log \left(\frac{x}{4}\right) \qquad \frac{\log x}{\log 4} \qquad \log(x - 4) \qquad \log x - \log 4$$

5. Which of the following expressions are equivalent?

$$\log(x + 2) \qquad \log x + \log 2 \qquad \log(2x) \qquad (\log 2)(\log x)$$

6. Which of the following expressions are equivalent?

$$\log x^3 \qquad (\log 3)(\log x) \qquad 3 \log x \qquad \log 3^x$$

SKILLS   *Practice each skill in the Homework problems listed.*

1. Use the properties of logarithms to simplify expressions: #1–24, #45–52

2. Solve exponential equations using logs base 10: #25–36

3. Solve problems about exponential models: #37–44

4. Solve problems about compound interest: #53–58

5. Solve formulas involving exponential expressions: #59–64

ThomsonNOW™ Test yourself on key content at www.thomsonedu.com/login.

1. a. Simplify $10^2 \cdot 10^6$.
   b. Compute $\log 10^2$, $\log 10^6$, and $\log(10^2 \cdot 10^6)$. How are they related?

2. a. Simplify $\frac{10^9}{10^6}$.
   b. Compute $\log 10^9$, $\log 10^6$, and $\log\left(\frac{10^9}{10^6}\right)$. How are they related?

3. a. Simplify $\frac{b^8}{b^5}$.
   b. Compute $\log_b b^8$, $\log_b b^5$, and $\log_b\left(\frac{b^8}{b^5}\right)$. How are they related?

4. a. Simplify $b^4 \cdot b^3$.
   b. Compute $\log b^4$, $\log b^3$, and $\log(b^4 \cdot b^3)$. How are they related?

5. a. Simplify $(10^3)^5$.
   b. Compute $\log\left((10^3)^5\right)$ and $\log(10^3)$. How are they related?

6. a. Simplify $(b^2)^6$.
   b. Compute $\log_b(b^2)^6$ and $\log_b(b^2)$. How are they related?

Use the properties of logarithms to expand each expression in terms of simpler logarithms. Assume that all variable expressions denote positive numbers.

7. a. $\log_b 2x$
   b. $\log_b \dfrac{x}{2}$

8. a. $\log_b \dfrac{2x}{x-2}$
   b. $\log_b x(2x+3)$

9. a. $\log_3 3x^4$
   b. $\log_5 1.1^{1/t}$

10. a. $\log_b(4b)^t$
   b. $\log_2 5(2^x)$

11. a. $\log_b \sqrt{bx}$
   b. $\log_3 \sqrt[3]{x^2+1}$

12. a. $\log_{10} \sqrt{\dfrac{2L}{R^2}}$
   b. $\log_{10} 2\pi \sqrt{\dfrac{l}{g}}$

13. a. $\log P_0(1-m)^t$
   b. $\log_4 \left(1+\dfrac{r}{4}\right)^{4t}$

14. a. $\log_3 \dfrac{a^2-2}{a^5}$
   b. $\log \dfrac{a^3 b^2}{(a+b)^{3/2}}$

Combine into one logarithm and simplify. Assume all expressions are defined.

15. a. $\log_b 8 - \log_b 2$
   b. $2\log_4 x + 3\log_4 y$

16. a. $\log_b 5 + \log_b 2$
   b. $\dfrac{1}{4}\log_5 x - \dfrac{3}{4}\log_5 y$

17. a. $\log 2x + 2\log x - \log \sqrt{x}$
   b. $\log(t^2-16) - \log(t+4)$

18. a. $\log x^2 + \log x^3 - 5\log x$
   b. $\log(x^2-x) - \log \sqrt{x^3}$

19. a. $3 - 3\log 30$
   b. $\dfrac{1}{3}\log_6(8w^6)$

20. a. $2 - \log_4(16z^2)$
   b. $1 - 2\log_3 x$

Use the three logs below to find the value of each expression.

$\log_b 2 = 0.6931, \quad \log_b 3 = 1.0986, \quad \log_b 5 = 1.6094$

(Hint: For example, $\log_b 15 = \log_b 3 + \log_b 5$.)

21. a. $\log_b 6$
   b. $\log_b \dfrac{2}{5}$

22. a. $\log_b 10$
   b. $\log_b \dfrac{3}{2}$

23. a. $\log_b 9$
   b. $\log_b \sqrt{50}$

24. a. $\log_b 25$
   b. $\log_b 75$

Solve each equation by using logarithms base 10.

25. $2^x = 7$

26. $3^x = 4$

27. $3^{x+1} = 8$

28. $2^{x-1} = 9$

29. $4^{x^2} = 15$ **30.** $3^{x^2} = 21$ **31.** $4.26^{-x} = 10.3$ **32.** $2.13^{-x} = 8.1$

33. $25 \cdot 3^{2.1x} = 47$ **34.** $12 \cdot 5^{1.5x} = 85$ **35.** $3600 = 20 \cdot 8^{-0.2x}$ **36.** $0.06 = 50 \cdot 4^{-0.6x}$

37. If raw meat is allowed to thaw at 50°F, *Salmonella* grows at a rate of 9% per hour.
  a. Write a formula for the amount of *Salmonella* present after t hours, if the initial amount is S_0.
  b. Health officials advise that the amount of *Salmonella* initially present in meat should not be allowed to increase by more than 50%. How long can meat be left to thaw at 50°F?

38. Starting in 1998, the demand for electricity in Ireland grew at a rate of 5.8% per year. In 1998, 20,500 gigawatts were used. (Source: Electricity Supply Board of Ireland)
  a. Write a formula for electricity demand in Ireland as a function of time.
  b. If demand continues to grow at the same rate, when would it reach 30,000 gigawatts?

39. The concentration of a certain drug injected into the bloodstream decreases by 20% each hour as the drug is eliminated from the body. The initial dose creates a concentration of 0.7 milligrams per milliliter.
  a. Write a function for the concentration of the drug as a function of time.
  b. The minimum effective concentration of the drug is 0.4 milligrams per milliliter. When should the second dose be administered?
  c. Verify your answer with a graph.

40. A small pond is tested for pollution and the concentration of toxic chemicals is found to be 80 parts per million. Clean water enters the pond from a stream, mixes with the polluted water, then leaves the pond so that the pollution level is reduced by 10% each month.
  a. Write a function for the concentration of toxic chemicals as a function of time.
  b. How long will it be before the concentration of toxic chemicals reaches a safe level of 25 parts per million?
  c. Verify your answer with a graph.

41. According to the National Council of Churches, the fastest growing denomination in the United States in 2004 was the Jehovah's Witnesses, with an annual growth rate of 1.82%.
  a. The Jehovah's Witnesses had 1,041,000 members in 2004. Write a formula for the membership in the Jehovah's Witnesses as a function of time, assuming that the church continues to grow at the same rate.
  b. When will the Jehovah's Witnesses have 2,000,000 members?

42. In 2004, the Presbyterian Church had 3,241,000 members, but membership was declining by 4.87% annually.
  a. Write a formula for the membership in the Presbyterian Church as a function of time, assuming that the membership continues to decline at the same rate.
  b. When will the Presbyterian Church have 2,000,000 members?

43. Sodium-24 is a radioactive isotope that is used in diagnosing circulatory disease. It decays into stable isotopes of sodium at a rate of 4.73% per hour.
  a. Technicians inject a quantity of sodium-24 into a patient's bloodstream. Write a formula for the amount of sodium-24 present in the bloodstream as a function of time.
  b. How long will it take for 75% of the isotope to decay?

44. The population of Afghanistan is growing at 2.6% per year.
  a. Write a formula for the population of Afghanistan as a function of time.
  b. In 2005, the population of Afghanistan was 29.9 million. At the current rate of growth, how long will it take the population to reach 40 million?

Evaluate each expression. Which (if any) are equal?

45. **a.** $\log_2(4 \cdot 8)$
    b. $(\log_2 4)(\log_2 8)$
    c. $\log_2 4 + \log_2 8$

46. **a.** $\log_2(16 + 16)$
    b. $\log_2 16 + \log_2 16$
    c. $\log_2 2 + \log_2 16$

47. **a.** $\log_3(27^2)$
    b. $(\log_3 27)^2$
    c. $\log_3 27 + \log_3 27$

48. **a.** $\log_3(3 \cdot 27)$
    b. $\log_3 3 + \log_3 27$
    c. $\log_3 3 \cdot \log_3 27$

49. a. $\log_{10}\left(\dfrac{240}{10}\right)$ **50. a.** $\log_{10}\left(\dfrac{1}{2}\cdot 80\right)$ **51. a.** $\log_{10}(75-15)$ **52. a.** $\log_{10}(8\cdot 25)$

b. $\dfrac{\log_{10} 240}{\log_{10} 10}$ **b.** $\dfrac{1}{2}\log_{10} 80$ **b.** $\log_{10} 75 - \log_{10} 15$ **b.** $\log_{10}(25^8)$

c. $\log_{10}(240-10)$ **c.** $\log_{10}\sqrt{80}$ **c.** $\dfrac{\log_{10} 75}{\log_{10} 15}$ **c.** $\log_{10} 8 + 25$

For Problems 53–58, use the formula for compound interest,

$$A = P\left(1+\frac{r}{n}\right)^{nt}$$

53. What rate of interest is required so that $1000 will yield $1900 after 5 years if the interest rate is compounded monthly?

54. What rate of interest is required so that $400 will yield $600 after 3 years if the interest rate is compounded quarterly?

55. How long will it take a sum of money to triple if it is invested at 10% compounded daily?

56. How long will it take a sum of money to increase by a factor of 5 if it is invested at 10% compounded quarterly?

57. a. Suppose you invest $1000 at 12% annual interest for 5 years. In this problem, we will investigate how the number of compounding periods, n, affects the amount, A. Write A as a function of n, with $P = 1000$, $r = 0.12$, and $t = 5$.
b. Use your calculator to make a table of values for A as a function of n. What happens to A as n increases?
c. What value of n is necessary to produce an amount $A > 1818$? To produce $A > 1820$? To produce $A > 1822$?
d. Graph the function $A(n)$ in the window

Xmin $= 0$ Xmax $= 52$
Ymin $= 1750$ Ymin $= 1850$

Describe the graph: Is it increasing or decreasing? Concave up or down? Does it appear to have an asymptote? Give your best estimate for the asymptote.

58. a. In this problem we will repeat Problem 49 for 4% interest. Write A as a function of n, with $P = 1000$, $r = 0.04$, and $t = 5$.
b. Use your calculator to make a table of values for A as a function of n. What happens to A as n increases?
c. What value of n is necessary to produce an amount $A > 1218$? To produce $A > 1220$? To produce $A > 1221.40$?
d. Graph the function $A(n)$ in the window

Xmin $= 0$ Xmax $= 52$
Ymin $= 1210$ Ymin $= 1225$

Describe the graph: Is it increasing or decreasing? Concave up or down? Does it appear to have an asymptote? Give your best estimate for the asymptote.

Solve each formula for the specified variable.

59. $N = N_0 a^{kt}$, for k **60.** $Q = Q_0 b^{t/2}$, for t **61.** $A = A_0(10^{kt} - 1)$, for t

62. $B = B_0(1 - 10^{-kt})$, for t **63.** $w = pv^q$, for q **64.** $l = p^a q^b$, for b

In Problems 65–68 we use the laws of exponents to prove the properties of logarithms.

65. We will use the first law of exponents, $a^p \cdot a^q = a^{p+q}$, to prove the first property of logarithms.
  a. Let $m = \log_b x$ and $n = \log_b y$. Rewrite these equations in exponential form:

  $x = $ _____ and $y = $ _____

  b. Now consider the expression $\log_b (xy)$. Replace x and y by your answers to part (a).
  c. Apply the first law of exponents to your expression in part (b).
  d. Use the definition of logarithm to simplify your answer to part (c).
  e. Refer to the definitions of m and n in part (a) to finish the proof.

66. We will use the second law of exponents, $\frac{a^p}{a^q} = a^{p-q}$, to prove the second property of logarithms.
  a. Let $m = \log_b x$ and $n = \log_b y$. Rewrite these equations in exponential form:

  $x = $ _____ and $y = $ _____

  b. Now consider the expression $\log_b \left(\frac{x}{y}\right)$. Replace x and y by your answers to part (a).
  c. Apply the second law of exponents to your expression in part (b).
  d. Use the definition of logarithm to simplify your answer to part (c).
  e. Refer to the definitions of m and n in part (a) to finish the proof.

67. We will use the third law of exponents, $(a^p)^q = a^{pq}$, to prove the third property of logarithms.
  a. Let $m = \log_b x$. Rewrite this equation in exponential form:

  $x = $ _____

  b. Now consider the expression $\log_b (x^k)$. Replace x by your answer to part (a).
  c. Apply the third law of exponents to your expression in part (b).
  d. Use the definition of logarithm to simplify your answer to part (c).
  e. Refer to the definition of m in part (a) to finish the proof.

68. a. Use the laws of exponents to explain why $\log_b 1 = 0$.
  b. Use the laws of exponents to explain why $\log_b b^x = x$.
  c. Use the laws of exponents to explain why $b^{\log_b x} = x$.

4.5 Exponential Models

Fitting an Exponential Function through Two Points

To write a formula for an exponential function, we need to know the initial value, a, and the growth or decay factor, b. We can find these two parameters if we know any two function values.

EXAMPLE 1 Find an exponential function that has the values $f(2) = 4.5$ and $f(5) = 121.5$.

Solution We would like to find values of a and b so that the given function values satisfy $f(x) = ab^x$. By substituting the function values into the formula, we can write two equations.

$$f(2) = 4.5 \text{ means } x = 2, \ y = 4.5, \qquad \text{so } ab^2 = 4.5$$
$$f(5) = 121.5 \text{ means } x = 5, \ y = 121.5, \qquad \text{so } ab^5 = 121.5$$

This is a system of equations in the two unknowns, a and b, but it is not a linear system. We can solve the system by the method of elimination, but we will *divide* one of the equations by the other.

$$\frac{ab^5}{ab^2} = \frac{121.5}{4.5}$$
$$b^3 = 27$$

Note that by dividing the two equations, we eliminated a, and we can now solve for b.

$$b^3 = 27$$
$$b = \sqrt[3]{27} = 3$$

Next we substitute $b = 3$ into either of the two equations and solve for a.

$$a(3)^2 = 4.5$$
$$a = \frac{4.5}{9} = 0.5$$

Thus, $a = 0.5$ and $b = 3$, so the function is $f(x) = 0.5(3^x)$.

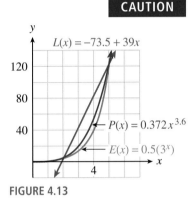

FIGURE 4.13

CAUTION

Knowing only two points on the graph of f is not enough to tell us what *kind* of function f is. Through the two points in Example 1, we can also fit a linear function or a power function. You can check that the three functions below all satisfy $f(2) = 4.5$ and $f(5) = 121.5$. The graphs of the functions are shown in Figure 4.13.

$$L(x) = -73.5 + 39x$$
$$P(x) = 0.372x^{3.6}$$
$$E(x) = 0.5(3^x)$$

However, if we already know that we are looking for an exponential function, we can follow the steps below to find its formula. This method is sometimes called the *ratio method*. (Of course, if one of the known function values is the initial value, we can find b without resorting to the ratio method.)

To find an exponential function $f(x) = ab^x$ through two points:

1. Use the coordinates of the points to write two equations in a and b.
2. Divide one equation by the other to eliminate a.
3. Solve for b.
4. Substitute b into either equation and solve for a.

EXERCISE 1

Use the ratio method to find an exponential function whose graph includes the points $(1, 20)$ and $(3, 125)$.

We can use the ratio method to find an exponential growth or decay model if we know two function values.

EXAMPLE 2

EXAMPLE 2

The unit of currency in Ghana is the cedi, denoted by ¢. Beginning in 1986, the cedi underwent a period of exponential inflation. In 1993, one U.S. dollar was worth ¢720, and in 1996, the dollar was worth about ¢1620. Find a formula for the number of cedi to the dollar as a function of time since 1986. What was the annual inflation rate?

Solution

We want to find a function $C(t) = ab^t$ for the number of cedi to the dollar, where $t = 0$ in 1986. We have two function values, $C(7) = 720$, and $C(10) = 1620$, and with these values we can write two equations.

$$ab^7 = 720$$
$$ab^{10} = 1620$$

We divide the second equation by the first to find

$$\frac{ab^{10}}{ab^7} = \frac{1620}{720}$$
$$b^3 = 2.25$$

Now we can solve this last equation for b to get $b = \sqrt[3]{2.25} \approx 1.31$. Finally, we substitute $b = 1.31$ into the first equation to find a.

$$a(1.31)^7 = 720$$
$$a = \frac{720}{1.31^7} = 108.75$$

Thus, $C(t) = 108.75(1.31)^t$, and the annual inflation rate was 31%.

EXERCISE 2

The number of earthquakes that occur worldwide is a decreasing exponential function of their magnitude on the Richter scale. Between 2000 and 2005, there were 7480 earthquakes of magnitude 5 and 793 earthquakes of magnitude 6. (Source: National Earthquake Information Center, U.S. Geological Survey)

a. Find a formula for the number of earthquakes, $N(m)$, in terms of their magnitude.

b. It is difficult to keep an accurate count of small earthquakes. Use your formula to estimate the number of magnitude 1 earthquakes that occurred between 2000 and 2005. How many earthquakes of magnitude 8 occurred?

Doubling Time

Instead of giving the rate of growth of a population, we can specify its rate of growth by giving the time it takes for the population to double.

EXAMPLE 3

In 2005, the population of Egypt was 74 million and was growing by 2% per year.

a. If it continues to grow at the same rate, how long will it take the population of Egypt to double?

b. How long will it take the population to double again?

c. Illustrate the results on a graph.

Solutions **a.** The population of Egypt is growing according to the formula $P(t) = 74(1.02)^t$, where t is in years and $P(t)$ is in millions. We would like to know when the population will reach 148 million (twice 74 million), so we solve the equation

$$74(1.02)^t = 148 \qquad \text{Divide both sides by 74.}$$
$$1.02^t = 2 \qquad \text{Take the log of both sides.}$$
$$t \log 1.02 = \log 2 \qquad \text{Divide both sides by log 1.02.}$$
$$t = \frac{\log 2}{\log 1.02} \approx 35 \text{ years}$$

It will take the population about 35 years to double.

b. Twice 148 million is 296 million, so we solve the equation

$$148(1.02)^t = 296 \qquad \text{Divide both sides by 128.}$$
$$1.02^t = 2 \qquad \text{Take the log of both sides.}$$
$$t \log 1.02 = \log 2 \qquad \text{Divide both sides by log 1.02.}$$
$$t = \frac{\log 2}{\log 1.02} \approx 35 \text{ years}$$

It will take the population about 35 years to double again.

c. A graph of $P(t) = 74(1.02)^t$ is shown in Figure 4.14. Note that the population doubles every 35 years.

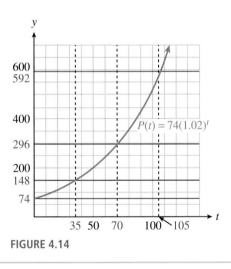

FIGURE 4.14

In Example 3, it took the population 35 years to double. Notice that the calculations in parts (a) and (b) are identical after the first step. In fact, we can start at any point, and it will take the population 35 years to double. We say that 35 years is the doubling time for this population. In the Homework problems, you will show that any increasing exponential function has a constant doubling time.

EXERCISE 3

In 2005, the population of Uganda was 26.9 million people and was growing by 3.2% per year.

  a. Write a formula for the population of Uganda as a function of years since 2005.

  b. How long will it take the population of Uganda to double?

  c. Use your formula from part (a) to verify the doubling time for three doubling periods.

     If we know the doubling time for a population, we can immediately write down its growth law. Because the population of Egypt doubles in 35 years, we can write

$$P(t) = 74 \cdot 2^{t/35}$$

In this form, the growth factor for the population is $2^{1/35}$, and you can check that, to five decimal places, $2^{1/35} = 1.02$.

Doubling Time

If D is the doubling time for an exponential function $P(t)$, then

$$P(t) = P_0 2^{t/D}$$

     So, from knowing the doubling time, we can easily find the growth rate of a population.

EXAMPLE 4   At its current rate of growth, the population of the United States will double in 115.87 years.

  a. Write a formula for the population of the United States as a function of time.

  b. What is the annual percent growth rate of the population?

Solutions   **a.** The current population of the United States is not given, so we represent it by P_0. With t expressed in years, the formula is then

$$P(t) = P_0 2^{t/115.87}$$

  b. We write $2^{t/115.87}$ in the form $(2^{1/115.87})^t$ to see that the growth factor is $b = 2^{1/115.87}$, or 1.006. For exponential growth, $b = 1 + r$, so $r = 0.006$, or 0.6%.

EXERCISE 4

At its current rate of growth, the population of Mexico will double in 36.8 years. What is its annual percent rate of growth?

Half-Life

The **half-life** of a decreasing exponential function is the time it takes for the output to decrease to half its original value. For example, the half-life of a radioactive isotope is the time it takes for half of the substance to decay. The half-life of a drug is the time it takes for half of the drug to be eliminated from the body. Like the doubling time, the half-life is constant for a particular function; no matter where you start, it takes the same amount of time to reach half that value.

EXAMPLE 5

If you take 200 mg of ibuprofen to relieve sore muscles, the amount of the drug left in your body after t hours is $Q(t) = 200(0.73)^t$.

  a. What is the half-life of ibuprofen?

  b. When will 50 mg of ibuprofen remain in your body?

  c. Use the half-life to sketch a graph of $Q(t)$.

Solutions
  a. To find the half-life, we calculate the time elapsed when only half the original amount, or 100 mg, is left.

$$200(0.73)^t = 100 \qquad \text{Divide both sides by 200.}$$
$$0.73^t = 0.5 \qquad \text{Take the log of both sides.}$$
$$t \log 0.73 = \log 0.5 \qquad \text{Divide both sides by } \log 0.73.$$
$$t = \frac{\log 0.5}{\log 0.73} = 2.2$$

The half-life is 2.2 hours.

  b. After 2.2 hours, 100 mg of ibuprofen is left in the body. After another 2.2 hours, half of that amount, or 50 mg, is left. Thus, 50 mg remain after 4.4 hours.

  c. We locate multiples of 2.2 hours on the horizontal axis. After each interval of 2.2 hours, the amount of ibuprofen is reduced to half its previous value. The graph is shown in Figure 4.15.

t	0	2.2	4.4	6.6	8.8
$Q(t)$	200	100	50	25	12.5

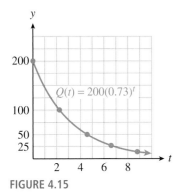

FIGURE 4.15

EXERCISE 5

Alcohol is eliminated from the body at a rate of 15% per hour.

  a. Write a decay formula for the amount of alcohol remaining in the body.

  b. What is the half-life of alcohol in the body?

Just as we can write an exponential growth law in terms of its doubling time, we can use the half-life to write a formula for exponential decay. For example, the half-life of ibuprofen is 2.2 hours, so every 2.2 hours the amount remaining is reduced by a factor of 0.5. After t hours a 200-mg dose will be reduced to

$$Q(t) = 200(0.5)^{t/2.2}$$

Once again, you can check that this formula is equivalent to the decay function given in Example 5.

> **Half-Life**
>
> If H is the half-life for an exponential function $Q(t)$, then
> $$Q(t) = Q_0(0.5)^{t/H}$$

Radioactive isotopes are molecules that decay into more stable molecules, emitting radiation in the process. Although radiation in large doses is harmful to living things, radioactive isotopes are useful as tracers in medicine and industry, and as treatment against cancer. The decay laws for radioactive isotopes are often given in terms of their half-lives.

EXAMPLE 6   Cobalt-60 is used in cold pasteurization to sterilize certain types of food. Gamma rays emitted by the isotope during radioactive decay kill any bacteria present without damaging the food. The half-life of cobalt-60 is 5.27 years.

a. Write a decay law for cobalt-60.

b. What is the annual decay rate for cobalt-60?

Solutions   **a.** We let $Q(t)$ denote the amount of cobalt-60 left after t years, and let Q_0 denote the initial amount. Every 5.27 years, $Q(t)$ is reduced by a factor of 0.5, so

$$Q(t) = Q_0(0.5)^{t/5.27}$$

b. We rewrite the decay law in the form $Q(t) = Q_0(1 + r)^t$ as follows:

$$Q(t) = Q_0(0.5)^{t/5.27} = Q_0((0.5)^{1/5.27})^t = Q_0(0.8768)^t$$

Thus, $1 - r = 0.8768$, so $r = 0.1232$, or 12.32%.

> **EXERCISE 6**
>
> Cesium-137, with a half-life of 30 years, is one of the most dangerous by-products of nuclear fission. What is the annual decay rate for cesium-137?

Annuities and Amortization

An **annuity** is a sequence of equal payments or deposits made at equal time intervals. A retirement fund is an example of an annuity. For ordinary annuities, payments are made at the end of each compounding period. The **future value** of an annuity is the sum of all the payments plus all the interest earned.

Future Value of an Annuity

If you make n payments per year for t years into an annuity that pays interest rate r compounded n times per year, the future value, FV, of the annuity is

$$FV = \frac{P\left[\left(1+\frac{r}{n}\right)^{nt} - 1\right]}{\frac{r}{n}}$$

where each payment is P dollars.

EXAMPLE 7

Greta plans to contribute $200 a month to a retirement fund that pays 5% interest compounded monthly.

a. What is the future value of Greta's retirement fund after 15 years?

b. For how many years must she contribute in order to accumulate $100,000?

Solutions

a. We evaluate the formula for FV when $P = 200$, $r = 0.05$, $n = 12$, and $t = 15$. Substituting these values into the formula, we find

$$FV = \frac{200\left[\left(1+\frac{0.05}{12}\right)^{12(15)} - 1\right]}{\frac{0.05}{12}}$$

$$= \frac{200[(1.004\overline{16})^{180} - 1]}{0.004\overline{16}} = 53{,}457.79$$

In 15 years, Greta's retirement fund will be worth $53,457.79.

b. We would like to find the value of t when $P = 200$, $r = 0.05$, $n = 12$, and $FV = 100{,}000$, so we must solve the equation

$$100{,}000 = \frac{200\left[\left(1+\frac{0.05}{12}\right)^{12t} - 1\right]}{\frac{0.05}{12}}$$ Isolate the expression in brackets.

$$\frac{1}{200}\left(\frac{0.05}{12}\right)100{,}000 = \left(1+\frac{0.05}{12}\right)^{12t} - 1$$ Simplify. Add 1 to both sides.

$$2.08\overline{3} + 1 = (1.004\overline{16})^{12t}$$ Take the log of both sides.

$$\log 3.08\overline{3} = 12t \log 1.004\overline{16}$$ Solve for t.

$$t = \frac{\log 3.08\overline{3}}{12 \log 1.004\overline{16}} \approx 22.6$$

Greta must contribute for over 22 years in order to accumulate $100,000.

Rufus is saving for a new car. He puts $2500 a year into an account that pays 4% interest compounded annually. How many years will it take him to accumulate $20,000? (Round up to the next whole year.)

In Example 7a, we knew the monthly deposits into the annuity and calculated how much the sum of all the deposits (plus interest) would be in the future. Now imagine that you have just retired and you want to begin drawing monthly payments from your retirement fund. The total amount accumulated in your fund is now its **present value,** and that amount must cover your future withdrawal payments.

Present Value of an Annuity

If you wish to receive n payments per year for t years from a fund that earns interest rate r compounded n times per year, the present value, PV, of the annuity must be

$$PV = \frac{P\left[1-\left(1+\frac{r}{n}\right)^{-nt}\right]}{\frac{r}{n}}$$

where each payment is P dollars.

EXAMPLE 8

Candace Welthy is setting up a college fund for her nephew Delbert that will provide $400 a month for the next 5 years. If the interest rate is 4% compounded monthly, how much money should she deposit now to cover the fund?

Solution   We would like to find the present value of an annuity in which $P = 400$, $r = 0.04$, $n = 12$, and $t = 5$. Substituting these values into the formula gives

$$PV = \frac{400\left[1 - \left(1 + \frac{0.04}{12}\right)^{-(12)(5)}\right]}{\frac{0.04}{12}}$$

$$= \frac{400[1 - (1.00\overline{3})]^{-60}}{0.00\overline{3}} = 21{,}719.63$$

Delbert's Aunt Welthy should deposit $21,719.63.

Payments on a loan, such as a home mortgage, are also an annuity, but in this case the monthly payments do not collect interest; instead, we must pay interest on the present value of the loan. Repaying a loan (plus interest) by making a sequence of equal payments is called **amortizing** the loan.

ANSWERS TO 4.5 EXERCISES

1. $f(x) = 8(2.5)^x$ **2. a.** $N(m) = 558{,}526{,}329(0.106)^m$ **b.** 59,212,751; 9

3. a. $P(t) = 26.9(1.032)^t$ million   **b.** 22 years

    c. $P(0) = 26.9$; $P(22) \approx 53.8$, $P(44) \approx 107.6$, $P(66) \approx 215.1$ **4.** 1.9%

5. a. $A(t) = A_0(0.85)^t$ **b.** 4.3 hours   **6.** 2.28%   **7.** 8 years   **8.** $1498.88

VOCABULARY *Look up the definitions of new terms in the Glossary.*

Doubling time   Amortization   Half-life   Annuity

CONCEPTS

1. We can use the ratio method to fit an exponential function through two points.

To find an exponential function $f(x) = ab^x$ through two points:

1. Use the coordinates of the points to write two equations in a and b.

2. Divide one equation by the other to eliminate a.

3. Solve for b.

4. Substitute b into either equation and solve for a.

2. Every increasing exponential has a fixed **doubling time.** Every decreasing exponential function has a fixed **half-life.**

3. If D is the doubling time for a population, its growth law can be written as $P(t) = P_0 2^{t/D}$.

4. If H is the half-life for a quantity, its decay law can be written as $Q(t) = Q_0(0.5)^{t/H}$.

5.

Future Value of an Annuity

If you make n payments per year for t years into an annuity that pays interest rate r compounded n times per year, the future value, FV, of the annuity is

$$FV = \frac{P\left[\left(1 + \dfrac{r}{n}\right)^{nt} - 1\right]}{\dfrac{r}{n}}$$

where each payment is P dollars.

6.

Present Value of an Annuity

If you wish to receive n payments per year for t years from a fund that earns interest rate r compounded n times per year, the present value, PV, of the annuity must be

$$PV = \frac{P\left[1 - \left(1 + \dfrac{r}{n}\right)^{-nt}\right]}{\dfrac{r}{n}}$$

where each payment is P dollars.

STUDY QUESTIONS

1. Compare the methods for fitting a line through two points and fitting an exponential function through two points.
2. A population of 3 million people has a doubling time of 15 years. What is the population 15 years from now? 30 years from now? 60 years from now?
3. Francine says that because the half-life of radium-223 is 11.7 days, after 23.4 days it will have all decayed. Is she correct? Why or why not?

4. Which is larger: the sum of all the deposits you make into your retirement fund, or the future value of the fund? Why?
5. Which is larger: the sum of all the payments you make towards your mortgage, or the amount of the loan? Why?

SKILLS *Practice each skill in the Homework problems listed.*

1. Fit an exponential function through two points: #1–18
2. Find the doubling time or half-life: #19–26
3. Write an exponential function, given the doubling time or half-life: #27–34, #39–42

4. Use the formula for future value of an annuity: #43 and 44
5. Use the formula for present value of an annuity: #45 and 46

HOMEWORK 4.5

ThomsonNOW™ Test yourself on key content at **www.thomsonedu.com/login.**

Find an exponential function that has the given values.

1. $A(0) = 0.14$, $A(3) = 7$

2. $B(0) = 28$, $B(5) = 0.25$

3. $f(7) = 12$, $f(8) = 9$

4. $g(2) = 2.6$, $g(3) = 3.9$

5. $M(4) = 100$, $M(7) = 0.8$

6. $N(12) = 512{,}000$, $N(14) = 1{,}024{,}000$

7. $s(3.5) = 16.2$, $s(6) = 3936.6$

8. $T(1.2) = 15$, $T(1.8) = 1.875$

Find a formula for the exponential function shown.

9.
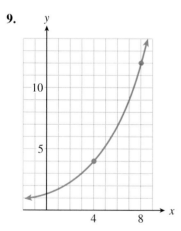

10.
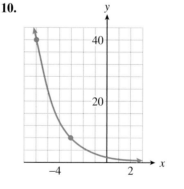

11.

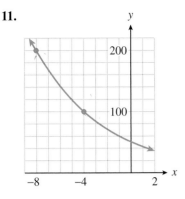

12.

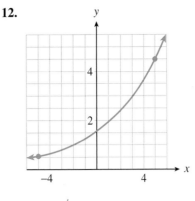

For Problems 13–18,
(a) Fit a linear function to the points.
(b) Fit an exponential function to the points.
(c) Graph both functions in the same window.

13. $(0, 2.6), (1, 1.3)$

14. $(0, 0.48), (1, 0.16)$

15. $(-6, 60), (-3, 12)$

16. $(2, 1.5), (4, 4.5)$

17. $(-2, 0.75), (4, 6)$

18. $(-1, 0.5), (1, 1)$

19. Nevada was the fastest growing state in the nation between 1990 and 2000, with an annual growth rate of over 5.2%.
   a. Write a function for the population of Nevada as a function of time. Let the initial population be P_0.
   b. How long will it take for the population to double?
   c. In 1990, the population of Nevada was 12 hundred thousand. Graph your function in the window Xmin = 0, Xmax = 47, Ymin = 0, Ymax = 100.
   d. Use **intersect** to verify that the population doubles from 12 to 24, from 24 to 48, and from 48 to 96 hundred thousand people in equal periods of time.

20. In 1986, the inflation rate in Bolivia was 8000% annually. The unit of currency in Bolivia is the boliviano.
   a. Write a formula for the price of an item as a function of time. Let P_0 be its initial price.
   b. How long did it take for prices to double? Give both an exact value and a decimal approximation rounded to two decimal places.
   c. Suppose $P_0 = 5$ bolivianos. Graph your function in the window Xmin = 0, Xmax = 0.94, Ymin = 0, Ymax = 100.
   d. Use **intersect** to verify that the price of the item doubles from 5 to 10 bolivianos, from 10 to 20, and from 20 to 40 in equal periods of time.

21. The gross domestic product (GDP) of the United Kingdom was 1 million pounds in the year 2000 and is growing at a rate of 2.8% per year. (The unit of currency in the U.K. is the pound, denoted by £.)
   a. Write a formula for the GDP as a function of years since 2000.
   b. How long will it take for the GDP to grow to 2 million pounds? Give both an exact value and a decimal approximation rounded to two decimal places.
   c. How long should it take for the GDP to 4 million pounds?
   d. Using your answers to (b) and (c), make a rough sketch of the function.

22. The number of *phishing* Web sites (fraudulent Web sites designed to trick victims into revealing personal financial information) is growing by 15% each month. In June 2005, there were 4000 phishing Web sites.
   (Source: www.itnews.com.au/newsstory)
   a. Write a formula for the number of phishing Web sites as a function of months since June 2005.
   b. How long will it take for the number of sites to reach 8000? Give both an exact value and a decimal approximation rounded to two decimal places.
   c. How long should it take for the number of sites to reach 16,000?
   d. Using your answers to (b) and (c), make a rough sketch of the function.

23. Radioactive potassium-42, which is used by cardiologists as a tracer, decays at a rate of 5.4% per hour.
  a. Find the half-life of potassium-42.
  b. How long will it take for three-fourths of the sample to decay? For seven-eighths of the sample?
  c. Suppose you start with 400 milligrams of potassium-42. Using your answers to (a) and (b), make a rough sketch of the decay function.

24. In October 2005, the *Los Angeles Times* published an article about efforts to save the endangered Channel Island foxes. "Their population declined by 95% to about 120 between 1994 and 2000, according to the park service."
  a. What was the fox population in 1994?
  b. Write a formula for the fox population as a function of time since 1994, assuming that their numbers declined exponentially.
  c. How long did it take for the fox population to be reduced to half its 1994 level? To one-quarter of the 1994 level?
  d. Using your answers to part (c), make a rough sketch of the decay function.

25. Caffeine leaves the body at a rate of 15.6% each hour. Your first cup of coffee in the morning has 100 mg of caffeine.
  a. How long will it take before you have 50 mg of that caffeine in your body?
  b. How long will it take before you have 25 mg of that caffeine in your body?
  c. Using your answers to (a) and (b), make a rough sketch of the decay function.

26. Pregnant women should monitor their intake of caffeine, because it leaves the body more slowly during pregnancy and can be absorbed by the unborn child through the bloodstream. Caffeine leaves a pregnant woman's body at a rate of 6.7% each hour.
  a. How long will it take before the 100 mg of caffeine in a cup of coffee is reduced to 50 mg?
  b. How long will it take before the 100 mg of caffeine in a cup of coffee is reduced to 25 mg?
  c. Make a rough sketch of the decay function, and compare with the graph in Problem 25.

For Problems 27–30,
(a) Write a growth or decay formula for the exponential function.
(b) Find the percent growth or decay rate.

27. A population starts with 2000 and has a doubling time of 5 years.

28. You have 10 grams of a radioactive isotope whose half-life is 42 years.

29. A certain medication has a half-life of 18 hours in the body. You are given an initial dose of D_0 mg.

30. The doubling time of a certain financial investment is 8 years. You invest an amount M_0.

31. The half-life of radium-226 is 1620 years.
  a. Write a decay law for radium-226.
  b. What is the annual decay rate for radium-226?

32. Dichloro-diphenyl-trichloroethane (DDT) is a pesticide that was used in the middle decades of the twentieth century to control malaria. After 1945, it was also widely used on crops in the United States, and as much as one ton might be sprayed on a single cotton field. However, after the toxic effects of DDT on the environment began to appear, the chemical was banned in 1972.
  a. A common estimate for the half-life of DDT in the soil is 15 years. Write a decay law for DDT in the soil.
  b. In 1970, many soil samples in the United States contained about 0.5 mg of DDT per kg of soil. The NOAA (National Oceanic and Atmospheric Administration) safe level for DDT in the soil is 0.008 mg/kg. When will DDT content in the soil be reduced to a safe level?

33. In 1798, the English political economist Thomas R. Malthus claimed that human populations, unchecked by environmental or social constraints, double every 25 years, regardless of the initial population size.

   a. Write a growth law for human populations under these conditions.

   b. What is the growth rate in unconstrained conditions?

35. Let $y = f(t) = ab^t$ be an exponential growth function, with $a > 0$ and $b > 1$.

   a. Suppose that the value of y doubles from $t = 0$ to $t = D$, so that $f(D) = 2 \cdot f(0)$. Rewrite this fact as an equation in terms of a, b, and D.

   b. What does your answer to (a) tell you about the value of b^D?

   c. Use the first law of exponents and your result from (b) to rewrite $f(t + D)$ in terms of $f(t)$.

   d. Explain why your result from (c) shows that the doubling time is constant.

37. Let $y = g(t) = ab^t$ be an exponential decay function, with $a > 0$ and $0 < b < 1$. In this problem, we will show that there is a fixed value R such that y is decreased by a factor of $\frac{1}{3}$ every R units.

   a. Suppose that $g(R) = \frac{1}{3} \cdot g(0)$. Rewrite this fact as an equation in terms of a, b, and R.

   b. What does your answer to (a) tell you about the value of b^R?

   c. Use the first law of exponents and your result from (b) to rewrite $g(t + R)$ in terms of $g(t)$.

   d. Explain why your result from (c) shows that an exponential decay function has a constant "one-third-life."

In Problems 39–42,
(a) Write a decay law for the isotope.
(b) Use the decay law to answer the question.

39. Carbon-14 occurs in living organisms with a fixed ratio to nonradioactive carbon-12. After a plant or animal dies, the carbon-14 decays into stable carbon with a half-life of 5730 years. When samples from the Shroud of Turin were analyzed in 1988, they were found to have 91.2% of their original carbon-14. How old were those samples in 1988?

41. Americium-241 (Am-241) is used in residential smoke detectors. Particles emitted as Am-241 decays cause the air in a smoke alarm to ionize, allowing current to flow between two electrodes. If smoke absorbs the particles, the current changes and sets off the alarm. The half-life of Am-241 is 432 years. How long will it take for 30% of the Am-241 to decay?

34. David Sifry observed in 2005 that over the previous two years, the number of Weblogs, or *blogs,* was doubling every 5 months. (Source: www.sifry.com/alerts/archives)

   a. Write a formula for the number of blogs t years after January 2005, assuming it continues to grow at the same rate.

   b. What is the growth rate for the number of blogs?

36. Let $y = g(t) = ab^t$ be an exponential decay function, with $a > 0$ and $0 < b < 1$.

   a. Suppose that the value of y is halved from $t = 0$ to $t = H$, so that $g(H) = \frac{1}{2} \cdot f(0)$. Rewrite this fact as an equation in terms of a, b, and H.

   b. What does your answer to (a) tell you about the value of b^H?

   c. Use the first law of exponents and your result from (b) to rewrite $g(t + H)$ in terms of $g(t)$.

   d. Explain why your result from (c) shows that the half-life is constant.

38. Let $y = f(t) = ab^t$ be an exponential growth function, with $a > 0$ and $b > 1$. In this problem, we will show that there is a fixed value T such that y triples every T units.

   a. Suppose that $f(T) = 3 \cdot f(0)$. Rewrite this fact as an equation in terms of a, b, and T.

   b. What does your answer to (a) tell you about the value of b^T?

   c. Use the first law of exponents and your result from (b) to rewrite $f(t + T)$ in terms of $f(t)$.

   d. Explain why your result from (c) shows that an exponential growth function has a constant tripling time.

40. Rubidium-strontium radioactive dating is used in geologic studies to measure the age of minerals. Rubidium-87 decays into strontium-87 with a half-life of 48.8 billion years. Several meteors were found to have 93.7% of their original rubidium. How old are the meteors?

42. Doctors can measure the amount of blood in a patient by injecting a known volume of red blood cells tagged with chromium-51. After allowing the blood to mix, they measure the percentage of tagged cells in a sample of the patient's blood and use a proportion to compute the original blood volume. Chromium-51 has a half-life of 27.7 days. How much of the original chromium-51 will still be present after 2 days?

For Problems 43 and 44, use the formula for future value of an annuity.

43. You want to retire with a nest egg of one million dollars. You plan to make fixed monthly payments of $1000 into a savings account until then. How long will you need to make payments if the account earns 6% interest compounded monthly? What if the annual interest rate is 5%?

44. Francine plans to make monthly payments into an account to save up for a cruise vacation. She wants to save $25,000 for the trip. How many $200 payments will she need if the account pays 3% interest compounded monthly? What if the rate is 4%?

For Problems 45 and 46, use the formula for present value of an annuity.

45. You want to finance $25,000 to purchase a new car, and your financing institution charges an annual interest rate of 2.7%, compounded monthly. How large will your monthly payment be to pay off the loan in 5 years? In 6 years?

46. Delbert has accumulated $5000 in credit card debt. The account charges an annual interest rate of 17%, compounded monthly. Delbert decides not to make any further charges to his account and to pay it off in equal monthly payments. What will the payment be if Delbert decides to pay off the entire amount in 5 years? In 10 years?

47. Moore's law predicts that the number of transistors per computer chip will continue to grow exponentially, with a doubling time of 18 months.
   a. Write a formula for Moore's law, with t in years and $M_0 = 2200$ in 1970.
   b. From 1970 to 1999, the number of transistors per chip was actually modeled approximately by $N(t) = 2200(1.356)^t$. How does this function compare with your answer to part (a)?
   c. Complete the table showing the number of transistors per chip in recent years, the number predicted by Moore's law, and the number predicted by $N(t)$.

Name of chip	Year	Moore's law	$N(t)$	Actual number
Pentium IV	2000			42,000,000
Pentium M (Banias)	2003			77,000,000
Pentium M (Dothan)	2004			140,000,000

   d. What is the doubling time for $N(t)$?

48. If the population of a particular animal is very small, inbreeding will cause a loss of genetic diversity. In a population of N individuals, the percent of the species' original genetic variation that remains after t generations is given by

$$V = V_0 \left(1 - \frac{1}{2N}\right)^t$$

(Source: Chapman and Reiss, 1992)

   a. Assuming $V_0 = 100$, graph V as a function of t for three different values of N: $N = 1000$, 100, and 10.
   b. Fill in the table to compare the values of V after 5, 50, and 100 generations.

Population size	Number of generations		
	5	50	100
1000			
100			
10			

   c. Studies of the cheetah have revealed variation at only 3.2% of its genes. (Other species show variation at 10 to 43% of their genes.) The population of cheetah may be less than 5000. Assuming the population can be maintained at its current level, how many generations will it take before the cheetah's genetic variation is reduced to 1%?

KEY CONCEPTS

1. If a quantity is multiplied by a constant factor, b, in each time period, we say that it undergoes **exponential growth** or **decay**. The constant b is called the **growth factor** if $b > 1$ and the **decay factor** if $0 < b < 1$.

2. Quantities that increase or decrease by a constant percent in each time period grow or decay exponentially.

3.

Exponential Growth and Decay

The function

$$P(t) = P_0 b^t$$

models exponential growth and decay.

$P_0 = P(0)$ is the initial value of P;
b is the growth or decay factor.

1. If $b > 1$, then $P(t)$ is increasing, and $b = 1 + r$, where r represents percent increase.

2. If $0 < b < 1$, then $P(t)$ is decreasing, and $b = 1 - r$, where r represents percent decrease.

4.

Interest Compounded Annually

The **amount $A(t)$** accumulated (principal plus interest) in an account bearing interest compounded annually is

$$A(t) = (1 + r)^t$$

where P is the principal invested,
        r is the interest rate,
        t is the time period, in years.

5. In linear growth, a constant amount is *added* to the output for each unit increase in the input. In exponential growth, the output is *multiplied* by a constant factor for each unit increase in the input.

6. An **exponential function** has the form

$$f(x) = ab^x, \quad \text{where} \quad b > 0 \quad \text{and} \quad b \neq 1, \ a \neq 0$$

7.

Properties of Exponential Functions, $f(x) = ab^x, a > 0$

1. Domain: all real numbers.

2. Range: all positive numbers.

3. If $b > 1$, the function is increasing and concave up; if $0 < b < 1$, the function is decreasing and concave up.

4. The y-intercept is $(a, 0)$. There is no x-intercept.

8. The graphs of exponential functions can be transformed by shifts, stretches, and reflections.

9.

Reflections of Graphs

1. The graph of $y = -f(x)$ is the reflection of the graph of $y = f(x)$ about the x-axis.

2. The graph of $y = f(-x)$ is the reflection of the graph of $y = f(x)$ about the y-axis.

10. Exponential functions $f(x) = ab^x$ have different properties than power functions $f(x) = kx^p$.

11. We can solve **exponential equations** by writing both sides with the same **base** and equating the exponents.

12. We can use graphs to find approximate solutions to exponential equations.

13. We use logrithms to help us solve exponential equations.

14. The **base b logarithm of x**, written $\log_b x$, is the exponent to which b must be raised in order to yield x.

15. If $b > 0$, $b \neq 1$, and $x > 0$,

$$y = \log_b x \qquad \text{if and only if} \qquad x = b^y$$

16. The operation of taking a base b logarithm is the *inverse* operation for raising the base b to a power.

17. Base 10 logarithms are called **common logarithms,** and $\log x$ means $\log_{10} x$.

18.

Steps for Solving Base 10 Exponential Equations

1. Isolate the power on one side of the equation.
2. Rewrite the equation in logarithmic form.
3. Use a calculator, if necessary, to evaluate the logarithm.
4. Solve for the variable.

19.

Properties of Logarithms

If x, y, and $b > 0$, then

1. $\log_b (xy) = \log_b x + \log_b y$
2. $\log_b \dfrac{x}{y} = \log_b x - \log_b y$
3. $\log_b x^k = k \log_b x$

20. We can use the properties of logarithms to solve exponential equations with any base.

21.

Compound Interest

The **amount** $A(t)$ accumulated (principal plus interest) in an account bearing interest compounded n times annually is

$$A(t) = P \left(1 + \frac{r}{n}\right)^{nt}$$

where P is the principal invested,

         r is the interest rate,

         t is the time period, in years.

22. We can use the ratio method to fit an exponential function through two points.

To find an exponential function $f(x) = ab^x$ through two points:

1. Use the coordinates of the points to write two equations in a and b.
2. Divide one equation by the other to eliminate a.
3. Solve for b.
4. Substitute b into either equation and solve for a.

23. Every increasing exponential has a fixed **doubling time.** Every decreasing exponential function has a fixed **half-life.**

24. If D is the doubling time for a population, its growth law can be written as $P(t) = P_0 2^{t/D}$.

25. If H is the half-life for a quantity, its decay law can be written as $Q(t) = Q_0 (0.5)^{t/H}$.

26.

Future Value of an Annuity

If you make n payments per year for t years into an annuity that pays interest rate r compounded n times per year, the future value, FV, of the annuity is

$$FV = \frac{P \left[\left(1 + \dfrac{r}{n}\right)^{nt} - 1\right]}{\dfrac{r}{n}}$$

where each payment is P dollars.

27.

Present Value of an Annuity

If you wish to receive n payments per year for t years from a fund that earns interest rate r compounded n times per year, the present value, PV, of the annuity must be

$$PV = \frac{P \left[1 - \left(1 + \dfrac{r}{n}\right)^{-nt}\right]}{\dfrac{r}{n}}$$

where each payment is P dollars.

REVIEW PROBLEMS

For Problems 1–4,
(a) Write a function that describes exponential growth or decay.
(b) Evaluate the function at the given values.

1. The number of computer science degrees awarded by Monroe College has increased by a factor of 1.5 every 5 years since 1984. If the college granted 8 degrees in 1984, how many did it award in 1994? In 2005?

2. The price of public transportation has been rising by 10% per year since 1975. If it cost $0.25 to ride the bus in 1975, how much did it cost in 1985? How much will it cost in the year 2010 if the current trend continues?

3. A certain medication is eliminated from the body at a rate of 15% per hour. If an initial dose of 100 milligrams is taken at 8 a.m., how much is left at 12 noon? At 6 p.m.?

4. After the World Series, sales of T-shirts and other baseball memorabilia decline 30% per week. If $200,000 worth of souvenirs were sold during the Series, how much will be sold 4 weeks later? After 6 weeks?

Use the laws of exponents to simplify.

5. $(4n^{x+5})^2$

6. $9^x \cdot 3^{x-3}$

7. $\dfrac{m^{x+2}}{m^{2x+4}}$

8. $\sqrt[3]{8^{2x+1} \cdot 8^{x-2}}$

Find a growth or decay law for the function.

9.

t	0	1	2	3
$g(t)$	16	13.6	11.56	9.83

10.

t	0	1	2	3
$f(t)$	12	19.2	30.72	49.15

11.

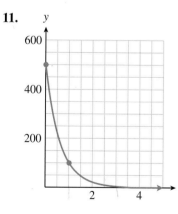

12.

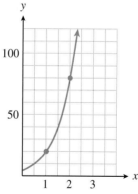

13. The president's approval rating increased by 12% and then decreased by 15%. What was the net change in his approval rating?

14. The number of students at Salt Creek Elementary School fell by 18% last year but increased by 26% this year. What was the net change in the number of students?

15. Enviroco's stock is growing exponentially in value and increased by 33.8% over the past 5 years. What was its annual rate of increase?

16. Sales of the software package Home Accountant 3.0 fell exponentially when the new version came out, decreasing by 60% over the past 3 months. What was the monthly rate of decrease?

For Problems 17–20,
(a) Graph the function.
(b) List all intercepts and asymptotes.
(c) Give the range of the function on the domain [−3, 3].

17. $f(t) = 6(1.2)^t$ **18.** $g(t) = 35(0.6)^{-t}$ **19.** $P(x) = 2^x - 3$ **20.** $R(x) = 2^{x+3}$

Solve each equation.

21. $3^{x+2} = 9^{1/3}$ **22.** $2^{x-1} = 8^{-2x}$ **23.** $4^{2x+1} = 8^{x-3}$ **24.** $3^{x^2-4} = 27$

For Problems 25–28,
(a) Graph both functions in the same window. Are they equivalent?
(b) Justify your answer to part (a) algebraically.

25. $P(t) = 5(2^{t/8}), \quad Q(t) = 5(1.0905)^t$

26. $M(x) = 4(3^{x/5}), \quad N(x) = 4(1.2457)^x$

27. $H(x) = \left(\dfrac{1}{3}\right)^{x-2}, G(x) = 9\left(\dfrac{1}{3}\right)^x$

28. $F(x) = \left(\dfrac{1}{2}\right)^{2x-3}, L(x) = 8\left(\dfrac{1}{4}\right)^x$

For Problems 29–32, $f(x) = 2^x$.
(a) Write a formula for the function.
(b) Use transformations to sketch the graph, indicating any intercepts and asymptotes.

29. $y = 4 + f(x+1)$ **30.** $y = -3 + f(x-2)$ **31.** $y = 6 - 3f(x)$ **32.** $y = 10 - 4f(x)$

In Problems 33–36, we compare power and exponential functions. Let

$$f(x) = 4x^{1.5}, \quad g(x) = 4(1.5)^x$$

33. Graph both functions in the window Xmin = 0, Xmax = 10, Ymin = 0, Ymax = 120. Which function grows more rapidly for large values of x?

34. Estimate the solutions of $f(x) = g(x)$. For what values of x is $f(x) > g(x)$?

35. When x doubles from 2 to 4, $f(x)$ grows by a factor of _____, and $g(x)$ grows by a factor of _____.

36. What is the range of $f(x)$ on the domain [0, 100]? What is the range of $g(x)$ on the same domain?

37. "Within belts of uniform moisture conditions and comparable vegetation, the organic matter content of soil decreases exponentially with increasing temperature." Data indicate that the organic content doubles with each 10°C decrease in temperature. Write a formula for this function, stating clearly what each variable represents. (Source: Leopold, Wolman, Gordon, and Miller, 1992)

38. In 1951, a study of barley yields under diverse soil conditions led to the formula

$$Y = cV^a G^b$$

where V is a soil texture rating, G is a drainage rating, and a, b, and c are constants. In fields with similar drainage systems, the formula gives barley yields, Y, as a function of V, the soil texture. What type of function is it? If it is an increasing function, what can you say about a? (Source: Briggs and Courtney, 1985)

Find each logarithm.

39. $\log_2 16$ **40.** $\log_4 2$ **41.** $\log_3 \frac{1}{3}$ **42.** $\log_7 7$

43. $\log_{10} 10^{-3}$ **44.** $\log_{10} 0.0001$

Write each equation in logarithmic form.

45. $0.3^{-2} = x + 1$

46. $4^{0.3t} = 3N_0$

Solve.

47. $4 \cdot 10^{1.3x} = 20.4$

48. $127 = 2(10^{0.5x}) - 17.3$

49. $3(10^{-0.7x}) + 6.1 = 9$

50. $40(1 - 10^{-1.2x}) = 30$

Write each expression in terms of simpler logarithms. (Assume that all variables and variable expressions denote positive real numbers.)

51. $\log_b \left(\dfrac{xy^{1/3}}{z^2} \right)$

52. $\log_b \sqrt{\dfrac{L^2}{2R}}$

53. $\log_{10} \left(x \sqrt[3]{\dfrac{x}{y}} \right)$

54. $\log_{10} \sqrt{(s-a)(s-g)^2}$

Write each expression as a single logarithm with coefficient 1.

55. $\dfrac{1}{3}(\log_{10} x - 2\log_{10} y)$

56. $\dfrac{1}{2}\log_{10}(3x) - \dfrac{2}{3}\log_{10} y$

57. $\dfrac{1}{3}\log_{10} 8 - 2(\log_{10} 8 - \log_{10} 2)$

58. $\dfrac{1}{2}(\log_{10} 9 + 2\log_{10} 4) + 2\log_{10} 5$

Solve each equation by using base 10 logarithms.

59. $3^{x-2} = 7$

60. $4 \cdot 2^{1.2x} = 64$

61. $1200 = 24 \cdot 6^{-0.3x}$

62. $0.08 = 12 \cdot 3^{-1.5x}$

63. Solve $N = N_0(10^{kt})$ for t.

64. Solve $Q = R_0 + R\log_{10} kt$ for t.

65. The population of Dry Gulch has been declining according to the function

$$P(t) = 3800 \cdot 2^{-t/20}$$

where t is the number of years since the town's heyday in 1910.
a. What was the population of Dry Gulch in 1990?
b. In what year will the population dip below 120 people?

66. The number of compact discs produced each year by Delta Discs is given by the function

$$N(t) = 8000 \cdot 3^{t/4}$$

where t is the number of years since discs were introduced in 1980.
a. How many discs did Delta produce in 1989?
b. In what year will Delta first produce over 2 million discs?

67. a. Write a formula for the cost of a camera t years from now if it costs $90 now and the inflation rate is 6% annually.
b. How much will the camera cost 10 months from now?
c. How long will it be before the camera costs $120?

68. a. Write a formula for the cost of a sofa t years from now if it costs $1200 now and the inflation rate is 8% annually.
b. How much will the sofa cost 20 months from now?
c. How long will it be before the sofa costs $1500?

69. Francine inherited $5000 and plans to deposit the money in an account that compounds interest monthly.
   a. If she can get 5.5% interest, how long will it take for the money to grow to $7500?
   b. What interest rate will she need if she would like the money to grow to $6000 in 3 years?

70. Delbert received a signing bonus of $2500 and wants to invest the money in a certificate of deposit (CD) that compounds interest quarterly.
   a. If the CD pays 4.8% interest, how long will it take his money to grow to $3000?
   b. What interest rate will he need if he would like the money to grow to $3000 in 1 year?

Find an exponential growth or decay function that fits the data.

71. $f(2) = 1714$, $f(4) = 1836$

72. $g(1) = 10{,}665$, $g(6) = 24{,}920$

73. $g(1) = 45$, $g(5) = 0.00142$

74. $f(2) = 17.464$, $f(5) = 16.690$

75. The population of Sweden is growing at 0.1% annually.
   a. What is the doubling time for Sweden's population?
   b. In 2005, the population of Sweden was 9 million. At the current rate of growth, how long will it take the population to reach 10 million?

76. The bacteria *E. sakazakii* is found in powdered infant formula and can has a doubling time of 4.98 hours even if kept chilled to 50°F.
   a. What is the hourly growth rate for *E. sakazakii*?
   b. How long would it take a colony of *E. sakazakii* to increase by 50%?

77. Manganese-53 decays to chromium-53 with a half-life of 3.7 million years and is used to estimate the age of meteorites. What is the decay rate of manganese-53, with time expressed in millions of years?

78. The cold medication pseudoephedrine decays at a rate of 5.95% per hour in the body. What is the half-life of pseudoephedrine?

79. You would like to buy a house with a 20-year mortgage for $300,000, at an interest rate of 6.25%, compounded monthly. Use the formula for the present value of an annuity to calculate your monthly payment.

80. Rosalie's retirement fund pays 7% interest compounded monthly. Use the formula for the future value of an annuity to calculate how much should she contribute monthly in order to have $500,000 in 25 years.

81. An eccentric millionaire offers you a summer job for the month of June. She will pay you 2 cents for your first day of work and will double your wages every day thereafter. (Assume that you work every day, including weekends.)
   a. Make a table showing your wages on each day. Do you see a pattern?
   b. Write a function that gives your wages in terms of the number of days you have worked.
   c. How much will you make on June 15? On June 30?

82. The king of Persia offered one of his subjects anything he desired in return for services rendered. The subject requested that the king give him an amount of grain calculated as follows: Place one grain of wheat on the first square of a chessboard, two grains on the second square, four grains on the third square, and so on, until the entire chessboard is covered.
   a. Make a table showing the number of grains of wheat on each square of the chessboard.
   b. Write a function for the amount of wheat on each square.
   c. How many grains of wheat should be placed on the last (64th) square?

PROJECTS FOR CHAPTER 4

1. In 1772, the astronomer Johann Bode promoted a formula for the orbital radii of the six planets known at the time. This formula calculated the orbital radius, r, as a function of the planet's position, n, in line from the Sun. (Source: Bolton, 1974)

   a. Evaluate Bode's law, $r(n) = 0.4 + 0.3(2^{n-1})$, for the values in the table. (Use a large negative number, such as $n = -100$, to approximate $r(-\infty)$.)

n	$-\infty$	1	2	3	4	5	6
$r(n)$							

   b. How do the values of $r(n)$ compare with the actual orbital radii of the planets shown in the table? (The radii are given in astronomical units (AU). One AU is the distance from the Earth to the Sun, about 149.6×10^6 kilometers.) Assign values of n to each of the planets so that they conform to Bode's law.

Planet	Mercury	Venus	Earth
Orbital radius (AU)	0.39	0.72	1.00
n			

Planet	Mars	Jupiter	Saturn
Orbital radius (AU)	1.52	5.20	9.54
n			

   c. In 1781, William Herschel discovered the planet Uranus at a distance of 19.18 AU from the Sun. If $n = 7$ for Uranus, what does Bode's law predict for the orbital radius of Uranus?

   d. None of the planets' orbital radii corresponds to $n = 2$ in Bode's law. However, in 1801 the first of a group of asteroids between the orbits of Mars and Jupiter was discovered. The asteroids have orbital radii between 2.5 and 3.0 AU. If we consider the asteroids as one planet, what orbital radius does Bode's law predict?

   e. In 1846, Neptune was discovered 30.6 AU from the Sun, and Pluto was discovered in 1930 39.4 AU from the Sun. What orbital radii does Bode's law predict for these planets?

2. In 1665, there was an outbreak of the plague in London. The table shows the number of people who died of plague during each week of the summer that year. (Source: Bolton, 1974)

Week	Deaths	Week	Deaths
0, May 9	9	12, August 1	2010
1, May 16	3	13, August 8	2817
2, May 23	14	14, August 15	3880
3, May 30	17	15, August 22	4237
4, June 6	43	16, August 29	6102
5, June 13	112	17, September 5	6988
6, June 20	168	18, September 12	6544
7, June 27	267	19, September 19	7165
8, July 4	470	20, September 26	5533
9, July 11	725	21, October 3	4929
10, July 18	1089	22, October 10	4327
11, July 25	1843		

   a. Scale horizontal and vertical axes for the entire data set, but plot only the data for the first 8 weeks of the epidemic, from May 9 through July 4. On the same axes, graph the function $f(x) = 2.18 (1.83)^x$.

   b. By what weekly percent rate did the number of victims increase during the first eight weeks?

   c. Add data points for July 11 through October 10 to your graph. Describe the progress of the epidemic relative to the function f and offer an explanation.

   d. Make a table showing the total number of plague victims at the end of each week and plot the data. Describe the graph.

3. The Koch snowflake is an example of a fractal. It is named in honor of the Swiss mathematician Niels Fabian Helge von Koch (1870–1924). To construct a Koch snowflake, draw an equilateral triangle with sides of length 1 unit. This is stage $n = 0$. Divide each side into 3 equal segments and draw a smaller equilateral triangle on each middle segment, as shown in the figure. The new figure (stage $n = 1$) is a 6-pointed star with 12 sides. Repeat the process to obtain stage $n = 2$: Trisect each of the 12 sides and draw an equilateral triangle on each middle third. If you continue this process forever, the resulting figure is the Koch snowflake.

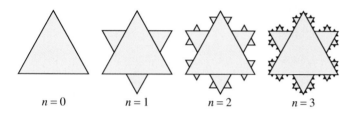

$n = 0$ $n = 1$ $n = 2$ $n = 3$

a. We will consider several functions related to the Koch snowflake:

$S(n)$ is the length of each side in stage n

$N(n)$ is the number of sides in stage n

$P(n)$ is the perimeter of the snowflake at stage n

Fill in the table describing the snowflake at each stage.

Stage n	$S(n)$	$N(n)$	$P(n)$
0			
1			
2			
3			

b. Write an expression for $S(n)$.
c. Write an expression for $N(n)$.
d. Write an expression for $P(n)$.
e. What happens to the perimeter as n gets larger and larger?
f. As n increases, the area of the snowflake increases also. Is the area of the completed Koch snowflake finite or infinite?

4. The Sierpinski carpet is another fractal. It is named for the Polish mathematician Waclaw Sierpinski (1882–1969). To build a Sierpinski carpet, start with a unit square (sides of length 1 unit.) For stage $n = 1$, trisect each side and partition the square into 9 smaller squares. Remove the center square, leaving a hole surrounded by 8 squares, as shown in the figure. For stage $n = 2$, repeat the process on each of the remaining 8 squares. If you continue this process forever, the resulting is the Sierpinski carpet.

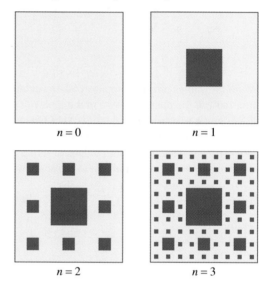

$n = 0$ $n = 1$

$n = 2$ $n = 3$

a. We will consider several functions related to the Sierpinski carpet:

$S(n)$ is the side of a new square at stage n

$A(n)$ is the area of a new square at stage n

$N(n)$ is the number of new squares removed at stage n

$R(n)$ is the total area removed at stage n

$T(n)$ is the total area remaining at stage n

Fill in the table describing the carpet at each stage.

Stage n	$S(n)$	$A(n)$	$N(n)$	$R(n)$	$T(n)$
1					
2					
3					

b. Write an expression for $S(n)$.
c. Write an expression for $A(n)$.
d. Write an expression for $N(n)$.
e. Write an expression for $R(n)$.
f. Write an expression for $T(n)$.
g. What happens to the area remaining as n approaches infinity?

5. The *order* of a stream or river is a measure of its relative size. A first-order stream is the smallest, one that has no tributaries. Second-order streams have only first-order streams as tributaries. Third-order streams may have first- and second-order streams as tributaries, and so on. The Mississippi River is an example of a tenth-order stream, and the Columbia River is ninth order. Both the number of streams of a given order and their average length are exponential functions of their order. In this problem, we consider all streams in the United States. (Source: Leopold, Luna, Gordon, and Miller, 1992)

a. Using the given values, find a function $N(x) = ab^{x-1}$ for the number of streams of a given order.

b. Complete the column for number of streams of each order. (Round to the nearest whole number of streams for each order.)

c. Find a function $L(x) = ab^{x-1}$ for the average length of streams of a given order, then complete that column.

d. Find the total length of all streams of each order, hence estimating the total length of all stream channels in the United States.

Order	Number	Average length	Total length
1	1,600,000	1	
2	339,200	2.3	
3			
4			
5			
6			
7			
8			
9			
10			

6. Related species living in the same area often evolve in different sizes to minimize competition for food and habitat. Here are the masses of eight species of fruit pigeon found in New Guinea, ranked from smallest to largest. (Source: Burton, 1998)

Size rank	1	2	3	4
Mass (grams)	49	76	123	163

Size rank	5	6	7	8
Mass (grams)	245	414	592	802

a. Plot the masses of the pigeons against their order of increasing size. What kind of function might fit the data?

b. Compute the ratios of the masses of successive sizes of fruit pigeons. Are the ratios approximately constant? What does this information tell you about your answer to part (a)?

c. Compute the average ratio to two decimal places. Using this ratio, estimate the mass of a hypothetical fruit pigeon of size rank 0.

d. Using your answers to part (c), write an exponential function that approximates the data. Graph this function on top of the data and evaluate the fit.

In Projects 7 and 8, we will prove the formulas in Section 4.5 for the present and future values of an annuity.
The future value of $M of money is its value in the future: its current value plus the interest it will accrue in the interval.
The present value of $M of money is the amount you would need to deposit now so that it will grow to $M in the future.

7. Suppose you deposit $100 at the end of every 6 months into an account that pays 4% compounded annually. How much money will be in the account at the end of 3 years?

   a. During the 3 years, you will make 6 deposits. Use the formula $F = P\left(1 + \frac{r}{n}\right)^{nt}$ to write an expression for the future value (principal plus interest) of each deposit. (Do not evaluate the expression!)

Deposit number	Amount deposited	Time in account	Future value
1	100	2.5	$100(1.02)^5$
2	100	2	
3	100	1.5	
4	100	1	
5	100	0.5	
6	100	0	

   b. Let S stand for the sum of the future values of all the deposits. Write out the sum, without evaluating the terms you found in part (a).

$$S =$$

   c. You could find S by working out all the terms and adding them up, but what if there were 100 terms, or more? We will use a trick to find the sum in an easier way. Multiply both sides of the equation in part (b) by 1.02. (Use the distributive law on the right side!)

$$1.02S =$$

   d. Now subtract the equation in part (b) from the equation in part (c). Be sure to line up like terms on the right side.

$$1.02S =$$
$$-S =$$
$$0.02S =$$

   e. Finally, solve for S. If you factor 100 from the numerator on the right side, your expression should look a lot like the formula for the future value of an annuity. (To help you see this, note that, for this example, $\frac{r}{n} = ?$ and $nt = ?$)

   f. Try to repeat the argument above, using letters for the parameters instead of numerical values.

8. You would like to set up an account that pays 4% interest compounded semiannually so that you can withdraw $100 at the end of every 6 months for the next 3 years. How much should you deposit now?

   a. During the next 3 years, you will make 6 withdrawals. Use the formula $P = A\left(1 + \frac{r}{n}\right)^{-nt}$ to write an expression for the present value of each of those withdrawals. (Do not evaluate the expression!)

Withdrawal number	Amount withdrawn	Time in account	Present value
1	100	0.5	$100(1.02)^{-1}$
2	100	1	
3	100	1.5	
4	100	2	
5	100	2.5	
6	10	3	

   b. Let S stand for the sum of the present values of all the withdrawals. Write out the sum, without evaluating the terms you found in part (a).

$$S =$$

   c. We will use a trick to evaluate the sum. Multiply both sides of the equation in part (b) by 1.02. (Use the distributive law on the right side!)

$$1.02S =$$

   d. Now subtract the equation in part (b) from the equation in part (c). Be sure to line up like terms on the right side.

$$1.02S =$$
$$-S =$$
$$0.02S =$$

   e. Finally, solve for S. If you factor 100 from the numerator on the right side, your expression should look a lot like the formula for the future value of an annuity. (To help you see this, note that, for this example, $\frac{r}{n} = ?$ and $nt = ?$)

   f. Try to repeat the argument above, using letters for the parameters instead of numerical values.

Logarithmic Functions

In Chapter 4, we used logarithms to solve exponential equations. In this chapter, we consider logarithmic functions as models in their own right. We also study another base for exponential and logarithmic functions, the natural base e, which is the most useful base for many scientific applications.

In 1885, the German philosopher Hermann Ebbinghaus conducted one of the first experiments on memory, using himself as a subject. He memorized lists of nonsense syllables and then tested his memory of the syllables at intervals ranging from 20 minutes to 31 days. After one hour, he remembered less than 50% of the items, but he found that the rate of forgetting leveled off over time. He modeled his data by the function

$$y = \frac{184}{2.88 \log t + 1.84}$$

Time elapsed	Percent remembered
20 minutes	58.2%
1 hour	44.2%
9 hours	35.8%
1 day	33.7%
2 days	27.8%
6 days	25.4%
31 days	21.1%

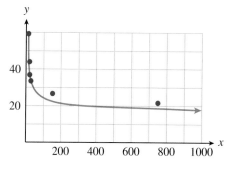

ThomsonNOW™

Throughout this chapter, the ThomsonNOW logo indicates an opportunity for online self-study, linking you to interactive tutorials and videos based on your level of understanding.

Ebbinghaus's model uses a logarithmic function. The graph of the data is called the "forgetting curve." Ebbinghaus's work, including his application of the scientific method to his research, provides part of the foundation of modern psychology.

3D4Medical.com/Getty Images

See Section 4.5 to read about the formula for computing mortgage payments.

When you buy a house, your monthly mortgage payment is a function of the size of the loan. Table 5.1 shows mortgage payments on 30-year loans of various sizes at 6% interest.

Loan amount, L	150,000	175,000	200,000	225,000	250,000
Mortgage payment, M	899.33	1049.21	1199.10	1348.99	1498.88

TABLE 5.1

For the function $M = f(L)$, the input value is the amount of the loan, and the output is the mortgage payment. However, when you are shopping for a house, you may think of the mortgage payment as the input variable: If you can afford a certain monthly mortgage payment, how large a loan can you finance? Now the mortgage payment is the input value, and the loan amount is the output. By interchanging the inputs and outputs, we define a new function, $L = g(M)$, shown in Table 5.2.

Mortgage payment, M	899.33	1049.21	1199.10	1348.99	1498.88
Loan amount, L	150,000	175,000	200,000	225,000	250,000

TABLE 5.2

This new function gives the same information as the original function, f, but from a different point of view. We call the function g the **inverse function** for f. The elements of the *range* of f are used as the input values for g, and the output values of g are the corresponding domain elements of f. For example, from the tables you can verify that $f(200,000) = 1199.10$, and $g(1199.10) = 200,000$. In fact, this property defines the inverse function.

Inverse Functions

Suppose g is the **inverse function** for f. Then

$$g(b) = a \quad \text{if and only if} \quad f(a) = b$$

EXAMPLE 1

Suppose g is the inverse function for f, and we know the following function values for f:

$$f(-3) = 5, \quad f(2) = 1, \quad f(5) = 0$$

Find $g(5)$ and $g(0)$.

Solution   We know that $g(5) = -3$ because $f(-3) = 5$, and $g(0) = 5$ because $f(5) = 0$. Tables may be helpful in visualizing the two functions, as shown below.

$y = f(x)$	
x	y
-3	5
2	1
5	0

→ Interchange the columns →

$x = g(y)$	
y	x
5	-3
1	2
0	5

For the function f, the input variable is x and the output variable is y. For the inverse function g, the roles of the variables are interchanged: y is now the input and x is the output.

Finding a Formula for the Inverse Function

If a function is given by a table of values, we can interchange the columns (or rows) of the table to obtain the inverse function. Swapping the columns works because we are really interchanging the input and output variables. If a function is defined by an equation, we can find a formula for its inverse function in the same way: Interchange the roles of the variables in the equation so that the old *output* variable becomes the new *input* variable.

EXAMPLE 2

a. The function $H = f(t) = 6 + 2t$ gives the height of corn seedlings, in inches, t days after they are planted. Find a formula for the inverse function and explain its meaning in this context.

b. Make a table of values for $f(t)$ and a table for its inverse function.

Solutions

a. Write the equation for f in the form

$$H = 6 + 2t$$

In this equation, t is the input and H is the output. We interchange the roles of the variables by solving for t to obtain

$$t = \frac{H - 6}{2}$$

In this equation, H is the input and t is the output. The formula for the inverse function is $t = g(H) = \frac{H-6}{2}$. The function g gives the number of days it will take the corn seedlings to grow to a height of H inches.

b. To make a table for f, we choose values for t and evaluate $f(t) = 6 + 2t$ at those t-values, as shown in Table 5.3a.

$H = f(t)$	
t	H
0	6
1	8
2	10
3	12

(a)

$t = g(H)$	
H	t
6	0
8	1
10	2
12	3

(b)

TABLE 5.3

To make a table for g, we could choose values for H and evaluate $\frac{H-6}{2}$, but because g is the inverse function for f, we can simply interchange the columns in our table for f, as shown in Table 5.3b. You can check that the values in the second table do satisfy the formula for the inverse function, $g(H) = \frac{H-6}{2}$. Note once again that the two tables show the *same relationship* between t and H, but the roles of input and output have been interchanged. The function f tells us the height of the seedlings after t days, and g tells us how long it will take the seedlings to grow to height H.

EXERCISE 2

Carol can burn 600 calories per hour bicycling and 400 calories per hour swimming. She would like to lose 5 pounds, which is equivalent to 16,000 calories.

a. Write an equation relating the number of hours of cycling, x, and the number of hours swimming, y, that Carol must spend to lose 5 pounds.

b. Write y as a function of x, $y = f(x)$. What does $f(10)$ tell you?

c. Find the inverse function, $x = g(y)$. What does $g(10)$ tell you?

Inverse Function Notation

If the inverse of a function f is also a function, we often denote the inverse by the symbol f^{-1}, read "f inverse." This notation makes it clear that the two functions are related in a special way. For example, the function $f(t) = 6 + 2t$ in Example 2 has inverse function $f^{-1}(H) = \frac{H-6}{2}$.

EXAMPLE 3   If $y = f(x) = x^3 + 2$, find $f^{-1}(10)$.

Solution   We first find the inverse function for $y = x^3 + 2$. Solve for x:

$$x^3 = y - 2 \qquad \text{Substract 2 from both sides.}$$
$$x = \sqrt[3]{y - 2} \qquad \text{Take cube roots.}$$

The inverse function is $x = f^{-1}(y) = \sqrt[3]{y - 2}$. Now we evaluate the inverse function at $y = 10$:

$$f^{-1}(10) = \sqrt[3]{10 - 2} = 2$$

CAUTION   Although the same symbol, $^{-1}$, is used for both reciprocals and inverse functions, the two notions are *not* equivalent. That is, the inverse of a given function is usually not the same as the reciprocal of that function. In Example 3, note that $f^{-1}(y)$ is not the same as the reciprocal of $f(y)$, because

$$\frac{1}{f(y)} = \frac{1}{y^3 + 2} \quad \text{but} \quad f^{-1}(y) = \sqrt[3]{y - 2}$$

To avoid confusion, we use the notation $\frac{1}{f}$ to refer to the reciprocal of the function f.

In Example 3, you can check that $f(2) = 10$. In fact, the two statements

$$f^{-1}(10) = 2 \quad \text{and} \quad f(2) = 10$$

are equivalent; they convey the same information. This fact is a restatement of our earlier observation about inverse functions, this time using inverse function notation.

Inverse Functions

Suppose the inverse of f is a function, denoted by f^{-1}. Then

$$f^{-1}(y) = x \quad \textbf{if and only if} \quad f(x) = y$$

EXERCISE 3

a. If $z = f(w) = \dfrac{1}{w+3}$, find $f^{-1}(1)$.

b. Write two equations about the value of $f^{-1}(1)$, one using f^{-1} and one using f.

c. Show that $f^{-1}(1)$ is not equal to $\dfrac{1}{f(1)}$.

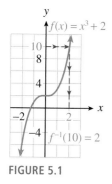

FIGURE 5.1

We can use a graph of a function $y = f(x)$ to find values of the inverse function $x = f^{-1}(y)$. Figure 5.1 shows the graph of $f(x) = x^3 + 2$. You already know how to evaluate a function from its graph: We start with the horizontal axis. For instance, to evaluate $f(-2)$, we find -2 on the x-axis, move vertically to the point on the graph with $x = -2$, in this case $(-2, -6)$, and read the y-coordinate of the point. We see that $f(-2) = -6$.

To evaluate the inverse function, we start with the *vertical axis*. For example, to find $f^{-1}(10)$, we find 10 on the vertical axis and move horizontally to the point on the graph with $y = 10$. In this case, the point is $(2, 10)$, so $f^{-1}(10) = 2$.

EXAMPLE 4

The function $C = h(F)$ gives Celsius temperature as a function of Fahrenheit temperature. The graph of the function is shown in Figure 5.2. Use the graph to evaluate $h(68)$ and $h^{-1}(10)$, and then explain their meaning in this context.

Solution   To evaluate $h(68)$, we find the input $F = 68$ on the horizontal axis, then find the point on the graph with $F = 68$ and read its vertical coordinate. We see that the point $(68, 20)$ lies on the graph, so $h(68) = 20$. When the Fahrenheit temperature is 68°, the Celsius temperature is 20°.

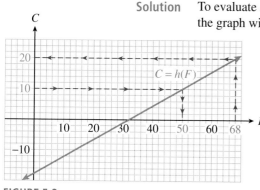

FIGURE 5.2

The inverse function reverses the roles of input and output. Because $C = h(F)$, $F = h^{-1}(C)$, so the inverse function gives us the Fahrenheit temperature if we know the Celsius temperature. In particular, $h^{-1}(10)$ is the Fahrenheit temperature when the Celsius temperature is 10°. To use the graph of h to find values of h^{-1}, we start with the *vertical* axis and find the point on the graph with $C = 10$. This point is $(50, 10)$, so $F = 50$ when $C = 10$, or $h^{-1}(10) = 50$. When the Celsius temperature is 10°, the Fahrenheit temperature is 50°.

a. Use the graph of h in Figure 5.2 to find $h^{-1}(-10)$.

b. Does $h^{-1}(-10) = -h^{-1}(10)$?

c. Write two equations, one using h and one using h^{-1}, stating the Fahrenheit temperature when the Celsius temperature is $0°$.

Graph of the Inverse Function

In Example 4, we used a graph of h to read values of h^{-1}. But we can also plot the graph of h^{-1} itself. Because C is the input variable for h^{-1}, we plot C on the horizontal axis and F on the vertical axis. To find some points on the graph of h^{-1}, we interchange the coordinates of points on the graph of h. The graph of h^{-1} is shown in Figure 5.3.

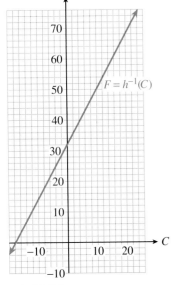

FIGURE 5.3

$C = h(F)$	
F	C
14	-10
32	0
50	10
68	20

$F = h^{-1}(C)$	
C	F
-10	14
0	32
10	50
20	68

EXAMPLE 5

The Park Service introduced a flock of 12 endangered pheasant into a wildlife preserve. After t years, the population of the flock was given by

$$P = f(t) = 12 + 2t^3$$

a. Graph the function on the domain $[0, 5]$.

b. Find a formula for the inverse function, $t = f^{-1}(P)$. What is the meaning of the inverse function in this context?

c. Sketch a graph of the inverse function.

Solutions

a. The graph of f is shown in Figure 5.4a, with t on the horizontal axis and P on the vertical axis.

b. We solve $P = 12 + 2t^3$ for t in terms of P.

$$2t^3 = P - 12 \qquad \text{Substract 12 from both sides.}$$

$$t^3 = \frac{P - 12}{2} \qquad \text{Divide both sides by 12.}$$

$$t = \sqrt[3]{\frac{P - 12}{2}} \qquad \text{Take cube roots.}$$

The inverse function is $t = f^{-1}(P) = \sqrt[3]{\frac{P-12}{2}}$. It tells us the number of years it takes for the pheasant population to grow to size P.

c. The graph of f^{-1} is shown in Figure 5.4b, with P on the horizontal axis and t on the vertical axis.

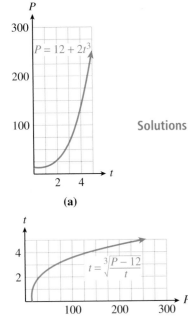

FIGURE 5.4

The formula $T = f(L) = 2\pi\sqrt{\frac{L}{32}}$ gives the period in seconds, T, of a pendulum as a function of its length in feet, L.

a. Graph the function on the domain $[0, 5]$.

b. Find a formula for the inverse function, $L = f^{-1}(T)$. What is the meaning of the inverse function in this context?

c. Sketch a graph of the inverse function.

When Is the Inverse a Function?

See Section 1.3 to review the vertical line test.

We can always find the inverse of a function simply by interchanging the input and output variables. In the preceding examples, interchanging the variables created a new function. However, *the inverse of a function does not always turn out to be a function* itself. For example, to find the inverse of $y = f(x) = x^2$, we solve for x to get $x = \pm\sqrt{y}$. When we regard y as the input and x as the output, the relationship does not describe a function. The graphs of f and its inverse are shown in Figure 5.5. (Note that for the graph of the inverse, we plot y on the horizontal axis and x on the vertical axis.) Since the graph of the inverse does not pass the vertical line test, it is *not* a function.

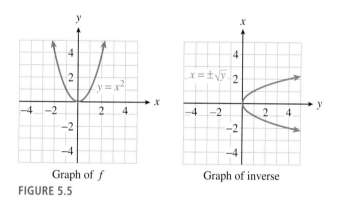

Graph of f Graph of inverse

FIGURE 5.5

For many applications, it is important to know whether or not the inverse of f is a function. This can be determined from the graph of f. When we interchange the roles of the input and output variables, horizontal lines of the form $y = k$ become vertical lines. Thus, if the graph of the *inverse* is going to pass the vertical line test, the graph of the *original function* must pass the **horizontal line test,** namely, that no horizontal line should intersect the graph in more than one point. Notice that the graph of $f(x) = x^2$ does not pass the horizontal line test, so we would not expect its inverse to be a function.

Horizontal Line Test

If no horizontal line intersects the graph of a function more than once, then the inverse is also a function.

EXAMPLE 6   Which of the functions in Figure 5.6 have inverses that are also functions?

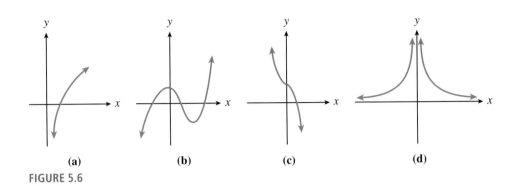

FIGURE 5.6

Solution   In each case, apply the horizontal line test to determine whether the inverse is a function. Because no horizontal line intersects their graphs more than once, the functions pictured in Figures 5.6a and 5.6c have inverses that are also functions. The functions in Figures 5.6b and 5.6d do not have inverses that are functions.

EXERCISE 6

Which of the functions whose graphs are shown in Figure 5.7 have inverses that are also functions?

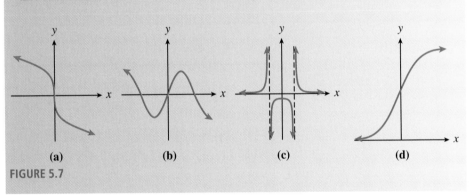

FIGURE 5.7

A function that passes the horizontal line test is called **one-to-one,** because each input has only one output and each output has only one input. A one-to-one function passes the horizontal line test as well as the vertical line test. With this terminology, we can state the following theorem.

One-to-one Functions

The inverse of a function f is also a function if and only if f is one-to-one.

CAUTION   A function may have an inverse function even if we cannot find its formula. The function $f(x) = x^5 + x + 1$ shown in Figure 5.8a is one-to-one, so it has an inverse function. We can even graph the inverse function, as shown in Figure 5.8b, by interchanging the coordinates of points on the graph of f. However, we cannot find a formula for the inverse function because we cannot solve the equation $y = x^5 + x + 1$ for x in terms of y.

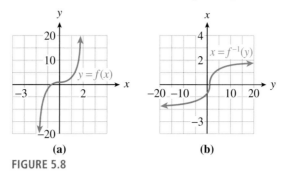

(a)   (b)

FIGURE 5.8

Mathematical Properties of the Inverse Function

The inverse function f^{-1} undoes the effect of the function f. In Example 2, the function $f(t) = 6 + 2t$ multiplies the input by 2 and then adds 6 to the result. The inverse function $f^{-1}(H) = \frac{H-6}{2}$ undoes those operations in reverse order: It subtracts 6 from the input and then divides the result by 2. If we apply the function f to a given input value and then apply the function f^{-1} to the output from f, the end result will be the original input value. For example, if we choose $t = 5$ as an input value, we find that

$$f(5) = 6 + 2(5) = 16 \quad \text{Multiply by 2, then add 6.}$$

$$\text{and} \quad f^{-1}(16) = \frac{16-6}{2} = 5 \quad \text{Subtract 6, then divide by 2.}$$

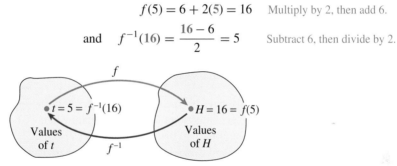

FIGURE 5.9

We return to the original input value, 5, as illustrated in Figure 5.9.

Example 7 illustrates the fact that if f^{-1} is the inverse function for f, then f is also the inverse function for f^{-1}.

EXAMPLE 7

Consider the function $f(x) = x^3 + 2$ and its inverse, $f^{-1}(y) = \sqrt[3]{y - 2}$.

a. Show that the inverse function undoes the effect of f on $x = 2$.

b. Show that f undoes the effect of the inverse function on $y = -25$.

Solutions

a. First evaluate the function f for $x = 2$:

$$f(2) = 2^3 + 2 = 10$$

Then evaluate the inverse function f^{-1} at $y = 10$:

$$f^{-1}(10) = \sqrt[3]{10 - 2} = \sqrt[3]{8} = 2$$

We started and ended with 2.

b. First evaluate the function f^{-1} for $y = -25$:

$$f^{-1}(-25) = \sqrt[3]{-25 - 2} = -3$$

Then evaluate the function f for $x = -3$:

$$f(-3) = (-3)^3 + 2 = -25$$

We started and ended with -25.

EXERCISE 7

a. Find a formula for the inverse of the function $f(x) = \dfrac{2}{x - 1}$.

b. Show that f^{-1} undoes the effect of f on $x = 3$.

c. Show that f undoes the effect of f^{-1} on $y = -2$.

We can state this property of inverse functions in the following way.

Functions and Inverse Functions

Suppose f^{-1} is the inverse function for f. Then

$$f^{-1}(f(x)) = x \quad \text{and} \quad f(f^{-1}(y)) = y$$

as long as x is in the domain of f, and y is in the domain of f^{-1}.

Symmetry

So far we have been careful to keep track of the input and output variables when we work with inverse functions. This is important when we are dealing with applications; the names of the variables are usually chosen because they have a meaning in the context of the application, and it would be confusing to change them.

However, we can also study inverse functions purely as mathematical objects. There is a relationship between the graph of a function and the graph of its inverse that is easier to see if we plot them both on the same set of axes. A graph does not change if we change the names of the variables, so we can let x represent the input for both functions, and let y represent the output. Consider the function $C = h(F)$ from Example 4, and its inverse function, $F = h^{-1}(C)$. The formulas for these functions are

$$C = h(F) = \frac{5}{9}(F - 32)$$

$$F = h^{-1}(C) = 32 + \frac{9}{5}C$$

But their graphs are the same if we write them as

$$y = h(x) = \frac{5}{9}(x - 32)$$

$$y = h^{-1}(x) = 32 + \frac{9}{5}x$$

The graphs are shown in Figure 5.10.

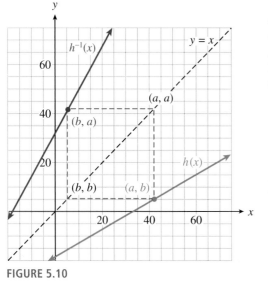

FIGURE 5.10

Now, for every point (a, b) on the graph of f, the point (b, a) is on the graph of the inverse function. Observe in Figure 5.10 that the points (a, b) and (b, a) are always located symmetrically across the line $y = x$. The graphs are **symmetric about the line $y = x$**, which means that if we were to place a mirror along the line $y = x$, each graph would be the reflection of the other.

EXAMPLE 8

Graph the function $f(x) = 2\sqrt{x + 4}$ on the domain $[-4, 12]$. Graph its inverse function f^{-1} on the same grid.

Solution

The graph of f has the same shape as the graph of $y = \sqrt{x}$, shifted 4 units to the left and stretched vertically by a factor of 2. Figure 5.11a shows the graph of f, along with a table of values. By interchanging the rows of the table, we obtain points on the graph of the inverse function, shown in Figure 5.11b.

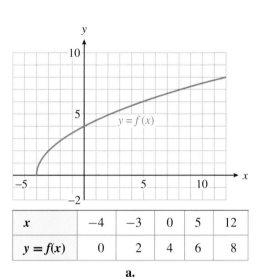

x	-4	-3	0	5	12
$y = f(x)$	0	2	4	6	8

a.

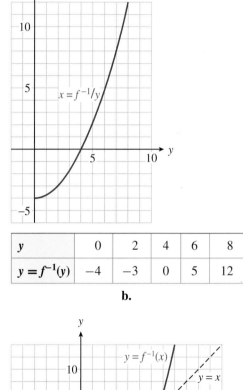

y	0	2	4	6	8
$y = f^{-1}(y)$	-4	-3	0	5	12

b.

FIGURE 5.11

If we use x as the input variable for both functions, and y as the output, we can graph f and f^{-1} on the same grid, as shown in Figure 5.12. The two graphs are symmetric about the line $y = x$.

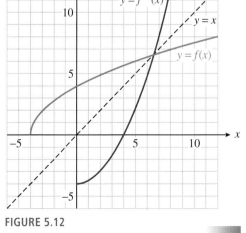

FIGURE 5.12

Graph the function $f(x) = x^3 + 2$ and its inverse $f^{-1}(x) = \sqrt[3]{x - 2}$ on the same set of axes, along with the line $y = x$.

Domain and Range

When we interchange the input and output variables to obtain the inverse function, we interchange the domain and range of the function. For the functions graphed in Example 8, you can see that

$$\text{Domain}(f) = [-4, 12] \quad \text{and} \quad \text{Domain}(f^{-1}) = [0, 8]$$
$$\text{Range}(f) = [0, 8] \qquad\qquad \text{Range}(f^{-1}) = [-4, 12]$$

This relationship between the domain and range of a function and its inverse holds in general.

Domain and Range of the Inverse Function

If $f^{-1}(x)$ is the inverse function for $f(x)$, then

$$\textbf{Domain}\,(f^{-1}) = \textbf{Range}\,(f)$$

$$\textbf{Range}\,(f^{-1}) = \textbf{Domain}\,(f)$$

EXAMPLE 9

a. Graph the function $y = f(x) = \frac{1}{x+3}$ in the window

$$\text{Xmin} = -6 \quad \text{Xmax} = 3.4$$
$$\text{Ymin} = -6 \quad \text{Ymax} = 3$$

b. Graph the inverse function in the same window, along with the line $y = x$.

c. State the domain and range of f, and of f^{-1}.

Solutions

a. The graph of f is shown in Figure 5.13. It looks like the graph of $y = \frac{1}{x}$, shifted 3 units to the left.

b. To find the inverse function, we solve for x. Take the reciprocal of both sides of the equation.

$$\frac{1}{y} = x + 3 \quad \text{Subtract 3 from both sides.}$$

$$x = \frac{1}{y} - 3$$

The inverse function is $x = f^{-1}(y) = \frac{1}{y} - 3$, or, using x for the input variable, $f^{-1}(x) = \frac{1}{x} - 3$. The graph of f^{-1} looks like the graph of $y = \frac{1}{x}$, shifted down 3 units, as shown in Figure 5.13.

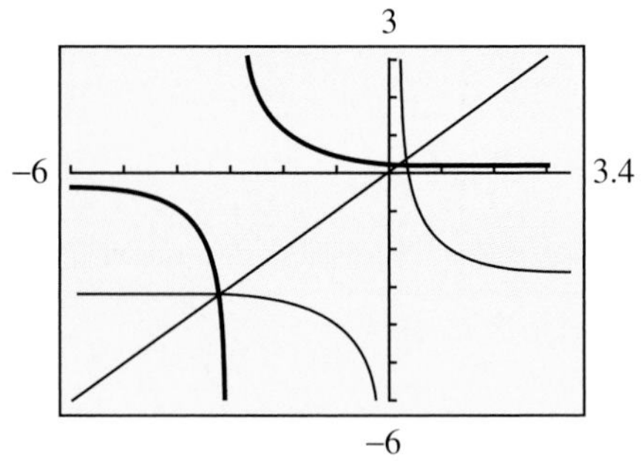

FIGURE 5.13

c. Because f is undefined at $x = -3$, the domain of f is all real numbers except -3. The graph has a horizontal asymptote at $y = 0$, so the range is all real numbers except 0. The inverse function $f^{-1}(x) = \frac{1}{x} - 3$ is undefined at $x = 0$, so its domain is all real numbers except 0. The graph of f^{-1} has a horizontal asymptote at $y = -3$, so its range is all real numbers except -3.

EXERCISE 9

a. Graph the function $f(x) = \frac{2}{x-1}$ and its inverse function, f^{-1} (which you found in Exercise 7), on the same set of axes, along with the line $y = x$.

b. State the domain and range of f, and of f^{-1}.

ANSWERS TO 5.1 EXERCISES

1. $g(0) = -1$, $g(1) = 0$ **2. a.** $600x + 400y = 16{,}000$

b. $y = f(x) = 40 - 1.5x$; $f(10) = 25$; If Carol cycles for 10 hrs, she must swim for 25 hrs.

c. $x = g(y) = 26\frac{2}{3} - \frac{2}{3}y$; $g(10) = 20$;   If Carol swims for 10 hrs, she must cycle for 20 hrs.   **3. a.** -2 **b.** $f^{-1}(1) = -2$, $f(-2) = 1$

c. $f^{-1}(1) = -2$, but $\frac{1}{f(1)} = 4$

4. a. -14 **b.** No   **c.** $h(32) = 0$, $h^{-1}(0) = 32$

5. a., c.

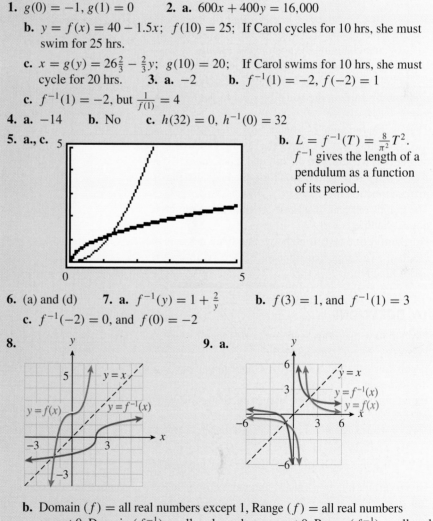

b. $L = f^{-1}(T) = \frac{8}{\pi^2}T^2$. f^{-1} gives the length of a pendulum as a function of its period.

6. (a) and (d)   **7. a.** $f^{-1}(y) = 1 + \frac{2}{y}$ **b.** $f(3) = 1$, and $f^{-1}(1) = 3$

c. $f^{-1}(-2) = 0$, and $f(0) = -2$

8.

9. a.

b. Domain (f) = all real numbers except 1, Range (f) = all real numbers except 0, Domain (f^{-1}) = all real numbers except 0, Range (f^{-1}) = all real numbers except 1

VOCABULARY *Look up the definitions of new terms in the Glossary.*

Inverse function   Horizontal line test   One-to-one

CONCEPTS

1. The **inverse** of a function describes the same relationship between two variables but interchanges the roles of the input and output.

2.
> ### Inverse Functions
>
> If the inverse of a function f is also a function, then the inverse is denoted by the symbol f^{-1}, and
>
> $$f^{-1}(b) = a \quad \text{if and only if} \quad f(a) = b$$

3. We can make a table of values for the inverse function, f^{-1}, by interchanging the columns of a table for f.

4. If a function is defined by a formula in the form $y = f(x)$, we can find a formula for its inverse function by solving the equation for x to get $x = f^{-1}(y)$.

5. The inverse function f^{-1} undoes the effect of the function f, that is, if we apply the inverse function to the output of f, we return to the original input value.

6. If f^{-1} is the inverse function for f, then f is also the inverse function for f^{-1}.

7. The graphs of f and its inverse function are **symmetric about the line $y = x$.**

8. **Horizontal line test:** If no horizontal line intersects the graph of a function more than once, then the inverse is also a function.

9. A function that passes the horizontal line test is called **one-to-one.**

10. The inverse of a function f is also a function if and only if f is one-to-one.

11.
> ### Functions and Inverse Functions
>
> Suppose f^{-1} is the inverse function for f. Then
>
> $$f^{-1}(f(x)) = x \quad \text{and} \quad f(f^{-1}(y)) = y$$
>
> as long as x is in the domain of f, and y is in the domain of f^{-1}.

12.
> ### Domain and Range of the Inverse Function
>
> If $f^{-1}(x)$ is the inverse function for $f(x)$, then
>
> $$\text{Domain } (f^{-1}) = \text{Range } (f)$$
> $$\text{Range } (f^{-1}) = \text{Domain } (f)$$

STUDY QUESTIONS

1. Explain how the terms *inverse function, one-to-one*, and *horizontal line test* are related.

2. If you know that $f^{-1}(3) = -7$, what can you say about the values of f?

3. Explain how to use a graph of the function g to evaluate $g^{-1}(2)$.

4. Evaluate $f(f^{-1}(5))$.

5. Delbert says that if $f(x) = x^{3/5}$, then $f^{-1}(x) = x^{-3/5}$. Is he correct? Why or why not?

SKILLS   *Practice each skill in the Homework problems listed.*

1. Given certain function values, find values of the inverse function: #1–4

2. Interpret values of the inverse function: #5–12

3. Find a formula for the inverse function: #9–22, 27–34

4. Graph the inverse function: #15 and 16, 23–34

5. Find the domain and range of the inverse function: #33 and 34

6. Use the horizontal line test to identify one-to-one functions: #35–42

1. Let $f(-1) = 0$, $f(0) = 1$, $f(1) = -2$, and $f(2) = -1$.
   a. Make a table of values for $f(x)$ and another table for its inverse function.
   b. Find $f^{-1}(1)$.
   c. Find $f^{-1}(-1)$.

2. Let $f(-2) = 1$, $f(-1) = -2$, $f(0) = 0$, and $f(1) = -1$.
   a. Make a table of values for $f(x)$ and another table for its inverse function.
   b. Find $f^{-1}(-1)$.
   c. Find $f^{-1}(1)$.

3. $f(x) = x^3 + x + 1$
   a. Make a table of values for $f(x)$ and another table for its inverse function.
   b. Find $f^{-1}(1)$.
   c. Find $f^{-1}(3)$.

4. $f(x) = x^5 + x^3 + 7$
   a. Make a table of values for $f(x)$ and another table for its inverse function.
   b. Find $f^{-1}(7)$.
   c. Find $f^{-1}(5)$.

Use the graph to evaluate each expression.

5. An insurance investigator measures the length, d, of the skid marks at an accident scene, in feet. The graph shows the function $v = f(d)$, which gives the velocity, v (mph), at which a car was traveling when it hit the brakes.

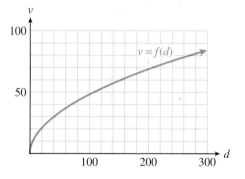

   a. Use the graph to estimate $f(60)$ and explain its meaning in this context.
   b. Use the graph to estimate $f^{-1}(60)$ and explain its meaning in this context.

6. The weight, m, of a missile launched from a catapult is a function of the distance, d, to the target. The graph shows the function $m = f(d)$, where d is in meters and m is in kilograms.

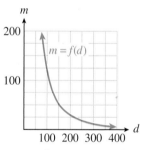

   a. Use the graph to estimate $f(100)$ and explain its meaning in this context.
   b. Use the graph to estimate $f^{-1}(100)$ and explain its meaning in this context.

7. After eating, the weight of a vampire bat drops steadily until its next meal. The graph shows the function $W = f(t)$, which gives the weight, W, of the bat in grams t hours since its last meal.

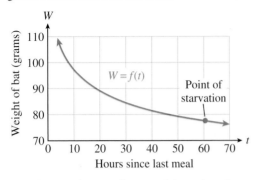

a. What are the coordinates of the point of starvation? Include units in your answer.
b. Use the graph to estimate $f^{-1}(90)$ and explain its meaning in this context.

8. The amount of money, A, in an interest-bearing savings account is a function of the number of years, t, it remains in the account. The graph shows $A = f(t)$, where A is in thousands of dollars.

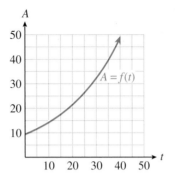

a. Use the graph to estimate $f(30)$ and explain its meaning in this context.
b. Use the graph to estimate $f^{-1}(30)$ and explain its meaning in this context.

9. The function $I = g(r) = (1 + r)^5 - 1$ gives the interest, I, that a dollar earns in 5 years in terms of the interest rate, r.
a. Evaluate $g(0.05)$ and explain its meaning in this context.
b. Find the interest rate needed to earn \$0.50 by substituting $I = 0.50$ in the formula and solving for r.
c. Find a formula for the inverse function.
d. Write your answer to part (b) with inverse function notation.

10. The function $C = h(F) = \frac{5}{9}(F - 32)$ gives the Celsius temperature C in terms of the Fahrenheit temperature F.
a. Evaluate $h(104)$ and explain its meaning in this context.
b. Find the Fahrenheit temperature of $37°$ Celsius by substituting $C = 37$ in the formula and solving for F.
c. Find a formula for the inverse function.
d. Write your answer to part (b) with inverse function notation.

11. If you are flying in an airplane at an altitude of h miles, on a clear day you can see a distance of d miles to the horizon, where $d = f(h) = \sqrt{7920h}$.
a. Evaluate $f(0.5)$ and explain its meaning in this context.
b. Find the altitude needed in order to see a distance of 10 mile by substituting $d = 10$ in the formula and solving for h.
c. Find a formula for the inverse function.
d. Write your answer to part (b) with inverse function notation.

12. A moving ship creates waves that impede its own speed. The function $v = f(L) = 1.3\sqrt{L}$ gives the ship's maximum speed in knots in terms of its length, L, in feet.
a. Evaluate $f(400)$ and explain its meaning in this context.
b. Find the length needed for a maximum speed of 35 knots by substituting $v = 35$ in the formula and solving for L.
c. Find a formula for the inverse function.
d. Write your answer to part (b) with inverse function notation.

13. a. Use the graph of $h(x) = \sqrt{5 - x}$ to find $h^{-1}(3)$.
b. Find a formula for $h^{-1}(x)$ and evaluate $h^{-1}(3)$.

14. a. Use the graph of $g(x) = \frac{1}{3 - x}$ to find $g^{-1}(-2)$.
b. Find a formula for $g^{-1}(x)$ and evaluate $g^{-1}(-2)$.

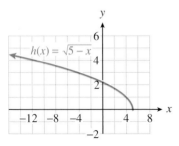

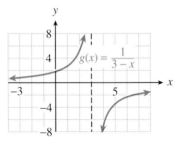

15. a. Find f^{-1} for the function $f(x) = (x-2)^3$.
   b. Show that f^{-1} undoes the effect of f on $x = 4$.
   c. Show that f undoes the effect of f^{-1} on $x = -8$.
   d. Graph the function and its inverse on the same grid, along with the graph of $y = x$.

16. a. Find f^{-1} for the function $f(x) = \frac{2}{x+1}$.
   b. Show that f^{-1} undoes the effect of f on $x = 3$.
   c. Show that f undoes the effect of f^{-1} on $x = -1$.
   d. Graph the function and its inverse on the same grid, along with the graph of $y = x$.

17. If $F(t) = \frac{2}{3}t + 1$, find $F^{-1}(5)$.

18. If $G(s) = \frac{s-3}{4}$, find $G^{-1}(-2)$.

19. If $m(v) = 6 - \frac{2}{v}$, find $m^{-1}(-3)$.

20. If $p(z) = 1 - 2z^3$, find $p^{-1}(7)$.

21. If $f(x) = \frac{x+2}{x-1}$, find $f^{-1}(2)$.

22. If $g(n) = \frac{3n+1}{n-3}$, find $g^{-1}(-2)$.

For Problems 23–26,
(a) Use the graph to make a table of values for the function $y = f(x)$.
(b) Make a table of values and a graph of the inverse function.

23.

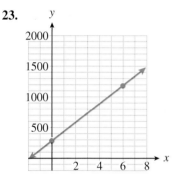

24.

25.

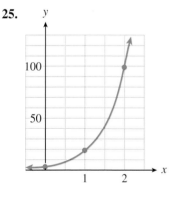

26.

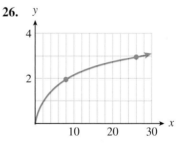

For Problems 27–32,
(a) Find a formula for the inverse of the function.
(b) Graph the function and its inverse on the same set of axes, along with the graph of $y = x$.

27. $f(x) = 2x - 6$

28. $f(x) = 3x - 1$

29. $f(x) = x^3 + 1$

30. $f(x) = \sqrt[3]{x+1}$

31. $f(x) = \dfrac{1}{x-1}$

32. $f(x) = \dfrac{1}{x} - 3$

33. a. Find the domain and range of the function
$g(x) = \sqrt{4 - x}$.
  b. Find a formula for $g^{-1}(x)$.
  c. State the domain and range of $g^{-1}(x)$.
  d. Graph g and g^{-1} on the same grid.

34. a. Find the domain and range of the function
$g(x) = 8 - \sqrt{x}$.
  b. Find a formula for $g^{-1}(x)$.
  c. State the domain and range of $g^{-1}(x)$.
  d. Graph g and g^{-1} on the same grid.

Which of the functions in Problems 35–42 have inverses that are also functions?

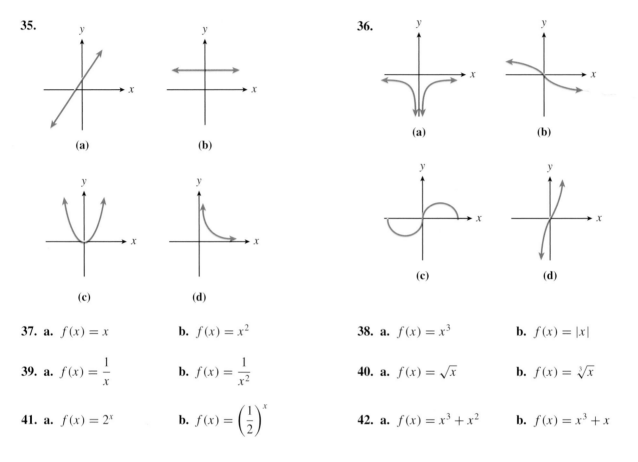

35. (a) (b) (c) (d)

36. (a) (b) (c) (d)

37. a. $f(x) = x$ **b.** $f(x) = x^2$

38. a. $f(x) = x^3$ **b.** $f(x) = |x|$

39. a. $f(x) = \dfrac{1}{x}$ **b.** $f(x) = \dfrac{1}{x^2}$

40. a. $f(x) = \sqrt{x}$ **b.** $f(x) = \sqrt[3]{x}$

41. a. $f(x) = 2^x$ **b.** $f(x) = \left(\dfrac{1}{2}\right)^x$

42. a. $f(x) = x^3 + x^2$ **b.** $f(x) = x^3 + x$

43. Find a formula for each function shown in (a)–(d). Then match each function with its inverse from I–IV.

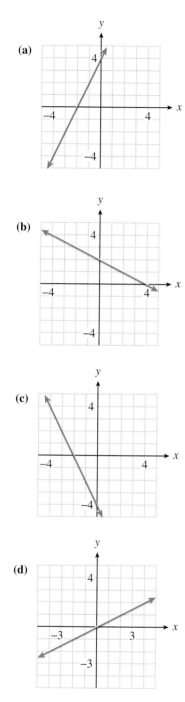

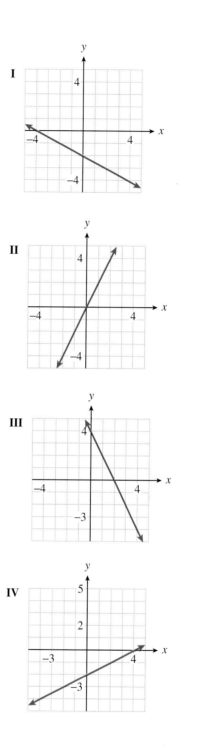

44. Find a formula for each transformation of a basic
function shown in (a)–(d). Then match each function
with its inverse from I–IV.

(a)

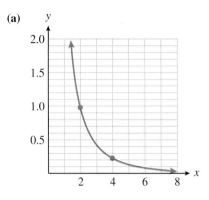

I

(b)

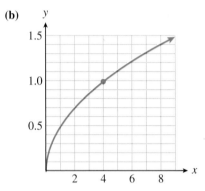

II

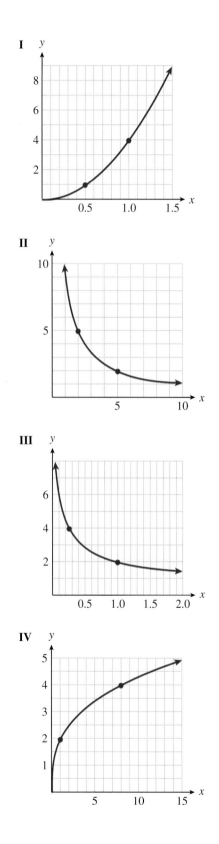

(c)

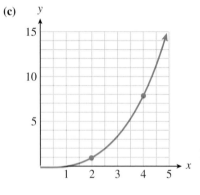

III

(d)

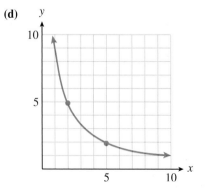

IV

For Problems 45 and 46, use the graph of *f* to match the other graphs with the appropriate function. (*Hint:* Look at the coordinates of some specific points.)

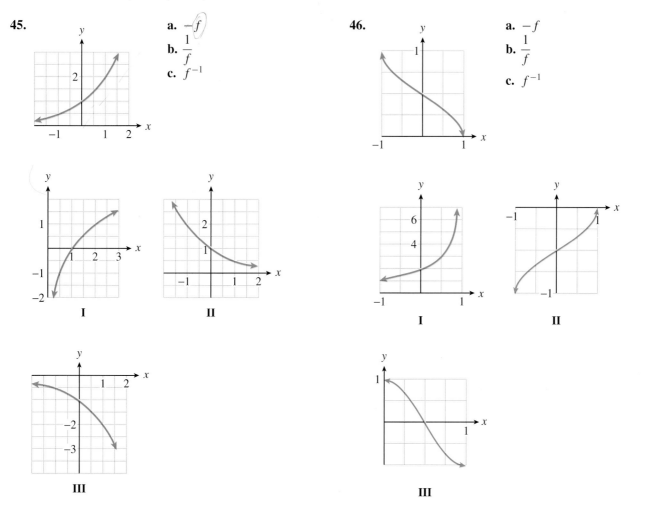

45.

a. $-f$

b. $\dfrac{1}{f}$

c. f^{-1}

I   II   III

46.

a. $-f$

b. $\dfrac{1}{f}$

c. f^{-1}

I   II   III

5.2   Logarithmic Functions

Inverse of the Exponential Function

Inverse functions are really a generalization of inverse operations. For example, raising to the *n*th power and taking *n*th roots are inverse operations. In fact, we use the following rule to define cube roots:

$$b = \sqrt[3]{a} \quad \text{if and only if} \quad a = b^3$$

Compare this rule to the definition of inverse functions from Section 5.1. In this case, if $f(x) = x^3$ and $g(x) = \sqrt[3]{x}$, we see that

$$b = g(a) \quad \text{if and only if} \quad a = f(b)$$

We have shown that the two functions $f(x) = x^3$ and $g(x) = \sqrt[3]{x}$ are inverse functions.

In Chapter 4, we saw that a similar rule relates the operations of raising a base b to a power and taking a base b logarithm, because they are inverse operations.

▶▶▶
See Section 4.3 to review logarithms.

Conversion Formulas for Logarithms

For any base $b > 0$,

$$y = \log_b x \quad \text{if and only if} \quad x = b^y$$

We can now define the **logarithmic function,** $g(x) = \log_b x$, that takes the log base b of its input values. The conversion formulas tell us that the log function, $g(x) = \log_b x$, is the inverse of the exponential function, $f(x) = b^x$.

Logarithmic Function

The **logarithmic function** base b, $g(x) = \log_b x$, is the inverse of the exponential function of the same base, $f(x) = b^x$.

For example, the function $g(x) = \log_2 x$ is the inverse of $f(x) = 2^x$. Each function undoes the effect of the other. So, if we start with $x = 3$, apply f, and then apply g to the result, we return to the original number, 3.

$$x = 3 \quad \rightarrow \quad f(3) = 2^3 = 8 \quad \rightarrow \quad g(8) = \log_2 8 = 3$$

Apply the exponential function.   Apply the log function.   Original number

We can write both calculations together as

$$\log_2(2^3) = 3$$

A similar equation holds for any value of x and for any base $b > 0$. In other words, applying first the exponential function and then the log function returns the original input value, so that

$$\log_b b^x = x$$

EXAMPLE 1 Simplify each expression.

a. $\log_4 4^6$ **b.** $\log_8 8^{2a+3}$

Solutions **a.** In this expression, we start with 6, apply the exponential function with base 4, and then take a logarithm base 4. Because the logarithm is the inverse of the exponential function, we return to the original number, 6.

$$6 \quad \rightarrow \quad 4^6 \quad \rightarrow \quad \log_4 4^6 = 6$$

Apply the exponential function.   Apply the log function.   Original number

b. The input of the exponential function is the expression $2a + 3$. Because the bases of the log and the exponential function are both 8, they are inverse functions, and applying them in succession returns us to the original input. Thus,
$\log_8 8^{2a+3} = 2a + 3$.

EXERCISE 1

Simplify each expression.

a. $\log_{10} 10^6$ **b.** $\log_w w^{x+1}$, for $w > 0$

We can also apply the two functions in the opposite order. For example,

$$2^{\log_2 8} = 8$$

To see that this equation is true, we simplify the exponent first. We start with 8, and apply the log base 2 function. Because $\log_2 8 = 3$, we have

$$8 \quad \rightarrow \quad \log_2 8 = 3 \quad \rightarrow \quad 2^{(\log_2 8)} = 2^3 = 8$$

Apply the   Apply the   Original number
log function.   exponential function.

Of course, a similar equation holds for any positive value of x and any base $b > 0$:

$$b^{\log_b x} = x$$

EXAMPLE 2   Simplify each expression.

a. $10^{\log_{10} 1000}$ **b.** $Q^{\log_Q 25}$, for $Q > 0$

Solutions   **a.** In this expression, we start with 1000, take the logarithm base 10, and then apply the exponential function base 10 to the result. We return to the original input, so $10^{\log_{10} 1000} = 1000$.

b. The log function, $\log_Q x$, and the exponential function, Q^x, are inverse functions, so $Q^{\log_Q 25} = 25$.

EXERCISE 2

Simplify each expression.

a. $4^{\log_4 64}$ **b.** $2^{\log_2 (x^2+1)}$

We summarize these relationships as follows.

Exponential and Logarithmic Functions

Because $f(x) = b^x$ and $g(x) = \log_b x$ are inverse functions for $b > 0$,

$$\log_b b^x = x, \quad \text{for all } x \quad \text{and} \quad b^{\log_b x} = x, \quad \text{for} \quad x > 0$$

Graphs of Logarithmic Functions

We can obtain a table of values for $g(x) = \log_2 x$ by making a table for $f(x) = 2^x$ and then interchanging the columns, as shown in Table 5.4. You can see that the graphs of $f(x) = 2^x$ and $g(x) = \log_2 x$, shown in Figure 5.14, are symmetric about the line $y = x$.

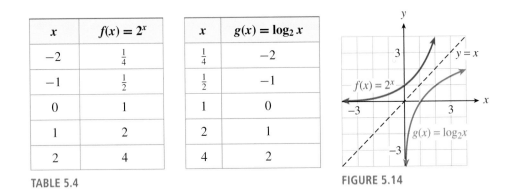

x	$f(x) = 2^x$
-2	$\frac{1}{4}$
-1	$\frac{1}{2}$
0	1
1	2
2	4

x	$g(x) = \log_2 x$
$\frac{1}{4}$	-2
$\frac{1}{2}$	-1
1	0
2	1
4	2

TABLE 5.4   FIGURE 5.14

The same procedure works for graphing log functions with any base: If we want to find values for the function $y = \log_b x$, we can find values for the exponential function $y = b^x$, and then interchange the x- and y-values in each ordered pair.

EXAMPLE 3   Graph the function $f(x) = 10^x$ and its inverse $g(x) = \log_{10} x$ on the same axes.

Solution   Make a table of values for the function $f(x) = 10^x$. A table of values for the inverse function, $g(x) = \log_{10} x$, can be obtained by interchanging the components of each ordered pairs in the table for f.

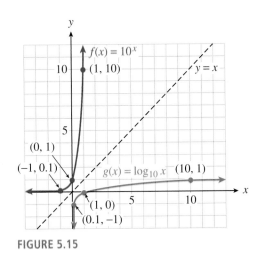

x	$f(x)$
-2	0.01
-1	0.1
0	1
1	10
2	100

x	$g(x)$
0.01	-2
0.1	-1
1	0
10	1
100	2

FIGURE 5.15

Plot each set of points and connect them with smooth curves to obtain the graphs shown in Figure 5.15.

EXERCISE 3

Make a table of values and graph the function $h(x) = \log_4 x$.

While an exponential growth function increases very rapidly for positive values, its inverse, the logarithmic function, grows extremely slowly, as you can see in Figure 5.15. In addition, the logarithmic function $y = \log_b x$ for any base $b > 0$, $b \neq 1$, has the following properties.

Logarithmic Functions $y = \log_b x$

1. Domain: all positive real numbers
2. Range: all real numbers
3. x-intercept: $(1, 0)$
   y-intercept: none
4. Vertical asymptote at $x = 0$
5. The graphs of $y = \log_b x$ and $y = b^x$ are symmetric about the line $y = x$.

Because the domain of a logarithmic function includes only the *positive* real numbers, *the logarithm of a negative number or zero is undefined.*

EXAMPLE 4

a. Find the inverse of the function $f(x) = 2^{x-3} - 4$.

b. Graph f and f^{-1} on the same grid.

c. State the domain and range of f and of f^{-1}.

Solutions

a. Write the function as $y = 2^{x-3} - 4$, and solve for x in terms of y. First, isolate the power:

$$y + 4 = 2^{x-3}$$

$$\log_2(y + 4) = \log_2 2^{x-3} = x - 3 \quad \text{Take logs base 2.}$$

$$x = 3 + \log_2(y + 4)$$

The inverse function is $f^{-1}(x) = 3 + \log_2(x + 4)$. However, to graph both f and f^{-1} on the same grid, we write the inverse function as $f^{-1}(x) = 3 + \log_2(x + 4)$.

b. To graph f, we translate the graph of $y = 2^x$ by 3 units to the right and 4 units down. The graph of f^{-1} looks like the graph of $y = \log_2 x$ shown in Figure 5.14, but shifted 4 units to the left and 3 units up. The graphs are shown in Figure 5.16, along with the line $y = x$.

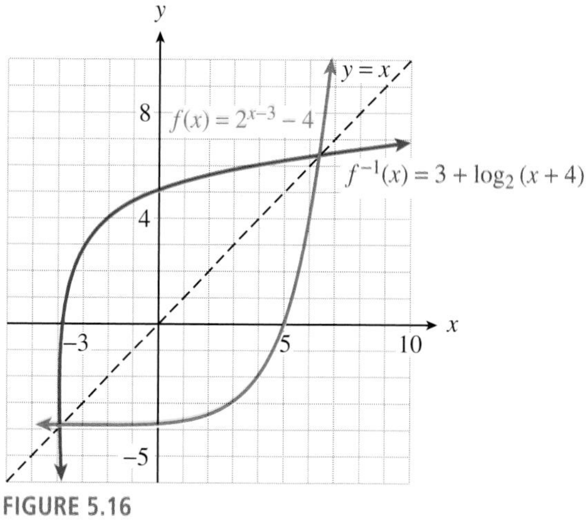

FIGURE 5.16

c. The function f is exponential, and its domain consists of all real numbers. Because the graph is shifted 4 units down, the range of f is $(-4, \infty)$. Because the log of a negative number or zero is undefined, for $f^{-1}(x) = 3 + \log_2(x + 4)$, we must have $x + 4 > 0$, or $x > -4$. We can verify on the graph that the range of f^{-1} includes all real numbers. Thus,

$$\text{Domain}(f) = \text{all real numbers} = \text{Range}(f^{-1})$$

$$\text{Range}(f) = (-4, \infty) = \text{Domain}(f^{-1})$$

EXERCISE 4

a. Find the inverse function for $f(x) = 2\log(x + 1)$.

b. Graph f and f^{-1} in the window

$$\text{Xmin} = -6 \quad \text{Xmax} = 6$$
$$\text{Ymin} = -4 \quad \text{Ymax} = 4$$

c. State the domain and range of f and f^{-1}.

Evaluating Logarithmic Functions

We can use the $\boxed{\text{LOG}}$ key on a calculator to evaluate the function $f(x) = \log_{10} x$.

EXAMPLE 5   Let $f(x) = \log_{10} x$. Evaluate the following expressions.

a. $f(35)$ **b.** $f(-8)$ **c.** $2f(16) + 1$

Solutions   **a.** $f(35) = \log_{10} 35 \approx 1.544$

b. Because -8 is not in the domain of f, $f(-8)$, or $\log_{10}(-8)$, is undefined.

> To evaluate log functions to other bases, we can use the change of base formula. See the Projects for this chapter.

c. $2f(16) + 1 = 2(\log_{10} 16) + 1$
$$\approx 2(1.204) + 1 = 3.408$$

EXERCISE 5

The formula $T = \dfrac{\log 2 \cdot t_i}{3\log(D_f/D_0)}$ is used by X-ray technicians to calculate the doubling time of a malignant tumor. D_0 is the diameter of the tumor when first detected, D_f is its diameter at the next reading, and t_i is the time interval between readings, in days. Calculate the doubling time of a tumor if its diameter when first detected was 1 cm, and 1.05 cm 7 days later.

Logarithmic Equations

A **logarithmic equation** is one in which the variable appears inside of a logarithm. For example,

$$\log_4 x = 3$$

is a log equation. To solve a log equation, remember that logarithms and exponentials with the same base are inverse functions. Therefore,

$$y = \log_b x \quad \text{if and only if} \quad x = b^y$$

Thus, we can rewrite a logarithmic equation in exponential form.

EXAMPLE 6 Solve for x.

a. $2(\log_3 x) - 1 = 4$ **b.** $\log_{10}(2x + 100) = 3$

Solutions **a.** Isolate the logarithm, then rewrite in exponential form:

$$2(\log_3 x) = 5 \qquad \text{Divide both sides by 5.}$$
$$\log_3 x = \frac{5}{2} \qquad \text{Convert to exponential form.}$$
$$x = 3^{5/2}$$

b. First, convert the equation to exponential form:

$$2x + 100 = 10^3 = 1000$$

Solve for x to find $2x = 900$, or $x = 450$.

EXERCISE 6

Solve for the unknown value in each equation.

a. $\log_b 2 = \frac{1}{2}$ **b.** $\log_3(2x - 1) = 4$

EXAMPLE 7 If $f(x) = \log_{10} x$, find x so that $f(x) = -3.2$.

Solution We must solve the equation $\log_{10} x = -3.2$. Rewriting the equation in exponential form yields

$$x = 10^{-3.2} \approx 0.00063$$

Evaluating 10^x

In Example 7, the expression $10^{-3.2}$ can be evaluated in two different ways with a calculator. We can use the $\boxed{\wedge}$ key and press

$$10 \;\boxed{\wedge}\; \boxed{(-)} \; 3.2 \; \boxed{\text{ENTER}}$$

which gives 6.30957344 **E** $- 4$, or approximately 0.00063. Alternatively, because 10^x is the inverse function for $\log x$, we can press

$$\boxed{\text{2nd}} \; \boxed{\text{LOG}} \; \boxed{(-)} \; 3.2$$

which gives the same answer as before.

Imagine the graph of $f(x) = \log_{10} x$. How far must you travel along the x-axis until the y-coordinate reaches a height of 5.25?

Using the Properties of Logarithms

> To review the properties of logarithms, see Section 4.4.

The properties of logarithms are useful in solving both exponential and logarithmic equations. To solve logarithmic equations, we first combine any expressions involving logs into a single logarithm.

EXAMPLE 8   Solve $\log_{10}(x + 1) + \log_{10}(x - 2) = 1$

Solution   Use Property (1) of logarithms to rewrite the left-hand side as a single logarithm:

$$\log_{10}(x + 1)(x - 2) = 1$$

Once the left-hand side is expressed as a *single* logarithm, we can rewrite the equation in exponential form as

$$(x + 1)(x - 2) = 10^1$$

from which

$$x^2 - x - 2 = 10 \qquad \text{Subtract 10 from both sides.}$$
$$x^2 - x - 12 = 0 \qquad \text{Factor the left side.}$$
$$(x - 4)(x + 3) = 0 \qquad \text{Apply the zero-factor principle.}$$

Thus, $x = 4$ or $x = -3$. The number -3 is not a solution of the original equation, because neither $\log_{10}(x + 1)$ nor $\log_{10}(x - 2)$ is defined for $x = -3$. The solution of the original equation is 4.

In Example 8, the apparent solution $x = -3$ is called **extraneous** because it does not solve the original equation. We should always check for extraneous solutions when solving log equations. The following steps give a rough outline for solving log equations.

Steps for Solving Logarithmic Equations

1. Use the properties of logarithms to combine all logs into one log.

2. Isolate the log on one side of the equation.

3. Convert the equation to exponential form.

4. Solve for the variable.

5. Check for extraneous solutions.

ANSWERS TO 5.2 EXERCISES

1. a. 6   **b.** $x+1$ **2. a.** 64   **b.** x^2+1

3.

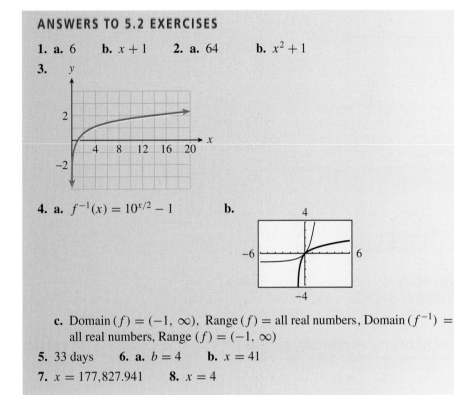

4. a. $f^{-1}(x) = 10^{x/2} - 1$ **b.**

c. Domain $(f) = (-1, \infty)$, Range (f) = all real numbers, Domain (f^{-1}) = all real numbers, Range $(f) = (-1, \infty)$

5. 33 days   **6. a.** $b = 4$ **b.** $x = 41$

7. $x = 177,827.941$ **8.** $x = 4$

SECTION 5.2 SUMMARY

VOCABULARY   *Look up the definitions of new terms in the Glossary.*

Logarithmic function   Logarithmic equation   Extraneous solution

CONCEPTS

1. We define the **logarithmic function,** $g(x) = \log_b x$, which takes the log base b of its input values. The log function $g(x) = \log_b x$ is the inverse of the exponential function $f(x) = b^x$.

2.

Because $f(x) = b^x$ and $g(x) = \log_b x$ are inverse functions for $b > 0$,

$$\log_b b^x = x, \quad \text{for all } x \quad \text{and} \quad b^{\log_b x} = x, \quad \text{for } x > 0$$

3.

Logarithmic Functions $y = \log_b x$

1. Domain: all positive real numbers $(0, \infty)$
2. Range: all real numbers
3. x-intercept: $(1, 0)$
   y-intercept: none
4. Vertical asymptote at $x = 0$
5. The graphs of $y = \log_b x$ and $y = b^x$ are symmetric about the line $y = x$.

4. A **logarithmic equation** is one in which the variable appears inside of a logarithm. We can solve logarithmic equations by converting to exponential form.

5.

> **Steps for Solving Logarithmic Equations**
>
> 1. Use the properties of logarithms to combine all logs into one log.
> 2. Isolate the log on one side of the equation.
> 3. Convert the equation to exponential form.
> 4. Solve for the variable.
> 5. Check for extraneous solutions.

STUDY QUESTIONS

1. Can the output of the function $y = \log_b x$ be negative?

2. Francine says that $\log_2 \frac{1}{x} = -\log_2 x$. Is she correct? Why or why not?

3. Sketch a typical logarithmic function.

4. Simplify: **a.** $10^{\log 13}$ **b.** $7^{\log_7 13}$

5. Why is the following attempt to solve the equation incorrect?

$$\text{Solve}: \quad \log x + \log(x + 1) = 2$$
$$x + x + 1 = 10^2$$

SKILLS   *Practice each skill in the Homework problems listed.*

1. Evaluate log functions: #1–16, 27 and 28

2. Simplify expressions involving logs: #15 and 16, 19 and 20

3. Graph logarithmic functions and transformations of log functions: #1–4, 25–28

4. Find formulas for inverse functions: #17–24

5. Solve logarithmic equations: #29–54

6. Solve formulas involving logs: #55–60

HOMEWORK 5.2

ThomsonNOW™ Test yourself on key content at **www.thomsonedu.com/login.**

In Problems 1–4,
(a) Make tables of values for each exponential function and its inverse logarithmic function.
(b) Graph both functions on the same set of axes.

1. $f(x) = 2^x$

2. $f(x) = 3^x$

3. $f(x) = \left(\dfrac{1}{3}\right)^x$

4. $f(x) = \left(\dfrac{1}{2}\right)^x$

5. a. How large must x be before the graph of $y = \log_{10} x$ reaches a height of 4?
   b. How large must x be before the graph of $y = \log_{10} x$ reaches a height of 8?

6. a. How large must x be before the graph of $y = \log_2 x$ reaches a height of 5?
   b. How large must x be before the graph of $y = \log_2 x$ reaches a height of 10?

7. For what values of x is $y = \log_{10} x < -2$?

8. For what values of x is $y = \log_2 x < -3$?

In Problems 9–14, $f(x) = \log_{10} x$. Evaluate.

9. a. $f(487) + f(206)$ **b.** $f(487 + 206)$

10. a. $f(93) + f(1500)$ **b.** $f(93 + 1500)$

11. a. $f(-7)$ **b.** $6f(28)$

12. a. $f(0)$ **b.** $3f(41)$

13. a. $18 - 5f(3)$ **b.** $\dfrac{2}{5 + f(0.6)}$

14. a. $15 - 4f(7)$ **b.** $\dfrac{3}{2 + f(0.2)}$

15. Let $f(x) = 3^x$ and $g(x) = \log_3 x$.
  a. Compute $f(4)$.
  b. Compute $g[f(4)]$.
  c. Explain why $\log_3 3^x = x$ for any x.
  d. Compute $\log_3 3^{1.8}$.
  e. Simplify $\log_3 3^a$.

16. Let $f(x) = \log_2 x$ and $g(x) = 2^x$.
  a. Compute $f(32)$.
  b. Compute $g[f(32)]$.
  c. Explain why $2^{\log_2 x} = x$ for $x > 0$.
  d. Compute $2^{\log_2 6}$.
  e. Simplify $2^{\log_2 Q}$.

17. a. If $h(r) = \log_2 r$, find $h^{-1}(8)$.
  b. If $H(w) = 3^w$, find $H^{-1}\left(\dfrac{1}{9}\right)$.

18. a. If $g(z) = \log_3 z$, find $g^{-1}(-3)$.
  b. If $G(q) = 2^q$, find $G^{-1}(1)$.

Simplify.

19. a. $10^{\log 2k}$ **b.** $10^{3 \log x}$
  c. $(\sqrt{10})^{\log x}$ **d.** $\log 100^m$

20. a. $\log 10^{(1-x)}$ **b.** $100^{\log 2x}$
  c. $(0.1)^{\log(x-1)}$ **d.** $\log 10^{\log 10}$

21. a. What is the domain of the function
$$f(x) = 4 + \log_3(x - 9)?$$
  b. Find a formula for $f^{-1}(x)$.

22. a. What is the domain of the function
$$f(x) = 1 - \log_2(16 - 4x)?$$
  b. Find a formula for $f^{-1}(x)$.

23. a. Find the inverse of the function $f(x) = 100 - 4^{x+2}$.
  b. Show that f^{-1} undoes the effect of f on $x = 1$.
  c. Show that f undoes the effect of f^{-1} on $x = 84$.

24. a. Find the inverse of the function $f(x) = 5 + 2^{-x}$.
  b. Show that f^{-1} undoes the effect of f on $x = -2$.
  c. Show that f undoes the effect of f^{-1} on $x = 6$.

Match each graph to its equation.

25. a. $y = \log_2(x - 3)$ **b.** $y = 3 + \log_2 x$ **c.** $y = 2 - \log_2 x$ **d.** $y = \log_2(x + 4) - 1$

26. a. $y = 5 \log x$ **b.** $y = \log \dfrac{x}{2}$ **c.** $y = \log \dfrac{1}{x}$ **d.** $y = \log(-x)$

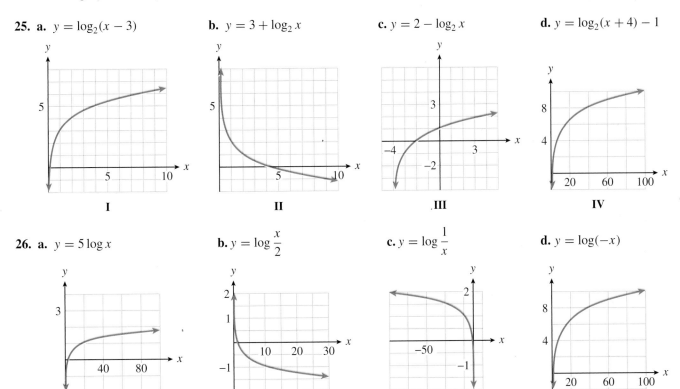

27. In a psychology experiment, volunteers were asked to memorize a list of nonsense words, then 24 hours later were tested to see how many of the words they recalled. On average, the subjects had forgotten 20% of the words. The researchers found that the more lists their volunteers memorized, the larger the fraction of words they were unable to recall. (Source: Underwood, *Scientific American,* vol. 210, no. 3)

Number of lists, n	1	4	8	12	16	20
Percent forgotten, F	20	40	55	66	74	80

   a. Plot the data. What sort of function seems to fit the data points?

   b. Psychologists often describe rates of forgetting by logarithmic functions. Graph the function

$$f(n) = 16.6 + 46.3 \log n$$

on the same graph with your data. Comment on the fit.

   c. What happens to the function $f(n)$ as n grows increasingly large? Does this behavior accurately reflect the situation being modeled?

28. The water velocity at any point in a stream or river is related to the logarithm of the depth at that point. For the Hoback River near Bondurant, Wyoming,

$$v = 2.63 + 1.03 \log d$$

where v is the velocity of the water, in feet per second, and d is the vertical distance from the stream bed, in feet, at that point. For Pole Creek near Pinedale, Wyoming,

$$v = 1.96 + 0.65 \log d$$

Both streams are 1.2 feet deep at the locations mentioned. (Source: Leopold, Luna, Wolman, and Gordon, 1992)

   a. Complete the table of values for each stream.

Distance from bed (feet)	0.2	0.4	0.6	0.8	1.0	1.2
Velocity, Hoback River (ft/sec)						
Velocity, Pole Creek (ft/sec)						

   b. If you double the distance from the bed, by how much does the velocity increase in each stream?

   c. Plot both functions on the same graph.

   d. The average velocity of the entire stream can be closely approximated as follows: Measure the velocity at 20% of the total depth of the stream from the surface and at 80% of the total depth, then average these two values. Find the average velocity for the Hoback River and for Pole Creek.

In Problems 29–30, $f(x) = \log_{10} x$. Solve for x.

29. **a.** $f(x) = 1.41$ **b.** $f(x) = -1.69$
     c. $f(x) = 0.52$

30. **a.** $f(x) = 2.3$ **b.** $f(x) = -1.3$
     c. $f(x) = 0.8$

Convert each logarithmic equation to exponential form.

31. $\log_{16} 256 = w$

32. $\log_9 729 = y$

33. $\log_b 9 = -2$

34. $\log_b 8 = -3$

35. $\log_{10} A = -2.3$

36. $\log_{10} C = -4.5$

37. $\log_u v = w$

38. $\log_m n = p$

Solve for the unknown value.

39. $\log_b 8 = 3$

40. $\log_b 625 = 4$

41. $\log_b 10 = \dfrac{1}{2}$

42. $\log_b 0.1 = -1$

43. $\log_2 (3x - 1) = 5$

44. $\log_5 (9 - 4x) = 3$

45. $3(\log_7 x) + 5 = 7$

46. $5(\log_2 x) + 6 = -14$

Solve each logarithmic equation.

47. $\log_{10} x + \log_{10}(x + 21) = 2$

48. $\log_{10}(x + 3) + \log_{10} x = 1$

49. $\log_8(x + 5) - \log_8 2 = 1$

50. $\log_{10}(x - 1) - \log_{10} 4 = 2$

51. $\log_{10}(x + 2) + \log_{10}(x - 1) = 1$

52. $\log_4(x + 8) + \log_4(x + 2) = 2$

53. $\log_3(x - 2) - \log_3(x + 1) = 3$

54. $\log_{10}(x + 3) - \log_{10}(x - 1) = 1$

Solve for the indicated variable.

55. $t = T \log_{10}\left(1 + \dfrac{A}{k}\right)$, for A

56. $\log_{10} R = \log_{10} R_0 + kt$, for R

57. $N = N_0 \log_b(ks)$, for s

58. $T = \dfrac{H \log_{10} \dfrac{N}{N_0}}{\log_{10} \dfrac{1}{2}}$, for N

59. $M = \sqrt{\dfrac{\log_{10} H}{k \log_{10} H_0}}$, for H

60. $h = a - \sqrt{\dfrac{\log_{10} B}{t}}$, for B

61. Choose the graph for each function described below.
   a. The area, A, of a pentagon is a quadratic function of the length, l, of its side.
   b. The strength, F, of a hurricane varies inversely with its speed, s.
   c. The price of food has increased by 3% every year for a decade.
   d. The magnitude, M, of a star is a logarithmic function of its brightness, I.
   e. The speed of the train increased at a constant rate.
   f. If you do not practice a foreign language, you lose $\frac{1}{8}$ of the words in your working vocabulary, V, each year.

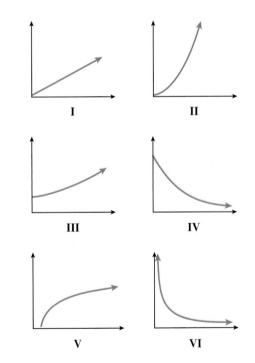

62. For each of the functions listed below, select the graph of its inverse function, if possible, from the figures labeled I–VI. (The inverse of one of the functions is not shown.)

a. $f(x) = 2^x$

b. $f(x) = x^2, \ x \geq 0$

c. $f(x) = \dfrac{2}{x}$

d. $f(x) = \sqrt{x}$

e. $f(x) = \log_2 x$

f. $f(x) = \left(\dfrac{1}{2}\right)^x$

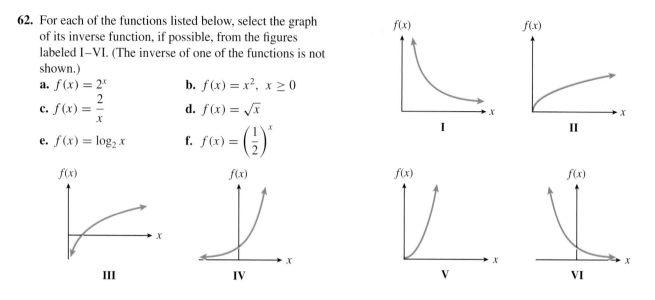

I   II   III   IV   V   VI

Graph each function on the domain [−4, 4] and a suitable range. Which have inverses that are also functions?

63. a. $f(x) = 5(2^{-x^2})$
   b. $f(x) = 2^x + 2^{-x}$

64. a. $f(x) = 5\,(\log(x))^2 + 1$
   b. $f(x) = 5\log(x^2 + 1)$

Graph each pair of functions on your calculator. Explain the result.

65. $f(x) = \log(2x), \ g(x) = \log 2 + \log x$

66. $f(x) = \log\left(\dfrac{x}{3}\right), \ g(x) = \log x - \log 3$

67. $f(x) = \log\left(\dfrac{1}{x}\right), \ g(x) = -\log x$

68. $f(x) = \log(x^3), \ g(x) = 3\log x$

69. a. Complete the following table.

x	x^2	$\log_{10} x$	$\log_{10} x^2$
1			
2			
3			
4			
5			
6			

b. Do you notice a relationship between $\log_{10} x$ and $\log_{10} x^2$? State the relationship as an equation.

70. a. Complete the following table.

x	$\dfrac{1}{x}$	$\log_{10} x$	$\log_{10} \dfrac{1}{x}$
1			
2			
3			
4			
5			
6			

b. Do you notice a relationship between $\log_{10} x$ and $\log_{10} \dfrac{1}{x}$? State the relationship as an equation.

In Problems 69 and 70, you found relationships between $\log_{10} x$ and $\log_{10} x^2$, and between $\log_{10} x$ and $\log_{10} 1/x$. Assuming that those relationships hold for any base, complete the following tables and use them to graph the given functions.

71.

x	$y = \log_e x$
1	0
2	0.693
4	
16	
$\frac{1}{2}$	
$\frac{1}{4}$	
$\frac{1}{16}$	

72.

x	$y = \log_f x$
1	0
2	0.431
4	
16	
$\frac{1}{2}$	
$\frac{1}{4}$	
$\frac{1}{16}$	

Investigation 9

Interest Compounded Continuously

We learned in Section 4.4 that the amount, A (principal plus interest), accumulated in an account with interest compounded n times annually is

$$A = P\left(1 + \frac{r}{n}\right)^{nt}$$

where P is the principal invested, r is the interest rate, and t is the time period, in years.

1. Suppose you keep $1000 in an account that pays 8% interest. How much is the amount A after 1 year if the interest is compounded twice a year? Four times a year?

$$n = 2:\ A = 1000\left(1 + \frac{0.08}{2}\right)^{2(1)} =$$

$$n = 4:\ A = 1000\left(1 + \frac{0.08}{4}\right)^{4(1)} =$$

2. What happens to A as we increase n, the number of compounding periods per year? Fill in the table showing the amount in the account for different values of n.

n	A
1 (annually)	1080
2 (semiannually)	
4 (quarterly)	
6 (bimonthly)	
12 (monthly)	
365 (daily)	
1000	
10,000	

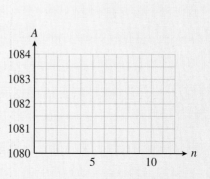

3. Plot the values in the table from $n = 1$ to $n = 12$, and connect them with a smooth curve. Describe the curve: What is happening to the value of A?

4. In part (2), as you increased the value of n, the other parameters in the formula stayed the same. In other words, A is a function of n, given by $A = 1000\,(1 + \frac{0.08}{n})^n$. Use your calculator to graph A on successively larger domains:

   a. Xmin $= 0$, Xmax $= 12$; Ymin $= 1080$, Ymax $= 1084$
   b. Xmin $= 0$, Xmax $= 50$; Ymin $= 1080$, Ymax $= 1084$
   c. Xmin $= 0$, Xmax $= 365$; Ymin $= 1080$, Ymax $= 1084$

5. Use the **Trace** feature or the **Table** feature to evaluate A for very large values of n. Rounded to the nearest penny, what is the largest value of A that you can find?

6. As n increases, the values of A approach a *limiting value*. Although A continues to increase, it does so by smaller and smaller increments and will never exceed $1083.29. When the number of compounding periods increases without bound, we call the limiting result *continuous compounding*.

7. Is there an easier way to compute A under continues compounding? Yes! Compute $1000e^{0.08}$ on your calculator. (Press $\boxed{\text{2nd}}$ $\boxed{\text{LN}}$ to enter e^x.) Compare the value to your answer in part (5) for the limiting value. The number e is called the **natural base.** We'll compute its value shortly.

8. Repeat your calculations for two other interest rates, 15% and (an extremely unrealistic) 100%, again for an investment of $1000 for 1 year. In each case, compare the limiting value of A, and compare to the value of $1000e^r$.

a.

r = 0.15	
n	*A*
1	115
2	
4	
6	
12	
365	
1000	
10,000	

$1000\,e^{0.15} =$

b.

r = 1	
n	*A*
1	200
2	
4	
6	
12	
365	
1000	
10,000	

$1000\,e^1 =$

9. In part (8b), you have computed an approximation for $1000e$. What is the value of e, rounded to 5 decimal places?

10. Complete the table of values. What does $(1 + \frac{1}{n})^n$ appear to approach as n increases?

n	100	1000	10,000	100,000
$(1 + \frac{1}{n})^n$				

There is another base for logarithms and exponential functions that is often used in applications. This base is an irrational number called e, where

$$e \approx 2.71828182845$$

The number e is essential for many advanced topics, and it is often called the **natural base.**

The Natural Exponential Function

The **natural exponential function** is the function $f(x) = e^x$. Values for e^x can be obtained with a calculator using the $\boxed{e^x}$ key ($\boxed{\text{2nd}}$ $\boxed{\text{LN}}$ on most calculators). For example, you can evaluate e^1 by pressing

$$\boxed{\text{2nd}} \quad \boxed{\text{LN}} \quad 1$$

to confirm the value of e given above.

Because e is a number between 2 and 3, the graph of $f(x) = e^x$ lies between the graphs of $y = 2^x$ and $y = 3^x$. Compare the tables of values (Table 5.5) and the graphs of the three functions in Figure 5.17. As with other exponential functions, the domain of the natural exponential function includes all real numbers, and its range is the set of positive numbers.

x	$y = 2^x$	$y = e^x$	$y = 3^x$
-3	0.125	0.050	0.037
-2	0.250	0.135	0.111
-1	0.500	0.368	0.333
0	1	1	1
1	2	2.718	3
2	4	7.389	9
3	8	20.086	27

TABLE 5.5

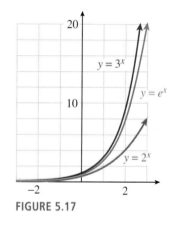

FIGURE 5.17

EXAMPLE 1

Graph each function. How does each graph differ from the graph of $y = e^x$?

a. $g(x) = e^{x+2}$ **b.** $h(x) = e^x + 2$

Solutions   If $f(x) = e^x$, then $g(x) = f(x + 2)$, so the graph of g is shifted 2 units to the left of $y = e^x$. Also, $h(x) = f(x) + 2$, so the graph of h is shifted 2 units up from $y = e^x$. The graphs are shown in Figure 5.18.

FIGURE 5.18

EXERCISE 1

Use your calculator to evaluate the following powers.

a. e^2

b. $e^{3.5}$

c. $e^{-0.5}$

The Natural Logarithmic Function

▶▶▶ Your calculator has keys for computing $\log_{10} x$ and $\ln x$. To compute logs to other bases, we use the change of base formula. See the Projects for Chapter 5.

The base e logarithm of a number x, or $\log_e x$, is called the **natural logarithm** of x and is denoted by **$\ln x$**.

The Natural Logarithm

The natural logarithm is the logarithm base e.

$$\ln x = \log_e x \quad x > 0$$

The natural logarithm of x is the exponent to which e must be raised to produce x. For example, the natural logarithm of 10, or $\ln 10$, is the solution of the equation

$$e^y = 10$$

You can verify on your calculator that

$$e^{2.3} \approx 10 \qquad \text{or} \qquad \ln 10 \approx 2.3$$

In general, natural logs obey the same conversion formulas that work for logs to other bases.

Conversion Formulas for Natural Logs

$$y = \ln x \quad \text{if and only if} \quad e^y = x$$

In particular,

$$\ln e = 1 \qquad \text{because} \qquad e^1 = e$$
$$\ln 1 = 0 \qquad \text{because} \qquad e^0 = 1$$

The conversion formulas tell us that the **natural log function,** $g(x) = \ln x$, is the inverse function for the natural exponential function, $f(x) = e^x$.

EXAMPLE 2

a. Graph $f(x) = e^x$ and $f^{-1}(x) = \ln x$ on the same grid.

b. Give the domain and range of the natural log function.

Solutions

a. We can make a table of values for $f^{-1}(x) = \ln x$ by interchanging the columns in Table 5.4 for $f(x) = e^x$. Plotting the points gives us the graph in Figure 5.19.

x	$y = \ln x$
0.050	-3
0.135	-2
0.368	-1
1	0
2.718	1
7.389	2
20.086	3

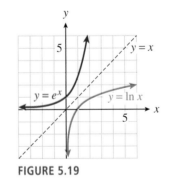

FIGURE 5.19

b. The domain of the natural log function is the same as the range of $y = e^x$, or all positive numbers. The range of $y = \ln x$ is the same as the domain of $y = e^x$, or all real numbers. These results are confirmed by the graph of $y = \ln x$.

Observe that the natural log of a number greater than 1 is positive, while the logs of fractions between 0 and 1 are negative. In addition, *the natural logs of negative numbers and zero are undefined.*

EXERCISE 2

Use your calculator to evaluate each logarithms. Round your answers to four decimal places.

a. $\ln 100$ **b.** $\ln 0.01$ **c.** $\ln e^3$

Properties of the Natural Logarithm

We use natural logarithms in the same way that we use logs to other bases. The properties of logarithms that we studied in Section 4.4 also apply to logarithms base e.

Properties of Natural Logarithms

If $x, y > 0$, then

1. $\ln(xy) = \ln x + \ln y$
2. $\ln \dfrac{x}{y} = \ln x - \ln y$
3. $\ln x^m = m \ln x$

Because the functions $y = e^x$ and $y = \ln x$ are inverse functions, the following properties are also true.

The Natural log and e^x

$$\ln e^x = x, \quad \text{for all } x \text{ and} \quad e^{\ln x} = x, \quad \text{for } x > 0$$

EXAMPLE 3

Simplify each expression.

a. $\ln e^{0.3x}$ **b.** $e^{2\ln(x+3)}$

Solutions

a. The natural log is the log base e, and hence the inverse of e^x. Therefore,

$$\ln e^{0.3x} = 0.3x$$

b. First, simplify the exponent using the third property of logs to get

$$2\ln(x+3) = \ln(x+3)^2$$

Then $e^{2\ln(x+3)} = e^{\ln(x+3)^2} = (x+3)^2$.

Solving Equations

We use the natural logarithm to solve exponential equations with base e. The techniques we've learned for solving other exponential equations also apply to equations with base e.

EXAMPLE 4 Solve each equation for x.

a. $e^x = 0.24$

b. $\ln x = 3.5$

Solutions **a.** Convert the equation to logarithmic form and evaluate using a calculator.

$$x = \ln 0.24 \approx -1.427$$

b. Convert the equation to exponential form and evaluate.

$$x = e^{3.5} \approx 33.1155$$

To solve more complicated exponential equations, we isolate the power on one side of the equation before converting to logarithmic form.

EXAMPLE 5 Solve $140 = 20\,e^{0.4x}$.

Solution First, divide each side by 20 to obtain

$$7 = e^{0.4x}$$

Then convert the equation to logarithmic form.

$$0.4x = \ln 7 \qquad \text{Divide both sides by 0.4.}$$

$$x = \frac{\ln 7}{0.4}$$

Rounded to four decimal places, $x \approx 4.8648$.

We can also solve the equation in Example 5,

$$7 = e^{0.4x}$$

by taking the natural logarithm of both sides. This gives us

$$\ln 7 = \ln e^{0.4x} \qquad \text{Simplify the right side.}$$

$$\ln 7 = 0.4x$$

because $\ln e^a = a$ for any number a. We then proceed with the solution as before.

EXERCISE 5

Solve $80 - 16e^{-0.2x} = 70.3$

Subtract 80 from both sides and divide by -16.
Take the natural log of both sides.
Divide by -0.2.

EXAMPLE 6

Solve $P = \dfrac{a}{1 + be^{-kt}}$ for t.

Solution Multiply both sides of the equation by the denominator, $1 + be^{-kt}$, to get

$$P(1 + be^{-kt}) = a$$

Then isolate the power, e^{-kt}, as follows:

$$1 + be^{-kt} = \frac{a}{P} \qquad \text{Subtract 1 from both sides and simplify.}$$

$$be^{-kt} = \frac{a}{P} - 1 = \frac{a - P}{P} \qquad \text{Divide both sides by } b.$$

$$e^{-kt} = \frac{a - P}{bP}$$

Take the natural logarithm of both sides to get

$$\ln e^{-kt} = \ln\frac{a - P}{bP}$$

and recall that $\ln e^x = x$ to simplify the left side.

$$-kt = \ln\frac{a - P}{bP}$$

Finally, divide both sides by $-k$ to solve for t.

$$t = \frac{-1}{k}\ln\frac{a - P}{bP}$$

EXERCISE 6

Solve $N = Ae^{-kt}$ for k.

Divide both sides by A.
Take the natural log of both sides.
Divide both sides by $-t$.

Exponential Growth and Decay

In Section 4.1, we considered functions of the form

$$P(t) = P_0 \cdot b^t \tag{1}$$

which describe exponential growth when $b > 1$ and exponential decay when $0 < b < 1$. Exponential growth and decay can also be modeled by functions of the form

$$\boldsymbol{P(t) = P_0 \cdot e^{kt}} \tag{2}$$

where we have substituted e^k for the growth factor b in Equation (1), so that

$$P(t) = P_0 \cdot b^t$$
$$= P_0 \cdot (e^k)^t = P_0 \cdot e^{kt}$$

We can find the value of k by solving the equation $b = e^k$ for k, to get $k = \ln b$.

For instance, in Example 1 in Section 4.1 we found that a colony of bacteria grew according to the formula

$$P(t) = 100 \cdot 3^t$$

We can express this function in the form $P(t) = 100 \cdot e^{kt}$ if we set

$$3 = e^k \quad \text{or} \quad k = \ln 3 \approx 1.0986$$

Thus, the growth law for the colony of bacteria can be written

$$P(t) \approx 100 \cdot e^{1.0986t}$$

By graphing both functions on your calculator, you can verify that $P(t) = 100 \cdot 3^t$ and $P(t) = 100 \cdot e^{1.0986t}$ are just two ways of writing the same function.

EXAMPLE 7

From 1990 to 2000, the population of Clark County, Nevada, grew by 6.4% per year.

a. What was the growth factor for the population of Clark County from 1990 to 2000? If the population of Clark County was 768,000 in 1990, write a formula for the population t years later.

b. Write a growth formula for Clark County using base e.

Solutions   a. The growth factor was $b = 1 + r = 1.064$. The population t years later was

$$P(t) = 768,000 (1.064)^t$$

b. We use the formula $P(t) = P_0 \cdot e^{kt}$, where $e^k = 1.064$. Solving for k, we find

$$k = \ln 1.064 = 0.062$$

so $P(t) = 768,000 \, e^{0.062t}$.

EXERCISE 7

From 1994 to 1998, the number of personal computers connected to the Internet grew according to the formula $N(t) = 2.8e^{0.85t}$, where $t = 0$ in 1994 and N is in millions.

a. Evaluate $N(1)$. By what percent did the number of Internet users grow in one year?

b. Express the growth law in the form $N(t) = N_0 (1 + r)^t$. (*Hint:* $e^k = 1 + r$.)

(Source: *Los Angeles Times*, September 6, 1999)

If k is negative, then e^k is a fraction less than 1. For example, if $k = -2$,

$$e^{-2} = \frac{1}{e^2} = \frac{1}{7.3891} = 0.1353$$

Thus, for negative values of k, the function $P(t) = P_0 e^{kt}$ describes exponential decay.

Exponential Growth and Decay

The function

$$P(t) = P_0 e^{kt}$$

describes exponential growth if $k > 0$,
and exponential decay if $k < 0$.

EXAMPLE 8 Express the decay law $N(t) = 60\,(0.8)^t$ in the form $N(t) = N_0 e^{kt}$.

Solution For this decay law, $N_0 = 60$ and $b = 0.8$. We would like to find a value for k so that $e^k = b = 0.8$, that is, we must solve the equation

$$e^k = 0.8 \qquad \text{Take natural log of both sides.}$$
$$\ln e^k = \ln 0.8 \qquad \text{Simplify.}$$
$$k = \ln 0.8 \approx -0.2231$$

Replacing b with e^k, we find that the decay law is

$$N(t) \approx 60 e^{-0.2231t}$$

EXERCISE 8

A scientist isolates 25 grams of krypton-91, which decays according to the formula $N(t) = 25 e^{-0.07t}$, where t is in seconds.

a. Complete the table of values showing the amount of krypton-91 left at 10-second intervals over the first minute.

t	0	10	20	30	40	50	60
$N(t)$							

b. Use the table to choose a suitable window and graph the function $N(t)$.

c. Write and solve an equation to answer the question: How long does it take for 60% of the krypton-91 to decay? (*Hint:* If 60% of the krypton-91 has decayed, 40% of the original 25 grams remains.)

Continuous Compounding

Some savings institutions offer accounts on which the interest is **compounded continuously.** The amount accumulated in such an account after t years at interest rate r is given by the function

$$A(t) = P e^{rt}$$

where P is the principal invested.

EXAMPLE 9 Suppose you invest $500 in an account that pays 8% interest compounded continuously. You leave the money in the account without making any additional deposits or withdrawals.

a. Write a formula that gives the value of your account $A(t)$ after t years.

b. Make a table of values showing $A(t)$ for the first 5 years.

c. Graph the function $A(t)$.

d. How much will the account be worth after 10 years?

e. How long will it be before the account is worth $1000?

Solutions   a. Substitute 500 for P, and 0.08 for r to find

$$A(t) = 500e^{0.08t}$$

b. Evaluate the formula for $A(t)$ to obtain a table.

t	$A(t)$
0	500
1	541.64
2	586.76
3	635.62
4	688.56
5	745.91

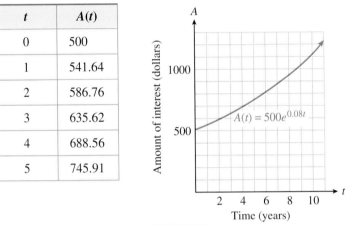

FIGURE 5.20

c. The graph of $A(t)$ is shown in Figure 5.20.

d. Evaluate $A(t)$ for $t = 10$.

$$A(10) = 500\,e^{0.08(10)}$$
$$= 500\,e^{0.8}$$
$$\approx 500(2.2255) = 1112.77$$

The account will be worth $1112.77 after 10 years.

e. Substitute 1000 for $A(t)$ and solve the equation.

$$1000 = 500\,e^{0.08t} \qquad \text{Divide both sides by 500.}$$
$$2 = e^{0.08t} \qquad \text{Take natural log of both sides.}$$
$$\ln 2 = \ln e^{0.08t} = 0.08t \qquad \text{Divide both sides by 0.08.}$$
$$t = \frac{\ln 2}{0.08} \approx 8.6643$$

The account will be worth $1000 after approximately 8.7 years.

Zelda invested $1000 in an account that pays 4.5% interest compounded continuously. How long will it be before the account is worth $2000?

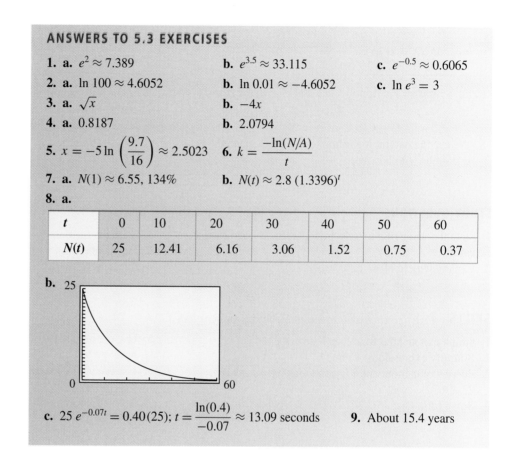

ANSWERS TO 5.3 EXERCISES

1. **a.** $e^2 \approx 7.389$ **b.** $e^{3.5} \approx 33.115$ **c.** $e^{-0.5} \approx 0.6065$

2. **a.** $\ln 100 \approx 4.6052$ **b.** $\ln 0.01 \approx -4.6052$ **c.** $\ln e^3 = 3$

3. **a.** $\sqrt{x}$ **b.** $-4x$

4. **a.** 0.8187 **b.** 2.0794

5. $x = -5\ln\left(\dfrac{9.7}{16}\right) \approx 2.5023$ 6. $k = \dfrac{-\ln(N/A)}{t}$

7. **a.** $N(1) \approx 6.55$, 134% **b.** $N(t) \approx 2.8\,(1.3396)^t$

8. **a.**

t	0	10	20	30	40	50	60
$N(t)$	25	12.41	6.16	3.06	1.52	0.75	0.37

b.

c. $25\,e^{-0.07t} = 0.40(25)$; $t = \dfrac{\ln(0.4)}{-0.07} \approx 13.09$ seconds   9. About 15.4 years

---

VOCABULARY   *Look up the definitions of new terms in the Glossary.*

Natural exponential function   Natural logarithm   Continuous compounding

CONCEPTS

1. The **natural base** is an irrational number called e, where

$$e \approx 2.71828182845$$

2. The **natural exponential function** is the function $f(x) = e^x$. The **natural log function** is the function $g(x) = \ln x = \log_e x$.

3.
Conversion Formulas for Natural Logs

$$y = \ln x \quad \text{if and only if} \quad e^y = x$$

4.

> ## Properties of Natural Logarithms
>
> If $x,\ y > 0$, then
>
> **1.** $\ln(xy) = \ln x + \ln y$
>
> **2.** $\ln\left(\dfrac{x}{y}\right) = \ln x - \ln y$
>
> **3.** $\ln x^m = m \ln x$
>
> also
>
> $\ln e^x = x,\quad$ for all x and $\quad e^{\ln x} = x$, for $(x > 0)$

5. We use the natural logarithm to solve exponential equations with base e.

6.

> ## Exponential Growth and Decay
>
> The function
>
> $$P(t) = P_0 e^{kt}$$
>
> describes and   exponential growth if $k > 0$, exponential decay if $k < 0$.

7. Continuous compounding: The amount accumulated in an account after t years at interest rate r compounded continuously is given by

$$A(t) = Pe^{rt}$$

where P is the principal invested.

STUDY QUESTIONS

1. State the value of e to 3 decimal places. Memorize this value.

2. Explain why $\ln e^x = x$.

3. State the formula for exponential growth using base e.

4. How is the formula for exponential decay in base e different from the formula for exponential growth?

SKILLS   *Practice each skill in the Homework problems listed.*

1. Graph exponential functions base e: #1–4

2. Simplify expressions: #5 and 6

3. Solve exponential and log equations base e: #7–10, 23–30

4. Solve formulas: #31–36

5. Use the properties of logs and exponents with the natural base: #19–22, 37–40

6. Use the natural exponential function in applications: #11–14, 47–58

7. Convert between $P(t) = P_0(1 + r)^t$ and $P(t) = P_0 e^{kt}$: #15–18, 41–46

HOMEWORK 5.3

ThomsonNOW™ Test yourself on key content at www.thomsonedu.com/login.

Use your calculator to complete the table for each function. Then choose a suitable window and graph the function.

x	-10	-5	0	5	10	15	20
$f(x)$							

1. $f(x) = e^{0.2x}$ **2.** $f(x) = e^{0.6x}$ **3.** $f(x) = e^{-0.3x}$ **4.** $f(x) = e^{-0.1x}$

Simplify.

5. a. $\ln e^2$ **b.** $e^{\ln 5t}$
     c. $e^{-\ln x}$ **d.** $\ln \sqrt{e}$

6. a. $\ln e^{x^4}$ **b.** $e^{3\ln x}$
     c. $e^{\ln x - \ln y}$ **d.** $\ln\left(\dfrac{1}{e^{2t}}\right)$

Solve for *x*. Round your answers to two decimal places.

7. a. $e^x = 1.9$ **b.** $e^x = 45$ **c.** $e^x = 0.3$ **8. a.** $e^x = 2.1$ **b.** $e^x = 60$ **c.** $e^x = 0.9$

9. a. $\ln x = 1.42$ **b.** $\ln x = 0.63$ **10. a.** $\ln x = 2.03$ **b.** $\ln x = 0.59$
    c. $\ln x = -2.6$ **c.** $\ln x = -3.4$

11. The number of bacteria in a culture grows according to the function

$$N(t) = N_0 e^{0.04t}$$

where N_0 is the number of bacteria present at time $t = 0$ and t is the time in hours.
  a. Write a growth law for a sample in which 6000 bacteria were present initially.
  b. Make a table of values for $N(t)$ in 5-hour intervals over the first 30 hours.
  c. Graph $N(t)$.
  d. How many bacteria were present at $t = 24$ hours?
  e. How much time must elapse (to the nearest tenth of an hour) for the original 6000 bacteria to increase to 100,000?

12. Hope invests $2000 in a savings account that pays $5\frac{1}{2}\%$ annual interest compounded continuously.
  a. Write a formula that gives the amount of money $A(t)$ in Hope's account after t years.
  b. Make a table of values for $A(t)$ in 2-year intervals over the first 10 years.
  c. Graph $A(t)$.
  d. How much will Hope's account be worth after 7 years?
  e. How long will it take for the account to grow to $5000?

13. The intensity, I (in lumens), of a light beam after passing through t centimeters of a filter having an absorption coefficient of 0.1 is given by the function

$$I(t) = 1000e^{-0.1t}$$

  a. Graph $I(t)$.
  b. What is the intensity (to the nearest tenth of a lumen) of a light beam that has passed through 0.6 centimeters of the filter?
  c. How many centimeters (to the nearest tenth) of the filter will reduce the illumination to 800 lumens?

14. X-rays can be absorbed by a lead plate so that

$$I(t) = I_0 e^{-1.88t}$$

where I_0 is the X-ray count at the source and $I(t)$ is the X-ray count behind a lead plate of thickness t inches.
  a. Graph $I(t)$.
  b. What percent of an X-ray beam will penetrate a lead plate $\frac{1}{2}$ inch thick?
  c. How thick should the lead plate be in order to screen out 70% of the X-rays?

Express each exponential function in the form $P(t) = P_0 b^t$. Is the function increasing or decreasing? What is its initial value?

15. $P(t) = 20e^{0.4t}$ **16.** $P(t) = 0.8e^{1.3t}$ **17.** $P(t) = 6500e^{-2.5t}$ **18.** $P(t) = 1.7e^{-0.02t}$

19. a. Fill in the table, rounding your answers to four decimal places.

x	0	0.5	1	1.5	2	2.5
e^x						

  b. Compute the ratio of each function value to the previous one. Explain the result.

20. a. Fill in the table, rounding your answers to four decimal places.

x	0	2	4	6	8	10
e^x						

  b. Compute the ratio of each function value to the previous one. Explain the result.

21. a. Fill in the table, rounding your answers to the nearest integer.

x	0	0.6931	1.3863	2.0794	2.7726	3.4657	4.1589
e^x							

    b. Subtract each x-value from the next one. Explain the result.

22. a. Fill in the table, rounding your answers to the nearest integer.

x	0	1.0986	2.1972	3.2958	4.3944	5.4931	6.5917
e^x							

    b. Subtract each x-value from the next one. Explain the result.

Solve. Round your answers to two decimal places.

23. $6.21 = 2.3e^{1.2x}$

24. $22.26 = 5.3e^{0.4x}$

25. $6.4 = 20e^{0.3x} - 1.8$

26. $4.5 = 4e^{2.1x} + 3.3$

27. $46.52 = 3.1e^{1.2x} + 24.2$

28. $1.23 = 1.3e^{2.1x} - 17.1$

29. $16.24 = 0.7e^{-1.3x} - 21.7$

30. $55.68 = 0.6e^{-0.7x} + 23.1$

Solve each equation for the specified variable.

31. $y = e^{kt}$, for t

32. $\dfrac{T}{R} = e^{t/2}$, for t

33. $y = k(1 - e^{-t})$, for t

34. $B - 2 = (A + 3)e^{-t/3}$, for t

35. $T = T_0 \ln(k + 10)$, for k

36. $P = P_0 + \ln 10k$, for k

37. a. Fill in the table, rounding your answers to three decimal places.

n	0.39	3.9	39	390
$\ln n$				

    b. Subtract each natural logarithm in your table from the next one. (For example, compute $\ln 3.9 - \ln 0.39$.) Explain the result.

38. a. Fill in the table, rounding your answers to three decimal places.

n	0.64	6.4	64	640
$\ln n$				

    b. Subtract each natural logarithm in your table from the next one. (For example, compute $\ln 6.4 - \ln 0.64$.) Explain the result.

39. a. Fill in the table, rounding your answers to three decimal places.

n	2	4	8	16
$\ln n$				

    b. Divide each natural logarithm in your table by $\ln 2$. Explain the result.

40. a. Fill in the table, rounding your answers to three decimal places.

n	5	25	125	625
$\ln n$				

    b. Divide each natural logarithm in your table by $\ln 5$. Explain the result.

For Problems 41–46,
(a) Express each growth or decay law in the form $N(t) = N_0 e^{kt}$.
(b) Check your answer by graphing both forms of the function on the same axes.
   Do they have the same graph?

41. $N(t) = 100 \cdot 2^t$

42. $N(t) = 50 \cdot 3^t$

43. $N(t) = 1200(0.6)^t$

44. $N(t) = 300(0.8)^t$

45. $N(t) = 10(1.15)^t$

46. $N(t) = 1000(1.04)^t$

47. The population of Citrus Valley was 20,000 in 1980. In 1990, it was 35,000.
   a. What is P_0 if $t = 0$ in 1980?
   b. Use the population in 1990 to find the growth factor e^k.
   c. Write a growth law of the form $P(t) = P_0 e^{kt}$ for the population of Citrus Valley.
   d. If it continues at the same rate of growth, what will the population be in 2010?

48. A copy of *Time* magazine cost $1.50 in 1981. In 1988, the cover price was $2.00
   a. What is P_0 if $t = 0$ in 1981?
   b. Use the price in 1988 to find the growth factor e^k.
   c. Find a growth law of the form $P(t) = P_0 e^{kt}$ for the price of *Time*.
   d. In 1999, a copy of *Time* cost $3.50. Did the price of the magazine continue to grow at the same rate from 1981 to 1999?

49. Cobalt-60 is a radioactive isotope used in the treatment of cancer. A 500-milligram sample of cobalt-60 decays to 385 milligrams after 2 years.
   a. Using $P_0 = 500$, find the decay factor e^k for cobalt-60.
   b. Write a decay law $N(t) = N_0 e^{kt}$ for cobalt-60.
   c. How much of the original sample will be left after 10 years?

50. Weed seeds can survive for a number of years in the soil. An experiment on cultivated land found 155 million weed seeds per acre, and in the following years the experimenters prevented the seeds from coming to maturity and producing new weeds. Four years later, there were 13.6 million seeds per acre. (Source: Burton, 1998)
   a. Find the annual decay factor e^k for the number of weed seeds in the soil.
   b. Write an exponential formula with base e for the number of weed seeds that survived after t years.

Problems 51–58 are about doubling time and half-life.

51. Delbert invests $500 in an account that pays 9.5% interest compounded continuously.
   a. Write a formula for $A(t)$ that gives the amount of money in Delbert's account after t years.
   b. How long will it take Delbert's investment to double to $1000?
   c. How long will it take Delbert's money to double again, to $2000?

   d. Graph $A(t)$ and illustrate the doubling time on your graph.
   e. Choose any point (t_1, A_1) on the graph, then find the point on the graph with vertical coordinate $2A_1$. Verify that the difference in the t-coordinates of the two points is the doubling time.

52. The growth of plant populations can be measured by the amount of pollen they produce. The pollen from a population of pine trees that lived more than 9500 years ago in Norfolk, England, was deposited in the layers of sediment in a lake basin and dated with radiocarbon techniques. The figure shows the rate of pollen accumulation plotted against time, and the fitted curve $P(t) = 650e^{0.00932t}$. (Source: Burton, 1998)

  a. What was the annual rate of growth in pollen accumulation?

  b. Find the doubling time for the pollen accumulation, that is, the time it took for the accumulation rate to double.

  c. By what factor did the pollen accumulation rate increase over a period of 500 years?

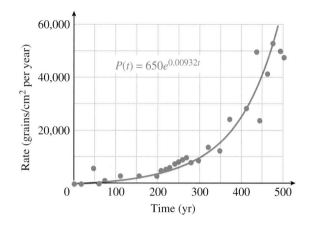

53. Technetium-99m (Tc-99m) is an artificially produced radionuclide used as a tracer for producing images of internal organs such as the heart, liver, and thyroid. A solution of Tc-99m with initial radioactivity of 10,000 becquerels (Bq) decays according to the formula

$$N(t) = 10,000e^{-0.1155t}$$

where t is in hours.

  a. How long will it take the radioactivity to fall to half its initial value, or 5000 Bq?

  b. How long will it take the radioactivity to be halved again?

  c. Graph $N(t)$ and illustrate the half-life on your graph.

  d. Choose any point (t_1, N_1) on the graph, then find the point on the graph with vertical coordinate $0.5N_1$. Verify that the difference in the t-coordinates of the two points is the half-life.

54. All living things contain a certain amount of the isotope carbon-14. When an organism dies, the carbon-14 decays according to the formula

$$N(t) = N_0e^{-0.000124t}$$

where t is measured in years. Scientists can estimate the age of an organic object by measuring the amount of carbon-14 remaining.

  a. When the Dead Sea scrolls were discovered in 1947, they had 78.8% of their original carbon-14. How old were the Dead Sea scrolls then?

  b. What is the half-life of carbon-14, that is, how long does it take for half of an object's carbon-14 to decay?

55. The half-life of iodine-131 is approximately 8 days.

  a. If a sample initially contains N_0 grams of iodine-131, how much will it contain after 8 days? How much will it contain after 16 days? After 32 days?

  b. Use your answers to part (a) to sketch a graph of $N(t)$, the amount of iodine-131 remaining, versus time. (Choose an arbitrary height for N_0 on the vertical axis.)

  c. Calculate k, and hence find a decay law of the form $N(t) = N_0e^{kt}$, where $k < 0$, for iodine-131.

56. The half-life of hydrogen-3 is 12.5 years.

  a. If a sample initially contains N_0 grams of hydrogen-3, how much will it contain after 12.5 years? How much will it contain after 25 years?

  b. Use your answers to part (a) to sketch a graph of $N(t)$, the amount of hydrogen-3 remaining, versus time. (Choose an arbitrary height for N_0 on the vertical axis.)

  c. Calculate k, and hence find a decay law of the form $N(t) = N_0e^{kt}$, where $k < 0$, for hydrogen-3.

57. A Geiger counter measures the amount of radioactive material present in a substance. The table shows the count rate for a sample of iodine-128 as a function of time. (Source: Hunt and Sykes, 1984)

Time (min)	0	10	20	30	40
Counts/sec	120	90	69	54	42

Time (min)	50	60	70	80	90
Counts/sec	33	25	19	15	13

a. Graph the data and use your calculator's exponential regression feature to fit a curve to them.
b. Write your equation in the form $G(t) = G_0 e^{kt}$.
c. Calculate the half-life of iodine-128.

58. The table shows the count rate for sodium-24 registered by a Geiger counter as a function of time. (Source: Hunt and Sykes, 1984)

Time (min)	0	10	20	30	40
Counts/sec	180	112	71	45	28

Time (min)	50	60	70	80	90
Counts/sec	18	11	7	4	3

a. Graph the data and use your calculator's exponential regression feature to fit a curve to them.
b. Write your equation in the form $G(t) = G_0 e^{kt}$.
c. Calculate the half-life of sodium-24.

5.4 Logarithmic Scales

Because logarithmic functions grow very slowly, they are useful for modeling phenomena that take on a very wide range of values. For example, biologists study how metabolic functions such as heart rate are related to an animal's weight, or mass. Table 5.6 shows the mass in kilograms of several mammals.

Animal	Shrew	Cat	Wolf	Horse	Elephant	Whale
Mass, kg	0.004	4	80	300	5400	70,000

TABLE 5.6

Imagine trying to scale the *x*-axis to show all of these values. If we set tick marks at intervals of 10,000 kg, as shown in Figure 5.21a, we can plot the mass of the whale, and maybe the elephant, but the dots for the smaller animals will be indistinguishable.

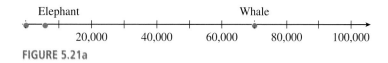

FIGURE 5.21a

On the other hand, we can plot the mass of the cat if we set tick marks at intervals of 1 kg, as shown in Figure 5.21b, but the axis will have to be extremely long to include even the wolf. We cannot show the masses of all these animals on the same scale.

FIGURE 5.21b

To get around this problem, we can plot the log of the mass, instead of the mass itself. Table 5.7 shows the base 10 log of each animal's mass, rounded to 2 decimal places.

Animal	Shrew	Cat	Wolf	Horse	Elephant	Whale
Mass, kg	0.004	4	80	300	5400	70,000
Log (mass)	−2.40	0.60	1.90	2.48	3.73	4.85

TABLE 5.7

The logs of the masses range from −2.40 to 4.85. We can easily plot these values on a single scale, as shown in Figure 5.22.

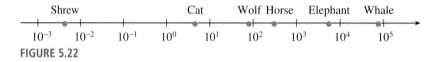

FIGURE 5.22

The scale in Figure 5.22 is called a **logarithmic scale,** or log scale. The tick marks are labeled with powers of 10, because, as you recall, a logarithm is actually an exponent. For example, the mass of the horse is 300 kg, and

$$\log 300 = 2.48 \quad \text{so} \quad 10^{2.48} = 300$$

When we plot 2.48 for the horse, we are really plotting the power of 10 that gives its mass, because $10^{2.48} = 300$ kg. The exponents on base 10 are evenly spaced on a log scale, so we plot $10^{2.48}$ about halfway between 10^2 and 10^3.

EXAMPLE 1   Plot the values on a log scale.

x	0.0007	0.2	3.5	1600	72,000	4×10^8

Solution   We actually plot the logs of the values, so we first compute the base 10 logarithm of each number.

x	0.0007	0.2	3.5	1600	72,000	4×10^8
log x	−3.15	−0.70	0.54	3.20	4.86	8.60

Then we plot each logarithm, estimating its position between integer exponents. For example, we plot the first value, −3.15, closer to −3 than to −4. The finished plot is shown in Figure 5.23.

FIGURE 5.23

EXERCISE 1

Complete the table by estimating the logarithm of each point plotted on the log scale in Figure 5.24. Then give a decimal value for each point.

log x				
x				

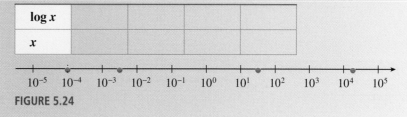

FIGURE 5.24

Using Log Scales

By now, you have noticed that the values represented by points on a log scale increase rapidly as we move to the right along the scale. Also notice that $10^0 = 1$, so the "middle" of a log scale represents 1 (not zero, as on a linear scale). Points to the left of 10^0 represent fractions between 0 and 1, because powers of 10 with negative exponents are numbers less than 1. Their values decrease toward 0 as we move to the left, but they never become negative. We cannot plot negative numbers or zero on a log scale, because the log of a negative number or zero is undefined.

EXAMPLE 2   Figure 5.25 shows a timeline for life on Earth, in units of Mya (million years ago).

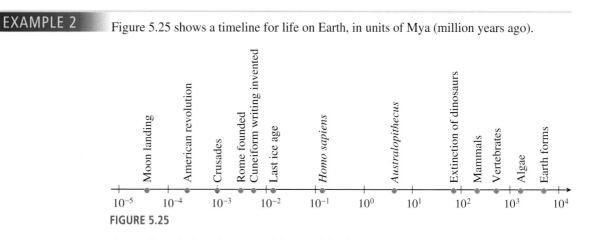

FIGURE 5.25

Approximately how long ago did each of the following events occur?

a. Formation of Earth   **b.** Dinosaurs became extinct

c. The last ice age   **d.** The Crusades

Solutions   **a.** Reading from the timeline, the Earth was formed between 10^3 and 10^4, or between 1000 and 10,000 million years ago. We estimate that Earth formed 5000 million years ago.

b. The extinction of the dinosaurs is plotted between 10^1 and 10^2, or between 10 and 100 million years ago. Because the point is closer to 10^2, we estimate their extinction at 70 million years ago.

c. The last ice age is plotted just after 10^{-2}, or 0.01 million years ago. One-hundredth of a million is 10,000, so we estimate that the ice age occurred a little more than 10,000 years ago.

d. The Crusades occurred about 10^{-3}, or about 0.001 million years ago. One-thousandth of a million is 1000, so the Crusades occurred about 1000 years ago, or about 1000 A.D.

Plot the following dollar values on a log scale.

Postage stamp	0.39
Medium cappuccino	3.15
Notebook computer	679
One year at Harvard	37,928
2005 Lamborghini	282,300
Kobe Bryant's salary	15,946,875
Bill Gates's financial worth	46,600,000,000
National debt	8,146,222,469,143

Equal Increments on a Log Scale

Log scales allow us to plot a wide range of values, but there is a trade-off. Equal increments on a log scale do not correspond to equal differences in value, as they do on a linear scale. You can see why in Figure 5.26: The difference between 10^1 and 10^0 is $10 - 1 = 9$, but the difference between 10^2 and 10^1 is $100 - 10 = 90$.

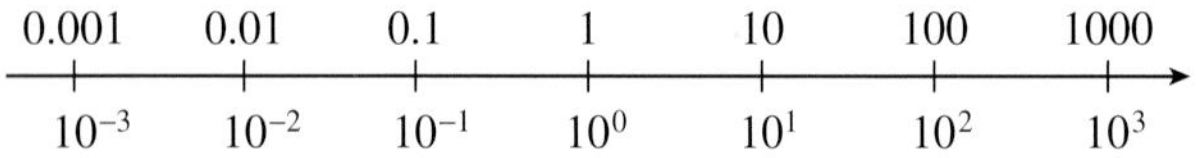

FIGURE 5.26

If we include tick marks for intermediate values on the log scale, they look like Figure 5.27.

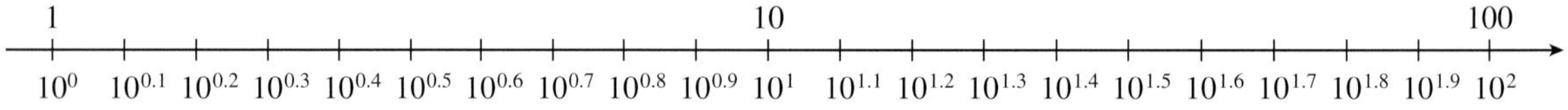

FIGURE 5.27

Once again, the difference between, say, $10^{0.1}$ and $10^{0.2}$ is not the same as the difference between $10^{0.2}$ and $10^{0.3}$. The decimal values of the powers $10^{0.1}$ through $10^{0.9}$, rounded to two places, are shown in Figure 5.28.

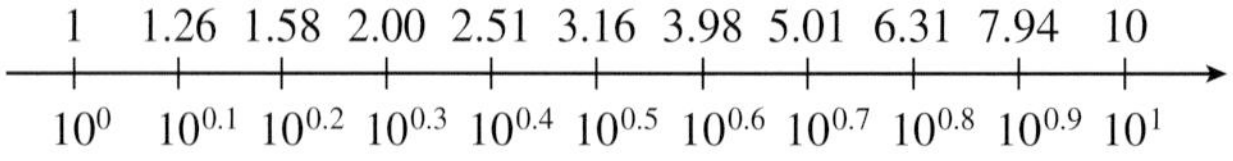

FIGURE 5.28

As we move from left to right on this scale, we *multiply* the value at the previous tick mark by $10^{0.1}$, or about 1.258. For example,

$$10^{0.2} = 1.258 \times 10^{0.1} = 1.585$$

$$10^{0.3} = 1.585 \times 10^{0.1} = 1.995$$

and so on. Moving up by equal increments on a log scale does not add equal amounts to the values plotted; it *multiplies* the values by equal *factors*.

EXAMPLE 3 What number is halfway between 10 and 100 on a log scale?

Solution   On a log scale, the number $10^{1.5}$ is halfway between 10^1 and 10^2, as shown in Figure 5.29.

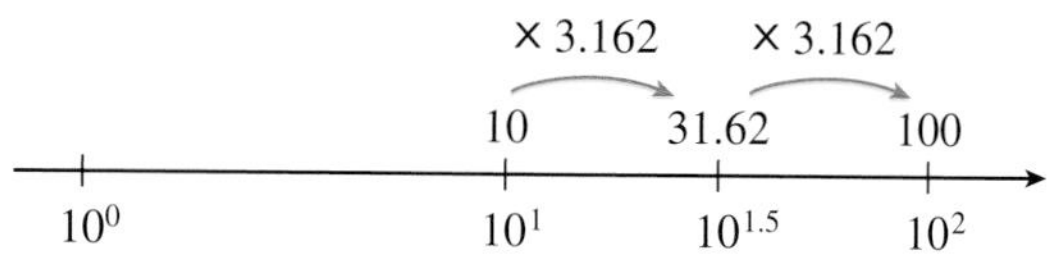

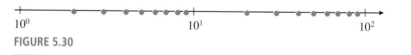

FIGURE 5.29

Now, $10^{1.5} = 10\sqrt{10}$, or approximately 31.62. Note how equal increments of 0.5 on the log scale correspond to equal factors of $10^{0.5}$ in the values plotted:

$$10 \times 3.162 = 31.62 \quad \text{and} \quad 31.62 \times 3.162 = 100$$
$$10^1 \times 10^{0.5} = 10^{1.5} \quad \text{and} \quad 10^{1.5} \times 10^{0.5} = 10^2$$

EXERCISE 3

What number is halfway between $10^{1.5}$ and 10^2 on a log scale?

If we would like to label the log scale with integers, we get a very different-looking scale, one in which the tick marks are not evenly spaced.

EXAMPLE 4 Plot the integer values 2 through 9 and 20 through 90 on a log scale.

Solution   Compute the logarithm of each integer value.

x	2	3	4	5	6	7	8	9
$\log x$	0.301	0.477	0.602	0.699	0.778	0.845	0.903	0.954

x	20	30	40	50	60	70	80	90
$\log x$	1.301	1.477	1.602	1.699	1.778	1.845	1.903	1.954

Plot on a log scale, as shown in Figure 5.30.

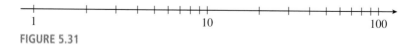

FIGURE 5.30

On the log scale in Figure 5.30, notice how the integer values are spaced: They get closer together as they approach the next power of 10. You will often see log scales labeled not with powers of 10, but with integer values, as in Figure 5.31.

FIGURE 5.31

In fact, **log-log graph paper** scales both axes with logarithmic scales.

The opening page of Chapter 3 shows the "mouse-to-elephant" curve, a graph of the metabolic rate of mammals as a function of their mass. (The elephant does not appear on that graph, because its mass is too big.) Figure 5.32 shows the same function, graphed on log-log paper.

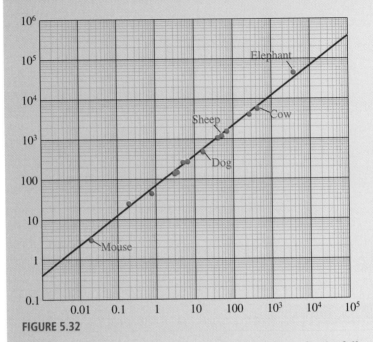

FIGURE 5.32

Use this graph to estimate the mass and metabolic rate for the following animals, labeled on the graph.

Animal	Mouse	Dog	Sheep	Cow	Elephant
Mass (kg)					
Metabolic rate (kcal/day)					

Acidity and the pH Scale

You may have already encountered log scales in some everyday applications. A simple example is the pH scale, used by chemists to measure the acidity of a substance or chemical compound. This scale is based on the concentration of hydrogen ions in the substance, denoted by $[H^+]$. The pH value is defined by the formula

$$pH = -\log_{10}[H^+]$$

Values for pH fall between 0 and 14, with 7 indicating a neutral solution. The lower the pH value, the more acidic the substance. Some common substances and their pH values are shown in Table 5.8.

Substance	pH	H⁺
Battery acid	1	0.1
Lemon juice	2	0.01
Vinegar	3	0.001
Milk	6.4	$10^{-6.4}$
Baking soda	8.4	$10^{-8.4}$
Milk of magnesia	10.5	$10^{-10.5}$
Lye	13	10^{-13}

TABLE 5.8

EXAMPLE 5

a. Calculate the pH of a solution with a hydrogen ion concentration of 3.98×10^{-5}.

b. The water in a swimming pool should be maintained at a pH of 7.5. What is the hydrogen ion concentration of the water?

Solutions   a. Use a calculator to evaluate the function with $[H^+] = 3.98 \times 10^{-5}$.

$$pH = -\log_{10}(3.98 \times 10^{-5}) \approx 4.4$$

b. Solve the equation

$$7.5 = -\log_{10}[H^+]$$

for $[H^+]$. First, write

$$-7.5 = \log_{10}[H^+]$$

Then convert the equation to exponential form to get

$$[H^+] = 10^{-7.5} \approx 3.2 \times 10^{-8}$$

The hydrogen ion concentration of the water is 3.2×10^{-8}.

EXERCISE 5

The pH of the water in a tide pool is 8.3. What is the hydrogen ion concentration of the water?

A decrease of 1 on the pH scale corresponds to an increase in acidity by a factor of 10. Thus, lemon juice is 10 times more acidic than vinegar, and battery acid is 100 times more acidic than vinegar.

Decibels

The decibel scale, used to measure the loudness or intensity of a sound, is another example of a logarithmic scale. The loudness of a sound is measured in decibels, D, by

$$D = 10 \log_{10}\left(\frac{I}{10^{-12}}\right)$$

where I is the intensity of its sound waves (in watts per square meter). Table 5.9 shows the intensity of some common sounds, measured in watts per square meter.

Sound	Intensity (watts/m^2)	Decibels
Whisper	10^{-10}	20
Background music	10^{-8}	40
Loud conversation	10^{-6}	60
Heavy traffic	10^{-4}	80
Jet airplane	10^{-2}	100
Thunder	10^{-1}	110

TABLE 5.9

Consider the ratio of intensities of thunder to a whisper:

$$\frac{\text{Intensity of thunder}}{\text{Intensity of a whisper}} = \frac{10^{-1}}{10^{-10}} = 10^9$$

Thunder is 10^9, or one billion times more intense than a whisper. It would be impossible to show such a wide range of values on a graph and still maintain reasonable precision. When we use a log scale, however, there is a difference of only 90 decibels between a whisper and thunder.

EXAMPLE 6

a. Normal breathing generates about 10^{-11} watts per square meter at a distance of 3 feet. Find the number of decibels for a breath 3 feet away.

b. Normal conversation registers at about 40 decibels. How many times more intense than breathing is normal conversation?

Solutions

a. Evaluate the decibel formula with $I = 10^{-11}$ to find

$$D = 10 \log_{10}\left(\frac{10^{-11}}{10^{-12}}\right) = 10 \log_{10} 10^1$$

$$= 10(1) = 10 \ \text{decibels}$$

b. Let I_b stand for the sound intensity of breathing, and let I_c stand for the intensity of normal conversation. We are looking for the ratio I_c/I_b. From part (a), we know that

$$I_W = 10^{-11}$$

and from the formula for decibels, we have

$$40 = 10 \log_{10}\left(\frac{I_c}{10^{-12}}\right)$$

which we can solve for I_c. Dividing both members of the equation by 10 and rewriting in exponential form, we have

$$\frac{I_c}{10^{-12}} = 10^4 \qquad\qquad \text{Multiply both sides by } 10^{-12}.$$
$$I_c = 10^4(10^{-12}) = 10^{-8}$$

Finally, compute the ratio $\frac{I_c}{I_b}$:

$$\frac{I_c}{I_b} = \frac{10^{-8}}{10^{-11}} = 10^3$$

Normal conversation is 1000 times more intense than breathing.

EXERCISE 6

The noise of city traffic registers at about 70 decibels.

a. What is the intensity of traffic noise, in watts per square meter?

b. How many times more intense is traffic noise than conversation?

CAUTION

Both the decibel model and the Richter scale in the next example use expressions of the form $\log\left(\frac{a}{b}\right)$. Be careful to follow the order of operations when using these models. We must compute the quotient $\frac{a}{b}$ before taking a logarithm. In particular, it is *not* true that $\log\left(\frac{a}{b}\right)$ can be simplified to $\frac{\log a}{\log b}$.

The Richter Scale

One method for measuring the magnitude of an earthquake compares the amplitude A of its seismographic trace with the amplitude A_0 of the smallest detectable earthquake. The log of their ratio is the Richter magnitude, M. Thus,

$$M = \log_{10}\left(\frac{A}{A_0}\right)$$

EXAMPLE 7

a. The Northridge earthquake of January 1994 registered 6.9 on the Richter scale. What would be the magnitude of an earthquake 100 times as powerful as the Northridge quake?

b. How many times more powerful than the Northridge quake was the San Francisco earthquake of 1989, which registered 7.1 on the Richter scale?

Solutions

a. Let A_N represent the amplitude of the Northridge quake and let A_H represent the amplitude of a quake 100 times more powerful. From the Richter model, we have

$$6.9 = \log_{10}\left(\frac{A_N}{A_0}\right)$$

or, rewriting in exponential form,

$$\frac{A_N}{A_0} = 10^{6.9}$$

Now, $A_H = 100A_N$, so

$$\frac{A_H}{A_0} = \frac{100A_N}{A_0}$$

$$= 100\left(\frac{A_N}{A_0}\right) = 10^2(10^{6.9})$$

$$= 10^{8.9}$$

Thus, the magnitude of the more powerful quake is

$$\log_{10}\left(\frac{A_H}{A_0}\right) = \log_{10} 10^{8.9}$$

$$= 8.9$$

b. Let A_S stand for the amplitude of the San Francisco earthquake. We are looking for the ratio A_S/A_N. First, we will use the Richter formula to compute values for A_S and A_N.

$$6.9 = \log_{10}\left(\frac{A_N}{A_0}\right) \quad \text{and} \quad 7.1 = \log_{10}\left(\frac{A_S}{A_0}\right)$$

Rewriting each equation in exponential form, we have

$$\frac{A_N}{A_0} = 10^{6.9} \quad \text{and} \quad \frac{A_S}{A_0} = 10^{7.1}$$

or

$$A_N = 10^{6.9}A_0 \quad \text{and} \quad A_S = 10^{7.1}A_0$$

Now we can compute the ratio we want:

$$\frac{A_S}{A_N} = \frac{10^{7.1}A_0}{10^{6.9}A_0} = 10^{0.2}$$

The San Francisco earthquake was $10^{0.2}$, or approximately 1.58 times as powerful as the Northridge quake.

EXERCISE 7

In October 2005, a magnitude 7.6 earthquake struck Pakistan. How much more powerful was this earthquake than the 1989 San Francisco earthquake of magnitude 7.1?

An earthquake 100, or 10^2, times as strong is only two units greater in magnitude on the Richter scale. In general, a *difference* of K units on the Richter scale (or any logarithmic scale) corresponds to a *factor* of 10^K units in the intensity of the quake.

| EXAMPLE 8 | On a log scale, the weights of two animals differ by 1.6 units. What is the ratio of their actual weights? |

Solution   A difference of 1.6 on a log scale corresponds to a factor of $10^{1.6}$ in the actual weights. Thus, the heavier animal is $10^{1.6}$, or 39.8 times heavier than the lighter animal.

EXERCISE 8

Two points, labeled A and B, differ by 2.5 units on a log scale. What is the ratio of their decimal values?

ANSWERS TO 5.4 EXERCISES

1.

$\log x$	-4	-2.5	1.5	4.25
x	0.0001	0.0032	31.6	17,782.8

2.

$10^{-1}\ \ 10^{0}\ \ 10^{1}\ \ 10^{2}\ \ 10^{3}\ \ 10^{4}\ \ 10^{5}\ \ 10^{6}\ \ 10^{7}\ \ 10^{8}\ \ 10^{9}\ \ 10^{10}\ \ 10^{11}\ \ 10^{12}\ \ 10^{13}$

3. 56.23

4.

Animal	Mouse	Dog	Sheep	Cow	Elephant
Mass (kg)	0.02	15	50	500	4000
Metabolic rate (kcal/day)	3.5	500	1500	6000	50,000

5. 5.01×10^{-9} **6. a.** $I = 10^{-5}$ watts/m^2 **b.** 1000   **7.** 3.16

8. 316.2

SECTION 5.4 SUMMARY

VOCABULARY   *Look up the definitions of new terms in the Glossary.*

Log scale   Log-log paper

CONCEPTS

1. A **log scale** is useful for plotting values that vary greatly in magnitude. We plot the log of the variable instead of the variable itself.
2. A log scale is a **multiplicative scale:** Each increment of equal length on the scale indicates that the value is multiplied by an equal amount.
3. The pH value of a substance is defined by the formula

$$\text{pH} = -\log_{10}[\text{H}^+]$$

where $[\text{H}^+]$ denotes $[\text{H}^+]$ the concentration of hydrogen ions in the substance.

4. The loudness of a sound is measured in decibels, D, by

$$D = 10\log_{10}\left(\frac{I}{10^{-12}}\right)$$

where I is the intensity of its sound waves (in watts per square meter).

5. The Richter magnitude, M, of an earthquake is given by

$$M = \log_{10}\left(\frac{A}{A_0}\right)$$

where A is the amplitude of its seismographic trace and A_0 is the amplitude of the smallest detectable earthquake.

6. A *difference* of K units on a logarithmic scale corresponds to a *factor* of 10^K units in the value of the variable.

STUDY QUESTIONS

1. What numbers are used to label the axis on a log scale?
2. What does it mean to say that a log scale is a multiplicative scale?

3. Delbert says that 80 decibels is twice as loud as 40 decibels. Is he correct? Why or why not?
4. Which is farther on a log scale, the distance between 5 and 15, or the distance between 0.5 and 1.5?

SKILLS   *Practice each skill in the Homework problems listed.*

1. Plot values on a log scale: #1–4, 9 and 10
2. Read values from a log scale: #5–8, 11–14, 19 and 20

3. Compare values on a log scale: #15–18
4. Use log scales in applications: #21–40

HOMEWORK 5.4

ThomsonNOW™ Test yourself on key content at www.thomsonedu.com/login.

1. a. The log scale is labeled with powers of 10. Finish labeling the tick marks in the figure with their corresponding decimal values.

```
    10                                      100
  ──┼──┼──┼──┼──┼──┼──┼──┼──┼──┼──┼──→
   10¹ 10¹·¹ 10¹·² 10¹·³ 10¹·⁴ 10¹·⁵ 10¹·⁶ 10¹·⁷ 10¹·⁸ 10¹·⁹ 10²
```

b. The log scale is labeled with integer values. Label the tick marks in the figure with the corresponding powers of 10.

```
    10          20        30    40   50  60 70 80 90 100
  ──┼───────────┼─────────┼─────┼────┼───┼──┼──┼─┼─┼──→
   10¹                                            10²
```

2. a. The log scale is labeled with powers of 10. Finish labeling the tick marks in the figure with their corresponding decimal values.

```
    0.1                                       1
  ──┼──┼──┼──┼──┼──┼──┼──┼──┼──┼──┼──→
  10⁻¹ 10⁻⁰·⁹ 10⁻⁰·⁸ 10⁻⁰·⁷ 10⁻⁰·⁶ 10⁻⁰·⁵ 10⁻⁰·⁴ 10⁻⁰·³ 10⁻⁰·² 10⁻⁰·¹ 10⁰
```

b. The log scale is labeled with integer values. Label the tick marks in the figure with the corresponding powers of 10.

```
    0.1         0.2       0.3   0.4  0.5 0.6 0.7 0.8 0.9 1
  ──┼───────────┼─────────┼─────┼────┼───┼──┼──┼─┼─┼──→
  10⁻¹                                            10⁰
```

3. Plot the values on a log scale.

x	0.075	1.3	4200	87,000	6.5×10^7

4. Plot the values on a log scale.

x	4×10^{-4}	0.008	0.9	27	90

5. Estimate the decimal value of each point on the log scale.

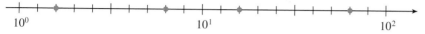

6. Estimate the decimal value of each point on the log scale.

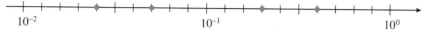

7. The log scale shows various temperatures in Kelvins. Estimate the temperatures of the events indicated.

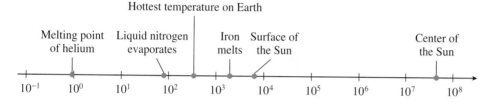

8. The log scale shows the size of various objects, in meters. Estimate the sizes of the objects indicated.

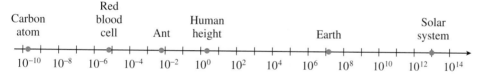

9. Plot the values of [H^+] in Table 5.8 on a log scale.

10. Plot the values of sound intensity in Table 5.9 on a log scale.

11. The magnitude of a star is a measure of its brightness. It is given by the formula

$$m = 4.83 - 2.5 \log L$$

where L is the luminosity of the star, measured in solar units. Calculate the magnitude of the stars whose luminosities are given in the figure.

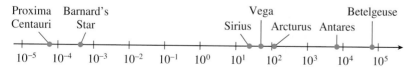

12. Estimate the wavelength, in meters, of the types of electromagnetic radiation shown in the figure.

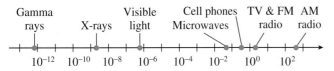

13. The *risk magnitude* of an event is defined by $R = 10 + \log p$, where p is the probability of the event occurring. Calculate the probability of each event.
  a. The sun will rise tomorrow, $R = 10$.
  b. The next child born in Arizona will be a boy, $R = 9.7$.
  c. A major hurricane will strike North Carolina this year, $R = 9.1$.
  d. A 100-meter asteroid will collide with Earth this year, $R = 8.0$.
  e. You will be involved in an automobile accident during a 10-mile trip, $R = 5.9$.
  f. A comet will collide with Earth this year, $R = 3.5$.
  g. You will die in an automobile accident on a 1000-mile trip, $R = 2.3$.
  h. You will die in a plane crash on a 1000-mile trip, $R = 0.9$

14. Have you ever wondered why time seems to pass more quickly as we grow older? One theory suggests that the human mind judges the length of a long period of time by comparing it with its current age. For example, a year is 20% of a 5-year-old's lifetime, but only 5% of a 20-year-old's, so a year feels longer to a 5-year-old. Thus, psychological time follows a log scale, like the one shown in the figure.

  a. Label the tick marks with their base 10 logarithms, rounded to 3 decimal places. What do you notice about the values?
  b. By computing their logs, locate 18 and 22 on the scale.
  c. Four years of college seems like a long time to an 18-year-old. What length of time feels the same to a 40-year-old?
  d. How long will the rest of your life feel? Let A be your current age, and let L be the age to which you think you will live. Compute the difference of their logs. Now move backward on the log scale an equal distance from your current age. What is the age at that spot? Call that age B. The rest of your life will feel the same as your life from age B until now.
  e. Compute B using a proportion instead of logs.

15. a. What number is halfway between $10^{1.5}$ and 10^2 on a log scale?
  b. What number is halfway between 20 and 30 on a log scale?

16. a. What number is halfway between $10^{3.0}$ and $10^{3.5}$ on a log scale?
  b. What number is halfway between 500 and 600 on a log scale?

17. The distances to two stars are separated by 3.4 units on a log scale. What is the ratio of their distances?

18. The populations of two cities are separated by 2.8 units on a log scale. What is the ratio of their populations?

19. The probability of discovering an oil field increases with its *diameter*, defined to be the square root of its area. Use the graph to estimate the diameter of the oil fields at the labeled points, and their probability of discovery.
(Source: Deffeyes, 2001)

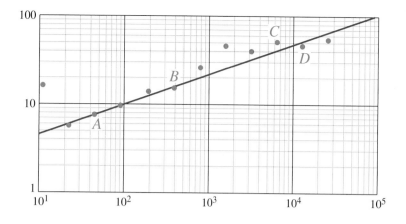

20. The *order* of a stream is a measure of its size. Use the graph to estimate the drainage area, in square miles, for streams of orders 1 through 4. (Source: Leopold, Wolman, and Miller)

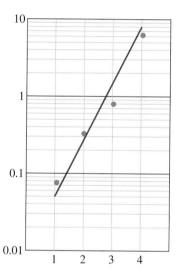

In Problems 21–40, use the appropriate formulas for logarithmic models.

21. The hydrogen ion concentration of vinegar is about 6.3×10^{-4}. Calculate the pH of vinegar.

22. The hydrogen ion concentration of spinach is about 3.2×10^{-6}. Calculate the pH of spinach.

23. The pH of lime juice is 1.9. Calculate its hydrogen ion concentration.

24. The pH of ammonia is 9.8. Calculate its hydrogen ion concentration.

25. A lawn mower generates a noise of intensity 10^{-2} watts per square meter. Find the decibel level of the sound of a lawn mower.

26. A jet airplane generates 100 watts per square meter at a distance of 100 feet. Find the decibel level for a jet airplane.

27. The loudest sound emitted by any living source is made by the blue whale. Its whistles have been measured at 188 decibels and are detectable 500 miles away. Find the intensity of the blue whale's whistle in watts per square meter.

28. The loudest sound created in a laboratory registered at 210 decibels. The energy from such a sound is sufficient to bore holes in solid material. Find the intensity of a 210-decibel sound.

29. At a concert by The Who in 1976, the sound level 50 meters from the stage registered 120 decibels. How many times more intense was this than a 90-decibel sound (the threshold of pain for the human ear)?

30. The loudest scientifically measured shouting by a human being registered 123.2 decibels. How many times more intense was this than normal conversation at 40 decibels?

31. The pH of normal rain is 5.6. Some areas of Ontario have experienced acid rain with a pH of 4.5. How many times more acidic is acid rain than normal rain?

32. The pH of normal hair is about 5, the average pH of shampoo is 8, and 4 for conditioner. Compare the acidity of normal hair, shampoo, and conditioner.

33. How much more acidic is milk than baking soda? (Refer to Table 5.8.)

34. Compare the acidity of lye and milk of magnesia. (Refer to Table 5.8.)

35. In 1964, an earthquake in Alaska measured 8.4 on the Richter scale. An earthquake measuring 4.0 is considered small and causes little damage. How many times stronger was the Alaska quake than one measuring 4.0?

36. On April 30, 1986, an earthquake in Mexico City measured 7.0 on the Richter scale. On September 21, a second earthquake occurred, this one measuring 8.1, hit Mexico City. How many times stronger was the September quake than the one in April?

37. A small earthquake measured 4.2 on the Richter scale. What is the magnitude of an earthquake three times as strong?

38. Earthquakes measuring 3.0 on the Richter scale often go unnoticed. What is the magnitude of a quake 200 times as strong as a 3.0 quake?

39. The sound of rainfall registers at 50 decibels. What is the decibel level of a sound twice as loud?

40. The magnitude, m, of a star is a function of its luminosity, L, given by

$$m = 4.83 - 2.5 \log L$$

If one star is 10 times as luminous as another star, what is the difference in their magnitudes?

CHAPTER 5 Summary and Review

KEY CONCEPTS

1.

> **Inverse Functions**
>
> If the inverse of a function f is also a function, then the inverse is denoted by the symbol f^{-1}, and
>
> $$f^{-1}(b) = a \quad \text{if and only if} \quad f(a) = b$$

2. We can make a table of values for the inverse function, f^{-1}, by interchanging the columns of a table for f.

3. If a function is defined by a formula in the form $y = f(x)$, we can find a formula for its inverse function by solving the equation for x to get $x = f^{-1}(y)$.

4. The inverse function f^{-1} undoes the effect of the function f, that is, if we apply the inverse function to the output of f, we return to the original input value.

5. If f^{-1} is the inverse function for f, then f is also the inverse function for f^{-1}.

6. The graphs of f and its inverse function are **symmetric about the line $y = x$**.

7. **Horizontal line test:** If no horizontal line intersects the graph of a function more than once, then the inverse is also a function.

8. A function that passes the horizontal line test is called **one-to-one.**

9. The inverse of a function f is also a function if and only if f is one-to-one.

10. We define the **logarithmic function,** $g(x) = \log_b x$, which takes the log base b of its input values. The log function $g(x) = \log_b x$ is the inverse of the exponential function $f(x) = b^x$.

11.

> Because $f(x) = b^x$ and $g(x) = \log_b x$ are inverse functions for $b > 0$,
>
> $\log_b b^x = x$, for all x and $b^{\log_b x} = x$, for $x > 0$

12.

> **Logarithmic Functions $y = \log_b x$**
>
> 1. Domain: all positive real numbers
> 2. Range: all real numbers
> 3. x-intercept: $(1, 0)$
> y-intercept: none
> 4. Vertical asymptote at $x = 0$
> 5. The graphs of $y = \log_b x$ and $y = b^x$ are symmetric about the line $y = x$.

13. A **logarithmic equation** is one where the variable appears inside of a logarithm. We can solve logarithmic equations by converting to exponential form.

14.

> **Steps for Solving Logarithmic Equations**
>
> 1. Use the properties of logarithms to combine all logs into one log.
> 2. Isolate the log on one side of the equation.
> 3. Convert the equation to exponential form.
> 4. Solve for the variable.
> 5. Check for extraneous solutions.

15. The **natural base** is an irrational number called e, where

$$e \approx 2.71828182845$$

16. The **natural exponential function** is the function $f(x) = e^x$. The **natural log function** is the function $g(x) = \ln x = \log_e x$.

17.

> **Conversion Formulas for Natural Logs**
>
> $y = \ln x$ if and only if $e^y = x$

18.

> **Properties of Natural Logarithms**
>
> If $x, y > 0$, then
>
> 1. $\ln(xy) = \ln x + \ln y$
> 2. $\ln \dfrac{x}{y} = \ln x - \ln y$
> 3. $\ln x^m = m \ln x$
>
> also
>
> $\ln e^x = x$,   for all x and $e^{\ln x} = x$, for $x > 0$

19. We use the natural logarithm to solve exponential equations with base e.

20.

> **Exponential Growth and Decay**
>
> The function
>
> $$P(t) = P_0 e^{kt}$$
>
> describes   exponential growth if $k > 0$,
> exponential decay if $k < 0$.

21. **Continuous compounding:** The amount accumulated in an account after t years at interest rate r compounded continuously is given by

$$A(t) = Pe^{rt}$$

where P is the principal invested.

22. A **log scale** is useful for plotting values that vary greatly in magnitude. We plot the log of the variable, instead of the variable itself.

23. A log scale is a **multiplicative scale:** Each increment of equal length on the scale indicates that the value is multiplied by an equal amount.

24. The pH value of a substance is defined by the formula

$$pH = -\log_{10}[H^+]$$

where $[H^+]$ denotes the concentration of hydrogen ions in the substance.

25. The loudness of a sound is measured in decibels, D, by

$$D = 10\log_{10}\left(\frac{I}{10^{-12}}\right)$$

where I is the intensity of its sound waves (in watts per square meter).

26. The Richter magnitude, M, of an earthquake is given by

$$M = \log_{10}\left(\frac{A}{A_0}\right)$$

where A is the amplitude of its seismographic trace and A_0 is the amplitude of the smallest detectable earthquake.

27. A *difference* of K units on a logarithmic scale corresponds to a *factor* of 10^K units in the value of the variable.

REVIEW PROBLEMS

Make a table of values for the *inverse* function.

1. $f(x) = x^3 + x + 1$ **2.** $g(x) = x + 6\sqrt[3]{x}$ **3.** $g(w) = \dfrac{1 + w}{w - 3}$ **4.** $f(n) = \dfrac{n}{1 + n}$

Use the graph to find the function values.

5.

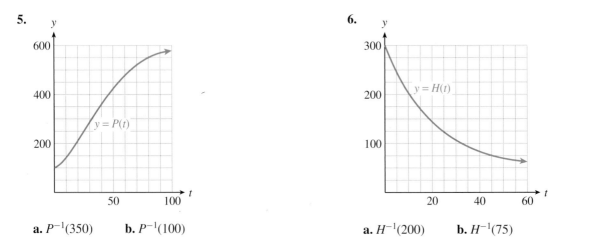

a. $P^{-1}(350)$ **b.** $P^{-1}(100)$

6.

a. $H^{-1}(200)$ **b.** $H^{-1}(75)$

For Problems 7–12,
(a) Find a formula for the inverse f^{-1} of each function.
(b) Graph the function and its inverse on the same set of axes, along with the graph of $y = x$.

7. $f(x) = x + 4$ **8.** $f(x) = \dfrac{x - 2}{4}$ **9.** $f(x) = x^3 - 1$

10. $f(x) = \dfrac{1}{x + 2}$ **11.** $f(x) = \dfrac{1}{x} + 2$ **12.** $f(x) = \sqrt[3]{x} - 2$

13. If $F(t) = \dfrac{3}{4}t + 2$, find $F^{-1}(2)$.   **14.** If $G(x) = \dfrac{1}{x} - 4$, find $G^{-1}(3)$.

15. The table shows the revenue, R, from sales of the Miracle Mop as a function of the number of dollars spent on advertising, A. Let f be the name of the function defined by the table, so $R = f(A)$.

A (thousands of dollars)	100	150	200	250	300
R (thousands of dollars)	250	280	300	310	315

  a. Evaluate $f^{-1}(300)$. Explain its meaning in this context.

  b. Write two equations to answer the following question, one using f and one using f^{-1}: How much should we spend on advertising to generate revenue of $250,000?

16. The table shows the systolic blood pressure, S, of a patient as a function of the dosage, d, of medication he receives. Let g be the name of the function defined by the table, so $S = g(d)$.

d (mg)	190	195	200	210	220
S (mm Hg)	220	200	190	185	183

  a. Evaluate $g^{-1}(200)$. Explain its meaning in this context.

  b. Write two equations to answer the following question, one using g and one using g^{-1}. What dosage results in systolic blood pressure of 220?

Write in exponential form.

17. $\log_{10} 0.001 = z$

18. $\log_3 20 = t$

19. $\log_2 3 = x - 2$

20. $\log_5 3 = 6 - 2p$

21. $\log_b(3x + 1) = 3$

22. $\log_m 8 = 4t$

23. $\log_n q = p - 1$

24. $\log_q(p + 2) = w$

Simplify.

25. $10^{\log 6n}$

26. $\log 100^x$

27. $\log_2 4^{x+3}$

28. $3^{2 \log_3 t}$

Solve.

29. $\log_3 \dfrac{1}{3} = y$

30. $\log_3 x = 4$

31. $\log_2 y = -1$

32. $\log_5 y = -2$

33. $\log_b 16 = 2$

34. $\log_b 9 = \dfrac{1}{2}$

35. $\log_4 \left(\dfrac{1}{2}t + 1 \right) = -2$

36. $\log_2(3x - 1) = 3$

Solve.

37. $\log_3 x + \log_3 4 = 2$

38. $\log_2(x + 2) - \log_2 3 = 6$

39. $\log_{10}(x - 1) + \log_{10}(x + 2) = 1$

40. $\log_{10}(x + 2) - \log_{10}(x - 3) = 1$

Solve.

41. $e^x = 4.7$

42. $e^x = 0.5$

43. $\ln x = 6.02$

44. $\ln x = -1.4$

45. $4.73 = 1.2\, e^{0.6x}$

46. $1.75 = 0.3\, e^{-1.2x}$

Simplify.

47. $e^{(\ln x)/2}$

48. $\ln\left(\dfrac{1}{e}\right)^{2m}$

49. $\ln\left(\dfrac{e^k}{e^3}\right)$

50. $e^{\ln(e+x)}$

51. In 1970, the population of New York City was 7,894,862. In 1980, the population had fallen to 7,071,639.
   a. Write an exponential function using base e for the population of New York over that decade.
   b. By what percent did the population decline annually?

52. In 1990, the population of New York City was 7,322,564. In 2000, the population was 8,008,278.
   a. Write an exponential function using base e for the population of New York over that decade.
   b. By what percent did the population increase annually?

53. You deposit $1000 in a savings account paying 5% interest compounded continuously.
   a. Find the amount in the account after 7 years.
   b. How long will it take for the original principal to double?
   c. Find a formula for the time t required for the amount to reach A.

54. The voltage, V, across a capacitor in a certain circuit is given by the function

$$V(t) = 100\left(1 - e^{-0.5t}\right)$$

where t is the time in seconds.
   a. Make a table of values and graph $V(t)$ for $t = 0$ to $t = 10$.
   b. Describe the graph. What happens to the voltage in the long run?
   c. How much time must elapse (to the nearest hundredth of a second) for the voltage to reach 75 volts?

55. Solve for t: $y = 12\,e^{-kt} + 6$

56. Solve for k: $N = N_0 + 4\ln(k + 10)$

57. Solve for M: $Q = \dfrac{1}{t}\left(\dfrac{\log M}{\log N}\right)$

58. Solve for t: $C_H = C_L \cdot 10^{kt}$

59. Express $P(t) = 750e^{0.32t}$ in the form $P(t) = P_0 b^t$.

60. Express $P(t) = 80e^{-0.6t}$ in the form $P(t) = P_0 b^t$.

61. Express $N(t) = 600(0.4)^t$ in the form $N(t) = N_0\,e^{kt}$

62. Express $N(t) = 100(1.06)^t$ in the form $N(t) = N_0 e^{kt}$

63. Plot the values on a log scale.

x	0.04	45	1200	560,000

64. Plot the values on a log scale.

x	0.0007	0.8	3.2	2500

65. The graph describes a network of streams near Santa Fe, New Mexico. It shows the number of streams of a given order, which is a measure of their size. Use the graph to estimate the number of streams of orders 3, 4, 8, and 9.
(Source: Leopold, Wolman, and Miller)

66. Large animals use oxygen more efficiently when running than small animals do. The graph shows the amount of oxygen various animals use, per gram of their body weight, to run 1 kilometer. Estimate the body mass and oxygen use for a kangaroo rat, a dog, and a horse.
(Source: Schmidt-Neilsen, 1972)

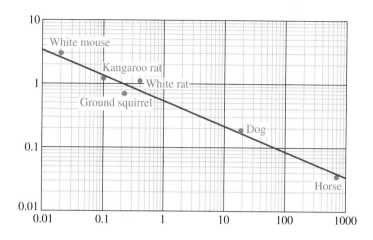

67. The pH of an unknown substance is 6.3. What is its hydrogen ion concentration?

68. The noise of a leaf blower was measured at 110 decibels. What was the intensity of the sound waves?

69. A refrigerator produces 50 decibels of noise, and a vacuum cleaner produces 85 decibels. How much more intense are the sound waves from a vacuum cleaner than those from a refrigerator?

70. In 2004, a magnitude 9.0 earthquake struck Sumatra in Indonesia. How much more powerful was this quake than the 1906 San Francisco earthquake of magnitude 8.3?

PROJECTS FOR CHAPTER 5

1. In this project, we investigate the graph of the logistic function.

   a. Graph the *sigmoid* function, $s(t) = \frac{1}{1+e^{-t}}$, in the window

$$\text{Xmin} = -4 \quad \text{Xmax} = 4$$
$$\text{Ymin} = -1 \quad \text{Ymax} = 2$$

   What are the domain and range of the function? List the intercepts of the graph, as well as any horizontal or vertical asymptotes. Estimate the coordinates of the *inflection point*, where the graph changes concavity.

   b. Graph the two functions $Y_1(t) = \frac{5}{1+4e^{-t}}$ and $Y_2 = \frac{10}{1+9e^{-t}}$ in the window

$$\text{Xmin} = -2 \quad \text{Xmax} = 10$$
$$\text{Ymin} = -1 \quad \text{Ymax} = 11$$

   How do the graphs of these functions differ from the sigmoid function? State the domain and range, intercepts, and asymptotes of Y_1 and Y_2. Estimate the coordinates of their inflection points.

   c. The function $P(t) = \frac{KP_0}{P_0 + (K - P_0)e^{-rt}}$ is called a *logistic* function. It is used to model population growth, among other things. It has three parameters, K, P_0, and r. The parameter K is called the *carrying capacity*. The functions Y_1 and Y_2 in part (b) are logistic functions with $P_0 = 1$ and $r = 1$. What does the value of K tell you about the graph? What do you notice about the vertical coordinate of the inflection point?

   d. Graph the function $P(t) = \frac{10P_0}{P_0 + (10 - P_0)e^{-t}}$ for $P_0 = 3, 4$, and 5. What does the value of P_0 tell you about the graph?

   e. Graph the function $P(t) = \frac{20}{2 + 8e^{-rt}}$ for $r = 0.5, 1$, and 2. What does the value of r tell you about the graph?

2. In this project, we investigate the normal or bell-shaped curve.

   a. Graph the function $f(x) = e^{-x^2}$ in the window

$$\text{Xmin} = -2 \quad \text{Xmax} = 2$$
$$\text{Ymin} = -1 \quad \text{Ymax} = 2$$

   What are the domain and range of the function? List the intercepts of the graph, and any horizontal or vertical asymptotes. Estimate the coordinates of the *inflection points*, where the graph changes concavity.

   b. Graph the function $f(x) = e^{-(x-m)^2}$ for $m = -1, 0, 1$, and 2. How does the value of m affect the graph?

   c. The function $N(x) = \frac{1}{s\sqrt{2\pi}}e^{-(x-m)^2/2s^2}$ is called the *normal* curve. It is used in statistics to describe the distribution of a variable, such as height, among a population. The parameter m gives the *mean* of the distribution, and s gives the *standard deviation*. For example, the distribution of height among American women has a mean of 64 inches and a standard deviation of 2.5 inches. Graph $N(x)$ for these values.

   d. Graph the function $N(x) = \frac{1}{s\sqrt{2\pi}}e^{-x^2/2s^2}$ for $s = 0.5, 0.8, 1$, and 1.2. (You may have to adjust the window to get a good graph.) How does the value of s affect the graph?

3. Do hedgerows planted at the boundaries of a field have a good or bad effect on crop yields? Hedges provide some shelter for the crops and retain moisture, but they may compete for nutrients or create too much shade. Results of studies on the microclimates produced by hedges are summarized in the figure, which shows how crop yields increase or decrease as a function of distance from the hedgerow. (Source: Briggs, David, and Courtney, 1985)

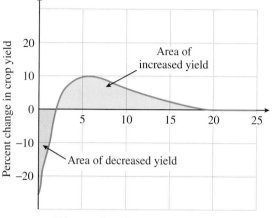

Distance from hedge ($\times$ height of hedge)

a. We will use trial-and-improvement to fit a curve to the graph. First, graph $y_1 = xe^{-x}$ in the window Xmin $= -2$, Xmax $= 5$, Ymin $= -1$, Ymax $= 1$ to see that it has the right shape.

b. Graph $y_2 = (x - 2)e^{-(x-2)}$ on the same axes. How is the graph of y_2 different from the graph of y_1?

c. Next we'll find the correct scale by trying functions of the form $y = a(x - 2)e^{-(x-2)/b}$. Experiment with different whole number values of a and b. How do the values of a and b affect the curve?

d. Graph $y = 5(x - 2)e^{(x-2)/4}$ in the window Xmin $= -5$, Xmax $= 25$, Ymin $= -20$, Ymax $= 25$. This function is a reasonable approximation for the curve in the figure. Compare the area of decreased yield (below the x-axis) with the area of increased yield (above the x-axis). Which area is larger? Is the overall effect of hedgerows on crop yield good or bad?

e. About how far from the hedgerow do the beneficial effects extend? If the average hedgerow is about 2.5 meters tall, how large should the field be to exploit their advantages?

4. Organic matter in the ground decomposes over time, and if the soil is cultivated properly, the fraction of its original organic carbon content is given by

$$C(t) = \frac{a}{b} - \frac{a - b}{b}e^{-bt}$$

where t is in years, and a and b are constants.
(Source: Briggs, David, and Courtney, 1985)

a. Write and simplify the formula for $C(t)$ if $a = 0.01$, $b = 0.028$.

b. Graph $C(t)$ in the window Xmin $= 0$, Xmax $= 200$, Ymin $= 0$, Ymax $= 1.5$.

c. What value does $C(t)$ approach as t increases? Compare this value to $\frac{a}{b}$.

d. The half-life of this function is the amount of time until $C(t)$ declines halfway to its limiting value, $\frac{a}{b}$. What is the half-life?

5. This project derives the *change of base* formula.

a. Follow the steps below to calculate $\log_8 20$.

Step 1 Let $x = \log_8 20$. Write the equation in exponential form.

Step 2 Take the logarithm base 10 of both sides of your new equation.

Step 3 Simplify and solve for x.

b. Follow the steps in part (a) to calculate $\log_8 5$.

c. Use part (a) to find a formula for calculating $\log_8 Q$, where Q is any positive number.

d. Find a formula for calculating $\log_b Q$, where $b > 1$ and Q is any positive number.

e. Find a formula for calculating $\ln Q$ in terms of $\log_{10} Q$.

f. Find a formula for calculating $\log_{10} Q$ in terms of $\ln Q$.

6. In this project, we solve logarithm equations with a graphing calculator. We have already used the **Intersect** feature to find approximate solutions for linear, exponential, and other types of equations in one variable. The same technique works for equations that involve common or natural logarithms.

a. Solve $\log_{10}(x + 1) + \log_{10}(x - 2) = 1$ using the **Intersect** feature by setting $Y_1 = \log(x + 1) + \log(x - 2)$ and $Y_2 = 1$. What about logarithmic equations with other bases? The calculator does not have a log key for bases other than 10 or e. However, by using the change of base formula from Project 5, we can rewrite any logarithm in terms of a common or natural logarithm.

b. Use the change of base formula to write $y = \log_2 x$ and $y = \log_2(x - 2)$ in terms of common logarithms.

c. Solve $\log_2 x + \log_2(x - 2) = 3$ by using the **Intersect** feature on your calculator.

d. Solve $\log_3(x - 2) - \log_3(x + 1) = 3$ by using the **Intersect** feature on your calculator.

Quadratic Functions

The models we have explored so far, namely, linear, exponential, logarithmic, and power, are **monotonic** functions, that is, always increasing or always decreasing on their domains. (Remember that we used power functions as models in the first quadrant only.) In this chapter, we investigate problems where the output variable may change from increasing to decreasing, or vice versa. The simplest sort of function that models this behavior is a *quadratic* function, one that involves the square of the variable.

Around 1600, Galileo began to study the motion of falling objects. He used a ball rolling down an inclined plane or ramp to slow down the motion, but he had no accurate way to measure time; clocks had not been invented yet. So he used water running into a jar to mark equal time intervals.

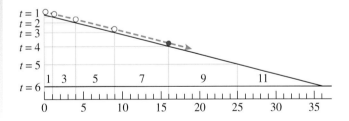

Time	Distance traveled	Total distance
1	1	1
2	3	4
3	5	9
4	7	16
5	9	25

ThomsonNOW

Throughout this chapter, the ThomsonNOW logo indicates an opportunity for online self-study, linking you to interactive tutorials and videos based on your level of understanding.

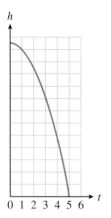

After many trials, Galileo found that the ball traveled 1 unit of distance down the plane in the first time interval, 3 units in the second time interval, 5 units in the third time interval, and so on, as shown in the figure, with the distances increasing through odd units of distance as time went on.

As you can see in the table on the preceding page, the total distance traveled by the ball is proportional to the square of the time elapsed, $s = kt^2$. Galileo found that this relationship held no matter how steep he made the ramp. Plotting the height of the ball as a function of time, we obtain a portion of the graph of a quadratic function.

Investigation 10

Height of a Baseball

Suppose a baseball player pops up, that is, hits the baseball straight up into the air. The height, h, of the baseball t seconds after it leaves the bat can be calculated using a formula from physics. This formula takes into account the initial speed of the ball (64 feet per second) and its height when it was hit (4 feet). The formula for the height of the ball (in feet) is

$$h = -16t^2 + 64t + 4$$

1. Evaluate the formula to complete the table of values for the height of the baseball.

t	0	1	2	3	4
h					

2. Graph the height of the baseball as a function of time. Plot data points from your table, then connect the points with a smooth curve.

3. What are the coordinates of the highest point on the graph? When does the baseball reach its maximum height, and what is that height?

4. Use the formula to find the height of the baseball after $\frac{1}{2}$ second.

5. Check that your answer to part (4) corresponds to a point on your graph. Approximate from your graph another time at which the baseball is at the same height as your answer to part (4).

6. Use your graph to find two times when the baseball is at a height of 64 feet.

7. Use your graph to approximate two times when the baseball is at a height of 20 feet. Then use the formula to find the actual heights at those times.

8. Suppose the catcher catches the baseball at a height of 4 feet, before it strikes the ground. At what time was the ball caught?

9. Use your calculator to make a table of values for the equation

   $$h = -16t^2 + 64t + 4$$

   with **TblStart** $= 0$ and Δ**Tbl** $= 0.5$.

10. Use your calculator to graph the equation for the height of the ball, with window settings

    $$\text{Xmin} = 0, \quad \text{Xmax} = 4.5, \quad \text{Xscl} = 0.5$$
    $$\text{Ymin} = 0, \quad \text{Ymax} = 70, \quad \text{Yscl} = 5$$

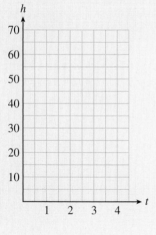

▶▶▶
See Appendix B for instructions on using your calculator.

11. Use the **intersect** command to verify your answer part (7): Estimate two times when the baseball is at a height of 20 feet.

12. Use the **intersect** command to verify your answer to part (8): At what time was the ball caught if it was caught at a height of 4 feet?

In Investigation 9, perhaps you recognized the graph of the baseball's height as a parabola. In this chapter, we shall see that the graph of any **quadratic function** is a parabola.

> **Quadratic Function**
>
> A **quadratic** function is one that can be written in the form
>
> $$f(x) = ax^2 + bx + c$$
>
> where a, b, and c are constants, and a is not equal to zero.

In the definition above, notice that if a is zero, there is no x-squared term, so the function is not quadratic.

In Investigation 10, the height of a baseball t seconds after being hit was given by

$$h = -16t^2 + 64t + 4$$

We used a graph to find two times when the baseball was 64 feet high. Can we solve the same problem algebraically?

We are looking for values of t that produce $h = 64$ in the height equation. So, if we substitute $h = 64$ into the height equation, we would like to solve the **quadratic equation**

$$64 = -16t^2 + 64t + 4$$

This equation cannot be solved by extraction of roots, because there are two terms containing the variable t, and they cannot be combined. To solve this equation, we will appeal to a property of our number system, called the zero-factor principle.

Zero-Factor Principle

Can you multiply two numbers together and obtain a product of zero? Only if one of the two numbers happens to be zero. This property of numbers is called the **zero-factor principle.**

> **Zero-Factor Principle**
>
> The product of two factors equals zero if and only if one or both of the factors equals zero. In symbols,
>
> $$ab = 0 \quad \text{if and only if} \quad a = 0 \quad \text{or} \quad b = 0$$

The principle is true even if the numbers a and b are represented by algebraic expressions, such as $x - 5$ or $2x + 1$. For example, if

$$(x - 5)(2x + 1) = 0$$

then it must be true that either $x - 5 = 0$ or $2x + 1 = 0$. Thus, we can use the zero-factor principle to solve equations.

EXAMPLE 1

a. Solve the equation $(x - 6)(x + 2) = 0$.

b. Find the x-intercepts of the graph of $f(x) = x^2 - 4x - 12$.

Solutions

a. Apply the zero-factor principle to the product $(x - 6)(x + 2)$.

$$(x - 6)(x + 2) = 0 \qquad \text{Set each factor equal to zero.}$$

$$x - 6 = 0 \quad \text{or} \quad x + 2 = 0 \qquad \text{Solve each equation.}$$

$$x = 6 \quad \text{or} \qquad x = -2$$

There are two solutions, 6 and -2. (You should check that both of these values satisfy the original equation.)

b. To find the x-intercepts of the graph, we set $y = 0$ and solve the equation

$$0 = x^2 - 4x - 12$$

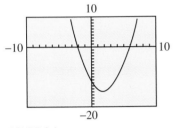

FIGURE 6.1

But this is the equation we solved in part (a), because $(x - 6)(x + 2) = x^2 - 4x - 12$. The solutions of that equation were 6 and -2, so the x-intercepts of the graph are 6 and -2. You can see this by graphing the equation on your calculator, as shown in Figure 6.1.

Example 1 illustrates an important fact about the x-intercepts of a graph.

---

x-Intercepts of a Graph

The x-intercepts of the graph of $y = f(x)$ are the solutions of the equation $f(x) = 0$.

EXERCISE 1

Graph the function

$$f(x) = (x - 3)(2x + 3)$$

on a calculator, and use your graph to solve the equation $f(x) = 0$. (Use Xmin $= -9.4$, Xmax $= 9.4$.) Check your answer with the zero-factor principle.

Solving Quadratic Equations by Factoring

Before we apply the zero-factor principle, we must first write the quadratic equation so that one side of the equation is zero. Let us introduce some terminology.

Forms for Quadratic Equations

1. A quadratic equation written

$$ax^2 + bx + c = 0$$

is in **standard form.**

2. A quadratic equation written

$$a(x - r_1)(x - r_2) = 0$$

is in **factored form.**

If you need to refresh your factoring skills, see Algebra Skills Refresher A.8.

Once we have written the equation in standard form, we factor the left side and set each variable factor equal to zero separately.

EXAMPLE 2

Solve $3x(x + 1) = 2x + 2$.

Solution

First, we write the equation in standard form.

$$3x(x + 1) = 2x + 2 \quad \text{Apply the distributive law to the left side.}$$
$$3x^2 + 3x = 2x + 2 \quad \text{Subtract } 2x + 2 \text{ from both sides.}$$
$$3x^2 + x - 2 = 0$$

Next, we factor the left side to obtain

$$(3x - 2)(x + 1) = 0$$

We then apply the zero-factor principle by setting each factor equal to zero.

$$3x - 2 = 0 \quad \text{or} \quad x + 1 = 0$$

Finally, we solve each equation to find

$$x = \frac{2}{3} \quad \text{or} \quad x = -1$$

The solutions are $\frac{2}{3}$ and -1.

CAUTION

When we apply the zero-factor principle, one side of the equation **must be zero.** For example, to solve the equation

$$(x - 2)(x - 4) = 15$$

it is *incorrect* to set each factor equal to 15! (There are *many* ways that the product of two numbers can equal 15; it is not necessary that one of the numbers be 15.) We must first simplify the left side and write the equation in standard form. (The correct solutions are 7 and -1; make sure you can find these solutions.)

We summarize the factoring method for solving quadratic equations as follows.

To Solve a Quadratic Equation by Factoring

1. Write the equation in standard form.
2. Factor the left side of the equation.
3. Apply the zero-factor principle: Set each factor equal to zero.
4. Solve each equation. There are two solutions (which may be equal).

EXERCISE 2

Solve by factoring: $(t - 3)^2 = 3(9 - t)$.

We can use factoring to solve the equation from Investigation 10.

EXAMPLE 3 The height, h, of a baseball t seconds after being hit is given by

$$h = -16t^2 + 64t + 4$$

When will the baseball reach a height of 64 feet?

Solution Substitute 64 for h in the formula, and solve for t.

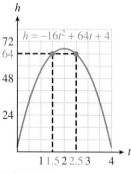

h

$h = -16t^2 + 64t + 4$

72
64

48

24

$1\ 1.5\ 2\ 2.5\ 3$ 4 t

FIGURE 6.2

$64 = -16t^2 + 64t + 4$	Write the equation in standard form.
$16t^2 - 64t + 60 = 0$	Factor 4 from the left side.
$4(4t^2 - 16t + 15) = 0$	Factor the quadratic expression.
$4(2t - 3)(2t - 5) = 0$	Set each variable factor equal to zero.
$2t - 3 = 0$ or $2t - 5 = 0$	Solve each equation.
$t = \dfrac{3}{2}$ or $t = \dfrac{5}{2}$	

There are two solutions to the quadratic equation. At $t = \frac{3}{2}$ seconds, the ball reaches a height of 64 feet on the way up, and at $t = \frac{5}{2}$ seconds, the ball is 64 feet high on its way down. (See Figure 6.2.)

In the solution to Example 3, the factor 4 does not affect the solutions of the equation at all. You can understand why this is true by looking at some graphs. First, check that the two equations

$$x^2 - 4x + 3 = 0 \quad \text{and} \quad 4(x^2 - 4x + 3) = 0$$

have the same solutions, $x = 1$ and $x = 3$. Then use your graphing calculator to graph the equation

$$Y_1 = X^2 - 4X + 3$$

in the window

$$\text{Xmin} = -2 \qquad \text{Xmax} = 8$$
$$\text{Ymin} = -5 \qquad \text{Ymax} = 10$$

Notice that when $y = 0$, $x = 3$ or $x = 1$. These two points are the x-intercepts of the graph. In the same window, now graph

$$Y_2 = 4(X^2 - 4X + 3)$$

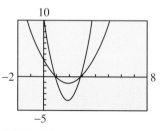

10

-2 8

-5

FIGURE 6.3

(See Figure 6.3.) This graph has the same x-values when $y = 0$. The factor of 4 stretches the graph vertically but does not change the location of the x-intercepts. The value of the constant factor a in the factored form of a quadratic function, $f(x) = a(x - r_1)(x - r_2)$, does not affect the location of the x-intercepts, because it does not affect the solutions of the equation $a(x - r_1)(x - r_2) = 0$.

EXERCISE 3

a. Solve $f(t) = 4t - t^2 = 0$ by factoring.

b. Solve $g(t) = 20t - 5t^2 = 0$ by factoring.

c. Graph $y = f(t)$ and $y = g(t)$ together in the window

$$\text{Xmin} = -2 \qquad \text{Xmax} = 6$$
$$\text{Ymin} = -20 \qquad \text{Ymax} = 25$$

and locate the horizontal intercepts of each graph.

Applications

Here is another example of how quadratic equations arise in applications.

EXAMPLE 4

The size of a rectangular computer monitor screen is given by the length of its diagonal. (See Figure 6.4.) If the length of the screen should be 3 inches greater than its width, what are the dimensions of a 15-inch monitor?

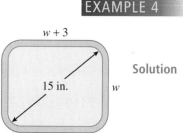

$w + 3$

15 in. w

FIGURE 6.4

See Algebra Skills Refresher A.11 if you would like to review the Pythagorean theorem.

Solution   We express the two dimensions of the screen in terms of a single variable:

$$\text{Width of screen:} \quad w$$
$$\text{Length of screen:} \quad w + 3$$

We can use the Pythagorean theorem to write an equation.

$$w^2 + (w + 3)^2 = 15^2$$

Solve the equation. Begin by simplifying the left side.

$w^2 + w^2 + 6w + 9 = 225$	Write the equation in standard form.
$2w^2 + 6w - 216 = 0$	Factor 2 from the left side.
$2(w^2 + 3w - 108) = 0$	Factor the quadratic expression.
$2(w - 9)(w + 12) = 0$	Set each variable factor equal to zero.
$w - 9 = 0 \quad \text{or} \quad w + 12 = 0$	Solve each equation.
$w = 9 \quad \text{or} \qquad w = -12$	

Because the width of the screen cannot be a negative number, we discard the solution $w = -12$. Thus, the width is $w = 9$ inches, and the length is $w + 3 = 12$ inches.

EXERCISE 4

Francine is designing the layout for a botanical garden. The plan includes a square herb garden, with a path 5 feet wide through the center of the garden, as shown in Figure 6.5. To include all the species of herbs, the planted area must be 300 square feet. Find the dimensions of the herb garden.

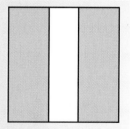

FIGURE 6.5

Solutions of Quadratic Equations

We have seen that the solutions of the quadratic equation

$$a(x - r_1)(x - r_2) = 0$$

are r_1 and r_2. Thus, if we know the two solutions of a quadratic equation, we can work backward and reconstruct the equation, starting from its factored form. We can then write the equation in standard form by multiplying together the factors.

EXAMPLE 5

Find a quadratic equation whose solutions are $\frac{1}{2}$ and -3.

Solution   The quadratic equation is

$$\left(x - \frac{1}{2}\right)[x - (-3)] = 0$$
$$\left(x - \frac{1}{2}\right)(x + 3) = 0$$

To write the equation in standard form, multiply the factors together.

$$x^2 + \frac{5}{2}x - \frac{3}{2} = 0$$

We can also find an equation with integer coefficients if we clear the equation of fractions. Multiply both sides by 2:

$$2\left(x^2 + \frac{5}{2}x - \frac{3}{2}\right) = 2(0)$$
$$2x^2 + 5x - 3 = 0$$

You can check that the solutions of this last equation are in fact $\frac{1}{2}$ and -3. Multiplying both sides of an equation by a constant factor does not change its solutions.

EXERCISE 5

Find a quadratic equation with integer coefficients whose solutions are $\frac{2}{3}$ and -5.

A quadratic equation in one variable always has two solutions. However, in some cases, the solutions may be equal. For example, the equation

$$x^2 - 2x + 1 = 0$$

can be solved by factoring as follows:

$$(x - 1)(x - 1) = 0 \quad \text{Apply the zero-factor principle.}$$
$$x - 1 = 0 \quad \text{or} \quad x - 1 = 0$$

Both of these equations have solution 1. We say that 1 is a solution of **multiplicity** two, meaning that it occurs twice as a solution of the quadratic equation.

Equations Quadratic in Form

The equation $x^6 - 4x^3 - 5 = 0$ is not quadratic, but if we make the substitution $u = x^3$, the equation becomes $u^2 - 4u - 5 = 0$. An equation is called **quadratic in form** if we can use a substitution to write it as $au^2 + bu + c = 0$, where u stands for an algebraic expression. Such equations can be solved by the same techniques we use to solve quadratic equations.

EXAMPLE 6   Use the substitution $u = x^3$ to solve the equation $x^6 - 4x^3 - 5 = 0$.

Solution   We set $u = x^3$, so that $u^2 = \left(x^3\right)^2 = x^6$. The original equation then becomes a quadratic equation in the variable u, which we can solve by factoring.

$$u^2 - 4u - 5 = 0 \qquad \text{Factor the left side.}$$
$$(u + 1)(u - 5) = 0 \qquad \text{Apply the zero-factor principle.}$$
$$u + 1 = 0 \quad \text{or} \quad u - 5 = 0 \quad \text{Solve each equation for } u.$$
$$u = -1 \quad \text{or} \quad u = 5$$

Finally, we replace u by x^3 and solve for x.

$$x^3 = -1 \quad \text{or} \quad x^3 = 5 \qquad \text{Take cube roots.}$$
$$u = \sqrt[3]{-1} = -1 \quad \text{or} \quad x = \sqrt[3]{5}$$

You can verify that the solutions of the original equation are -1 and $\sqrt[3]{5}$.

We say that the equation in Example 6, $x^6 - 4x^3 - 5 = 0$, is *quadratic in* x^3. We chose the substitution $u = x^3$ because $x^6 = u^2$.

EXERCISE 6

Use the substitution $u = x^2$ to solve the equation $x^4 - 5x^2 + 6 = 0$.

Usually, you can choose the simpler variable term in the equation for the u-substitution. For example, in Exercise 6 we chose $u = x^2$ because $u^2 = (x^2)^2 = x^4$, which is the first term of the equation. Once you have chosen the u-substitution, you should check that the other variable term is then a multiple of u^2; otherwise, the equation is not quadratic in form.

EXAMPLE 7   Solve the equation $e^{2x} - 7e^x + 12 = 0$.

Solutions   We use the substitution $u = e^x$, because $u^2 = (e^x)^2 = e^{2x}$. The original equation then becomes

$$u^2 - 7u + 12 = 0 \qquad \text{Factor the left side.}$$
$$(u - 3)(u - 4) = 0 \qquad \text{Apply the zero-factor principle.}$$
$$u - 3 = 0 \quad \text{or} \quad u - 4 = 0 \qquad \text{Solve each equation for } u.$$
$$u = 3 \quad \text{or} \qquad u = 4$$

Finally, we replace u by e^x and solve for x.

$$e^x = 3 \quad \text{or} \quad e^x = 4$$
$$x = \ln 3 \quad \text{or} \quad x = \ln 4$$

You should verify that the solutions of the original equation are $\ln 3$ and $\ln 4$.

EXERCISE 7

Solve the equation $10^{2x} - 3 \cdot 10^x + 2 = 0$, and check the solutions.

ANSWERS TO 6.1 EXERCISES

1. $x = -\frac{3}{2}, x = 3$ **2.** $x = -3, x = 6$ **3. a.** $t = 0, t = 4$

b. $t = 0, t = 4$ **c.** $(0, 0), (4, 0)$ **4.** 20 feet by 20 feet

5. $3x^2 + 13x - 10 = 0$ **6.** $x = \pm\sqrt{2}, x = \pm\sqrt{3}$ **7.** $x = 0, x = \log 2$

SECTION 6.1 SUMMARY

VOCABULARY *Look up the definitions of new terms in the Glossary.*

Quadratic function	Zero-factor principle	Standard form	Factored form
Multiplicity	Monotonic		

CONCEPTS

1. A **quadratic** function has the form $f(x) = ax^2 + bx + c$, where a, b, and c are constants and a is not equal to zero.

2. **Zero-factor principle:** The product of two factors equals zero if and only if one or both of the factors equals zero. In symbols,

$$ab = 0 \quad \text{if and only if} \quad a = 0 \quad \text{or} \quad b = 0$$

3. The x-intercepts of the graph of $y = f(x)$ are the solutions of the equation $f(x) = 0$.

4. A quadratic equation written as $ax^2 + bx + c = 0$ is in **standard form.**

   A quadratic equation written as $a(x - r_1)(x - r_2) = 0$ is in **factored form.**

5.
   > ### To Solve a Quadratic Equation by Factoring
   >
   > 1. Write the equation in standard form.
   > 2. Factor the left side of the equation.
   > 3. Apply the zero-factor principle: Set each factor equal to zero.
   > 4. Solve each equation. There are two solutions (which may be equal).

6. Every quadratic equation has two solutions, which may be the same.

7. The value of the constant a in the factored form of a quadratic equation does not affect the solutions.

8. Each solution of a quadratic equation corresponds to a factor in the factored form.

9. An equation is called **quadratic in form** if we can use a substitution to write it as $au^2 + bu + c = 0$, where u stands for an algebraic expression.

STUDY QUESTIONS

1. **a.** Find a pair of numbers whose product is 6. Now find a different pair of numbers whose product is 6. Can you find more such pairs?
   b. Find a pair of numbers whose product is 0. What is true about any such pair?

2. Before you begin factoring to solve a quadratic equation, what should you do?

3. How can you find the x-intercepts of the graph of $y = f(x)$ without looking at the graph?

4. How many solutions does a quadratic equation have?

5. **a.** Write a linear equation whose only solution is $x = 3$.
   b. Write a quadratic equation whose only solution is $x = 3$.

6. If you know the solutions of $ax^2 + bx + c = 0$, how can you find the solutions of $5(ax^2 + bx + c) = 0$?

7. Is the equation $x^9 - 6x^3 + 8 = 0$ quadratic in form? Why or why not?

8. Delbert says that he can solve the equation $x(x + 5) = 2(x + 5)$ by canceling the factor $(x + 5)$ to get $x = 2$. Comment on his method.

SKILLS *Practice each skill in the Homework problems listed.*

1. Use the zero-factor principle and find x-intercepts: #3–10

2. Solve quadratic equations by factoring: #11–24

3. Use the x-intercepts of the graph to factor a quadratic equation: #25–28, 37–40

4. Write a quadratic equation with given solutions: #29–36

5. Solve applied problems involving quadratic equations: #41–50

6. Solve equations that are quadratic in form: #51–62

1. Delbert stands at the top of a 300-foot cliff and throws his algebra book directly upward with a velocity of 20 feet per second. The height of his book above the ground t seconds later is given by the equation

$$h = -16t^2 + 20t + 300$$

where h is in feet.

   a. Use your calculator to make a table of values for the height equation, with increments of 0.5 second.

   b. Graph the height equation on your calculator. Use your table of values to help you choose appropriate window settings.

   c. What is the highest altitude Delbert's book reaches? When does it reach that height? Use the ⃞ TRACE feature to find approximate answers first. Then use the **Table** feature to improve your estimate.

   d. When does Delbert's book pass him on its way down? (Delbert is standing at a height of 300 feet.) Use the **intersect** command.

   e. How long will it take Delbert's book to hit the ground at the bottom of the cliff?

2. James Bond stands on top of a 240-foot building and throws a film canister upward to a fellow agent in a helicopter 16 feet above the building. The height of the film above the ground t seconds later is given by the formula

$$h = -16t^2 + 32t + 240$$

where h is in feet.

   a. Use your calculator to make a table of values for the height formula, with increments of 0.5 second.

   b. Graph the height formula on your calculator. Use your table of values to help you choose appropriate window settings.

   c. How long will it take the film canister to reach the agent in the helicopter? (What is the agent's altitude?) Use the ⃞ TRACE feature to find approximate answers first. Then use the **Table** feature to improve your estimate.

   d. If the agent misses the canister, when will it pass James Bond on the way down? Use the **intersect** command.

   e. How long will it take to hit the ground?

In Problems 3–10, use a graph to solve the equation $y = 0$. (Use Xmin $= -9.4$, Xmax $= 9.4$.) Check your answers with the zero-factor principle.

3. $y = (2x + 5)(x - 2)$

4. $y = (x + 1)(4x - 1)$

5. $y = x(3x + 10)$

6. $y = x(3x - 7)$

7. $y = (4x + 3)(x + 8)$

8. $y = (x - 2)(x - 9)$

9. $y = (x - 4)^2$

10. $y = (x + 6)^2$

Solve by factoring. (See Algebra Skills Refresher A.8 to review factoring.)

11. $2a^2 + 5a - 3 = 0$

12. $3b^2 - 4b - 4 = 0$

13. $2x^2 = 6x$

14. $5z^2 = 5z$

15. $3y^2 - 6y = -3$

16. $4y^2 + 4y = 8$

17. $x(2x - 3) = -1$

18. $2x(x - 2) = x + 3$

19. $t(t - 3) = 2(t - 3)$

20. $5(t + 2) = t(t + 2)$

21. $z(3z + 2) = (z + 2)^2$

22. $(z - 1)^2 = 2z^2 + 3z - 5$

23. $(v + 2)(v - 5) = 8$

24. $(w + 1)(2w - 3) = 3$

In Problems 25–28, graph each set of functions in the standard window. What do you notice about the x-intercepts? Generalize your observation, and test your idea with examples.

25. **a.** $y = x^2 - x - 20$
    b. $y = 2(x^2 - x - 20)$
    c. $y = 0.5(x^2 - x - 20)$

26. **a.** $y = x^2 + 2x - 15$
    b. $y = 3(x^2 + 2x - 15)$
    c. $y = 0.2(x^2 + 2x - 15)$

27. **a.** $y = x^2 + 6x - 16$
    b. $y = -2(x^2 + 6x - 16)$
    c. $y = -0.1(x^2 + 6x - 16)$

28. **a.** $y = x^2 - 16$
    b. $y = -1.5(x^2 - 16)$
    c. $y = -0.4(x^2 - 16)$

In Problems 29–36, write a quadratic equation whose solutions are given. The equation should be in standard form with integer coefficients.

29. -2 and 1
30. -4 and 3
31. 0 and -5
32. 0 and 5

33. -3 and $\frac{1}{2}$
34. $\frac{-2}{3}$ and 4
35. $\frac{-1}{4}$ and $\frac{3}{2}$
36. $\frac{-1}{3}$ and $\frac{-1}{2}$

Graph the function in the *ZInteger* window, and locate the x-intercepts of the graph. Use the x-intercepts to write the quadratic expression in factored form.

37. $y = 0.1(x^2 - 3x - 270)$
38. $y = 0.1(x^2 + 9x - 360)$

39. $y = -0.08(x^2 + 14x - 576)$
40. $y = -0.06(x^2 - 22x - 504)$

Use the Pythagorean theorem to solve Problems 41 and 42. (See Algebra Skills Refresher A.11 to review the Pythagorean theorem.)

41. One end of a ladder is 10 feet from the base of a wall, and the other end reaches a window in the wall. The ladder is 2 feet longer than the height of the window.
    a. Write a quadratic equation about the height of the window.
    b. Solve your equation to find the height of the window.

42. The diagonal of a rectangle is 20 inches. One side of the rectangle is 4 inches shorter than the other side.
    a. Write a quadratic equation about the length of the rectangle.
    b. Solve your equation to find the dimensions of the rectangle.

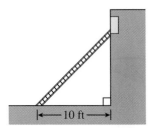

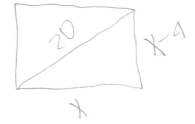

Use the following formula to answer Problems 43 and 44. If an object is thrown into the air from a height s_0 above the ground with an initial velocity v_0, its height t seconds later is given by the formula

$$h = -\frac{1}{2}gt^2 + v_0 t + s_0$$

where *g* is a constant that measures the force of gravity.

43. A tennis ball is thrown into the air with an initial velocity of 16 feet per second from a height of 8 feet. The value of g is 32.

   a. Write a quadratic equation that gives the height of the tennis ball at time t.

   b. Find the height of the tennis ball at $t = \frac{1}{2}$ second and at $t = 1$ second.

   c. Write and solve an equation to answer the question: At what time is the tennis ball 11 feet high?

   d. Use the **Table** feature on your calculator to verify your answers to parts (b) and (c). (What value of Δ **Tbl** is useful for this problem?)

   e. Graph your equation from part (a) on your calculator. Use your table to help you choose an appropriate window.

   f. If nobody hits the tennis ball, approximately how long will it be in the air?

44. A mountain climber stands on a ledge 80 feet above the ground and tosses a rope down to a companion clinging to the rock face below the ledge. The initial velocity of the rope is -8 feet per second, and the value of g is 32.

   a. Write a quadratic equation that gives the height of the rope at time t.

   b. What is the height of the rope after $\frac{1}{2}$ second? After 1 second?

   c. Write and solve an equation to answer the question: How long does it take the rope to reach the second climber, who is 17 feet above the ground?

   d. Use the **Table** feature on your calculator to verify your answers to parts (b) and (c). (What value of Δ **Tbl** is useful for this problem?)

   e. Graph your equation from part (a) on your calculator. Use your table to help you choose an appropriate window.

   f. If the second climber misses the rope, approximately how long will the rope take to reach the ground?

For Problems 45 and 46, you may want to review Investigation 3, Perimeter and Area, in Chapter 2.

45. A rancher has 360 yards of fence to enclose a rectangular pasture. If the pasture should be 8000 square yards in area, what should its dimensions be? We will use 3 methods to solve this problem: a table of values, a graph, and an algebraic equation.

   a. Make a table by hand that shows the areas of pastures of various widths, as shown here.

Width	Length	Area
10	170	1700
20	160	3200
⋮	⋮	⋮

   (To find the length of each pasture, ask yourself, What is the sum of the length plus the width if there are 360 yards of fence?) Continue the table until you find the pasture whose area is 8000 square yards.

   b. Write an expression for the length of the pasture if its width is x. Next, write an expression for the area, A, of the pasture if its width is x. Graph the equation for A on your calculator, and use the graph to find the pasture of area 8000 square yards.

   c. Write an equation for the area, A, of the pasture in terms of its width x. Solve your equation algebraically for $A = 8000$. Explain why there are two solutions.

46. If the rancher in Problem 45 uses a riverbank to border 1 side of the pasture as shown in the figure, he can enclose 16,000 square yards with 360 yards of fence. What will the dimensions of the pasture be then? We will use three methods to solve this problem: a table of values, a graph, and an algebraic equation.

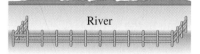

River

   a. Make a table by hand that shows the areas of pastures of various widths, as shown here.

Width	Length	Area
10	340	3400
20	320	6400
⋮	⋮	⋮

   (Be careful computing the length of the pasture: Remember that one side of the pasture does not need any fence!) Continue the table until you find the pasture whose area is 16,000 square yards.

   b. Write an expression for the length of the pasture if its width is x. Next, write an expression for the area, A, of the pasture if its width is x. Graph the equation for A, and use the graph to find the pasture of area 16,000 square yards.

   c. Write an equation for the area, A, of the pasture in terms of its width x. Solve your equation algebraically for $A = 16,000$.

For Problems 47 and 48, you will need the formula for the volume of a box.

47. A box is made from a square piece of cardboard by cutting 2-inch squares from each corner and turning up the edges.

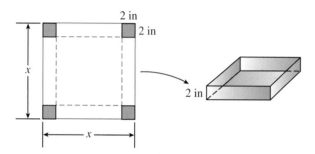

a. If the piece of cardboard is x inches square, write expressions for the length, width, and height of the box. Then write an expression for the volume, V, of the box in terms of x.
b. Use your calculator to make a table of values showing the volumes of boxes made from cardboard squares of side 4 inches, 5 inches, and so on.
c. Graph your expression for the volume on your calculator. What happens to V as x increases?
d. Use your table or your graph to find what size cardboard you need to make a box with volume 50 cubic inches.
e. Write and solve a quadratic equation to answer part (d).

48. A length of rain gutter is made from a piece of aluminum 6 feet long and 1 foot wide.
a. If a strip of width x is turned up along each long edge, write expressions for the length, width, and height of the gutter. Then write an expression for the volume, V, of the gutter in terms of x.

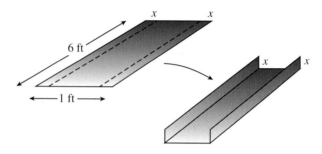

b. Use your calculator to make a table of values showing the volumes of various rain gutters formed by turning up edges of 0.1 foot, 0.2 foot, and so on.
c. Graph your expression for the volume. What happens to V as x increases?
d. Use your table or your graph to discover how much metal should be turned up along each long edge so that the gutter has a capacity of $\frac{3}{4}$ cubic foot of rainwater.
e. Write and solve a quadratic equation to answer part (d).

Problems 49 and 50 deal with wildlife management. The annual increase, I, in a population often depends on the size x of the population, according to the formula

$$I = kCx - kx^2$$

where k and C are constants related to the fertility of the population and the availability of food.

49. The annual increase, $f(x)$, in the deer population in a national park is given by

$$f(x) = 1.2x - 0.0002x^2$$

where x is the size of the population that year.
a. Make a table of values for $f(x)$ for $0 \le x \le 7000$. Use increments of 500 in x.
b. How much will a population of 2000 deer increase? A population of 5000 deer? A population of 7000 deer?
c. Use your calculator to graph the annual increase, $f(x)$, versus the size of the population, x, for $0 \le x \le 7000$.
d. What do the x-intercepts tell us about the deer population?
e. Estimate the population size that results in the largest annual increase. What is that increase?

50. Commercial fishermen rely on a steady supply of fish in their area. To avoid overfishing, they adjust their harvest to the size of the population. The function

$$g(x) = 0.4x - 0.0001x^2$$

gives the annual rate of growth, in tons per year, of a fish population of biomass x tons.
a. Make a table of values for $g(x)$ for $0 \le x \le 5000$. Use increments of 500 in x.
b. How much will a population of 1000 tons increase? A population of 3000 tons? A population of 5000 tons?
c. Use your calculator to graph the annual increase, $g(x)$, versus the size of the population, x, for $0 \le x \le 5000$.
d. What do the x-intercepts tell us about the fish population?
e. Estimate the population size that results in the largest annual increase. What is that increase?

Use a substitution to solve the equation.

51. $a^4 + a^2 - 2 = 0$

52. $t^6 - t^3 - 6 = 0$

53. $4b^6 - 3 = b^3$

54. $3x^4 + 1 = 4x^2$

55. $c^{2/3} + 2c^{1/3} - 3 = 0$

56. $y^{1/2} - 5y^{1/4} + 4 = 0$

57. $10^{2w} - 5 \cdot 10^w + 6 = 0$

58. $e^{2x} - 5e^x + 4 = 0$

59. $5^{2t} - 30 \cdot 5^t + 125 = 0$

60. $e^{4r} - 3e^{2r} + 2 = 0$

61. $\dfrac{1}{m^2} + \dfrac{5}{m} - 6 = 0$

62. $\dfrac{1}{s^2} - \dfrac{4}{s} - 5 = 0$

63. The sail in the figure is a right triangle of base and height x. It has a colored stripe along the hypotenuse and a white triangle of base and height y in the lower corner.

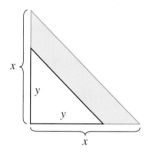

   a. Write an expression for the area of the colored stripe.
   b. Express the area of the stripe in factored form.
   c. If the sail is $7\frac{1}{2}$ feet high and the white strip is $4\frac{1}{2}$ feet high, use your answer to (b) to calculate mentally the area of the stripe.

64. An hors d'oeuvres tray has radius x, and the dip container has radius y, as shown in the figure.

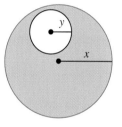

   a. Write an expression for the area for the chips (shaded region).
   b. Express the area in factored form.
   c. If the tray has radius $8\frac{1}{2}$ inches and the space for the dip has radius $2\frac{1}{2}$ inches, use your answer to part (b) to calculate mentally the area for chips. (Express your answer as a multiple of π.)

6.2   Solving Quadratic Equations

Not every quadratic equation can be solved by factoring or by extraction of roots. For example, the expression $x^2 + x - 1$ cannot be factored, so the equation $x^2 + x - 1 = 0$ cannot be solved by factoring. For other equations, factoring may be difficult. In this section we learn two methods that can be used to solve any quadratic equation.

Squares of Binomials

In Section 2.1 we used extraction of roots to solve equations of the form

$$a(px + q)^2 + r = 0$$

where the left side of the equation includes the square of a binomial, or a *perfect square*. We can write any quadratic equation in this form by completing the square.

Consider the following squares of binomials.

Square of binomial $(x + p)^2$	p	$2p$	p^2
1. $(x + 5)^2 = x^2 + 10x + 25$	5	$2(5) = 10$	$5^2 = 25$
2. $(x-3)^2 = x^2 - 6x + 9$	-3	$2(-3) = -6$	$(-3)^2 = 9$
3. $(x - 12)^2 = x^2 - 24x + 144$	-12	$2(-12) = -24$	$(-12)^2 = 144$

In each case, the square of the binomial is a **quadratic trinomial,**

$$(x + p)^2 = x^2 + 2px + p^2$$

The coefficient of the linear term, $2p$, is twice the constant in the binomial, and the constant term of the trinomial, p^2, is its square.

We would like to reverse the process and write a quadratic expression as the square of a binomial. For example, what constant term can we add to

$$x^2 - 16x$$

to produce a perfect square trinomial? Compare the expression to the formula above:

$$x^2 + 2px + p^2 = (x + p)^2$$
$$x^2 - 16x + \underline{?} = (x + ?)^2$$

We see that $2p = -16$, so $p = \frac{1}{2}(-16) = -8$, and $p^2 = (-8)^2 = 64$. Substitute these values for p^2 and p into the equation to find

$$x^2 - 16x + 64 = (x - 8)^2$$

Notice that in the resulting trinomial, the constant term is equal to *the square of one-half the coefficient of x.* In other words, we can find the constant term by taking one-half the coefficient of x and then squaring the result. Adding a constant term obtained in this way is called **completing the square.**

EXAMPLE 1

Complete the square by adding an appropriate constant; write the result as the square of a binomial.

a. $x^2 - 12x +$ _____   **b.** $x^2 + 5x +$ _____

Solutions

a. One-half of -12 is -6, so the constant term is $(-6)^2$, or 36. Add 36 to obtain

$$x^2 - 12x + 36 = (x - 6)^2 \qquad \begin{array}{l} p = \frac{1}{2}(-12) = -6 \\ p^2 = (-6)^2 = 36 \end{array}$$

b. One-half of 5 is $\frac{5}{2}$, so the constant term is $\left(\frac{5}{2}\right)^2$, or $\frac{25}{4}$. Add $\frac{25}{4}$ to obtain

$$x^2 + 5x + \frac{25}{4} = \left(x + \frac{5}{2}\right)^2 \qquad \begin{array}{l} p = \frac{1}{2}(5) = \frac{5}{2} \\ p^2 = \left(\frac{5}{2}\right)^2 = \frac{25}{4} \end{array}$$

You may find it helpful to visualize completing the square geometrically. We can think of the expression $x^2 + 2px$ as the area of a rectangle with dimensions x and $x + 2p$. For example, the rectangle with length $x + 10$ and width x has area $x(x + 10) = x^2 + 10x$, as shown in Figure 6.6a. We would like to cut the rectangle into pieces and rearrange them so that we can make a square. In Figure 6.6b, we move half of the x-term so that each side of the square has length $x + 5$ (note that $p = \frac{1}{2}(10) = 5$), and in Figure 6.6c we see that the missing corner piece has area $p^2 = 5^2 = 25$.

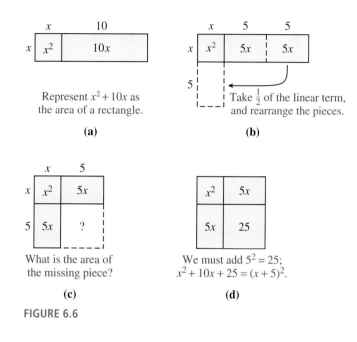

Represent $x^2 + 10x$ as the area of a rectangle.

(a)

Take $\frac{1}{2}$ of the linear term, and rearrange the pieces.

(b)

What is the area of the missing piece?

(c)

We must add $5^2 = 25$; $x^2 + 10x + 25 = (x + 5)^2$.

(d)

FIGURE 6.6

EXERCISE 1

Complete the square by adding an appropriate constant; write the result as the square of a binomial.

a. $x^2 - 18x +$ ___ $= (x$ ___ $)^2$ **b.** $x^2 + 9x +$ ___ $= (x$ ___ $)^2$

$p = \dfrac{1}{2}(-18) =$ ___, $p^2 =$ ___ $p = \dfrac{1}{2}(9) =$ ___, $p^2 =$ ___

Solving Quadratic Equations by Completing the Square

Now we will use completing the square to solve quadratic equations. First, we will solve equations in which the coefficient of the squared term is 1. Consider the equation

$$x^2 - 6x - 7 = 0$$

Step 1   Begin by moving the constant term to the other side of the equation, to get

$$x^2 - 6x \underline{\quad} = 7$$

Step 2   Now complete the square on the left. Because

$$p = \frac{1}{2}(-6) = -3 \quad \text{and} \quad p^2 = (-3)^2 = 9$$

we add 9 to *both* sides of our equation to get

$$x^2 - 6x + 9 = 7 + 9$$

Step 3 The left side of the equation is now the square of a binomial, namely $(x - 3)^2$. We write the left side in its square form and simplify the right side, which gives us

$$(x - 3)^2 = 16$$

(You can check that this equation is equivalent to the original one; if you expand the left side and collect like terms, you will return to the original equation.)

Step 4 We can now use extraction of roots to find the solutions. Taking square roots of both sides, we get

$$x - 3 = 4 \qquad \text{or} \qquad x - 3 = -4 \qquad \text{Solve each equation.}$$
$$x = 7 \qquad \text{or} \qquad x = -1$$

The solutions are 7 and -1.

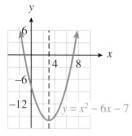

The graph of $y = x^2 - 6x - 7$ is shown in Figure 6.7. Note that the x-intercepts of the graph are $x = 7$ and $x = -1$, and the parabola is symmetric about the vertical line halfway between the intercepts, at $x = 3$.

We can also solve $x^2 - 6x - 7 = 0$ by factoring instead of completing the square. Of course, we get the same solutions by either method. In Example 2, we will solve an equation that cannot be solved by factoring.

FIGURE 6.7

EXAMPLE 2 Solve $x^2 - 4x - 3 = 0$ by completing the square.

Solution

Step 1 Write the equation with the constant term on the right side.

$$x^2 - 4x \underline{\hspace{1cm}} = 3$$

Step 2 Now complete the square on the left side. The coefficient of x is -4, so

$$p = \frac{1}{2}(-4) = -2, \qquad \text{and} \qquad p^2 = (-2)^2 = 4$$

We add 4 to both sides of the equation:

$$x^2 - 4x + 4 = 3 + 4$$

Step 3 Write the left side as the square of a binomial, and combine terms on the right side:

$$(x - 2)^2 = 7$$

Step 4 Finally, use extraction of roots to obtain

$$x - 2 = \sqrt{7} \qquad \text{or} \qquad x - 2 = -\sqrt{7} \qquad \text{Solve each equation.}$$
$$x = 2 + \sqrt{7} \qquad \qquad x = 2 - \sqrt{7}$$

These are the exact values of the solutions. We can use a calculator to find decimal approximations for each solution:

$$2 + \sqrt{7} \approx 4.646 \quad \text{and} \quad 2 - \sqrt{7} \approx -0.646$$

The graph of $y = x^2 - 4x - 3$ is shown in Figure 6.8.

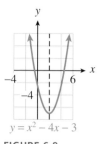

FIGURE 6.8

EXERCISE 2

a. Follow the steps to solve by completing the square: $x^2 - 1 = 3x$.

    Step 1   Write the equation with the constant on the right.

    Step 2   Complete the square on the left:

$$p = \frac{1}{2}(-3) = \underline{\quad}, \quad p^2 = \underline{\quad}$$

    Add p^2 to both sides.

    Step 3   Write the left side as a perfect square; simplify the right side.

    Step 4   Solve by extracting roots.

b. Find approximations to two decimal places for the solutions.

c. Graph the parabola $y = x^2 - 3x - 1$ in the window

$$\text{Xmin} = -4.7, \quad \text{Xmax} = 4.7$$
$$\text{Ymin} = -5, \quad \text{Ymax} = 5$$

The General Case

Our method for completing the square works only if the coefficient of x^2 is 1. If we want to solve a quadratic equation whose lead coefficient is *not* 1, we first divide each term of the equation by the lead coefficient.

EXAMPLE 3   Solve $2x^2 - 6x - 5 = 0$.

Solution

Step 1   Because the coefficient of x^2 is 2, we must divide each term of the equation by 2.

$$x^2 - 3x - \frac{5}{2} = 0$$

Now we proceed as before. Rewrite the equation with the constant on the right side.

$$x^2 - 3x \underline{\quad\quad} = \frac{5}{2}$$

Step 2   Complete the square:

$$p = \frac{1}{2}(-3) = \frac{-3}{2} \quad \text{and} \quad p^2 = \left(\frac{-3}{2}\right)^2 = \frac{9}{4}$$

Add $\frac{9}{4}$ to both sides of the equation:

$$x^2 - 3x + \frac{9}{4} = \frac{5}{2} + \frac{9}{4}$$

Step 3   Rewrite the left side as the square of a binomial and simplify the right side to get

$$\left(x - \frac{3}{2}\right)^2 = \frac{19}{4}$$

Step 4   Finally, extract roots and solve each equation for x.

$$x - \frac{3}{2} = \sqrt{\frac{19}{4}} \quad \text{or} \quad x - \frac{3}{2} = -\sqrt{\frac{19}{4}}$$

The solutions are $\frac{3}{2} + \sqrt{\frac{19}{4}}$ and $\frac{3}{2} - \sqrt{\frac{19}{4}}$. Using a calculator, we can find decimal approximations for the solutions: 3.679 and −0.679.

CAUTION

In Example 3, it is essential that we first divide each term of the equation by 2, the coefficient of x^2. The following attempt at a solution is *incorrect*.

$$2x^2 - 6x = 5$$
$$2x^2 - 6x + 9 = 5 + 9$$
$$(2x - 3)^2 = 14 \quad \leftarrow \text{Incorrect!}$$

You can check that $(2x - 3)^2$ is *not* equal to $2x^2 - 6x + 9$. We have not written the left side of the equation as a perfect square, so the solutions we obtain by extracting roots will not be correct.

EXERCISE 3

a. Follow the steps to solve by completing the square: $-4x^2 - 36x - 65 = 0$.

Step 1 Divide each term by −4. Write the equation with the constant on the right.

Step 2 Complete the square on the left:

$$p = \frac{1}{2}(9) = \_\_ , \quad p^2 = \_\_$$

Add p^2 to both sides.

Step 3 Write the left side as a perfect square; simplify the right side.

Step 4 Solve by extracting roots.

b. Graph $y = -4x^2 - 36x - 65$ in the window

$$\text{Xmin} = -9.4, \quad \text{Xmax} = 0$$
$$\text{Ymin} = -10, \quad \text{Ymax} = 20$$

Here is a summary of the steps for solving quadratic equations by completing the square.

To Solve a Quadratic Equation by Completing the Square:

1. a. Write the equation in standard form.

   b. Divide both sides of the equation by the coefficient of the quadratic term, and subtract the constant term from both sides.

2. Complete the square on the left side:

   a. Multiply the coefficient of the first-degree term by one-half, then square the result.

   b. Add the value obtained in (a) to both sides of the equation.

3. Write the left side of the equation as the square of a binomial. Simplify the right side.

4. Use extraction of roots to finish the solution.

Quadratic Formula

Instead of completing the square every time we solve a new quadratic equation, we can complete the square on the general quadratic equation,

$$ax^2 + bx + c = 0, \qquad a \neq 0$$

and obtain a formula for the solutions of any quadratic equation.

The Quadratic Formula

The solutions of the equation $ax^2 + bx + c = 0$, $a \neq 0$, are

$$x = \frac{-b \pm \sqrt{b^2 - 4ac}}{2a}$$

This formula expresses the solutions of a quadratic equation in terms of its coefficients. (The proof of the formula is considered in the Homework problems.) The symbol $\pm$, read *plus or minus*, is used to combine the two equations

$$x = \frac{-b + \sqrt{b^2 - 4ac}}{2a} \quad \text{and} \quad x = \frac{-b - \sqrt{b^2 - 4ac}}{2a}$$

into a single equation.

To solve a quadratic equation using the quadratic formula, all we have to do is substitute the coefficients a, b, and c into the formula.

EXAMPLE 4　　Solve $2x^2 + 1 = 4x$.

Solution　　Write the equation in standard form as

$$2x^2 - 4x + 1 = 0$$

Substitute 2 for a, -4 for b, and 1 for c into the quadratic formula, then simplify.

$$x = \frac{-(-4) \pm \sqrt{(-4)^2 - 4(2)(1)}}{2(2)}$$

$$= \frac{4 \pm \sqrt{8}}{4}$$

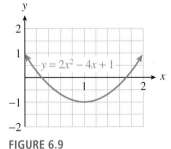

FIGURE 6.9

Using a calculator, we find that the solutions are approximately 1.707 and 0.293. (If you would like to review following the order of operations on a calculator, please see Appendix B.) We can also verify that the x-intercepts of the graph of $y = 2x^2 - 4x + 1$ are approximately 1.707 and 0.293, as shown in Figure 6.9.

EXERCISE 4

Use the quadratic formula to solve $x^2 - 3x = 1$.

Note that the solutions to this equation are the same as the solutions we found in Exercise 2.

Applications

We have now seen four different algebraic methods for solving quadratic equations:

1. Factoring
2. Extraction of roots
3. Completing the square
4. Quadratic formula

Factoring and extraction of roots are relatively fast and simple, but they do not work on all quadratic equations. The quadratic formula will work on any quadratic equation.

EXAMPLE 5

The owners of a day-care center plan to enclose a divided play area against the back wall of their building, as shown in Figure 6.10. They have 300 feet of picket fence and would like the total area of the playground to be 6000 square feet. Can they enclose the playground with the fence they have, and if so, what should the dimensions of the playground be?

Solution

Suppose the width of the play area is x feet. Because there are three sections of fence along the width of the play area, that leaves $300 - 3x$ feet of fence for its length. The area of the play area should be 6000 square feet, so we have the equation

$$x(300 - 3x) = 6000$$

This is a quadratic equation. In standard form,

$$3x^2 - 300x + 6000 = 0 \quad \text{Divide each term by 3.}$$
$$x^2 - 100x + 2000 = 0$$

The left side cannot be factored, so we use the quadratic formula with $a = 1$, $b = -100$, and $c = 2000$.

$$x = \frac{-(-100) \pm \sqrt{(-100)^2 - 4(1)(2000)}}{2(1)}$$

$$= \frac{100 \pm \sqrt{2000}}{2} \approx \frac{100 \pm 44.7}{2}$$

Simplifying the last fraction, we find that $x \approx 72.35$ or $x \approx 27.65$. Both values give solutions to the problem. If the width of the play area is 72.35 feet, then the length is $300 - 3(72.35)$, or 82.95 feet. If the width is 27.65 feet, the length is $300 - 3(27.65)$, or 217.05 feet.

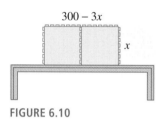

$300 - 3x$

x

FIGURE 6.10

EXERCISE 5

In Investigation 10, we considered the height of a baseball, given by the equation

$$h = -16t^2 + 64t + 4$$

Find two times when the ball is at a height of 20 feet. Give your answers to two decimal places.

Sometimes it is useful to solve a quadratic equation for one variable in terms of the others.

EXAMPLE 6   Solve $x^2 - xy + y = 2$ for x in terms of y.

Solution   We first write the equation in standard form as a quadratic equation in the variable x.

$$x^2 - yx + (y - 2) = 0$$

Expressions in y are treated as constants with respect to x, so that $a = 1$, $b = -y$, and $c = y - 2$. Substitute these expressions into the quadratic formula.

$$x = \frac{-(-y) \pm \sqrt{(-y)^2 - 4(1)(y - 2)}}{2(1)}$$

$$= \frac{y \pm \sqrt{y^2 - 4y + 8}}{2}$$

EXERCISE 6

Solve $2x^2 + kx + k^2 = 1$ for x in terms of k.

ANSWERS TO 6.2 EXERCISES

1. **a.** $x - 18x + 81 = (x - 9)^2$ **b.** $x^2 + 9x + \frac{81}{4} = \left(x + \frac{9}{2}\right)^2$

2. **a.** $x = \frac{3}{2} \pm \sqrt{\frac{13}{4}}$ **b.** $x \approx -0.30$ or $x \approx 3.30$

3. $x = \frac{-13}{2}$, $x = \frac{-5}{2}$ 4. $x = \frac{3 \pm \sqrt{13}}{2}$

5. 0.27 sec, 3.73 sec   6. $x = \frac{-k \pm \sqrt{8 - 7k^2}}{4}$

SECTION 6.2 SUMMARY

VOCABULARY   *Look up the definitions of new terms in the Glossary.*

Quadratic trinomial   Complete the square   Quadratic formula

CONCEPTS

1. The square of the binomial is a **quadratic trinomial,**

$$(x + p)^2 = x^2 + 2px + p^2$$

2.

To Solve a Quadratic Equation by Completing the Square:

1. **a.** Write the equation in standard form.
   b. Divide both sides of the equation by the coefficient of the quadratic term, and subtract the constant term from both sides.

2. Complete the square on the left side:
   a. Multiply the coefficient of the first-degree term by one-half, then square the result.
   b. Add the value obtained in (a) to both sides of the equation.

3. Write the left side of the equation as the square of a binomial. Simplify the right side.

4. Use extraction of roots to finish the solution.

3.

The Quadratic Formula

The solutions of the equation $ax^2 + bx + c = 0$, $a \neq 0$, are

$$x = \frac{-b \pm \sqrt{b^2 - 4ac}}{2a}$$

4. We have four methods for solving quadratic equations: extracting of roots, factoring, completing the square, and using the quadratic formula. The first two methods are faster, but they don't work on all equations. The last two methods work on any quadratic equation.

STUDY QUESTIONS

1. Name four algebraic methods for solving a quadratic equation.
2. Give an example of a quadratic trinomial that is the square of a binomial.
3. What number must be added to $x^2 - 26x$ to make it the square of a binomial?
4. After completing the square, how do we finish solving the quadratic equation?
5. What is the first step in solving the equation $2x^2 - 6x = 5$ by completing the square?

SKILLS *Practice each skill in the Homework problems listed.*

1. Solve quadratic equations by completing the square: #3–24
2. Solve quadratic equations by using the quadratic formula: #27–36
3. Solve problems by writing and solving quadratic equations: #37–44
4. Solve formulas: #45–64

HOMEWORK 6.2

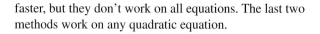 Test yourself on key content at www.thomsonedu.com/login.

Complete the square and write the result as the square of a binomial.

1. a. $x^2 + 8x$ **b.** $x^2 - 7x$
   c. $x^2 + \frac{3}{2}x$ **d.** $x^2 - \frac{4}{5}x$

2. a. $x^2 - 14x$ **b.** $x^2 + 3x$
   c. $x^2 - \frac{5}{2}x$ **d.** $x^2 + \frac{2}{3}x$

Solve by completing the square.

3. $x^2 - 2x + 1 = 0$ **4.** $x^2 + 4x + 4 = 0$ **5.** $x^2 + 9x + 20 = 0$ **6.** $x^2 - x - 20 = 0$

7. $x^2 = 3 - 3x$ **8.** $x^2 = 5 - 5x$ **9.** $2x^2 + 4x - 3 = 0$ **10.** $3x^2 + 12x + 2 = 0$

11. $3x^2 + x = 4$ **12.** $4x^2 + 6x = 3$ **13.** $4x^2 - 3 = 2x$ **14.** $2x^2 - 5 = 3x$

15. $3x^2 - x - 4 = 0$ **16.** $2x^2 - x - 3 = 0$ **17.** $5x^2 + 8x = 4$ **18.** $9x^2 - 12x - 5 = 0$

Solve by completing the square. Your answers will involve *a*, *b*, or *c*.

19. $x^2 + 2x + c = 0$ **20.** $x^2 - 4x + c = 0$ **21.** $x^2 + bx + 1 = 0$ **22.** $x^2 + bx - 4 = 0$

23. $ax^2 + 2x - 4 = 0$ **24.** $ax^2 - 4x + 9 = 0$

25. a. Write an expression for the area of the square in the figure.
   b. Express the area as a polynomial.
   c. Divide the square into four pieces whose areas are given by the terms of your answer to part (b).

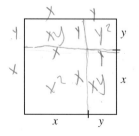

26. **a.** Write an expression for the area of the shaded region in the figure.
    b. Express the area in factored form.
    c. By making one cut in the shaded region, rearrange the pieces into a rectangle whose area is given by your answer to part (b).

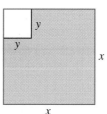

Solve using the quadratic formula. Round your answers to three decimal places.

27. $x^2 - x - 1 = 0$

28. $x^2 + x - 1 = 0$

29. $y^2 + 2y = 5$

30. $y^2 - 4y = 4$

31. $3z^2 = 4.2z + 1.5$

32. $2z^2 = 7.5z - 6.3$

33. $0 = x^2 - \dfrac{5}{3}x + \dfrac{1}{3}$

34. $0 = -x^2 + \dfrac{5}{2}x - \dfrac{1}{2}$

35. $-5.2z^2 + 176z + 1218 = 0$

36. $15z^2 - 18z - 2750 = 0$

37. A car traveling at s miles per hour on a dry road surface requires approximately d feet to stop, where d is given by

$$d = \frac{s^2}{24} + \frac{s}{2}$$

   a. Make a table showing the stopping distance, d, for speeds of 10, 20, . . . , 100 miles per hour. (Use the **Table** feature of your calculator.)
   b. Graph the equation for d in terms of s. Use your table values to help you choose appropriate window settings.
   c. Write and solve an equation to answer the question: If a car must be able to stop in 50 feet, what is the maximum safe speed it can travel? Verify your answer on your graph.

38. A car traveling at s miles per hour on a wet road surface requires approximately d feet to stop, where d is given by

$$d = \frac{s^2}{12} + \frac{s}{2}$$

   a. Make a table showing the stopping distance, d, for speeds of 10, 20, . . . , 100 miles per hour. (Use the **Table** feature of your calculator.)
   b. Graph the equation for d in terms of s. Use your table values to help you choose appropriate window settings.
   c. Insurance investigators at the scene of an accident find skid marks 100 feet long leading up to the point of impact. Write and solve an equation to discover how fast the car was traveling when it put on the brakes. Verify your answer on your graph.

39. A skydiver jumps out of an airplane at 11,000 feet. While she is in free-fall, her altitude in feet t seconds after jumping is given by

$$h = -16t^2 - 16t + 11{,}000$$

   a. Make a table of values showing the skydiver's altitude at 5-second intervals after she jumps from the airplane. (Use the **Table** feature of your calculator.)
   b. Graph the equation. Use your table of values to choose appropriate window settings.
   c. If the skydiver must open her parachute at an altitude of 1000 feet, how long can she free-fall? Write and solve an equation to find the answer.
   d. If the skydiver drops a marker just before she opens her parachute, how long will it take the marker to hit the ground? (*Hint:* The marker continues to fall according to the equation given above.)
   e. Find points on your graph that correspond to your answers to parts (c) and (d).

40. A high diver jumps from the 10-meter springboard. His height in meters above the water t seconds after leaving the board is given by

$$h = -4.9t^2 + 8t + 10$$

   a. Make a table of values showing the diver's altitude at 0.25-second intervals after he jumps from the springboard. (Use the **Table** feature of your calculator.)
   b. Graph the equation. Use your table of values to choose appropriate window settings.
   c. How long is it before the diver passes the board on the way down?
   d. How long is it before the diver hits the water?
   e. Find points on your graph that correspond to your answers to parts (c) and (d).

41. A dog trainer has 100 meters of chain link fence. She wants to enclose 250 square meters in three pens of equal size, as shown in the figure.

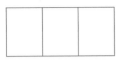

  a. Let l and w represent the length and width, respectively, of the entire area. Write an equation about the amount of chain link fence.
  b. Solve your equation for l in terms w.
  c. Write and solve an equation in w for the total area enclosed.
  d. Find the dimensions of each pen.

42. An architect is planning to include a rectangular window topped by a semicircle in his plans for a new house, as shown in the figure. In order to admit enough light, the window should have an area of 120 square feet. The architect wants the rectangular portion of the window to be 2 feet wider than it is tall.

  a. Let x stand for the horizontal width of the window. Write expressions for the height of the rectangular portion and for the radius of the semicircular portion.
  b. Write an expression for the total area of the window.
  c. Write and solve an equation to find the width and overall height of the window.

43. When you look down from a height, say a tall building or a mountain peak, your line of sight is tangent to the Earth at the horizon, as shown in the figure.
  a. Suppose you are standing on top of the Petronas Tower in Kuala Lumpur, 1483 feet high. How far can you see on a clear day? (You will need to use the Pythagorean theorem and the fact that the radius of the Earth is 3960 miles. Do not forget to convert the height of the Petronas Tower to miles.)
  b. How tall a building should you stand on in order to see 100 miles?

44. a. If the radius of the Earth is 6370 kilometers, how far can you see from an airplane at an altitude of 10,000 meters? (*Hint*: See Problem 43.)
  b. How high would the airplane have to be in order for you to see a distance of 10 kilometers?

Use the quadratic formula to solve each equation for the indicated variable.

45. $A = 2w^2 + 4lw$, for w **46.** $A = \pi r^2 + \pi rs$, for r **47.** $h = 4t - 16t^2$, for t **48.** $P = IE - RI^2$, for I

49. $s = vt - \dfrac{1}{2}at^2$, for t **50.** $S = \dfrac{n^2 + n}{2}$, for n

51. $3x^2 + xy + y^2 = 2$, for y **52.** $y^2 - 3xy + x^2 = 3$, for x

Solve for *y* in terms of *x*. Use whichever method of solution seems easiest.

53. $x^2y - y^2 = 0$ **54.** $x^2y^2 - y = 0$ **55.** $(2y + 3x)^2 = 9$ **56.** $(3y - 2x)^2 = 4$

57. $4x^2 - 9y^2 = 36$ **58.** $9x^2 + 4y^2 = 36$ **59.** $4x^2 - 25y^2 = 0$ **60.** $(2x - 5y)^2 = 0$

Solve each formula for the indicated variable.

61. $V = \pi(r - 3)^2 h$, for r **62.** $A = P(1 + r)^2$, for r **63.** $E = \dfrac{1}{2}mv^2 + mgh$, for v

64. $h = \dfrac{1}{2}gt^2 + dl$, for t **65.** $V = 2(s^2 + t^2)w$, for t **66.** $V = \pi(r^2 + R^2)h$, for R

67. What is the sum of the two solutions of the quadratic equation $ax^2 + bx + c = 0$? (*Hint:* The two solutions are given by the quadratic formula.)

68. What is the product of the two solutions of the quadratic equation $ax^2 + bx + c = 0$? (*Hint:* Do *not* try to multiply the two solutions given by the quadratic formula! Think about the factored form of the equation.)

In Problems 69 and 70, we prove the quadratic formula.

69. Complete the square to find the solutions of the equation $x^2 + bx + c = 0$. (Your answers will be expressions in b and c.)

70. Complete the square to find the solutions of the equation $ax^2 + bx + c = 0$. (Your answers will be expressions in a, b, and c.)

6.3 Graphing Parabolas

The graph of a quadratic function $f(x) = ax^2 + bx + c$ is called a **parabola.** Some parabolas are shown in Figure 6.11.

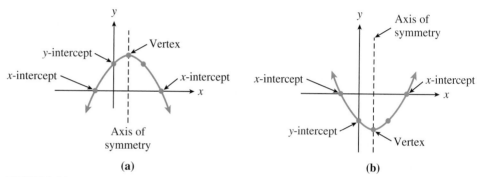

(a)　　　　　　　　　　**(b)**

FIGURE 6.11

All these parabolas share certain features. The graph has either a highest point (if the parabola opens downward, as in Figure 6.11a) or a lowest point (if the parabola opens upward, as in Figure 6.11b). This high or low point is called the **vertex** of the graph. The parabola is symmetric about a vertical line, called the **axis of symmetry,** that runs through the vertex. The **y-intercept** is the point where the parabola intersects the y-axis. The graph of a quadratic function always has exactly one y-intercept. However, the graph may cross the x-axis at one point, at two points, or not at all. Points where the parabola intersects the x-axis are called the **x-intercepts.** If there are two x-intercepts, they are equidistant from the axis of symmetry.

The values of the constants a, b, and c determine the location and orientation of the parabola. We will begin by considering each of these constants separately.

The Graph of $y = ax^2$

In Chapter 2, we saw that the graph of $y = af(x)$ is a transformation of the graph of $y = f(x)$. The scale factor, a, stretches or compresses the graph vertically, and if a is negative, the graph is reflected about the x-axis.

EXAMPLE 1　　Sketch a graph of each quadratic equation by hand.

　　a. $y = 2x^2$　　　　**b.** $y = -\frac{1}{2}x^2$

Solutions   Both equations are of the form $y = ax^2$. The graph of $y = 2x^2$ opens upward because $a = 2 > 0$, and the graph of $y = -\frac{1}{2}x^2$ opens downward because $a = -\frac{1}{2} < 0$. To make a reasonable sketch by hand, it is enough to plot a few *guidepoints;* the points with x-coordinates 1 and −1 are easy to compute.

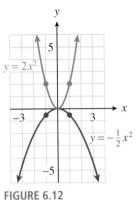

FIGURE 6.12

x	$y = 2x^2$	$y = -\frac{1}{2}x^2$
-1	2	$-\frac{1}{2}$
0	0	0
1	2	$-\frac{1}{2}$

Sketch parabolas through each set of guidepoints, as shown in Figure 6.12.

EXERCISE 1

Match each parabola in Figure 6.13 with its equation. The basic parabola is shown in black.

a. $y = -\frac{3}{4}x^2$

b. $y = \frac{1}{4}x^2$

c. $y = \frac{5}{2}x^2$

d. $y = -\frac{5}{4}x^2$

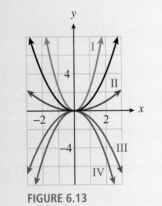

FIGURE 6.13

The Graph of $y = x^2 + c$

Next, we consider the effect of the constant term, c, on the graph. Adding a constant c to the formula for $y = f(x)$ causes a vertical translation of the graph.

EXAMPLE 2   Sketch graphs for the following quadratic equations.

a. $y = x^2 - 2$ **b.** $y = -x^2 + 4$

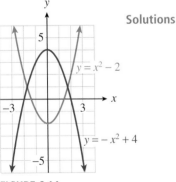

FIGURE 6.14

Solutions   **a.** The graph of $y = x^2 - 2$ is shifted downward by two units, compared to the basic parabola. The vertex is the point $(0, -2)$ and the x-intercepts are the solutions of the equation

$$0 = x^2 - 2$$

or $\sqrt{2}$ and $-\sqrt{2}$. The graph is shown in Figure 6.14.

b. The graph of $y = -x^2 + 4$ opens downward and is shifted 4 units up, compared to the basic parabola. Its vertex is the point $(0, 4)$. Its x-intercepts are the solutions of the equation

$$0 = -x^2 + 4$$

or 2 and −2. (See Figure 6.14.) You can verify both graphs with your graphing calculator.

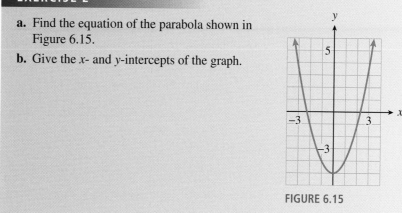

EXERCISE 2

a. Find the equation of the parabola shown in Figure 6.15.

b. Give the *x*- and *y*-intercepts of the graph.

FIGURE 6.15

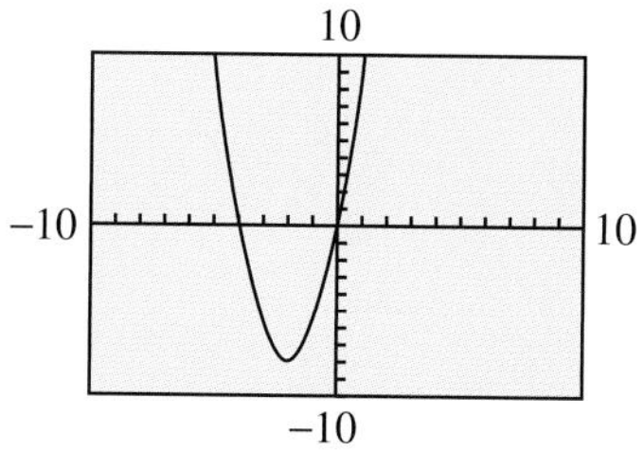

The Graph of $y = ax^2 + bx$

How does the linear term, bx, affect the graph? Let us begin by considering an example. Graph the equation

$$y = 2x^2 + 8x$$

on your calculator. The graph is shown in Figure 6.16. Note that $a = 2$ and that $2 > 0$, so the parabola opens upward. We can find the *x*-intercepts of the graph by setting y equal to zero:

$$0 = 2x^2 + 8x$$
$$= 2x(x + 4)$$

FIGURE 6.16

The solutions of this equation are 0 and -4, so the *x*-intercepts are the points $(0, 0)$ and $(-4, 0)$.

Recall that the parabola is symmetric about a vertical line through its vertex. (We will prove that this is true in the Homework problems.) The two *x*-intercepts are equidistant from this line of symmetry, so the *x*-coordinate of the vertex lies exactly halfway between the *x*-intercepts. We can average their values to find

$$x = \frac{1}{2}[0 + (-4)] = -2$$

To find the *y*-coordinate of the vertex, substitute $x = -2$ into the equation for the parabola:

$$y = 2(-2)^2 + 8(-2)$$
$$= 8 - 16 = -8$$

Thus, the vertex is the point $(-2, -8)$.

EXERCISE 3

a. Find the *x*-intercepts and the vertex of the parabola $y = 6x - x^2$.

b. Verify your answers by graphing the equation in the window

$$\text{Xmin} = -9.4 \qquad \text{Xmax} = 9.4$$
$$\text{Ymin} = -10 \qquad \text{Ymax} = 10$$

Finding the Vertex

We can use the same method to find a formula for the vertex of any parabola of the form

$$y = ax^2 + bx$$

We proceed as we did in the previous example. First, find the x-intercepts of the graph by setting y equal to zero and solving for x.

$$0 = ax^2 + bx \qquad \text{Factor.}$$
$$= x(ax + b)$$

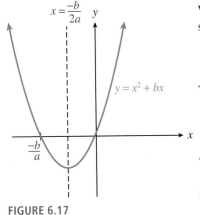

$$x = \frac{-b}{2a}$$

$$y = x^2 + bx$$

$$\frac{-b}{a}$$

FIGURE 6.17

Thus,

$$x = 0 \quad \text{or} \quad ax + b = 0$$
$$x = 0 \quad \text{or} \quad x = \frac{-b}{a}$$

The x-intercepts are the points $(0, 0)$ and $(\frac{-b}{a}, 0)$.

Next, find the x-coordinate of the vertex by taking the average of the two x-intercepts found above:

$$x = \frac{1}{2}\left[0 + \left(\frac{-b}{a}\right)\right] = \frac{-b}{2a}$$

This gives us a formula for the x-coordinate of the vertex.

Vertex of a Parabola

For the graph of $y = ax^2 + bx$, the x-coordinate of the vertex is

$$x_v = \frac{-b}{2a}$$

Also, the axis of symmetry is the vertical line $x = \frac{-b}{2a}$ (see Figure 6.17). Finally, we find the y-coordinate of the vertex by substituting its x-coordinate into the equation for the parabola.

EXAMPLE 3

a. Find the vertex of the graph of $f(x) = -1.8x^2 - 16.2x$.

b. Find the x-intercepts of the graph.

Solutions　**a.** The x-coordinate of the vertex is

$$x_v = \frac{-b}{2a} = \frac{-(-16.2)}{2(-1.8)} = -4.5$$

To find the y-coordinate of the vertex, evaluate $f(x)$ at $x = -4.5$.

$$y_v = -1.8(-4.5)^2 - 16.2(-4.5) = 36.45$$

The vertex is $(-4.5, 36.45)$.

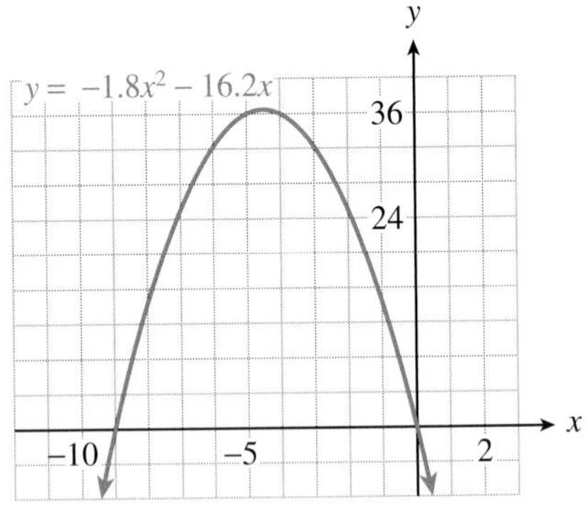

$$y = -1.8x^2 - 16.2x$$

FIGURE 6.18

b. To find the x-intercepts of the graph, set $f(x) = 0$ and solve.

$$-1.8x^2 - 16.2x = 0 \qquad \text{Factor.}$$
$$-x(1.8x + 16.2) = 0 \qquad \text{Set each factor equal to zero.}$$
$$-x = 0 \quad 1.8x + 16.2 = 0 \qquad \text{Solve each equation.}$$
$$x = 0 \qquad\qquad x = -9$$

The x-intercepts of the graph are $(0, 0)$ and $(-9, 0)$. The graph is shown in Figure 6.18.

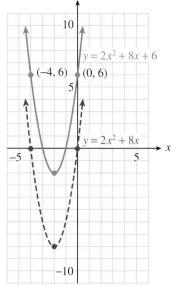

FIGURE 6.19

The Graph of $y = ax^2 + bx + c$

Now we will see that the vertex formula holds for any parabola. Consider the function

$$y = 2x^2 + 8x + 6$$

Adding 6 to $2x^2 + 8x$ shifts each point on the graph 6 units upward, as shown in Figure 6.19. The x-coordinate of the vertex will not be affected by an upward shift. Thus, the formula

$$x_v = \frac{-b}{2a}$$

for the x-coordinate of the vertex still holds. We have

$$x_v = \frac{-8}{2(2)} = -2$$

We find the y-coordinate of the vertex by substituting $x_v = -2$ into the equation for the parabola.

$$y_v = 2(-2)^2 + 8(-2) + 6 \qquad \text{Substitute } -2 \text{ for } x.$$
$$= 8 - 16 + 6 = -2$$

So the vertex is the point $(-2, -2)$. (Notice that this point is shifted 6 units upward from the vertex of $y = 2x^2 + 8x$.)

We find the x-intercepts of the graph by setting y equal to zero.

$$
\begin{aligned}
0 &= 2x^2 + 8x + 6 && \text{Factor the right side.} \\
&= 2(x + 1)(x + 3) && \text{Set each factor equal to zero.} \\
x + 1 &= 0 \quad \text{or} \quad x + 3 = 0 \\
x &= -1 \qquad\qquad x = -3
\end{aligned}
$$

The x-intercepts are the points $(-1, 0)$ and $(-3, 0)$.

The y-intercept of the graph is found by setting x equal to zero:

$$y = 2(0)^2 + 8(0) + 6 = 6$$

You can see that the y-intercept, 6, is just the constant term of the quadratic equation. The completed graph is shown in Figure 6.18.

EXAMPLE 4

Find the vertex of the graph of $y = -2x^2 + x + 1$.

Solution For this equation, $a = -2$, $b = 1$, and $c = 1$. The x-coordinate of the vertex is given by

$$x_v = \frac{-b}{2a} = \frac{-1}{2(-2)} = \frac{1}{4}$$

To find the y-coordinate of the vertex, we substitute $x = \frac{1}{4}$ into the equation. We can do this by hand to find

$$
\begin{aligned}
y_v &= -2\left(\frac{1}{4}\right)^2 + \frac{1}{4} + 1 \\
&= -2\left(\frac{1}{16}\right) + \frac{4}{16} + \frac{16}{16} = \frac{18}{16} = \frac{9}{8}
\end{aligned}
$$

So the coordinates of the vertex are $(\frac{1}{4}, \frac{9}{8})$. Alternatively, we can use the calculator to evaluate $-2x^2 + x + 1$ for $x = 0.25$. The calculator returns the y-value 1.125. Thus, the vertex is the point $(0.25, 1.125)$, which is the decimal equivalent of $(\frac{1}{4}, \frac{9}{8})$.

EXERCISE 4

Find the vertex of the graph of $y = 3x^2 - 6x + 4$. Decide whether the vertex is a maximum point or a minimum point of the graph.

Number of *x*-Intercepts

The graph of the quadratic function $y = ax^2 + bx + c$ may have two, one, or no x-intercepts, according to the number of distinct real-valued solutions of the equation $ax^2 + bx + c = 0$. Consider the three equations graphed in Figure 6.20. The graph of

$$y = x^2 - 4x + 3$$

has two x-intercepts, because the equation

$$x^2 - 4x + 3 = 0$$

has two real-valued solutions, $x = 1$ and $x = 3$. The graph of

$$y = x^2 - 4x + 4$$

has only one x-intercept, because the equation

$$x^2 - 4x + 4 = 0$$

has only one (repeated) real-valued solution, $x = 2$. The graph of $y = x^2 - 4x + 6$ has no x-intercepts, because the equation $x^2 - 4x + 6 = 0$ has no real-valued solutions.

A closer look at the quadratic formula reveals useful information about the solutions of quadratic equations. For the three equations above, we have the following:

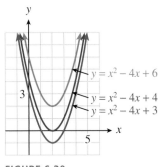

$y = x^2 - 4x + 6$
$y = x^2 - 4x + 4$
$y = x^2 - 4x + 3$

FIGURE 6.20

$y = x^2 - 4x + 3$	$y = x^2 - 4x + 4$	$y = x^2 - 4x + 6$
two x-intercepts	one x-intercept	no x-intercepts

$$x = \frac{4 \pm \sqrt{(-4)^2 - 4(1)(3)}}{2} \qquad x = \frac{4 \pm \sqrt{(-4)^2 - 4(1)(4)}}{2} \qquad x = \frac{4 \pm \sqrt{(-4)^2 - 4(1)(6)}}{2}$$

$$= \frac{4 \pm \sqrt{4}}{2} \text{ (two solutions)} \qquad = \frac{4 \pm \sqrt{0}}{2} \text{ (one repeated solution)} \qquad = \frac{4 \pm \sqrt{-12}}{2} \text{ (no solutions)}$$

The expression $b^2 - 4ac$, which appears under the radical in the quadratic formula, is called the **discriminant, D,** of the equation. The value of the discriminant determines the nature of the solutions of the equation. In particular, if the discriminant is negative, the solutions of the quadratic equation are **complex numbers.** (We will study complex numbers in Section 7.3.)

The Discriminant

The **discriminant** of a quadratic equation is $D = b^2 - 4ac$.

1. If $D > 0$, there are two unequal real solutions.
2. If $D = 0$, there is one real solution of multiplicity two.
3. If $D < 0$, there are two complex solutions.

We can also use the discriminant to decide whether a quadratic equation can be solved by factoring. First, clear the equation of fractions. If the discriminant is a perfect square, that is, the square of an integer, the solutions are rational numbers. This in turn means that the equation can be solved by factoring. If the discriminant is not a perfect square, the solutions will be irrational. Irrational solutions always occur in **conjugate pairs,**

$$\frac{-b}{2a} + \frac{\sqrt{b^2 - 4ac}}{2a} \quad \text{and} \quad \frac{-b}{2a} - \frac{\sqrt{b^2 - 4ac}}{2a}$$

The only difference between the two solutions is the sign between the terms. For example, if we know that one solution of a particular quadratic equation is $3 + \sqrt{2}$, then the other solution must be $3 - \sqrt{2}$.

EXAMPLE 5

Use the discriminant to determine the nature of the solutions of each equation. Can the equation be solved by factoring?

a. $x^2 - x - 3 = 0$ **b.** $2x^2 + x + 1 = 0$ **c.** $x^2 - 2x - 3 = 0$

Solutions Compute the discriminant for each equation.

a. $D = b^2 - 4ac = (-1)^2 - 4(1)(-3) = 13 > 0$. The equation has two real, unequal solutions. Because 13 is not a perfect square, the solutions will be irrational numbers, so the equation cannot be solved by factoring.

b. $D = b^2 - 4ac = 1^2 - 4(2)(1) = -7 < 0$. The equation has two complex solutions, which cannot be found by factoring.

c. $D = b^2 - 4ac = (-2)^2 - 4(1)(-3) = 16 > 0$. The equation has two real, unequal solutions. Because $16 = 4^2$, the solutions are rational numbers and can be found by factoring.

(You can verify the conclusions above by solving each equation.)

EXERCISE 5

Use the discriminant to discover how many x-intercepts the graph of each equation has.

a. $y = x^2 + 5x + 7$
b. $y = -\frac{1}{2}x^2 + 4x - 8$

In Exercise 5b, you should check that the single x-intercept is also the vertex of the parabola.

Sketching a Parabola

Once we have located the vertex of the parabola, the x-intercepts, and the y-intercept, we can sketch a reasonably accurate graph. Recall that the graph should be symmetric about a vertical line through the vertex. We summarize the procedure as follows.

To Graph the Quadratic Function $y = ax^2 + bx + c$:

1. Determine whether the parabola opens upward (if $a > 0$) or downward (if $a < 0$).
2. Locate the vertex of the parabola.
   a. The x-coordinate of the vertex is $x_v = \dfrac{-b}{2a}$.
   b. Find the y-coordinate of the vertex by substituting x_v into the equation of the parabola.
3. Locate the x-intercepts (if any) by setting $y = 0$ and solving for x.
4. Locate the y-intercept by evaluating y for $x = 0$.
5. Locate the point symmetric to the y-intercept across the axis of symmetry.

EXAMPLE 6   Sketch a graph of the equation $y = x^2 + 3x + 1$, showing the significant points.

Solution   We follow the steps outlined above.

Step 1   Because $a = 1 > 0$, we know that the parabola opens upward.

Step 2   Compute the coordinates of the vertex:

$$x_v = \frac{-b}{2a} = \frac{-3}{2(1)} = -1.5$$

$$y_v = (-1.5)^2 + 3(-1.5) + 1 = -1.25$$

The vertex is the point $(-1.5, -1.25)$.

Step 3   Set y equal to zero to find the x-intercepts.

$$0 = x^2 + 3x + 1 \qquad \text{Use the quadratic formula.}$$

$$x = \frac{-3 \pm \sqrt{3^2 - 4(1)(1)}}{2(1)}$$

$$= \frac{-3 \pm \sqrt{5}}{2}$$

Rounding to the nearest tenth, we find that the x-intercepts are approximately $(-2.6, 0)$ and $(-0.4, 0)$.

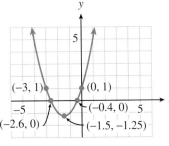

FIGURE 6.21

Step 4   Substitute $x = 0$ to find the y-intercept, $(0, 1)$.

Step 5   The axis of symmetry is the vertical line $x = -1.5$, so the y-intercept lies 1.5 units to the right of the axis of symmetry. There must be another point on the parabola with the same y-coordinate as the y-intercept but 1.5 units to the left of the axis of symmetry. The coordinates of this point are $(-3, 1)$.

Finally, plot the x-intercepts, the vertex, and the y-intercept and its symmetric point, and draw a parabola through them. The finished graph is shown in Figure 6.21.

ANSWERS TO 6.3 EXERCISES

1. **a.** III   **b.** II   **c.** I   **d.** IV
2. **a.** $y = x^2 - 5$ **b.** $(-\sqrt{5}, 0), (\sqrt{5}, 0), (0, -5)$
3. x-intercepts: $(0, 0)$ and $(6, 0)$; vertex: $(3, 9)$
4. $(1, 1)$, minimum
5. **a.** None   **b.** One
6. **a.** $(0, 4)$; $(1, 0), (4, 0)$; vertex $(\frac{5}{2}, \frac{-9}{4})$ **b.**

SECTION 6.3 SUMMARY

VOCABULARY   *Look up the definitions of new terms in the Glossary.*

Vertex   Conjugate pair   Axis of symmetry   Discriminant

CONCEPTS

1. The graph of a quadratic function $f(x) = ax^2 + bx + c$ is called a **parabola.** The values of the constants a, b, and c determine the location and orientation of the parabola.
2. For the graph of $y = ax^2 + bx + c$, the x-coordinate of the **vertex** is $x_v = \frac{-b}{2a}$. To find the y-coordinate of the vertex, we substitute x_v into the formula for the parabola.
3. The graph of the quadratic function $y = ax^2 + bx + c$ may have two, one, or no x-intercepts, according to the number of distinct real-valued solutions of the equation $ax^2 + bx + c = 0$.

4.

The Discriminant

The **discriminant** of a quadratic equation is $D = b^2 - 4ac$.

1. If $D > 0$, there are two unequal real solutions.
2. If $D = 0$, there is one real solution of multiplicity two.
3. If $D < 0$, there are two complex solutions.

5.

> **To graph the quadratic function $y = ax^2 + bx + c$:**
>
> 1. Determine whether the parabola opens upward (if $a > 0$) or downward (if $a < 0$).
> 2. Locate the vertex of the parabola.
>   a. The x-coordinate of the vertex is $x_v = \frac{-b}{2a}$.
>   b. Find the y-coordinate of the vertex by substituting x_v into the equation of the parabola.
>
> 3. Locate the x-intercepts (if any) by setting $y = 0$ and solving for x.
> 4. Locate the y-intercept by evaluating y for $x = 0$.
> 5. Locate the point symmetric to the y-intercept across the axis of symmetry.

STUDY QUESTIONS

1. Sketch a parabola that opens downward. Show the location of the x-intercepts, the y-intercept, the vertex, and the axis of symmetry.
2. Describe how the value of a in $y = ax^2$ alters the graph of the basic parabola.
3. Describe how the value of c in $y = x^2 + c$ alters the graph of the basic parabola.
4. Suppose you know that the x-intercepts of a parabola are $(-8, 0)$ and $(2, 0)$. What is the equation of the parabola's axis of symmetry?

5. State a formula for the x-coordinate of the vertex of a parabola. How can you find the y-coordinate of the vertex?
6. Suppose that a given parabola has only one x-intercept. What can you say about the vertex of the parabola?
7. Explain why a quadratic equation has one (repeated) solution if its discriminant is zero, and none if the discriminant is negative.

SKILLS *Practice each skill in the Homework problems listed.*

1. Graph transformations of the basic parabola: #1 and 2, 7 and 8
2. Locate the x-intercepts of a parabola: #3–6
3. Locate the vertex of a parabola: #3–6, 13 and 14

4. Sketch the graph of a quadratic function: #15–24, 41 and 42
5. Use the discriminant to describe the solutions of a quadratic equation: #25–40

HOMEWORK 6.3

Describe what each graph will look like compared to the basic parabola. Then sketch a graph by hand and label the coordinates of three points on the graph.

1. a. $y = 2x^2$ b. $y = 2 + x^2$ 2. a. $y = -4x^2$ b. $y = (x - 4)^2$
   c. $y = (x + 2)^2$ d. $y = x^2 - 2$ c. $y = 4 - x^2$ d. $y = x^2 - 4$

Find the vertex and the x-intercepts (if there are any) of the graph. Then sketch the graph by hand.

3. a. $y = x^2 - 16$ b. $y = 16 - x^2$ 4. a. $y = x^2 - 1$ b. $y = 1 - x^2$
   c. $y = 16x - x^2$ d. $y = x^2 - 16x$ c. $y = x^2 - x$ d. $y = x - x^2$

5. a. $y = 3x^2 + 6x$ b. $y = 3x^2 - 6x$ 6. a. $y = 12x - 2x^2$ b. $y = 12 - 2x^2$
   c. $y = 3x^2 + 6$ d. $y = 3x^2 - 6$ c. $y = 12 + 2x^2$ d. $y = 12x + 2x^2$

7. Match each function with its graph. In each equation, $a > 0$.
   a. $y = x^2 + a$ b. $y = x^2 + ax$ c. $y = ax^2$
   d. $y = ax$ e. $y = x + a$ f. $y = x^2 - a$

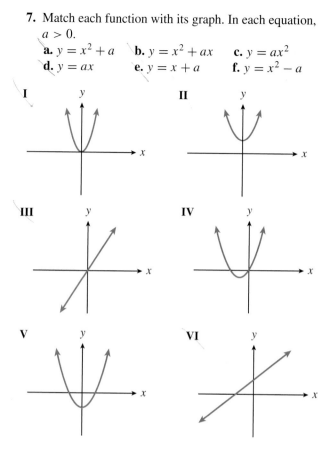

I   II   III   IV   V   VI

8. Match each function with its graph. In each equation, $b > 0$.
   a. $y = -bx$ b. $y = -bx^2$ c. $y = b - x^2$
   d. $y = x - b$ e. $y = b - x$ f. $y = x^2 - bx$

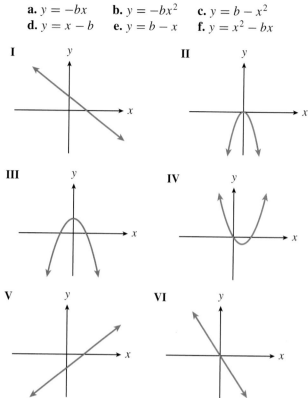

I   II   III   IV   V   VI

9. Commercial fishermen rely on a steady supply of fish in their area. To avoid overfishing, they adjust their harvest to the size of the population. The function
$$g(x) = 0.4x - 0.0001x^2$$
gives the annual rate of growth, in tons per year, of a fish population of biomass x tons.
   a. Find the vertex of the graph. What does it tell us about the fish population?
   b. Sketch the graph for $0 \le x \le 5000$.
   c. For what values of x does the fish population decrease rather than increase? Suggest a reason why the population might decrease.

10. The annual increase, I, in the deer population in a national park depends on the size, x, of the population that year, according to the formula
$$I = 1.2x - 0.0002x^2$$
   a. Find the vertex of the graph. What does it tell us about the deer population?
   b. Sketch the graph for $0 \le x \le 7000$.
   c. For what values of x does the deer population decrease rather than increase? Suggest a reason why the population might decrease.

11. Many animals live in groups. A species of marmot found in Colorado lives in harems composed of a single adult male and several females with their young. The number of offspring each female can raise depends on the number of females in the harem. On average, if there are x females in the harem, each female can raise $y = 2 - 0.4x$ young marmots each year.
   a. Complete the table of values for the average number of offspring per female, and the total number of young marmots, A, produced by the entire harem in one year.

x	1	2	3	4	5
y					
A					

   b. Write a formula for A in terms of x.
   c. Graph A as a function of x.
   d. What is the maximum number of young marmots a harem can produce (on average)? What is the optimal number of female marmots per harem?

12. Greenshield's model for traffic flow assumes that the average speed, u, of cars on a highway is a linear function of the traffic density, k, in vehicles per mile, given by

$$u = u_f\left(1 - \frac{k}{k_j}\right)$$

where u_f is the free-flow speed and k_j is the maximum density (the point when traffic jams). Then the traffic flow, q, in vehicles per hour, is given by $q = uk$.
a. Write a formula for q as a function of k.
b. If the free-flow speed is 70 mph and the maximum density is 240 vehicles per mile, graph q as a function of k.
c. What value of k gives the maximum traffic flow? What is the average speed of vehicles at that density?

13. After touchdown, the distance the space shuttle travels is given by

$$d = vT + \frac{v^2}{2a}$$

where v is the shuttle's velocity in ft/sec at touchdown, T is the pilot's reaction time before the brakes are applied, and a is the shuttle's deceleration.
a. Graph $d = f(v)$ for $T = 0.5$ seconds and $a = 12$ ft/sec^2. Find the coordinates of the vertex and the horizontal intercepts. Explain their meaning, if any, in this context.
b. The runway at Edwards Air Force base is 15,000 feet long. What is the maximum velocity the shuttle can have at touchdown and still stop on the runway?

14. When setting the pump pressure at the engine, firefighters must take into account the pressure loss due to friction inside the fire hose. For every 100 feet of hoseline, a hose of diameter 2.5 inches loses pressure according to the formula

$$L = \begin{cases} 2Q^2 + Q, & Q \geq 1 \\ 2Q^2 + \frac{1}{2}Q, & Q < 1 \end{cases}$$

where Q is the water flow in hundreds of gallos per minute. The friction loss, L, is measured in pounds per square inch (psi)
(Source: www.hcc.hawaii.edu/~jkemmer)
a. Graph $L = g(Q)$ on the domain $[0, 5]$.
b. The firefighters have unrolled 600 feet of 2.5-inch-diameter hose, and they would like to deliver water at a rate of 200 gallons per minute, with nozzle pressure at 100 psi. They must add the friction loss to the nozzle pressure to calculate the engine pressure required. What should the engine pressure be?

Find the coordinates of the vertex. Decide whether the vertex is a maximum point or a minimum point on the graph.

15. a. $y = 2 + 3x - x^2$

b. $y = \frac{1}{2}x^2 - \frac{2}{3}x + \frac{1}{3}$

c. $y = 2.3 - 7.2x - 0.8x^2$

16. a. $y = 3 - 5x + x^2$

b. $y = \frac{-3}{4}x^2 + \frac{1}{2}x - \frac{1}{4}$

c. $y = 5.1 - 0.2x + 4.6x^2$

In Problems 17–26,
(a) Find the coordinates of the intercepts and the vertex.
(b) Sketch the graph by hand.
(c) Use your calculator to verify your graph.

17. $y = -2x^2 + 7x + 4$

18. $y = -3x^2 + 2x + 8$

19. $y = 0.6x^2 + 0.6x - 1.2$

20. $y = 0.5x^2 - 0.25x - 0.75$

21. $y = x^2 + 4x + 7$

22. $y = x^2 - 6x + 10$

23. $y = x^2 + 2x - 1$

24. $y = x^2 - 6x + 2$

25. $y = -2x^2 + 6x - 3$

26. $y = -2x^2 - 8x - 5$

27. a. Graph the three functions $y = x^2 - 6x + 5$, $y = x^2 - 6x + 9$, and $y = x^2 - 6x + 12$ in the window

$$\text{Xmin} = -2, \quad \text{Xmax} = 7.4$$
$$\text{Ymin} = -5, \quad \text{Ymax} = 15$$

Use the **Trace** to locate the x-intercepts of each graph.

b. Set $y = 0$ for each of the equations in part (a) and calculate the discriminant. What does the discriminant tell you about the solutions of the equation? How does your answer relate to the graphs in part (a)?

28. a. Graph the three functions $y = 3 - 2x - x^2$, $y = -1 - 2x - x^2$, and $y = -4 - 2x - x^2$ in the window

$$\text{Xmin} = -6.4, \quad \text{Xmax} = 3$$
$$\text{Ymin} = -10, \quad \text{Ymax} = 5$$

Use the **Trace** to locate the x-intercepts of each graph.

b. Set $y = 0$ for each of the equations in part (a) and calculate the discriminant. What does the discriminant tell you about the solutions of the equation? How does your answer relate to the graphs in part (a)?

Use the discriminant to determine the nature of the solutions of each equation.

29. $3x^2 + 26 = 17x$

30. $4x^2 + 23x = 19$

31. $16x^2 - 712x + 7921 = 0$

32. $121x^2 + 1254x + 3249 = 0$

33. $65.2x = 13.2x^2 + 41.7$

34. $0.03x^2 = 0.05x - 0.12$

Use the discriminant to decide if we can solve the equation by factoring.

35. $3x^2 - 7x + 6 = 0$

36. $6x^2 - 11x - 7 = 0$

37. $15x^2 - 52x - 32 = 0$

38. $17x^2 + 65x - 12 = 0$

For Problems 39–42,
(a) Given one solution of a quadratic equation with rational coefficients, find the other solution.
(b) Write a quadratic equation that has those solutions.

39. $2 + \sqrt{5}$

40. $3 - \sqrt{2}$

41. $4 - 3\sqrt{2}$

42. $5 + 2\sqrt{3}$

For Problems 43 and 44, match each equation with one of the eight graphs shown.

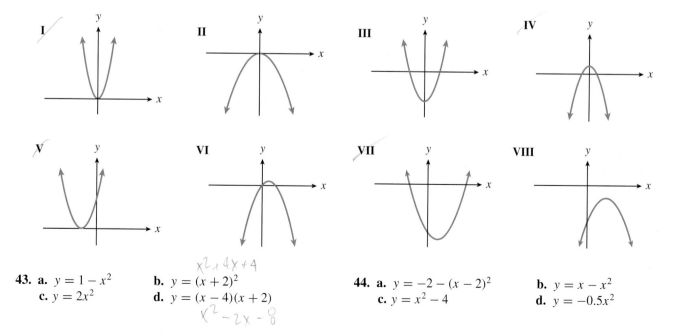

43. a. $y = 1 - x^2$ **b.** $y = (x + 2)^2$
     c. $y = 2x^2$ **d.** $y = (x - 4)(x + 2)$

44. a. $y = -2 - (x - 2)^2$ **b.** $y = x - x^2$
     c. $y = x^2 - 4$ **d.** $y = -0.5x^2$

45. a. Write an equation for a parabola that has x-intercepts at $(2, 0)$ and $(-3, 0)$. What is the equation of the parabola's axis of symmetry?
   b. Write an equation for another parabola that has the same x-intercepts. What is the equation of the parabola's axis of symmetry?

46. a. Write an equation for a parabola that opens upward and has x-intercepts $(-1, 0)$ and $(4, 0)$. What is the equation of its axis of symmetry?
   b. Write an equation for a parabola that opens downward and has x-intercepts $(-1, 0)$ and $(4, 0)$. What is the equation of its axis of symmetry?

47. a. Graph the functions in the same window on your calculator:

$$y = x^2 + 2x, \ y = x^2 + 4x,$$
$$y = x^2 + 6x, \ y = x^2 + 8x$$

   b. Find the vertex of each graph in part (a) and plot the points.
   c. Find the equation of the curve in part (b).
   d. Show that the vertex of $y = x^2 + 2kx$ lies on the curve for any value of k.

48. a. Graph the functions in the same window on your calculator:

$$y = x - \frac{1}{2}x^2, \ y = 3x - \frac{1}{2}x^2,$$
$$y = 5x - \frac{1}{2}x^2, \ y = 7x - \frac{1}{2}x^2$$

   b. Find the vertex of each graph in part (a) and plot the points.
   c. Find the equation of the curve in part (b).
   d. Show that the vertex of $y = kx - \frac{1}{2}x^2$ lies on the curve for any value of k.

49. Because of air resistance, the path of a kicked soccer ball is not actually parabolic. However, both the horizontal and vertical coordinates of points on its trajectory can be approximated by quadratic functions. For a soccer ball kicked from the ground, these functions are

$$x = f(t) = 12.8t - 1.3t^2$$
$$y = g(t) = 17.28t - 4.8t^2$$

where x and y are given in meters and t is the number of seconds since the ball was kicked.
   a. Fill in the table.

t	0	0.5	1.0	1.5	2.0	2.5	3.0	3.5
x								
y								

   b. Plot the points (x, y) from your table and connect them with a smooth curve to represent the path of the ball.
   c. Use your graph to estimate the maximum height of the ball.
   d. Estimate the horizontal distance traveled by the ball before it strikes the ground.
   e. Using the formula given for y, determine how long the ball is in the air.
   f. Use your answer from part (e) and the formula for x to find the horizontal distance traveled by the ball before it strikes the ground.
   g. Use the formula given for y to find the maximum height for the ball.

50. How far can you throw a baseball? The distance depends on the initial speed of the ball, v, and on the angle at which you throw it. For maximum range, you should throw the ball at $45°$.
   a. If there were no air resistance, the height, h, of the ball t seconds after its release would be given in meters by the formula

$$h = \frac{vt}{\sqrt{2}} - \frac{gt^2}{2}$$

where g is the acceleration due to gravity. Find an expression for the total time the ball is in the air. (*Hint:* Set $h = 0$ and solve for t in terms of the other variables.)
   b. At time t, the ball has traveled a horizontal distance d given by

$$d = \frac{vt}{\sqrt{2}}$$

Find an expression for the range of the ball in terms of its velocity, v. (*Hint:* In part (a), you found an expression for t when $h = 0$. Use that value of t to calculate d when $h = 0$.)
   c. The fastest baseball pitch on record was 45 meters per second, or about 100 miles per hour. Use your formula from part (b) to calculate the theoretical range of such a pitch. The value of g is 9.8.
   d. The maximum distance a baseball has actually been thrown is 136 meters. Can you explain the discrepancy between this figure and your answer to part (c)?

Many quadratic models arise as the product of two variables, one of which increases while the other decreases. For example, the area of a rectangle is the product of its length and its width, or $A = lw$. If we require that the rectangle have a certain perimeter, then as we increase its length, we must also decrease its width. (We analyzed this problem in Investigation 3 of Chapter 2.)

From Investigation 10, recall the formula for the revenue from sales of an item:

Revenue = (price of one item)(number of items sold)

Usually, when the price of an item increases, the number of items sold decreases.

Investigation 11   Revenue from Theater Tickets

The local theater group sold tickets to its opening night performance for $5 and drew an audience of 100 people. The next night, the group reduced the ticket price by $0.25 and 10 more people attended; that is, 110 people bought tickets at $4.75 apiece. In fact, for each $0.25 reduction in ticket price, 10 additional tickets can be sold.

1. Complete the table.

2. Use your table to make a graph. Plot *total revenue* on the vertical axis versus *number of price reductions* on the horizontal axis. (Use a grid like the one on page 516.)

3. Let x represent the *number of price reductions*, as in the first column of the table. Write algebraic expressions in terms of x for each quantity.

   The *price of a ticket* after x price reductions:

   Price =

   The *number of tickets sold* at that price:

   Number =

   The *total revenue* from ticket sales:

   Revenue =

Number of price reductions	Price of ticket	Number of tickets sold	Total revenue
0	5.00	100	500
1	4.75	110	522.50
2			
3			
4			
5			
6			
7			
8			
9			
10			
11			

4. Enter your expressions for the price of a ticket, the number of tickets sold, and the total revenue into the calculator as Y_1, Y_2, and Y_3. Use the **Table** feature to verify that your algebraic expressions agree with your table from part (1).

5. Use your calculator to graph your expression for total revenue in terms of x. Use your table to choose appropriate window settings that show the high point of the graph and both x-intercepts.

6. What is the maximum revenue possible from ticket sales? What price should the theater group charge for a ticket to generate that revenue? How many tickets will the group sell at that price?

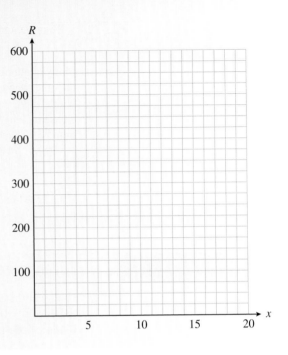

Maximum or Minimum Values

Finding the maximum or minimum value for a variable expression is a common problem in many applications. For example, if you own a company that manufactures blue jeans, you might like to know how much to charge for your jeans in order to maximize your revenue. As you increase the price of the jeans, your revenue may increase for a while. But if you charge too much for the jeans, consumers will not buy as many pairs, and your revenue may actually start to decrease. Is there some optimum price you should charge for a pair of jeans in order to achieve the greatest revenue?

EXAMPLE 1 Late Nite Blues finds that it can sell $600 - 15x$ pairs of jeans per week if it charges x dollars per pair. (Notice that as the price increases, the number of pairs of jeans sold decreases.)

a. Write an equation for the revenue as a function of the price of a pair of jeans.

b. Graph the function.

c. How much should Late Nite Blues charge for a pair of jeans in order to maximize its revenue?

Solutions   a. Using the formula for revenue stated above, we find

$$\text{Revenue} = (\text{price of one item})(\text{number of items sold})$$
$$R = x(600 - 15x)$$
$$R = 600x - 15x^2$$

b. We recognize the equation as quadratic, so the graph is a parabola. You can use your calculator to verify the graph in Figure 6.22.

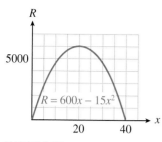

FIGURE 6.22

c. The maximum value of R occurs at the vertex of the parabola. Thus,

$$x_v = \frac{-b}{2a} = \frac{-600}{2(-15)} = 20$$

$$R_v = 600(20) - 15(20)^2 = 6000$$

The revenue takes on its maximum value when $x = 20$, and the maximum value is $R = 6000$. This means that Late Nite Blues should charge \$20 for a pair of jeans in order to maximize revenue at \$6000 a week.

If the equation relating two variables is quadratic, then the maximum or minimum value is easy to find: It is the value at the vertex. If the parabola opens downward, as in Example 1, there is a maximum value at the vertex. If the parabola opens upward, there is a minimum value at the vertex.

> ### EXERCISE 1
>
> The Metro Rail service sells $1200 - 80x$ tickets each day when it charges x dollars per ticket.
>
> **a.** Write an equation for the revenue, R, as a function of the price of a ticket.
> **b.** What ticket price will return the maximum revenue? What is the maximum revenue?

The Vertex Form for a Parabola

Consider the quadratic equation

$$y = 2(x - 3)^2 - 8$$

By expanding the squared expression and collecting like terms, we can rewrite the equation in standard form as

$$y = 2(x^2 - 6x + 9) - 8$$
$$y = 2x^2 - 12x + 10$$

FIGURE 6.23

The vertex of this parabola is

$$x_v = \frac{-(-12)}{2(2)} = 3$$

$$y_v = 2(3)^2 - 12(3) + 10 = -8$$

and its graph is shown in Figure 6.23.

Notice that the coordinates of the vertex, $(3, -8)$, are apparent in the original equation; we don't need to do any computation to find the vertex.

$$y = 2\underbrace{(x - 3)}_{x_v}{}^2 - \underbrace{8}_{y_v}$$

This equation is an example of the **vertex form** for a quadratic function.

> **Vertex Form for a Quadratic Function**
>
> A quadratic function $y = ax^2 + bx + c$, $a \neq 0$, can be written in the **vertex form**
>
> $$y = a(x - x_v)^2 + y_v$$
>
> where the vertex of the graph is (x_v, y_v).

EXAMPLE 2   Find the vertex of the graph of $y = -3(x - 4)^2 + 6$. Is the vertex a maximum or a minimum point of the graph?

Solution   Compare the equation to the vertex form to see that the coordinates of the vertex are $(4, 6)$. For this equation, $a = -3 < 0$, so the parabola opens downward. The vertex is the maximum point of the graph.

To understand why the vertex form works, substitute $x_v = 4$ into $y = -3(x - 4)^2 + 6$ from Example 4 to find

$$y = -3(4 - 4)^2 + 6 = 6$$

which confirms that when $x = 4$, $y = 6$. Next, notice that if x is any number except 4, the expression $-3(x - 4)^2$ is negative, so $y < 6$. Therefore, 6 is the maximum value for y on the graph, so $(4, 6)$ is the high point or vertex.

You can also rewrite $y = -3(x - 4)^2 + 6$ in standard form and use the formula $x_v = \frac{-b}{2a}$ to confirm that the vertex is the point $(4, 6)$.

> **EXERCISE 2**
>
> **a.** Find the vertex of the graph of $y = 5 - \frac{1}{2}(x + 2)^2$.
> **b.** Write the equation of the parabola in standard form.

Any quadratic equation in vertex form can be written in standard form by expanding, and any quadratic equation in standard form can be put into vertex form by completing the square.

EXAMPLE 3   Write the equation $y = 3x^2 - 6x - 1$ in vertex form and find the vertex of its graph.

Solution   We factor the lead coefficient, 3, from the variable terms, leaving a space to complete the square.

$$y = 3(x^2 - 2x\underline{\quad\quad}) - 1$$

Next, we complete the square inside parentheses. Take half the coefficient of x and square the result: $p = \frac{1}{2}(-2) = -1$, and $p^2 = (-1)^2 = 1$. We must add 1 to complete the square. However, we are really adding $3(1)$ to the right side of the equation, so we must also subtract 3 to compensate:

$$y = 3(x^2 - 2x + 1) - 1 - 3$$

The expression inside parentheses is now a perfect square, and the vertex form is

$$y = 3(x - 1)^2 - 4$$

The vertex of the parabola is $(1, -4)$.

EXERCISE 3

Write the equation $y = 2x^2 + 12x + 13$ in vertex form, and find the vertex of its graph.

Step 1 Factor 2 from the variable terms.

Step 2 Complete the square inside parentheses.

Step 3 Subtract $2p^2$ outside parentheses.

Step 4 Write the vertex form.

Graphing with the Vertex Form

We can also use the vertex form to sketch a graph, using what we know about transformations.

EXAMPLE 4 Use transformations to graph $f(x) = -3(x - 4)^2 + 6$.

Solution We can graph $f(x)$ by applying transformations to the basic parabola, $y = x^2$. To identify the transformations, think of evaluating the function for a specific value of x. What operations would we perform on x, besides squaring?

1. Subtract 4: $y = (x - 4)^2$ Shift 4 units right.

2. Multiply by -3: $y = -3(x - 4)^2$ Stretch by a factor of 3, and reflect about the x-axis.

3. Add 6: $y = -3(x - 4)^2 + 6$ Shift up 6 units.

We perform the same transformations on the graph of $y = x^2$, as shown in Figure 6.24.

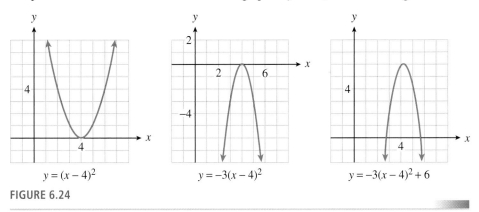

$y = (x - 4)^2$ $y = -3(x - 4)^2$ $y = -3(x - 4)^2 + 6$

FIGURE 6.24

EXERCISE 4

a. List the transformations of $y = x^2$ needed to graph $g(x) = 5 - \frac{1}{2}(x + 2)^2$.

b. Use transformations to sketch the graph.

Systems Involving Quadratic Equations

Recall that the solution to a 2×2 system of linear equations is the intersection point of the graphs of the equations. This is also true of systems in which one or both of the equations is quadratic. Such a system may have either one solution, two solutions, or no solutions. Figure 6.25 shows the three cases for systems of one quadratic and one linear equation.

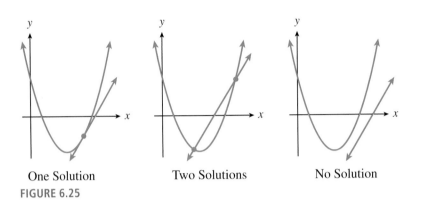

One Solution   Two Solutions   No Solution

FIGURE 6.25

In Example 5, we use both graphical and algebraic techniques to solve the system.

EXAMPLE 5

The Pizza Connection calculates that the cost, in dollars, of producing x pizzas per day is given by

$$C = 0.15x^2 + 0.75x + 180$$

The Pizza Connection charges $15 per pizza, so the revenue from selling x pizzas is

$$R = 15x$$

How many pizzas per day must the Pizza Connection sell in order to break even?

Solution

To break even means to make zero profit. Because

$$Profit = Revenue - Cost$$

the break-even points occur when revenue equals cost. In mathematical terms, we would like to find any values of x for which $R = C$. If we graph the revenue and cost equations on the same axes, these values correspond to points where the two graphs intersect. Use the window settings

$$Xmin = 0, \quad Xmax = 94,$$
$$Ymin = 0, \quad Ymax = 1400$$

on your calculator to obtain the graph shown in Figure 6.25. You can verify that the two intersection points are (15, 225) and (80, 1200).

Thus, the Pizza Connection must sell either 15 or 80 pizzas in order to break even. On the graph we see that revenue is greater than cost for x-values between 15 and 80, so the Pizza Connection will make a profit if it sells between 15 and 80 pizzas.

We can also solve algebraically for the break-even points. The intersection points of the two graphs correspond to the solutions of the system of equations

$$y = 0.15x^2 + 0.75x + 180$$
$$y = 15x$$

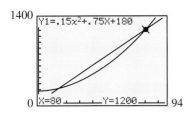

FIGURE 6.26

We equate the two expressions for y and solve for x:

$$0.15x^2 + 0.75x + 180 = 15x \qquad \text{Subtract } 15x \text{ from both sides.}$$

$$0.15x^2 - 14.25x + 180 = 0 \qquad \text{Use the quadratic formula.}$$

$$x = \frac{14.25 \pm \sqrt{14.25^2 - 4(0.15)(180)}}{2(-0.05)} \qquad \text{Simplify.}$$

$$= \frac{14.25 \pm 9.75}{0.3}$$

The solutions are 15 and 80, as we found from the graph in Figure 6.26.

EXERCISE 5

a. Solve the system algebraically:

$$y = x^2 - 6x - 7$$
$$y = 13 - x^2$$

b. Graph both equations, and show the solutions on the graph.

Solving Systems with the Graphing Calculator

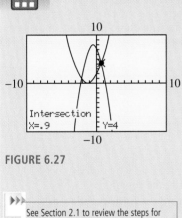

FIGURE 6.27

See Section 2.1 to review the steps for using the **intersect** feature.

We can use the **intersect** feature of the graphing calculator to solve systems of quadratic equations. Consider the system

$$y = (x + 1.1)^2$$
$$y = 7.825 - 2x - 2.5x^2$$

We will graph these two equations in the standard window. The two intersection points are visible in the window, but we do not find their exact coordinates when we trace the graphs. Use the **intersect** command to locate one of the solutions, as shown in Figure 6.27. You can check that the point (0.9, 4) is an exact solution to the system by substituting $x = 0.9$ and $y = 4$ into *each* equation of the system. (The calculator is not always able to find the exact coordinates, but it usually gives a very good approximation.)

You can find the other solution of the system by following the same steps and moving the bug close to the other intersection point. You should verify that the other solution is the point $(-2.1, 1)$.

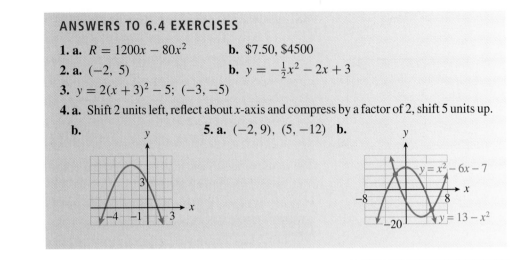

ANSWERS TO 6.4 EXERCISES

1. a. $R = 1200x - 80x^2$ **b.** $7.50, $4500

2. a. $(-2, 5)$ **b.** $y = -\frac{1}{2}x^2 - 2x + 3$

3. $y = 2(x + 3)^2 - 5$; $(-3, -5)$

4. a. Shift 2 units left, reflect about x-axis and compress by a factor of 2, shift 5 units up.

b.

5. a. $(-2, 9)$, $(5, -12)$ **b.**

SECTION 6.4 SUMMARY

VOCABULARY *Look up the definitions of new terms in the Glossary.*

Maximum value   Minimum value   Vertex form

CONCEPTS

1. Quadratic models may arise as the product of two variables.

2. The maximum or minimum of a quadratic function occurs at the vertex.

3.

> **Vertex Form for a Quadratic Function**
>
> A quadratic function $y = ax^2 + bx + c$, $a \neq 0$, can be written in the **vertex form**
> $$y = a(x - x_v)^2 + y_v$$
> where the vertex of the graph is (x_v, y_v).

4. We can convert a quadratic equation to vertex form by completing the square.

5. We can graph a quadratic equation in vertex form using transformations.

6. A 2×2 system involving quadratic equations may have one, two, or no solutions.

STUDY QUESTIONS

1. How can you tell whether a variable given by a quadratic equation has a maximum value or a minimum value?

2. What is wrong with this statement: The maximum or minimum value given by a quadratic equation is the average of the x-intercepts?

3. Explain why -4 is the smallest function value for $f(x) = 2(x - 3)^2 - 4$.

4. In the equation $y = \frac{1}{3}(x + 5)^2 - 2$, what does each of the constants tell you about the graph?

5. Francine attempts to write the equation $g(x) = 2x^2 - 6x + 1$ in vertex form as follows: $g(x) = (2x^2 - 6x + 9) + 1 - 9$. What is wrong with her work?

6. Without doing any calculations, solve the system $y = x^2 + 4$, $y = 2 - 3x^2$. (*Hint:* Visualize the graphs.)

SKILLS   *Practice each skill in the Homework problems listed.*

1. Find the maximum or minimum value of a quadratic function: #1–14

2. Convert a quadratic equation from vertex form to standard form: #19–22

3. Convert a quadratic equation from standard form to vertex form: #23–28

4. Use transformations to graph a quadratic equation: #15–28

5. Solve a system involving quadratic equations: #31–50

1. The owner of a motel has 60 rooms to rent. She finds that if she charges $20 per room per night, all the rooms will be rented. For every $2 that she increases the price of a room, 3 rooms will stand vacant.

   a. Complete the table. The first two rows are filled in for you.

No. of price increases	Price of room	No. of rooms rented	Total revenue
0	20	60	1200
1	22	57	1254
2			
3			
4			
5			
6			
7			
8			
10			
12			
16			
20			

2. The owner of a video store sells 96 blank tapes per week if he charges $6 per tape. For every $0.50 he increases the price, he sells 4 fewer tapes per week.

   a. Complete the table. The first two rows are filled in for you.

No. of price increases	Price of tape	No. of tapes sold	Total revenue
0	6	96	576
1	6.50	92	598
2			
3			
4			
5			
6			
7			
8			
12			
16			
20			
24			

   b. Let x stand for the number of $2 price increases the owner makes. Write algebraic expressions for the price of a room, the number of rooms that will be rented, and the total revenue earned at that price.

   c. Use your calculator to make a table of values for your algebraic expressions. Let Y_1 stand for the price of a room, Y_2 for the number of rooms rented, and Y_3 for the total revenue. Verify the values you calculated in part (a).

   d. Use your table to find a value of x that causes the total revenue to be zero.

   e. Use your graphing calculator to graph your formula for total revenue.

   f. What is the lowest price that the owner can charge for a room if she wants her revenue to exceed $1296 per night? What is the highest price she can charge to obtain this revenue?

   g. What is the maximum revenue the owner can earn in one night? How much should she charge for a room to maximize her revenue? How many rooms will she rent at that price?

   b. Let x stand for the number of $0.50 price increases the owner makes. Write algebraic expressions for the price of a tape, the number of tapes sold, and the total revenue.

   c. Use your calculator to make a table of values for your algebraic expressions. Let Y_1 stand for the price of a tape, Y_2 for the number of tapes sold, and Y_3 for the total revenue. Verify the values you calculated in part (a).

   d. Use your table to find a value of x for which the total revenue is zero.

   e. Use your graphing calculator to graph your formula for total revenue.

   f. How much should the owner charge for a tape in order to bring in $630 per week from tapes? (You should have two answers.)

   g. What is the maximum revenue the owner can earn from tapes in one week? How much should he charge for a tape to maximize his revenue? How many tapes will he sell at that price?

3. **a.** Give the dimensions of two different rectangles with perimeter 60 meters. Compute the areas of the two rectangles.
   b. A rectangle has a perimeter of 60 meters. If the length of the rectangle is x meters, write an expression for its width.
   c. Write an expression for the area of the rectangle.

4. **a.** Give the dimensions of two different rectangles with perimeter 48 inches. Compute the areas of the two rectangles.
   b. A rectangle has a perimeter of 48 inches. If the width of the rectangle is w inches, write an expression for its length.
   c. Write an expression for the area of the rectangle.

For Problems 5–8,
(a) Find the maximum or minimum value algebraically.
(b) Obtain a good graph on your calculator and verify your answer. (Use the coordinates of the vertex and the vertical intercept to help you choose an appropriate window for the graph.)

5. Delbert launches a toy water rocket from ground level. Its distance above the ground t seconds after launch is given, in feet, by $d = 96t - 16t^2$. When will the rocket reach its greatest height, and what will that height be?

6. Francine throws a wrench into the air from the bottom of a trench 12 feet deep. Its height t seconds later is given, in feet, by $h = -12 + 32t - 16t^2$. When will the wrench reach its greatest height, and what will that height be?

7. The owners of a small fruit orchard decide to produce gift baskets as a sideline. The cost per basket for producing x baskets is $C = 0.01x^2 - 2x + 120$. How many baskets should they produce in order to minimize the cost per basket? What will their total cost be at that production level?

8. A new electronics firm is considering marketing a line of telephones. The cost per phone for producing x telephones is $C = 0.001x^2 - 3x + 2270$. How many telephones should the firm produce in order to minimize the cost per phone? What will the firm's total cost be at that production level?

9. As part of a collage for her art class, Sheila wants to enclose a rectangle with 100 inches of yarn.
   a. Let w represent the width of the rectangle, and write an expression for its length. Then write an expression in terms of w for the area, A, of the rectangle.
   b. What is the area of the largest rectangle that Sheila can enclose with 100 inches of yarn?

10. Gavin has rented space for a booth at the county fair. As part of his display, he wants to rope off a rectangular area with 80 yards of rope.
    a. Let w represent the width of the roped-off rectangle, and write an expression for its length. Then write an expression in terms of w for the area, A, of the roped-off space.
    b. What is the largest area that Gavin can rope off? What will the dimensions of the rectangle be?

11. A farmer plans to fence a rectangular grazing area along a river with 300 yards of fence as shown in the figure.

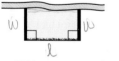

   a. Write an expression for the area A of the grazing land in terms of the width w of the rectangle.
   b. What is the largest area the farmer can enclose?

12. A breeder of horses wants to fence two rectangular grazing areas along a river with 600 meters of fence as shown in the figure.

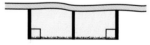

    a. Write an expression for the total area, A, of the grazing land in terms of the width, w, of the rectangles.
    b. What is the largest area the breeder can enclose?

13. A travel agent offers a group rate of $2400 per person for a week in London if 16 people sign up for the tour. For each additional person who signs up, the price per person is reduced by $100.

   a. Let x represent the number of additional people who sign up. Write expressions for the total number of people signed up, the price per person, and the total revenue.

   b. How many people must sign up for the tour in order for the travel agent to maximize her revenue?

14. An entrepreneur buys an apartment building with 40 units. The previous owner charged $240 per month for a single apartment and on the average rented 32 apartments at that price. The entrepreneur discovers that for every $20 he raises the price, another apartment stands vacant.

   a. Let x represent the number of $20 price increases. Write expressions for the new price, the number of rented apartments, and the total revenue.

   b. What price should the entrepreneur charge for an apartment in order to maximize his revenue?

15. During a statistical survey, a public interest group obtains two estimates for the average monthly income of young adults aged 18 to 25. The first estimate is $860 and the second estimate is $918. To refine its estimate, the group will take a weighted average of these two figures:

$$I = 860a + 918(1 - a) \quad \text{where } 0 \le a \le 1$$

To get the best estimate, the group must choose a to minimize the expression

$$V = 576a^2 + 5184(1 - a)^2$$

(The numbers that appear in this expression reflect the *variance* of the data, which measures how closely the data cluster around the *mean,* or average.) Find the value of a that minimizes V, and use this value to get a refined estimate for the average income.

16. The rate at which an antigen precipitates during an antigen-antibody reaction depends upon the amount of antigen present. For a fixed quantity of antibody, the time required for a particular antigen to precipitate is given in minutes by

$$t = 2w^2 - 20w + 54$$

where w is the quantity of antigen present, in grams. For what quantity of antigen will the reaction proceed most rapidly, and how long will the precipitation take?

Use transformations to graph each parabola. What is the vertex of each graph?

17. **a.** $y = (x - 3)^2$
    b. $y = -(x - 3)^2$
    c. $y = -(x - 3)^2 + 4$

18. **a.** $y = (x + 1)^2$
    b. $y = 2(x + 1)^2$
    c. $y = 2(x + 1)^2 - 4$

19. **a.** $y = (x + 4)^2$
    b. $y = \frac{1}{2}(x + 4)^2$
    c. $y = 3 + \frac{1}{2}(x + 4)^2$

20. **a.** $y = (x - 2)^2$
    b. $y = -(x - 2)^2$
    c. $y = -3 - (x - 2)^2$

In Problems 21–24,
(a) Find the vertex of the parabola.
(b) Use transformations to sketch the graph.
(c) Write the equation in standard form.

21. $y = 2(x - 3)^2 + 4$ **22.** $y = -3(x + 1)^2 - 2$ **23.** $y = -\frac{1}{2}(x + 4)^2 - 3$ **24.** $y = 4(x - 2)^2 - 6$

For Problems 25–30,
(a) Write each equation in the form $y = a(x - p)^2 + q$ by completing the square.
(b) Using horizontal and vertical translations, sketch the graph by hand.

25. $y = x^2 - 4x + 7$ **26.** $y = x^2 - 2x - 1$ **27.** $y = 3x^2 + 6x - 2$

28. $y = \frac{1}{2}x^2 + 2x + 5$ **29.** $y = -2x^2 - 8x + 3$ **30.** $y = -x^2 + 5x + 2$

31. A system of two quadratic equations may have no solution, one solution, or two solutions. Sketch a system illustrating each case. In your sketches, one of the parabolas should open up, and the other down.

32. A system of two quadratic equations may have no solution, one solution, or two solutions. Sketch a system illustrating each case. In your sketches, both parabolas should open up.

Solve each system algebraically. Use your calculator to graph both equations and verify your solutions.

33. $y = x^2 - 4x + 7$
$y = 11 - x$

34. $y = x^2 + 6x + 4$
$y = 3x + 8$

35. $y = -x^2 - 2x + 7$
$y = 2x + 11$

36. $y = x^2 - 8x + 17$
$y + 4x = 13$

37. $y = x^2 + 8x + 8$
$3y + 2x = -36$

38. $y = -x^2 + 4x + 2$
$4y - 3x = 24$

39. $y = x^2 - 9$
$y = -2x^2 + 9x + 21$

40. $y = 4 - x^2$
$y = 3x^2 - 12x - 12$

41. $y = x^2 - 0.5x + 3.5$
$y = -x^2 + 3.5x + 1.5$

42. $y = x^2 + 10x + 22$
$y = -0.5x^2 - 8x - 32$

43. $y = x^2 - 4x + 4$
$y = x^2 - 8x + 16$

44. $y = 0.5x^2 + 3x + 5.5$
$y = 2x^2 + 12x + 4$

Problems 45–48 deal with wildlife management and sustainable yield.

45. In Problem 9 of Section 6.3, you graphed the annual growth rate of a population of fish, $R = 0.4x - 0.0001x^2$, where x is the current biomass of the population, in tons.
  a. Suppose that fishermen harvest 300 tons of fish each year. Sketch the graph of $H = 300$ on the same axes with your graph of y.
  b. If the biomass is currently 2500 tons and 300 tons are harvested, will the population be larger or smaller next year? By how much? What if the biomass is currently 3500 tons?
  c. What sizes of biomass will remain stable from year to year if 300 tons are harvested annually?
  d. If the biomass ever falls below 1000 tons, what will happen after several years of harvesting 300 tons annually?

46. In Problem 10 of Section 6.3, you graphed the annual increase, I, in the deer population in a national park, $I = 1.2x - 0.0002x^2$, where x is the current population.
  a. Suppose hunters are allowed to kill 1000 deer per year. Sketch the graph of $H = 1000$ on the same axes with a graph of y.
  b. What sizes of deer populations will remain stable from year to year if 1000 deer are hunted annually?
  c. Suppose 1600 deer are killed annually. What sizes of deer populations will remain stable?
  d. What is the largest annual harvest that still allows for a stable population? (This harvest is called the maximum sustainable yield.) What is the stable population?
  e. What eventually happens if the population falls below the stable value but hunting continues at the maximum sustainable yield?

47. The annual increase, N, in a bear population of size x is given by

$$N = 0.0002x(2000 - x)$$

if the bears are not hunted. The number of bears killed each year by hunters is related to the bear population by the equation $K = 0.2x$. (Notice that in this model, hunting is adjusted to the size of the bear population.)

a. Sketch the graphs of N and K on the same axes.

b. When the bear population is 1200, which is greater, N or K? Will the population increase or decrease in the next year? By how many bears?

c. When the bear population is 900, will the population increase or decrease in the next year? By how many bears?

d. What sizes of bear population will remain stable after hunting?

e. What sizes of bear populations will increase despite hunting? What sizes of populations will decrease?

f. Toward what size will the population tend over time?

g. Suppose hunting limits are raised so that $K = 0.3x$. Toward what size will the population tend over time?

48. The annual increase in the biomass of a whale population is given in tons by

$$w = 0.001x(1000 - x)$$

where x is the current population, also in tons.

a. Sketch a graph of w for $0 \leq x \leq 1100$. What size biomass remains stable?

b. Each year hunters are allowed to harvest a biomass given by $H = 0.6x$. Sketch H on the same graph with w. What is the stable biomass with hunting?

c. What sizes of populations will increase despite hunting? What sizes will decrease?

d. What size will the population approach over time? What biomass are hunters allowed to harvest for that size population?

e. Find a value of k so that the graph of $H = kx$ will pass through the vertex of $w = 0.001x(1000 - x)$.

f. For the value of k found in part (e), what size will the population approach over time? What biomass are hunters allowed to harvest for that size population?

g. Explain why the whaling industry should prefer hunting quotas of kx rather than $0.6x$ for a long-term strategy, even though $0.6x > kx$ for any positive value of x.

For Problems 49–52,
(a) Find the break-even points by solving a system of equations.
(b) Graph the equations for Revenue and Cost in the same window and verify your solutions on the graph.
(c) Use the fact that

$$Profit = Revenue - Cost$$

to find the value of *x* for which profit is maximum.

49. Writewell, Inc. makes fountain pens. It costs Writewell $C = 0.0075x^2 + x + 2100$ dollars to manufacture x pens, and the company receives $R = 13x$ dollars in revenue from the sale of the pens.

50. It costs The Sweetshop $C = 0.01x^2 + 1836$ dollars to produce x pounds of chocolate creams. The company brings in $R = 12x$ dollars revenue from the sale of the chocolates.

51. It costs an appliance manufacturer $C = 1.625x^2 + 33{,}150$ dollars to produce x front-loading washing machines, which will then bring in revenues of $R = 650x$ dollars.

52. A company can produce x lawn mowers for a cost of $C = 0.125x^2 + 100{,}000$ dollars. The sale of the lawn mowers will generate $R = 300x$ dollars in revenue.

Problems 53 and 54 prove that the vertical line $x = \frac{-b}{2a}$ is the axis of symmetry of the graph of $y = ax^2 + bx + c$. A graph is *symmetric about the line $x = h$* if the point $(h + d, v)$ lies on the graph whenever the point $(h - d, v)$ lies on the graph.

53. a. Sketch a parabola $y = a(x - h)^2 + k$ and the line $x = h$. We will show that the parabola is symmetric about the line $x = h$.

   b. Label a point on the parabola with x-coordinate $x = h + d$, where $d > 0$. What is the y-coordinate of that point?

   c. Label the point on the parabola with x-coordinate $x = h - d$. What is the y-coordinate of that point?

   d. Explain why your answers to parts (b) and (c) prove that the line $x = h$ is the axis of symmetry for the graph of $y = a(x - h)^2 + k$.

54. To find the axis of symmetry for the graph of $y = ax^2 + bx + c$, we will use the results of Problem 51 and the technique of completing the square.

   a. Write the equation $y = ax^2 + bx + c$ in vertex form by completing the square. (Follow the steps in Example 3.)

   b. Your answer to part (a) has the form $y = a(x - h)^2 + k$. What is your value of h? What is your value of k?

   c. What is the axis of symmetry for the parabola $y = ax^2 + bx + c$?

6.5   Quadratic Inequalities

Solving Inequalities Graphically

In Chapter 1, we used graphs to solve equations and inequalities. The graphing technique is especially helpful for solving quadratic inequalities.

EXAMPLE 1

The Chamber of Commerce in River City plans to put on a Fourth of July fireworks display. City regulations require that fireworks at public gatherings explode higher than 800 feet above the ground. The mayor particularly wants to include the Freedom Starburst model, which is launched from the ground. Its height after t seconds is given by

$$h = 256t - 16t^2$$

When should the Starburst explode in order to satisfy the safety regulation?

Solution   We can get an approximate answer to this question by looking at the graph of the rocket's height, shown in Figure 6.28.

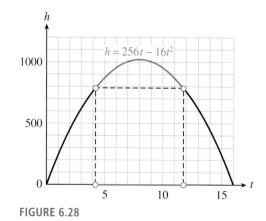

FIGURE 6.28

When is the rocket's height greater than 800 feet, or, in mathematical terms, for what values of t is $h > 800$? The answer to this question is the solution of the inequality

$$256t - 16t^2 > 800$$

Points on the graph with $h > 800$ are shown in color, and the t-coordinates of those points are marked on the horizontal axis. If the Freedom Starburst explodes at any of these times, it will satisfy the safety regulation. From the graph, the safe time interval runs from approximately 4.25 seconds to 11.75 seconds after launch. The solution of the inequality is the set of all t-values greater than 4.25 but less than 11.75

The solution set in Example 1 is called a **compound inequality,** because it involves more than one inequality symbol. We write this set as

$$4.25 < t < 11.75$$

and read "t greater than 4.25 but less than 11.75."

See Algebra Skills Refresher A.2 to review compound inequalities.

Solving an Inequality With a Graphing Calculator

You can use your graphing calculator to solve the problem in Example 1. Graph the two equations

$$Y_1 = 256X - 16X^2$$
$$Y_2 = 800$$

on the same screen. (See Figure 6.29.) Use ⌈ **WINDOW** ⌉ settings to match the graph in Figure 4.0. Then use the **intersect** feature to find the x-coordinates of the points where the two graphs intersect (there are two of them). These points will have y-coordinates of 800. The parabola is above the line, so $h > 800$ when t is between these two x-values. To two decimal places, you can see that $h > 800$ when $4.26 < t < 11.74$.

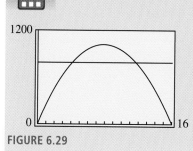

FIGURE 6.29

EXERCISE 1

a. Graph the function $y = x^2 - 2x - 9$ in the window

$$\text{Xmin} = -9.4 \qquad \text{Xmax} = 9.4$$
$$\text{Ymin} = -10 \qquad \text{Ymax} = 10$$

b. Use the graph to solve the inequality $x^2 - 2x - 9 \geq 6$.

In Example 1, we solved the inequality $256t - 16t^2 > 800$ by comparing points on the graph of $h = 256 - 16t^2$ with points on the line $h = 800$. If one side of an inequality is zero, we can compare points on the graph with the line $y = 0$, which is the x-axis.

EXAMPLE 2

Consider the graph of $y = x^2 - 4$. Find the solutions of the following equations and inequalities.

a. $x^2 - 4 = 0$ **b.** $x^2 - 4 < 0$ **c.** $x^2 - 4 > 0$

Solutions   Look at the graph of $y = x^2 - 4$ shown in Figure 6.30. When we substitute a value of x into the expression $x^2 - 4$, the result is either positive, negative, or zero. (You can see this more clearly if you compute a few values yourself to complete the table below. Your table should agree with the coordinates of points on the graph.)

x	-3	-2	-1	0	1	2	3
y							

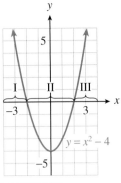

FIGURE 6.30

We will use the graph to solve the given equation and inequalities.

a. First, locate the two points on the graph where $y = 0$. These points are $(-2, 0)$ and $(2, 0)$. Their x-coordinates, -2 and 2, are the solutions of the equation $x^2 - 4 = 0$. The two points divide the x-axis into three sections, which are labeled on the graph. On each of these sections, the value of $x^2 - 4$ is either always positive or always negative.

b. To solve $x^2 - 4 < 0$, find the points on the graph where $y < 0$, that is, below the x-axis. These points have x-coordinates between -2 and 2 (labeled section II on the figure). Thus, the solution to the inequality $x^2 - 4 < 0$ is $-2 < x < 2$.

c. To solve $x^2 - 4 > 0$, locate the points on the graph where $y > 0$, or above the x-axis. Points with positive y-values correspond to two sections of the x-axis, labeled I and III on the figure. In section I, $x < -2$, and in section III, $x > 2$. Thus, the solution to the inequality $x^2 - 4 > 0$ includes all values of x for which either $x < -2$ or $x > 2$.

CAUTION

In Example 2c, the solution of the inequality $x^2 - 4 > 0$ is the set

$$x < -2 \quad \text{or} \quad x > 2.$$

This set is another type of compound inequality, and its graph consists of two pieces, as shown in Figure 6.31.

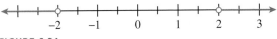

FIGURE 6.31

The left piece of the set is $x < -2$, and the right piece is $x > 2$. It would be incorrect to describe the solution set as $-2 > x > 2$, because this notation implies that $-2 > 2$. We must write the solution as two parts: $x < -2$ or $x > 2$.

EXERCISE 2

Use a graph of $y = 6 - x^2$ to solve the inequalities.

a. $6 - x^2 > 0$ **b.** $6 - x^2 \leq 0$

Because it is relatively easy to decide whether the y-coordinate of a point on a graph is positive or negative (the point lies above the x-axis or below the x-axis), we often rewrite a given inequality so that one side is zero.

EXAMPLE 3

Use a graph to solve $x^2 - 2x - 3 \leq 12$.

Solution We first write the inequality with zero on one side: $x^2 - 2x - 15 \leq 0$. We would like to find points on the graph of $y = x^2 - 2x - 15$ that have y-coordinates less than or equal to zero. A graph of the equation is shown in Figure 6.32. You can check that the x-intercepts of the graph are -3 and 5. The points shown in red on the graph lie below the x-axis and have $y \leq 0$, so the x-coordinates of these points are the solutions of the inequality. All of these points have x-coordinates between -3 and 5. Thus, the solution is $-3 \leq x \leq 5$, or in interval notation, $[-3, 5]$.

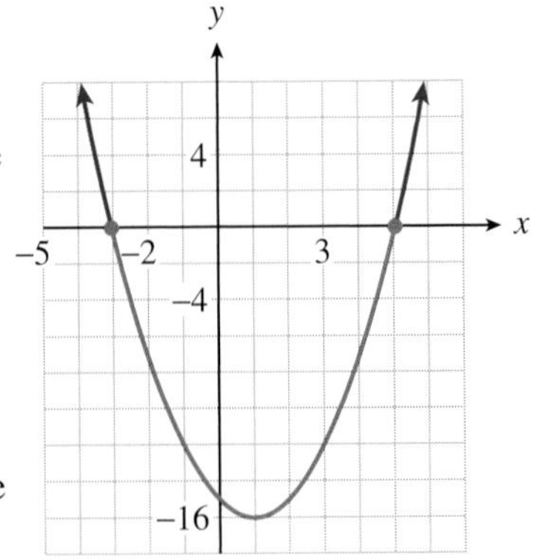

See Algebra Skills Refresher A.2 to review interval notation.

FIGURE 6.32

Follow the steps below to solve the inequality $36 + 6x - x^2 \leq 20$.

a. Rewrite the inequality so that the right side is zero.

b. Graph the equation $y = 16 + 6x - x^2$.

c. Locate the points on the graph with y-coordinate less than zero, and mark the x-coordinates of the points on the x-axis.

d. Write the solution with interval notation.

Solving Quadratic Inequalities Algebraically

Although a graph is very helpful in solving inequalities, it is not completely necessary. Every quadratic inequality can be put into one of the forms

$$ax^2 + bx + c < 0, \qquad ax^2 + bx + c > 0$$
$$ax^2 + bx + c \leq 0, \qquad ax^2 + bx + c \geq 0$$

All we really need to know is whether the corresponding parabola $y = ax^2 + bx + c$ opens upward or downward. Consider the parabolas shown in Figure 6.33.

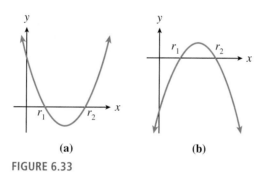

(a)   **(b)**

FIGURE 6.33

The parabola in Figure 6.33a opens upward. It crosses the x-axis at two points, $x = r_1$ and $x = r_2$. At these points, $y = 0$. The graph lies below the x-axis between r_1 and r_2, so the solutions to the inequality $y < 0$ lie between r_1 and r_2. The graph lies above the x-axis for x-values less than r_1 or greater than r_2, so the solutions to the inequality $y > 0$ are $x < r_1$ or $x > r_2$.

If the parabola opens downward, as in Figure 6.33b, the situation is reversed. The solutions to the inequality $y > 0$ lie between the x-intercepts, and the solutions to $y < 0$ lie outside the x-intercepts.

From the graphs in Figure 6.33, we see that the x-intercepts are the boundary points between the portions of the graph with positive y-coordinates and the portions with negative y-coordinates. To solve a quadratic inequality, we need only locate the x-intercepts of the corresponding graph and then decide which intervals of the x-axis produce the correct sign for y.

To Solve a Quadratic Inequality Algebraically:

1. Write the inequality in standard form: One side is 0, and the other has the form $ax^2 + bx + c$.

2. Find the x-intercepts of the graph of $y = ax^2 + bx + c$ by setting $y = 0$ and solving for x.

3. Make a rough sketch of the graph, using the sign of a to determine whether the parabola opens upward or downward.

4. Decide which intervals on the x-axis give the correct sign for y.

EXAMPLE 4   Solve the inequality $36 + 6x - x^2 \le 20$ algebraically.

Solution

Step 1   We subtract 20 from both sides of the inequality so that we have 0 on the right side.

$$16 + 6x - x^2 \le 0$$

Step 2   Consider the equation $y = 16 + 6x - x^2$. To locate the x-intercepts, we set $y = 0$ and solve for x.

$$\begin{aligned} 16 + 6x - x^2 &= 0 &&\text{Multiply each term by } -1. \\ x^2 - 6x - 16 &= 0 &&\text{Factor the left side.} \\ (x - 8)(x + 2) &= 0 &&\text{Apply the zero-factor principle.} \end{aligned}$$

$$\begin{aligned} x - 8 &= 0 \quad \text{or} \quad x + 2 = 0 \\ x &= 8 \quad \text{or} \qquad x = -2 \end{aligned}$$

The x-intercepts are $x = -2$ and $x = 8$.

Step 3   Make a rough sketch of the graph of $y = 16 + 6x - x^2$, as shown in Figure 6.34. Because $a = -1 < 0$, the graph is a parabola that opens downward.

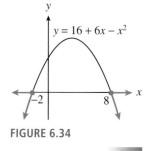

FIGURE 6.34

Step 4   We are interested in points on the graph for which $y \le 0$. The points with negative y-coordinates (that is, points below the x-axis) lie outside the x-intercepts of the graph, so the solution of the inequality is $x \le -2$ or $x \ge 8$. Or, using interval notation, the solution is $(-\infty, -2] \cup [8, \infty)$.

CAUTION   Many people think that the inequality signs in the solution should point in the same direction as the sign in the original problem, and hence would incorrectly write the solution to Example 1 as $x \le -2$ or $x \le 8$. However, you can see from the graph that this is incorrect. Remember that the graph of a quadratic equation is a parabola, not a straight line!

If we cannot find the x-intercepts of the graph by factoring or extraction of roots, we can use the quadratic formula.

EXAMPLE 5   TrailGear, Inc. manufactures camping equipment. The company finds that the profit from producing and selling x alpine parkas per month is given, in dollars, by

$$P = -0.8x^2 + 320x - 25,200$$

How many parkas should the company produce and sell each month if it must keep the profits above $2000?

Solution

Step 1   We would like to solve the inequality

$$-0.8x^2 + 320x - 25,200 > 2000$$

or, subtracting 2000 from both sides,

$$-0.8x^2 + 320x - 27,200 > 0$$

Step 2   Consider the equation

$$y = -0.8x^2 + 320x - 27,200$$

We locate the x-intercepts of the graph by setting $y = 0$ and solving for x. We will use the quadratic formula to solve the equation

$$-0.8x^2 + 320x - 27,200 = 0$$

so $a = -0.8$, $b = 320$, and $c = -27,200$.

$$
\begin{aligned}
x &= \frac{-(320) \pm \sqrt{(320)^2 - 4(-0.8)(-27,200)}}{2(-0.8)} \\
&= \frac{-320 \pm \sqrt{102,400 - 87,040}}{-1.6} \\
&= \frac{-320 \pm \sqrt{15,360}}{-1.6}
\end{aligned}
$$

To two decimal places, the solutions to the equation are 122.54 and 277.46.

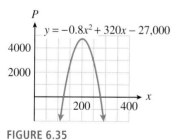

P

$y = -0.8x^2 + 320x - 27,000$

4000

2000

200   400

x

FIGURE 6.35

Step 3   The graph of the equation is a parabola that opens downward, because the coefficient of x^2 is negative. (See Figure 6.35.)

Step 4   The graph lies above the x-axis, and hence $y > 0$, for x-values between the two x-intercepts, that is, for $122.54 < x < 277.46$. Because we cannot produce a fraction of a parka, we restrict the interval to the closest whole number x-values included, namely 123 and 277. Thus, TrailGear can produce as few as 123 parkas or as many as 277 parkas per month to keep its profit above $2000.

Solve the inequality $10 - 8x + x^2 > 4$.

Step 1 Write the inequality in standard form.

Step 2 Find the x-intercepts of the corresponding graph. Use the quadratic formula.

Step 3 Make a rough sketch of the graph.

Step 4 Decide which intervals on the x-axis give the correct sign for y. Write your answer with interval notation.

ANSWERS TO 6.5 EXERCISES

1. $x \le -3$ or $x \ge 5$

3. a. $16 + 6x - x^2 \le 0$

   b., c.

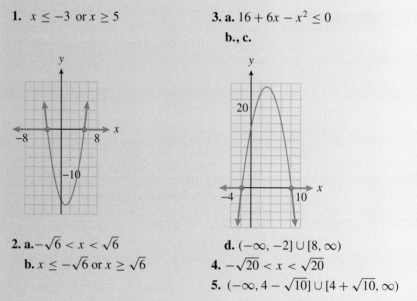

2. a. $-\sqrt{6} < x < \sqrt{6}$

   b. $x \le -\sqrt{6}$ or $x \ge \sqrt{6}$

d. $(-\infty, -2] \cup [8, \infty)$

4. $-\sqrt{20} < x < \sqrt{20}$

5. $(-\infty, 4 - \sqrt{10}] \cup [4 + \sqrt{10}, \infty)$

SECTION 6.5 SUMMARY

VOCABULARY *Look up the definitions of new terms in the Glossary.*

Compound inequality   Interval notation

CONCEPTS

1. We can use a graphical technique to solve quadratic inequalities.

2.

> **To Solve a Quadratic Inequality Algebraically:**
>
> **1.** Write the inequality in standard form: One side is 0, and the other has the form $ax^2 + bx + c$.
>
> **2.** Find the x-intercepts of the graph of $y = ax^2 + bx + c$ by setting $y = 0$ and solving for x.
>
> **3.** Make a rough sketch of the graph, using the sign of a to determine whether the parabola opens upward or downward.
>
> **4.** Decide which intervals on the x-axis give the correct sign for y.

1. If $ax^2 + bx > c$ for a particular value of x, what can you say about the graph of $y = ax^2 + bx - c$ at that x-value?
2. What are the only x-values at which the graph of $y = ax^2 + bx + c$ can change sign?
3. Explain the difference between an open interval and a closed interval.
4. Explain what is wrong with the following "solution" to a quadratic inequality: $2 > x > 8$.
5. The parabola $y = x^2 + bx + c$ has x-intercepts at r_1 and r_2, with $r_1 < r_2$. What are the solutions of the inequality $x^2 + bx + c > 0$?

6. The parabola $y = x^2 + bx + c$ has x-intercepts at r_1 and r_2, with $r_1 < r_2$. What are the solutions of the inequality $x^2 + bx + c \leq 0$?
7. The parabola $y = -x^2 + bx + c$ has x-intercepts at r_1 and r_2, with $r_1 < r_2$. What are the solutions of the inequality $-x^2 + bx + c \leq 0$?
8. The parabola $y = -x^2 + bx + c$ has x-intercepts at r_1 and r_2, with $r_1 < r_2$. What are the solutions of the inequality $-x^2 + bx + c > 0$?

SKILLS *Practice each skill in the Homework problems listed.*

1. Solve a quadratic inequality graphically: #1–30
2. Solve a quadratic inequality algebraically: #31–50
3. Solve problems involving quadratic inequalities: #51–60

HOMEWORK 6.5

1. **a.** Graph the equation $y = x^2$ by hand.
   b. Darken the portion of the x-axis for which $y > 9$.
   c. Solve the inequality $x^2 > 9$. Explain why $x > 3$ is incorrect as an answer.

2. **a.** Graph the equation $y = x^2$ by hand on graph paper.
   b. Darken the portion of the x-axis for which $y < 1$.
   c. Solve the inequality $x^2 < 1$. Explain why $x < 1$ is incorrect as an answer.

3. **a.** Graph the equation $y = x^2 - 2x - 3$ by hand on graph paper.
   b. Darken the portion of the x-axis for which $y > 0$.
   c. Solve the inequality $x^2 - 2x - 3 > 0$.

4. **a.** Graph the equation $y = x^2 + 2x - 8$ by hand on graph paper.
   b. Darken the portion of the x-axis for which $y < 0$.
   c. Solve the inequality $x^2 + 2x - 8 < 0$.

Use the graphs provided to estimate the solutions to each equation and inequality.

5. **a.** $x^2 - 3x - 180 = 0$
   b. $x^2 - 3x - 180 > 0$

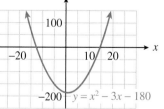

6. **a.** $175 - 18x - x^2 = 0$
   b. $175 - 18x - x^2 < 0$

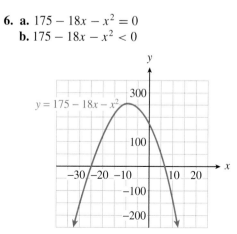

7. a. $-6x^2 + 4.8x - 0.9 = 0$
  b. $-6x^2 + 4.8x - 0.9 \geq 0$

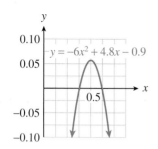

8. a. $5x^2 + 7.5x + 1.8 = 0$
  b. $5x^2 + 7.5x + 1.8 \leq 0$

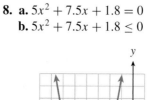
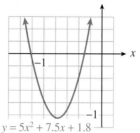

Graph the parabola in the window

$$\text{Xmin} = -9.4, \quad \text{Xmax} = 9.4, \quad \text{Ymax} = -25, \quad \text{Ymax} = 25$$

Use the graph to solve the inequalities. Write your answers in interval notation.

9. $y = x^2 - 3x - 18$
(For parts (c) and (d), it may be helpful to graph $Y_2 = -8$ as well.)
a. $x^2 - 3x - 18 > 0$ **b.** $x^2 - 3x - 18 < 0$
c. $x^2 - 3x - 18 \leq -8$ **d.** $x^2 - 3x - 18 \geq -8$

10. $y = 16 - 6x - x^2$
(For parts (c) and (d), it may be helpful to graph $Y_2 = 21$ as well.)
a. $16 - 6x - x^2 \geq 0$ **b.** $16 - 6x - x^2 \leq 0$
c. $16 - 6x - x^2 < 21$ **d.** $16 - 6x - x^2 > 21$

11. $y = 16 - x^2$
(For parts (c) and (d), it may be helpful to graph $Y_2 = 7$ as well.)
a. $16 - x^2 > 0$ **b.** $16 - x^2 < 0$
c. $16 - x^2 \leq 7$ **d.** $16 - x^2 \geq 7$

12. $y = x^2 - 9$
(For parts (c) and (d), it may be helpful to graph $Y_2 = 16$ as well.)
a. $x^2 - 9 \geq 0$ **b.** $x^2 - 9 \leq 0$
c. $x^2 - 9 > 16$ **d.** $x^2 - 9 < 16$

Solve each inequality by graphing. Use the following window settings:

$$\text{Xmin} = -9.4, \quad \text{Xmax} = 9.4, \quad \text{Ymin} = -15, \quad \text{Ymax} = 15$$

13. $(x - 3)(x + 2) > 0$

14. $(x + 1)(x - 5) > 0$

15. $k(4 - k) \geq 0$

16. $-m(7 + m) \geq 0$

17. $6 + 5p - p^2 < 0$

18. $q^2 + 9q + 18 < 0$

Solve each inequality by graphing. Use the following window settings:

$$\text{Xmin} = -9.4, \quad \text{Xmax} = 9.4, \quad \text{Ymin} = -62, \quad \text{Ymax} = 62$$

19. $x^2 - 1.4x - 20 < 9.76$

20. $-x^2 + 3.2x + 20 > 6.56$

21. $5x^2 + 39x + 27 \geq 5.4$

22. $-6x^2 - 36x - 20 \leq 25.36$

23. $-8x^2 + 112x - 360 < 6.08$

24. $10x^2 + 96x + 180 > 17.2$

Solve each inequality by graphing. Choose a suitable window for each problem. Use the **intersect** feature to estimate your solutions accurate to one decimal place.

25. $x^2 > 12.2$

26. $x^2 \le 45$

27. $-3x^2 + 7x - 25 \le 0$

28. $2.4x^2 - 5.6x + 18 \le 0$

29. $0.4x^2 - 54x < 620$

30. $-0.05x^2 - 3x > 76$

Solve each inequality algebraically. Write your answers in interval notation, rounding to two decimal places if necessary.

31. $(x+3)(x-4) < 0$

32. $(x+2)(x+5) > 0$

33. $28 - 3x - x^2 \ge 0$

34. $32 + 4x - x^2 \le 0$

35. $2z^2 - 7z > 4$

36. $6h^2 + 13h < 15$

37. $64 - t^2 > 0$

38. $121 - y^2 < 0$

39. $v^2 < 5$

40. $t^2 \ge 7$

41. $5a^2 - 32a + 12 \ge 0$

42. $6b^2 + 16b - 9 < 0$

43. $4x^2 + x \ge -2x^2 + 2$

44. $2x^2 + 8x \le -x^2 + 3$

45. $x^2 - 4x + 1 \ge 0$

46. $x^2 + 4x + 2 \le 0$

47. $-3 - m^2 < 0$

48. $11 + n^2 < 0$

49. $w^2 - w + 4 \le 0$

50. $-z^2 + z - 1 \le 0$

In Problems 51–58,
(a) Solve each problem by writing and solving an inequality.
(b) Graph the equation and verify your solution on the graph.

51. A fireworks rocket is fired from ground level. Its height in feet t seconds after launch is given by $h = 320t - 16t^2$. During what time interval is the rocket higher than 1024 feet?

52. A baseball thrown vertically reaches a height, h, in feet given by $h = 56t - 16t^2$, where t is measured in seconds. During what time intervals is the ball between 40 and 48 feet high?

53. The cost, in dollars, of manufacturing x pairs of garden shears is given by

$$C = -0.02x^2 + 14x + 1600$$

for $0 \le x \le 700$. How many pairs of shears can be produced if the total cost must be kept under \$2800?

54. The cost, in dollars, of producing x cashmere sweaters is given by

$$C = x^2 + 4x + 90$$

How many sweaters can be produced if the total cost must be kept under \$1850?

55. The Locker Room finds that it sells $1200 - 30p$ sweatshirts each month when it charges p dollars per sweatshirt. It would like its revenue from sweatshirts to be over \$9000 per month. In what range should it keep the price of a sweatshirt?

56. Green Valley Nursery sells $120 - 10p$ boxes of rose food per month at a price of p dollars per box. It would like to keep its monthly revenue from rose food over \$350. In what range should it price a box of rose food?

57. A group of cylindrical storage tanks must be 20 feet tall. If the volume of each tank must be between 500π and 2880π cubic feet, what are the possible values for the radius of a tank?

58. The volume of a cylindrical can should be between 21.2 and 21.6 cubic inches. If the height of the can is 5 inches, what values for the radius (to the nearest hundredth of an inch) will produce an acceptable can?

59. A travel agency offers a group rate of $600 per person for a weekend in Lake Tahoe if 20 people sign up. For each additional person who signs up, the price for all participants is reduced by $10 per person.

   a. Write algebraic expressions for the size of the group and the price per person if x additional people sign up.

   b. Write a formula for the travel agency's total income as a function of x.

   c. What is the maximum income the travel agency can earn on the Lake Tahoe weekend? How many people should the agency enroll to achieve this income?

   d. How many people must sign up in order for the agency to bring in at least $15,750?

   e. Graph the income function and use the graph to verify your answers to parts (c) and (d).

60. A farmer inherits an apple orchard on which 60 trees are planted per acre. Each tree yields 12 bushels of apples. Experimentation has shown that for each tree removed per acre, the yield per tree increases by $\frac{1}{2}$ bushel.

   a. Write algebraic expressions for the number of trees per acre and for the yield per tree if x trees per acre are removed.

   b. Write a formula for the total yield per acre as a function of x.

   c. What is the maximum yield per acre that can be achieved by removing trees? How many trees per acre should be removed to achieve this yield?

   d. How many trees should be removed per acre in order to harvest at least 850 bushels per acre?

   e. Graph the yield function and use the graph to verify your answers to parts (c) and (d).

6.6   Curve Fitting

In Section 1.6, we used linear regression to fit a line to a collection of data points. If the data points do not cluster around a line, it does not make sense to describe them by a linear function. Compare the scatterplots in Figure 6.36. The points in Figure 6.36a are roughly linear in appearance, but the points in Figure 6.36b are not. However, we can visualize a parabola that would approximate the data we will. In this section, we will see how to fit a quadratic function to a collection of data points.

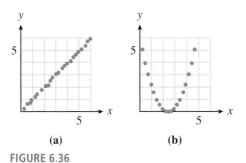

(a)   **(b)**

FIGURE 6.36

▶▶▶

See Algebra Skills Refresher A.5 if you would like to review the elimination method for solving linear systems.

We will need to solve a special type of 3×3 linear system, that is, a linear system of three equations in three variables. We can solve these systems using the **elimination** method.

EXAMPLE 1   Use elimination to solve the system of equations.

$$3a + 2b + c = -1 \qquad (1)$$
$$a - 2b + c = -3 \qquad (2)$$
$$2a + 3b + c = 4 \qquad (3)$$

Solution We first eliminate c from the system by combining the equations in pairs. We can add -1 times Equation (2) to Equation (1) to get a new equation in two variables:

$$
\begin{array}{ll}
3a + 2b + c = -1 & (1) \\
\underline{-a + 2b - c = 3} & (2) \\
2a + 4b \phantom{{}+ c} = 2 & (4)
\end{array}
$$

Next, we add -1 times Equation (2) to Equation (3) to get a second equation in two variables:

$$
\begin{array}{ll}
2a + 3b + c = 4 & (3) \\
\underline{-a + 2b - c = 3} & (2) \\
a + 5b \phantom{{}+ c} = 7 & (5)
\end{array}
$$

By combining Equations (4) and (5), we have a 2×2 linear system, which we can solve as usual.

$$
\begin{array}{ll}
2a + 4b = 2 & (4) \\
a + 5b = 7 & (5)
\end{array}
$$

To eliminate a, we add -2 times Equation (5) to Equation (4):

$$
\begin{array}{l}
2a + 4b = 2 \\
\underline{-2a - 10b = -14} \\
{-6b} = -12
\end{array}
$$

Solving this last equation gives us $b = 2$. Then we substitute $b = 2$ into either of Equations (4) or (5) to find $a = -3$. Finally, we substitute both values into one of the three original equations to find $c = 4$. The solution of the system is $a = -3, b = 2, c = 4$.

EXERCISE 1

Follow the steps to solve the system

$$
\begin{array}{ll}
a + b + c = 3 & (1) \\
4a - b + c = -4 & (2) \\
-3a + 2b + c = 4 & (3)
\end{array}
$$

Step 1 Eliminate c from Equations (1) and (2) to obtain a new Equation (4).

Step 2 Eliminate c from Equations (2) and (3) to obtain a new Equation (5).

Step 3 Solve the system of Equations (4) and (5).

Step 4 Substitute the values of a and b into one of the original equations to find c.

Finding a Quadratic Function through Three Points

Every linear function can be written in the form

$$y = mx + b$$

To find a specific line, we must find values for the two parameters (constants) m and b. We need two data points in order to find those two parameters. A quadratic function, however, has three parameters, a, b, and c:

$$y = ax^2 + bx + c$$

To find these parameters, we need three data points. We then use the method of elimination to solve a system of three linear equations.

EXAMPLE 2   Find values for a, b, and c so that the points $(1, 3)$, $(3, 5)$, and $(4, 9)$ lie on the graph of $y = ax^2 + bx + c$.

Solution   Substitute the coordinates of each of the three points into the equation of the parabola to obtain three equations:

$$3 = a(1)^2 + b(1) + c$$
$$5 = a(3)^2 + b(3) + c$$
$$9 = a(4)^2 + b(4) + c$$

or, equivalently,

$$a + b + c = 3 \qquad (1)$$
$$9a + 3b + c = 5 \qquad (2)$$
$$16a + 4b + c = 9 \qquad (3)$$

This is a system of three equations in the three unknowns a, b, and c. To solve the system, we first eliminate c. Add -1 times Equation (1) to Equation (2) to obtain

$$8a + 2b = 2 \qquad (4)$$

and add -1 times Equation (1) to Equation (3) to get

$$15a + 3b = 6 \qquad (5)$$

We now have a system of two linear equations in two variables:

$$8a + 2b = 2 \quad (4)$$
$$15a + 3b = 6 \quad (5)$$

We eliminate b from Equations (4) and (5): Add -3 times Equation (4) to 2 times Equation (5) to get

$$
\begin{array}{ll}
-24a - 6b = -6 & \quad -3 \text{ times Equation (4)} \\
\underline{30a + 6b = 12} & \quad 2 \text{ times Equation (5)} \\
6a = 6 &
\end{array}
$$

or $a = 1$. Substitute 1 for a in Equation (4) to find

$$8(1) + 2b = 2 \qquad \text{Solve for } b.$$
$$b = -3$$

Finally, substitute -3 for b and 1 for a in Equation (1) to find

$$1 + (-3) + c = 3 \quad \text{Solve for } c.$$
$$c = 5$$

Thus, the equation of the parabola is

$$y = x^2 - 3x + 5$$

The parabola and the three points are shown in Figure 6.37.

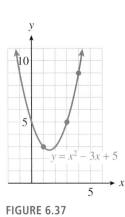

FIGURE 6.37

a. Find the equation of a parabola

$$y = ax^2 + bx + c$$

that passes through the points (0, 80), (15, 95), and (25, 55).

b. Plot the data points and sketch the parabola.

The simplest way to fit a parabola to a set of data points is to pick three of the points and find the equation of the parabola that passes through those three points.

EXAMPLE 3

Major Motors Corporation is testing a new car designed for in-town driving. The data below show the cost of driving the car at different speeds. The speeds, v, are given in miles per hour, and the cost, C, includes fuel and maintenance for driving the car 100 miles at that speed.

v	30	40	50	60	70
C	6.50	6.00	6.20	7.80	10.60

Find a possible quadratic model for C as a function of v, $C = av^2 + bv + c$.

Solution When we plot the data, it is clear that the relationship between v and C is not linear, but it may be quadratic (see Figure 6.38). We will use the last three data points, (50, 6.20), (60, 7.80), and (70, 10.60), to fit a parabola to the data. We would like to find the coefficients a, b, and c of a parabola $C = av^2 + bv + c$ that includes the three data points. This gives us a system of equations:

$$2500a + 50b + c = 6.20 \quad (1)$$
$$3600a + 60b + c = 7.80 \quad (2)$$
$$4900a + 70b + c = 10.60 \quad (3)$$

Eliminating c from Equations (1) and (2) yields Equation (4), and eliminating c from Equations (2) and (3) yields Equation (5).

$$1100a + 10b = 1.60 \quad (4)$$
$$1300a + 10b = 2.80 \quad (5)$$

Eliminating b from Equations (4) and (5) gives us

$$200a = 1.20$$
$$a = 0.006$$

Substitute this value into Equation (4) to find $b = -0.5$, then substitute both values into Equation (1) to find $c = 16.2$. Thus, our quadratic model is

$$C = 0.006v^2 - 0.5v + 16.2$$

The graph of this function, along with the data points, is shown in Figure 6.39.

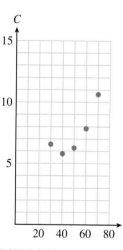

FIGURE 6.38

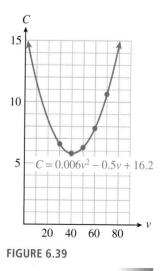

$C = 0.006v^2 - 0.5v + 16.2$

FIGURE 6.39

Sara plans to start a side business selling eggs. She finds that the total number of eggs produced each day depends on the number of hens confined in the henhouse, as shown in the table. Use the first three data points to find a quadratic model $E = an^2 + bn + c$. Plot the data and sketch the curve on the same axes.

Number of hens, n	15	20	25	30	36	39
Number of eggs, E	15	18	20	21	21	20

Finding an Equation in Vertex Form

It is easier to find a quadratic model if one of the points we know happens to be the vertex of the parabola. In that case, we need only one other point, and we can use the vertex form to find its equation.

EXAMPLE 4

When Andre practices free-throws at the park, the ball leaves his hands at a height of 7 feet and reaches the vertex of its trajectory 10 feet away at a height of 11 feet, as shown in Figure 6.40.

a. Find a quadratic equation for the ball's trajectory.

b. Do you think Andre's free-throw will score on a basketball court where the hoop is 15 feet from the shooter and 10 feet high?

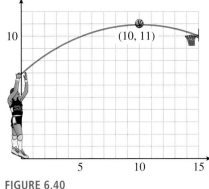

FIGURE 6.40

Solutions **a.** If Andre's feet are at the origin, then the vertex of the ball's trajectory is the point (10, 11), and its y-intercept is (0, 7). Start with the vertex form for a parabola:

$$y = a(x - x_v)^2 + y_v$$
$$y = a(x - 10)^2 + 11$$

We still need to know the value of a. We can substitute the coordinates of any point on the parabola for x and y and solve for a. We will use the point (0, 7):

$$7 = a(0 - 10)^2 + 11$$
$$7 = 100a + 11$$
$$a = -0.04$$

The equation of the trajectory is $y = -0.04(x - 10)^2 + 11$.

b. We would like to know if the point (15, 10) is on the trajectory of Andre's free-throw. Substitute $x = 15$ into the equation:

$$y = -0.04(15 - 10)^2 + 11$$
$$= -0.04(25) + 11 = 10$$

Andre's shot will score.

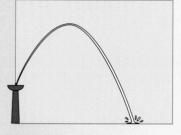

Using a Calculator for Quadratic Regression

We can use a graphing calculator to find an approximate quadratic fit for a set of data. The procedure is similar to the steps for linear regression outlined in Section 1.6.

EXAMPLE 5

a. Use your calculator to find a quadratic fit for the data in Example 3.

b. How many of the given data points actually lie on the graph of the quadratic approximation?

Solution

a. We press $\boxed{\text{STAT}}$ $\boxed{\text{ENTER}}$ and enter the data under columns $\mathbf{L_1}$ and $\mathbf{L_2}$, as shown in Figure 6.42a. Next, we calculate the quadratic regression equation and store it in $\mathbf{Y_1}$ by pressing $\boxed{\text{STAT}}$ $\boxed{\triangleright}$ $\boxed{5}$ $\boxed{\text{VARS}}$ $\boxed{\triangleright}$ $\boxed{1}$ $\boxed{1}$ $\boxed{\text{ENTER}}$. The regression equation has the form $y = ax^2 + bx + c$, where $a = 0.0057$, $b = -0.47$, and $c = 15.56$. Notice that a, b, and c are all close to the values we computed in Example 3.

b. Next, we will graph the data and the regression equation. We press $\boxed{\text{Y=}}$ and select **Plot1**, then press $\boxed{\text{ZOOM}}$ $\boxed{9}$ to see the graph shown in Figure 6.43a. The parabola seems to pass close to all the data points. However, we can use either the **value** feature or a table to find the y-coordinates of points on the regression curve. By comparing these y-coordinates with our original data points, we find that none of the given data points lies precisely on the parabola. (See Figure 6.43b.)

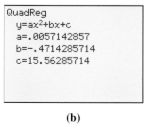

(a)

(b)

FIGURE 6.42

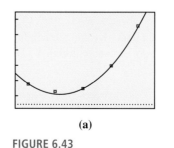

(a)

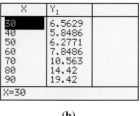

(b)

FIGURE 6.43

To test the effects of radiation, a researcher irradiated male mice with various dosages and bred them with unexposed female mice. The table below shows the fraction of fertilized eggs that survived, as a function of the radiation dosage.

Radiation (rems)	100	300	500	700	900	1100	1500
Relative survival of eggs	0.94	0.700	0.544	0.424	0.366	0.277	0.195

(Source: Strickberger, Monroe W., 1976)

a. Enter the data into your calculator and create a scatterplot. Does the graph appear to be linear? Does it appear to be quadratic?

b. Fit a quadratic regression equation to the data and graph the equation on the scatterplot.

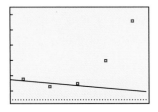

FIGURE 6.44

We must be careful that our data set gives a complete picture of the situation we want to model. A regression equation may fit a particular collection of data and still be a poor model if the rest of the data diverge from the regression graph. In Example 3, suppose Major Motors had collected only the first three data points and fit a line through them, as shown in Figure 6.44. This regression line gives poor predictions for the cost of driving at 60 or 70 miles per hour.

EXAMPLE 6

Francine records the height of the tip of the minute hand on the classroom's clock at different times. The data are shown in the table, where time is measured in minutes since noon. (A negative time indicates a number of minutes before noon.) Find a quadratic regression equation for the data and use it to predict the height of the minute hand's tip at 40 minutes past noon. Do you believe this prediction is valid?

Time (minutes)	−25	−20	−15	−10	−5	0	5	10	15	20	25
Height (feet)	7.13	7.50	8.00	8.50	8.87	9.00	8.87	8.50	8.00	7.50	7.13

Solution

We enter the time data under L_1 and the height data under L_2. Then we calculate and store the quadratic regression equation in Y_1, as we did in Example 5. The regression equation is

$$y = -0.00297x^2 + 0x + 8.834$$

From either the graph of the regression equation or from the table (see Figure 6.45), we can see that the fit is not perfect, although the curve certainly fits the data better than any straight line could.

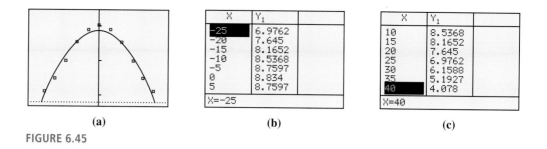

(a)   **(b)**   **(c)**

FIGURE 6.45

If we scroll down the table, we find that this equation predicts a height of approximately 4.08 feet at time 40 minutes. (See Figure 6.45c.) This is a preposterous estimate! The position of the minute hand at 40 minutes after noon should be the same as it was exactly one hour earlier (at 20 minutes before noon), when it was 7.50 feet.

Using the wrong type of equation to fit the data is a common error in making predictions. We know that the minute hand of a clock repeats its position every 60 minutes. The graph of the height of its tip oscillates up and down, repeating the same pattern over and over. We cannot describe such a graph using either a linear or a quadratic equation. The graph of the height is shown in Figure 6.46, along with the graph of our quadratic regression equation. You can see that the regression equation fits the actual curve only on a small interval.

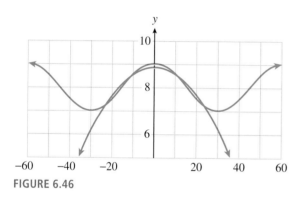

FIGURE 6.46

Your calculator can always compute a regression equation, but that equation is not necessarily appropriate for your data. Choosing a reasonable type of regression equation for a particular data set requires knowledge of different kinds of models and the physical or natural laws that govern the situation at hand.

EXERCISE 6

A speeding motorist slams on the brakes when she sees an accident directly ahead of her. The distance she has traveled t seconds after braking is shown in the table.

Time (seconds)	0	0.5	1.0	1.5	2.0	2.5
Distance (feet)	0	51	95	131	160	181

a. Enter the data into your calculator and create a scatterplot. Fit a quadratic regression equation to the data and graph the equation on the scatterplot.

b. Use your regression equation to find the vertex of the parabola. What do the coordinates represent in terms of the problem?

ANSWERS TO 6.6 EXERCISES

1. $a = 1$, $b = 5$, $c = -3$ **2.** $y = \frac{-1}{5}x^2 + 4x + 80$

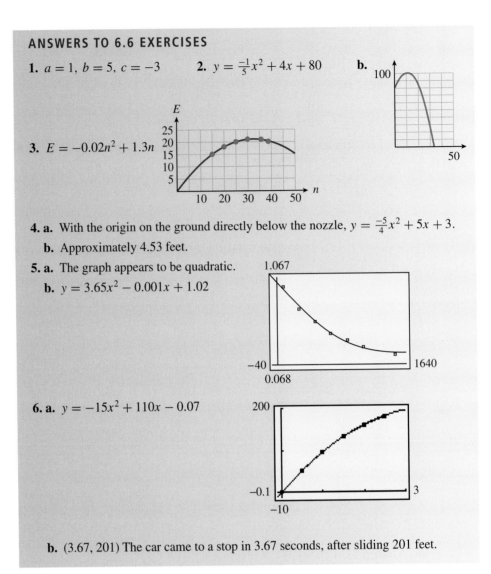

3. $E = -0.02n^2 + 1.3n$

4. a. With the origin on the ground directly below the nozzle, $y = \frac{-5}{4}x^2 + 5x + 3$.

 b. Approximately 4.53 feet.

5. a. The graph appears to be quadratic.

 b. $y = 3.65x^2 - 0.001x + 1.02$

6. a. $y = -15x^2 + 110x - 0.07$

 b. $(3.67, 201)$ The car came to a stop in 3.67 seconds, after sliding 201 feet.

SECTION 6.6 SUMMARY

VOCABULARY *Look up the definitions of new terms in the Glossary.*

Elimination method   Quadratic regression

CONCEPTS

1. We need three points to determine a parabola.

2. We can use the method of elimination to find the equation of a parabola through three points.

3. If we know the vertex of a parabola, we need only one other point to find its equation.

4. We can use quadratic regression to fit a parabola to a collection of data points.

1. How many points are necessary to determine a parabola?
2. Why do we need a second point to find the equation of a parabola if we know its vertex?

3. How can you decide whether linear regression, quadratic regression, or neither one is appropriate for a collection of data?

SKILLS *Practice each skill in the Homework problems listed.*

1. Fit a quadratic equation through three points: #5–12
2. Find a quadratic model in vertex form: #13–30

3. Use quadratic regression to fit a parabola to data: #31–34

HOMEWORK 6.6

ThomsonNOW™ Test yourself on key content at www.thomsonedu.com/login.

Solve the system by elimination. Begin by eliminating *c*.

1.
$$a + b + c = -3$$
$$a - b + c = -9$$
$$4a + 2b + c = -6$$

2.
$$a + b + c = 10$$
$$4a + 2b + c = 19$$
$$9a + 3b + c = 38$$

3.
$$a - b + c = 12$$
$$4a - 2b + c = 19$$
$$9a + 3b + c = 4$$

4.
$$4a + 2b + c = 14$$
$$9a - 3b + c = -41$$
$$16a - 4b + c = -70$$

Find a quadratic equation that fits the data points.

5. Find values for a, b, and c so that the graph of the parabola $y = ax^2 + bx + c$ includes the points $(-1, 0)$, $(2, 12)$, and $(-2, 8)$.

6. Find values for a, b, and c so that the graph of the parabola $y = ax^2 + bx + c$ includes the points $(-1, 2)$, $(1, 6)$, and $(2, 11)$.

7. A survey to determine what percent of different age groups regularly use marijuana collected the following data.

Age	15	20	25	30
Percent	4	13	11	7

   a. Use the percentages for ages 15, 20, and 30 to fit a quadratic equation to the data, $P = ax^2 + bx + c$, where x represents age.

   b. What does your equation predict for the percentage of 25-year-olds who use marijuana?

   c. Sketch the graph of your quadratic equation and the given data on the same axes.

8. The following data show the number of people of certain ages who were the victims of homicide in a large city last year.

Age	10	20	30	40
Number of victims	12	62	72	40

   a. Use the first three data points to fit a quadratic equation to the data, $N = ax^2 + bx + c$, where x represents age.

   b. What does your equation predict for the number of 40-year-olds who were the victims of homicide?

   c. Sketch the graph of your quadratic equation and the given data on the same axes.

9. The data below show Americans' annual per capita consumption of chicken for several years since 1985.

Year	1986	1987	1988	1989	1990
Pounds of chicken	51.3	55.5	57.4	60.8	63.6

   a. Use the values for 1987 through 1989 to fit a quadratic equation to the data, $C = at^2 + bt + c$, where t is measured in years since 1985.

   b. What does your equation predict for per capita chicken consumption in 1990?

   c. Sketch the graph of your equation and the given data.

10. The data show sales of in-line skates at a sporting goods store at the beach.

Year	1990	1991	1992	1993	1994
Skates sold	54	82	194	446	726

   a. Use the values for 1991 through 1993 to fit a quadratic equation to the data, $S = at^2 + bt + c$, where t is measured in years since 1990.

   b. What does your equation predict for the number of pairs of skates sold in 1994?

   c. Sketch the graph of your equation and the given data.

11. Find a quadratic formula for the number of diagonals that can be drawn in a polygon of n sides. Some data are provided.

Sides	4	5	6	7
Diagonals	2	5	9	14

12. Your are driving at 60 miles per hour when you step on the brakes. Find a quadratic formula for the distance in feet that your car travels in t seconds after braking. Some data are provided.

Seconds	1	2	3	4
Feet	81	148	210	267

13. a. Write an equation for a parabola whose vertex is the point $(-2, 6)$. (Many answers are possible.)

   b. Find the value of a if the y-intercept of the parabola in part (a) is 18.

14. a. Write an equation for a parabola whose vertex is the point $(5, -10)$. (Many answers are possible.)

   b. Find the value of a if the y-intercept of the parabola in part (a) is -5.

15. a. Write an equation for a parabola with vertex at $(0, -3)$ and one of its x-intercepts at $(2, 0)$.

   b. Write an equation for a parabola with vertex at $(0, -3)$ and no x-intercepts.

16. Write an equation for a parabola with vertex at $(4, 0)$ and y-intercept at $(0, 4)$. How many x-intercepts does the parabola have?

17. Find the equation for a parabola that has a vertex of $(30, 280)$ and passes through the point $(20, 80)$.

18. Find the equation for a parabola that has a vertex of $(-12, -40)$ and passes through the point $(6, 68)$.

Find an equation for each parabola. Use the vertex form or the factored form of the equation, whichever is more appropriate.

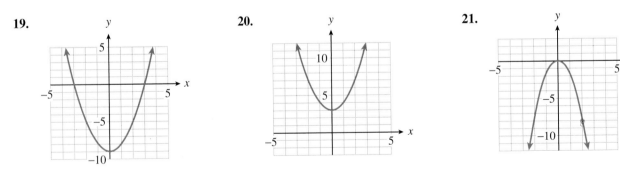

19.

20.

21.

22.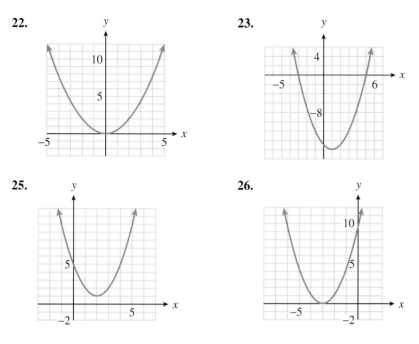

23.

24.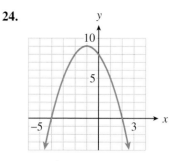

25.

26.

27. In skeet shooting, the clay pigeon is launched from a height of 4 feet and reaches a maximum height of 164 feet at a distance of 80 feet from the launch site.
   a. Write an equation for the height of the clay pigeon in terms of the horizontal distance it has traveled.
   b. If the shooter misses the clay pigeon, how far from the launch site will it hit the ground?

28. The batter in a softball game hits the ball when it is 4 feet above the ground. The ball reaches the greatest height on its trajectory, 35 feet, directly above the head of the left-fielder, who is 200 feet from home plate.
   a. Write an equation for the height of the softball in terms of its horizontal distance from home plate.
   b. Will the ball clear the left field wall, which is 10 feet tall and 375 feet from home plate?

The cables on a suspension bridge hang in the shape of parabolas. For Problems 29 and 30, imagine a coordinate system superimposed on a diagram of the bridge, as shown in the figure.

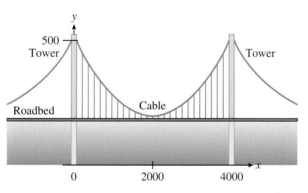

29. The Akashi Kaikyo bridge in Japan is the longest suspension bridge in the world, with a main span of 1991 meters. Its main towers are 297 meters tall. The roadbed of the bridge is 14 meters thick and clears the water below by 65 meters.
   a. Find the coordinates of the vertex and one other point on the cable.
   b. Use the points from part (a) to find an equation for the shape of the cable in vertex form.

30. A suspension bridge joining Sicily to the tip of Italy over the Straits of Messina is planned for completion in 2012. The main span of the bridge will be 3300 meters, and its main towers will be 376 meters tall. The roadbed will be 3 meters thick, clearing the water below by 65 meters.
   a. Find the coordinates of the vertex and one other point on the cable.
   b. Use the points from part (a) to find an equation for the shape of the cable in vertex form.

31. The Square Kilometre Array (SKA) is an international radio telescope project. Project members plan to build a telescope 30 times larger than the largest one currently available. The Australia Telescope National Facility held a workshop in 2005 to design an appropriate antenna. The antenna should be a parabolic dish with diameter from 12 to 20 meters, and the ratio of the focal length to the diameter should be 0.4. The figure shows a cross section of the dish. (Source: www.atnf.csiro.au/projects/ska/)

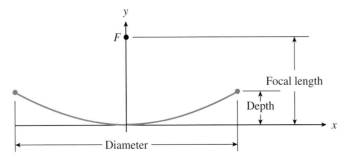

a. You want to design a 20-meter-diameter parabolic antenna for the project. What will the focal length of your antenna be?

b. The equation of the dish has the form $y = \frac{x^2}{4F}$, where F is the focal length. What is the equation of the parabola for your antenna?

c. What is the depth of your parabolic antenna?

32. Some comets move about the sun in parabolic orbits. In 1973, the comet Kohoutek passed within 0.14 AU (astronomical units), or 21 million kilometers, of the Sun. Imagine a coordinate system superimposed on a diagram of the comet's orbit, with the Sun at the origin, as shown in the figure. The units on each axis are measured in AU.

a. The comet's closest approach to the Sun (called *perihelion*) occurred at the vertex of the parabola. What were the comet's coordinates at perihelion?

b. When the comet was first discovered, its coordinates were (1.68, −4.9). Find an equation for comet Kohoutek's orbit in vertex form.

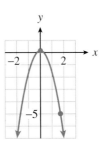

Use your calculator's statistics features for Problems 33–38.

33. The table shows the height of a projectile at different times after it was fired.

Time (seconds)	2	4	6	8	10	12	14
Height (meters)	39.2	71.8	98.0	117.8	131.0	137.8	138.0

a. Find the equation of the least-squares regression line for height in terms of time.

b. Use the linear regression equation to predict the height of the projectile 15 seconds after it was fired.

c. Make a scatterplot of the data and draw the regression line on the same axes.

d. Find the quadratic regression equation for height in terms of time.

e. Use the quadratic regression equation to predict the height of the projectile 15 seconds after it was fired.

f. Draw the quadratic regression curve on the graph from part (c).

g. Which model is more appropriate for the height of the projectile, linear or quadratic? Why?

34. The table shows the height of a star-flare at different times after it exploded from the surface of a star.

Time (seconds)	0.2	0.4	0.6	0.8	1.0	1.2
Height (kilometers)	6.8	12.5	17.1	20.5	22.8	23.9

a. Find the equation of the least-squares regression line for height of the flare in terms of time.

b. Use the linear regression equation to predict the height of the flare 1.4 seconds after it exploded.

c. Make a scatterplot of the data and draw the regression line on the same axes.

d. Find the quadratic regression equation for height in terms of time.

e. Use the quadratic regression equation to predict the height of the flare 1.4 seconds after it exploded.

f. Draw the quadratic regression curve on the graph from part (c).

g. Which model is more appropriate for the height of the star-flare, linear or quadratic? Why?

35. In the 1990s, an outbreak of mad cow disease (Creutzfeldt-Jakob disease) alarmed health officials in England. The table shows the number of deaths each year from the disease.

Year	'94	'95	'96	'97	'98	'99	2000	'01	'02	'03	'04
Deaths	0	3	10	10	18	15	28	20	17	19	9

(Source: www.cjd.ed.ac.uk/vcjdqsep05)

a. The Health Protection Agency determined that a quadratic model was the best-fitting model for the data. Find a quadratic regression equation for the data.

b. Use your model to estimate when the peak of the epidemic occurred and how many deaths from mad cow disease were expected in 2005.

37. The number of daylight hours increases each day from the beginning of winter until the beginning of summer, and then begins to decrease. The table below gives the number of daylight hours in Delbert's hometown last year in terms of the number of days since January 1.

Days since January 1	0	50	100	150	200	250	300
Hours of daylight	9.8	10.9	12.7	14.1	13.9	12.5	10.7

a. Find the equation of the least-squares regression line for the number of daylight hours in terms of the number of days since January 1.

b. Use the linear regression equation to predict the number of daylight hours 365 days after January 1.

c. Make a scatterplot of the data and draw the regression line on the same axes.

d. Find the quadratic regression equation for the number of daylight hours in terms of the number of days since January 1.

e. Use the quadratic regression equation to predict the number of daylight hours 365 days after January 1.

f. Draw the quadratic regression curve on the graph from part (c).

g. Predict the number of daylight hours 365 days since January 1 without using any regression equation. What does this tell you about the linear and quadratic models you found?

36. The table shows the amount of nitrogen fertilizer applied to a crop of soybeans per hectare of land in a trial in Thailand and the resulting yield.

Nitrogen (kg)	0	15	30	60	120
Yield (tons)	2.12	2.46	2.65	2.80	2.60

(Source: www.arc-avrdc.org)

a. Fit a quadratic regression equation to the data.

b. Use your model to predict the maximum yield and the amount of nitrogen needed.

38. To observers on Earth, the Moon looks like a disk that is completely illuminated at full moon and completely dark at new moon. The table below shows what fraction of the Moon is illuminated at 5-day interval after the last full moon.

Days since full moon	0	5	10	15	20	25
Fraction illuminated	1.000	0.734	0.236	0.001	0.279	0.785

a. Find the equation of the least-squares regression line for the fraction illuminated in terms of days.

b. Use the linear regression equation to predict the fraction illuminated 30 days after the full moon.

c. Make a scatterplot of the data and draw the regression line on the same axes.

d. Find the quadratic regression equation for the fraction illuminated in terms of days.

e. Use the quadratic regression equation to predict the fraction illuminated 30 days after the full moon.

f. Draw the quadratic regression curve on the graph from part (c).

g. Predict the fraction of the disk that is illuminated 30 days after the full moon without using any regression equation. What does this tell you about the linear and quadratic models you found?

KEY CONCEPTS

1. A **quadratic** function has the form $f(x) = ax^2 + bx + c$, where a, b, and c are constants and a is not equal to zero.

2. **Zero-factor principle:** The product of two factors equals zero if and only if one or both of the factors equals zero. In symbols,

$$ab = 0 \quad \text{if and only if} \quad a = 0 \quad \text{or} \quad b = 0$$

3. The x-intercepts of the graph of $y = f(x)$ are the solutions of the equation $f(x) = 0$.

4. A quadratic equation written as $ax^2 + bx + c = 0$ is in **standard form.**
   A quadratic equation written as $a(x - r_1)(x - r_2) = 0$ is in **factored form.**

5.

> **To Solve a Quadratic Equation by Factoring:**
>
> 1. Write the equation in standard form.
> 2. Factor the left side of the equation.
> 3. Apply the zero-factor principle: Set each factor equal to zero.
> 4. Solve each equation. There are two solutions.

6. Every quadratic equation has two solutions, which may be the same.

7. The value of the constant a in the factored form of a quadratic equation does not affect the solutions.

8. Each solution of a quadratic equation corresponds to a factor in the factored form.

9. An equation is called **quadratic in form** if we can use a substitution to write it as $au^2 + bu + c = 0$, where u stands for an algebraic expression.

10. The square of the binomial is a **quadratic trinomial,**

$$(x + p)^2 = x^2 + 2px + p^2$$

11.

> **To Solve a Quadratic Equation by Completing the Square:**
>
> 1. a. Write the equation in standard form.
>   b. Divide both sides of the equation by the coefficient of the quadratic term, and subtract the constant term from both sides.
>
> 2. Complete the square on the left side:
>   a. Multiply the coefficient of the first-degree term by one-half, then square the result.
>   b. Add the value obtained in (a) to both sides of the equation.
> 3. Write the left side of the equation as the square of a binomial. Simplify the right side.
> 4. Use extraction of roots to finish the solution.

12.

> **The Quadratic Formula**
>
> The solutions of the equation $ax^2 + bx + c = 0$, $a \neq 0$, are
>
> $$x = \frac{-b \pm \sqrt{b^2 - 4ac}}{2a}$$

13. We have four methods for solving quadratic equations: extracting roots, factoring, completing the square, and using the quadratic formula. The first two methods are faster, but they do not work on all equations. The last two methods work on any quadratic equation.

14. The graph of a quadratic function $f(x) = ax^2 + bx + c$ is called a **parabola.** The values of the constants a, b, and c determine the location and orientation of the parabola.

15. For the graph of $y = ax^2 + bx + c$, the x-coordinate of the vertex is $x_v = \dfrac{-b}{2a}$.
    To find the y-coordinate of the vertex, we substitute x_v into the formula for the parabola.

16. The graph of the quadratic function $y = ax^2 + bx + c$ may have two, one, or no x-intercepts, according to the number of distinct real-valued solutions of the equation $ax^2 + bx + c = 0$.

17.

The Discriminant

The **discriminant** of a quadratic equation is $D = b^2 - 4ac$.

1. If $D > 0$, there are two unequal real solutions.
2. If $D = 0$, there is one real solution of multiplicity two.
3. If $D < 0$, there are two complex solutions.

18.

To graph the quadratic function $y = ax^2 + bx + c$:

1. Determine whether the parabola opens upward (if $a > 0$) or downward (if $a < 0$).
2. Locate the vertex of the parabola.
   a. The x-coordinate of the vertex is $x_v = \dfrac{-b}{2a}$.
   b. Find the y-coordinate of the vertex by substituting x_v into the equation of the parabola.
3. Locate the x-intercepts (if any) by setting $y = 0$ and solving for x.
4. Locate the y-intercept by evaluating y for $x = 0$.
5. Locate the point symmetric to the y-intercept across the axis of symmetry.

19. Quadratic models may arise as the product of two variables.

20. The maximum or minimum of a quadratic function occurs at the vertex.

21.

Vertex Form for a Quadratic Function

A quadratic function $y = ax^2 + bx + c$, $a \neq 0$, can be written in the **vertex form**

$$y = a(x - x_v)^2 + y_v$$

where the vertex of the graph is (x_v, y_v).

22. We can convert a quadratic equation to vertex form by completing the square.

23. We can graph a quadratic equation in vertex form using transformations.

24. A 2×2 system involving quadratic equations may have one, two, or no solutions.

25. We can use a graphical technique to solve quadratic inequalities.

26.

To Solve a Quadratic Inequality Algebraically:

1. Write the inequality in standard form: One side is 0, and the other has the form $ax^2 + bx + c$.
2. Find the x-intercepts of the graph of $y = ax^2 + bx + c$ by setting $y = 0$ and solving for x.
3. Make a rough sketch of the graph, using the sign of a to determine whether the parabola opens upward or downward.
4. Decide which intervals on the x-axis give the correct sign for y.

27. We need three points to determine a parabola.

28. We can use the method of elimination to find the equation of a parabola through three points.

29. If we know the vertex of a parabola, we need only one other point to find its equation.

30. We can use quadratic regression to fit a parabola to a collection of data points.

REVIEW PROBLEMS

Solve by factoring.

1. $x^2 + x = 4 - (x + 2)^2$

2. $(n - 3)(n + 2) = 6$

3. $x(3x + 2) = (x + 2)^2$

4. $6y = (y + 1)^2 + 3$

5. $4x - (x + 1)(x + 2) = -8$

6. $3(x + 2)^2 = 15 + 12x$

Write a quadratic equation with integer coefficients and with the given solutions.

7. $\dfrac{-3}{4}$ and 8

8. $\dfrac{5}{3}$ and $\dfrac{5}{3}$

Graph each equation using the *ZDecimal* setting. Locate the *x*-intercepts and use them to write the quadratic expression in factored form.

9. $y = x^2 - 0.6x - 7.2$

10. $y = -x^2 + 0.7x + 2.6$

Use a substitution to solve.

11. $2^{2p} - 6 \cdot 2^p + 8 = 0$

12. $3^{2r} - 6 \cdot 3^r + 5 = 0$

13. $\left(\dfrac{1}{b}\right)^2 - 3\left(\dfrac{1}{b}\right) - 4 = 0$

14. $\left(\dfrac{1}{q}\right)^2 + \dfrac{1}{q} - 2 = 0$

Solve by completing the square.

15. $x^2 - 4x - 6 = 0$

16. $x^2 + 3x = 3$

17. $2x^2 + 3 = 6x$

18. $3x^2 = 2x + 3$

Solve by using the quadratic formula.

19. $\dfrac{1}{2}x^2 + 1 = \dfrac{3}{2}x$

20. $x^2 - 3x + 1 = 0$

21. $x^2 - 4x + 2 = 0$

22. $2x^2 + 2x = 3$

Solve each formula for the indicated variable.

23. $K = \dfrac{1}{2}mv^2$, for v

24. $a^2 + b^2 = c^2$, for b

25. $h = 6t - 3t^2$, for t

26. $D = \dfrac{n^2 - 3n}{2}$, for n

27. In a tennis tournament among n competitors, $\dfrac{n(n-1)}{2}$ matches must be played. If the organizers can schedule 36 matches, how many players should they invite?

28. The formula $S = \dfrac{n(n+1)}{2}$ gives the sum of the first n positive integers. How many consecutive integers must be added to make a sum of 91?

29. Irene wants to enclose 2 adjacent chicken coops of equal size against the henhouse wall. She has 66 feet of chicken wire fencing and would like the total area of the two coops to be 360 square feet. What should the dimensions of the chicken coops be?

30. The base of an isosceles triangle is one inch shorter than the equal sides, and the altitude of the triangle is 2 inches shorter than the equal sides. What is the length of the equal sides?

31. A car traveling at 50 feet per second (about 34 miles per hour) can stop in 2.5 seconds after applying the brakes hard. The distance the car travels, in feet, t seconds after applying the brakes is $d = 50t - 10t^2$. How long does it take the car to travel 40 feet?

32. You have 300 feet of wire fence to mark off a rectangular Christmas tree lot with a center divider, using a brick wall as one side of the lot. If you would like to enclose a total area of 7500 square feet, what should be the dimensions of the lot?

33. The height, h, of an object t seconds after being thrown from ground level is given by

$$h = v_0 t - \frac{1}{2} g t^2$$

where v_0 is its starting velocity and g is a constant that depends on gravity. On the Moon, the value of g is approximately 5.6. Suppose you hit a golf ball on the Moon with an upward velocity of 100 feet per second.

a. Write an equation for the height of the golf ball t seconds after you hit it.

b. Graph your equation in the window

$$\text{Xmin} = 0, \quad \text{Xmax} = 47,$$
$$\text{Ymin} = 0, \quad \text{Ymax} = 1000$$

c. Use the **Trace** to estimate the maximum height the golf ball reaches.

d. Use your equation to calculate when the golf ball will reach a height of 880 feet.

34. An acrobat is catapulted into the air from a springboard at ground level. Her height, h, in meters is given by the formula $h = -4.9t^2 + 14.7t$, where t is the time in seconds from launch. Use your calculator to graph the acrobat's height versus time. Use the window

$$\text{Xmin} = 0, \quad \text{Xmax} = 4.7$$
$$\text{Ymin} = 0, \quad \text{Ymax} = 12$$

a. Use the **Trace** to find the coordinates of the highest point on the graph. When does the acrobat reach her maximum height, and what is that height?

b. Use the formula to find the height of the acrobat after 2.4 seconds.

c. Use the **Trace** to verify your answer to part (b). Find another time when the acrobat is at the same height.

d. Use the formula to find two times when the acrobat is at a height of 6.125 meters. Verify your answers on the graph.

e. What are the coordinates of the horizontal intercepts of your graph? What do these points have to do with the acrobat?

Show that the shaded areas are equal.

35.

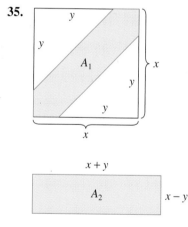

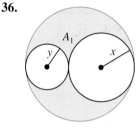

36.

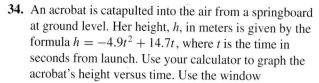

(a) Find the coordinates of the vertex and the intercepts.
(b) Sketch the graph.

37. $y = \dfrac{1}{2}x^2$

38. $y = x^2 - 4$

39. $y = x^2 - 9x$

40. $y = -2x^2 - 4x$

41. $y = x^2 + x - 6$

42. $y = x^2 - 3x + 4$

43. $y = 8 - x - 2x^2$

44. $y = -2x^2 + x - 4$

45. $y = x^2 - x - 9$

46. $y = -x^2 + 2x + 4$

Use the discriminant to determine how many *x*-intercepts the graph has.

47. $y = -2x^2 + 5x - 1$

48. $y = -12 - 3x + 4x^2$

Use the discriminant to determine the nature of the solution of each equation.

49. $4x^2 - 12x + 9 = 0$

50. $2t^2 + 6t + 5 = 0$

51. $2y^2 = 3y - 4$

52. $\dfrac{x^2}{4} = x + \dfrac{5}{4}$

53. The total profit Kiyoshi makes from producing and selling *x* floral arrangements is

$$P(x) = -0.4x^2 + 36x - 400$$

   a. How many floral arrangements should Kiyoshi produce and sell to maximize his profit? What is his maximum profit?

   b. Verify your answers on a graph.

54. Lightning does about one billion dollars damage annually in the United States and kills 85 people. To study lightning, meteorologists fire small rockets at passing thunderclouds to induce lightning bolts. The rocket trails a thin copper wire that is vaporized by the lightning, leaving a plasma channel that carries the current to the grounding point. The rocket boosts the wire to a height of 250 meters, and *t* seconds later, its height is given in meters by $h(t) = -4.9t^2 + 32t + 250$.

   a. When does the rocket reach its maximum height? What is the maximum height?

   b. Verify your answers on a graph.

55. A beekeeper has beehives distributed over 60 square miles of pastureland. When she places 4 hives per square mile, each hive produces about 32 pints of honey per year. For each additional hive per square mile, honey production drops by 4 pints per hive.

   a. Write an equation for the total production of honey, in pints, in terms of the number of additional hives per square mile.

   b. How many additional hives per square mile should the beekeeper install in order to maximize honey production?

56. A small company manufactures radios. When it charges $20 for a radio, it sells 500 radios per month. For each dollar the price is increased, 10 fewer radios are sold per month.

   a. Write an equation for the monthly revenue in terms of the price increase over $20.

   b. What should the company charge for a radio in order to maximize its monthly revenue?

For Problems 55–58,
(a) Find all values of *x* for which $f(x) = 0$.
(b) Find all values of *x* for which $g(x) = 0$.
(c) Find all values of *x* for which $f(x) = g(x)$.
(d) Graph each pair of functions in the same window, then sketch the graph on paper. Illustrate your answers to (a)–(d) as points on the graph.

57. $f(x) = 2x^2 + 3x, \ g(x) = 5 - 6x$

58. $f(x) = 3x^2 - 6x, \quad g(x) = 8 + 4x$

59. $f(x) = 2x^2 - 2x, \ g(x) = x^2 + 3$

60. $f(x) = x^2 + 4x + 6, \ g(x) = 4 - x^2$

Solve each inequality algebraically, and give your answers in interval notation. Verify your solutions by graphing.

61. $(x - 3)(x + 2) > 0$

62. $y^2 - y - 12 \leq 0$

63. $2y^2 - y \leq 3$

64. $3z^2 - 5z > 2$

65. $s^2 \leq 4$

66. $4t^2 > 12$

67. The Sub Station sells $220 - \frac{1}{4}p$ submarine sandwiches at lunchtime if it sells them at p cents each.
   a. Write a formula for the Sub Station's daily revenue in terms of p.
   b. What range of prices can the Sub Station charge if it wants to keep its daily revenue from subs over \$480? (Remember to convert \$480 to cents.)

68. When it charges p dollars for an electric screwdriver, Handy Hardware will sell $30 - \frac{1}{2}p$ screwdrivers per month.
   a. Write a formula in terms of p for Handy Hardware's monthly revenue from electric screwdrivers.
   b. How much should Handy charge per screwdriver if it wants the monthly revenue from the screwdrivers to be over \$400?

Solve each system algebraically, and verify your solution with a graph.

69. $y + x^2 = 4$
$y = 3$

70. $y = 3 - x^2$
$5x + y = 7$

71. $y = x^2 - 5$
$y = 4x$

72. $y = x^2 - 2x + 1$
$y = 3 - x$

73. $y = x^2 - 6x + 20$
$y = 2x^2 - 2x - 25$

74. $y = x^2 - 5x - 28$
$y = -x^2 + 4x + 28$

75. $y = \frac{1}{2}x^2 - \frac{3}{2}x$
$y = -\frac{1}{2}x^2 + \frac{1}{2}x + 3$

76. $y = 2x^2 + 5x - 3$
$y = x^2 + 4x - 1$

77. Find values of a, b, and c so that the graph of the parabola $y = ax^2 + bx + c$ contains the points $(-1, -4)$, $(0, -6)$, and $(4, 6)$.

78. a. Find the equation of a parabola $y = ax^2 + bx + c$ that passes through the points $(0, -2)$, $(-6, 1)$ and $(4, 6)$.
   b. Plot the data points and sketch the graph on the grid.

79. Find a parabola that fits the following data points.

x	-8	-4	2	4
y	10	18	0	-14

80. Find a parabola that fits the following data points.

x	-3	0	2	4
y	-46	8	-6	-60

81. Find the equation for a parabola that has a vertex of $(15, -6)$ and passes through the point $(3, 22.8)$.

82. Find the equation for a parabola that has a vertex of $(-3, -8)$ and passes through the point $(6, 12.25)$.

For Problems 83–86,
(a) Write the equation in vertex form.
(b) Use transformations to sketch the graph.

83. $f(x) = x^2 - 24x + 44$

84. $g(x) = x^2 + 30x + 300$

85. $y = \frac{1}{3}x^2 + 2x + 1$

86. $y = -2x^2 + 4x + 3$

87. The height of a cannonball was observed at 0.2-second intervals after the cannon was fired, and the data were recorded in the table.

Time (seconds)	0.2	0.4	0.6	0.8	1.0	1.2	1.4	1.6	1.8	2.0
Height (meters)	10.2	19.2	27.8	35.9	43.7	51.1	58.1	64.7	71.0	76.8

   a. Find the equation of the least-squares regression line for height in terms of time.
   b. Use the linear regression equation to predict the height of the cannonball at 3 seconds and at 4 seconds after it was fired.
   c. Make a scatterplot of the data and draw the regression line on the same axes.
   d. Find the quadratic regression equation for height in terms of time.
   e. Use the quadratic regression equation to predict the height of the cannonball at 3 seconds and at 4 seconds after it was fired.
   f. Draw the quadratic regression curve on the graph from part (c).
   g. Which model is more appropriate for the height of the cannonball, linear or quadratic? Why?

88. Max took a sequence of photographs of an explosion spaced at equal time intervals. From the photographs, he was able to estimate the height and vertical velocity of some debris from the explosion, as shown in the table. (Negative velocities indicate that the debris is falling back to Earth.)

Velocity (meters/second)	67	47	27	8	−12	−31
Height (meters)	8	122	196	232	228	185

a. Enter the data into your calculator and create a scatterplot. Fit a quadratic regression equation to the data, then graph the equation on the scatterplot.
b. Use your regression equation to find the vertex of the parabola. What do the coordinates represent, in terms of the problem? What should the velocity of the debris be at the maximum height of the debris?

PROJECTS FOR CHAPTER 6

1. Starlings often feed in flocks, and their rate of feeding depends on the size of the flock. If the flock is too small, the birds are nervous and spend a lot of time watching for predators. If the flock is too large, the birds become overcrowded and fight each other, which interferes with feeding. Here are some data gathered at a feeding station. The data show the number of starlings in the flock and the total number of pecks per minute recorded at the station while the flock was feeding.
(Source: Chapman & Reiss, 1992)

Number of starlings	Pecks per minute	Pecks per starling per minute
1	9	
2	26	
3	48	
4	80	
5	120	
6	156	
7	175	
8	152	
9	117	
10	180	
12	132	

a. For each flock size, calculate the number of pecks per starling per minute. For purposes of efficient feeding, what flock size appears to be optimum? How many pecks per minute would each starling make in a flock of optimal size?
b. Plot the number of pecks per starling per minute against flock size. Do the data points appear to lie on (or near) a parabola?
c. The quadratic regression equation for the data is $y = -0.45x^2 + 5.8x + 3.9$. Graph this parabola on the same axes with the data points.
d. What are the optimum flock size and the maximum number of pecks per starling per minute predicted by the regression equation?

2. Biologists conducted a four-year study of the nesting habits of the species *Parus major* in an area of England called Wytham Woods. The bar graph shows the clutch size (the number of eggs) in 433 nests. (Source: Perrins and Moss, 1975)

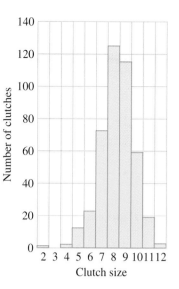

a. Which clutch size was observed most frequently? Fill in the table, showing the total number of eggs produced in each clutch size.

Clutch size	2	3	4	5	6	7	8	9	10	11	12
Number of clutches	1	0	2	12	23	73	126	116	59	19	3
Number of eggs											

b. The average weight of the nestlings declines as the size of the brood increases, and the survival of individual nestlings is linked to their weight. A hypothetical (and simplified) model of this phenomenon is described by the table below. Calculate the number of surviving nestlings for each clutch size. Which clutch size produces the largest average number of survivors?

Clutch size	1	2	3	4	5	6	7	8	9	10
Percent survival	100	90	80	70	60	50	40	30	20	10
Number of survivors										

c. The figure shows the number of survivors for each clutch size in Wytham Woods, along with the curve of best fit. The equation for the curve is $y = -0.0105x^2 + 0.2x - 0.035$.

Find the optimal clutch size for maximizing the number of surviving nestlings. How does this optimum clutch size compare with the most frequently observed clutch size in part (a)?

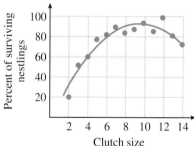

3. In this project, we minimize a quadratic expression to find the line of best fit. The figure shows a set of three data points and a line of best fit. For this example, the regression line passes through the origin, so its equation is $y = mx$ for some positive value of m. How shall we choose m to give the best fit for the data? We want the data points to lie as close to the line as possible. One way to achieve this is to minimize the sum of the squares of the vertical distances shown in the figure.

a. The data points are $(1, 2)$, $(2, 6)$, and $(3, 7)$. Verify that the sum S we want to minimize is

$$S = (2 - m)^2 + (6 - 2m)^2 + (7 - 3m)^2$$
$$= 14m^2 - 70m + 89$$

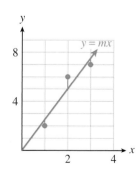

b. Graph the formula for S in the window

$$\text{Xmin} = 0 \quad \text{Xmax} = 9.4$$
$$\text{Ymin} = 0 \quad \text{Ymax} = 100$$

c. Find the vertex of the graph of S.
d. Use the value of m to write the equation of the regression line $y = mx$.
e. Graph the three data points and your regression line on the same axes.

4. The figure shows the typical weight of two species of birds each day after hatching. (Source: Perrins, 1979)

a. Describe the rate of growth for each species over the first 15 days of life. How are the growth rates for the two species similar, and how are they different?

b. Complete the tables showing the weight and the daily rate of growth for each species.

Parus major

Day	1	2	3	4	5	6	7	8	9	10	11	12	13	14	15
Weight															
Growth rate															

Parus caeruleus

Day	1	2	3	4	5	6	7	8	9	10	11	12	13	14	15
Weight															
Growth rate															

c. Plot the rate of growth against weight in grams for each species. What type of curve does the growth rate graph appear to be?

d. For each species, at what weight did the maximum growth rate occur? Locate the corresponding point on each original curve in Figure 6.43.

5. a. Find a quadratic regression equation for the growth rate of *Parus major* in terms of its weight using the data from Project 4.

b. Make a scatterplot of the data and draw the regression curve on the same axes.

c. Find the vertex of the graph of the regression equation. How does this estimate for the maximum growth rate compare with your estimate in Project 4?

6. a. Find a quadratic regression equation for the growth rate of *Parus caeruleus* in terms of its weight using the data from Project 4.

b. Make a scatterplot of the data and draw the regression curve on the same axes.

c. Find the vertex of the graph of the regression equation. How does this estimate for the maximum growth rate compare with your estimate in Project 4?

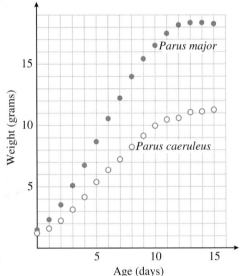

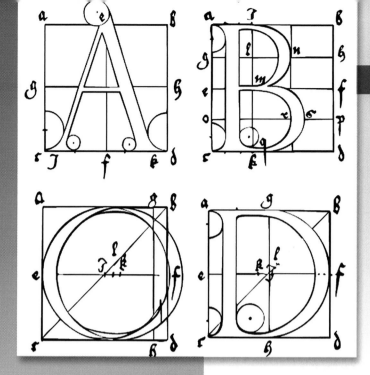

Polynomial and Rational Functions

The graphs of linear functions, quadratic functions, power functions, and exponentials all have a characteristic shape. On the other hand, the family of polynomial functions has graphs that represent a huge variety of different shapes.

Ever since Gutenberg's invention of movable type in 1455, artists and printers have been interested in the design of pleasing and practical fonts. In 1525, Albrecht Durer published *On the Just Shaping of Letters,* which set forth a system of rules for the geometric construction of Roman capitals. The letters shown above are examples of Durer's font. Until the twentieth century, a ruler and compass were the only practical design tools, and consequently straight lines and circular arcs were the only geometric objects that could be accurately reproduced.

With the advent of computers, however, complex curves and surfaces, such as the smooth contours of modern cars, can be defined precisely and reproduced in metal or plastic. In fact, it was the French automobile engineer Pierre Bézier who in the 1960s developed a new design tool based on polynomials. *Bézier curves* are widely used today in all fields of design, from technical plans and blueprints to the most creative artistic endeavors.

The study of Bézier curves falls under the general topic of curve fitting, but these curves do not really have a scientific purpose. A scientist would not use Bézier curves to fit a function to data. Rather, Bézier curves have more of an artistic purpose. Computer

ThomsonNOW™

Throughout this chapter, the ThomsonNOW logo indicates an opportunity for online self-study, linking you to interactive tutorials and videos based on your level of understanding.

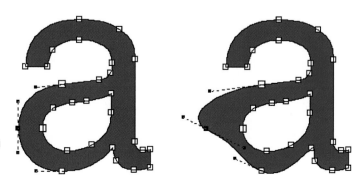

programs like Illustrator, Freehand, and CorelDraw™ use cubic Bézier curves. The PostScript printer language and Type 1 fonts also use cubic Bézier curves, and TrueType fonts use quadratic Bézier curves.

Investigation 12

Bézier Curves

A Bézier curve is actually a sequence of short curves pieced together. Each piece has two *endpoints* and (for nonlinear curves) at least one *control point*. The control points do not lie on the curve itself, but they determine its shape. Two polynomials define the curve, one for the *x*-coordinate and one for the *y*-coordinate.

Part I

A. Linear Bézier Curves

The linear Bézier curve for two endpoints, (x_1, y_1) and (x_2, y_2), is the straight line segment joining those two points. The curve is defined by the two functions

$$x = f(t) = x_1 \cdot (1 - t) + x_2 \cdot t$$
$$y = g(t) = y_1 \cdot (1 - t) + y_2 \cdot t$$

for $0 \leq t \leq 1$.

1. Find the functions f and g defining the linear Bézier curve joining the two points $(-4, 7)$ and $(2, 0)$. Simplify the formulas defining each function.

2. Fill in the table of values and plot the curve.

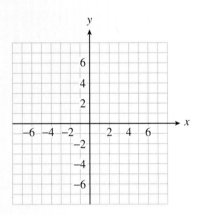

t	0	0.25	0.5	0.75	1
x					
y					

B. Quadratic Bézier Curves: Drawing a Simple Numeral 7

The quadratic Bézier curve is defined by two endpoints, (x_1, y_1) and (x_3, y_3), and a control point, (x_2, y_2).

$$x = f(t) = x_1 \cdot (1 - t)^2 + 2x_2 \cdot t(1 - t) + x_3 \cdot t^2$$
$$y = g(t) = y_1 \cdot (1 - t)^2 + 2y_2 \cdot t(1 - t) + y_3 \cdot t^2$$

for $0 \leq t \leq 1$.

3. Find the functions f and g for the quadratic Bézier curve defined by the endpoints $(4, 7)$ and $(0, -7)$, and the control point $(0, 5)$. Simplify the formulas defining each function.

4. Fill in the table of values and plot the curve.

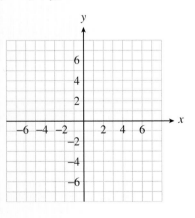

t	0	0.25	0.5	0.75	1
x					
y					

5. Draw a line segment from $(-4, 7)$ to $(4, 7)$ on the grid on the bottom of page 562 to complete the numeral **7**.

6. We can adjust the curvature of the diagonal stroke of the **7** by moving the control point. Find the functions f and g for the quadratic Bézier curve defined by the endpoints $(4, 7)$ and $(0, -7)$, and the control point $(0, -3)$. Simplify the formulas defining each function.

7. Fill in the table of values and plot the curve.

t	0	0.25	0.5	0.75	1
x					
y					

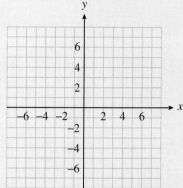

8. Draw a line segment from $(-4, 7)$ to $(4, 7)$ on the grid above to complete the numeral **7**.

9. On your graphs in steps (5) and (8), plot the three points that defined the curved section of the numeral **7**, then connect them in order with line segments. How does the position of the control point change the curve?

C. Cubic Bézier Curves: Drawing a Letter *y*

A cubic Bézier curve is defined by two endpoints, (x_1, y_1) and (x_4, y_4), and two control points, (x_2, y_2) and (x_3, y_3).

$$x = f(t) = x_1 \cdot (1 - t)^2 + 3x_2 \cdot t(1 - t)^2 + 3x_3 \cdot t^2(1 - t) + x_4 \cdot t^3$$
$$y = g(t) = y_1 \cdot (1 - t)^2 + 3y_2 \cdot t(1 - t)^2 + 3y_3 \cdot t^2(1 - t) + y_4 \cdot t^3$$

for $0 \le t \le 1$.

10. Find the functions f and g for the cubic Bézier curve defined by the endpoints $(4, 7)$ and $(-4, -5)$, and the control points $(3, 3)$ and $(0, -8)$. Simplify the formulas defining each function.

11. Fill in the table of values and plot the curve.

t	0	0.25	0.5	0.75	1
x					
y					

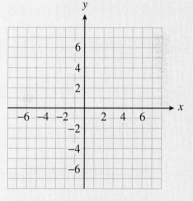

12. Connect the four given points in order using three line segments. How does the position of the control points affect the curve? Finish the letter **y** by including the linear Bézier curve you drew for step (2) in Part 1.

Part II

A. Bézier Curves on the Graphing Calculator

We can draw Bézier curves on the graphing calculator using the **parametric** mode. First, press the **MODE** key and highlight **PAR,** as shown in Figure 7.1a. To remove the *x*- and *y*-axes from the display, press **2nd** **ZOOM** to get the **Format** menu, then choose **AxesOff** as shown in Figure 7.1b. Finally, we set the viewing window: Press $\boxed{\textbf{WINDOW}}$ and set

$$\text{Tmin} = 0 \qquad \text{Tmax} = 1 \qquad \text{TStep} = 0.05$$
$$\text{Xmin} = -10 \qquad \text{Xmax} = 10 \qquad \text{Ymin} = -10 \qquad \text{Ymax} = 10$$

as shown in Figure 7.1c.

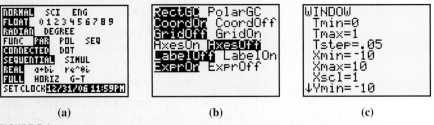

(a)   (b)   (c)

FIGURE 7.1

As an example, we will graph the linear curve from part (A). Press $\boxed{\textbf{Y=}}$ and enter the definitions for $x(t)$ and $y(t)$, as shown in Figure 7.2a. Press $\boxed{\textbf{GRAPH}}$ and the calculator displays the line segment joining $(-4, 7)$ and $(2, 0)$, as shown in Figure 7.2b. Experiment by modifying the endpoints to see how the graph changes.

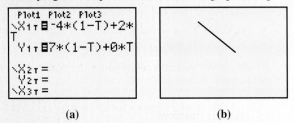

(a)   (b)

FIGURE 7.2

B. Designing a Numeral 7

13. Press $\boxed{\textbf{2nd}}$ $\boxed{\textbf{Y=}}$ and enter the formulas for the quadratic Bézier curve defined by the endpoints $(4, 7)$ and $(0, -7)$, and the control point $(0, 5)$ under X_{1T} and Y_{1T}. (These are the same formulas you found in step (3) of Part I.)

14. Find the functions f and g for the linear Bézier curve joining the points $(-4, 7)$ and $(4, 7)$. Simplify the formulas for those functions and enter them into your calculator under X_{2T} and Y_{2T}. Press $\boxed{\textbf{GRAPH}}$ to see the graph.

15. Now we will alter the image slightly: Go back to X_{1T} and Y_{1T} and change the control point to $(0, -3)$. (These are the formulas you found in step (6).) How does the graph change?

16. We can see exactly how the control point affects the graph by connecting the three data points with line segments. Press $\boxed{\textbf{STAT}}$ $\boxed{\textbf{ENTER}}$ and enter the coordinates of $(4, 7)$, $(0, -3)$, and $(0, -7)$ in L_1 and L_2, as shown in Figure 7.3a. Press $\boxed{\textbf{2nd}}$ $\boxed{\textbf{Y=}}$ $\boxed{\textbf{ENTER}}$, turn on **Plot1,** and select the second plot type, as shown in Figure 7.3b. You should see the line segments superimposed on your numeral **7.** How are those segments related to the curve?

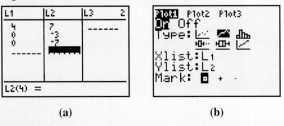

(a)   (b)

FIGURE 7.3

17. Now edit L_2 so that the control point is $(0, 5)$, and again define X_{1T} and Y_{1T} as in step (13) to see how the graph is altered. Experiment by using different coordinates for the control point and entering the appropriate new equations in X_{1T} and Y_{1T}. Describe how the point (x_2, y_2) affects the direction of the quadratic Bézier curve. (*Hint:* Consider the direction of the curve starting off from either endpoint.)

C. Designing a Letter *y* with Cubic Bézier Curves

18. First, turn off **Plot1** by pressing [Y=] [△] [ENTER].

19. Enter the formulas for the cubic Bézier curve from part (10) under X_{1T} and Y_{1T}. Modify the expressions in X_{2T} and Y_{2T} to the linear Bézier curve joining the points $(-4, 7)$ and $(2, 0)$ from step (1).

20. How do the control points affect the direction of the Bézier curve? (*Suggestion:* Enter the coordinates of the four points under L_1 and L_2 and turn **Plot1** back on.)

When you finish experimenting with Bézier curves, set the calculator **MODE** ([2nd] [QUIT]) back to **FUNC**, and reset the **FORMAT** ([2nd] [ZOOM]) to **AxesOn.**

7.1 Polynomial Functions

We have already encountered some examples of polynomial functions. Linear functions,

$$f(x) = ax + b$$

and quadratic functions

$$f(x) = ax^2 + bx + c$$

are special cases of polynomial functions. In general, we make the following definition.

> **Polynomial Function**
>
> A **polynomial function** has the form
>
> $$f(x) = a_n x^n + a_{n-1} x^{n-1} + a_{n-2} x^{n-2} + \cdots + a_2 x^2 + a_1 x + a_0$$
>
> where $a_0, a_1, a_2, \ldots, a_n$ are constants and $a_n \neq 0$. The coefficient a_n of the highest power term is called the **lead coefficient.**

Some examples of polynomials are

$$f(x) = 6x^3 - 4x^2 + x - 2 \qquad g(x) = 9x^5 - 2$$
$$p(x) = x^4 + x^2 + 1 \qquad q(x) = 2x^{10} - x^7 + 3x^6 + 5x^3 + 3x$$

Each of the polynomials above is written in **descending powers,** which means that the highest-degree term comes first, and the degrees of the terms decrease from largest to smallest. Sometimes it is useful to write a polynomial in **ascending powers,** so that the degrees of the terms increase. For example, the polynomial $f(x)$ above would be written as

$$f(x) = -2 + x - 4x^2 + 6x^3$$

in ascending powers.

▶▶▶ Before studying this chapter, you may want to review Algebra Skills Refresher A.7, Polynomials and Factoring.

Products of Polynomials

When we multiply two or more polynomials together, we get another polynomial of higher degree. (See Algebra Skills Refresher A.7 for the definition of **degree**.)

EXAMPLE 1　Compute the products.

　　　　a. $(x + 2)(5x^3 - 3x^2 + 4)$　　　　**b.** $(x - 3)(x + 2)(x - 4)$

Solutions　**a.** $(x + 2)(5x^3 - 3x^2 + 4)$　　　　　　　　　　Apply the distributive law.

$$= x(5x^3 - 3x^2 + 4) + 2(5x^3 - 3x^2 + 4) \qquad \text{Apply the distributive law again.}$$
$$= 5x^4 - 3x^3 + 4x + 10x^3 - 6x^2 + 8 \qquad \text{Combine like terms.}$$
$$= 5x^4 + 7x^3 - 6x^2 + 4x + 8$$

b. $(x - 3)(x + 2)(x - 4)$　　　　　　　　　　　Multiply two of the factors first.

$$= (x - 3)(x^2 - 2x - 8) \qquad\qquad\qquad \text{Apply the distributive law.}$$

$$= x(x^2 - 2x - 8) - 3(x^2 - 2x - 8) \qquad \text{Apply the distributive law again.}$$
$$= x^3 - 2x^2 - 8x - 3x^2 + 6x + 24 \qquad \text{Combine like terms.}$$
$$= x^3 - 5x^2 - 2x + 24$$

EXERCISE 1

Multiply $(y + 2)(y^2 - 2y + 3)$.

In Example 1a, we multiplied a polynomial of degree 1 by a polynomial of degree 3, and the product was a polynomial of degree 4. In Example 1b, the product of three first-degree polynomials is a third-degree polynomial.

Degree of a Product

The degree of a product of nonzero polynomials is the sum of the degrees of the factors. That is,

If $P(x)$ has degree m and $Q(x)$ has degree n, then their product $P(x)Q(x)$ has degree $n + m$.

EXAMPLE 2　Let $P(x) = 5x^4 - 2x^3 + 6x^2 - x + 2$, and $Q(x) = 3x^3 - 4x^2 + 5x + 3$.

　　a. What is the degree of their product? What is the coefficient of the lead term?

　　b. Find the coefficient of the x^3-term of the product.

Solutions　**a.** The degree of P is 4, and the degree of Q is 3, so the degree of their product is $4 + 3 = 7$. The only degree 7 term of the product is $(5x^4)(3x^3) = 15x^7$, which has coefficient 15.

b. In the product, each term of $P(x)$ is multiplied by each term of $Q(x)$. We get degree 3 terms by multiplying together terms of degree 0 and 3, or 1 and 2. For these polynomials, the possible combinations are:

$P(x)$	$Q(x)$	Product
2	$3x^3$	$6x^3$
$-2x^3$	3	$-6x^3$
$-x$	$-4x^2$	$4x^3$
$6x^2$	$5x$	$30x^3$

The sum of the third-degree terms of the product is $34x^3$, with coefficient 34.

EXERCISE 2

Find the coefficient of the fourth-degree term of the product of
$f(x) = 2x^6 + 2x^4 - x^3 + 5x^2 + 1$ and $g(x) = x^5 - 3x^4 + 2x^3 + x^2 - 4x - 2$.

Special Products

In the Algebra Skills Refresher A.8, you can review the following special products involving quadratic expressions.

Special Products of Binomials

$(a + b)^2 = (a + b)(a + b) = a^2 + 2ab + b^2$
$(a - b)^2 = (a - b)(a - b) = a^2 - 2ab + b^2$
$(a + b)(a - b) = a^2 - b^2$

There are also special products resulting in cubic polynomials. In the Homework problems, you will be asked to verify the following products.

Cube of a Binomial

1. $(x + y)^3 = x^3 + 3x^2y + 3xy^2 + y^3$
2. $(x - y)^3 = x^3 - 3x^2y + 3xy^2 - y^3$

If you become familiar with these general forms, you can use them as patterns to find specific examples of such products.

EXAMPLE 3   Write $(2w - 3)^3$ as a polynomial.

Solution   Use product (2) on the preceding page, with x replaced by $2w$ and y replaced by 3.

$$(x - y)^3 = x^3 - 3x^2y + 3xy^2 - y^3$$
$$(2w - 3)^3 = (2w)^3 - 3(2w)^2(3) + 3(2w)(3)^2 - 3^3 \quad \text{Simplify.}$$
$$= 8w^3 - 36w^2 + 54w - 27$$

Of course, we can also expand the product in Example 3 simply by polynomial multiplication and arrive at the same answer.

EXERCISE 3

Write $(5 + x^2)^3$ as a polynomial.

Factoring Cubics

Another pair of products is useful for factoring cubic polynomials. In the Homework problems, you will be asked to verify the following products:

$$(x + y)(x^2 - xy + y^2) = x^3 + y^3$$
$$(x - y)(x^2 + xy + y^2) = x^3 - y^3$$

Viewing these products from right to left, we have the following special factorizations for the sum and difference of two cubes.

Factoring the Sum or Difference of Two Cubes

3. $x^3 + y^3 = (x + y)(x^2 - xy + y^2)$

4. $x^3 - y^3 = (x - y)(x^2 + xy + y^2)$

When we recognize a polynomial as a sum or difference of two perfect cubes, we then identify the two cubed expressions and apply the formula.

EXAMPLE 4   Factor each polynomial.

a. $8a^3 + b^3$

b. $1 - 27h^6$

Solutions   **a.** This polynomial is a sum of two cubes. The cubed expressions are $2a$, because $(2a)^3 = 8a^3$, and b. Use formula (3) as a pattern, replacing x with $2a$, and y with b.

$$x^3 + y^3 = (x + y)(x^2 - xy + y^2)$$
$$(2a)^3 + b^3 = (2a + b)((2a)^2 - (2a)b + b^2) \quad \text{Simplify.}$$
$$= (2a + b)(4a^2 - 2ab + b^2)$$

b. This polynomial is a difference of two cubes. The cubed expressions are 1, because $1^3 = 1$, and $3h^2$, because $(3h^2)^3 = 27h^6$. Use formula (4) on the preceding page as a pattern, replacing x by 1, and y by $3h^2$:

$$x^3 - y^3 = (x - y)\ (x^2 + \ xy \ + \ y^2)$$
$$1^3 - (3h^2)^3 = (1 - 3h^2)(1^2 + 1(3h^2) + (3h^2)^2) \quad \text{Simplify.}$$
$$= (1 - 3h^2)(1 \ + 3h^2 \ + 9\,h^4)$$

EXERCISE 4

Factor $125n^3 - p^3$.

Modeling with Polynomials

Polynomials model many variable relationships, including volume and surface area.

EXAMPLE 5

A closed box has a square base of length and width x inches and a height of 8 inches, as shown in Figure 7.4.

a. Write a polynomial function $S(x)$ that gives the surface area of the box in terms of the dimensions of the base.

b. What is the surface area of a box of length and width 18 inches?

Solutions   **a.** The surface area of a box is the sum of the areas of its six faces,

$$S = 2lh + 2wh + 2lw$$

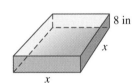

8 in

x

x

FIGURE 7.4

Substituting x for l and w, and 8 for h gives us

$$S(x) = 2(8)x + 2(8)x + 2x^2 = 2x^2 + 32x$$

b. We evaluate the polynomial for $x = 18$ to find

$$S(18) = 2(18)^2 + 32(18) = 1224 \text{ square inches}$$

EXERCISE 5

An empty reflecting pool is 3 feet deep. It is 8 feet longer than it is wide, as illustrated in Figure 7.5.

a. Write a polynomial function $S(x)$ that gives the surface area of the empty pool.

b. Write a polynomial function $V(x)$ for the volume of the pool.

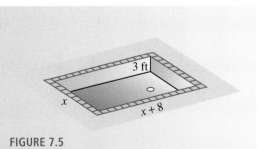

FIGURE 7.5

3 ft

x

$x + 8$

Cubic polynomials are often used in economics to model cost functions. The cost of producing x items is an increasing function of x, but its rate of increase is usually not constant.

EXAMPLE 6

Pegasus Printing, Ltd. is launching a new magazine. The cost of printing x thousand copies is given by

$$C(x) = x^3 - 24x^2 + 195x + 250$$

a. What are the *fixed costs,* that is, the costs incurred before any copies are printed?

b. Graph the cost function in the window below and describe the graph.

$$\text{Xmin} = 0 \quad \text{Xmax} = 20$$
$$\text{Ymin} = 0 \quad \text{Ymax} = 1500$$

c. How many copies can be printed for $1200?

d. What does the concavity of the graph tell you about the cost function?

Solutions
a. Fixed costs are given by $C(0) = 250$, or $250. The fixed costs include expenses like utility bills that must be paid even if no magazines are produced.

b. The graph is shown in Figure 7.6a. It is increasing from a vertical intercept of 250. The graph is concave down for $x < 8$ approximately, and concave up for $x > 8$.

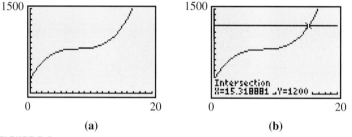

(a)   **(b)**

FIGURE 7.6

c. We must solve the equation

$$x^3 - 24x^2 + 195x + 250 = 1200$$

We will solve the equation graphically, as shown in Figure 7.6b. Graph $y = 1200$ along with the cost function, and use the **intersect** command to find the intersection point of the graphs, $(15.319, 1200)$. $C(x) = 1200$ when x is about 15.319, so 15,319 copies can be printed for $1200.

d. Although the cost is always increasing, it increases very slowly from about $x = 5$ to about $x = 11$. The flattening of the graph in this interval is a result of *economy of scale:* By buying supplies in bulk and using time efficiently, the cost per magazine can be minimized. However, if the production level is too large, costs begin to rise rapidly again.

In Example 5c, we solved a cubic equation graphically. There is a *cubic formula,* analogous to the quadratic formula, that allows us to solve cubic equations algebraically, but it is complicated and not often used. See the Projects for Chapter 7 if you would like to know more about the cubic formula.

Cubic polynomials are also used to model smooth curves connecting given points. Such a curve is called a *cubic spline.*

EXERCISE 6

Leon is flying his plane to Au Gres, Michigan. He maintains a constant altitude until he passes over a marker just outside the neighboring town of Omer, when he begins his descent for landing. During the descent, his altitude, in feet, is given by

$$A(x) = 128x^3 - 960x^2 + 8000$$

where x is the number of miles Leon has traveled since passing over the marker in Omer.

a. What is Leon's altitude when he begins his descent?

b. Graph $A(x)$ in the window

Xmin = 0	Xmax = 5
Ymin = 0	Ymax = 8000

c. Use the **Trace** feature to discover how far from Omer Leon will travel before landing. (In other words, how far is Au Gres from Omer?)

d. Verify your answer to part (c) algebraically.

ANSWERS TO 7.1 EXERCISES

1. $y^3 - y + 6$ **2.** 2   **3.** $125 + 75x^2 + 15x^4 + x^6$

4. $(5n - p)(25n^2 + 5np + p^2)$

5. a. $S(x) = x^2 + 20x + 48$ **6. a.** 8000 ft   **b.**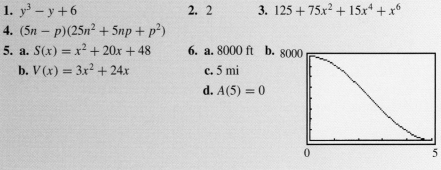

      b. $V(x) = 3x^2 + 24x$ **c.** 5 mi

                                  d. $A(5) = 0$

SECTION 7.1 SUMMARY

VOCABULARY *Look up the definitions of new terms in the Glossary.*

Polynomial function   Degree   Ascending powers   Lead coefficient
Descending powers

CONCEPTS

1. The degree of a product of nonzero polynomials is the sum of the degrees of the factors.

2.

> **Cube of a Binomial**
>
> **1.** $(x + y)^3 = x^3 + 3x^2y + 3xy^2 + y^3$
> **2.** $(x - y)^3 = x^3 - 3x^2y + 3xy^2 - y^3$

3.

> **Factoring the Sum or Difference of Two Cubes**
>
> **3.** $x^3 + y^3 = (x + y)(x^2 - xy + y^2)$
> **4.** $x^3 - y^3 = (x - y)(x^2 + xy + y^2)$

STUDY QUESTIONS

1. If you add two polynomials of degree 3, what can you say about the degree of the sum?
2. If you multiply a polynomial of degree 3 and a polynomial of degree 4, what can you say about the degree of the product?
3. If you multiply together

$$(2x - 1)(2x - 2)(2x - 3) \cdots (2x - 8)$$

what will be the degree of the product? What will be the lead coefficient?
4. What are the two middle terms in the expansion of $(x + y)^3$?
5. Is it possible to factor the sum of two cubes? What about the sum of two squares?

SKILLS *Practice each skill in the Homework problems listed.*

1. Multiply polynomials: #1–8
2. Find specific terms of polynomial products: #9–22
3. Use the formula for the cube of a binomial: #23–34
4. Factor the sum or difference of two cubes: #35–46
5. Write and analyze polynomial models: #47–64

HOMEWORK 7.1

ThomsonNOW™ Test yourself on key content at **www.thomsonedu.com/login**.

Multiply.

1. $(3x - 2)(4x^2 + x - 2)$
2. $(2x + 3)(3x^2 - 4x + 2)$
3. $(x - 2)(x - 1)(x - 3)$
4. $(z - 5)(z + 6)(z - 1)$
5. $(2a^2 - 3a + 1)(3a^2 + 2a - 1)$
6. $(b^2 - 3b + 5)(2b^2 - b + 1)$
7. $(y - 2)(y + 2)(y + 4)(y + 1)$
8. $(z + 3)(z + 2)(z - 1)(z + 1)$

Find the first three terms of the product in ascending powers. (Do not compute the entire product!)

9. $(2 - x + 3x^2)(3 + 2x - x^2 + 2x^4)$
10. $(1 + x - 2x^2)(-3 + 2x - 4x^3)$
11. $(1 - 2x^2 - x^4)(4 + x^2 - 2x^4)$
12. $(3 + 2x)(5 - 2x^2 - 3x^3 - x^5 + 2x^6)$

Find the indicated term in each product. (Do not compute the entire product!)

13. $(4 + 2x - x^2)(2 - 3x + 2x^2);\quad x^2$
14. $(1 - 2x + 3x^2)(6 - x - x^3);\quad x^3$
15. $(3x + x^3 - 7x^5)(1 + 4x - 3x^2);\quad x^3$
16. $(2 + 3x^2 + 2x^4)(2 - x - x^2 - x^4);\quad x^4$

Without performing the multiplication, give the degree of each product.

17. **a.** $(x^2 - 4)(3x^2 - 6x + 2)$
    b. $(x - 3)(2x - 5)(x^3 - x + 2)$
    c. $(3x^2 + 2x)(x^3 + 1)(-2x^2 + 8)$
18. **a.** $(6x^2 - 1)(4x^2 - 9)$
    b. $(3x + 4)(3x + 1)(2x^3 + x^2 - 7)$
    c. $(x^2 - 3)(2x^3 - 5x^2 + 2)(-x^3 - 5x)$

Verify the following products discussed in the text.

19. $(x + y)^3 = x^3 + 3x^2y + 3xy^2 + y^3$
20. $(x - y)^3 = x^3 - 3x^2y + 3xy^2 - y^3$
21. $(x + y)(x^2 - xy + y^2) = x^3 + y^3$
22. $(x - y)(x^2 + xy + y^2) = x^3 - y^3$

23. a. As if you were addressing a classmate, explain how to remember the formula for expanding $(x + y)^3$. In particular, mention the exponents on each term and the numerical coefficients.
   b. Explain how to remember the formula for expanding $(x - y)^3$, assuming your listener already knows the formula for $(x + y)^3$.

24. a. As if you were addressing a classmate, explain how to remember the formula for factoring a sum of two cubes. Pay particular attention to the placement of the variables and the signs of the terms.
   b. Explain how to remember the formula for factoring a difference of two cubes, assuming your listener already knows how to factor a sum of two cubes.

Use the formulas for the cube of a binomial to expand the products.

25. $(1 + 2z)^3$

26. $(1 - x^2)^3$

27. $(1 - 5\sqrt{t})^3$

28. $\left(1 - \dfrac{3}{a}\right)^3$

Write each product as a polynomial and simplify.

29. $(x - 1)(x^2 + x + 1)$

30. $(x + 2)(x^2 - 2x + 4)$

31. $(2x + 1)(4x^2 - 2x + 1)$

32. $(3x - 1)(9x^2 + 3x + 1)$

33. $(3a - 2b)(9a^2 + 6ab + 4b^2)$

34. $(2a + 3b)(4a^2 - 6ab + 9b^2)$

Factor completely.

35. $x^3 + 27$

36. $y^3 - 1$

37. $a^3 - 8b^3$

38. $27a^3 + b^3$

39. $x^3y^6 - 1$

40. $8 + x^{12}y^3$

41. $27a^3 + 64b^3$

42. $8a^3 - 125b^3$

43. $125a^3b^3 - 1$

44. $64a^3b^3 + 1$

45. $64t^9 + w^6$

46. $w^{15} - 125t^9$

47. a. Write a polynomial function, $A(x)$, that gives the area of the front face of the speaker frame (the region in color) in the figure.
   b. If $x = 8$ inches, find the area of the front face of the frame.

48. a. A Norman window is shaped like a rectangle whose length is twice its width, with a semicircle at the top (see the figure). Write a polynomial, $A(x)$, that gives its area.
   b. If $x = 3$ feet, find the area of the window.

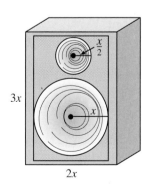

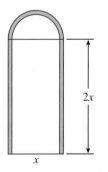

49. a. A grain silo is built in the shape of a cylinder with a hemisphere on top (see the figure). Write an expression for the volume of the silo in terms of the radius and height of the cylindrical portion of the silo.

b. If the total height of the silo is five times its radius, write a polynomial function $V(r)$ in one variable for its volume.

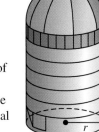

50. a. A cold medication capsule is shaped like a cylinder with a hemispherical cap on each end (see the figure). Write an expression for the volume of the capsule in terms of the radius and length of the cylindrical portion.

b. If the radius of the capsule is one-fourth of its overall length, write a polynomial function $V(r)$ in one variable for its volume.

51. Jack invests $500 in an account bearing interest rate r, compounded annually. This means that each year his account balance is increased by a factor of $1 + r$.

a. Write expressions for the amount of money in Jack's account after 2 years, after 3 years, and after 4 years.

b. Expand the expressions you found in part (a) as polynomials.

c. How much money will be in Jack's account at the end of 2 years, 3 years, and 4 years if the interest rate is 8%?

52. A small company borrows $800 for start-up costs and agrees to repay the loan at interest rate r, compounded annually. This means that each year the debt is increased by a factor of $1 + r$.

a. Write expressions for the amount of money the company will owe if it repays the loan after 2 years, after 3 years, or after 4 years.

b. Expand the expressions you found in part (a) as polynomials.

c. How much money will the company owe after 2 years, after 3 years, or after 4 years at an interest rate of 12%?

53. A paper company plans to make boxes without tops from sheets of cardboard 12 inches wide and 16 inches long. The company will cut out four squares of side x inches from the corners of the sheet and fold up the edges as shown in the figure.

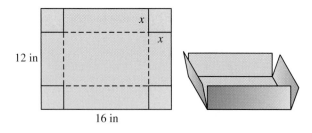

a. Write expressions in terms of x for the length, width, and height of the resulting box.

b. Write a formula for the volume, V, of the box as a function of x.

c. What is the domain of the function V? (What are the largest and smallest reasonable values for x?)

d. Make a table of values for $V(x)$ on its domain.

e. Graph your function V in a suitable window.

f. Use your graph to find the value of x that will yield a box with maximum possible volume. What is the maximum possible volume?

54. The paper company also plans to make boxes with tops from 12-inch by 16-inch sheets of cardboard by cutting out the shaded areas shown in the figure and folding along the dotted lines.

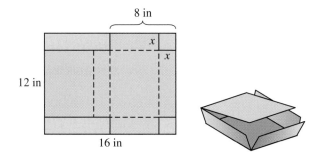

a. Write expression in terms of x for the length, width, and height of the resulting box.

b. Write a formula for the volume V of the box as a function of x.

c. What is the domain of the function V? (What are the largest and smallest reasonable values for x?)

d. Make a table of values for $V(x)$ on its domain.

e. Graph your function V in a suitable window.

f. Use your graph to find the value of x that will yield a box with maximum possible volume. What is the maximum possible volume?

55. A doctor who is treating a heart patient wants to prescribe medication to lower the patient's blood pressure. The body's reaction to this medication is a function of the dose administered. If the patient takes x milliliters of the medication, his blood pressure should decrease by $R = f(x)$ points, where

$$f(x) = 3x^2 - \frac{1}{3}x^3$$

a. For what values of x is $R = 0$?
b. Find a suitable domain for the function and explain why you chose this domain.
c. Graph the function f on its domain.
d. How much should the patient's blood pressure drop if he takes 2 milliliters of medication?
e. What is the maximum drop in blood pressure that can be achieved with this medication?
f. There may be risks associated with a large change in blood pressure. How many milliliters of the medication should be administered to produce half the maximum possible drop in blood pressure?

57. The population, $P(t)$, of Cyberville has been growing according to the formula

$$P(t) = t^3 - 63t^2 + 1403t + 900$$

where t is the number of years since 1970.
a. Graph $P(t)$ in the window

$$\text{Xmin} = 0, \quad \text{Xmax} = 47$$
$$\text{Ymin} = 0, \quad \text{Ymax} = 20{,}000$$

b. What was the population in 1970? In 1985? In 2004?
c. By how much did the population grow from 1970 to 1971? From 1985 to 1986? From 2004 to 2005?
d. Approximately when was the population growing at the slowest rate, that is, when is the graph the least steep?

56. A soup bowl has the shape of a hemisphere of radius 6 centimeters. The volume of the soup in the bowl, $V = f(x)$, is a function of the depth, x, of the soup.

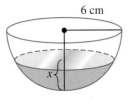

6 cm

a. What is the domain of f? Why?
b. The function f is given by

$$f(x) = 6\pi x^2 - \frac{\pi}{3}x^3$$

Graph the function on its domain.
c. What is the volume of the soup if it is 3 centimeters deep?
d. What is the maximum volume of soup that the bowl can hold?
e. Find the depth of the soup (to within 2 decimal places of accuracy) when the bowl is filled to half its capacity.

58. The annual profit, $P(t)$, of the Enviro Company, in thousands of dollars, is given by

$$P(t) = 2t^3 - 152t^2 + 3400t + 30$$

where t is the number of years since 1960, the first year that the company showed a profit.
a. Graph $P(t)$ on the window

$$\text{Xmin} = 0, \quad \text{Xmax} = 94$$
$$\text{Ymin} = 0, \quad \text{Ymax} = 50{,}000$$

b. What was the profit in 1960? In 1980? In 2000?
c. How did the profit change from 1960 to 1961? From 1980 to 1981? From 2000 to 2001?
d. During which years did the profit decrease from one year to the next?

59. The total annual cost of educating postgraduate research students at an Australian university, in thousands of dollars, is given by the function

$$C(x) = 0.0173x^3 - 0.647x^2 + 9.587x + 195.366$$

where x is the number of students, in hundreds. (Source: Creedy, Johnson, and Valenzuela, 2002)

a. Graph the function in a suitable window for up to 3500 students.

b. Describe the concavity of the graph. For what value of x is the cost growing at the slowest rate?

c. Approximately how many students can be educated for $350,000?

60. It has been proposed that certain cubic functions model the response of wheat and barley to nitrogen fertilizer. These functions exhibit a "plateau" that fits observations better than the standard quadratic model. (See Problem 36 of Section 6.6.) In trials in Denmark, the yield per acre was a function of the amount of nitrogen applied. A typical response function is

$$Y(x) = 54.45 + 0.305x - 0.001655x^2 + 2.935 \times 10^{-6}x^3$$

where x is the amount of fertilizer, in kilograms per acre. (Source: Beattie, Mortensen, and Knudsen, 2005)

a. Graph the function on the domain $[0, 400]$.

b. Describe the concavity of the graph. In reality, the yield does not increase after reaching its plateau. Give a suitable domain for the model in this application.

c. Estimate the maximum yield attainable and the optimum application of fertilizer.

61. During an earthquake, Nordhoff Street split in two, and one section shifted up several centimeters. Engineers created a ramp from the lower section to the upper section. In the coordinate system shown in the figure below, the ramp is part of the graph of

$$y = f(x) = -0.00004x^3 - 0.006x^2 + 20$$

a. By how much did the upper section of the street shift during the earthquake?

b. What is the horizontal distance from the bottom of the ramp to the raised part of the street?

62. The off-ramp from a highway connects to a parallel one-way road. The accompanying figure shows the highway, the off-ramp, and the road. The road lies on the x-axis, and the off-ramp begins at a point on the y-axis. The off-ramp is part of the graph of the polynomial

$$y = f(x) = 0.00006x^3 - 0.009x^2 + 30$$

a. How far east of the exit does the off-ramp meet the one-way road?

b. How far apart are the highway and the road?

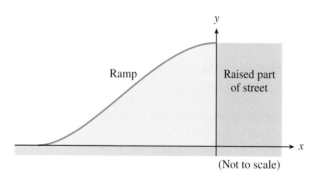

(Not to scale)

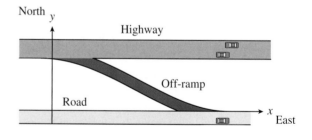

63. The number of minutes of daylight per day in Chicago is approximated by the polynomial

$$H(t) = 0.000\ 000\ 525t^4 - 0.0213t^2 + 864$$

where t is the number of days since the summer solstice. The approximation is valid for $-74 < t < 74$. (A negative value of t corresponds to a number of days before the summer solstice.)

a. Use a table of values with increments of 10 days to estimate the range of the function on its domain.

b. Graph the polynomial on its domain.

c. How many minutes of daylight are there on the summer solstice?

d. How much daylight is there two weeks before the solstice?

e. When are the days more than 14 hours long?

f. When are the days less than 13 hours long?

64. The water level (in feet) at a harbor is approximated by the polynomial

$$W(t) = 0.00733t^4 - 0.332t^2 + 9.1$$

where t is the number of hours since the high tide. The approximation is valid for $-4 \le t \le 4$. (A negative value of t corresponds to a number of hours before the high tide.)

a. Use a table of values to estimate the range of the function on its domain.

b. Graph the polynomial on its domain.

c. What is the water level at high tide?

d. What is the water level 3 hours before high tide?

e. When is the water level below 8 feet?

f. When is the water level above 7 feet?

7.2 Graphing Polynomial Functions

In Section 7.1, we considered applications of polynomial functions. Although most applications use only a portion of the graph of a particular polynomial, we can learn a lot about these functions by taking a more global view of their behavior.

Classifying Polynomials by Degree

The graph of a polynomial function depends first of all on its degree. We have already studied the graphs of polynomials of degrees 0, 1, and 2. A polynomial of degree 0 is a constant, and its graph is a horizontal line. An example of such a polynomial function is $f(x) = 3$ (see Figure 7.7a). A polynomial of degree 1 is a linear function, and its graph is a straight line. The function $f(x) = 2x - 3$ is an example of a polynomial of degree 1. (See Figure 7.7b.)

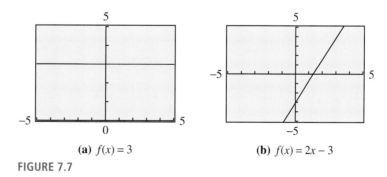

(a) $f(x) = 3$ **(b)** $f(x) = 2x - 3$

FIGURE 7.7

Quadratic functions, such as $f(x) = -2x^2 + 6x + 8$, are polynomials of degree 2. The graph of every quadratic function is a parabola, with the same basic shape as the standard parabola, $y = x^2$. (See Figure 7.8 on page 578.) It has one **turning point,** where the graph changes from increasing to decreasing or vice versa. The turning point of a parabola is the same as its vertex.

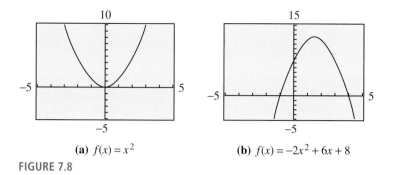

(a) $f(x) = x^2$ **(b)** $f(x) = -2x^2 + 6x + 8$

FIGURE 7.8

Cubic Polynomials

Do the graphs of all **cubic,** or third-degree, polynomials have a basic shape in common? We can graph a few examples and find out. Unlike the basic parabola, the graph of $y = x^3$ is always increasing. At the origin, however, it changes from concave down to concave up. A point where the graph changes concavity is called an **inflection point.**

EXAMPLE 1   Graph the cubic polynomial $P(x) = x^3 - 4x$ and compare its graph with that of the basic cubic, $y = x^3$.

Solution   The graph of the basic cubic is shown in Figure 7.9a. To help us understand the graph of the polynomial $P(x) = x^3 - 4x$, we will evaluate the function to make a table of values. We can do this by hand or use the **Table** feature on the graphing calculator.

x	-3	-2	-1	0	1	2	3
$P(x)$	-15	0	3	0	-3	0	15

The graph of $P(x) = x^3 - 4x$ is shown in Figure 7.9b. It is not exactly the same shape as the basic cubic—it has two turning points—but it is similar, especially at the edges of the graphs.

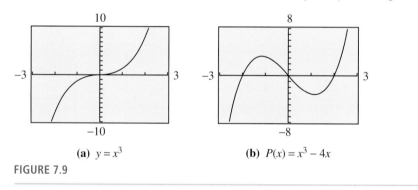

(a) $y = x^3$ **(b)** $P(x) = x^3 - 4x$

FIGURE 7.9

     Despite the differences in the central portions of the two graphs, they exhibit similar *long-term* behavior. For very large and very small values of x, both graphs look like the power function $y = x^3$. The y-values increase from $-\infty$ toward zero in the third quadrant, and they increase from zero toward $+\infty$ in the first quadrant. Or we might say that the graphs start at the lower left and extend to the upper right. All cubic polynomials display this behavior when their lead coefficients (the coefficient of the x^3 term) are positive.

Both of the graphs in Example 1 are smooth curves without any breaks or holes. This smoothness is a feature of the graphs of all polynomial functions. The domain of any polynomial function is the entire set of real numbers.

EXERCISE 1

a. Complete the table of values below for $C(x) = -x^3 - 2x^2 + 4x + 4$.

x	-4	-3	-2	-1	0	1	2	3	4
y									

b. Graph $y = C(x)$ in the standard window. Compare the graph to the graphs in Example 1: What similarities do you notice? What differences?

Quartic Polynomials

Now let's compare the long-term behavior of two **quartic**, or fourth-degree, polynomials.

EXAMPLE 2

Graph the polynomials $f(x) = x^4 - 10x^2 + 9$ and $g(x) = x^4 + 2x^3$, and compare.

Solution For each function we make a table of values.

x	-4	-3	-2	-1	0	1	2	3	4
$f(x)$	105	0	-15	0	9	0	-15	0	105
$g(x)$	128	27	0	-1	0	3	32	135	384

The graphs are shown in Figure 7.10. All the essential features of the graphs are shown in these viewing windows. The graphs continue forever in the directions indicated, without any additional twists or turns. You can see that the graph of f has three turning points, and the graph of g has one turning point.

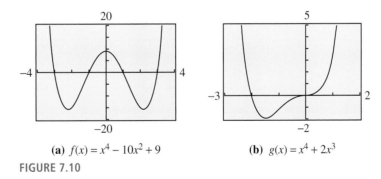

(a) $f(x) = x^4 - 10x^2 + 9$ **(b)** $g(x) = x^4 + 2x^3$

FIGURE 7.10

As in Example 1, both graphs have similar long-term behavior. The y-values decrease from $+\infty$ toward zero as x increases from $-\infty$, and the y-values increase toward $+\infty$ as x increases to $+\infty$. This long-term behavior is similar to that of the power function $y = x^4$. Its graph also starts at the upper left and extends to the upper right.

In Examples 1 and 2, we have seen polynomials of degree 3, whose graphs have a characteristic shape, illustrated in Figure 7.11a, and polynomials of degree 4, whose graphs are illustrated by Figure 7.11b. In the exercises, you will consider more graphs to help you verify the following observations.

Long-Term Behavior of Polynomial Functions

1. A polynomial of **odd** degree (with positive lead coefficient) has negative y-values for large negative x and positive y-values for large positive x.

2. A polynomial of **even** degree (with positive lead coefficient) has positive y-values for both large positive and large negative x.

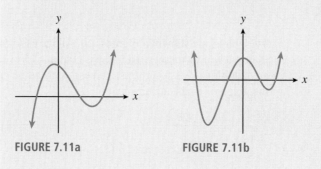

FIGURE 7.11a   FIGURE 7.11b

Another way to describe the long-term behavior of a polynomial graph is to note that for large values of $|x|$, the shape is similar to a power function of the same degree. It is the presence of the lower-degree terms in the polynomial that are responsible for any extra wiggles or turning points in the graph.

x-Intercepts and the Factor Theorem

In Chapter 6, we saw that the x-intercepts of a quadratic polynomial, $f(x) = ax^2 + bx + c$, occur at values of x for which $f(x) = 0$, that is, at the real-valued solutions of the equation $ax^2 + bx + c = 0$. The same holds true for polynomials of higher degree.

Solutions of the equation $P(x) = 0$ are called **zeros** of the polynomial P. In Example 1, we graphed the cubic polynomial $P(x) = x^3 - 4x$. Its x-intercepts are the solutions of the equation $x^3 - 4x = 0$, which we can solve by factoring the polynomial $P(x)$.

$$x^3 - 4x = 0$$
$$x(x - 2)(x + 2) = 0$$

The zeros of P are 0, 2, and -2. Each zero of P corresponds to a factor of $P(x)$. This result suggests the following theorem, which holds for any polynomial P.

Factor Theorem

Let $P(x)$ be a polynomial with real number coefficients. Then $(x - a)$ is a factor of $P(x)$ if and only if $P(a) = 0$.

▶▶▶
If you would like to review polynomial division, see Algebra Skills Refresher A.7.

The factor theorem follows from the division algorithm for polynomials. We will consider both of these results in more detail in the Homework problems.

Because a polynomial function of degree n can have at most n linear factors of the form $(x - a)$, it follows that P can have at most n distinct zeros. Another way of saying this is that if $P(x)$ is a polynomial of nth degree, the equation $P(x) = 0$ can have at most n distinct solutions, some of which may be complex numbers. (We consider complex numbers in Section 7.3.) Because only real-valued solutions appear on the graph as x-intercepts, we have the following corollary to the factor theorem.

x-Intercepts of Polynomials

A polynomial of degree n can have at most n x-intercepts.

If some of the zeros of P are complex numbers, they will not appear on the graph, so a polynomial of degree n may have *fewer* than n x-intercepts.

EXAMPLE 3   Find the real-valued zeros of each of the following polynomials, and list the x-intercepts of its graph.

a. $f(x) = x^3 + 6x^2 + 9x$ **b.** $g(x) = x^4 - 3x^2 - 4$

Solutions   **a.** Factor the polynomial to obtain

$$f(x) = x(x^2 + 6x + 9)$$
$$= x(x + 3)(x + 3)$$

By the factor theorem, the zeros of f are 0, -3, and -3. (We say that f has a zero *of multiplicity two* at -3.) Because all of these are real numbers, all will appear as x-intercepts on the graph. Thus, the x-intercepts occur at $(0, 0)$ and at $(-3, 0)$.

b. Factor the polynomial to obtain

$$g(x) = (x^2 - 4)(x^2 + 1)$$
$$= (x - 2)(x + 2)(x^2 + 1)$$

Because $x^2 + 1$ cannot be factored in real numbers, the graph has only two x-intercepts, at $(-2, 0)$ and $(2, 0)$. The graphs of both polynomials are shown in Figure 7.12 on page 582.

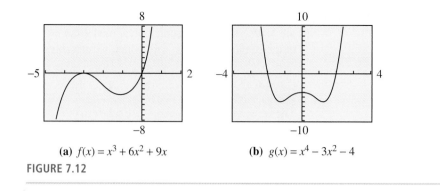

(a) $f(x) = x^3 + 6x^2 + 9x$ **(b)** $g(x) = x^4 - 3x^2 - 4$

FIGURE 7.12

EXERCISE 3

a. Find the real-valued zeros of $P(x) = -x^4 + x^3 + 2x^2$ by factoring.

b. Sketch a rough graph of $y = P(x)$ by hand.

Zeros of Multiplicity Two or Three

The appearance of the graph near an x-intercept is determined by the multiplicity of the zero there. For example, both real zeros of the polynomial $g(x) = x^4 - 3x^2 - 4$ in Example 3b are of multiplicity one, and the graph *crosses* the x-axis at each intercept. However, the polynomial $f(x) = x^3 + 6x^2 + 9x$ in Example 3a has a zero of multiplicity two at $x = -3$. The graph of f just *touches* the x-axis and then reverses direction without crossing the axis.

To understand what happens in general, compare the graphs of the three polynomials in Figure 7.13.

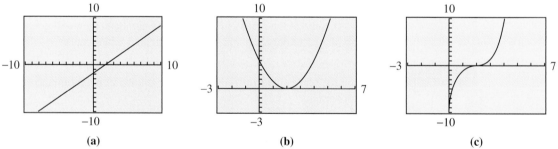

(a)   **(b)**   **(c)**

FIGURE 7.13

1. In Figure 7.13a, $L(x) = x - 2$ has a zero of *multiplicity one* at $x = 2$, and its graph crosses the x-axis there.

2. In Figure 7.13b, $Q(x) = (x - 2)^2$ has a zero of *multiplicity two* at $x = 2$, and its graph touches the x-axis there but changes direction without crossing.

3. In Figure 7.13c, $C(x) = (x - 2)^3$ has a zero of *multiplicity three* at $x = 2$. In this case, the graph makes an S-shaped curve at the intercept, like the graph of $y = x^3$.

Near its x-intercepts, the graph of a polynomial takes one of the characteristic shapes illustrated in Figure 7.13. (Although we will not consider zeros of multiplicity greater than three, they correspond to similar behavior in the graph: At a zero of odd multiplicity, the graph has an inflection point at the intercept; its graph makes an S-shaped curve. At a zero of even multiplicity, the graph has a turning point; it changes direction without crossing the x-axis.)

EXAMPLE 4   Graph the polynomial

$$f(x) = (x + 2)^3(x - 1)(x - 3)^2$$

Solution   The polynomial has degree six, an even number, so its graph starts at the upper left and extends to the upper right. Its y-intercept is

$$f(0) = (2)^3(-1)(-3)^2 = -72$$

f has a zero of multiplicity three at $x = -2$, a zero of multiplicity one at $x = 1$, and a zero of multiplicity two at $x = 3$. The graph has an S-shaped curve at $x = -2$, crosses the x-axis at $x = 1$, touches the x-axis at $x = 3$, and then changes direction, as shown in Figure 7.14.

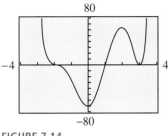

FIGURE 7.14

EXERCISE 4

Sketch a rough graph of $f(x) = (x + 3)(x - 1)^2$ by hand. Label the x- and y-intercepts.

ANSWERS TO 7.2 EXERCISES

1. a.

x	−4	−3	−2	−1	0	1	2	3	4
y	20	1	−4	−1	4	5	−4	−29	−76

b.

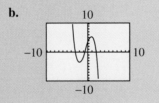

Both graphs have three x-intercepts, but the function in Example 1 has long-term behavior like $y = x^3$, and this function has long-term behavior like $y = -x^3$.

2. a.

x	−4	−3	−2	−1	0	1	2	3	4
y	−286	−106	−30	−4	2	−6	−46	−160	−414

b.

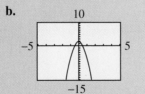

The graphs all have long-term behavior like a fourth degree power function, $y = ax^4$. The long-term behavior of the graphs in Example 2 is the same as that of $y = x^4$, but the graph here has long-term behavior like $y = -x^4$.

3. a. −1, 0, 2   **b.**   **4.**

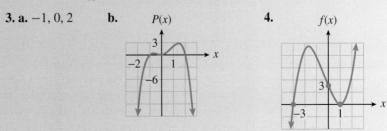

VOCABULARY   *Look up the definitions of new terms in the Glossary.*

Degree	Multiplicity	Inflection point	Quartic
Turning point	Cubic	Corollary	Zero

CONCEPTS

1. The graphs of all polynomials are smooth curves without breaks or holes.
2. The graph of a polynomial of degree n (with positive leading coefficient) has the same long-term behavior as the power function of the same degree.
3. **Factor theorem:** $(x - a)$ is a factor of a polynomial $P(x)$ if and only if $P(a) = 0$.

4. A polynomial of degree n can have at most n x-intercepts.
5. At a zero of multiplicity two, the graph of a polynomial has a turning point. At a zero of multiplicity three, the graph of a polynomial has an inflection point.

STUDY QUESTIONS

1. Describe the graphs of polynomials of degrees 0, 1, and 2.
2. What does the degree of a polynomial tell you about its long-term behavior?
3. What is a zero of a polynomial?
4. How are zeros related to the factors of a polynomial?

5. What do the zeros tell you about the graph of a polynomial?
6. Explain the difference between a turning point and an inflection point.

SKILLS   *Practice each skill in the Homework problems listed.*

1. Identify x-intercepts, turning points, and inflection points: #1–8, 11–18
2. Use a graph to factor a polynomial: #21–28
3. Sketch the graph of a polynomial: #29–46

4. Find a possible formula for a polynomial whose graph is shown: #47–52
5. Graph translations of polynomials: #53–56

HOMEWORK 7.2

ThomsonNOW™ Test yourself on key content at **www.thomsonedu.com/login.**

In Problems 1–8, use your calculator to graph each cubic polynomial.
(a) Describe the long-term behavior of each graph. How does this behavior compare to that of the basic cubic? How does the sign of the lead coefficient affect the graph?
(b) How many x-intercepts does each graph have? How many turning points? How many inflection points?

1. $y = x^3 + 4$

2. $y = x^3 - 8$

3. $y = -2 - 0.05x^3$

4. $y = 5 - 0.02x^3$

5. $y = x^3 - 3x$

6. $y = 9x - x^3$

7. $y = x^3 + 5x^2 - 4x - 20$

8. $y = -x^3 - 2x^2 + 5x + 6$

Use a calculator to graph each cubic polynomial. Which graphs are the same?

9. **a.** $y = x^3 - 2$
   b. $y = (x - 2)^3$
   c. $y = x^3 - 6x^2 + 12x - 8$

10. **a.** $y = x^3 + 3$
    b. $y = (x + 3)^3$
    c. $y = x^3 + 9x^2 + 27x + 27$

In Problems 11–18, use your calculator to graph each quartic polynomial.
(a) Describe the long-term behavior of each graph. How does this behavior compare to that of the basic quartic? How does the sign of the lead coefficient affect the graph?
(b) How many x-intercepts does each graph have? How many turning points? How many inflection points?

11. $y = 0.5x^4 - 4$

12. $y = 0.3x^4 + 1$

13. $y = -x^4 + 6x^2 - 10$

14. $y = x^4 - 8x^2 - 8$

15. $y = x^4 - 3x^3$

16. $y = -x^4 - 4x^3$

17. $y = -x^4 - x^3 - 2$

18. $y = x^4 + 2x^3 + 4x^2 + 10$

19. From your answers to Problems 1–8, what you can conclude about the graphs of cubic polynomials? Consider the long-term behavior, x-intercepts, turning points, and inflection points.

20. From your answers to Problems 11–18, what you can conclude about the graphs of quartic polynomials? Consider the long-term behavior, x-intercepts, turning points, and inflection points.

For Problems 21–26,
(a) Use your calculator to graph each polynomial and locate the x-intercepts. Set Xmin = −4.7, Xmax = 4.7, and adjust Ymin and Ymax to get a good graph.
(b) Write the polynomial in factored form.
(c) Expand the factored form of the polynomial (that is, multiply the factors together). Do you get the original polynomial?

21. $P(x) = x^3 - 7x - 6$

22. $Q(x) = x^3 + 3x^2 - x - 3$

23. $R(x) = x^4 - x^3 - 4x^2 + 4x$

24. $S(x) = x^4 + 3x^3 - x^2 - 3x$

25. $p(x) = x^3 - 3x^2 - 6x + 8$

26. $q(x) = x^3 + 6x^2 - x - 30$

27. $r(x) = x^4 - x^3 - 10x^2 + 4x + 24$

28. $s(x) = x^4 - x^3 - 12x^2 - 4x + 16$

Sketch a rough graph of each polynomial function by hand, paying attention to the shape of the graph near each x-intercept. Check by graphing with a calculator.

29. $q(x) = (x + 4)(x + 1)(x - 1)$

30. $p(x) = x(x + 2)(x + 4)$

31. $G(x) = (x - 2)^2(x + 2)^2$

32. $F(x) = (x - 1)^2 (x - 3)^2$

33. $h(x) = x^3(x + 2)(x - 2)$

34. $H(x) = (x + 1)^3(x - 2)^2$

35. $P(x) = (x + 4)^2(x + 1)^2(x - 1)^2$

36. $Q(x) = x^2(x - 5)(x - 1)^2(x + 2)$

For Problems 37–46,
(a) Find the zeros of each polynomial by factoring.
(b) Sketch a rough graph by hand.

37. $P(x) = x^4 + 4x^2$

38. $P(x) = x^3 + 3x$

39. $f(x) = x^4 + 4x^3 + 4x^2$

40. $g(x) = x^4 + 4x^3 + 3x^2$

41. $g(x) = 4x - x^3$

42. $f(x) = 8x - x^4$

43. $k(x) = x^4 - 10x^2 + 16$

44. $m(x) = x^4 - 15x^2 + 36$

45. $r(x) = (x^2 - 1)(x + 3)^2$

46. $s(x) = (x^2 - 9)(x - 1)^2$

Find a possible equation for the polynomial whose graph is shown.

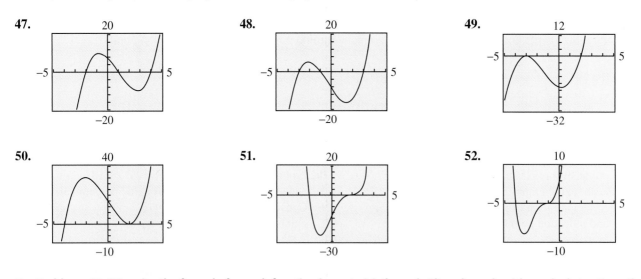

For Problems 53–56, write the formula for each function in parts (a) through (d) and graph with a calculator. Describe how the graph differs from the graph of $y = f(x)$.

53. $f(x) = x^3 - 4x$
    a. $y = f(x) + 3$
    b. $y = f(x) - 5$
    c. $y = f(x - 2)$
    d. $y = f(x + 3)$

54. $f(x) = x^3 - x^2 + x - 1$
    a. $y = f(x) + 4$
    b. $y = f(x) - 4$
    c. $y = f(x - 3)$
    d. $y = f(x + 5)$

55. $f(x) = x^4 - 4x^2$
    a. $y = f(x) + 6$
    b. $y = f(x) - 2$
    c. $y = f(x - 1)$
    d. $y = f(x + 2)$

56. $f(x) = x^4 + 3x^3$
    a. $y = f(x) + 5$
    b. $y = f(x) - 3$
    c. $y = f(x - 2)$
    d. $y = f(x + 1)$

> ### Division Algorithm for Polynomials
>
> If $f(x)$ and $g(x)$ are nonconstant polynomials with real coefficients, then there exist unique polynomials $q(x)$ and $r(x)$ such that
>
> $$f(x) = g(x)q(x) + r(x),$$
>
> where deg $r(x) <$ deg $g(x)$.

In Problems 57–60, use polynomial division to divide $f(x)$ by $g(x)$, and hence find the quotient, $q(x)$, and remainder, $r(x)$. (See Algebra Skills Refresher A.7 to review polynomial division.)

57. $f(x) = 2x^3 - 2x^2 - 19x - 11$, $g(x) = x - 3$

58. $f(x) = 3x^3 + 12x^2 - 13x - 32$, $g(x) = x + 4$

59. $f(x) = x^5 + 2x^4 - 7x^3 - 12x^2 + 5$, $g(x) = x^2 + 2x - 1$

60. $f(x) = x^5 - 4x^4 + 11x^3 - 12x^2 + 5x + 2$, $g(x) = x^2 - x + 3$

61. The remainder theorem states, If $P(x)$ is a polynomial and a is any real number, there is a unique polynomial $Q(x)$ such that

$$P(x) = (x - a)Q(x) + P(a)$$

Follow the steps below to prove the remainder theorem.
    a. State the division algorithm applied to the polynomials $P(x)$ and $x - a$.
    b. What must be the degree of $r(x)$ in this case?
    c. Evaluate your expression from part (a) at $x = a$. What does this tell you about the remainder, $r(x)$?

62. Verify the remainder theorem for the following:
   a. $P(x) = x^3 - 4x^2 + 2x - 1, \ a = 2$
   b. $P(x) = 3x^2 + x - 5, \ a = -3$

63. Use the remainder theorem to prove the factor theorem, stated earlier in this section. You will need to justify two statements:
   a. If $P(a) = 0$, show that $x - a$ is a factor of $P(x)$.
   b. If $x - a$ is a factor of $P(x)$, show that $P(a) = 0$.

64. Verify the factor theorem for the following:
   a. $P(x) = x^4 - 4x^3 - 11x^2 + 3x + 2, \ a = -2$
   b. $P(x) = x^3 + 2x^2 - 31x - 20, \ a = 5$

For Problems 65–68,
(a) Verify that the given value is a zero of the polynomial.
(b) Find the other zeros. (*Hint:* Use polynomial division to write $P(x) = (x - a)Q(x)$, then factor $Q(x)$.)

65. $P(x) = x^3 - 2x^2 + 1; \quad a = 1$

66. $P(x) = x^3 + 2x^2 - 1; \quad a = -1$

67. $P(x) = x^4 - 3x^3 - 10x^2 + 24x; \quad a = -3$

68. $P(x) = x^4 + 5x^3 - x^2 - 5x; \quad a = -5$

In Problems 69–70, we use polynomials to approximate other functions.

69. **a.** Graph the functions $f(x) = e^x$ and
   $$p(x) = 1 + x + \frac{1}{2}x^2 + \frac{1}{6}x^3 \text{ in the standard window.}$$
   For what values of x does it appear that $p(x)$ would be a good approximation for $f(x)$?
   b. Change the window settings to
   $$\text{Xmin} = -4.7 \quad \text{Xmax} = 4.7$$
   $$\text{Ymin} = 0 \quad \text{Ymax} = 20$$
   and fill in the table of values below. (You can use the **value** feature on your calculator.)

x	-1	-0.5	0	0.5	1	1.5	2
$f(x)$							
$p(x)$							

   c. The *error* in the approximation is the difference $f(x) - p(x)$. We can reduce the error by using a polynomial of higher degree. The nth degree polynomial for approximating e^x is
   $$P_n(x) = 1 + x + \frac{1}{2!}x^2 + \frac{1}{3!}x^3 + \cdots + \frac{1}{n!}x^n$$
   where $n! = n(n - 1)(n - 2) \cdots 3 \cdot 2 \cdot 1$. Graph $f(x)$ and $P_5(x)$ in the same window as in part (b). What is the error in approximating $f(2)$ by $P_5(2)$?
   d. Graph $f(x) - P_5(x)$ in the same window as in part (b). What does the graph tell you about the error in approximating $f(x)$ by $P_5(x)$?

70. In the Projects for Chapter 2, we investigated periodic functions. The *sine function*, $f(x) = \sin(x)$, is a useful periodic function.
   a. Graph the functions $f(x) = \sin(x)$ and $p(x) = x - \frac{1}{6}x^3$ in the standard window. (Check that your calculator is set in **Radian** mode.) For what values of x does it appear that $p(x)$ would be a good approximation for $f(x)$?
   b. Change the window settings to
   $$\text{Xmin} = -4.7 \quad \text{Xmax} = 4.7$$
   $$\text{Ymin} = -2 \quad \text{Ymax} = 2$$
   and fill in the table of values below. (You can use the **value** feature on your calculator.)

x	-1	-0.5	0	0.5	1	1.5	2
$f(x)$							
$p(x)$							

   c. Two more polynomials for approximating $f(x) = \sin(x)$ are
   $$P_5(x) = x - \frac{1}{3!}x^3 + \frac{1}{5!}x^5$$
   $$P_7(x) = x - \frac{1}{3!}x^3 + \frac{1}{5!}x^5 - \frac{1}{7!}x^7$$
   (See Problem 69 for the definition of $n!$.) Graph $f(x)$ and $P_5(x)$ in the same window as in part (b). What is the error in approximating $f(2)$ by $P_5(2)$?
   d. Graph $f(x) - P_5(x)$ in the same window as in part (b). What does the graph tell you about the error in approximating $f(x)$ by $P_5(x)$?

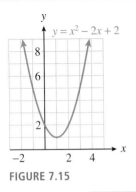

FIGURE 7.15

You know that not all quadratic equations have real solutions. For example, the graph of

$$f(x) = x^2 - 2x + 2$$

has no x-intercepts (as shown in Figure 7.15), and the equation

$$x^2 - 2x + 2 = 0$$

has no real solutions.

We can still use completing the square or the quadratic formula to solve the equation.

EXAMPLE 1   Solve the equation $x^2 - 2x + 2 = 0$ by using the quadratic formula.

Solution   We substitute $a = 1$, $b = -2$, and $c = 2$ into the quadratic formula to get

$$x = \frac{-(-2) \pm \sqrt{(-2)^2 - 4(1)(2)}}{2} = \frac{2 \pm \sqrt{-4}}{2}$$

Because $\sqrt{-4}$ is not a real number, the equation $x^2 - 2x + 2 = 0$ has no real solutions.

EXERCISE 1

Solve the equation $x^2 - 6x + 13 = 0$ by using the quadratic formula.

Imaginary Numbers

See Algebra Skills Refresher A.12 if you would like to review the properties of real numbers.

Although square roots of negative numbers such as $\sqrt{-4}$ are not real numbers, they occur often in mathematics and its applications. Mathematicians began working with square roots of negative numbers in the sixteenth century, in their attempts to solve quadratic and cubic equations. René Descartes gave them the name *imaginary numbers,* which reflected the mistrust with which mathematicians regarded them at the time. Today, however, such numbers are well understood and used routinely by scientists and engineers.

We begin by defining a new number, i, whose square is -1.

Imaginary Unit

We define the **imaginary unit** i by

$$i^2 = -1 \qquad \text{or} \qquad i = \sqrt{-1}$$

The letter i used in this way is not a variable; it is the name of a specific number and hence is a constant. The square root of any negative number can be written as the product of a real number and i. For example,

$$\sqrt{-4} = \sqrt{-1 \cdot 4}$$
$$= \sqrt{-1} \ \sqrt{4} = i \cdot 2$$

or $\sqrt{-4} = 2i$. Any number that is the product of i and a real number is called an **imaginary number.**

> **Imaginary Numbers**
>
> For $a > 0$,
> $$\sqrt{-a} = \sqrt{-1}\sqrt{a} = i\sqrt{a}$$

Examples of imaginary numbers are

$$3i, \quad \frac{7}{8}i, \quad -38i, \quad \text{and} \quad i\sqrt{5}$$

EXAMPLE 2   Write each radical as an imaginary number.

a. $\sqrt{-25}$ **b.** $2\sqrt{-3}$

Solutions   **a.** $\sqrt{-25} = \sqrt{-1}\sqrt{25}$ **b.** $2\sqrt{-3} = 2\sqrt{-1}\sqrt{3}$
$$= i\sqrt{25} = 5i \qquad\qquad\qquad\qquad = 2i\sqrt{3}$$

> **EXERCISE 2**
>
> Write each radical as an imaginary number.
>
> **a.** $\sqrt{-18}$ **b.** $-6\sqrt{-5}$

Every negative real number has two imaginary square roots, $i\sqrt{a}$ and $-i\sqrt{a}$, because

$$(i\sqrt{a})^2 = i^2(\sqrt{a})^2 = -1 \cdot a = -a$$

and

$$(-i\sqrt{a})^2 = (-i)^2(\sqrt{a})^2 = i^2 \cdot a = -a$$

For example, the two square roots of -9 are $3i$ and $-3i$.

Complex Numbers

Consider the quadratic equation

$$x^2 - 2x + 5 = 0$$

Using the quadratic formula to solve the equation, we find

$$x = \frac{-(-2) \pm \sqrt{(-2)^2 - 4(1)(5)}}{2} = \frac{2 \pm \sqrt{-16}}{2}$$

If we now replace $\sqrt{-16}$ with $4i$, we have

$$x = \frac{2 \pm 4i}{2} = 1 \pm 2i$$

The two solutions are $1 + 2i$ and $1 - 2i$. These numbers are examples of **complex numbers.**

Complex Numbers

A **complex number** can be written in the form $a + bi$, where a and b are real numbers.

Examples of complex numbers are

$$3 - 5i, \quad 2 + \sqrt{7}\,i, \quad \frac{4 - i}{3}, \quad 6i, \quad \text{and} \quad -9$$

In a complex number $a + bi$, a is called the **real part,** and b is called the **imaginary part.** All real numbers are also complex numbers (with the imaginary part equal to zero). A complex number whose real part equals zero is called a **pure imaginary** number.

EXAMPLE 3   Write the solutions to Example 1, $\dfrac{2 \pm \sqrt{-4}}{2}$, as complex numbers.

Solution   Because $\sqrt{-4} = \sqrt{-1}\sqrt{4} = 2i$, we have $\dfrac{2 \pm \sqrt{-4}}{2} = \dfrac{2 \pm 2i}{2}$, or $1 \pm i$. The solutions are $1 + i$ and $1 - i$.

EXERCISE 3

Use extraction of roots to solve $(2x + 1)^2 + 9 = 0$. Write your answers as complex numbers.

Arithmetic of Complex Numbers

All the properties of real numbers listed in Algebra Skills Refresher A.12 are also true of complex numbers. We can carry out arithmetic operations with complex numbers.

We add and subtract complex numbers by combining their real and imaginary parts separately. For example,

$$(4 + 5i) + (2 - 3i) = (4 + 2) + (5 - 3)i$$
$$= 6 + 2i$$

Sums and Differences of Complex Numbers

$$(a + bi) + (c + di) = (a + c) + (b + d)i$$

$$(a + bi) - (c + di) = (a - c) + (b - d)i$$

EXAMPLE 4   Subtract: $(8 - 6i) - (5 + 2i)$.

Solution   Combine the real and imaginary parts.

$$(8 - 6i) - (5 + 2i) = (8 - 5) + (-6 - 2)i$$
$$= 3 + (-8)i = 3 - 8i$$

Products of Complex Numbers

To find the product of two imaginary numbers, we use the fact that $i^2 = -1$. For example,

$$(3i) \cdot (4i) = 3 \cdot 4\, i^2$$
$$= 12(-1) = -12$$

To find the product of two complex numbers, we use the FOIL method, as if the numbers were binomials. For example,

$$(2 + 3i)(3 - 5i) = 6 - 10i + 9i - 15i^2$$

Because $i^2 = -1$, the last term, $-15i^2$, can be replaced by $-15\,(-1)$, or 15, to obtain

$$6 - 10i + 9i + 15$$

Finally, we combine the real parts and imaginary parts to obtain

$$(6 + 15) + (-10i + 9i) = 21 - i$$

EXAMPLE 5   Multiply $(7 - 4i)(-2 - i)$.

Solution

$$(7 - 4i)(-2 - i) = -14 - 7i + 8i + 4i^2 \quad \text{Replace } i^2 \text{ by } -1.$$
$$= -14 - 7i + 8i - 4 \quad \text{Combine real parts and imaginary}$$
$$= -18 + i \quad \text{parts.}$$

You can verify that in general the following rule holds.

Product of Complex Numbers

$$(a + bi)(c + di) = (ac - bd) + (ad + bc)i$$

CAUTION   One property of real numbers that is *not* true of complex numbers is $\sqrt{ab} = \sqrt{a} \cdot \sqrt{b}$. This identity fails when a and b are both negative. For example, if $a = b = -2$, we have

$$\sqrt{ab} = \sqrt{(-2)(-2)} = \sqrt{4} = 2$$

but

$$\sqrt{a} \cdot \sqrt{b} = \sqrt{-2} \cdot \sqrt{-2} = \sqrt{-1 \cdot 2} \cdot \sqrt{-1 \cdot 2} = i\sqrt{2} \cdot i\sqrt{2} = i^2(\sqrt{2})^2 = -2$$

so $\sqrt{ab} \neq \sqrt{a} \cdot \sqrt{b}$. We can avoid possible errors by writing square roots of negative numbers as imaginary numbers.

Quotients of Complex Numbers

▶▶ See Algebra Skills Refresher A.11 to review rationalizing the denominator.

To find the quotient of two complex numbers, we use the technique of rationalizing the denominator. For example, consider the quotient

$$\frac{3 + 4i}{2i}$$

Because i is really a radical (remember that $i = \sqrt{-1}$), we multiply the numerator and denominator of the quotient by i to obtain

$$\frac{(3 + 4i) \cdot i}{2i \cdot i} = \frac{3i + 4i^2}{2i^2} \qquad \text{Apply the distributive law to the numerator.}$$
$$\text{Recall that } i^2 = -1.$$
$$= \frac{3i - 4}{-2}$$

To write the quotient in the form $a + bi$, we divide -2 into each term of the numerator to get

$$\frac{3i}{-2} - \frac{4}{-2} = \frac{-3}{2}i + 2 = 2 + \frac{-3}{2}i$$

EXAMPLE 6   Divide $\dfrac{10 - 15i}{5i}$.

Solution   We multiply numerator and denominator by i.

$$\frac{10 - 15i}{5i} = \frac{(10 - 15i) \cdot i}{5i \cdot i}$$

$$= \frac{10i - 15i^2}{5i^2} = \frac{10i + 15}{-5} \qquad \text{Replace } i^2 \text{ by } -1.$$

$$= \frac{10i}{-5} + \frac{15}{-5} = -2i - 3 \qquad \text{Divide } -5 \text{ into each term of numerator.}$$

The quotient is $-3 - 2i$.

EXERCISE 6

Divide $\dfrac{8 + 9i}{3i}$.

If $z = a + bi$ is any nonzero complex number, then the number $\bar{z} = a - bi$ is called the **complex conjugate** of z. The product of a nonzero complex number and its conjugate is always a positive real number.

$$z\bar{z} = (a + bi)(a - bi) = a^2 - b^2i^2 = a^2 - b^2(-1) = a^2 + b^2$$

We use this fact to find the quotient of complex numbers. If the divisor has both a real and an imaginary part, we multiply numerator and denominator by the conjugate of the denominator.

EXAMPLE 7   Divide $\dfrac{2 + 3i}{4 - 2i}$.

Solution   We multiply numerator and denominator by $4 + 2i$, the conjugate of the denominator.

$$\frac{2 + 3i}{4 - 2i} = \frac{(2 + 3i)(4 + 2i)}{(4 - 2i)(4 + 2i)} \qquad \text{Expand numerator and denominator.}$$

$$= \frac{8 + 4i + 12i + 6i^2}{16 + 8i - 8i - 4i^2} \qquad \text{Replace } i^2 \text{ by } -1.$$

$$= \frac{8 + 16i - 6}{16 - (-4)} \qquad \text{Combine like terms.}$$

$$= \frac{2 + 16i}{20} \qquad \text{Divide 20 into each term of numerator.}$$

$$= \frac{2}{20} + \frac{16i}{20} = \frac{1}{10} + \frac{4}{5}i$$

EXERCISE 7

Write the quotient $\dfrac{4 - 2i}{1 + i}$ in the form $a + bi$.

Zeros of Polynomials

Because we can add, subtract, and multiply any two complex numbers, we can use a complex number as an input for a polynomial function. Thus, we can extend the domain of any polynomial to include all complex numbers.

EXAMPLE 8   Evaluate the polynomial $f(x) = x^2 - 2x + 2$ for $x = 1 + i$, then simplify.

Solution   We substitute $x = 1 + i$ to find

$$f(1 + i) = (1 + i)^2 - 2(1 + i) + 2$$
$$= 1^2 + 2i + i^2 - 2 - 2i + 2$$
$$= 1 + 2i + (-1) - 2 - 2i + 2$$
$$= 0$$

Thus, $f(1 + i) = 0$, so $1 + i$ is a solution of $x^2 - 2x + 2 = 0$.

EXERCISE 8

If $f(x) = x^2 - 6x + 13$, evaluate $f(3 + 2i)$.

In Chapter 6, we learned that irrational solutions of quadratic equations occur in conjugate pairs,

$$x = \frac{-b}{2a} + \frac{\sqrt{b^2 - 4ac}}{2a} \quad \text{and} \quad x = \frac{-b}{2a} - \frac{\sqrt{b^2 - 4ac}}{2a}$$

If the discriminant $D = b^2 - 4ac$ is negative, the two solutions are complex conjugates,

$$z = \frac{-b}{2a} + \frac{i\sqrt{|D|}}{2a} \quad \text{and} \quad \bar{z} = \frac{-b}{2a} - \frac{i\sqrt{|D|}}{2a}$$

Thus, if we know that z is a complex solution of a quadratic equation, we know that $\bar{z}$ is the other solution. The quadratic equation with solutions z and $\bar{z}$ is

$$(x - z)(x - \bar{z}) = 0$$
$$x^2 - (z + \bar{z})x + z\bar{z} = 0$$

EXAMPLE 9

a. Let $z = 7 - 5i$. Compute $z\bar{z}$.

b. Find a quadratic equation with one solution being $z = 7 - 5i$.

Solutions   **a.** The conjugate of $z = 7 - 5i$ is $\bar{z} = 7 + 5i$, so

$$z\bar{z} = (7 - 5i)(7 + 5i)$$
$$= 49 - 25i^2 = 49 + 25 = 74$$

b. The other solution of the equation is $\bar{z} = 7 + 5i$, and the equation is
$(x - z)(x - \bar{z}) = 0$. We expand the product to find

$$(x - z)(x - \bar{z}) = x^2 - (z + \bar{z})x + z\bar{z}$$
$$= x^2 - (7 - 5i + 7 + 5i)x + 74$$
$$= x^2 - 14x + 74$$

The equation is $x^2 - 14x + 74 = 0$.

EXERCISE 9

a. Let $z = -3 + 4i$. Compute $z\bar{z}$.

b. Find a quadratic equation with one solution being $z = -3 + 4i$.

One of the most important results in mathematics is the **fundamental theorem of algebra,** which says that if we allow complex numbers as inputs, then every polynomial $p(x)$ of degree $n \geq 1$ has exactly n complex number zeros.

Fundamental Theorem of Algebra

Let $p(x)$ be a polynomial of degree $n \geq 1$. Then $p(x)$ has exactly n complex zeros.

▶▶▶
See the Glossary or Section 7.2 for the definition of the zero of a polynomial.

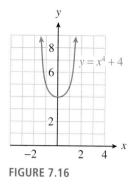

FIGURE 7.16

As a result, the factor theorem tells that every polynomial of degree n can be factored as the product of n linear terms. For example, although the graph of $y = x^4 + 4$ shown in Figure 7.16 has no x-intercepts, the fundamental theorem tells us that there are four complex solutions to $x^4 + 4 = 0$, and that $x^4 + 4$ can be factored. You can check that the four solutions to $x^4 + 4 = 0$ are $1 + i$, $-1 + i$, $-1 - i$, and $1 - i$. For example, if $x = 1 + i$, then

$$x^2 = (1 + i)^2 = 1 + 2i + i^2 = 2i$$

and

$$x^4 = (x^2)^2 = (2i)^2 = -4$$

Because each zero corresponds to a factor of the polynomial, the factored form of $x^4 + 4$ is

$$x^4 + 4 = [x - (1 + i)][x - (-1 + i)][x - (-1 - i)][x - (1 - i)]$$

The four solutions to $x^4 + 4 = 0$ form two complex conjugate pairs, namely $1 \pm i$ and $-1 \pm i$. In fact, for every polynomial with real coefficients, the nonreal zeros always occur in complex conjugate pairs.

EXAMPLE 10   Find a fourth-degree polynomial with real coefficients, two of whose zeros are $3i$ and $2 + i$.

Solution   The other two zeros are $-3i$ and $2 - i$. The factored form of the polynomial is

$$(x - 3i)(x + 3i)[x - (2 + i)][x - (2 - i)]$$

We multiply together the factors to find the polynomial. The product of $(x - 3i)(x + 3i)$ is $x^2 + 9$, and

$$\begin{aligned}[x - (2 + i)][x - (2 - i)] &= x^2 - (2 + i + 2 - i)x + (2 + i)(2 - i)\\ &= x^2 - 4x + 5\end{aligned}$$

Finally, we multiply these two partial products to find the polynomial we seek,

$$(x^2 + 9)(x^2 - 4x + 5) = x^4 - 4x^3 + 14x^2 - 36x + 45$$

EXERCISE 10

a. Find the zeros of the polynomial $f(x) = x^4 + 15x^2 - 16$.

b. Write the polynomial in factored form.

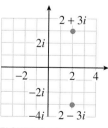

FIGURE 7.17

Graphing Complex Numbers

Real numbers can be plotted on a number line, but to graph a complex number we use a plane, called the **complex plane.** In the complex plane, the real numbers lie on the horizontal or **real axis,** and pure imaginary numbers lie on the vertical or **imaginary axis.** To plot a complex number $a + bi$, we move a units from the origin in the horizontal direction and b units in the vertical direction. The numbers $2 + 3i$ and $2 - 3i$ are plotted in Figure 7.17.

EXAMPLE 11 Plot the numbers z, $\bar{z}$, $-z$, and $-\bar{z}$ as points on the complex plane, for $z = 2 - 2i$.

Solution To plot $z = 2 - 2i$, we move from the origin 2 units to the right and 2 units down. To plot $\bar{z} = 2 + 2i$, we move from the origin 2 units to the right and 2 units up. To plot $-z = -2 + 2i$, we move from the origin 2 units to the left and 2 units up. To plot $-\bar{z} = -2 - 2i$, we move from the origin 2 units to the left and 2 units down. All four points are plotted in Figure 7.18.

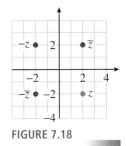

FIGURE 7.18

EXERCISE 11

Plot the numbers $z = 1 + i$, $iz = i + i^2$, $i^2z = i^2 + i^3$ and $i^3z = i^3 + i^4$ as points on the complex plane.

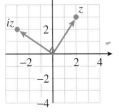

FIGURE 7.19

If we draw an arrow from the origin to the point $a + bi$ in the complex plane, we can see that multiplication by i corresponds to rotating a point around the origin by $90°$ in the counterclockwise direction. For example, Figure 7.19 shows the graphs of $z = 2 + 3i$ and $iz = 2i - 3$.

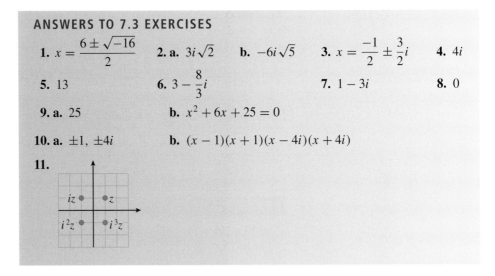

ANSWERS TO 7.3 EXERCISES

1. $x = \dfrac{6 \pm \sqrt{-16}}{2}$ 2. **a.** $3i\sqrt{2}$ **b.** $-6i\sqrt{5}$ 3. $x = \dfrac{-1}{2} \pm \dfrac{3}{2}i$ 4. $4i$

5. 13 6. $3 - \dfrac{8}{3}i$ 7. $1 - 3i$ 8. 0

9. **a.** 25 **b.** $x^2 + 6x + 25 = 0$

10. **a.** ± 1, $\pm 4i$ **b.** $(x - 1)(x + 1)(x - 4i)(x + 4i)$

11.

SECTION 7.3 SUMMARY

VOCABULARY *Look up the definitions of new terms in the Glossary.*

Imaginary unit	Imaginary number	Complex number	Imaginary axis
Complex conjugate	Real part	Imaginary part	
Complex plane	Real axis		

CONCEPTS

1. The square root of a negative number is an imaginary number.
2. A complex number is the sum of a real number and an imaginary number.
3. We can perform the four arithmetic operations on complex numbers.
4. The product of a nonzero complex number and its conjugate is always a positive real number.

5. **Fundamental theorem of algebra:** A polynomial of degree $n \geq 1$ has exactly n complex zeros.
6. The nonreal zeros of a polynomial with real coefficients always occur in conjugate pairs.
7. We can graph complex numbers in the complex plane.
8. Multiplying a complex number by i rotates its graph by $90°$ around the origin.

STUDY QUESTIONS

1. What are imaginary numbers, and why were they invented?
2. Simplify the following powers of i:
$$i^2, \ i^3, \ i^4, \ i^5, \ i^6, \ i^7, \ i^8$$
What do you notice?
3. Explain how the complex conjugate is used in dividing complex numbers.

4. If one solution of a quadratic equation is $3 + i\sqrt{2}$, what is the other solution?
5. If $P(x)$ is a polynomial of degree 7, how many zeros does $P(x)$ have? How many x-intercepts could its graph have? How many complex zeros could $P(x)$ have?

SKILLS   *Practice each skill in the Homework problems listed.*

1. Write and simplify complex numbers: #1–10
2. Perform arithmetic operations on complex numbers: #11–36
3. Evaluate polynomials at complex numbers, expand polynomials: #37–48

4. Find a polynomial with given zeros: #53–56, 59–62
5. Graph complex numbers: #63–70

HOMEWORK 7.3

ThomsonNOW™ Test yourself on key content at **www.thomsonedu.com/login.**

Write each complex number in the form $a + bi$, where a and b are real numbers.

1. $\sqrt{-25} - 4$

2. $\sqrt{-9} + 3$

3. $\dfrac{-8 + \sqrt{-4}}{2}$

4. $\dfrac{6 - \sqrt{-36}}{2}$

5. $\dfrac{-5 - \sqrt{-2}}{6}$

6. $\dfrac{7 + \sqrt{-3}}{4}$

Find the zeros of each quadratic polynomial. Write each in the form $a + bi$, where a and b are real numbers.

7. $x^2 + 6x + 13$

8. $x^2 - 2x + 10$

9. $3x^2 - x + 1$

10. $5x^2 + 2x + 2$

Add or subtract.

11. $(11 - 4i) - (-2 - 8i)$

12. $(7i - 2) + (6 - 4i)$

13. $(2.1 + 5.6i) + (-1.8i - 2.9)$

14. $\left(\dfrac{1}{5}i - \dfrac{2}{5}\right) - \left(\dfrac{4}{5} - \dfrac{3}{5}i\right)$

Multiply.

15. $5i(2 - 4i)$

16. $-7i(-1 + 4i)$

17. $(4 - i)(-6 + 7i)$

18. $(2 - 3i)(2 - 3i)$

19. $(7 + i\sqrt{3})(7 + i\sqrt{3})$

20. $(5 - i\sqrt{2})^2$

21. $(7 + i\sqrt{3})(7 - i\sqrt{3})$

22. $(5 - i\sqrt{2})(5 + i\sqrt{2})$

23. $(1 - i)^3$

24. $(2 + i)^3$

Divide.

25. $\dfrac{12 + 3i}{-3i}$

26. $\dfrac{12 + 4i}{8i}$

27. $\dfrac{10 + 15i}{2 + i}$

28. $\dfrac{4 - 6i}{1 - i}$

29. $\dfrac{5i}{2 - 5i}$

30. $\dfrac{-2i}{7 + 2i}$

31. $\dfrac{\sqrt{3}}{\sqrt{3} + i}$

32. $\dfrac{2\sqrt{2}}{1 - i\sqrt{2}}$

33. $\dfrac{1 + i\sqrt{5}}{1 - i\sqrt{5}}$

34. $\dfrac{\sqrt{2} - i}{\sqrt{2} + i}$

35. $\dfrac{3 + 2i}{2 - 3i}$

36. $\dfrac{4 - 6i}{-3 - 2i}$

Evaluate each polynomial for the given value of the variable.

37. $z^2 + 9$
   a. $z = 3i$
   b. $z = -3i$

38. $2y^2 - y - 2$
   a. $y = 2 - i$
   b. $y = -2 - i$

39. $x^2 - 2x + 2$
   a. $x = 1 - i$
   b. $x = 1 + i$

40. $3w^2 + 5$
   a. $w = 2i$
   b. $w = -2i$

41. $q^2 + 4q + 13$
   a. $q = -2 + 3i$
   b. $q = -2 - 3i$

42. $v^2 + 2v + 3$
   a. $v = 1 + i$
   b. $v = -1 + i$

Expand each product of polynomials.

43. $(2z + 7i)(2z - 7i)$

44. $(5w + 3i)(5w - 3i)$

45. $[x + (3 + i)][x + (3 - i)]$

46. $[s - (1 + 2i)][s - (1 - 2i)]$

47. $[v - (4 + i)][v - (4 - i)]$

48. $[Z + (2 + i)][Z + (2 - i)]$

49. For what values of x will $\sqrt{x - 5}$ be real? Imaginary?

50. For what values of x will $\sqrt{x + 3}$ be real? Imaginary?

51. Simplify.
   a. i^6
   b. i^{12}
   c. i^{15}
   d. i^{102}

52. Express with a positive exponent and simplify.
   a. i^{-1}
   b. i^{-2}
   c. i^{-3}
   d. i^{-6}

In Problems 53–56,
(a) Given one solution of a quadratic equation with rational coefficients, find the other solution.
(b) Write a quadratic equation that has those solutions.

53. $2 + \sqrt{5}$ **54.** $3 - \sqrt{2}$ **55.** $4 - 3i$ **56.** $5 + i$

Every polynomial factors into a product of a constant and linear factors of the form $(x - a)$, where a can be either real or complex. How many linear factors are in the factored form of the given polynomial?

57. a. $x^4 - 2x^3 + 4x^2 + 8x - 6$ **58. a.** $x^6 - 6x$
     b. $2x^5 - x^3 + 6x - 4$ **b.** $x^3 + 3x^2 - 2x + 1$

Find a fourth-degree polynomial with real coefficients that has the given complex numbers as two of its zeros.

59. $1 + 3i, \ 2 - i$ **60.** $5 - 4i, \ -i$

61. $\dfrac{1}{2} - \dfrac{\sqrt{3}}{2}i, \ 3 + 2i$ **62.** $-\dfrac{\sqrt{2}}{2} + \dfrac{\sqrt{2}}{2}i, \ 4 - i$

Plot each number and its complex conjugate in the complex plane. What is the geometric relationship between complex conjugates?

63. $z = -3 + 2i$ **64.** $z = 4 - 3i$

65. $z = \dfrac{\sqrt{3}}{2} - \dfrac{1}{2}i$ **66.** $z = -\dfrac{\sqrt{2}}{2} - \dfrac{\sqrt{2}}{2}i$

Simplify and plot each complex number as a point on the complex plane.

67. $1, \ i, \ i^2, \ i^3$ and i^4 **68.** $-1, \ -i, \ -i^2, \ -i^3,$ and $-i^4$

69. $1 + 2i$ and $i(1 + 2i)$ **70.** $3 - 4i$ and $i(3 - 4i)$

Problems 71 and 72 show that multiplication by i results in a rotation of 90°.

71. Suppose that $z = a + bi$ and that the real numbers a and b are both nonzero.
     a. What is the slope of the segment in the complex plane joining the origin to z?
     b. What is the slope of the segment in the complex plane joining the origin to zi?
     c. What is the product of the slopes of the two segments from parts (a) and (b)? What can you conclude about the angle between the two segments?

72. Suppose that $z = a + bi$ and that a and b are both real numbers.
     a. If $a \neq 0$ and $b = 0$, then what is the slope of the segment in the complex plane joining the origin to z? What is the slope of the segment joining the origin to iz?
     b. If $a = 0$ and $b \neq 0$, then what is the slope of the segment in the complex plane joining the origin to z? What is the slope of the segment joining the origin to iz?
     c. What can you conclude about the angle between the two segments from parts (a) and (b)?

A rational function is the quotient of two polynomials. (As with rational numbers, the word *rational* refers to a ratio.)

> **Rational Function**
>
> A **rational function** is one of the form
>
> $$f(x) = \frac{P(x)}{Q(x)}$$
>
> where $P(x)$ and $Q(x)$ are polynomials and $Q(x)$ is not the zero polynomial.

The graphs of rational functions can be quite different from the graphs of polynomials.

EXAMPLE 1

Francine is planning a 60-mile training flight through the desert on her cycle-plane, a pedal-driven aircraft. If there is no wind, she can pedal at an average speed of 15 miles per hour, so she can complete the flight in 4 hours.

a. If there is a headwind of x miles per hour, it will take Francine longer to fly 60 miles. Express the time it will take to complete the training flight as a function of x.

b. Make a table of values for the function.

c. Graph the function and explain what it tells you about the time Francine should allot for the flight.

Solutions

a. If there is a headwind of x miles per hour, Francine's ground speed will be $15 - x$ miles per hour. Using the fact that $time = \frac{distance}{rate}$, we find that the time needed for the flight will be

$$t = f(x) = \frac{60}{15 - x}$$

b. Evaluate the function for several values of x, as shown in the table below.

x	0	3	5	7	9	10
t	4	5	6	7.5	10	12

For example, if the headwind is 5 miles per hour, then

$$t = \frac{60}{15 - 5} = \frac{60}{10} = 6$$

Francine's effective speed is only 10 miles per hour, and it will take her 6 hours to fly the 60 miles. The table shows that as the speed of the headwind increases, the time required for the flight increases also.

c. The graph of the function is shown in Figure 7.20. You can use your calculator with the window

$$\text{Xmin} = -8.5, \quad \text{Xmax} = 15$$
$$\text{Ymin} = 0, \qquad \text{Ymax} = 30$$

to verify the graph. In particular, the point (0, 4) lies on the graph. This point tells us that if there is no wind, Francine can fly 60 miles in 4 hours, as we calculated earlier.

The graph is increasing, as indicated by the table of values. In fact, as the speed of the wind gets close to 15 miles per hour, Francine's flying time becomes extremely large. In theory, if the wind speed were exactly 15 miles per hour, Francine would never complete her flight. On the graph, the time becomes infinite at $x = 15$.

What about negative values for x? If we interpret a negative headwind as a tailwind, Francine's flying time should decrease for negative x-values. For example, if $x = -5$, there is a tailwind of 5 miles per hour, so Francine's effective speed is 20 miles per hour, and she can complete the flight in 3 hours. As the tailwind gets stronger (that is, as we move farther to the left in the x-direction), Francine's flying time continues to decrease, and the graph approaches the x-axis.

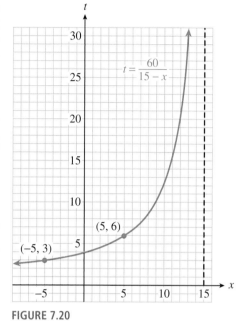

FIGURE 7.20

The vertical dashed line at $x = 15$ on the graph of $t = \dfrac{60}{15 - x}$ is a *vertical asymptote* for the graph. We first encountered asymptotes in Section 2.2 when we studied the graph of $y = \frac{1}{x}$. Locating the vertical asymptotes of a rational function is an important part of determining the shape of the graph.

EXERCISE 1

Queueing theory is used to predict your waiting time in a line, or queue. For example, suppose the attendant at a toll booth can process 6 vehicles per minute. The average total time spent by a motorist negotiating the toll booth depends on the rate, r, at which vehicles arrive, according to the formula

$$T = g(r) = \frac{12 - r}{12(6 - r)}$$

a. What is the average time spent at the toll booth if vehicles arrive at a rate of 3 vehicles per minute?

b. Graph the function on the domain [0, 6].

c. What is the vertical asymptote of the graph? What does it tell you about the queue?

EXAMPLE 2

EarthCare decides to sell T-shirts to raise money. The company makes an initial invest-ment of $100 to pay for the design of the T-shirt and to set up the printing process. After that, the T-shirts cost $5 each for labor and materials.

a. Express the average cost per T-shirt as a function of the number of T-shirts EarthCare produces.

b. Make a table of values for the function.

c. Graph the function and explain what it tells you about the cost of the T-shirts.

Solutions

a. If EarthCare produces x T-shirts, the total costs will be $100 + 5x$ dollars. To find the average cost per T-shirt, we divide the total cost by the number of T-shirts produced, to get

$$C = g(x) = \frac{100 + 5x}{x}$$

b. Evaluate the function for several values of x, as shown in the table.

x	1	2	4	5	10	20
C	105	55	40	25	15	10

If EarthCare makes only one T-shirt, its cost is $105. But if more than one T-shirt is made, the cost of the original $100 investment is distributed among them. For example, the average cost per T-shirt for 2 T-shirts is

$$\frac{100 + 5(2)}{2} = 55$$

and the average cost for 5 T-shirts is

$$\frac{100 + 5(5)}{5} = 25$$

c. The graph is shown in Figure 7.21. You can use your calculator with the window

$$\text{Xmin} = 0 \qquad \text{Xmax} = 470$$
$$\text{Ymin} = 0 \qquad \text{Ymax} = 30$$

to verify the graph. Use the **Trace** to locate on the graph several points from the table of values. For example, the point $(5, 25)$ indicates that if EarthCare makes 5 T-shirts, the cost per shirt is $25.

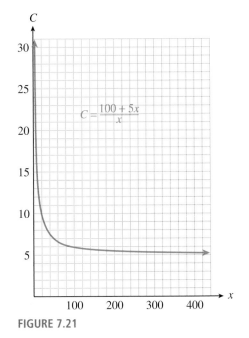

FIGURE 7.21

The graph shows that as the number of T-shirts increases, the average cost per shirt continues to decrease, but not as rapidly as at first. Eventually the average cost levels off and approaches $5 per T-shirt. For example, if EarthCare produces 400 T-shirts, the average cost per shirt is

$$\frac{100 + 5(400)}{400} = 5.25$$

The horizontal line $C = 5$ on the graph of $C = \dfrac{100 + 5x}{x}$ is a *horizontal asymptote*. As x increases, the graph approaches the line $C = 5$ but never actually meets it. The average price per T-shirt will always be slightly more than \$5. Horizontal asymptotes are also important in sketching the graphs of rational functions.

EXERCISE 2

Delbert prepares a 20% glucose solution of by mixing 2 mL of glucose with 8 mL of water. If he adds x ml of glucose to the solution, its concentration is given by

$$C(x) = \frac{2 + x}{8 + x}$$

a. How many milliliters of glucose should Delbert add to increase the concentration to 50%?

b. Graph the function on the domain [0, 100].

c. What is the horizontal asymptote of the graph? What does it tell you about the solution?

Domain of a Rational Function

Most applications of rational functions have restricted domains, that is, they make sense for only a subset of the real numbers on the x-axis. Consequently, only a portion of the graph is useful for analyzing the application. However, a knowledge of the general shape and properties of the whole graph can be very helpful in understanding a rational function.

As we stated earlier, a rational function is a quotient of two polynomials. Some examples of rational functions are shown below.

$$f(x) = \frac{2}{(x - 3)^2} \qquad g(x) = \frac{x}{x + 1}$$

$$h(x) = \frac{2x^2}{x^2 + 4} \qquad k(x) = \frac{x^2 - 1}{x^2 - 9}$$

Because we cannot divide by zero, a rational function $f(x) = \dfrac{P(x)}{Q(x)}$ is undefined for any value $x = a$ where $Q(a) = 0$. These x-values are not in the domain of the function.

EXAMPLE 3 Find the domains of the rational functions f, g, h, and k defined above.

Solution The domain of f is the set of all real numbers except 3, because the denominator, $(x - 3)^2$, equals 0 when $x = 3$. The domain of g is the set of all real numbers except -1, because $x + 1$ equals zero when $x = -1$. The denominator of the function h, $x^2 + 4$, is never equal to zero, so the domain of h is all the real numbers. The domain of k is the set of all real numbers except 3 and -3, because $x^2 - 9$ equals 0 when $x = 3$ or $x = -3$.

We only need to exclude the zeros of the *denominator* from the domain of a rational function. We do not exclude the zeros of the numerator. In fact, the zeros of the numerator include the zeros of the rational function itself, because a fraction is equal to 0 when its numerator is 0 but its denominator is not 0.

a. Find the domain of $F(x) = \dfrac{x-2}{x+4}$.

b. Find the zeros of $F(x)$.

Vertical Asymptotes

As we saw in Section 7.2, a polynomial function is defined for all values of x, and its graph is a smooth curve without any breaks or holes. The graph of a rational function, on the other hand, will have breaks or holes at those x-values where it is undefined.

EXAMPLE 4 Investigate the graph of $f(x) = \dfrac{2}{(x-3)^2}$ near $x = 3$.

Solution This function is undefined for $x = 3$, so there is no point on the graph with x-coordinate 3. However, we can make a table of values for other values of x. Plotting the ordered pairs in the table results in the points shown in Figure 7.22.

x	y
0	$\frac{2}{9}$
1	$\frac{1}{2}$
2	2
3	undefined
4	2
5	$\frac{1}{2}$
6	$\frac{2}{9}$

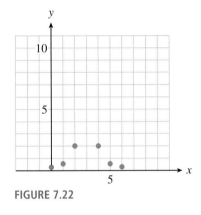

FIGURE 7.22

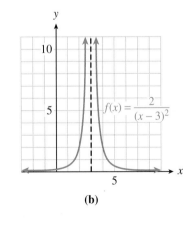

(b)

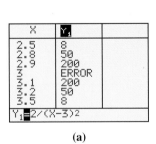

(a)

FIGURE 7.23

Next, we make a table showing x-values close to 3, as in Figure 7.23a. As we choose x-values closer and closer to 3, $(x-3)^2$ gets closer to 0, so the fraction $\dfrac{2}{(x-3)^2}$ gets very large.

This means that the graph approaches, but never touches, the vertical line $x = 3$. In other words, the graph has a vertical asymptote at $x = 3$. We indicate the vertical asymptote by a dashed line, as shown in Figure 7.23b.

In general, we have the following result.

Vertical Asymptotes

If $Q(a) = 0$ but $P(a) \neq 0$, then the graph of the rational function $f(x) = \dfrac{P(x)}{Q(x)}$ has a **vertical asymptote** at $x = a$.

If $P(a)$ and $Q(a)$ are both zero, then the graph of the rational function $\dfrac{P(x)}{Q(x)}$ may have a hole at $x = a$ rather than an asymptote. (This possibility is considered in Problems 51–54 of Homework 7.4.)

EXERCISE 4

Find the vertical asymptotes of $G(x) = \dfrac{4x^2}{x^2 - 4}$.

Near a vertical asymptote, the graph of a rational function has one of the four characteristic shapes, illustrated in Figure 7.24. Locating the vertical asymptotes can help us make a quick sketch of a rational function.

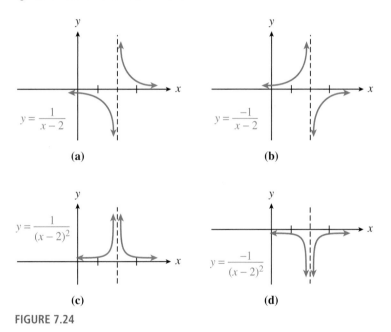

FIGURE 7.24

EXAMPLE 5

Locate the vertical asymptotes and sketch the graph of $g(x) = \dfrac{x}{x + 1}$.

Solution   The denominator, $x + 1$, equals zero when $x = -1$. Because the numerator does not equal zero when $x = -1$, there is a vertical asymptote at $x = -1$. The asymptote separates the graph into two pieces. Use the **Table** feature of your calculator to evaluate $g(x)$ for several values of x on either side of the asymptote, as shown in Figure 7.25a. Plot the points found in this way; then connect the points on either side of the asymptote to obtain the graph shown in Figure 7.25b.

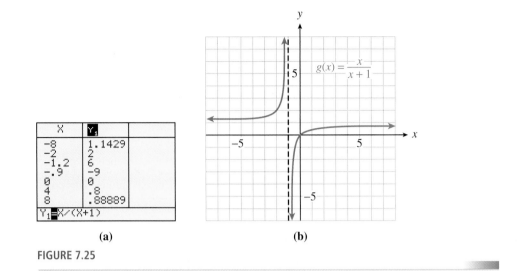

(a)   (b)

FIGURE 7.25

Horizontal Asymptotes

Look again at the graph of $g(x) = \dfrac{x}{x + 1}$ in Example 5. As $|x|$ gets large—that is, as we move away from the origin along the x-axis in either direction—the corresponding y-values get closer and closer to 1. The graph approaches, but never coincides with, the line $y = 1$. We say that the graph has a **horizontal asymptote** at $y = 1$.

When does a rational function $f(x) = \dfrac{P(x)}{Q(x)}$ have a horizontal asymptote? It depends on the degrees of the two polynomials $P(x)$ and $Q(x)$. The degree of the numerator of $g(x)$ is equal to the degree of the denominator. Equivalently, the highest power of x in the numerator (1, in this case) is the same as the highest power in the denominator.

Consider the three rational functions whose graphs are shown in Figure 7.26.

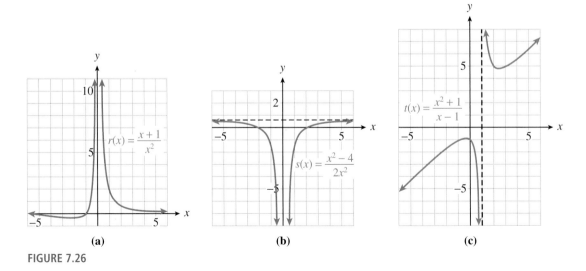

(a)   (b)   (c)

FIGURE 7.26

The graph of $r(x) = \dfrac{x+1}{x^2}$ in Figure 7.26a has a horizontal asymptote at $y = 0$, the x-axis, because the degree of the denominator is larger than the degree of the numerator. Higher powers of x grow much more rapidly than smaller powers. Thus, for large values of $|x|$, the denominator is much larger in absolute value than the numerator of $r(x)$, so the function values approach 0.

The graph of $s(x) = \dfrac{x^2 - 4}{2x^2}$ in Figure 7.26b has a horizontal asymptote at $y = \frac{1}{2}$, because the numerator and denominator of the fraction have the same degree. For large values of $|x|$, the terms of lower degree are negligible compared to the squared terms. As x increases, $s(x)$ is approximately equal to $\frac{x^2}{2x^2}$, or $\frac{1}{2}$. Thus, the function values approach a constant value of $\frac{1}{2}$.

The graph of $t(x) = \dfrac{x^2 + 1}{x - 1}$ in Figure 7.26c does not have a horizontal asymptote, because the degree of the numerator is larger than the degree of the denominator. As $|x|$ increases, $x^2 + 1$ grows much faster than $x - 1$, so their ratio does not approach a constant value. The function values increase without bound.

We summarize our discussion as follows.

Horizontal Asymptotes

Suppose $f(x) = \dfrac{P(x)}{Q(x)}$ is a rational function, where the degree of $P(x)$ is m and the degree of $Q(x)$ is n.

1. If $m < n$, the graph of f has a horizontal asymptote at $y = 0$.

2. If $m = n$, the graph of f has a horizontal asymptote at $y = \frac{a}{b}$, where a is the lead coefficient of $P(x)$ and b is the lead coefficient of $Q(x)$.

3. If $m > n$, the graph of f does not have a horizontal asymptote.

EXAMPLE 6

Locate the horizontal asymptotes and sketch the graph of $h(x) = \dfrac{2x^2}{x^2 + 4}$.

Solution The numerator and denominator of the fraction are both second-degree polynomials, so the graph does have a horizontal asymptote. The lead coefficients of $P(x)$ and $Q(x)$ are 2 and 1, respectively, so the horizontal asymptote is $y = \frac{2}{1}$, or $y = 2$. The function h does not have a vertical asymptote because the denominator, $x^2 + 4$, is never equal to zero. The y-intercept of the graph is the point $(0, 0)$. We can plot several points by evaluating the function at convenient x-values, and use the asymptote to help us sketch the graph, as shown in Figure 7.27.

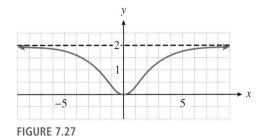

FIGURE 7.27

EXERCISE 6

Locate the horizontal and vertical asymptotes and sketch the graph of $k(x) = \dfrac{x^2 - 1}{x^2 - 9}$. Label the x- and y-intercepts with their coordinates.

▶▶▶
See Algebra Skills Refresher A.9 to review operations on algebraic fractions.

Applications

It is often useful to simplify the formula for a rational function before using it.

EXAMPLE 7

When estimating their travel time, pilots must take into account the prevailing winds. A tailwind adds to the plane's ground speed, while a headwind decreases the ground speed. Skyhigh Airlines is setting up a shuttle service from Dallas to Phoenix, a distance of 800 miles.

  a. Express the time needed for a one-way trip, without wind, as a function of the speed of the plane.

  b. Suppose there is a prevailing wind of 30 miles per hour blowing from the west. Write expressions for the flying time from Dallas to Phoenix and from Phoenix to Dallas.

  c. Write an expression for the round-trip flying time, excluding stops, with a 30-mile-per-hour wind from the west, as a function of the plane's speed. Simplify your expression.

Solutions a. Recall that time $= \frac{\text{distance}}{\text{rate}}$. If we let r represent the speed of the plane in still air, then the time required for a one-way trip is

$$f(r) = \frac{800}{r}$$

  b. On the trip from Dallas to Phoenix, the plane encounters a headwind of 30 miles per hour, so its actual ground speed is $r - 30$. On the return trip, the plan enjoys a tailwind of 30 miles per hour, so its actual ground speed is $r + 30$. Therefore, the flying times are

$$\text{Dallas to Phoenix:} \quad \frac{800}{r - 30}$$

and

$$\text{Phoenix to Dallas:} \quad \frac{800}{r + 30}$$

  c. The round-trip flying time from Dallas to Phoenix and back is

$$F(r) = \frac{800}{r - 30} + \frac{800}{r + 30}$$

The LCD for these fractions is $(r - 30)(r + 30)$. Thus,

$$\frac{800}{r - 30} + \frac{800}{r + 30} = \frac{800(r + 30)}{(r - 30)(r + 30)} + \frac{800(r - 30)}{(r + 30)(r - 30)}$$

$$= \frac{(800r + 24000) + (800r - 24000)}{(r + 30)(r - 30)} = \frac{1600r}{r^2 - 900}$$

EXERCISE 7

Navid took his outboard motorboat 20 miles upstream to a fishing site, returning downstream later that day. His boat travels 10 miles per hour in still water. Write an expression for the time Navid spent traveling, as a function of the speed of the current.

ANSWERS TO 7.4 EXERCISES

1. a. 0.25 min   **b.** 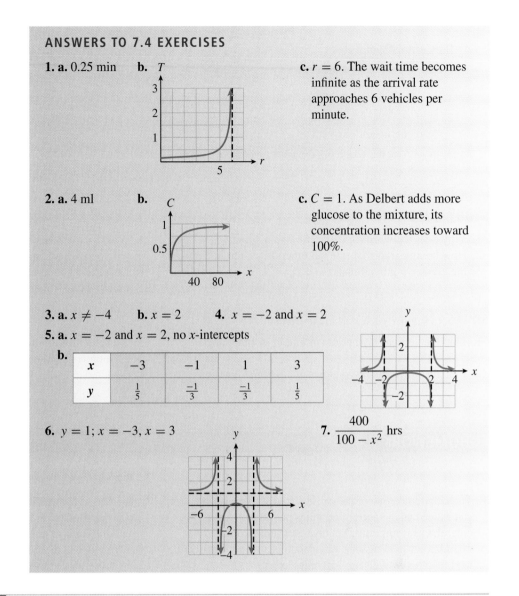   **c.** $r = 6$. The wait time becomes infinite as the arrival rate approaches 6 vehicles per minute.

2. a. 4 ml   **b.**   **c.** $C = 1$. As Delbert adds more glucose to the mixture, its concentration increases toward 100%.

3. a. $x \neq -4$ **b.** $x = 2$ **4.** $x = -2$ and $x = 2$

5. a. $x = -2$ and $x = 2$, no x-intercepts

b.

x	-3	-1	1	3
y	$\frac{1}{5}$	$\frac{-1}{3}$	$\frac{-1}{3}$	$\frac{1}{5}$

6. $y = 1$; $x = -3$, $x = 3$ **7.** $\dfrac{400}{100 - x^2}$ hrs

---

SECTION 7.4 SUMMARY

VOCABULARY   *Look up the definitions of new terms in the Glossary.*

Rational function   Vertical asymptote   Horizontal asymptote

CONCEPTS

1.

> **Rational Function**
>
> A **rational function** is one of the form
>
> $$f(x) = \frac{P(x)}{Q(x)}$$
>
> where $P(x)$ and $Q(x)$ are polynomials and $Q(x)$ is not the zero polynomial.

2. A rational function $f(x) = \dfrac{P(x)}{Q(x)}$ is undefined for any value $x = a$ where $Q(a) = 0$. These x-values are not in the domain of the function.

3.

> **Vertical Asymptotes**
>
> If $Q(a) = 0$ but $P(a) \neq 0$, then the graph of the rational function $f(x) = \frac{P(x)}{Q(x)}$ has a **vertical asymptote** at $x = a$.

4.

> ### Horizontal Asymptotes
>
> Suppose $f(x) = \frac{P(x)}{Q(x)}$ is a rational function, where the degree of $P(x)$ is m and the degree of $Q(x)$ is n.
>
> **1.** If $m < n$, the graph of f has a horizontal asymptote at $y = 0$.
> **2.** If $m = n$, the graph of f has a horizontal asymptote at $y = \frac{a}{b}$, where a is the lead coefficient of $P(x)$ and b is the lead coefficient of $Q(x)$.
> **3.** If $m > n$, the graph of f does not have a horizontal asymptote.

STUDY QUESTIONS

1. Why does the word *rational* refer to a quotient?
2. How are the graphs of rational functions different from the graphs of polynomials?
3. What do the zeros of the numerator of a rational function tell you? What about the zeros of the denominator?
4. Under what circumstances can the graph of a rational function have a horizontal asymptote?

SKILLS *Practice each skill in the Homework problems listed.*

1. Find the vertical asymptotes of a rational function: #13–32
2. Find the horizontal asymptotes of a rational function: #13–32
3. Interpret the significance of horizontal and vertical asymptotes in context: #1–10
4. Sketch the graph of a rational function: #13–36, 51–54
5. Write a rational function to model a situation: #37–42

HOMEWORK 7.4

1. The eider duck, one of the world's fastest flying birds, can exceed an airspeed of 65 miles per hour. A flock of eider ducks is migrating south at an average airspeed of 50 miles per hour against a moderate headwind. Their next feeding grounds are 150 miles away.
   a. Express the ducks' travel time, t, as a function of the windspeed, v.
   b. Complete the table showing the travel time for various windspeeds.

v	0	5	10	15	20	25	30	35	40	45	50
t											

   What happens to the travel time as the headwind increases?
   c. Use the table to choose an appropriate window and graph your function $t(v)$. Give the equations of any horizontal or vertical asymptotes. What does the vertical asymptote signify in the context of the problem?

2. The fastest fish in the sea may be the bluefin tuna, which has been clocked at 43 miles per hour in short sprints. A school of tuna is migrating a distance of 200 miles at an average speed of 36 miles per hour in still water, but they have run into a current flowing against their direction of travel.
   a. Express the tuna's travel time, t, as a function of the current speed, v.
   b. Complete the table showing the travel time for various current speeds.

v	0	4	8	12	16	20	24	28	32	36
t										

   What happens to the travel time as the current increases?
   c. Use the table to choose an appropriate window and graph your function $t(v)$. Give the equations of any horizontal or vertical asymptotes. What does the vertical asymptote signify in the context of the problem?

3. The cost, in thousands of dollars, for immunizing p percent of the residents of Emporia against a dangerous new disease is given by the function

$$C(p) = \frac{72p}{100 - p}$$

a. What is the domain of C?

b. Complete the table showing the cost of immunizing various percentages of the population.

p	0	15	25	40	50	75	80	90	100
C									

c. Graph the function C. (Use Xmin = 6, Xmax = 100, and appropriate values of Ymin and Ymax.) What percentage of the population can be immunized if the city is able to spend $108,000?

d. For what values of p is the total cost more than $1,728,000?

e. The graph has a vertical asymptote. What is it? What is its significance in the context of this problem?

4. The cost, in thousands of dollars, for extracting p percent of a precious ore from a mine is given by the equation

$$C(p) = \frac{360p}{100 - p}$$

a. What is the domain of C?

b. Complete the table showing the cost of extracting various percentages of the ore.

p	0	15	25	40	50	75	80	90	100
C									

c. Graph the function C. (Use Xmin = 6, Xmax = 100 and appropriate values of Ymin and Ymax.) What percentage of the ore can be extracted if $540,000 can be spent on the extraction?

d. For what values of p is the total cost less than $1,440,000?

e. The graph has a vertical asymptote. What is it? What is its significance in the context of this problem?

5. The total cost in dollars of producing n calculators is approximately $20,000 + 8n$.

a. Express the cost per calculator, C, as a function of the number n of calculators produced.

b. Complete the table showing the cost per calculator for various production levels.

n	100	200	400	500	1000	2000	4000	5000	8000
C									

c. Graph the function $C(n)$ for the cost per calculator. Use the window

$$\text{Xmin} = 0, \quad \text{Xmax} = 9400$$
$$\text{Ymin} = 0, \quad \text{Ymax} = 50$$

d. How many calculators should be produced so that the cost per calculator is $18?

e. For what values of n is the cost less than $12 per calculator?

f. Find the horizontal asymptote of the graph. What does it represent in this context?

6. The number of loaves of Mom's Bread sold each day is approximated by the demand function

$$D(p) = \frac{100}{1 + (p - 1.10)^4}$$

where p is the price per loaf in dollars.

a. Complete the table showing the demand for Mom's Bread at various prices per loaf. Round the values of $D(p)$ to the nearest whole number.

p	0.25	0.50	1	1.25	1.50	1.75	2	2.25	2.50	2.75	3.00
Demand											

b. Graph the demand function in the window

$$\text{Xmin} = 0, \ \text{Xmax} = 3.74$$
$$\text{Ymin} = 0, \ \text{Ymax} = 170$$

What happens to the demand for Mom's Bread as the price increases?

c. Add a row to your table to show the daily revenue from Mom's Bread at various prices.

p	0.25	0.50	1	1.25	1.50	1.75	2	2.25	2.50	2.75	3.00
Demand											
Revenue											

d. Using the formula for $D(p)$, write an expression $R(p)$ that approximates the total daily revenue as a function of the price, p.

e. Graph the revenue function $R(p)$ in the same window with $D(p)$. Estimate the maximum possible revenue. Does the maximum for $D(p)$ occur at the same value of p as the maximum for $R(p)$?

f. Find the horizontal asymptote of the graphs. What does it represent in this context?

7. A computer store sells approximately 300 of its most popular model per year. The manager would like to minimize her annual inventory cost by ordering the optimal number of computers, x, at regular intervals. If she orders x computers in each shipment, the cost of storage will be $6x$ dollars, and the cost of reordering will be $\dfrac{300}{x}(15x + 10)$ dollars. The inventory cost is the sum of the storage cost and the reordering cost.

a. Use the distributive law to simplify the expression for the reordering cost. Then express the inventory cost, C, as a function of x.

b. Complete the table of values for the inventory cost for various reorder sizes.

x	10	20	30	40	50	60	70	80	90	100
C										

c. Graph your function C in the window

$$\text{Xmin} = 0, \qquad \text{Xmax} = 150$$
$$\text{Ymin} = 4500, \qquad \text{Ymax} = 5500$$

Estimate the minimum possible value for C.

d. How many computers should the manager order in each shipment so as to minimize the inventory cost? How many orders will she make during the year?

e. Graph the function $y = 6x + 4500$ in the same window with the function C. What do you observe?

8. A chain of electronics stores sells approximately 500 portable phones every year. The owner would like to minimize his annual inventory cost by ordering the optimal number of phones, x, at regular intervals. The cost of storing the phones will then be $2x$ dollars, and the cost of reordering will be $\dfrac{500}{x}(4x + 10)$. The total annual inventory cost is the sum of the storage cost and the reordering cost.

    a. Use the distributive law to simplify the expression for the reordering cost. Then express the inventory cost C as a function of x.

    b. Complete the following table of values for the inventory cost for various reorder sizes.

x	10	20	30	40	50	60	70	80	90	100
C										

    c. Graph your function C in the window

$$\text{Xmin} = 0, \qquad \text{Xmax} = 150$$
$$\text{Ymin} = 2000, \qquad \text{Ymax} = 2500$$

    Estimate the minimum possible value for C.

    d. How many portable phones should the owner order in each shipment so as to minimize the inventory cost? How many orders will he make during the year?

    e. Graph the function $y = 2x + 2000$ in the same window with the function C. What do you observe?

9. Francine wants to make a rectangular box. In order to simplify construction and keep her costs down, she plans for the box to have a square base and a total surface area of 96 square centimeters. She would like to know the largest volume that such a box can have.

    a. If the square base has length x centimeters, show that the height of the box is $h = \dfrac{24}{x} - \dfrac{x}{2}$ centimeters. (*Hint:* The surface area of the box is the sum of the areas of the six sides of the box.)

    b. Write an expression for the volume, V, of the box as a function of the length, x, of its base.

    c. Complete the table showing the heights and volumes of the box for various base lengths.

x	1	2	3	4	5	6	7
h							
V							

    Explain why the values of h and V are negative when $x = 7$.

    d. Graph your expression for volume $V(x)$ in an appropriate window. Approximate the maximum possible volume for a box of surface area 96 square centimeters.

    e. What value of x gives the maximum volume?

    f. Graph the height, $h(x)$, in the same window with $V(x)$. What is the height of the box with greatest volume? (Find the height directly from your graph and verify by using the formula given for $h(x)$.)

10. Delbert wants to make a box with a square base and a volume of 64 cubic centimeters. He would like to know the smallest surface area that such a box can have.
    a. If the square base has length x centimeters, show that the height of the box must be $h = \dfrac{64}{x^2}$ centimeters.
    b. Write an expression for the surface area, S, of the box as a function of the length x of its base. (*Hint:* The surface area of the box is the sum of the areas of the six sides of the box.)
    c. Complete the table showing the heights and surface areas of the box for various base lengths.

x	1	2	3	4	5	6	7	8
h								
S								

    d. Graph your expression for surface area $S(x)$ in an appropriate window. Approximate the minimum possible surface area for Delbert's box.
    e. What value of x gives the minimum surface area?
    f. Graph the height, $h(x)$, in the same window with $S(x)$. What is the height of the box with the smallest surface area? (Find the height directly from your graph and verify by using the formula given for $h(x)$.)

11. A train whistle sounds higher when the train is approaching you than when it is moving away from you. This phenomenon is known as the Doppler effect. If the actual pitch of the whistle is 440 hertz (this is the A note below middle C), then the note you hear will have the pitch

$$P(v) = \frac{440 \cdot (332)}{332 - v}$$

where the velocity, v, in meters per second is positive as the train approaches and is negative when the train is moving away. (The number 332 that appears in this expression is the speed of sound in meters per second.)
    a. Complete the table of values showing the pitch of the whistle at various train velocities.

v	-100	-75	-50	-25	0	25	50	75	100
P									

    b. Graph the function P. (Use the window Xmin = −94, Xmax = 94, and appropriate values of Ymin and Ymax.)
    c. What is the velocity of the train if the note you hear has a pitch of 415 hertz (corresponding to the note A-flat)? A pitch of 553.$\overline{3}$ hertz (C-sharp)?
    d. For what velocities will the pitch you hear be greater than 456.5 hertz?
    e. The graph has a vertical asymptote (although it is not visible in the suggested window). Where is it and what is its significance in this context?

12. The maximum altitude (in meters) attained by a projectile shot from the surface of the Earth is

$$h(v) = \frac{6.4 \times 10^6 v^2}{19.6 \cdot 6.4 \times 10^6 - v^2}$$

where v is the speed (in meters per second) at which the projectile was launched. (The radius of the Earth is 6.4×10^6 meters, and the constant 19.6 is related to the Earth's gravitational constant.)
    a. Complete the table of values showing the maximum altitude for various launch velocities.

v	100	200	300	400	500	600	700	800	900	1000
h										

    b. Graph the function h. (Use the window Xmin = 0, Xmax = 940, and appropriate values of Ymin and Ymax.)
    c. Approximately what speed is needed to attain an altitude of 4000 meters? An altitude of 16 kilometers?
    d. For what velocities will the projectile attain an altitude exceeding 32 kilometers?
    e. The graph has a vertical asymptote (although it is not visible in the suggested window). Where is it and what is its significance in this context?

For Problems 13–30,
(a) Sketch the horizontal and vertical asymptotes for each function.
(b) Use the asymptotes to help you sketch the rest of the graph.

13. $y = \dfrac{1}{x+3}$

14. $y = \dfrac{1}{x-3}$

15. $y = \dfrac{2}{x^2 - 5x + 4}$

16. $y = \dfrac{4}{x^2 - x - 6}$

17. $y = \dfrac{x}{x+3}$

18. $y = \dfrac{x}{x-2}$

19. $y = \dfrac{x+1}{x+2}$

20. $y = \dfrac{x-1}{x-3}$

21. $y = \dfrac{2x}{x^2 - 4}$

22. $y = \dfrac{x}{x^2 - 9}$

23. $y = \dfrac{x-2}{x^2 + 5x + 4}$

24. $y = \dfrac{x+1}{x^2 - x - 6}$

25. $y = \dfrac{x^2 - 1}{x^2 - 4}$

26. $y = \dfrac{2x^2}{x^2 - 1}$

27. $y = \dfrac{x+1}{(x-1)^2}$

28. $y = \dfrac{2(x^2 - 1)}{x^2 + 4}$

29. $y = \dfrac{x}{x^2 + 3}$

30. $y = \dfrac{x^2 + 2}{x^2 + 4}$

31. Graph the curve known as Newton's Serpentine:
$$y = \frac{4x}{x^2 + 1}.$$

32. Graph the curve known as the Witch of Agnesi:
$$y = \frac{8}{x^2 + 4}.$$

For Problems 33–38,
(a) Use polynomial division to write the fraction in the form $y = \dfrac{k}{p(x)} + c$, where k and c are constants.
(b) Use transformations to sketch the graph.

33. $y = \dfrac{2x + 2}{x}$

34. $y = \dfrac{4x^2 + 3}{x^2}$

35. $y = \dfrac{x+2}{x+1}$

36. $y = \dfrac{7 - 2x}{x - 3}$

37. $y = \dfrac{3x^2 - 12x + 13}{(x-2)^2}$

38. $y = \dfrac{-4x^2 + 8x - 3}{(x-1)^2}$

To review operations on algebraic fractions, see Algebra Skills Refresher A.9.

39. River Queen Tours offers a 50-mile round-trip excursion on the Mississippi River on a paddle wheel boat. The current in the Mississippi is 8 miles per hour.
   a. Express the time required for the downstream journey as a function of the speed of the paddle wheel boat in still water.
   b. Write a function for the time required for the return trip upstream.
   c. Write and simplify an expression for the time needed for the round trip as a function of the boat's speed.

40. A rowing team can maintain a speed of 15 miles per hour in still water. The team's daily training session includes a 5-mile run up the Red Cedar River and the return downstream.
   a. Express the team's time on the upstream leg as a function of the speed of the current.
   b. Write a function for the team's time on the downstream leg.
   c. Write and simplify an expression for the total time for the training run as a function of the current's speed.

41. Two pilots for the Flying Express parcel service receive packages simultaneously. Orville leaves Boston for Chicago at the same time Wilbur leaves Chicago for Boston. Each selects an airspeed of 400 miles per hour for the 900-mile trip. The prevailing winds blow from east to west.
   a. Express Orville's flying time as a function of the windspeed.
   b. Write a function for Wilbur's flying time.
   c. Who reaches his destination first? By how much time (in terms of windspeed)?

42. On New Year's Day, a blimp leaves its berth in Carson, California, and heads north for the Rose Bowl, 23 miles away. There is a breeze from the north at 6 miles per hour.
   a. Express the time required for the trip as a function of the blimp's airspeed.
   b. Write a function for the time needed for the return trip.
   c. Which trip takes longer? By how much time (in terms of the blimp's airspeed)?

43. The focal length of a lens is given by the formula

$$\frac{1}{f} = \frac{1}{p} + \frac{1}{q}$$

where f stands for the focal length, p is the distance from the object viewed to the lens, and q is the distance from the image to the lens. Suppose you estimate that the distance from your cat (the object viewed) to your camera lens is 60 inches greater than the distance from the lens to the film inside the camera, where the image forms.
   a. Express $1/f$ as a single fraction in terms of q.
   b. Write an expression for f as a function of q.

44. If two resistors, R_1 and R_2, in an electrical circuit are connected in parallel, the total resistance R in the circuit is given by

$$\frac{1}{R} = \frac{1}{R_1} + \frac{1}{R_2}$$

   a. Suppose that the second resistor, R_2, is 10 ohms greater than the first. Express $1/R$ as a single fraction in terms of R_1.
   b. Write an expression for R as a function of R_1.

45. a. Show that the equation $\dfrac{1}{y} - \dfrac{1}{x} = \dfrac{1}{k}$ is equivalent to $y = \dfrac{kx}{x + k}$ on their common domain.
   b. Graph the functions $y = \dfrac{kx}{x + k}$ for $k = 1, 2,$ and 3 in the window

$$\text{Xmin} = 0, \quad \text{Xmax} = 30$$
$$\text{Ymin} = 0, \quad \text{Ymax} = 4$$

   Describe the graphs.

46. Consider the graph of $y = \dfrac{ax}{x + k}$, where a and k are positive constants.
   a. What is the horizontal asymptote of the graph?
   b. Show that for $x = k$, $y = \frac{a}{2}$.
   c. Sketch the graph of $y = \dfrac{ax}{x + k}$ for $a = 4$ and $k = 10$ in the window

$$\text{Xmin} = 0, \quad \text{Xmax} = 60$$
$$\text{Ymin} = 0, \quad \text{Ymax} = 5$$

   Illustrate your answers to parts (a) and (b) on the graph.

For Problems 47–48,
(a) Use your answers to Problem 46 to find equations of the form $y = \dfrac{ax}{x + k}$ for the graphs shown.
(b) Check your answer with a graphing calculator.

47.

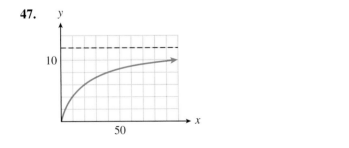

48.

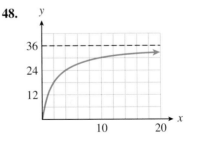

49. The Michaelis-Menten equation is the rate equation for chemical reactions catalyzed by enzymes. The speed of the reaction v is a function of the initial concentration of the reactant s and is given by

$$v = f(s) = \frac{Vs}{s + K}$$

where V is the maximum possible reaction rate and K is called the Michaelis constant. (Source: Holme and Peck, 1993)

a. What value does v approach as s increases?

b. What is the value of v when $s = K$?

c. The table gives data from reactions of the enzyme D-amino acid oxidase.

s	0.33	0.66	1.00	1.66	2.50	3.33	6.66
v	0.08	0.14	0.20	0.30	0.39	0.46	0.58

Plot the data and estimate the values of V and K from your graph.

d. Graph the function $v = \dfrac{0.88s}{s + 3.34}$ on top of your data points.

50. Show that

$$\frac{1}{v} = \frac{1}{V} + \frac{K}{Vs}$$

is another form of the Michaelis-Menten equation. (See Problem 49.)

51. a. Refer to the Michaelis-Menten equation in Problem 49. Solve for $\frac{1}{v}$, then write your new equation in the form $\dfrac{1}{v} = a \cdot \dfrac{1}{s} + b$. Express a and b in terms of V and K.

b. Use the data from part (c) of Problem 49 to make a table of values for $\left(\frac{1}{s}, \frac{1}{v}\right)$.

c. Plot the points $\left(\frac{1}{s}, \frac{1}{v}\right)$, then use linear regression to find the line of best fit.

d. Use your values for a and b to solve for V and K.

52. a. Refer to the Michaelis-Menten equation in Problem 49. Write an equation for $\frac{s}{v}$ of the form $\dfrac{s}{v} = cs + d$. Express c and d in terms of V and K.

b. Use the data from part (c) of Problem 49 to make a table of values for $\left(s, \frac{s}{v}\right)$.

c. Plot the points $\left(s, \frac{s}{v}\right)$, then use linear regression to find the line of best fit.

d. Use your values for a and b to solve for V and K.

The following are examples of functions whose graphs have holes.
(a) Find the domain of the function.
(b) Reduce the fraction to lowest terms.
(c) Graph the function. (*Hint:* The graph of the original function is identical to the graph of the function in part (b) except that certain points are excluded from the domain.) Indicate a hole in the graph by an open circle.

53. $y = \dfrac{x^2 - 4}{x - 2}$ **54.** $y = \dfrac{x^2 - 1}{x + 1}$ **55.** $y = \dfrac{x + 1}{x^2 - 1}$ **56.** $y = \dfrac{x - 3}{x^2 - 9}$

7.5   Equations That Include Algebraic Fractions

When working with rational functions, we often need to solve equations that involve algebraic fractions. In Example 1 of Section 7.4, we wrote a function that gave the time Francine needs for a 60-mile training run on her cycle-plane in terms of the windspeed, x:

$$t = f(x) = \frac{60}{15 - x}$$

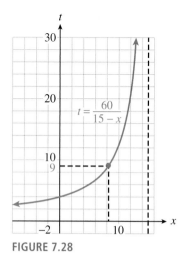

FIGURE 7.28

If it takes Francine 9 hours to cover 60 miles, what is the speed of the wind? We can answer this question by reading values from the graph of f, as shown in Figure 7.28. When $t = 9$, the value of x is between 8 and 9, so the windspeed is between 8 and 9 miles per hour.

Solving Equations with Fractions Algebraically

If we need a more accurate value for the windspeed, we can solve the equation

$$\frac{60}{15 - x} = 9$$

To solve an equation involving an algebraic fraction, we multiply each side of the equation by the denominator of the fraction. This has the effect of clearing the fraction, giving us an equivalent equation without fractions.

EXAMPLE 1   Solve the equation $\dfrac{60}{15 - x} = 9$.

Solution   Multiply both sides of the equation by $15 - x$ to obtain

$$(15 - x)\frac{60}{15 - x} = 9(15 - x)$$

$$60 = 9(15 - x) \qquad \text{Apply the distributive law.}$$

From here we can proceed as usual.

$$60 = 135 - 9x \qquad \text{Subtract 135 from both sides.}$$
$$-75 = -9x \qquad \text{Devide by } -9.$$
$$8.\bar{3} = x$$

The windspeed was $8.\bar{3}$, or $8\frac{1}{3}$ miles per hour.

EXERCISE 1

Solve $\dfrac{x^2}{x + 4} = 2$.

If the equation contains more than one fraction, we can clear all the denominators at once by multiplying both sides by the LCD of the fractions.

EXAMPLE 2   Rani times herself as she kayaks 30 miles down the Derwent River with the help of the current. Returning upstream against the current, she manages only 18 miles in the same amount of time. Rani knows that she can kayak at a rate of 12 miles per hour in still water. What is the speed of the current?

Solution   If we let x represent the speed of the current, we can use the formula $\text{time} = \frac{\text{distance}}{\text{rate}}$ to fill in the following table.

	Distance	**Rate**	**Time**
Downstream	30	$12 + x$	$\frac{30}{12 + x}$
Upstream	18	$12 - x$	$\frac{18}{12 - x}$

b. To solve the equation graphically, graph the two functions

$$Y_1 = \frac{6}{x} + 1 \quad \text{and} \quad Y_2 = \frac{1}{x+2}$$

in the window

$$Xmin = -4.7, \quad Xmax = 4.7$$
$$Ymin = -10, \quad Ymax = 10$$

as shown in Figure 7.30a.

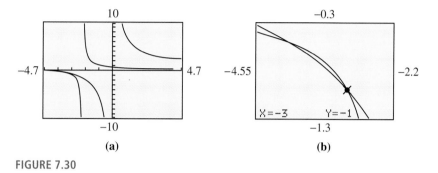

(a)   **(b)**

FIGURE 7.30

We see that the first graph has an asymptote at $x = 0$, and the second graph has one at $x = -2$. It appears that the two graphs may intersect in the third quadrant, around $x = -3$. To investigate further, we change the window settings to

$$Xmin = -4.55 \quad Xmax = -2.2$$
$$Ymin = -1.3 \quad Ymax = -0.3$$

to obtain the close-up view shown in Figure 7.30b. In this window, we can see that the graphs intersect in two distinct points, and by using the **Trace** we find that their x-coordinates are $x = -3$ and $x = -4$.

EXERCISE 3

Solve $\dfrac{9}{x^2 + x - 2} + \dfrac{1}{x^2 - x} = \dfrac{4}{x - 1}$.

Formulas

Algebraic fractions may appear in formulas that relate several variables. If we want to solve for one variable in terms of the others, we may need to clear the fractions.

EXAMPLE 4

Solve the formula $p = \dfrac{v}{q + v}$ for v.

Solution

Because the variable we want appears in the denominator, we must first multiply both sides of the equation by that denominator, $q + v$.

$$(q + v)p = \frac{v}{q + v}(q + v)$$

$$(q + v)p = v$$

Apply the distributive law on the left side, then collect all terms that involve v on one side of the equation.

$$qp + vp = v \qquad \text{Subtract } vp \text{ from both sides.}$$
$$qp = v - vp$$

We cannot combine the two terms containing v because they are not like terms. However, we can factor out v, so that the right side is written as a single term containing the variable v. We can then complete the solution.

$$qp = v(1-p) \qquad \text{Divide both sides by } 1-p.$$
$$\frac{qp}{1-p} = v$$

EXERCISE 4

Solve for a: $\dfrac{2ab}{a+b} = H$.

ANSWERS TO 7.5 EXERCISES

1. $x = -2, x = 4$ **2.** $x = 2$ **3.** $x = \frac{-1}{2}$ **4.** $a = \dfrac{bH}{2b - H}$

SECTION 7.5 SUMMARY

VOCABULARY *Look up the definitions of the new term in the Glossary.*

Extraneous solution

CONCEPTS

1. To solve an equation involving an algebraic fraction, we multiply each side of the equation by the denominator of the fraction. This has the effect of clearing the fraction, giving us an equivalent equation without fractions.

2. Whenever we multiply an equation by an expression containing the variable, we should check that the solutions obtained are not extraneous.

STUDY QUESTIONS

1. What is the first step in solving an equation that includes algebraic fractions?

2. If the equation also contains terms without fractions, should you multiply those terms by the LCD?

3. What are extraneous solutions, and when might they arise?

4. If you are solving a formula and two or more terms contain the variable you are solving for, what should you do?

SKILLS *Practice each skill in the Homework problems listed.*

1. Solve a fractional equation by clearing denominators: #1–14, 47–54

2. Write and solve proportions: #25–36

3. Solve equations by graphing: #15–22

4. Solve formulas that involve fractions: #39–48

5. Solve problems that involve algebraic fractions: #55–58

Solve each equation algebraically.

1. $\dfrac{6}{w+2} = 4$

2. $\dfrac{12}{r-7} = 3$

3. $9 = \dfrac{h-5}{h-2}$

4. $-3 = \dfrac{v+1}{v-6}$

5. $\dfrac{15}{s^2} = 8$

6. $\dfrac{3}{m^2} = 5$

7. $4.3 = \sqrt{\dfrac{18}{y}}$

8. $6.5 = \dfrac{52}{\sqrt{z}}$

9. The total weight, S, that a beam can support is given in pounds by

$$S = \frac{182.6wh^2}{l}$$

where w is the width of the beam in inches, h is its height in inches, and l is the length of the beam in feet. A beam over the doorway in an interior wall of a house must support 1600 pounds. If the beam is 4 inches wide and 9 inches tall, how long can it be?

10. If two appliances are connected in parallel in an electrical circuit, the total resistance, R, in the circuit is given by

$$R = \frac{ab}{a+b}$$

where a and b are the resistances of the two appliances. If one appliance has a resistance of 18 ohms, and the total resistance in the circuit is measured at 12 ohms, what is the resistance of the second appliance?

11. A flock of eider ducks is making a 150-mile flight at an average airspeed of 50 miles per hour against a moderate headwind.
   a. Express the ducks' travel time, t, as a function of the windspeed, v, and graph the function in the window

$$\text{Xmin} = 0, \qquad \text{Xmax} = 50$$
$$\text{Ymin} = 0, \qquad \text{Ymax} = 20$$

   (See Problem 1 of Homework 7.4.)
   b. Write and solve an equation to find the windspeed if the flock makes its trip in 4 hours. Label the corresponding point on your graph.

12. Bluefin tuna swim at average speed of 36 miles per hour in still water. A school of tuna is making a 200-mile trip against a current.
   a. Express the tuna's travel time, t, as a function of the current speed, v, and graph the function on the window

$$\text{Xmin} = 0, \qquad \text{Xmax} = 36$$
$$\text{Ymin} = 0, \qquad \text{Ymax} = 50$$

   (See Problem 2 of Homework 7.4.)
   b. Write and solve an equation to find the current speed if the school makes its trip in 8 hours. Label the corresponding point on your graph.

13. The cost, in thousands of dollars, for immunizing p percent of the residents of Emporia against a dangerous new disease is given by the function

$$C(p) = \frac{72p}{100-p}$$

Write and solve an equation to determine what percent of the population can be immunized for $168,000.

14. The cost, in thousands of dollars, for extracting $p\%$ of a precious ore from a mine, is given by the equation

$$C(p) = \frac{360p}{100-p}$$

Write and solve an equation to determine what percentage of the ore can be extracted for $390,000.

For Problems 15–18,
(a) Solve the equation graphically by graphing two functions, one for each side of the equation.
(b) Solve the equation algebraically.

15. $\dfrac{2x}{x+1} = \dfrac{x+1}{2}$

16. $\dfrac{3}{2x+1} = \dfrac{2x-3}{x}$

17. $\dfrac{2}{x+1} = \dfrac{x}{x+1} + 1$

18. $\dfrac{5}{x-3} = \dfrac{x+2}{x-3} + 3$

19. The manager of Joe's Burgers discovers that he will sell $\frac{160}{x}$ burgers per day if the price of a burger is x dollars. On the other hand, he can afford to make $6x + 49$ burgers if he charges x dollars apiece for them.
  a. Graph the *demand function*, $D(x) = \frac{160}{x}$, and the *supply function*, $S(x) = 6x + 49$, in the same window. At what price x does the demand for burgers equal the number that Joe can afford to supply? This value for x is called the *equilibrium price*.
  b. Write and solve an equation to verify your equilibrium price.

20. A florist finds that she will sell $\frac{300}{x}$ dozen roses per week if she charges x dollars for a dozen. Her suppliers will sell her $5x - 55$ dozen roses if she sells them at x dollars per dozen.
  a. Graph the demand function, $D(x) = \frac{300}{x}$, and the supply function, $S(x) = 5x - 55$, in the same window. At what equilibrium price x will the florist sell all the roses she purchases?
  b. Write and solve an equation to verify your equilibrium price.

21. Francine wants to fence a rectangular area of 3200 square feet to grow vegetables for her family of three.
  a. Express the length of the garden as a function of its width.
  b. Express the perimeter, P, of the garden as a function of its width.
  c. Graph your function for perimeter and find the coordinates of the lowest point on the graph. Interpret those coordinates in the context of the problem.
  d. Francine has 240 feet of chain link to make a fence for the garden, and she would like to know what the width of the garden should be. Write an equation that describes this situation.
  e. Solve your equation and find the dimensions of the garden.

22. The cost of wire fencing is $7.50 per foot. A rancher wants to enclose a rectangular pasture of 1000 square feet with this fencing.
  a. Express the length of the pasture as a function of its width.
  b. Express the cost of the fence as a function of its width.
  c. Graph your function for the cost and find the coordinates of the lowest point on the graph. Interpret those coordinates in the context of the problem.
  d. The rancher has $1050 to spend on the fence, and she would like to know what the width of the pasture should be. Write an equation to describe this situation.
  e. Solve your equation and find the dimensions of the pasture.

23. A proportion is an equation in which each side is a ratio: $\dfrac{a}{b} = \dfrac{c}{d}$. Show that this equation may be rewritten as $ad = bc$.

24. Suppose that y varies directly with x, and (a, b) and (c, d) are two points on the graph of y in terms of x. Show that $\dfrac{b}{a} = \dfrac{d}{c}$.

Solve each proportion using your result from Problem 23.

25. $\dfrac{3}{4} = \dfrac{y+2}{12-y}$

26. $\dfrac{-3}{4} = \dfrac{y-7}{y+14}$

27. $\dfrac{50}{r} = \dfrac{75}{r+20}$

28. $\dfrac{30}{r} = \dfrac{20}{r-10}$

29. Property taxes on a house vary directly with the value of the house. If the taxes on a house worth $120,000 are $2700, what would the taxes be on a house assessed at $275,000?

30. The cost of electricity varies directly with the number of units (BTUs) consumed. If a typical household in the Midwest uses 83 million BTUs of electricity annually and pays $1236, how much will a household that uses 70 million BTUs annually spend for energy?

31. Distances on a map vary directly with actual distances. The scale on a map of Michigan uses $\frac{3}{8}$ inch to represent 10 miles. If Isle Royale is $1\frac{11}{16}$ inches long on the map, what is the actual length of the island?

32. The dimensions of an enlargement vary directly with the dimensions of the original. A photographer plans to enlarge a photograph that measures 8.3 centimeters by 11.2 centimeters to produce a poster that is 36 centimeters wide. How long will the poster be?

33. The Forest Service tags 200 perch and releases them into Spirit Lake. One month later, it captures 80 perch and finds that 18 of them are tagged. What is the Forest Service's estimate of the original perch population of the lake?

34. The Wildlife Commission tags 30 Canada geese at one of its migratory feeding grounds. When the geese return, the commission captures 45 geese, of which 4 are tagged. What is the commission's estimate of the number of geese that use the feeding ground?

35. The highest point on Earth is Mount Everest in Tibet, with an elevation of 8848 meters. The deepest part of the ocean is the Challenger Deep in the Mariana Trench, near Indonesia, 11,034 meters below sea level.
    a. What is the total height variation in the surface of the Earth?
    b. What percentage of the Earth's radius, 6400 kilometers, is this variation?
    c. If the Earth were shrunk to the size of a basketball, with a radius of 4.75 inches, what would be the corresponding height of Mount Everest?

36. Shortly after the arrival of human beings at the Hawaiian islands around 400 A.D., many species of birds became extinct. Fossils of 29 different species have been found, but some species may have left no fossils for us to find. We can estimate the total number of extinct species using a proportion. Of 9 species that are still alive, biologists have found fossil evidence of 7. (Source: Burton, 1998)
    a. Assuming that the same fraction of extinct species have left fossil records, calculate the total number of extinct species.
    b. Give two reasons why this estimate may not be completely accurate.

37. In the figure, the rectangle *ABCD* is divided into a square and a smaller rectangle, *CDEF*. The two rectangles *ABCD* and *CDEF* are similar (their corresponding sides are proportional.) A rectangle *ABCD* with this property is called a *golden rectangle,* and the ratio of its length to its width is called the golden ratio. The golden ratio appears frequently in art and nature, and it is considered to give the most pleasing proportions to many figures. We will compute the golden ratio as follows.

    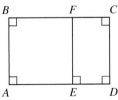

    a. Let *AB* = 1 and *AD* = x. What are the lengths of *AE*, *ED*, and *CD*?
    b. Write a proportion in terms of *x* for the similarity of rectangles *ABCD* and *CDEF*. Be careful to match up the corresponding sides.
    c. Solve your proportion for *x*. Find the golden ratio,
    $$\frac{AD}{AB} = \frac{x}{1}.$$

38. The figure shows the graphs of two equations, $y = x$ and $y = \frac{1}{x} + 1$.
    a. Find the *x*-coordinate of the intersection point of the two graphs.
    b. Compare your answer to the golden ratio you computed in Problem 37.

    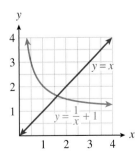

Solve each formula for the specified variable.

39. $S = \dfrac{a}{1 - r}$, for r

40. $I = \dfrac{E}{r + R}$, for R

41. $H = \dfrac{2xy}{x + y}$, for x

42. $M = \dfrac{ab}{a + b}$, for b

43. $F = \dfrac{Gm_1m_2}{d^2}$, for d

44. $F = \dfrac{kq_1q_2}{r^2}$, for r

45. $\dfrac{1}{Q} + \dfrac{1}{I} = \dfrac{2}{r}$, for r

46. $\dfrac{1}{R} = \dfrac{1}{A} + \dfrac{1}{B}$, for B

47. The sidereal period of a planet is the time for the planet to make one trip around the Sun (as seen from the Sun itself). The synodic period is the time between two successive conjunctions of the planet and the Sun, as seen from Earth. The relationship among the sidereal period, P, of a planet, the synodic period, S, of the planet, and the sidereal period of Earth, E, is given by

$$\frac{1}{P} = \frac{1}{S} + \frac{1}{E}$$

when the planet is closer to the Sun than the Earth is. Solve for P in terms of S and E.

48. When a planet is farther from the Sun than Earth is,

$$\frac{1}{P} = \frac{1}{E} - \frac{1}{S}$$

where P, E, and S are as defined in Problem 47. Solve for P in terms of S and E.

Solve the equation algebraically.

49. $\dfrac{3}{x - 2} = \dfrac{1}{2} + \dfrac{2x - 7}{2x - 4}$

50. $\dfrac{2}{x + 1} + \dfrac{1}{3x + 3} = \dfrac{1}{6}$

51. $\dfrac{4}{x + 2} - \dfrac{1}{x} = \dfrac{2x - 1}{x^2 + 2x}$

52. $\dfrac{1}{x - 1} + \dfrac{2}{x + 1} = \dfrac{x - 2}{x^2 - 1}$

53. $\dfrac{x}{x + 2} - \dfrac{3}{x - 2} = \dfrac{x^2 + 8}{x^2 - 4}$

54. $\dfrac{4}{2x - 3} + \dfrac{4x}{4x^2 - 9} = \dfrac{1}{2x + 3}$

55. $\dfrac{4}{3x} + \dfrac{3}{3x + 1} + 2 = 0$

56. $-3 = \dfrac{-10}{x + 2} + \dfrac{10}{x + 5}$

57. A chartered sightseeing flight over the Grand Canyon is scheduled to return to its departure point in 3 hours. The pilot would like to cover a distance of 144 miles before turning around, and he hears on the Weather Service that there will be a headwind of 20 miles per hour on the outward journey.
   a. Express the time it takes for the outward journey as a function of the airspeed of the plane.
   b. Express the time it takes for the return journey as a function of the speed of the plane
   c. Graph the sum of the two functions and find the point on the graph with y-coordinate 3. Interpret the coordinates of the point in the context of the problem.
   d. The pilot would like to know what airspeed to maintain in order to complete the tour in 3 hours. Write an equation to describe this situation.
   e. Solve your equation to find the appropriate airspeed.

58. Two student pilots leave the airport at the same time. They both fly at an airspeed of 180 miles per hour, but one flies with the wind and the other flies against the wind.
   a. Express the time it takes the first pilot to travel 500 miles as a function of the windspeed.
   b. Express the time it takes the second pilot to travel 400 miles as a function of the windspeed.
   c. Graph the two functions in the same window, and find the coordinates of the intersection point. Interpret those coordinates in the context of the problem.
   d. Both pilots check in with their instructors at the same time, and the first pilot has traveled 500 miles while the second pilot has gone 400 miles. Write an equation to describe this situation.
   e. Solve your equation to find the speed of the wind.

59. Andy drives 300 miles to Lake Tahoe at 70 miles per hour and returns home at 50 miles per hour. What is his average speed for the round trip? (It is not 60 miles per hour!)

  a. Write expressions for the time it takes for each leg of the trip if Andy drives a distance, d, at speed r_1 and returns at speed r_2.

  b. Write expressions for the total distance and total time for the trip.

  c. Write an expression for the average speed for the entire trip.

  d. Write your answer to part (c) as a simple fraction.

  e. Use your formula to answer the question stated in the problem.

60. The owner of a print shop volunteers to produce flyers for his candidate's campaign. His large printing press can complete the job in 4 hours, and the smaller model can finish the flyers in 6 hours. How long will it take to print the flyers if he runs both presses simultaneously?

  a. Suppose that the large press can complete a job in t_1 hours and the smaller press takes t_2 hours. Write expressions for the fraction of a job that each press can complete in 1 hour.

  b. Write an expression for the fraction of a job that can be completed in 1 hour with both presses running simultaneously.

  c. Write an expression for the amount of time needed to complete the job with both presses running.

  d. Write your answer to part (c) as a simple fraction.

  e. Use your formula to answer the question stated in the problem.

CHAPTER 7 Summary and Review

KEY CONCEPTS

1. The degree of a product of nonzero polynomials is the sum of the degrees of the factors.

2.

> ### Cube of a Binomial
>
> **1.** $(x + y)^3 = x^3 + 3x^2y + 3xy^2 + y^3$
>
> **2.** $(x - y)^3 = x^3 - 3x^2y + 3xy^2 - y^3$

3.

> ### Factoring the Sum or Difference of Two Cubes
>
> **3.** $x^3 + y^3 = (x + y)(x^2 - xy + y^2)$
>
> **4.** $x^3 - y^3 = (x - y)(x^2 + xy + y^2)$

4. The graphs of all polynomials are smooth curves without breaks or holes.

5. The graph of a polynomial of degree n (with positive lead coefficient) has the same long-term behavior as the power function of the same degree.

6. **Factor theorem:** $(x - a)$ is a factor of a polynomial $P(x)$ if and only if $P(a) = 0$.

7. A polynomial of degree n can have at most n x-intercepts.

8. At a zero of multiplicity 2, the graph of a polynomial has a turning point. At a zero of multiplicity 3, the graph of a polynomial has an inflection point.

9. The square root of a negative number is an imaginary number.

10. A complex number is the sum of a real number and an imaginary number.

11. We can perform the four arithmetic operations on complex numbers.

12. The product of a nonzero complex number and its conjugate is always a positive real number.

13. We can graph complex numbers in the complex plane.

14. Multiplying a complex number by i rotates its graph by 90° around the origin.

15. **Fundamental theorem of algebra:** A polynomial of degree $n \geq 1$ has exactly n complex zeros.

16. We can graph complex numbers in the complex plane.

17.

> ### Rational Function
>
> A **rational function** is one of the form
> $$f(x) = \frac{P(x)}{Q(x)}$$
> where $P(x)$ and $Q(x)$ are polynomials and $Q(x)$ is not the zero polynomial.

18. A rational function $f(x) = \frac{P(x)}{Q(x)}$ is undefined for any value $x = a$ where $Q(a) = 0$. These x-values are not in the domain of the function.

19.

> ### Vertical Asymptotes
>
> If $Q(a) = 0$ but $P(a) \neq 0$, then the graph of the rational function $f(x) = \frac{P(x)}{Q(x)}$ has a **vertical asymptote** at $x = a$.

21. To solve an equation involving an algebraic fraction, we multiply each side of the equation by the denominator of the fraction. This has the effect of clearing the fraction, giving us an equivalent equation without fractions.

22. Whenever we multiply an equation by an expression containing the variable, we should check that the solutions obtained are not extraneous.

20.

> ### Horizontal Asymptotes
>
> Suppose $f(x) = \frac{P(x)}{Q(x)}$ is a rational function, where the degree of $P(x)$ is m and the degree of $Q(x)$ is n.
>
> **1.** If $m < n$, the graph of f has a horizontal asymptote at $y = 0$.
>
> **2.** If $m = n$, the graph of f has a horizontal asymptote at $y = \frac{a}{b}$, where a is the lead coefficient of $P(x)$ and b is the lead coefficient of $Q(x)$.
>
> **3.** If $m > n$, the graph of f does not have a horizontal asymptote.

REVIEW PROBLEMS

Multiply.

1. $(2x - 5)(x^2 - 3x + 2)$

2. $(b^2 - 2b - 3)(2b^2 + b - 5)$

3. $(t + 4)(t^2 - t - 1)$

4. $(b + 3)(2b - 1)(2b + 5)$

Find the indicated term.

5. $(1 - 3x + 5x^2)(7 + x - x^2); \quad x^2$

6. $(-3 + x - 4x^2)(4 + 3x - 2x^3); \quad x^3$

7. $(4x - x^2 + 3x^3(1 + 4x - 3x^2); \quad x^3$

8. $(3 - 2x + 2x^3)(5 + 3x - 2x^2 + 4x^4); \quad x^4$

Factor.

9. $8x^3 - 27z^3$

10. $1 + 125a^3b^3$

11. $y^3 + 27x^3$

12. $x^9 - 8$

Write as a polynomial.

13. $(v - 10)^3$

14. $(a + 2b^2)^3$

15. The expression $\frac{n}{6}(n-1)(n-2)$ gives the number of different 3-item pizzas that can be created from a list of n toppings.
   a. Write the expression as a polynomial.
   b. If Mitch's Pizza offers 12 different toppings, how many different combinations for 3-item pizzas can be made?
   c. Use a table or graph to determine how many different toppings are needed in order to be able to have more than 1000 possible combinations for 3-item pizzas.

16. The expression $n(n-1)(n-2)$ gives the number of different triple-scoop ice cream cones that can be created from a list of n flavors.
   a. Write the expression as a polynomial.
   b. If Zanner's Ice Cream Parlor offers 21 flavors, how many different triple-scoop ice cream cones can be made?
   c. Use a table or graph to determine how many different toppings are needed in order to be able to have more than 10,000 possible triple-scoop ice cream cones.

For Problems 17 and 18,
(a) Graph each polynomial in the standard window.
(b) Find the range of the function on the domain $[-10, 10]$.

17. $f(x) = x^3 - 3x + 2$

18. $g(x) = -0.1(x^4 - 6x^3 + x^2 + 24x + 16)$

For Problems 19–28,
(a) Find the zeros of the polynomial.
(b) Sketch the graph by hand.

19. $f(x) = (x-2)(x+1)^2$

20. $g(x) = (x-3)^2(x+2)$

21. $G(x) = x^2(x-1)(x+3)$

22. $F(x) = (x+1)^2(x-2)^2$

23. $V(x) = x^3 - x^5$

24. $H(x) = x^4 - 9x^2$

25. $P(x) = x^3 + x^2 - x - 1$

26. $y = x^3 + x^2 - 2x$

27. $y = x^4 - 7x^2 + 6$

28. $y = x^4 + x^3 - 3x^2 - 3x$

Find a possible formula for the polynomial, in factored form.

29.

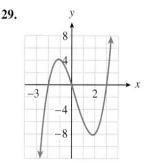

30.

31.

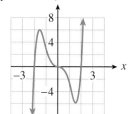

32.

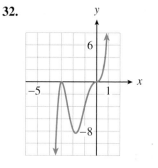

33.

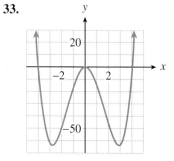

34.

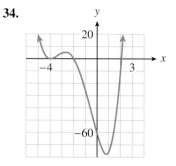

For Problems 35–36,
(a) Verify that the given value is a zero of the polynomial.
(b) Find the other zeros. (*Hint:* Use polynomial division to write $P(x) = (x - a)Q(x)$, then factor $Q(x)$.)

35. $P(x) = x^3 - x^2 - 7x - 2;\quad a = -2$

36. $P(x) = 3x^3 - 11x^2 - 5x + 4;\quad a = 4$

For Problems 37–40,
(a) Solve the quadratic equation, and write the solutions in the form $a + bi$.
(b) Check your solutions.

37. $x^2 + 4x + 10 = 0$

38. $x^2 - 2x + 7 = 0$

39. $3x^2 - 6x + 5 = 0$

40. $2x^2 + 5x + 4 = 0$

Evaluate each polynomial for the given value of the variable.

41. $z^2 - 6z + 5$
    a. $z = 3 + 2i$
    b. $z = 3 - 2i$

42. $w^2 + 4w + 7$
    a. $w = -1 - 3i$
    b. $w = -1 + 3i$

Find the quotient.

43. $\dfrac{2 - 5i}{3 - i}$

44. $\dfrac{1 + i}{1 - i}$

Find a fourth-degree polynomial with the given zeros.

45. $3i, 1 - 2i$

46. $2 - \sqrt{3}, 2 + 3i$

Plot each complex number as a point on the complex plane.

47 a. $z = 4 + i, \ \overline{z}, \ -z, \ -\overline{z}$
  b. $iz, i\overline{z}, -iz, -i\overline{z}$

48. a. $w = -2 + 3i, \ \overline{w}, \ -w, \ -\overline{w}$
  b. $iw, i\overline{w}, -iw, -i\overline{w}$

49. The radius, r, of a cylindrical can should be one-half its height, h.
  a. Express the volume, V, of the can as a function of its height.
  b. What is the volume of the can if its height is 2 centimeters? 4 centimeters?
  c. Graph the volume as a function of the height and verify your results of part (b) graphically. What is the approximate height of the can if its volume is 100 cubic centimeters?

50. The Twisty-Freez machine dispenses soft ice cream in a cone-shaped peak with a height 3 times the radius of its base. The ice cream comes in a round bowl with base diameter d.
  a. Express the volume V of Twisty-Freez in the bowl as a function of d.
  b. How much Twisty-Freez comes in a 3-inch diameter dish? A 4-inch dish?
  c. Graph the volume as a function of the diameter and verify your results of part (b) graphically. What is the approximate diameter of a Twisty-Freez if its volume is 5 cubic inches?

51. A new health club opened up, and the manager kept track of the number of active members over its first few months of operation. The equation below gives the number, N, of active members, in hundreds, t months after the club opened.
$$N = \frac{44t}{40 + t^2}$$
  a. Use your calculator to graph the function N on a suitable domain.
  b. How many active members did the club have after 8 months?
  c. In which months did the club have 200 active members?
  d. When does the health club have the largest number of active members? What happens to the number of active members as time goes on?

52. A small lake in a state park has become polluted by runoff from a factory upstream. The cost for removing p percent of the pollution from the lake is given, in thousands of dollars, by
$$C = \frac{25p}{100 - p}$$
  a. Use your calculator to graph the function C on a suitable domain.
  b. How much will it cost to remove 40% of the pollution?
  c. How much of the pollution can be removed for $100,000?
  d. What happens to the cost as the amount of pollution to be removed increases? How much will it cost to remove all the pollution?

State the domain of each function.

53. $h(x) = \dfrac{x^2 - 9}{x(x^2 - 4)}$

54. $j(x) = \dfrac{x - 3x + 10}{x^2(x^2 + 1)}$

For Problems 55–56,
(a) Sketch the horizontal and vertical asymptotes for each function.
(b) Use the asymptotes to help you sketch the graph.

55. $F(x) = \dfrac{2x}{x^2 - 1}$

56. $G(x) = \dfrac{2}{x^2 - 1}$

For Problems 25–30,
(a) Identify all asymptotes and intercepts.
(b) Sketch the graph.

57. $y = \dfrac{1}{x - 4}$

58. $y = \dfrac{2}{x^2 - 3x - 10}$

59. $y = \dfrac{x - 2}{x + 3}$

60. $y = \dfrac{x-1}{x^2 - 2x - 3}$ **61.** $y = \dfrac{3x^2}{x^2 - 4}$ **62.** $y = \dfrac{2x^2 - 2}{x^2 - 9}$

For Problems 63–66,
(a) Use polynomial division to write the fraction in the form $y = \dfrac{k}{p(x)} + c$, where k and c are constants.
(b) Use transformations to sketch the graph.

63. $y = \dfrac{3x+4}{x+3}$ **64.** $y = \dfrac{5x+1}{x-2}$

65. $y = \dfrac{x^2 + 2x + 3}{(x+1)^2}$ **66.** $y = \dfrac{x^2 - 4x + 3}{(x-2)^2}$

67. The Explorer's Club is planning a canoe trip to travel 90 miles up the Lazy River and return in 4 days. Club members plan to paddle for 6 hours each day, and they know that the current in the Lazy River is 2 miles per hour.
  a. Express the time it will take for the upstream journey as a function of their paddling speed in still water.
  b. Express the time it will take for the downstream journey as a function of their paddling speed in still water.
  c. Graph the sum of the two functions and find the point on the graph with y-coordinate 24. Interpret the coordinates of the point in the context of the problem.
  d. The Explorer's Club would like to know what average paddling speed members must maintain in order to complete their trip in 4 days. Write an equation to describe this situation.
  e. Solve your equation to find the required paddling speed.

68. Pam lives on the banks of the Cedar River and makes frequent trips in her outboard motorboat. The boat travels at 20 miles per hour in still water.
  a. Express the time it takes Pam to travel 8 miles upstream to the gas station as a function of the speed of the current.
  b. Express the time it takes Pam to travel 12 miles downstream to Marie's house as a function of the speed of the current.
  c. Graph the two functions in the same window, then find the coordinates of the intersection point. Interpret those coordinates in the context of the problem.
  d. Pam traveled to the gas station in the same time it took her to travel to Marie's house. Write an equation to describe this situation.
  e. Solve your equation to find the speed of the current in the Cedar River.

69. Mikala sells $\frac{320}{x}$ bottles of bath oil per week if she charges x dollars per bottle. Her supplier can manufacture $\frac{1}{2}x + 6$ bottles per week if she sells it at x dollars per bottle.
  a. Graph the demand function, $D(x) = \frac{320}{x}$, and the supply function, $S(x) = \frac{1}{2}x + 6$, in the same window.
  b. Write and solve an equation to find the equilibrium price, that is, the price at which the supply equals the demand for bath oil. Label this point on your graph.

70. Tomoko sells $\frac{4800}{x}$ exercise machines each month if the price of a machine is x dollars. On the other hand, her supplier can manufacture $2.5x + 20$ machines if she charges x dollars apiece for them.
  a. Graph the demand function, $D(x) = \frac{4800}{x}$, and the supply function, $S(x) = 2.5x + 20$, in the same window.
  b. Write and solve an equation to find the equilibrium price, that is, the price at which the supply equals the demand for exercise machines. Label this point on your graph.

Write and solve a proportion for each problem.

71. A polling firm finds that 78 of the 300 randomly selected students at Citrus College play some musical instrument. Based on the poll, how many of the college's 1150 students play a musical instrument?

72. Claire wants to make a scale model of Salem College. The largest building on campus, Lausanne Hall, is 60 feet tall, and her model of Lausanne Hall will be 8 inches tall. How tall should she make the model of Willamette Hall, which is 48 feet tall?

Solve.

73. $\dfrac{y+3}{y+5} = \dfrac{1}{3}$ **74.** $\dfrac{z^2 + 2}{z^2 - 2} = 3$

75. $\dfrac{x}{x-2} = \dfrac{2}{x-2} + 7$

76. $\dfrac{3x}{x+1} - \dfrac{2}{x^2+x} = \dfrac{4}{x}$

77. $\dfrac{2}{a+1} + \dfrac{1}{a-1} = \dfrac{3a-1}{a^2-1}$

78. $\dfrac{2b-1}{b^2+2b} = \dfrac{4}{b+2} - \dfrac{1}{b}$

79. $\dfrac{-10}{u-2} = \dfrac{u-4}{u^2-u-2} + \dfrac{3}{u+1}$

80. $\dfrac{1}{t^2+t} + \dfrac{1}{t} = \dfrac{3}{t+1}$

Solve for the indicated variable.

81. $V = C\left(1 - \dfrac{t}{n}\right)$, for n

82. $r = \dfrac{dc}{1-ec}$, for c

83. $\dfrac{p}{q} = \dfrac{r}{q+r}$, for q

84. $I = \dfrac{E}{R + \dfrac{r}{n}}$, for R

PROJECTS FOR CHAPTER 7

1. In this project, we solve cubic equations of the form

$$x^3 + mx = n$$

Note that there is no quadratic term. This special form was first solved by the Italian mathematicians Scipione del Ferro and Niccolò Fontana Tartaglia early in the sixteenth century. Tartaglia revealed the secret to solving the special cubic equation in a poem. He first found values u and v to satisfy the system

$$u - v = n$$
$$uv = \left(\dfrac{m}{3}\right)^3$$

   a. We will use Tartaglia's method to solve

$$x^3 + 6x = 7$$

   What are the values of m and n?
   b. Substitute the values of m and n into Tartaglia's system, then use substitution to solve for u and v. You should find two possible solutions.
   c. For each solution of the system, compute $x = \sqrt[3]{u} - \sqrt[3]{v}$. You should get the same value of x for each (u, v).
   d. Check that your value for x is a solution of $x^3 + 6x = 7$.

2. Tartaglia's method always works to solve the special cubic equation, even when u and v are not convenient values. We will show why in this project.
   a. Expand the expression $(a-b)^3 + 3ab(a-b)$ and complete the identity.

$$(a-b)^3 + 3ab(a-b) = \underline{\hspace{2cm}}$$

   b. Your answer to part (a) is actually Tartaglia's special cubic in disguise. Substitute $x = a - b$, $m = 3ab$,

and $n = a^3 - b^3$ to see this. Therefore, if we can find numbers a and b that satisfy

$$3ab = m$$
$$a^3 - b^3 = n$$

   then the solution to Tartaglia's cubic is $x = a - b$.
   c. Compare the system in part (b) to the system from Project 1,

$$u - v = n$$
$$uv = \left(\dfrac{m}{3}\right)^3$$

   to show that $u = a^3$ and $v = b^3$.
   d. Use your answer to part (c) to show that Tartaglia's value, $x = \sqrt[3]{u} - \sqrt[3]{v}$, is a solution of $x^3 + mx = n$.

3. Use Tartaglia's method to solve the equation

$$x^3 + 3x = 2$$

by carrying out the following steps.
   a. Identify the values of m and n from Equation (1) and write two equations for u and v.
   b. Solve for values of u and v. You will need to use the quadratic formula.
   c. Take the positive values of u and v. Write the solution $x = \sqrt[3]{u} - \sqrt[3]{v}$. Do not try to simply the radical expression; instead, use your calculator to check the solution numerically.

4. We can solve any cubic equation by first using a substitution to put the equation in Tartaglia's special form.
   a. Consider the equation $X^3 + bX^2 + cX + d = 0$. Make the substitution $X = x - \dfrac{b}{3}$, and expand the left side of the equation.
   b. What is the coefficient of x^2 in the resulting equation? What are the values of m and n?
   c. If you solve the special form in part (a) for x, how can you find the value of X that solves the original equation?

5. The time, T, it takes for the Moon to eclipse the Sun totally is given (in minutes) by the formula

$$T = \frac{1}{v}\left(\frac{rD}{R} - d\right)$$

where d is the diameter of the Moon, D is the diameter of the Sun, r is the distance from the Earth to the Moon, R is the distance from the Earth to the Sun, and v is the speed of the Moon.

a. Solve the formula for v in terms of the other variables.

b. It takes 2.68 minutes for the Moon to eclipse the Sun. Calculate the speed of the Moon given the following values:

$$d = 3.48 \times 10^3 \, \text{km} \qquad D = 1.41 \times 10^6 \, \text{km}$$
$$r = 3.82 \times 10^5 \, \text{km} \qquad R = 1.48 \times 10^8 \, \text{km}$$

6. The stopping distance, s, for a car traveling at speed v meters per second is given (in meters) by

$$s = vT + \frac{v^2}{2a}$$

where T is the reaction time of the driver and a is the average deceleration as the car brakes. Suppose that all the cars on a crowded motorway maintain the appropriate spacing determined by the stopping distance for their speed. What speed allows the maximum flow of cars along the road per unit time? Using the formula time $= \frac{\text{distance}}{\text{speed}}$, we see that the time interval, t, between cars is

$$t = \frac{s}{v} + \frac{L}{v}$$

where L is the length of the car. To achieve the maximum flow of cars, we would like t to be as small as possible. (Source: Bolton, 1974)

a. Substitute the expression for s into the formula for t, then simplify.

b. A typical reaction time is $T = 0.7$ seconds, a typical car length is $L = 5$ meters, and $a = 7.5$ meters per second squared. With these values, graph t as a function of v in the window

$$\text{Xmin} = 0 \qquad \text{Xmax} = 20$$
$$\text{Ymin} = 0 \qquad \text{Ymax} = 3$$

c. To one decimal place, what value of v gives the minimum value of t? Convert your answer to miles per hour.

7. Many endangered species have fewer than 1000 individuals left. To preserve the species, captive breeding programs must maintain a certain effective population, N, given by

$$N = \frac{4FM}{F + M}$$

where F is the number of breeding females and M the number of breeding males. (Source: Chapman and Reiss, 1992)

a. What is the effective population if there are equal numbers of breeding males and females?

b. In 1972, a breeding program for Speke's gazelle was established with just three female gazelle. Graph the effective population, N, as a function of the number of males.

c. What is the largest effective population that can be created with three females? How many males are needed to achieve the maximum value?

d. With three females, for what value of M is $N = M$?

e. The breeding program for Speke's gazelle began with only one male. What was the effective population?

8. When a drug or chemical is injected into a patient, biological processes begin removing that substance. If no more of the substance in introduced, the body removes a fixed fraction of the substance each hour. The amount of substance remaining in the body at time t is an exponential decay function, so there is a *biological half-life* to the substance denoted by T_b. If the substance is a radioisotope, it undergoes radioactive decay and so has a physical half-life as well, denoted T_p. The *effective half-life*, denoted by T_e, is related to the biological and physical half-lives by the equation

$$\frac{1}{T_e} = \frac{1}{T_b} + \frac{1}{T_p}$$

The radioisotope ^{131}I is used as a label for the human serum albumin. The physical half-life of ^{131}I is 8 days. (Source: Pope, 1989)

a. If ^{131}I is cleared from the body with a half-life of 21 days, what is the effective half-life of ^{131}I?

b. The biological half-life of a substance varies considerably from person to person. If the biological half-life of ^{131}I is x days, what is the effective half life?

c. Let $f(x)$ represent the effective half-life of ^{131}I when the biological half-life is x days. Graph $y = f(x)$.

d. What would the biological half-life of ^{131}I need to be to produce an effective half-life of 6 days? Label the corresponding point on your graph.

e. For what possible biological half-lives of ^{131}I will the effective half-life be less than 4 days?

9. Animals spend most of their time hunting or foraging for food to keep themselves alive. Knowing the rate at which an animal (or population of animals) eats can help us determine its metabolic rate or its impact on its habitat. The rate of eating is proportional to the availability of food in the area, but it has an upper limit imposed by mechanical considerations, such as how long it takes the animal to capture and ingest its prey. (Source: Burton, 1998)

   a. Sketch a graph of eating rate as a function of quantity of available food. This will be a qualitative graph only; you do not have enough information to put scales on the axes.

   b. Suppose that the rate at which an animal catches its prey is proportional to the number of prey available, or $r_c = ax$, where a is a constant and x is the number of available prey. The rate at which it handles and eats the prey is constant, $r_h = b$. Write expressions for T_c and T_h, the times for catching and handling N prey.

   c. Show that the rate of food consumption is given by

   $$y = \frac{abx}{b + ax} = \frac{bx}{b/a + x}$$

   Hint: $y = \frac{N}{T}$, where N is the number of prey consumed in a time interval T, where $T = T_c + T_h$.

   d. In a study of ladybirds, it was discovered that larvae in their second stage of development consumed aphids at a rate y_2 aphids per day, given by

   $$y_2 = \frac{20x}{x + 16}$$

   where x is the number of aphids available. Larvae in the third stage ate at rate y_3, given by

   $$y_3 = \frac{90x}{x + 79}$$

   Graph both of these functions on the domain $0 \le x \le 140$.

   e. What is the maximum rate at which ladybird larvae in each stage of development can consume aphids?

10. A person will float in fresh water if his or her density is less than or equal to 1 kilogram per liter, the density of water. (Density is given by the formula density $= \frac{\text{weight}}{\text{volume}}$.) Suppose a swimmer weighs $50 + F$ kilograms, where F is the amount of fat her body contains. (Source: Burton, 1998)

   a. Calculate the volume of her nonfat body mass if its density is 1.1 kilograms per liter.

   b. Calculate the volume of the fat if its density is 0.901 kilograms per liter.

   c. The swimmer's lungs hold 2.6 liters of air. Write an expression for the total volume of her body, including the air in her lungs.

   d. Write an expression for the density of the swimmer's body.

   e. Write an equation for the amount of fat needed for the swimmer to float in fresh water.

   f. Solve your equation. What percent of the swimmer's weight is fat?

   g. Suppose the swimmer's lungs can hold 4.6 liters of air. What percent body fat does she need to be buoyant?

Linear Systems

Linear programming is an application of linear systems developed in the 1940s to solve complex optimization problems during wartime operations. Today, linear programming is used in scheduling and allocation of resources, in shipping and telecommunications networks, and in the selection of stocks and bonds for portfolios. In 1980, the computer scientist Laszlo Lovasz said, "If one would take statistics about which mathematical problem is using up most of the computer time in the world, . . . the answer would probably be linear programming."

George Dantzig (1914–2005) invented the simplex method for solving linear programming problems in 1947. His first application of the method was to determine an adequate diet of least cost. This problem involved 9 equations in 77 unknowns and took 120 days to solve using the desk calculators available at the time. In this chapter, we use a graphing technique to solve problems in two unknowns. For example, the diet of the Columbian ground squirrel consists of two foods: grass and forb (a type of flowering weed). Small animals spend most of their time foraging for food, but they must also be alert for predators. Which foraging strategy favors survival: Should the squirrel try to satisfy its dietary requirements in minimum time, thus minimizing its exposure to predators and the elements, or should it try to maximize its intake of nutrients?

ThomsonNOW™

Throughout this chapter, the ThomsonNOW logo indicates an opportunity for online self-study, linking you to interactive tutorials and videos based on your level of understanding.

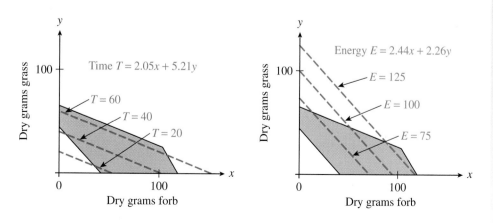

Using linear programming, Gary Belovsky of Notre Dame University identified the optimal amounts of forb and grass for each foraging strategy. We can compare those solutions to the squirrel's actual diet to see which strategy the squirrel favors.

Investigation 13

Interpolating Polynomials

In Chapter 6, we learned to fit a quadratic function through three points on its graph. A polynomial whose graph passes through a given set of points is called an *interpolating polynomial*. Because polynomials are easy to evaluate and manipulate, they are often used to approximate more complicated functions and to describe the shapes of curves.

In this Investigation, we find interpolating polynomials of degrees 1, 2, and 3 to approximate the function $f(x) = \frac{12}{x}$.

1. a. Graph the function $f(x) = \frac{12}{x}$ in the window

$$\text{Xmin} = -2 \quad \text{Xmax} = 7.4$$
$$\text{Ymin} = -5 \quad \text{Ymax} = 20$$

b. Complete the table of values.

x	1	2	3	4	5	6
$f(x)$						

2. First we will find a linear polynomial $P_1(x) = ax + b$ that matches $f(x)$ at $x = 1$ and $x = 6$. We must find constants a and b so that $P_1(1) = f(1)$ and $P_1(6) = f(6)$.

a. The two conditions above translate into equations about a and b. The constants a and b must satisfy the system

$$a \cdot 1 + b = f(1)$$
$$a \cdot 6 + b = f(6)$$

Solve the system and find the polynomial $P_1(x)$.

b. Graph $P_1(x)$ in the same window with $f(x)$.

3. Next, we will find a quadratic polynomial $P_2(x) = ax^2 + bx + c$ that matches $f(x)$ at $x = 1$, 2.5, and 6.

a. Write and solve a system of equations for the constants a, b, and c.

b. Graph $P_2(x)$ in the same window with $f(x)$.

4. We can also find a cubic polynomial $P_3(x) = ax^3 + bx^2 + cx + d$ that matches $f(x)$ at $x = 1$, 3, 4, and 6.

a. Write a system of equations for a, b, c, and d. We will see how to solve such a system later in this chapter. For now, we will use the calculator's cubic regression feature.

b. Enter the coordinates of $P_3(x)$ evaluated at $x = 1$, 3, 4, and 6 into $\mathbf{L_1}$ and $\mathbf{L_2}$ under the **STAT EDIT** menu. Then, from the **STAT CALC** menu, choose **6: CubicReg** and press $\boxed{\text{ENTER}}$.

c. Graph $P_3(x)$ in the same window with $f(x)$.

5. How well does each interpolating polynomial approximate the function $f(x)$? Graph the "error function," $E_n(x)$, for each polynomial on the interval $[-2, 7.4]$.

$$E_1(x) = f(x) - P_1(x)$$
$$E_2(x) = f(x) - P_2(x)$$
$$E_3(x) = f(x) - P_3(x)$$

(You will have to choose a suitable y-window for each error function.) What is the maximum error on the interval $[-2, 7.4]$ for each approximating polynomial?

8.1 Systems of Linear Equations in Two Variables

Systems of linear equations are some of the most useful and widely used mathematical tools for solving problems. Systems involving hundreds of variables and equations are not uncommon in applications such as scheduling airline flights or routing telephone calls.

Solving Systems by Graphing

A biologist wants to know the average weights of two species of birds in a wildlife preserve. She sets up a feeder whose platform is actually a scale and mounts a camera to monitor the feeder. She waits until the feeder is occupied only by members of the two species she is studying, robins and thrushes. Then she takes a picture, which records the number of each species on the scale and the total weight registered.

From her two best pictures, she obtains the following information. The total weight of three thrushes and six robins is 48 ounces, and the total weight of five thrushes and two robins is 32 ounces. Using these data, the biologist estimates the average weight of a thrush and of a robin. She begins by assigning variables to the two unknown quantities:

Average weight of a thrush: t

Average weight of a robin: r

Because there are two variables, the biologist must write two equations about the weights of the birds. In each of the two photos,

(weight of thrushes) + (weight of robins) = total weight

Thus,

$$3t + 6r = 48$$
$$5t + 2r = 32$$

This pair of equations is an example of a **linear system of two equations in two unknowns** (or a 2×2 linear system, for short). A **solution** to the system is an ordered pair of numbers, (t, r), that satisfies both equations in the system.

Recall that every point on the graph of an equation represents a solution to that equation. A solution to *both* equations corresponds to a point on *both* graphs. Therefore, a solution to the system is a point where the two graphs intersect. From Figure 8.1, it appears that the intersection point is $(4, 6)$, so we expect that the values $t = 4$ and $r = 6$ form the solution to the system. We can check by verifying that these values satisfy *both* equations in the system.

$$3(4) + 6(6) \overset{?}{=} 48 \quad \text{True}$$

$$5(4) + 2(6) \overset{?}{=} 32 \quad \text{True}$$

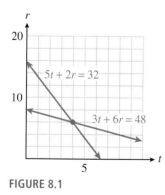

FIGURE 8.1

Both equations are true, so we conclude that the average weight of a thrush is 4 ounces, and the average weight of a robin is 6 ounces.

We can obtain graphs for the equations in a system quickly and easily using a calculator.

EXAMPLE 1   Use your calculator to solve the system by graphing.

$$y = 1.7x + 0.4$$
$$y = 4.1x + 5.2$$

Solution   Set the graphing window to

$$\text{Xmin} = -9.4 \quad \text{Xmax} = 9.4$$
$$\text{Ymin} = -10 \quad \text{Ymax} = 10$$

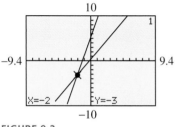

and enter the two equations We can see in Figure 8.2 that the two lines intersect in the third quadrant. Use the $\boxed{\textbf{TRACE}}$ key or the **intersect** feature to find the coordinates of the intersection point. Check that the point $(-2, -3)$ lies on both graphs. The solution to the system is $x = -2$, $y = -3$.

FIGURE 8.2

The values we obtain from a calculator may be only approximations, so it is a good idea to check the solution algebraically. In Example 1, we find that both equations are true when we substitute $x = -2$ and $y = -3$.

$$-3 = 1.7(-2) + 0.4 \quad \text{True}$$
$$-3 = 4.1(-2) + 5.2 \quad \text{True}$$

EXERCISE 1

a. Solve the system of equations

$$y = -0.7x + 6.9$$
$$y = 1.2x - 6.4$$

by graphing. Use the window

$$\text{Xmin} = -9.4 \quad \text{Xmax} = 9.4$$
$$\text{Ymin} = -10 \quad \text{Ymax} = 10$$

b. Verify algebraically that your solution satisfies both equations.

Using the Intersect Feature to Solve a System

Because the **Trace** feature does not show every point on a graph, we may not find the exact solution to a system by tracing the graphs.

EXAMPLE 2   Solve the system

$$3x - 2.8y = 21.06$$
$$2x + 1.2y = 5.3$$

Solution   We can graph this system in the standard window by solving each equation for y. Enter

$$Y_1 = (21.06 - 3X)/-2.8$$
$$Y_2 = (5.3 - 2X)/1.2$$

and then press **ZOOM** **6** . (Do not forget the parentheses around the numerator of each expression.) Trace along the first line to find the intersection point. It appears to be at $x = 4.468051$, $y = -2.734195$, as shown in Figure 8.3a. However, if we press the up or down arrow to read the coordinates off the second line, we see that for the same x-coordinate we obtain a different y-coordinate, as in Figure 8.3b. The different y-coordinates indicate that we have *not* found an intersection point, although we are close. However, the **intersect** feature can give us a better estimate, $x = 4.36$, $y = -2.85$.

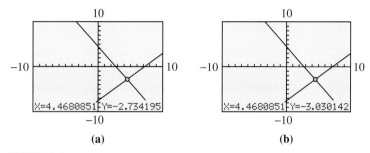

(a)   **(b)**

FIGURE 8.3

We can substitute these values into the original system to check that they satisfy both equations.

$$3(4.36) - 2.8(-2.85) = 21.06$$
$$2(4.36) + 1.2(-2.85) = 5.3$$

See Section 2.1 to review the **intersect** feature on your calculator.

EXERCISE 2

Solve the system of equations

$$y = 47x - 1930$$
$$y + 19x = 710$$

by graphing. Use the **intersect** feature in the window

$$Xmin = 0 \qquad Xmax = 94$$
$$Ymin = -2000 \quad Ymax = 1000$$

In previous algebra courses, you learned two algebraic techniques for solving 2×2 linear systems: **substitution** and **elimination.** See Algebra Skills Refresher A.5 if you would like to review these techniques.

EXAMPLE 3   Rani kayaks downstream for 45 minutes and travels a distance of 6000 meters. On the return journey upstream, she covers only 4800 meters in 45 minutes. How fast is the current in the river, and how fast would Rani kayak in still water? (Give your answers in meters per minute.)

Solution

Step 1 Rani's speed in still water: r

Speed of the current: s

Step 2 We must write two equations using the variables r and s. First, organize the information into a table. When Rani travels downstream, the current in the river helps her, so her effective speed is $r + s$. When she travels upstream she is fighting the current, so her speed is actually $r - s$.

	Rate	Time	Distance
Downstream	$r + s$	45	6000
Upstream	$r - s$	45	4800

Using the formula rate $\times$ time = distance, we write one equation describing Rani's journey downstream, and a second equation for the journey upstream.

$$(r + s) \cdot 45 = 6000$$
$$(r - s) \cdot 45 = 4800$$

Apply the distributive law to write each equation in standard form.

$$45r + 45s = 6000 \qquad (1)$$
$$45r - 45s = 4800 \qquad (2)$$

Step 3 To solve the system, we eliminate the variable s by adding the two equations vertically.

$$
\begin{aligned}
45r + 45s &= 6000 \\
+45r - 45s &= 4800 \\
\hline
90r &= 10{,}800
\end{aligned}
$$

We now have an equation in one variable only, which we can solve for r.

$$90r = 10{,}800 \qquad \text{\small Divide both sides by 90.}$$
$$r = 120$$

To solve for s we substitute $r = 120$ into any previous equation involving both r and s. We will use Equation (1).

$$45(\mathbf{120}) + 45s = 6000 \qquad \text{\small Simplify the left side.}$$
$$5400 + 45s = 6000 \qquad \text{\small Subtract 5400 from both sides.}$$
$$45s = 600 \qquad \text{\small Divide both sides by 45; reduce.}$$
$$s = \frac{40}{3}$$

Step 4 The speed of the current is $\frac{40}{3}$, or $13\frac{1}{3}$ meters per minute, and Rani's speed in still water is 120 meters per minute.

It took Leon 7 hours to fly the same distance that Marlene drove in 21 hours. Leon flies 120 miles per hour faster than Marlene drives. At what speed did each travel?

a. Choose variables for the unknown quantities, and fill in the table.

	Rate	Time	Distance
Leon			
Marlene			

b. Write one equation about the Leon's and Marlene's speeds.

c. Write a second equation about distances.

d. Solve the system and answer the question in the problem.

Inconsistent and Dependent Systems

Because two straight lines do not always intersect at a single point, a 2×2 system of linear equations does not always have a unique solution. In fact, there are three possibilities, as illustrated in Figure 8.4.

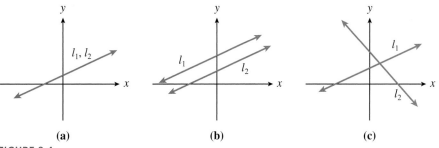

(a)   **(b)**   **(c)**

FIGURE 8.4

1. The graphs may be the same line, as shown in Figure 8.4a.

2. The graphs may be parallel but distinct lines, as shown in Figure 8.4b.

3. The graphs may intersect in one and only one point, as shown in Figure 8.4c.

EXAMPLE 4   Solve the system

$$y = -x + 5$$
$$2x + 2y = 3$$

Solution   We use the calculator to graph both equations on the same axes, as shown in Figure 8.5. First, rewrite the second equation in slope-intercept form by solving for y.

$$2x + 2y = 3 \qquad \text{Substract } 2x \text{ from both sides.}$$
$$2y = -2x + 3 \qquad \text{Divide both sides by 2.}$$
$$y = -x + 1.5$$

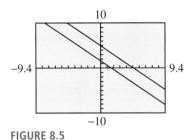

FIGURE 8.5

Now enter the equations as

$$Y_1 = -X + 5$$
$$Y_2 = -X + 1.5$$

The lines do not intersect within the viewing window; they appear to be parallel. If we look again at the equations of the lines, we recognize that both have slope -1 but different y-intercepts, so they *are* parallel. Since parallel lines never meet, there is no solution to the system.

A system with no solutions, such as the system in Example 2, is called **inconsistent.** A 2×2 system of linear equations is inconsistent when the two equations correspond to parallel lines. This situation occurs when the lines have the same slope but different y-intercepts.

EXERCISE 4

Without graphing, show that the following system is inconsistent:

$$3y = \frac{3}{2}x - 1$$
$$2x - 4y = 3$$

A linear system with infinitely many solutions is called **dependent.** A 2×2 system is dependent when the two equations actually describe the same line. This situation occurs when the two lines have the same slope *and* the same y-intercept.

EXAMPLE 5

Solve the system

$$x = \frac{2}{3}y + 3$$
$$3x - 2y = 9$$

Solution   We begin by putting each equation in slope-intercept form.

$$x = \frac{2}{3}y + 3 \qquad \text{Subtract 3.}$$
$$x - 3 = \frac{2}{3}y \qquad\qquad \text{Multiply by } \frac{3}{2}.$$
$$\frac{3}{2}x - \frac{9}{2} = y$$

For the second equation,

$$3x - 2y = 9 \qquad\qquad \text{Substract } 3x.$$
$$-2y = -3x + 9 \qquad \text{Divide by } -2.$$
$$y = \frac{3}{2}x - \frac{9}{2}$$

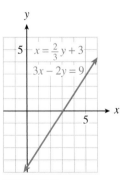

FIGURE 8.6

The two equations are actually different forms of the same equation. Because they are equivalent, they share the same line as a graph, as shown in Figure 8.6. Every point on the first line is also a point on the second line, so every solution to the first equation is also a solution of the second equation. Thus, the system has infinitely many solutions.

Here is a summary of the three cases for a 2×2 system of linear equations.

> ### Solutions of 2 × 2 Linear Systems
>
> 1. **Dependent system.** All the solutions of one equation are also solutions to the second equation and hence are solutions of the system. The graphs of the two equations are the same line. A dependent system has infinitely many solutions.
> 2. **Inconsistent system.** The graphs of the equations are parallel lines and hence do not intersect. An inconsistent system has no solutions.
> 3. **Consistent and independent system.** The graphs of the two lines intersect in exactly one point. The system has exactly one solution.

EXERCISE 5

a. Graph the system

$$y = -3x + 6$$
$$6x + 2y = 15$$

by hand, using either the intercept method or the slope-intercept method. (See Sections 1.1 and 1.4 to review these methods.)

b. Identify the system as dependent, inconsistent, or consistent and independent.

It is not always easy to tell from the equations themselves whether there is one solution, no solution, or infinitely many solutions. However, the method of elimination will reveal which of the three cases applies.

EXAMPLE 6 Solve the system

$$2x = 2 - 3y$$
$$6y = 7 - 4x$$

Solution First, rewrite the system in standard form as

$$2x + 3y = 2 \quad (1)$$
$$4x + 6y = 7 \quad (2)$$

Multiply Equation (1) by -2 and add the result to Equation (2) to obtain

$$
\begin{aligned}
-4x - 6y &= -4 \\
\underline{4x + 6y} &= \underline{7} \\
0x + 0y &= 3
\end{aligned}
$$

This equation has no solutions. The system is inconsistent. (Notice that both lines have slope $\frac{-2}{3}$, but they have different y-intercepts, so their graphs are parallel.)

We generalize the results from Example 6 as follows.

Inconsistent and Dependent Systems

1. If an equation of the form

$$0x + 0y = k \qquad (k \neq 0)$$

is obtained as a linear combination of the equations in a system, the system is **inconsistent.**

2. If an equation of the form

$$0x + 0y = 0$$

is obtained as a linear combination of the equations in a system, the system is **dependent.**

Exercise 6 illustrates a dependent system.

EXERCISE 6

a. Use the method of elimination to solve the system

$$3x - 4 = y$$
$$2y + 8 = 6x$$

b. Verify that both equations have the same graph.

Applications

Many practical problems involve two or more unknown quantities.

EXAMPLE 7 A cup of rolled oats provides 11 grams of protein. A cup of rolled wheat flakes provides 8.5 grams of protein. Francine wants to combine oats and wheat to make a cereal with 10 grams of protein per cup. How much of each grain will she need in one cup of her mixture?

Solution

Step 1 Fraction of a cup of oats needed: x

Fraction of a cup of wheat needed: y

Step 2 Because we have two variables, we must find two equations that describe the problem. It may be helpful to organize the information into a table.

	Cups	Grams of protein per cup	Grams of protein
Oats	x	11	$11x$
Wheat	y	8.5	$8.5y$
Mixture	1	—	10

The wheat and oats together will make one cup of mixture, so the first equation is

$$x + y = 1$$

The 10 grams of protein must come form the protein in the oats plus the protein in the wheat. This gives us a second equation:

$$11x + 8.5y = 10$$

We now have a system of equations.

Step 3   We will solve the system by graphing. First, solve each equation for y in terms of x to get

$$y = -x + 1$$
$$y = (10 - 11x)/8.5$$

Although we could simplify the second equation, the calculator can graph both equations as they are. We know that x and y represent fractions of one cup, so we set the window (as shown in Figure 8.7) with

$$\text{Xmin} = 0 \quad \text{Xmax} = 0.94$$
$$\text{Ymin} = 0 \quad \text{Ymax} = 1$$

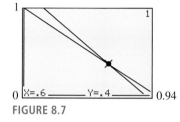

FIGURE 8.7

The lines intersect at $(0.6, 0.4)$, which we can verify by substituting these values into the original two equations of our system.

Step 4   Francine needs 0.6 cup of oats and 0.4 cup of wheat.

EXERCISE 7

Etienne plans to open a coffee house, and he has $7520 to spend on furniture. A table costs $460, and a chair costs $120. Etienne will buy four chairs for each table. How many tables can he buy?

a. Let x represent the number of tables Etienne should buy, and let y represent the number of chairs. Write an equation about the cost of the furniture.

b. Write a second equation about the number of tables and chairs.

c. Graph both equations, solve the system, and answer the question in the problem. (Find the intercepts of the graphs to help you choose a window.)

Being unable to read exact coordinates from a graph is not always a disadvantage. In many situations, fractional values of the unknowns are not acceptable.

EXAMPLE 8   The mathematics department has $40,000 to set up a new computer lab. The department will need one printer for every four terminals it purchases. If a printer costs $560 and a terminal costs $1520, how many of each should the department buy?

Solution

Step 1   Number of printers: p

Number of terminals: t

Step 2   Since the math department needs four times as many terminals as printers,

$$t = 4p$$

The total cost of the printers will be 560p dollars, and the total cost of the terminals will be 1520t dollars, so we have

$$560p + 1520t = 40{,}000$$

Step 3   Solve the second equation for t to get

$$t = (40{,}000 - 560p)/1520$$

Now we graph the equations

$$Y_1 = 4X$$
$$Y_2 = (40000 - 560X)/1520$$

on the same set of axes. The second graph is not visible in the standard graphing window, but with a little experimentation we can find an appropriate window setting. The **WINDOW** values used for Figure 8.8 are

$$Xmin = 0 \quad Xmax = 9.4$$
$$Ymin = 0 \quad Ymax = 30$$

The lines intersect at approximately (6, 24). These values satisfy the first equation, but not the second.

$$560(6) + 1520(24) \overset{?}{=} 40{,}000$$
$$39{,}840 \neq 40{,}000$$

Step 4   The exact solution to the system is $\left(\frac{500}{83}, \frac{2000}{83}\right)$. But this solution is not of practical use, since the math department cannot purchase fractions of printers or terminals. The department *can* purchase 6 printers and 24 terminals (with some money left over).

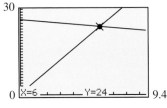

FIGURE 8.8

EXERCISE 8

The manager for Books for Cooks plans to spend $300 stocking a new diet cookbook. The paperback version costs her $5, and the hardback costs $10. She finds that she will sell three times as many paperbacks as hardbacks. How many of each should she buy?

a. Let x represent the number of hardbacks and y the number of paperbacks she should buy. Write an equation about the cost of the books.

b. Write a second equation about the number of each type of book.

c. Graph both equations and solve the system. (Find the intercepts of the graphs to help you choose a window.) Answer the question in the problem.

An Application from Economics

The owner of a retail business must try to balance the demand for his product from consumers with the supply he can obtain from manufacturers. Supply and demand both vary with the price of the product: Consumers usually buy fewer items if the price increases, but manufacturers will be willing to supply more units of the product if its price increases.

The **demand function** gives the number of units of the product that consumers will buy in terms of the price per unit. The **supply function** gives the number of units that the producer will supply in terms of the price per unit. The price at which the supply and demand are equal is called the **equilibrium price.** This is the price at which the consumer and the producer agree to do business.

EXAMPLE 9

A woolens mill can produce $400x$ yards of fine suit fabric if it can charge x dollars per yard. The mill's clients in the garment industry will buy $6000 - 100x$ yards of wool fabric at a price of x dollars per yard. Find the equilibrium price and the amount of fabric that will change hands at that price.

Solution

Step 1 Price per yard: x

Number of yards: y

Step 2 The supply equation tells us how many yards of fabric the mill will produce for a price of x dollars per yard.

$$y = 400x$$

The demand equation tells us how many yards of fabric the garment industry will buy at a price of x dollars per yard.

$$y = 6000 - 100x$$

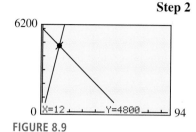

FIGURE 8.9

Step 3 Graph the two equations on the same set of axes, as shown in Figure 8.9. Set the window values to

$$\text{Xmin} = 0 \quad \text{Xmax} = 94$$
$$\text{Ymin} = 0 \quad \text{Ymax} = 6200$$

and use the $\boxed{\textbf{TRACE}}$ or the **intersect** command to locate the solution. the graphs intersect at the point (12, 4800).

Step 4 The equilibrium price is \$12 per yard, and the mill will sell 4800 yards of fabric at that price.

EXERCISE 9

Sanaz can afford to produce $35x$ pairs of hand-painted sunglasses if she can sell them at x dollars per pair, and the market will buy $1700 - 15x$ at x dollars a pair.

a. Write the supply and demand equations for the sunglasses.

b. Find the equilibrium price and the number of sunglasses Sanaz will produce and sell at that price.

ANSWERS FOR 8.1 EXERCISES

1. (7, 2)   **2.** (40, −50)   **3. a.**

	Rate	Time	Distance
Leon	x	7	$7x$
Marlene	y	21	$21y$

b. $x = y + 120$ or $x - y = 120$ **d.** (180, 60). Leon flies at 180 mph;
c. $7x = 21y$ Marlene drives at 60 mph.

4. Both lines have slope $\frac{1}{2}$.   **5. a.**

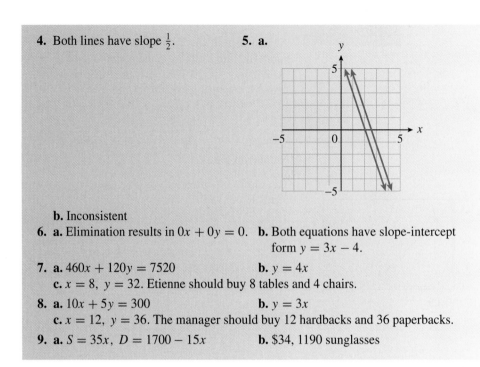

b. Inconsistent
6. a. Elimination results in $0x + 0y = 0$.   **b.** Both equations have slope-intercept form $y = 3x - 4$.
7. a. $460x + 120y = 7520$ **b.** $y = 4x$
   c. $x = 8$, $y = 32$. Etienne should buy 8 tables and 4 chairs.
8. a. $10x + 5y = 300$ **b.** $y = 3x$
   c. $x = 12$, $y = 36$. The manager should buy 12 hardbacks and 36 paperbacks.
9. a. $S = 35x$, $D = 1700 - 15x$ **b.** \$34, 1190 sunglasses

---

SECTION 8.1 SUMMARY

VOCABULARY   *Look up the definitions of new terms in the Glossary.*

Linear system	Solution of a system	Intersection point	Equilibrium point
Dependent	Inconsistent	Consistent and independent	
Demand equation	Supply equation		

CONCEPTS

1. We can solve a 2×2 linear system by graphing. The solution is the intersection point of the two graphs.

2. A linear system may be **inconsistent** (has no solution), **dependent** (has infinitely many solutions), or **consistent and independent** (has one solution).

4. We can use a system of equations to solve problems involving two unknown quantities.

5. In economics, the price at which the **supply** and **demand** are equal is called the **equilibrium price.**

3.

Inconsistent and Dependent Systems

1. If an equation of the form

$$0x + 0y = k \qquad (k \neq 0)$$

is obtained as a linear combination of the equations in a system, the system is **inconsistent.**

2. If an equation of the form

$$0x + 0y = 0$$

is obtained as a linear combination of the equations in a system, the system is **dependent.**

STUDY QUESTIONS

1. How can you test whether (a, b) is a solution to a system of two linear equations?
2. Do two lines always intersect in one point? Explain.
3. When is a system useful for solving an applied problem?
4. Name two algebraic methods for solving a 2×2 linear system.

5. What is the result of performing elimination on a dependent system?
6. Explain the terms *demand function*, *supply function*, and *equilibrium price*.

SKILLS   *Practice each skill in the Homework problems listed.*

1. Solve a 2×2 linear system by graphing: #1–14, 31–34
2. Identify inconsistent and dependent systems: #15–20

3. Write a system of two linear equations to solve a problem: #21–30

HOMEWORK 8.1

ThomsonNOW™ Test yourself on key content at **www.thomsonedu.com/login**.

In Problems 1–4, solve each system of equations using the graphs given. Verify algebraically that your solution satisfies both equations.

1. $2.3x - 3.7y = 6.9$
   $1.1x + 3.7y = 3.3$

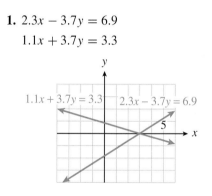

2. $-2.3x + 5.9y = 38.7$
   $9.3x + 7.4y = -0.2$

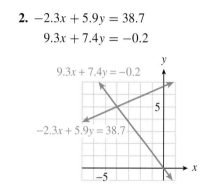

3. $35s - 17t = 560$
   $24s + 15t = 2250$

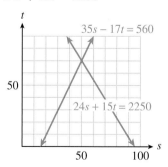

4. $56a + 32b = -880$
   $23a - 7b = 1250$

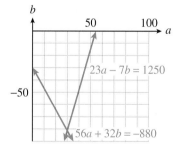

In Problems 5–8, solve each system of equations by graphing. Use the window

$$\text{Xmin} = -9.4 \quad \text{Xmax} = 9.4$$
$$\text{Ymin} = -10 \quad \text{Ymax} = 10$$

Verify algebraically that your solution satisfies both equations.

5. $y = 2.6x + 8.2$
    $y = 1.8 - 0.6x$

6. $y = 5.8x - 9.8$
    $y = 0.7 - 4.7x$

7. $y = 7.2 - 2.1x$
    $-2.8x + 3.7y = 5.5$

8. $y = -2.3x - 5.5$
    $3.1x + 2.4y = -1.1$

In Problems 9–14, graph each system by hand, using either the intercept method or the slope-intercept method. (See Sections 1.1 and 1.4 to review these methods.) Identify the system as dependent, inconsistent, or consistent and independent.

9. $2x = y + 4$
    $8x - 4y = 8$

10. $2t + 12 = -6s$
    $12s + 4t = 24$

11. $w - 3z = 6$
    $2w + z = 8$

12. $2u + v = 5$
    $u - 2v = 3$

13. $2L - 5W = 6$
    $\dfrac{15W}{2} + 9 = 3L$

14. $-3A = 4B + 12$
    $\dfrac{1}{2}A + 2 = \dfrac{-2}{3}B$

Use linear combinations to identify each system in Problems 15–20 as dependent, inconsistent, or consistent and independent. (See Algebra Skills Refresher A.5 to review linear combinations.)

15. $2m = n + 1$
    $8m - 4n = 3$

16. $6p = 1 - 2q$
    $12p + 4q = 2$

17. $r - 3s = 4$
    $2r + s = 6$

18. $2u + v = 4$
    $u - 3v = 2$

19. $2L - 5W = 6$
    $\dfrac{15W}{2} + 9 = 3L$

20. $-3A = 4B + 8$
    $\dfrac{1}{2}A + \dfrac{4}{3} = \dfrac{-2}{3}B$

Solve Problems 21–30 by graphing a system of equations. If you need to review writing equations, please see Algebra Skills Refresher A.3.

21. Dash Phone Company charges a monthly fee of $10, plus $0.09 per minute for long distance calls. Friendly Phone Company charges $15 per month, plus $0.05 per minute for long-distance calls.
    a. Write an equation for Dash Phone Company's monthly bill if you talk long distance for x minutes.
    b. Write an equation for Friendly Phone Company's monthly bill if you talk long distance for x minutes.
    c. Graph both equations in the window

$$\text{Xmin} = 0 \quad \text{Xmax} = 200$$
$$\text{Ymin} = 0 \quad \text{Ymax} = 30$$

and solve the system. How many minutes of long-distance calls would result in equal bills from the two companies?

22. The Olympus Health Club charges an initial fee of $230 and $13 monthly dues. The Valhalla Health Spa charges $140 initially and $16 per month.

   a. Write an equation for the cost of belonging to Olympus Health Club for x months.

   b. Write an equation for the cost of belonging to Valhalla Health Spa for x months.

   c. Graph both equations in the window

$$\text{Xmin} = 0 \quad \text{Xmax} = 50$$
$$\text{Ymin} = 0 \quad \text{Ymax} = 800$$

and solve the system. After how many months of membership would the costs of belonging to the two clubs be equal?

23. Yasuo can afford to produce $50x$ bushels of wheat if he can sell them at x cents per bushel, and the market will buy $2100 - 20x$ bushels at x cents per bushel.

   a. What is the supply equation?

   b. What is the demand equation?

   c. Graph both equations and solve the sustem. Find the equilibrium price and the number of bushels of wheat Yasuo can sell at that prince.

24. Mel's Pool Service can clean $1.5x$ pools per week if it charges x dollars per pool, and the public will book $120 - 2.5x$ pool cleanings at x dollars per pool.

   a. What is the supply equation?

   b. What is the demand equation?

   c. Graph both equations and solve the system. Find the equilibrium price and the number of pools Mel will clean at that price.

25. The Aquarius jewelry company determines that each production run to manufacture a pendant involves an initial set-up cost of $200 and $4 for each pendant produced. The pendants sell for $12 each.

   a. Express the cost C of production in terms of the number x of pendants produced.

   b. Express the revenue R in terms of the number x of pendants sold.

   c. Graph the revenue and cost on the same set of axes. How many pendants must be sold for the Aquarius company to break even on a particular production run?

26. The Bread Alone Bakery has a daily overhead of $90. It costs $0.60 to bake each loaf of bread, and the bread sells for $1.50 per loaf.

   a. Express the cost, C, in terms of the number x of loaves baked.

   b. Express the revenue, R, in terms of the number x of loaves sold.

   c. Graph the revenue and cost on the same set of axes. How many loaves must the bakery sell to break even on a given day?

27. The admissions at a Bengals' baseball game was $7.50 for adults and $4.25 for students. The ticket office took in $465.50 for 82 paid admissions. How many adults and how many students attended the game?

   a. Write algebraic expressions to fill in the table.

	Number of tickets	Cost per ticket	Revene
Adults	x		
Students	y		
Total			

   b. Write an equation about the number of tickets sold.

   c. Write a second equation about the revenue from the tickets.

   d. Graph both equations and solve the system.

28. There were 42 passengers on an airplane flight for which first-class fare was $400 and tourist fare was $320. If the revenue for the flight totaled $14,400, how many first-class and how many tourist passengers paid for the flight?

   a. Write algebraic expressions to fill in the table.

	Number of tickets	Cost per ticket	Revene
First-Class	x		
Tourist	y		
Total			

   b. Write an equation about the number of tickets sold.

   c. Write a second equation about the revenue from the tickets.

   d. Graph both equations and solve the system.

29. Earthquakes simultaneously send out two types of waves, called P waves and S waves, but the two types travel at different speeds. A seismograph records arrival of P waves from an earthquake, and 90 seconds later the seismograph receives S waves from the same earthquake. The P waves travel at 5.4 miles per second, and S waves at 3 miles per second. How far is the seismograph from the earthquake?

a. Let x represent the time in seconds for the P to arrive at the seismograph, and y the distance in miles between the earthquake and seismograph. Fill in the table.

	Rate	Time	Distance
P waves			
S waves			

b. Write an equation about how far the S waves travel.
c. Write a second equation about how far the P waves travel.
d. Solve the system and answer the question in the problem.

30. Thelma and Louise start together and drive in the same direction, Thelma driving twice as fast as Louise. At the end of 3 hours, they are 96 miles apart. How fast is each traveling?

a. Choose variables for the unknown quantities, and fill in the table.

	Rate	Time	Distance
Thelma			
Louise			

b. Write one equation about Thelma's and Louise's speeds.
c. Write a second equation about distances.
d. Solve the system and answer the question in the problem.

In Problems 31–34, solve each system of equations by graphing. Find the intercepts of each graph to help you choose a suitable window, then use the *intersect* feature to locate the solution.

31. $38x + 2.3y = -55.2$
   $y = 15x + 121$

32. $25x - 1.7y = 10.5$
   $y + 5x = 49$

33. $64x + 58y = 707$
   $82x - 21y = 496$

34. $35x - 76y = 293$
   $15x + 44y = -353$

8.2   Systems of Linear Equations in Three Variables

Some problems involve three (or more) unknown quantities, and efficient techniques for solving linear systems in many variables are available. In this section, we solve systems of three linear equations in three variables.

3 × 3 Linear Systems

A solution to an equation in three variables, such as

$$x + 2y - 3z = -4$$

is an **ordered triple** of numbers that satisfies the equation. For example, $(0, -2, 0)$ and $(-1, 0, 1)$ are solutions to the equation above, but $(1, 1, 1)$ is not. You can verify this by substituting the coordinates into the equation to see if a true statement results.

For $(0, -2, 0)$:	$0 + 2(-2) - 3(0) = -4$	True
For $(-1, 0, 1)$:	$-1 + 2(0) - 3(1) = -4$	True
For $(1, 1, 1)$:	$1 + 2(1) - 3(1) = -4$	Not true

As with the two-variable case, a single linear equation in three variables has infinitely many solutions.

An ordered triple (x, y, z) can be represented geometrically as a point in space using a three-dimensional Cartesian coordinate system (see Figure 8.10). In this coordinate system, the graph of a linear equation in three variables is a plane, and the fact that there are infinitely many solutions to the equation tells us that that there are infinitely many points in the corresponding plane.

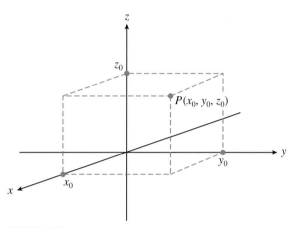

FIGURE 8.10

A solution to a *system* of three linear equations in three variables is an ordered triple that satisfies each equation in the system. That triple represents a point that must lie on all three graphs. Figure 8.11 shows the different ways in which three planes may intersect in space.

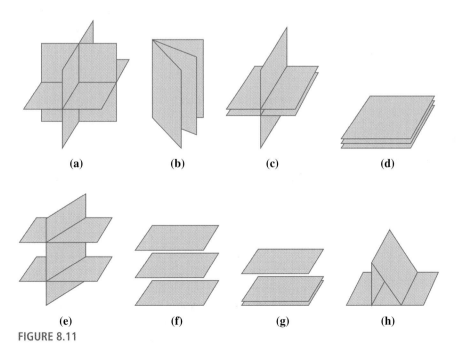

(a) (b) (c) (d)

(e) (f) (g) (h)

FIGURE 8.11

In Figure 8.11a, the three planes intersect in a single point, so the corresponding system of three equations has a unique solution. In Figures 8.11b, c, and d, the intersection is either a line or an entire plane, so the corresponding system has infinitely many solutions. Such a system is called **dependent.** In Figures 8.11e, f, g, and h, the three planes have no common intersection, so the corresponding system has no solution. In this case, the system is said to be **inconsistent.**

It is impractical to solve 3×3 systems by graphing. Even when technology for producing three-dimensional graphs is available, we cannot read coordinates on such graphs with any confidence. Thus, we will restrict our attention to algebraic methods of solving such systems.

Back-Substitution

When we review our algebraic methods for 2×2 systems—substitution and linear combinations—we find features common to both. In both methods, we obtain an equation in one variable. Once we have solved for that variable, we substitute the value into an earlier equation to find the other variable.

One strategy for solving a 3×3 system extends this idea to include a third variable and a third equation. The following special case illustrates the substitution part of the procedure.

EXAMPLE 1 Solve the system

$$x + 2y + 3z = 2$$
$$-2y - 4z = -2$$
$$3z = -3$$

Solution The third equation involves only the variable z, so we solve that equation to find $z = -1$. Then we substitute -1 for z in the second equation and solve for y.

$$-2y - 4(-1) = -2$$
$$-2y + 4 = -2$$
$$-2y = -6$$
$$y = 3$$

Finally, we substitute -1 for z and 3 for y into the first equation to find x.

$$x + 2(3) + 3(-1) = 2$$
$$x + 6 - 3 = 2$$
$$x = -1$$

The solution is the ordered triple $(-1, 3, -1)$. You should verify that this triple satisfies all three equations of the system.

The technique used in Example 1 is called **back-substitution.** It works in the special case where one of the equations involves exactly one variable, and a second equation involves that same variable and just one other variable. A 3×3 linear system with these properties is said to be in **triangular** form. If we can transform a system into triangular form, we can use back-substitution to complete the solution.

Use back-substitution to solve the system

$$2x + 2y + z = 10$$
$$y - 4z = 9$$
$$3z = -6$$

Gaussian Reduction

We can use linear combinations to reduce a 3×3 system to triangular form and then use back-substitution to find the solutions. Our strategy will be to eliminate one of the variables from each of the three equations by considering them in pairs. This results in a 2×2 system that we can solve using elimination. As an example, consider the system

$$x + 2y - 3z = -4 \qquad (1)$$
$$2x - y + z = 3 \qquad (2)$$
$$3x + 2y + z = 10 \qquad (3)$$

Step 1 This system is already in standard form. We can choose any one of the three variables to eliminate first. For this example, we will eliminate x.

Step 2 Choose two of the equations, say (1) and (2), and use a linear combination: We multiply Equation (1) by -2 and add the result to Equation (2) to produce Equation (4).

$$
\begin{array}{ll}
-2x - 4y + 6z = 8 & -2 \times (1) \\
\underline{2x - y + z = 3} & (2) \\
-5y + 7z = 11 & (4)
\end{array}
$$

Step 3 Now we have an equation involving only two variables. But we need *two* equations in two unknowns to find the solution. So we choose a different pair of equations, say (1) and (3), and eliminate x again. We multiply Equation (1) by -3 and add the result to Equation (3) to obtain Equation (5).

$$
\begin{array}{ll}
-3x - 6y + 9z = 12 & -3 \times (1) \\
\underline{3x + 2y + z = 10} & (3) \\
-4y + 10z = 22 & (5)
\end{array}
$$

Step 4 We now form a 2×2 system with our new Equations (4) and (5).

$$-5y + 7z = 11 \qquad (4)$$
$$-4y + 10z = 22 \qquad (5)$$

We eliminate either y or z to obtain an equation in a single variable. If we choose to eliminate y, we add 4 times Equation (4) to -5 times Equation (5) to obtain Equation (6).

$$
\begin{array}{ll}
-20y + 28z = 44 & 4 \times (4) \\
\underline{20y - 50z = -110} & -5 \times (5) \\
-22z = -66 & (6)
\end{array}
$$

Step 5 Now we can start solving for the variables. To keep things organized, we form a triangular system: Choose one of the original equations (in three variables), one of the equations from our 2 × 2 system, and our final equation in one variable. We choose Equations (1), (4), and (6).

$$x + 2y - 3z = -4 \qquad (1)$$
$$-5y + 7z = 11 \qquad (4)$$
$$-22z = -66 \qquad (6)$$

This new system is in triangular form, and it has the same solutions as the original system. We complete the solution by back-substitution. Solve Equation (6) to find $z = 3$. Substituting 3 for z in Equation (4), we find

$$-5y + 7(3) = 11$$
$$-5y + 21 = 11$$
$$-5y = -10$$
$$y = 2$$

Finally, we substitute 3 for z and 2 for y into Equation (1) to find

$$x + 2(2) - 3(3) = -4$$
$$x + 4 - 9 = -4$$
$$x = 1$$

The solution to the system is the ordered triple $(1, 2, 3)$. You can verify that this triple satisfies all three of the original equations.

The method described above for putting a linear system into triangular form is called **Gaussian reduction,** after the German mathematician Carl Gauss. We summarize our method for solving a 3 × 3 linear system as follows.

Steps for Solving a 3 × 3 Linear System

1. Clear each equation of fractions and put it in standard form.
2. Choose two of the equations and eliminate one of the variables by forming a linear combination.
3. Choose a different pair of equations and eliminate the *same* variable.
4. Form a 2 × 2 system with the equations found in steps (2) and (3). Eliminate one of the variables from this 2 × 2 system by using a linear combination.
5. Form a triangular system by choosing among the previous equations. Use back-substitution to solve the triangular system.

EXAMPLE 2   Solve the system

$$x + 2y - z = -3 \qquad (1)$$
$$\frac{1}{3}x - y + \frac{1}{3}z = 2 \qquad (2)$$
$$x + \frac{1}{2}y + z = \frac{5}{2} \qquad (3)$$

Solution   Follow the steps outlined above.

Step 1 Multiply each side of Equation (2) by 3, and multiply each side of Equation (3) by 2, to obtain the equivalent system

$$x + 2y - z = -3 \qquad (1)$$
$$x - 3y + z = 6 \qquad (2a)$$
$$2x + y + 2z = 5 \qquad (3a)$$

Step 2 Eliminate z from Equations (1) and (2a) by adding.

$$x + 2y - z = -3 \qquad (1)$$
$$\underline{x - 3y + z = 6} \qquad (2a)$$
$$2x - y \quad\ = 3 \qquad (4)$$

Step 3 Eliminate z from Equations (1) and (3a): Multiply Equation (1) by 2 and add the result to Equation (3a).

$$2x + 4y - 2z = -6 \qquad (1a)$$
$$\underline{2x + y + 2z = 5} \qquad (3a)$$
$$4x + 5y \qquad = -1 \qquad (5)$$

Step 4 Form the 2×2 system consisting of Equations (4) and (5).

$$2x - y = 3 \qquad (4)$$
$$4x + 5y = -1 \qquad (5)$$

Eliminate x by adding -2 times Equation (4) to equation (5).

$$-4x + 2y = -6 \qquad (4a)$$
$$\underline{4x + 5y = -1} \qquad (5)$$
$$7y = -7 \qquad (6)$$

Step 5 Form a triangular system using Equations (1), (4), and (6).

$$x + 2y - z = -3 \qquad (1)$$
$$2x - y \qquad = 3 \qquad (4)$$
$$7y \qquad = -7 \qquad (6)$$

Use back-substitution to find the solution. You can verify that the ordered triple $(1, -1, 2)$ satisfies all three of the original equations of the system.

EXERCISE 2

Use Gaussian reduction to solve the system

$$x - 2y + z = -1 \qquad (1)$$
$$\frac{2}{3}x + \frac{1}{3}y - z = 1 \qquad (2)$$
$$3x + 3y - 2z = 10 \qquad (3)$$

Follow the steps suggested below:

1. Clear the fractions from Equation (2).
2. Eliminate z from Equations (1) and (2).
3. Eliminate z from Equations (1) and (3).
4. Eliminate x from your new 2×2 system.
5. Form a triangular system and solve by back-substitution.

Inconsistent and Dependent Systems

The results in Section 8.1 for identifying dependent and inconsistent systems can be extended to 3×3 linear systems. If at any step in forming linear combinations we obtain an equation of the form

$$0x + 0y + 0z = k, \quad (k \neq 0)$$

then the system is inconsistent and has no solution. If we obtain an equation of the form

$$0x + 0y + 0z = 0$$

then the system is dependent and has infinitely many solutions.

EXAMPLE 3   Solve the system

$$
\begin{aligned}
3x + y - 2z &= 1 \qquad (1) \\
6x + 2y - 4z &= 5 \qquad (2) \\
-2x - y + 3z &= -1 \qquad (3)
\end{aligned}
$$

Solution   To eliminate y from Equations (1) and (2), multiply Equation (1) by -2 and add the result to Equation (2).

$$
\begin{aligned}
-6x - 2y + 4z &= -2 \\
\underline{6x + 2y - 4z = 5} \\
0x + 0y + 0z &= 3
\end{aligned}
$$

Since the resulting equation has no solution, the system is *inconsistent*.

EXERCISE 3

Decide whether the system is inconsistent, dependent, or consistent and independent.

$$
\begin{aligned}
x + 3y - z &= 4 \\
-2x - 6y + 2z &= 1 \\
x + 2y - z &= 3
\end{aligned}
$$

EXAMPLE 4   Solve the system

$$
\begin{aligned}
-x + 3y - z &= -2 \qquad (1) \\
2x + y - 4z &= 6 \qquad (2) \\
2x - 6y + 2z &= 4 \qquad (3)
\end{aligned}
$$

Solution   To eliminate x from Equations (1) and (3), multiply Equation (1) by 2 and add Equation (3).

$$
\begin{aligned}
-2x + 6y - 2z &= -4 \\
\underline{2x - 6y + 2z = 4} \\
0x + 0y + 0z &= 0
\end{aligned}
$$

Since the resulting equation vanishes, the system is *dependent* and has infinitely many solutions.

Decide whether the system is inconsistent, dependent, or consistent and independent.

$$a - c = 2$$
$$2a + b = 5$$
$$a + b + c = 3$$

Applications

Here are some problems that can be modeled by a system of three linear equations. When writing such systems, we must be careful to find three *independent* equations describing the conditions of the problem.

EXAMPLE 5

One angle of a triangle measures 4° less than twice the second angle, and the third angle is 20° greater than the sum of the first two. Find the measure of each angle.

Solution

Step 1 Represent the measure of each angle by a separate variable.

First angle: x

Second angle: y

Third angle: z

Step 2 Write the conditions stated in the problem as three equations.

$$x = 2y - 4$$
$$z = x + y + 20$$
$$x + y + z = 180$$

(The third equation states the fact that the sum of the angles of a triangle is 180°.)

Step 3 We follow the steps for solving a 3 × 3 linear system.

1. Write the three equations in standard form.

$$x - 2y \quad = -4 \qquad (1)$$
$$x + y - z = -20 \qquad (2)$$
$$x + y + z = 180 \qquad (3)$$

2–3. Since Equation (1) has no z-term, it will be most efficient to eliminate the variable z from Equations (2) and (3). Add these two equations.

$$x + y - z = -20 \qquad (2)$$
$$\underline{x + y + z = 180} \qquad (3)$$
$$2x + 2y \quad = 160 \qquad (4)$$

4. Form a 2 × 2 system from Equations (1) and (4). Add the two equations to eliminate the variable y, yielding

$$x - 2y = -4 \qquad (1)$$
$$\underline{2x + 2y = 160} \qquad (4)$$
$$3x \quad = 156 \qquad (5)$$

5. Form a triangular system using Equations (3), (1), and (5). Use back-substitution to complete the solution.

$$x + y + z = 180 \qquad (2)$$
$$x - 2y \quad\;\;\; = -4 \qquad (1)$$
$$3x \qquad\quad = 156 \qquad (5)$$

Divide both sides of Equation (5) by 3 to find $x = 52$. Substitute 52 for x in Equation (1) and solve for y to find

$$52 - 2y = -4$$
$$y = 28$$

Substitute 52 for x and 28 for y in Equation (3) to find

$$52 + 28 + z = 180$$
$$z = 100$$

Step 4   The angles measure $52°$, $28°$, and $100°$.

EXERCISE 5

A manufacturer of office supplies makes three types of file cabinet: two-drawer, four-drawer, and horizontal. The manufacturing process is divided into three phases: assembly, painting, and finishing. A two-drawer cabinet requires 3 hours to assemble, 1 hour to paint, and 1 hour to finish. The four-drawer model takes 5 hours to assemble, 90 minutes to paint, and 2 hours to finish. The horizontal cabinet takes 4 hours to assemble, 1 hour to paint, and 3 hours to finish. The manufacturer employs enough workers for 500 hours of assembly time, 150 hours of painting, and 230 hours of finishing per week. How many of each type of file cabinet should he make in order to use all the hours available?

Step 1   Represent the number of each model of file cabinet by a different variable.

Number of two-drawer cabinets: x

Number of four-drawer cabinets: y

Number of horizontal cabinets: z

Step 2   Organize the information into a table. (Convert all times to hours.)

	2-Drawer	**4-Drawer**	**Horizontal**	**Total available**
Assembly				
Painting				
Finishing				

Write three equations describing the time constraints in each of the three manufacturing phases. For example, the assembly phase requires $3x$ hours for the two-drawer cabinets, $5y$ hours for the four-drawer cabinets, and $4z$ hours for the horizontal cabinets, and the sum of these times should be the time available, 500 hours.

Assembly time: _____ (1)

Painting time: _____ (2)

Finishing time: _____ (3)

Step 3 Solve the system. Follow the steps suggested below.

1. Clear the fractions from the second equation.

2. Subtract Equation (1) from 3 times Equation (3) to obtain a new Equation (4).

3. Subtract Equation (2) from twice Equation (3) to obtain a new Equation (5).

4. Equations (4) and (5) form a 2×2 system in y and z. Subtract Equation (5) from Equation (4) to obtain a new Equation (6).

5. Form a triangular system with Equations (3), (4), and (6). Use back-substitution to complete the solution.

Step 4 You should have found the following solution: The manufacturer should make 60 two-drawer cabinets, 40 four-drawer cabinets, and 30 horizontal cabinets.

ANSWERS FOR 8.2 EXERCISES

1. $(5, 1, -2)$ **2.** $(2, 2, 1)$ **3.** Inconsistent   **4.** Dependent

5. $3x + 5y + 4z = 500$;

$\quad x + 1.5y + z = 150$

$\quad\; x + 2y + 3z = 230$

$\quad x = 60, \; y = 40, \; z = 30$

SECTION 8.2 SUMMARY

VOCABULARY *Look up the definitions of new terms in the Glossary.*

Ordered triple   Gaussian reduction   Triangular form   Back-substitution

CONCEPTS

1. The solution to a 3×3 linear system is an **ordered triple.**

2. A 3×3 system in **triangular form** can be solved by **back-substitution.**

3. **Gaussian reduction** is a generalized form of the elimination method that can be used to reduce a 3×3 linear system to triangular form.

4.

> **Steps for Solving a 3 × 3 Linear System**
>
> **1.** Clear each equation of fractions and put it in standard form.
>
> **2.** Choose two of the equations and eliminate one of the variables by forming a linear combination.

3. Choose a different pair of equations and eliminate the *same* variable.

4. Form a 2×2 system with the equations found in steps (2) and (3). Eliminate one of the variables from this 2×2 system by using a linear combination.

5. Form a triangular system by choosing among the previous equations. Use back-substitution to solve the triangular system.

5. 3×3 linear systems may be **inconsistent** or **dependent.**

STUDY QUESTIONS

1. How can you check whether an ordered triple (a, b, c) is a solution of a 3×3 system?
2. In order to solve by back-substitution, does the shortest equation in a triangular system have to be at the bottom?
3. After you have eliminated one variable from two of the equations in a 3×3 linear system, what is the next step?
4. How would you start Gaussian reduction on a 3×3 linear system if the first equation has only two variables?

SKILLS   *Practice each skill in the Homework problems listed.*

1. Solve a triangular system by back-substitution: #1–6
2. Solve a 3×3 linear system by Gaussian reduction: #7–20
3. Identify inconsistent and dependent systems: #21–30
4. Write a solve a 3×3 linear system to solve an applied problem: #31–40

HOMEWORK 8.2

ThomsonNOW™ Test yourself on key content at www.thomsonedu.com/login.

Use back-substitution to solve Problems 1–6.

1. $x + y + z = 2$
   $3y + z = 5$
   $-4y = -8$

2. $2x + 3y - z = -7$
   $y - 2z = -6$
   $5z = 15$

3. $2x - y - z = 6$
   $5y + 3z = -8$
   $13y = -13$

4. $x + y + z = 1$
   $x + 4y = 1$
   $3x = 3$

5. $2x + z = 5$
   $3y + 2z = 6$
   $5x = 20$

6. $3x - y = 6$
   $x - 2z = -7$
   $13x = 13$

Use Gaussian reduction to solve Problems 7–20.

7. $x + y + z = 0$
   $2x - 2y + z = 8$
   $3x + 2y + z = 2$

8. $x - 2y + 4z = -3$
   $3x + y - 2z = 12$
   $2x + y - 3z = 11$

9. $4x + z = 3$
   $2x - y = 2$
   $3y + 2x = 0$

10. $3y + z = 3$
    $-2x + 3y = 7$
    $3x + 2z = -6$

11. $2x + 3y - 2z = 5$
    $3x - 2y - 5z = 5$
    $5x + 2y + 3z = -9$

12. $3x - 4y + 2z = 20$
    $4x + 3y - 3z = -4$
    $2x - 5y + 5z = 24$

13. $4x + 6y + 3z = -3$
    $2x - 3y - 2z = 5$
    $-6x + 6y + 2z = -5$

14. $3x + 4y + 6z = 2$
    $-2x + 2y - 3z = 1$
    $4x - 10y + 9z = 0$

15. $x - \dfrac{1}{2}y - \dfrac{1}{2}z = 4$

    $x - \dfrac{3}{2}y - 2z = 3$

    $\dfrac{1}{4}x + \dfrac{1}{4}y - \dfrac{1}{4}z = 0$

16. $x + 2y + \dfrac{1}{2}z = 0$

    $x + \dfrac{3}{5}y - \dfrac{2}{5}z = \dfrac{1}{5}$

    $4x - 7y - 7z = 6$

17. $x + y - z = 2$

    $\dfrac{1}{2}x - y + \dfrac{1}{2}z = -\dfrac{1}{2}$

    $x + \dfrac{1}{3}y - \dfrac{2}{3}z = \dfrac{4}{3}$

18. $x + y - 2z = 3$

    $x - \dfrac{1}{3}y + \dfrac{1}{3}z = \dfrac{5}{3}$

    $\dfrac{1}{2}x - \dfrac{1}{2}y - z = \dfrac{3}{2}$

19.
$$x = -y$$
$$x + z = \frac{5}{6}$$
$$y - 2z = -\frac{7}{6}$$

20.
$$x = y + \frac{1}{2}$$
$$y = z + \frac{5}{4}$$
$$2z = x - \frac{7}{4}$$

Solve the systems in Problems 21–30. If the system is inconsistent or dependent, say so.

21.
$$3x - 2y + z = 6$$
$$2x + y - z = 2$$
$$4x + 2y - 2z = 3$$

22.
$$x - 2y + z = 5$$
$$-x + y = -2$$
$$y - z = -3$$

23.
$$2x + 3y - z = -2$$
$$x - y + \frac{1}{2}z = 2$$
$$4x - \frac{1}{3}y + 2z = 8$$

24.
$$3x + 6y + 2z = -2$$
$$\frac{1}{2}x - 3y - z = 1$$
$$4x + y + \frac{1}{3}z = -\frac{1}{3}$$

25.
$$x = 2y - 7$$
$$y = 4z + 3$$
$$z = 3x + y$$

26.
$$x = y + z$$
$$y = 2x - z$$
$$z = 3x - y$$

27.
$$\frac{1}{2}x + y = \frac{1}{2}z$$
$$x - y = -z - 2$$
$$-x - 2y = -z + \frac{4}{3}$$

28.
$$x = \frac{1}{2}y - \frac{1}{2}z + 1$$
$$x = 2y + z - 1$$
$$x = \frac{1}{2}y - \frac{1}{2}z + \frac{1}{4}$$

29.
$$x - y = 0$$
$$2x + 2y + z = 5$$
$$2x + y - \frac{1}{2}z = 0$$

30.
$$x + y = 1$$
$$2x - y + z = -1$$
$$x - 3y - z = -\frac{2}{3}$$

Solve Problems 31–40 by using a system of equations.
(a) Identify three unknown quantities and choose variables to represent them.
(b) If appropriate, make a table organizing the information in the problem.
(c) Write three equations about the variables in the problem.
(d) Solve the system and answer the question in the problem.

31. A box contains $6.25 in nickels, dimes, and quarters. There are 85 coins in all, with 3 times as many nickels as dimes. How many coins of each kind are there?

32. Vanita has $446 in 10-dollar, 5-dollar, and 1-dollar bills. There are 94 bills in all, with 10 more 5-dollar bills than 10-dollar bills. How many bills of each kind does she have?

33. The perimeter of a triangle is 155 inches. Side x is 20 inches shorter than side y, and side y is 5 inches longer than side z. Find the lengths of the sides of the triangle.

34. One angle of a triangle measures 10° more than a second angle, and the third angle is 10° more than six times the measure of the smallest angle. Find the measure of each angle.

35. Vegetable Medley is made of carrots, green beans, and cauliflower. The package says that 1 cup of Vegetable Medley provides 29.4 milligrams of vitamin C and 47.4 milligrams of calcium. One cup of carrots contains 9 milligrams of vitamin C and 48 milligrams of calcium. One cup of green beans contains 15 milligrams of vitamin C and 63 milligrams of calcium. One cup of cauliflower contains 69 milligrams of vitamin C and 26 milligrams of calcium. How much of each vegetable is in 1 cup of Vegetable Medley?

36. The Java Shoppe sells a house brand of coffee that is only 2.25% caffeine for $6.60 per pound. The house brand is a mixture of Colombian coffee that sells for $6 per pound and is 2% caffeine, French roast that sells for $7.60 per pound and is 4% caffeine, and Sumatran that sells for $6.80 per pound and is 1% caffeine. How much of each variety is in a pound of house brand?

37. The ABC Psychological Testing Service offers three types of reports on test results: score only, evaluation, and narrative report. Each score-only test takes 3 minutes to score using an optical scanner and 1 minute to print the interpretation. Each evaluation takes 3 minutes to score, 4 minutes to analyze, and 2 minutes to print. Each narrative report takes 3 minutes to score, 5 minutes to analyze, and 8 minutes to print. If ABC Services uses its optical scanner 7 hours per day, has 8 hours in which to analyze results, and has 12 hours of printer time available per day, how many of each type of report can it complete each day when it is using all its resources?

38. Reliable Auto Company wants to ship 1700 Status Sedans to three major dealers in Los Angeles, Chicago, and Miami. From past experience, Reliable figures that it will sell twice as many sedans in Los Angeles as in Chicago. It costs $230 to ship a sedan to Los Angeles, $70 to Chicago, and $160 to Miami. If Reliable Auto has $292,000 to pay for shipping costs, how many sedans should it ship to each city?

39. Ace, Inc. produces three kinds of wooden rackets: tennis rackets, Ping-Pong paddles, and squash rackets. After the pieces are cut, each racket goes through three phases of production: gluing, sanding, and finishing. A tennis racket takes 3 hours to glue, 2 hours to sand, and 3 hours to finish. A Ping-Pong paddle takes 1 hour to glue, 1 hour to sand, and 1 hour to finish. A squash racket takes 2 hours to glue, 2 hours to sand, and $2\frac{1}{2}$ hours to finish. Ace has available 95 labor-hours in its gluing department, 75 labor-hours in sanding, and 100 labor-hours in finishing per day. How many of each racket should it make in order to use all the available personnel?

40. A farmer has 1300 acres on which to plant wheat, corn, and soybeans. The seed costs $6 for an acre of wheat, $4 for an acre of corn, and $5 for an acre of soybeans. An acre of wheat requires 5 acre-feet of water during the growing season, while an acre of corn requires 2 acre-feet and an acre of soybeans requires 3 acre-feet. If the farmer has $6150 to spend on seed and can count on 3800 acre-feet of water, how many acres of each crop should he plant in order to use all his resources?

8.3 Solving Linear Systems Using Matrices

In this section, we consider a mathematical tool called a matrix (plural: matrices) that has wide application in mathematics, business, science, and engineering. A **matrix** is a rectangular array of numbers or **entries.** These entries are ordinarily displayed in rows and columns, and the entire matrix is enclosed in brackets or parentheses. Thus,

$$\begin{bmatrix} 1 & 2 & 3 \\ 4 & 5 & 6 \\ 7 & 8 & 9 \end{bmatrix}, \quad \begin{bmatrix} 2 & -1 & 3 \\ 4 & 0 & 2 \end{bmatrix}, \text{ and } \begin{bmatrix} 4 \\ 5 \\ 6 \end{bmatrix}$$

are matrices. A matrix of **order,** or **dimension,** $n \times m$ (read "n by m") has n (horizontal) rows and m (vertical) columns. The matrices above are 3×3, 2×3, and 3×1, respectively. The first matrix—in which the number of rows is equal to the number of columns—is an example of a **square matrix.**

Coefficient Matrix and Augmented Matrix of a System

We will use matrices to solve systems of linear equations. For a system of linear equations of the form

$$a_1 x + b_1 y + c_1 z = d_1$$
$$a_2 x + b_2 y + c_2 z = d_2$$
$$a_3 x + b_3 y + c_3 z = d_3$$

the matrices

$$\begin{bmatrix} a_1 & b_1 & c_1 \\ a_2 & b_2 & c_2 \\ a_3 & b_3 & c_3 \end{bmatrix} \quad \text{and} \quad \left[\begin{array}{ccc|c} a_1 & b_1 & c_1 & d_1 \\ a_2 & b_2 & c_2 & d_2 \\ a_3 & b_3 & c_3 & d_3 \end{array}\right]$$

are called the **coefficient matrix** and the **augmented matrix,** respectively. Each *row* of the augmented matrix represents one of the equations of the system. For example, the augmented matrix of the system

$$\begin{array}{rcl} 3x - 4y + z &=& 2 \\ -x + 2y &=& -1 \\ 2x - y - 3z &=& 4 \end{array} \quad \text{is} \quad \left[\begin{array}{ccc|c} 3 & -4 & 1 & 2 \\ -1 & 2 & 0 & -1 \\ 2 & -1 & -3 & 4 \end{array}\right]$$

and the augmented matrix of the system

$$\begin{array}{rcl} x - 3y + 2z &=& 5 \\ 2y - z &=& 4 \\ 4z &=& 8 \end{array} \quad \text{is} \quad \left[\begin{array}{ccc|c} 1 & -3 & 2 & 5 \\ 0 & 2 & -1 & 4 \\ 0 & 0 & 4 & 8 \end{array}\right]$$

The augmented matrix of this last system, which has all zero entries in the lower left corner (below the diagonal), is said to be in **upper triangular form.** As we saw in Section 8.2, it is easy to find the solution of such a system by back-substitution.

Elementary Row Operations

The method of elimination depends on two properties of linear systems that allow us to change one system into an equivalent one, that is, one that has the same solutions as the original system.

Properties of Linear Systems

1. Multiplying a linear equation by a (nonzero) constant does not change its solutions. That is, any solution of the equation

   $$ax + by = c$$

   is also a solution of the equation

   $$kax + kby = kc$$

2. Adding (or subtracting) two linear equations does not change their common solutions. That is, any solution of the system

   $$a_1 x + b_1 y = c_1$$
   $$a_2 x + b_2 y = c_2$$

   is also a solution of the equation

   $$(a_1 + a_2)x + (b_1 + b_2)y = c_1 + c_2$$

These properties apply to linear systems of any size. Thus, we can perform the following operations on the equations of a system:

1. Multiply both sides of an equation by a nonzero real number.

2. Add a constant multiple of one equation to another equation.

We can also perform the following operation without changing the solution of the system:

3. Interchange two equations.

Because each equation of the system corresponds to a row in the augmented matrix, the three operations above correspond to three **elementary row operations** for the augmented matrix. We can perform one or more of these operations on a matrix without changing the solution of the system it represents.

Elementary Row Operations

1. Multiply the entries of any row by a nonzero real number.

2. Add a constant multiple of one row to another row.

3. Interchange two rows.

EXAMPLE 1

a. $A = \begin{bmatrix} 1 & 3 & -1 & | & -1 \\ 2 & 1 & 4 & | & 5 \\ 6 & 2 & -1 & | & -12 \end{bmatrix}$ and $B = \begin{bmatrix} 1 & 3 & -1 & | & -1 \\ 6 & 3 & 12 & | & 15 \\ 6 & 2 & -1 & | & -12 \end{bmatrix}$

represent equivalent systems because we can multiply each entry in the second row of A by 3 to obtain B.

b. $A = \begin{bmatrix} 3 & -1 & 2 & | & 7 \\ 2 & 1 & 4 & | & -5 \\ 3 & 1 & 9 & | & -16 \end{bmatrix}$ and $B = \begin{bmatrix} 3 & 1 & 9 & | & -16 \\ 2 & 1 & 4 & | & -5 \\ 3 & -1 & 2 & | & 7 \end{bmatrix}$

represent equivalent systems because we can interchange the first and third rows of A to obtain B.

c. $A = \begin{bmatrix} 1 & 2 & 1 & | & -3 \\ 2 & 0 & -1 & | & 7 \\ 3 & 1 & 2 & | & 10 \end{bmatrix}$ and $B = \begin{bmatrix} 1 & 2 & 1 & | & -3 \\ 0 & -4 & -3 & | & 13 \\ 3 & 1 & 2 & | & 10 \end{bmatrix}$

represent equivalent systems because we can add -2 times each entry of the first row of A to the corresponding entry of the second row of A to obtain B.

EXERCISE 1

Subtract 4 times row 2 from row 3.

$$\begin{bmatrix} 3 & 2 & -5 & | & -3 \\ 2 & -3 & 2 & | & 6 \\ 8 & -4 & 2 & | & 12 \end{bmatrix}$$

EXAMPLE 2

Use row operations to form an equivalent matrix with the given elements:

$$\begin{bmatrix} 1 & -4 & -5 \\ 3 & 6 & 3 \end{bmatrix} \rightarrow \begin{bmatrix} 1 & -4 & -5 \\ 0 & ? & ? \end{bmatrix}$$

Solution   To obtain 0 as the first entry in the second row, we can multiply the first row by -3 and add the result to the second row. That is, we add -3(row 1) to row 2:

$$-3(\text{row } 1) + \text{row } 2 \quad \begin{bmatrix} 1 & -4 & -5 \\ 3 & 6 & 3 \end{bmatrix} \rightarrow \begin{bmatrix} 1 & -4 & -5 \\ 0 & 18 & 18 \end{bmatrix}$$

EXERCISE 2

Use row operations to form an equivalent matrix with the given elements:

$$\begin{bmatrix} 1 & -3 & 7 \\ -2 & 4 & -6 \end{bmatrix} \rightarrow \begin{bmatrix} 1 & -3 & 7 \\ 0 & ? & ? \end{bmatrix}$$

EXAMPLE 3

Use row operations on the first matrix to form an equivalent matrix in upper triangular form.

$$\begin{bmatrix} 1 & -3 & 1 & -4 \\ 3 & -1 & -1 & 8 \\ 2 & -2 & 3 & -1 \end{bmatrix} \rightarrow \begin{bmatrix} 1 & -3 & 1 & -4 \\ 0 & ? & ? & ? \\ 0 & 0 & ? & ? \end{bmatrix}$$

Solution   We perform the transformation in two steps. First, we obtain zeros in the lower two entries of the first column. We get these zeros by adding suitable multiples of the first row to the second and third rows:

$$\begin{matrix} -3(\text{row } 1) + \text{row } 2 \\ -2(\text{row } 1) + \text{row } 3 \end{matrix} \quad \begin{bmatrix} 1 & -3 & 1 & -4 \\ 3 & -1 & -1 & 8 \\ 2 & -2 & 3 & -1 \end{bmatrix} \rightarrow \begin{bmatrix} 1 & -3 & 1 & -4 \\ 0 & 8 & -4 & 20 \\ 0 & 4 & 1 & 7 \end{bmatrix}$$

Next we obtain a zero as the second entry of the third row by adding a suitable multiple of the *second* row. For this matrix, we add $-\frac{1}{2}$(row 2) to row 3:

$$-\frac{1}{2}(\text{row } 2) + \text{row } 3 \quad \begin{bmatrix} 1 & -3 & 1 & -4 \\ 0 & 8 & -4 & 20 \\ 0 & 4 & 1 & 7 \end{bmatrix} \rightarrow \begin{bmatrix} 1 & -3 & 1 & -4 \\ 0 & 8 & -4 & 20 \\ 0 & 0 & 3 & -3 \end{bmatrix}$$

The last matrix is in upper triangular form.

EXERCISE 3

Use the suggested row operations to form an equivalent matrix in upper triangular form.

$$\begin{bmatrix} 1 & -2 & 4 & 3 \\ 5 & -7 & 8 & 6 \\ -2 & 6 & -7 & 6 \end{bmatrix} \rightarrow \begin{bmatrix} 1 & -2 & 4 & 3 \\ 0 & ? & ? & ? \\ 0 & 0 & ? & ? \end{bmatrix}$$

1. Add -5 (row 1) to (row 2).

2. Add 2 (row 1) to (row 3).

3. Multiply (row 2) by $\frac{1}{3}$.

4. Add -2 (row 2) to (row 3).

Matrix Reduction

We use elementary row operations to apply Gaussian reduction to the augmented matrix of a linear system. This matrix method can be used to solve linear systems of any size and is well suited for implementation by a computer program. The method has three steps.

Solving a Linear System by Matrix Reduction

1. Write the augmented matrix for the system.
2. Using elementary row operations, transform the matrix into an equivalent one in upper triangular form.
3. Use back-substitution to find the solution to the system.

EXAMPLE 4 Use matrix reduction to solve the system.

$$x - 2y = -5$$
$$2x + 3y = 11$$

Solution The augmented matrix is

$$\begin{bmatrix} 1 & -2 & \bigm| & -5 \\ 2 & 3 & \bigm| & 11 \end{bmatrix}$$

We use row operations to obtain 0 in the first entry of the second row.

$$-2(\text{row } 1) + \text{row } 2 \quad \begin{bmatrix} 1 & -2 & \bigm| & -5 \\ 2 & 3 & \bigm| & 11 \end{bmatrix} \quad \rightarrow \quad \begin{bmatrix} 1 & -2 & \bigm| & -5 \\ 0 & 7 & \bigm| & 21 \end{bmatrix}$$

The last matrix is upper triangular and corresponds to the system

$$x - 2y = -5 \qquad (1)$$
$$7y = 21 \qquad (2)$$

Now we use back-substitution to solve the system. From Equation (2), $y = 3$. Substitute 3 for y in Equation (1) to find

$$x - 2(3) = -5$$
$$x = 1$$

The solution is the ordered pair $(1, 3)$.

EXERCISE 4

Use matrix reduction to solve the system

$$x + 4y = 3$$
$$3x + 8y = 1$$

Follow the suggested steps:

1. Write the augmented matrix for the system.
2. Add $-3(\text{row } 1)$ to (row 2).
3. Solve the resulting system by back-substitution.

Reducing a 3 × 3 Matrix

Although there are many ways to reduce an augmented 3 × 3 matrix to upper triangular form, the following two-step procedure may help you organize the row operations.

> **Strategy for Matrix Reduction**
>
> **1.** If the first entry in the first row is zero, interchange that row with another. Obtain zeros in the *first* entries of the second and third rows by adding suitable multiples of the *first* row to the second and third rows.
>
> **2.** If the second entry of the second row is zero, interchange the second and third rows. Obtain a zero in the *second* entry of the third row by adding a suitable multiple of the *second* row to the third row.

If you cannot obtain nonzero entries on the diagonal, then the system does not have a unique solution; the system is either inconsistent or dependent. See Homework problems 45–48.

EXAMPLE 5   Use matrix reduction to solve the system.

$$2x - 4y \quad\quad = 6$$
$$3x - 4y + \; z = 8$$
$$2x \quad\quad - 3z = -11$$

Solution   The augmented matrix is

$$\left[\begin{array}{ccc|c} 2 & -4 & 0 & 6 \\ 3 & -4 & 1 & 8 \\ 2 & 0 & -3 & -11 \end{array}\right]$$

Obtain 1 as the first entry of the first row by multiplying each entry in the first row by $\frac{1}{2}$. This will make it easier to obtain zeros in the first entries of the second and third rows.

$$\frac{1}{2}\text{(row 1)} \quad \left[\begin{array}{ccc|c} 2 & -4 & 0 & 6 \\ 3 & -4 & 1 & 8 \\ 2 & 0 & -3 & -11 \end{array}\right] \qquad \begin{array}{l} 2x - 4y + 0z = 6 \\ 3x - 4y + \; z = 8 \\ 2x + 0y - 3z = -11 \end{array}$$

$$\downarrow$$

$$\left[\begin{array}{ccc|c} 1 & -2 & 0 & 3 \\ 3 & -4 & 1 & 8 \\ 2 & 0 & -3 & -11 \end{array}\right] \qquad \begin{array}{l} x - 2y + 0z = 3 \\ 3x - 4y + \; z = 8 \\ 2x + 0y - 3z = -11 \end{array}$$

Next, obtain zeros in the first entries of the second and third rows by adding suitable multiples of the first row:

$$\begin{array}{l} -3\text{(row 1)} + 2 \\ -2\text{(row 1)} + 3 \end{array} \left[\begin{array}{ccc|c} 1 & -2 & 0 & 3 \\ 3 & -4 & 1 & 8 \\ 2 & 0 & -3 & -11 \end{array}\right] \qquad \begin{array}{l} x - 2y + 0z = 3 \\ 3x - 4y + z = 8 \\ 2x + 0y - 3z = -11 \end{array}$$

$$\downarrow$$

$$\left[\begin{array}{ccc|c} 1 & -2 & 0 & 3 \\ 0 & 2 & 1 & -1 \\ 0 & 4 & -3 & -17 \end{array}\right] \qquad \begin{array}{l} x - 2y + 0z = 3 \\ 0x + 2y + z = -1 \\ 0x + 4y - 3z = -17 \end{array}$$

Finally, obtain a zero in the second entry of the third row by adding -2(row 2) to row 3:

$$-2(\text{row 1}) + \text{row 2} \quad \begin{bmatrix} 1 & -2 & 0 & | & 3 \\ 0 & 2 & 1 & | & -1 \\ 0 & 4 & -3 & | & -17 \end{bmatrix} \qquad \begin{array}{l} x - 2y + 0z = 3 \\ 0x + 2y + z = -1 \\ 0x + 4y - 3z = -17 \end{array}$$

$$\downarrow$$

$$\begin{bmatrix} 1 & -2 & 0 & | & 3 \\ 0 & 2 & 1 & | & -1 \\ 0 & 0 & -5 & | & -15 \end{bmatrix} \qquad \begin{array}{l} x - 2y + 0z = 3 \\ 0x + 2y + z = -1 \\ 0x + 0y - 5z = -15 \end{array}$$

The system is now in upper triangular form, and we use back-substitution to find the solution. Solve the last equation to get $z = 3$ and substitute 3 for z in the second equation to find $y = -2$. Finally, substitute 3 for z and -2 for y in the first equation to find $x = -1$. The solution is the ordered triple $(-1, -2, 3)$.

EXERCISE 5

Use matrix reduction to solve the system

$$\begin{array}{r} x + 3z = -11 \\ 2x + y + z = 1 \\ -3x - 2y = 3 \end{array}$$

Follow the suggested steps:

1. Write the augmented matrix for the system.
2. Add -2(row 1) to (row 2).
3. Add 3(row 1) to (row 3).
4. Add 2(row 2) to (row 3).
5. Solve the resulting system by back-substitution.

It is a good idea to make the first entry in the first row equal to 1, as we did in Example 5, to simplify the calculations that follow. However, if that entry is zero we can interchange two rows. For example, to reduce the matrix

$$\begin{bmatrix} 0 & 1 & 2 & | & 3 \\ 1 & 2 & 3 & | & 4 \\ 2 & 5 & 1 & | & 4 \end{bmatrix}$$

we begin by interchanging the first and second rows to obtain

$$\begin{bmatrix} 1 & 2 & 3 & | & 4 \\ 0 & 1 & 2 & | & 3 \\ 2 & 5 & 1 & | & 4 \end{bmatrix}$$

We then follow the two-step procedure described above to find

$$\begin{bmatrix} 1 & 2 & 3 & | & 4 \\ 0 & 1 & 2 & | & 3 \\ 2 & 5 & 1 & | & 4 \end{bmatrix} \xrightarrow[-2(\text{row 1}) + \text{row 3}]{} \begin{bmatrix} 1 & 2 & 3 & | & 4 \\ 0 & 1 & 2 & | & 3 \\ 0 & 1 & -5 & | & -4 \end{bmatrix} \xrightarrow[-(\text{row 2}) + \text{row 3}]{} \begin{bmatrix} 1 & 2 & 3 & | & 4 \\ 0 & 1 & 2 & | & 3 \\ 0 & 0 & -7 & | & -7 \end{bmatrix}$$

Solving Larger Systems

Systems of linear equations in business and economics may involve hundreds of linear equations and hundreds of variables. The study of airflow for aircraft design can involve systems of millions of linear equations and variables. These large systems are solved using matrix methods.

EXAMPLE 6 Write the augmented matrix for the system

$$-w + x - y + z = 7$$
$$2w - 2x - 4y - 5z = -17$$
$$-3w + 9x + 3y + 6z = 30$$
$$4w + 8x - 2y + 5z = 17$$

Use matrix reduction and back-substitution to solve the system.

Solution The augmented matrix is

$$\left[\begin{array}{cccc|c} -1 & 1 & -1 & 1 & 7 \\ 2 & -2 & -4 & -5 & -17 \\ -3 & 9 & 3 & 6 & 30 \\ 4 & 8 & -2 & 5 & 17 \end{array}\right]$$

We use row operations to put the matrix in upper triangular form. First, we add multiples of the first row to the rows below it to get 0s in the first column.

$$\begin{array}{c} \\ 2(\text{row }1) + \text{row }2 \\ -3(\text{row }1) + \text{row }3 \\ 4(\text{row }1) + \text{row }4 \end{array} \left[\begin{array}{cccc|c} -1 & 1 & -1 & 1 & 7 \\ 0 & 0 & -6 & -3 & -3 \\ 0 & 6 & 6 & 3 & 9 \\ 0 & 12 & -6 & 9 & 45 \end{array}\right]$$

The second row has 0 as its second entry, so we interchange the second and third rows to put a nonzero entry in the second column.

$$\left[\begin{array}{cccc|c} -1 & 1 & -1 & 1 & 7 \\ 0 & 6 & 6 & 3 & 9 \\ 0 & 0 & -6 & -3 & -3 \\ 0 & 12 & -6 & 9 & 45 \end{array}\right]$$

Now we want 0s in the last two rows of the second column. The third row already has 0 in the second column, so we only need to add a multiple of the second row to the fourth row.

$$\begin{array}{c} \\ \\ \\ -2(\text{row }2) + \text{row }4 \end{array} \left[\begin{array}{cccc|c} -1 & 1 & -1 & 1 & 7 \\ 0 & 6 & 6 & 3 & 9 \\ 0 & 0 & -6 & -3 & -3 \\ 0 & 0 & -18 & 3 & 27 \end{array}\right]$$

Finally, we add a multiple of the third row to the fourth row to get 0 in the third column.

$$\begin{array}{c} \\ \\ \\ -3(\text{row }3) + \text{row }4 \end{array} \left[\begin{array}{cccc|c} -1 & 1 & -1 & 1 & 7 \\ 0 & 6 & 6 & 3 & 9 \\ 0 & 0 & -6 & -3 & -3 \\ 0 & 0 & 0 & 12 & 36 \end{array}\right]$$

This triangular matrix corresponds to the system

$$-w + x - y + z = 7$$
$$6x + 6y + 3z = 9$$
$$-6y - 3z = -3$$
$$12z = 36$$

which we solve by back-substitution. From the last equation, we find $z = 3$, and by substituting 3 for z in the third equation, we find $y = -1$. Next, we substitute -1 for y and 3 for z in the second equation to find $x = 1$, and finally, we substitute $x = 1$, $y = -1$, $z = 3$ into the first equation to find $w = -2$. The solution is the *ordered four-tuple* $(-2, 1, -1, 3)$. We can verify that these values satisfy all four equations of the original system.

Our strategy for reducing 3×4 augmented matrices can be generalized for larger matrices, as demonstrated in Example 6. Starting with the first row, we work our way along the diagonal, using row operations to obtain nonzero entries on the diagonal and zeros below the diagonal entry.

EXERCISE 6

Use matrix reduction to solve the system

$$w + x + y + z = -1$$
$$-w + x - y + z = 3$$
$$-2w + 10x + 4y + 7z = 23$$
$$3w + 9x - 3y + 6z = 0$$

Reduced Row Echelon Form

The triangular matrix we obtain by row reduction is an example of *row echelon form*. In a row echelon matrix, (1) the first nonzero entry in any row is farther to the right than the first nonzero entry in the rows above it, (2) the entries directly below the first nonzero entry in any row are all zeros, and (3) any row consisting entirely of zeros is below all rows with any nonzero entry. If we divide each row by its leading nonzero entry, so that the leading entry is 1, and if we use row reduction to obtain 0s above the leading 1 as well as below it, the resulting matrix is said to be in *reduced row echelon form*. This form is especially convenient because we do not have to use back-substitution to finish solving the system; we can read off the solutions directly from the matrix. For example, you can check that the matrix

$$\begin{bmatrix} 1 & 0 & 0 & 0 & | & -2 \\ 0 & 1 & 0 & 0 & | & 3 \\ 0 & 0 & 1 & 0 & | & -3 \\ 0 & 0 & 0 & 1 & | & 4 \end{bmatrix}$$

corresponds to the system

$$a + 0b + 0c + 0d = -2$$
$$0a + b + 0c + 0d = 3$$
$$0a + 0b + c + 0d = -3$$
$$0a + 0b + 0c + d = 4$$

From this system, it is easy to read the solutions $a = -2$, $b = 3$, $c = -3$, and $d = 4$.

TI-84 and TI-83 calculators have a command for finding the reduced row echelon form of a matrix.

EXAMPLE 7 Solve the system

$$a + 2b + 4c + 8d = 12$$
$$-2a + 2b - 2c + 2d = 1$$
$$6a + 6b + 6c + 6d = 19$$
$$4a + 20b - 8c + 14d = 41$$

by finding the reduced row echelon form of the augmented matrix.

Solution The augmented matrix is

$$\begin{bmatrix} 1 & 2 & 4 & 8 & 12 \\ -2 & 2 & -2 & 2 & 1 \\ 6 & 6 & 6 & 6 & 19 \\ 4 & 20 & -8 & 14 & 41 \end{bmatrix}$$

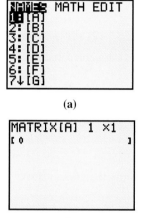

(a)

(b)

FIGURE 8.12

We enter this matrix into the calculator as follows: First access the **MATRIX** menu by pressing 2nd x⁻¹ on a TI-84 or MATRIX on a TI-83. You will see the menu shown in Figure 8.12a. We will use matrix **[A]**, which is already selected, and we presss ▷ ▷ ENTER to **EDIT** (or enter) the matrix, shown in Figure 8.12b.

We want to enter a 4×5 matrix, so we press 4 ENTER 5 ENTER, and we see the display in Figure 8.13a. Now type in the first row of the matrix, pressing ENTER after each entry. The calculator automatically moves to the second row. Continue filling in the rest of the augmented matrix, as shown in Figure 8.13b.

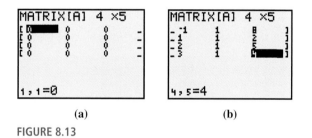

(a)   **(b)**

FIGURE 8.13

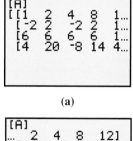

(a)

To make sure you have entered the values correctly, press 2nd MODE to quit to the home screen, then open the matrix menu again. Press 1 ENTER to retrieve matrix **[A]**; the calculator display should look like Figure 8.14a. To check the rest of the matrix, press the right arrow key until you see the last column, as in Figure 8.14b.

We are now ready to compute the reduced row echelon form of the matrix. Access the matrix menu again, but this time press the right arrow once to highlight **MATH** (Figure 8.15a). Scroll down until the **rref(** command is highlighted, as shown in Figure 8.15b, and press ENTER.

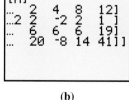

(b)

FIGURE 8.14

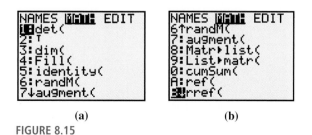

(a)   **(b)**

FIGURE 8.15

Finally, enter matrix **[A]** after the **rref(** command by pressing $\boxed{\text{2nd}}$ $\boxed{x^{-1}}$ $\boxed{1}$ $\boxed{)}$ $\boxed{\text{ENTER}}$ or $\boxed{\text{MATRX}}$ $\boxed{1}$ $\boxed{)}$ $\boxed{\text{ENTER}}$. The display should now look like Figure 8.16a. The last column of the matrix gives decimal approximations for the solutions.

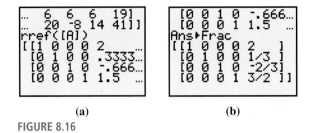

(a)   **(b)**

FIGURE 8.16

The calculator can also display the rational form of the solutions: Press $\boxed{\text{MATH}}$ $\boxed{\text{ENTER}}$ $\boxed{\text{ENTER}}$ to see Figure 8.16b. The reduced row echelon form of the augmented matrix is thus

$$\left[\begin{array}{cccc|c} 1 & 0 & 0 & 0 & 2 \\ 0 & 1 & 0 & 0 & 1/3 \\ 0 & 0 & 1 & 0 & -2/3 \\ 0 & 0 & 0 & 1 & 3/2 \end{array}\right]$$

and the solution is the ordered four-tuple $(2, \frac{1}{3}, \frac{-2}{3}, \frac{3}{2})$. You can verify that these values satisfy each of the original four equations.

EXERCISE 7

a. Find the row reduced echelon form of the augmented matrix for the system

$$-a + b - c + d = 8$$
$$a + b + c + d = 2$$
$$8a + 4b + 2c + d = 5$$
$$27a + 9b + 3c + d = 4$$

b. What third degree polynomial has a graph that passes through the points $(-1, 8)$, $(1, 2)$, $(2, 5)$, and $(3, 4)$?

ANSWERS FOR 8.3 EXERCISES

1. $\left[\begin{array}{ccc|c} 3 & 2 & -5 & -3 \\ 2 & -3 & 2 & 6 \\ 0 & 8 & -6 & -12 \end{array}\right]$ **2.** 2 (row 1) + row 2: $\left[\begin{array}{cc|c} 1 & -3 & 7 \\ 0 & -2 & 8 \end{array}\right]$

3. $\left[\begin{array}{ccc|c} 1 & -2 & 4 & 3 \\ 0 & 1 & -4 & -3 \\ 0 & 0 & 9 & 18 \end{array}\right]$ **4.** $(-5, 2)$ **5.** $(37, -57, -16)$

6. $(-3, 2, 1, -1)$ **7. a.** $\left[\begin{array}{cccc|c} 1 & 0 & 0 & 0 & -1 \\ 0 & 1 & 0 & 0 & 4 \\ 0 & 0 & 1 & 0 & -2 \\ 0 & 0 & 0 & 1 & 1 \end{array}\right]$ **b.** $f(x) = -x^3 + 4x^2 - 2x + 1$

VOCABULARY   *Look up the definitions of new terms in the Glossary.*

Matrix	Row echelon form	Upper triangular form	Coefficient matrix
Dimension	Entry	Reduced row echelon form	Elementary row operation
Augmented matrix	Square matrix	Order	

CONCEPTS

1. We can use a **matrix** to represent a system of linear equations. Each row of the matrix consists of the coefficients in one of the equations of the system.

2. We operate on a matrix by using the elementary row operations.

Elementary Row Operations

1. Multiply the entries of any row by a nonzero real number.

2. Add a constant multiple of one row to another row.

3. Interchange two rows.

3. We can solve a linear system by matrix reduction.

Solving a Linear System by Matrix Reduction

1. Write the augmented matrix for the system.

2. Using elementary row operations, transform the matrix into an equivalent one in upper triangular form.

3. Use back-substitution to find the solution to the system.

4.

Strategy for Matrix Reduction

1. If the first entry in the first row is zero, interchange that row with another. Obtain zeros in the *first* entries of the second and third rows by adding suitable multiples of the *first* row to the second and third rows.

2. If the second entry of the second row is zero, interchange the second and third rows. Obtain a zero in the *second* entry of the third row by adding a suitable multiple of the *second* row to the third row.

5. To reduce larger matrices, we start with the first row and work our way along the diagonal, using row operations to obtain nonzero entries on the diagonal and zeros below the diagonal entry.

STUDY QUESTIONS

1. What are the coefficient matrix and the augmented matrix for a system of equations?

2. What does it mean for a matrix to be in upper triangular form?

3. State the three elementary row operations.

4. Describe a two-step procedure for reducing a 3×3 matrix to upper triangular form.

5. Once the matrix is in upper triangular form, how do we complete the solution of the system?

6. Why is the reduced row echelon form of an augmented matrix convenient for solving a system?

SKILLS   *Practice each skill in the Homework problems listed.*

1. Perform elementary row operations: #1–8

2. Choose an elementry row operation to convert a matrix: #9–18

3. Solve a system by matrix reduction: #19–34

4. Solve a system using the reduced row echelon form of the augmented matrix: #35–40

ThomsonNOW™ Test yourself on key content at **www.thomsonedu.com/login.**

Perform the given elementary row operation on the matrices in Problems 1–8.

1. Multiply row 2 by -3:

$$\left[\begin{array}{cc|c} -2 & 1 & 0 \\ 3 & -1 & 2 \end{array}\right]$$

2. Multiply row 1 by $\dfrac{1}{4}$:

$$\left[\begin{array}{cc|c} 2 & 0 & 3 \\ -1 & 5 & 4 \end{array}\right]$$

3. Add 2 (row 1) to row 2:

$$\left[\begin{array}{cc|c} 1 & -3 & 6 \\ -2 & 4 & -1 \end{array}\right]$$

4. Add -3 (row 1) to row 2:

$$\left[\begin{array}{cc|c} 1 & -4 & 8 \\ 3 & -2 & 10 \end{array}\right]$$

5. Interchange row 1 and row 3:

$$\left[\begin{array}{ccc|c} 0 & -3 & 2 & -3 \\ 2 & 6 & -1 & 3 \\ 1 & 0 & -2 & 5 \end{array}\right]$$

6. Interchange row 2 and row 3:

$$\left[\begin{array}{ccc|c} 1 & 6 & 0 & -2 \\ 0 & 0 & 5 & -10 \\ 0 & 3 & -2 & 8 \end{array}\right]$$

7. Add -4(row 1) to row 3:

$$\left[\begin{array}{ccc|c} 1 & 2 & 1 & -5 \\ 0 & 4 & -2 & 3 \\ 4 & -1 & 6 & -8 \end{array}\right]$$

8. Add 2(row 2) to row 3:

$$\left[\begin{array}{ccc|c} 1 & -7 & 5 & 2 \\ 0 & 1 & -3 & -1 \\ 0 & -2 & -3 & 4 \end{array}\right]$$

In Problems 9–18, use row operations on the first matrix to form an equivalent matrix with the given entries.

9. $\left[\begin{array}{cc|c} 1 & -3 & 2 \\ 2 & 1 & 4 \end{array}\right] \rightarrow \left[\begin{array}{cc|c} 1 & -3 & 2 \\ 0 & ? & ? \end{array}\right]$

10. $\left[\begin{array}{cc|c} -2 & 3 & 0 \\ 4 & 1 & 6 \end{array}\right] \rightarrow \left[\begin{array}{cc|c} -2 & 3 & 0 \\ 0 & ? & ? \end{array}\right]$

11. $\left[\begin{array}{cc|c} 2 & 6 & -4 \\ 5 & 3 & 1 \end{array}\right] \rightarrow \left[\begin{array}{cc|c} 2 & 6 & -4 \\ ? & 0 & ? \end{array}\right]$

12. $\left[\begin{array}{cc|c} 6 & 4 & -2 \\ -1 & -2 & -3 \end{array}\right] \rightarrow \left[\begin{array}{cc|c} 6 & 4 & -2 \\ ? & 0 & ? \end{array}\right]$

13. $\left[\begin{array}{ccc|c} 1 & -2 & 2 & 1 \\ 2 & 3 & -1 & 6 \\ 4 & 1 & -3 & 3 \end{array}\right] \rightarrow \left[\begin{array}{ccc|c} 1 & -2 & 2 & 1 \\ 0 & ? & ? & ? \\ 0 & ? & ? & ? \end{array}\right]$

14. $\left[\begin{array}{ccc|c} 2 & -1 & 3 & -1 \\ -4 & 0 & 4 & 5 \\ 6 & 2 & -1 & -2 \end{array}\right] \rightarrow \left[\begin{array}{ccc|c} 2 & -1 & 3 & -1 \\ 0 & ? & ? & ? \\ 0 & ? & ? & ? \end{array}\right]$

15. $\left[\begin{array}{ccc|c} -1 & 4 & 3 & 2 \\ 2 & -2 & -4 & 6 \\ 1 & 2 & 3 & -3 \end{array}\right] \rightarrow \left[\begin{array}{ccc|c} -1 & 4 & 3 & 2 \\ ? & 0 & ? & ? \\ ? & 0 & ? & ? \end{array}\right]$

16. $\left[\begin{array}{ccc|c} 3 & -2 & 4 & -4 \\ 2 & 2 & 1 & 2 \\ -1 & 1 & 5 & -1 \end{array}\right] \rightarrow \left[\begin{array}{ccc|c} 3 & -2 & 4 & -4 \\ ? & ? & 0 & ? \\ ? & ? & 0 & ? \end{array}\right]$

17. $\left[\begin{array}{ccc|c} -2 & 1 & -3 & -2 \\ 4 & 2 & 0 & 2 \\ 6 & -1 & 2 & 0 \end{array}\right] \rightarrow \left[\begin{array}{ccc|c} -2 & 1 & -3 & -2 \\ 0 & ? & ? & ? \\ 0 & 0 & ? & ? \end{array}\right]$

18. $\left[\begin{array}{ccc|c} -1 & 2 & 3 & 3 \\ 4 & 0 & 1 & -6 \\ 2 & 2 & -3 & -2 \end{array}\right] \rightarrow \left[\begin{array}{ccc|c} -1 & 2 & 3 & 3 \\ 0 & ? & ? & ? \\ 0 & 0 & ? & ? \end{array}\right]$

Use matrix reduction on the augmented matrix to solve each system in Problems 19–26.

19. $x + 3y = 11$
$2x - y = 1$

20. $x - 5y = 11$
$2x + 3y = -4$

21. $x - 4y = -6$
$3x + y = -5$

22. $x + 6y = -14$
$5x + 3y = -4$

23. $2x + y = 5$
$3x - 5y = 14$

24. $3x - 2y = 16$
$4x + 2y = 12$

25. $x - y = -8$
$x + 2y = 9$

26. $4x - 3y = 16$
$2x + y = 8$

Use matrix reduction on the augmented matrix to solve each system in Problems 27–34.

27. $x + 3y - z = 5$
$3x - y + 2z = 5$
$x + y + 2z = 7$

28. $x - 2y + 3z = -11$
$2x + 3y - z = 6$
$3x - y - z = 3$

29. $2x - y + z = 5$
$x - 2y - 2z = 2$
$3x + 3y - z = 4$

30. $x - 2y - 2z = 4$
$2x + y - 3z = 7$
$x - y - z = 3$

31. $2x - y - z = -4$
$x + y + z = -5$
$x + 3y - 4z = 12$

32. $x - 2y - 5z = 2$
$2x + 3y + z = 11$
$3x - y - z = 11$

33. $2x - y = 0$
$3y + z = 7$
$2x + 3z = 1$

34. $3x - z = 7$
$2x + y = 6$
$3y - z = 7$

Solve the system by finding the reduced row echelon form of the augmented matrix.

35. $a + 2b - c + 4d = 20$
$-a + b + 4c + 5d = 7$
$2a + 2b + 6c + 7d = 22$
$3a + 2b - 3c + 4d = 28$

36. $2a - b + 3c - d = -6$
$3a - b + 2c - 2d = -37$
$a + 4c + d = 43$
$4a + 8b - 9c + 6d = 14$

37. $6a + 12b + 18c + 24d = 1$
$3a + 4c = 3$
$-2a + b + 3d = -4$
$18a + 6b + 30c = 23$

38. $a - b - c - d = 1$
$12a - 18b - 6d = 13$
$3a - 8c + 6d = 3$
$24a - 36b - 18c + 66d = 17$

39. $2a + 3b + 4c + d + 4e = 8$
$2a - 5b - 6c + 4d + 3e = 4$
$3a + 3b + 2c - 3d + 7e = -13$
$-3a + 4b - c + d - e = 6$
$4a - 4b + 3c + 2d - 2e = 1$

40. $3a + 2b + 5c + 2d + 4e = 6$
$4a + b + 5c + 2d + 8e = 13$
$6a + 3b + 9c + 3d + 4e = 17$
$7a + 8b + 4c + 7e = -2$
$8a + 9c + d + 9e = 26$

41. a. Find a third-degree polynomial $p(x)$ that satisfies $p(1) = 2$, $p(2) = 4$, $p(3) = 6$, and $p(4) = 1$.
   b. Find a third-degree polynomial $p(x)$ that satisfies $p(1) = 2$, $p(2) = 4$, $p(3) = 6$, and $p(4) = 5$.

42. a. Find a third-degree polynomial $p(x)$ that satisfies $p(1) = 1$, $p(2) = 4$, $p(3) = 9$, and $p(4) = 5$.
   b. Find a third-degree polynomial $p(x)$ that satisfies $p(1) = 2$, $p(2) = 4$, $p(3) = 6$, and $p(4) = 10$.

43. Find a fourth-degree polynomial $p(x)$ that satisfies $p(-2) = -45$, $p(-1) = -2$, $p(0) = 5$, $p(1) = 6$, and $p(2) = -17$.

44. Find a fourth-degree polynomial $p(x)$ that satisfies $p(-2) = 81$, $p(-1) = 9$, $p(0) = 3$, $p(1) = 15$, and $p(2) = 117$.

Problems 45–48 consider inconsistent and dependent systems.

45. Suppose that the reduced row echelon form of an augmented matrix is

$$\begin{bmatrix} 1 & 0 & 0 & 0 & | & 0 \\ 0 & 1 & 0 & 0 & | & 0 \\ 0 & 0 & 1 & 0 & | & 0 \\ 0 & 0 & 0 & 0 & | & 1 \end{bmatrix}$$

a. To what equation does the last row of the matrix correspond?
b. Is there a solution to the system?

46. Suppose that the reduced row echelon form of an augmented matrix is

$$\begin{bmatrix} 1 & 3 & 0 & 0 & | & 0 \\ 0 & 0 & 1 & 0 & | & 0 \\ 0 & 0 & 0 & 0 & | & 1 \\ 0 & 0 & 0 & 0 & | & 0 \end{bmatrix}$$

a. To what equation does the third row of the matrix correspond?
b. Is there a solution to the system?

47. Suppose that the reduced row echelon form of an augmented matrix is

$$\begin{bmatrix} 1 & 0 & 0 & 0 & | & 9 \\ 0 & 1 & 0 & 0 & | & 4 \\ 0 & 0 & 1 & 0 & | & 2 \\ 0 & 0 & 0 & 0 & | & 0 \end{bmatrix}$$

a. To what equation does the last row of the matrix correspond?
b. If the variables of the system are a, b, c, and d, what do you know about a, b, and c?
c. Is there more than one ordered four-tuple that makes all the equations true? (*Hint:* Does the value of d affect any of the equations?)

48. Suppose that the reduced row echelon form of an augmented matrix is

$$\begin{bmatrix} 1 & 2 & 0 & 0 & | & 9 \\ 0 & 0 & 1 & 0 & | & 4 \\ 0 & 0 & 0 & 1 & | & 2 \\ 0 & 0 & 0 & 0 & | & 0 \end{bmatrix}$$

a. To what equation does the last row of the matrix correspond?
b. If the variables of the system are a, b, c, and d, what do you know about c and d?
c. Is there more than one ordered four-tuple that makes all the equations true? (*Hint:* If you specify any value of b, will you be able to find values of a, c, and d that make all the equations true?)

Investigation 14   Input-Output Models

In order to understand the economy of a country or region, we need a model that describes how the various sectors of the economy interact. Each sector of the economy uses input from itself and other sectors to produce a product. The Leontief input-output model represents this interaction as a system of linear equations. The model was invented in the 1930s by Professor Wassily Leontief (1906–1999), who divided the U.S. economy into 500 sectors. Professor Leontief was awarded the Nobel Prize in economics in 1973 for his work on input-output analysis, which is widely used today in policy and economic planning.

In our model, we will partition the economy into just six sectors. We will use the variables u, v, w, x, y, and z to denote the output of each sector in billions of dollars, as shown in Table 8.1. The value of the goods and services demanded by the producing sectors themselves is called the **intermediate demand,** and the **final demand** is the value of the goods and services demanded by the **open sector,** that is, consumers who do not produce goods and services. The final demand for the products of each sector, in billions of dollars, is shown in Table 8.1.

	Production (10^9)	Final demand (10^9)
Agriculture	u	25
Manufacturing	v	110
Energy	w	80
Construction	x	15
Transportation	y	5
Services	z	200

TABLE 8.1

The dollar value of the intermediate demand for each sector, as a fraction of its output, is shown in Table 8.2. Each row lists how the output from that sector is distributed among the other sectors, and each column lists how much input a single sector needs from the other sectors. For example, the first column of the table shows the goods demanded by the agriculture sector from the other sectors. For each dollar's worth of goods it produces, the agriculture sector needs $0.20 worth of goods from its own sector, $0.15 worth of goods from manufacturing, $0.15 from energy, and so on.

Purchased from	Inputs consumed per unit of outputs					
	Agriculture	Manufact.	Energy	Construct.	Transport.	Con. serv.
Agriculture	0.20	0.05	0.05	0.05	0.10	0.05
Manufactuting	0.15	0.30	0.20	0.15	0.15	0.15
Energy	0.15	0.15	0.20	0.15	0.20	0.10
Construction	0.10	0.10	0.20	0.30	0.10	0.05
Transportation	0.20	0.15	0.10	0.15	0.20	0.10
Consumer services	0.10	0.10	0.05	0.10	0.15	0.30

TABLE 8.2

1. In order to provide goods and services for the open sector, what must be true of the entries in any column of Table 8.2?

   In order for the economy to function efficiently, we would like the production level for each sector to balance exactly the sum of the intermediate and final demands. To find the optimal production levels, we will write a system of equations in the variables u, v, w, x, y, and z.

2. **a.** We will with the construction industry: What should its output be to meet the needs of the entire economy? Look at the fourth row of the table. Because the total production of the agriculture sector is u billion dollars, agriculture will need $0.10u$ billion dollars worth of goods from the construction sector. In a similar manner, the manufacturing, energy, construction, transportation, and services sectors will require, respectively, $0.10v$, $0.20w$, $0.30x$, $0.10y$, and $0.05z$ billion dollars of construction goods. Write an expression for the intermediate demand on the construction sector.
   b. For each sector, we would like

   production = intermediate demand + final demand

   Write an equation for the optimal production level by the construction sector. Put the equation in standard form.

3. Repeat step (2) to write equations for the optimal production levels of the other sectors.

4. Write the augmented matrix for the system.

5. Solve the system by finding the row reduced echelon form of the augmented matrix. How much should each sector produce in order that the total production matches the sum of the intermediate and final demand?

In this section, we study linear inequalities in two variables and how they arise in applications.

Graphs of Inequalities in Two Variables

Ivana is investing in the hotel business. She has bought two hotels, and she will expand her investments when her total profit from the two hotels exceeds $10,000. If we let x represent the profit from one hotel and let Y represent the profit from the other, then Ivana will expand her investments when

$$x + y \geq 10{,}000$$

Notice that the equation $x + y = 10{,}000$ is not appropriate to model our situation, since Ivana will be delighted if her profits are not exactly equal to $10,000 but actually exceed that amount.

A **solution** to an inequality in two variables is an ordered pair of numbers that satisfies the inequality. The graph of the inequality must show all the points whose coordinates are solutions. As an example, let us graph the inequality above, $x + y \geq 10{,}000$.

Rewrite the inequality by subtracting x from both sides to get

$$y \geq -x + 10{,}000$$

This inequality says that for each x-value, we must choose points with y-values greater than or equal to $-x + 10{,}000$. For example, when $x = 2000$, we must choose points with y-values greater than or equal to 8000. Solutions for several choices of x are shown in Figure 8.17a.

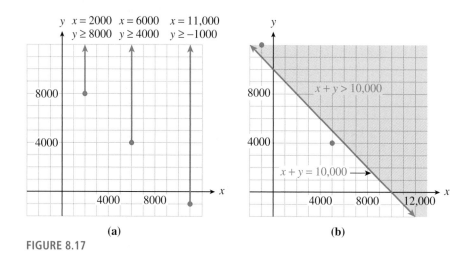

(a)   **(b)**

FIGURE 8.17

A more efficient way to find all the solutions of the inequality is to start with the graph of the corresponding equation

$$y = -x + 10{,}000$$

The graph is a straight line, as illustrated in Figure 8.17b. Observe that any point *above* this line has a y-coordinate greater than $-x + 10{,}000$ and hence satisfies the inequality. Thus, the graph of the inequality includes all the points on or above the line $y = -x + 10{,}000$, as shown by the shaded region in Figure 8.17b.

You can check that the shaded points are also solutions to the original inequality, $x + y \geq 10{,}000$. Consider the point $(-1000, 12{,}000)$, which lies in the shaded region above the line. This pair *does* satisfy $x + y \geq 10{,}000$, because

$$-1000 + 12{,}000 \geq 10{,}000$$

(Ivana will expand her investments if her first hotel loses \$1000 and her second has a profit of \$12,000.) On the other hand, the point $(5000, 4000)$ does *not* lie in the graph of $x + y \geq 10{,}000$, because the coordinates do not satisfy the inequality.

Linear Inequalities

A **linear inequality** can be written in the form

$$ax + by + c \leq 0 \qquad \text{or} \qquad ax + by + c \geq 0$$

The solutions consist of the line $ax + by + c = 0$ and a **half-plane** on one side of that line. We shade the half-plane to show that all its points are included in the solution set. If the inequality is strict, then the graph includes only the half-plane and not the line. In that case, we use a dashed line for the graph of the equation $ax + by + c = 0$ to show that it is not part of the solution.

To decide which side of the line to shade, we can solve the inequality for y in terms of x. If we obtain

$$y \geq mx + b \quad (\text{or} \quad y > mx + b)$$

then we shade the half-plane *above* the line. If the inequality is equivalent to

$$y \leq mx + b \quad (\text{or} \quad y < mx + b)$$

then we shade the half-plane *below* the line.

CAUTION

Be careful when isolating y: We must remember to reverse the direction of the inequality whenever we multiply or divide by a negative number. (See Algebra Skills Refresher A.2 if you would like to review solving inequalities.)

EXAMPLE 1   Graph $4x - 3y \geq 12$

Solution   Solve the inequality for y.

$$4x - 3y \geq 12 \qquad \text{Subtract } 4x \text{ from both sides.}$$
$$-3y \geq -4x + 12 \qquad \text{Divide both sides by } -3.$$
$$y \leq \frac{4}{3}x - 4$$

Graph the corresponding line

$$y = \frac{4}{3}x - 4$$

Note that the y-intercept is -4 and the slope is $\frac{4}{3}$. (See Section 1.4 to review the slope-intercept method of graphing.) Finally, shade the half-plane below the line. The completed graph is shown in Figure 8.18.

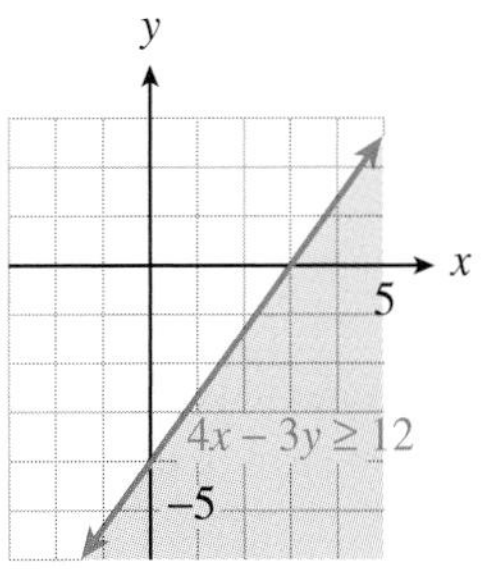

FIGURE 8.18

a. Find one y-value that satisfies the inequality $y - 3x < 6$ for each of the x-values in the table.

x	1	0	-2
y			

b. Graph the line $y - 3x = 6$. Then plot your solutions from part (a) on the same grid.

c. Graph the solutions of the inequality $y - 3x < 6$.

Using a Test Point

A second method for graphing inequalities does not require us to solve for y. Once we have graphed the boundary line, we can decide which half-plane to shade by using a *test point*. The test point can be any point that is not on the boundary line itself.

EXAMPLE 2

Graph the solutions of the inequality $3x - 2y < 6$.

Solution

First, graph the line $3x - 2y = 6$, as shown in Figure 8.19. We will use the intercept method. The intercepts are $(2, 0)$ and $(0, -3)$, so we sketch the boundary line through those points. Next, choose a test point. Since $(0, 0)$ does not lie on the line, we choose it as our test point. Substitute the coordinates of the test point into the inequality to obtain

$$3(0) - 2(0) < 6$$

Because this is a true statement, $(0, 0)$ *is* a solution of the inequality. Since *all* the solutions lie on the same side of the boundary line, we shade the half-plane that contains the test point. In this example, the boundary line is a dashed line because the original inequality was strict.

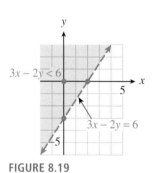

FIGURE 8.19

We can choose *any* point for the test point as long as it does not lie on the boundary line. We chose $(0, 0)$ in Example 2 because the coordinates are easy to substitute into the inequality. If the test point *is* a solution to the inequality, then the half-plane including that point should be shaded. If the test point is *not* a solution to the inequality, then the *other* half-plane should be shaded. For example, suppose we had chosen $(5, 0)$ as the test point in Example 2. When we substitute its coordinates into the inequality, we find

$$3(5) - 2(0) < 6$$

which is a *false* statement. This tells us that $(5, 0)$ is not a solution to the inequality, so the solutions must lie on the other side of the boundary line. Using $(5, 0)$ as the test point gives us the same solutions we found in Example 2.

Here is a summary of our test point method for graphing inequalities.

To Graph an Inequality Using a Test Point:

1. Graph the corresponding equation to obtain the boundary line.
2. Choose a test point that does not lie on the boundary line.
3. Substitute the coordinates of the test point into the inequality.
   a. If the resulting statement is true, shade the half-plane that includes the test point.
   b. If the resulting statement is false, shade the half-plane that does not include the test point.
4. If the inequality is strict, make the boundary line a dashed line.

EXERCISE 2

Graph the solutions of the inequality $y > \dfrac{-3}{2}x$.

1. Graph the line $y = \frac{-3}{2}x$. (Use the slope-intercept method.)
2. Choose a test point. (Do not choose $(0, 0)$!)
3. Decide which side of the line to shade.
4. Should the boundary line be dashed or solid?

Recall that the equation of a vertical line has the form

$$x = k$$

where k is a constant, and a horizontal line has an equation of the form

$$y = k$$

Similarly, the inequality $x \geq k$ may represent the inequality in two variables

$$x + 0y \geq k$$

Its graph is then a region in the plane.

EXAMPLE   Graph $x \geq 2$ in the plane.

Solution   First, graph the equation $x = 2$; its graph is a vertical line. Since the origin does not lie on this line, we can use it as a test point. Substitute 0 for x (there is no y) into the inequality to obtain

$$0 \geq 2$$

Since this is a false statement, shade the half-plane that does not contain the origin. We see in Figure 8.20 that the graph of the inequality contains all points whose x-coordinates are greater than or equal to 2.

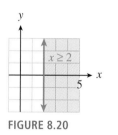

FIGURE 8.20

EXERCISE 3

Graph $-2 \leq y < 3$ in the plane.

Systems of Inequalities

Some applications are best described by a system of two or more inequalities. The solutions to a system of inequalities include all points that are solutions to each inequality in the system. The graph of the system is the intersection of the shaded regions for each inequality in the system. For example, Figure 8.21 shows the solutions of the system

$$y > x \quad \text{and} \quad y > 2$$

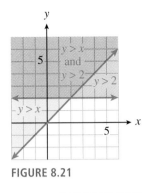

FIGURE 8.21

EXAMPLE 4

Laura is a finicky eater, and she dislikes most foods that are high in calcium. Her morning cereal satisfies some of her calcium requirements, but she needs an additional 500 milligrams of calcium, which she will get from a combination of broccoli, at 160 milligrams per serving, and zucchini, at 30 milligrams per serving. Draw a graph representing the possible combinations of broccoli and zucchini that fulfill Laura's calcium requirements.

Solution

Step 1 Number of servings of broccoli: x

Number of servings of zucchini: y

Step 2 To consume at least 500 milligrams of calcium, Laura must choose x and y so that

$$160x + 30y \geq 500$$

It makes no sense to consider negative values of x or of y, since Laura cannot eat a negative number of servings. Thus, we have two more inequalities to satisfy:

$$x \geq 0 \quad \text{and} \quad y \geq 0$$

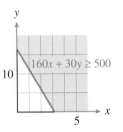

FIGURE 8.22

Step 3 Graph all three inequalities on the same axes. The inequalities $x \geq 0$ and $y \geq 0$ restrict the solutions to lie in the first quadrant. The solutions common to all three inequalities are shown in Figure 8.22.

Step 4 Laura can choose any combination of broccoli and zucchini represented by points in the shaded region. For example, the point $(3, 1)$ is a solution to the system of inequalities, so Laura could choose to eat 3 servings of broccoli and 1 serving of zucchini.

EXERCISE 4

Use the following steps to graph the solutions of the system

$$x + y \leq 12$$
$$3x - 4y \leq 8$$

1. Graph the boundary line $x + y = 12$.

2. Lightly shade the solutions of the inequality $x + y \leq 12$.

3. Graph the boundary line $3x - 4y = 8$.

4. Lightly shade the solutions of $3x - 4y \leq 8$.

5. Shade the intersection of the two solutions sets.

To describe the solutions of a system of inequalities, it is useful to locate the **vertices,** or corner points, of the boundary.

EXAMPLE 5 Graph the solution set of the system below and find the coordinates of its vertices.

$$x - y - 2 \le 0$$
$$x + 2y - 6 \le 0$$
$$x \ge 0, \quad y \ge 0$$

Solution The last two inequalities, $x \ge 0$ and $y \ge 0$, restrict the solutions to the first quadrant. Graph the line $x - y - 2 = 0$ and use the test point $(0, 0)$ to shade the half-plane, including the origin. Finally, graph the line $x - 2y - 6 = 0$ and again use the test point $(0, 0)$ to shade the half-plane below the line. The intersection of the shaded regions is shown in Figure 8.23.

To find the coordinates of the vertices A, B, C, and D, solve simultaneously the equations of the two lines that intersect at the vertex. Thus,

for A, solve the system $\quad\begin{aligned} x &= 0 \\ y &= 0 \end{aligned}\quad$ to find $(0, 0)$

for B, solve the system $\quad\begin{aligned} x &= 0 \\ x + 2y &= 6 \end{aligned}\quad$ to find $(0, 3)$

for C, solve the system $\quad\begin{aligned} x + 2y &= 6 \\ x - y &= 2 \end{aligned}\quad$ to find $\left(\dfrac{10}{3}, \dfrac{4}{3}\right)$

for D, solve the system $\quad\begin{aligned} y &= 0 \\ x - y &= 2 \end{aligned}\quad$ to find $(2, 0)$

The vertices are the points $(0, 0)$, $(0, 3)$, $\left(\frac{10}{3}, \frac{4}{3}\right)$, and $(2, 0)$.

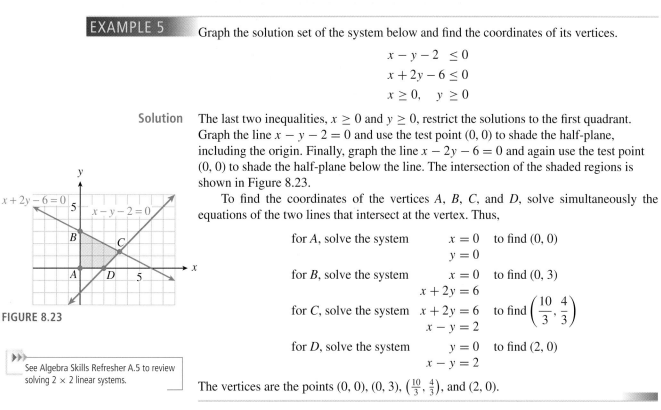

FIGURE 8.23

⟫ See Algebra Skills Refresher A.5 to review solving 2 × 2 linear systems.

EXERCISE 5

a. Graph the system of inequalities

$$5x + 4y < 40$$
$$-3x + 4y < 12$$
$$x < 6, \quad y > 2$$

b. Find the coordinates of the vertices of the solution set.

ANSWERS FOR 8.4 EXERCISES

1. a.

x	1	0	−2
y	5	0	−3

(Many answers are possible.)

b., c.

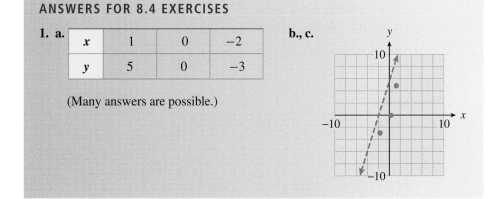

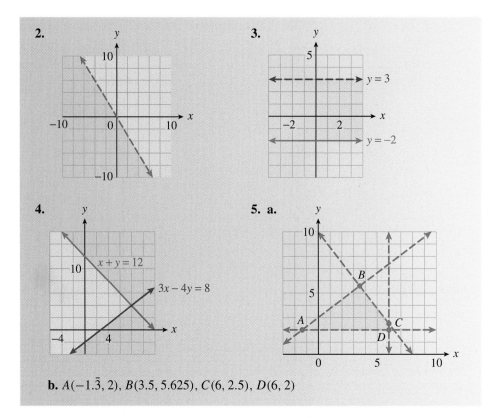

2.

3. $y = 3$; $y = -2$

4. $x + y = 12$; $3x - 4y = 8$

5. a.

b. $A(-1.\overline{3}, 2)$, $B(3.5, 5.625)$, $C(6, 2.5)$, $D(6, 2)$

SECTION 8.4 SUMMARY

VOCABULARY   *Look up the definitions of new terms in the Glossary.*

Half-plane   Test point   Vertices

CONCEPTS

1. The solutions of a linear inequality in two variables consist of a **half-plane** on one side of the line. The line itself is not included if the inequality is strict.

2. Once we have graphed the boundary line, we can decide which half-plane to shade by using a **test point.**

3.

> **To Graph an Inequality Using a Test Point:**
>
> **1.** Graph the corresponding equation to obtain the boundary line.
>
> **2.** Choose a test point that does not lie on the boundary line.
>
> **3.** Substitute the coordinates of the test point into the inequality.

> **a.** If the resulting statement is true, shade the half-plane that includes the test point.
>
> **b.** If the resulting statement is false, shade the half-plane that does not include the test point.
>
> **4.** If the inequality is strict, make the boundary line a dashed line.

4. The solutions to a system of inequalities include all points that are solutions to each inequality in the system. The graph of the system is the intersection of the shaded regions for each inequality in the system.

5. To describe the solutions of a system of inequalities, it is useful to locate the **vertices,** or corner points, of the boundary.

STUDY QUESTIONS

1. The solutions of a linear inequality in two variables form what sort of set?
2. How can you find the boundary of the solution set?

3. If your test point is not a solution of the inequality, which side of the line should you shade?
4. How can you find the vertices of the solution set of a system of inequalities?

SKILLS   *Practice each skill in the Homework problems listed.*

1. Graph the solutions of a linear inequality in two variables: #1–16

2. Graph the solutions of a system of inequalities: #17–36
3. Solve problems using inequalities: #37–42

HOMEWORK 8.4

Graph each inequality.

1. $y > 2x + 4$

2. $y < 9 - 3x$

3. $3x - 2y \leq 12$

4. $2x + 5y \geq 10$

5. $x + 4y \geq -6$

6. $3x - y \leq -2$

7. $x > -3y + 1$

8. $x > 2y - 5$

9. $x \geq -3$

10. $y < 4$

11. $y < \dfrac{1}{2}x$

12. $y > \dfrac{4}{3}x$

13. $0 \geq x - y$

14. $0 \geq x + 3y$

15. $-1 < y \leq 4$

16. $-2 \leq y < 0$

Graph each system of inequalities.

17. $y > 2$
    $x \geq -2$

18. $y \leq -1$
    $x > 2$

19. $y < x$
    $y \geq -3$

20. $y \geq -x$
    $y < 2$

21. $x + y \leq 6$
    $x + y \geq 4$

22. $x - y < 3$
    $x - y > -2$

23. $2x - y \leq 4$
    $x + 2y > 6$

24. $2y - x < 2$
    $x + y \leq 4$

25. $3y - 2x < 2$
    $y > x - 1$

26. $2x + y < 4$
    $y > 1 - x$

Graph each system of inequalities and find the coordinates of the vertices.

27. $2x + 3y - 6 < 0$
    $x \geq 0, \ y \geq 0$

28. $3x + 2y < 6$
    $x \geq 0, \ y \geq 0$

29. $5y - 3x \leq 15$
    $x + y \leq 11$
    $x \geq 0, \ y \geq 0$

30. $y - 2x \geq -4$
    $x + y \leq 5$
    $x \geq 0, \ y \geq 0$

31. $2y \leq x$
    $2x \leq y + 12$
    $x \geq 0, \ y \geq 0$

32. $y \geq 3x$
    $2y + x \leq 14$
    $x \geq 0, \ y \geq 0$

33. $x + y \geq 3$
    $2y \leq x + 8$
    $2y + 3x \leq 24$
    $x \geq 0, \ y \geq 0$

34. $2y + 3x \geq 6$
    $2y + x \leq 10$
    $y \geq 3x - 9$
    $x \geq 0, \ y \geq 0$

35. $3y - x \geq 3$
    $y - 4x \geq -10$
    $y - 2 \leq x$
    $x \geq 0, \ y \geq 0$

36. $2y + x \leq 12$
    $4y \leq 2x + 8$
    $x \leq 4y + 4$
    $x \geq 0, \ y \geq 0$

Graph the set of solutions to each problem. Two of the inequalities in each system are $x \geq 0$ and $y \geq 0$.

37. The math club is selling tickets for a show by a mathemagician. Student tickets will cost $1 and faculty tickets will cost $2. The ticket receipts must be at least $250 to cover the fee for the performer. Write a system of three inequalities for the number of student tickets and the number of faculty tickets that must be sold, and graph the solutions.

38. The math department is having a book sale of old textbooks to raise at least $300 for scholarships. Paperback textbooks will cost $2 and the hardcover textbooks will cost $5. Write a system of three inequalities for the number of paperback and hardback textbooks that must be sold, and graph the solutions.

39. Vassilis plans to invest at most $10,000 in two banks. One bank pays 6% annual interest and the other pays 5% annual interest. Vassilis wants at least $540 total annual interest from his two investments. Write a system of four inequalities for the amount Vassilis can invest in the two accounts, and graph the system.

40. Jeannette has 180 acres of farmland for growing wheat or soy. She can get a profit of $36 per acre for wheat and $24 per acre for soy. She wants to have a profit of at least $5400 from her crops. Write a system of four inequalities for the number of acres she can use for each crop, and graph the solutions.

41. Gary's pancake recipe includes corn meal and whole wheat flour. Corn meal has 2.4 grams of linoleic acid and 2.5 milligrams of niacin per cup. Whole wheat flour has 0.8 grams of linoleic acid and 5 milligrams of niacin per cup. These two ingredients should not exceed 3 cups total. The mixture should provide at least 3.2 grams of linoleic acid and at least 10 milligrams of niacin. Write a system of five inequalities for the amount of corn meal and the amount of whole wheat flour Gary can use, then graph the solutions.

42. Cho and his brother go into business making comic book costumes. They need 1 hour of cutting and 2 hours of sewing to make a Batman costume. They need 2 hours of cutting and 1 hour of sewing to make a Wonder Woman costume. They have available at most 10 hours per day for cutting and at most 8 hours per day for sewing. They must make at least one costume each day to stay in business. Write a system of five inequalities for the number of each type of costume Cho can make, then graph the solutions.

8.5   Linear Programming

The term **linear programming** was coined in the late 1940s. It describes a relatively young branch of mathematics, compared to subjects such as Euclidean geometry, where the major ideas were already well understood 23 centuries ago. (The Greek mathematician Euclid wrote what can be considered the first geometry textbook about 300 B.C.) Business managers routinely solve linear programming problems for purchasing and marketing strategy, so it is possible that linear programming affects your daily life as much as any other branch of mathematics.

The Objective Function and Constraints

TrailGear would like to maximize its profit from selling hiking boots. The company produces two kinds of hiking boots, a Weekender model, on which it makes $8 profit per pair, and a Sierra model, on which it makes $10 profit per pair. How many of each model should TrailGear produce each week in order to maximize its profit?

If we let x represent the number of Weekender boots and y the number of Sierra boots TrailGear produces, then the total weekly profit is given by

$$P = 8x + 10y$$

This expression for P is called the **objective function.** The goal of a linear programming problem is to maximize or minimize such an objective function, subject to one or more constraints.

If TrailGear had infinite resources and an infinite market, there would be no limit to the profit it could earn by producing more and more hiking boots. However, every business has to consider many factors, including its supplies of labor and materials, overhead and shipping costs, and the size of the market for its product. To keep things simple, we will concentrate on just two of these factors.

Each pair of Weekender boots requires 3 hours of labor to produce, and each pair of Sierra boots requires 6 hours. TrailGear has available 2400 hours of labor per week. Thus, x and y must satisfy the inequality

$$3x + 6y \leq 2400$$

In addition, suppose that TrailGear's suppliers can provide at most 1000 ounces of silicone gel each week, with each pair of Weekenders using 2 ounces and each Sierra model using 1 ounce. This means that

$$2x + y \leq 1000$$

Of course, we will also require that $x \geq 0$ and $y \geq 0$. These four inequalities are called the **constraints** of the problem.

Feasible Solutions

We have formulated the original problem into an objective function

$$P = 8x + 10y$$

and a system of inequalities called the constraints.

$$3x + 6y \leq 2400$$
$$2x + y \leq 1000$$
$$x \geq 0, \quad y \geq 0$$

Our goal is to find values for x and y that satisfy the constraints and produce the maximum value for P.

We begin by graphing the solutions to the constraint inequalities. These solutions are shown in the shaded region in Figure 8.24. The points in this region are called **feasible solutions** because they are the only values we can consider while looking for the maximum value of the objective function P.

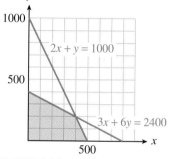

FIGURE 8.24

EXAMPLE 1

a. Verify that the points (300, 100) and (200, 300) represent feasible solutions for the problem above. Show that (300, 400) is not a feasible solution.

b. Find the values of the objective function $P = 8x + 10y$ at the two feasible solutions in part (a).

Solutions

a. The two points (300, 100) and (200, 300) lie within the shaded region in Figure 8.19, but (300, 400) does not. We can also verify that the coordinates of (300, 100) and (200, 300) satisfy each of the constraint inequalities.

b. For (300, 100), we have

$$P = 8(300) + 10(100) = 3400$$

For (200, 300), we have

$$P = 8(200) + 10(300) = 4600$$

a. Determine which of the points (0, 400), (400, 200), (500, 0), and (500, 400) represent feasible solutions for the TrailGear problem.

b. Find the values of the objective function $P = 8x + 10y$ at the feasible solutions in part (a).

The Optimum Solution

We cannot check all of the feasible solutions to see which one results in the largest profit. Fortunately, there is a simple way to find the optimal solution.

Consider the objective function,

$$P = 8x + 10y$$

If TrailGear would like to make $2000 on hiking boots, it could produce 200 pairs of Sierra boots, or 250 pairs of Weekenders. Or it could produce some of each; for example, 50 pairs of Weekenders and 160 pairs of Sierra boots. In fact, every point on the line

$$8x + 10y = 2000$$

represents a combination of Weekenders and Sierra boots that will yield a profit of $2000. This line is labeled $P = 2000$ in Figure 8.25.

If TrailGear would like to make $4000 on boots, it should choose a point on the line labeled $P = 4000$. Similarly, all the points on the line labeled $P = 6000$ will yield a profit of $6000, and so on. Different values of P correspond to parallel lines on the graph. Smaller values of P correspond to lines near the origin, and larger values of P have lines farther from the origin. Here is another example.

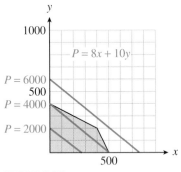

FIGURE 8.25

Figure 8.26 shows the feasible solutions for another linear programming problem. The objective function is $C = 3x + 5y$.

a. Find the value of C at the point (0, 3). Are there other feasible solutions that give the same value of C?

b. Find all feasible solutions that result in an objective value of 30.

c. How many feasible solutions result in an objective value of 39?

d. Is it possible for a feasible solution to result in an objective value of 45?

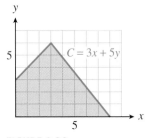

FIGURE 8.26

Solutions **a.** The objective value at the point (0, 3) is

$$C = 3(0) + 5(3) = 15$$

Another point with the same objective value is (5, 0). In fact, all points on the line $3x + 5y = 15$ have an objective value of 15. This line intersects the set of feasible solutions in a line segment, as shown in Figure 8.27. Thus, there are infinitely many feasible solutions with objective value 15.

b. Points that give an objective value of $C = 30$ lie on the line $3x + 5y = 30$, as shown in the figure. There are infinitely many feasible solutions that lie on this line; for example, one such point is (5, 3).

c. The line $3x + 5y = 39$ intersects the set of feasible solutions in only one point, the point (3, 6). This is the only feasible solution that yields an objective value of 39.

d. The line $3x + 5y = 45$ includes all points for which $C = 45$. This line does not intersect the set of feasible solutions, as we see in the figure. Thus, there are no feasible solutions that result in an objective value of 45.

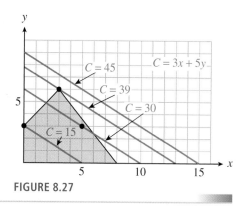

FIGURE 8.27

EXERCISE 2

a. Returning to the TrailGear problem of Example 1, find all feasible solutions for which the objective function $P = 8x + 10y$ has a value of 5200.

b. Find all feasible solutions that result in an objective value of 6000.

We are only allowed to choose points from the set of feasible solutions. Imagine the parallel lines representing different values of the objective function sweeping across the graph of the feasible solutions. The objective values increase as the lines sweep up across the graph. What is the last feasible solution the lines intersect before leaving the shaded region? If you study the preceding examples, perhaps you can see that the largest (and smallest) values of the objective function will occur at corner points of the set of feasible solutions. We have not proved this fact, but it is true.

Linear Programming

The maximum and the minimum values of the objective function always occur at vertices of the graph of feasible solutions.

Depending on the exact formula for the objective function, the maximum and minimum values may occur at *any* of the vertices of the shaded region.

EXAMPLE 3   Find TrailGear's maximum weekly profit.

Solution   Figure 8.28 shows the lines corresponding to the objective values $P = 2000$, $P = 4000$, and $P = 5200$. The maximum value of the profit, P, corresponds to the topmost line that intersects the region of feasible solutions. This is the line that passes through the vertex where the lines $3x + 6y = 2400$ and $2x + y = 1000$ intersect, namely the vertex at (400, 200). The profit for that point is

$$P = 8(400) + 10(200) = 5200$$

so the maximum weekly profit is $5200.

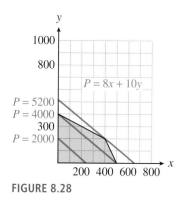

FIGURE 8.28

Figure 8.29 shows the feasible solutions for a linear programming problem. The objective function is $R = x + 5y$.

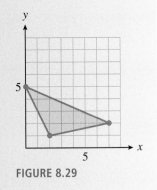
FIGURE 8.29

a. Sketch lines for objective values of $R = 5$, $R = 15$, $R = 25$, and $R = 35$.

b. Evaluate the objective function at each vertex of the shaded region.

c. Which vertex corresponds to the maximum value of the objective function? What is the maximum value?

d. Which vertex corresponds to the minimum value of the objective function? What is the minimum value?

We can now formulate a strategy for solving problems by linear programming.

To Solve a Linear Programming Problem:

1. Represent the unknown quantities by variables. Write the objective function and the constraints in terms of the variables.

2. Graph the solutions to the constraint inequalities.

3. Find the coordinates of each vertex of the solution set.

4. Evaluate the objective function at each vertex.

5. The maximum and minimum values of the objective function occur at vertices of the set of feasible solutions.

In Example 4, the set of feasible solutions is an unbounded region.

EXAMPLE 4   Each week, the Healthy Food Store buys both granola and muesli in bulk from two cereal companies. The store requires at least 12 kilograms of granola and 9 kilograms of muesli. Company A charges $15 for a package that contains 2 kilograms of granola and 1 kilogram of muesli. Company B charges $25 for a package of 3 kilograms of granola and 3 kilograms of muesli. How much should the Healthy Food Store purchase from each company in order to minimize its costs and still meet its needs for granola and muesli? What is the minimum cost?

Solution

Step 1 Number of packages purchased from Company A: x

Number of packages purchased from Company B: y

First write the objective function. The store would like to minimize its cost, so

$$C = 15x + 25y$$

Next, write the constraints. These will be a system of inequalities. It may help to organize the information into a table.

	Company A	Company B	Required
Granola	$2x$	$3y$	12
Muesli	x	$3y$	9

The Healthy Food Store will have $2x$ kilograms of granola and x kilograms of muesli from Company A, and $3y$ kilograms of granola and $3y$ kilograms of muesli from Company B. The store requires that

$$2x + 3y \geq 12$$
$$x + 3y \geq 9$$

Because the store cannot purchase negative quantities, we also have

$$x \geq 0, \quad y \geq 0$$

Step 2 Graph the solutions to the constraint system. The feasible solutions form the shaded region shown in Figure 8.30. Any ordered point on this graph corresponds to a way to purchase granola and muesli that meets the store's needs, but some of these choices cost more than others.

Step 3 We know that the minimum cost will occur at one of the vertex points, which are labeled in the figure. The coordinates of P and R are easy to see. To find the coordinates of Q, we notice that it is the intersection of the lines $2x + 3y = 12$ and $x + 3y = 9$. Thus, we must solve the system

$$2x + 3y = 12$$
$$x + 3y = 9$$

Subtracting the second equation from the first, we find that $x = 3$. Substituting this value into either of the original two equations, we find that $y = 2$. Thus the point Q has coordinates $(3, 2)$.

Step 4 Now we evaluate the objective function at each of the three vertices.

At P (0, 4): $C = 15(0) + 25(4) = 100$

At Q (3, 2): $C = 15(3) + 25(2) = 95$ Minimum cost

At R (9, 0): $C = 15(9) + 25(0) = 135$

The minimum cost occurs at point Q.

Step 5 The Healthy Food Store should buy three packages from Company A and two packages from Company B. It will pay $95 for its stock of granola and muesli.

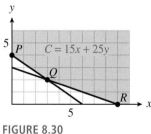

FIGURE 8.30

EXERCISE 4

Find the minimum value of the objective function, $O = 5x + 3y$, subject to the constraints $x \geq 0$, $y \geq 0$, $x + y \geq 7$, and $5x + 2y \geq 20$.

You can use your graphing calculator to solve the problem in Example 4. Set the window values at

$$\text{Xmin} = 0 \quad \text{Xmax} = 9.4$$
$$\text{Ymin} = 0 \quad \text{Ymax} = 6.2$$

Next, graph the set of feasible solutions. We have already taken care of the constraints $x \geq 0$ and $y \geq 0$ by setting Xmin and Ymin to zero. Solve each of the other constraints for y to get

$$y \geq (12 - 2x)/3$$
$$y \geq (9 - x)/3$$

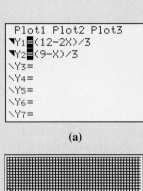

(a)

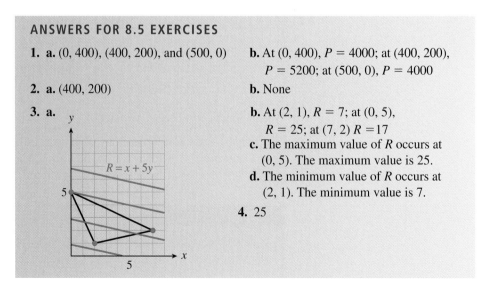

(b)

FIGURE 8.31

For each constraint, the set of feasible solutions lies above the boundary line, because y is *greater* than the expression in x. To shade the regions above the graphs of Y_1 and Y_2, move the cursor onto the backslash in front of the equations and press **ENTER** twice, as shown in Figure 8.31a. Then press **GRAPH**. Your display should look like Figure 8.31b.

The feasible solutions lie in the crosshatched region that is shaded with both the vertical and horizontal lines. We will use the calculator to evaluate the objective function at each vertex. First, use the **TRACE** (or **value** or **intersect** feature) to find the coordinates of one of the vertices, say $(0, 4)$. Then press **2nd** **QUIT** to get back to the **Home** screen; enter the formula for the objective function by keying in

$$15X + 25Y$$

(Enter Y by pressing **ALPHA** **1**.) Your calculator has stored the values $x = 0$ and $y = 4$ from the **TRACE** key, so when you press **ENTER**, the calculator returns 100 for the value of C at that point. Thus, when $x = 0$ and $y = 4$, $C = 100$.

Similarly, you can verify that $C = 135$ when $x = 9$ and $y = 0$, and that when $x = 3$ and $y = 2$, $C = 95$. Thus, the minimum cost of \$95 occurs when $x = 3$ and $y = 2$.

ANSWERS FOR 8.5 EXERCISES

1. a. $(0, 400)$, $(400, 200)$, and $(500, 0)$ **b.** At $(0, 400)$, $P = 4000$; at $(400, 200)$, $P = 5200$; at $(500, 0)$, $P = 4000$

2. a. $(400, 200)$ **b.** None

3. a.

b. At $(2, 1)$, $R = 7$; at $(0, 5)$, $R = 25$; at $(7, 2)$ $R = 17$
c. The maximum value of R occurs at $(0, 5)$. The maximum value is 25.
d. The minimum value of R occurs at $(2, 1)$. The minimum value is 7.

4. 25

VOCABULARY *Look up the definitions of new terms in the Glossary.*

Objective function   Constraint   Feasible solution

CONCEPTS

1. **Linear programming** is a technique for finding the maximum or minimum value of an **objective function,** subject to a system of **constraints.**
2. The optimum solution occurs at one of the vertices of the set of **feasible solutions.**

3.

> **To Solve a Linear Programming Problem:**
>
> 1. Represent the unknown quantities by variables. Write the objective function and the constraints in terms of the variables.
> 2. Graph the solutions to the constraint inequalities.
> 3. Find the coordinates of each vertex of the solution set.
> 4. Evaluate the objective function at each vertex.
> 5. The maximum and minimum values of the objective function occur at vertices of the set of feasible solutions.

STUDY QUESTIONS

1. Explain the terms *objective function, constraints,* and *feasible solution.*
2. Explain how to solve a linear programming problem by graphing.

3. How can you find the vertices of the set of feasible solutions?
4. Where do the maximum and minimum values of the objective function occur?

SKILLS *Practice each skill in the Homework problems listed.*

1. Find the maximum or minimum value for an objectice function and a given set of feasible solutions: #1–12
2. Solve a linear programming problem by graphing: #13–32

3. Write the objective function and the constraints for a linear programming problem: #19–24

ThomsonNOW™ Test yourself on key content at **www.thomsonedu.com/login.**

For Problems 1–4, find the minimum value of the cost function $C = 3x + 4y$ subject to following constraints:

$$x + y \geq 10, \quad x \leq 8, \quad y \leq 7, \quad x \geq 0, \quad y \geq 0$$

The graph of the feasible solutions is shown in in the figure.

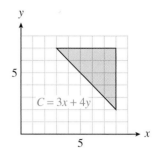

1. Use a graph to explain why it is impossible in this situation to have a cost as low as $12. (*Hint:* Draw the graph of $12 = 3x + 4y$ on the graph of the feasible solutions.)

2. Use a graph to explain why the cost will not be as great as $60. (*Hint:* Draw the graph of $60 = 3x + 4y$ together with the graph of the feasible solutions.)

3. Use a graph to determine which vertex of the shaded region will correspond to the minimum cost. What is the minimum cost?

4. Use a graph to determine which vertex of the shaded region will correspond to the maximum cost. What is the maximum cost?

For Problems 5–8, find the minimum value of the profit function $P = 4x - 2y$ subject to the following constraints:

$$5x - y \geq 22, \quad x + y \leq 8, \quad x \geq 0, \quad y \geq 0$$

The graph of the feasible solutions is shown in the figure.

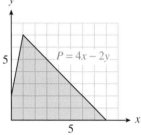

5. Graph the line that corresponds to a profit of $8. Find the coordinates of at least one feasible solution that gives a profit of $8.

6. Graph the line that corresponds to a profit of $22. Find the coordinates of at least one feasible solution that gives a profit of $22.

7. **a.** Which line is farther from the origin, the line for a profit of $8 or the line for the profit of $22?
   b. Use a graph to determine which vertex corresponds to a maximum profit.
   c. Find the maximum profit.

8. **a.** Use a graph to determine which vertex corresponds to a minimum profit.
   b. Find the minimum profit.

For Problems 9–12, objective functions and the graphs of the feasible solutions are given.
(a) Use the graph to find the vertex that yields the minimum value of the objective function, and find the minimum value.
(b) Use the graph to find the vertex that yields the maximum value of the objective function, and find the maximum value.

9. $C = 3x + y$ 10. $C = x + 4y$ 11. $C = 5x - 2y$ 12. $C = 2x - y$

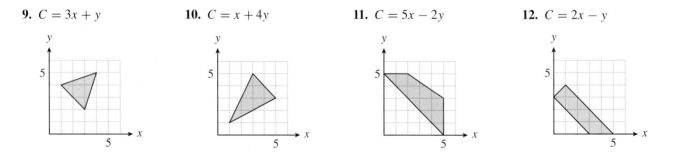

For Problems 13–18,
(a) Graph the set of feasible solutions.
(b) Find the vertex that gives the minimum of the objective function, and find the minimum value.
(c) Find the vertex that gives the maximum of the objective function, and find the maximum value.

13. Objective function $C = 3x + 2y$ with constraints $x \geq 0$, $y \geq 0, 2x + y \leq 8, 4x + 6y \leq 24$

14. Objective function $C = -2x + y$ with constraints $x \geq 0, y \geq 0, x - 2y \geq -10, 2x + y \leq 10$

15. Objective function $C = 3x - y$ with constraints $x \geq 0$, $y \geq 0, x + y \leq 14, 5x + y \leq 50$

16. Objective function $C = 5x + 4y$ with constraints $x \geq 0$, $y \geq 0, 2x + y \leq 10, x - 3y \geq -3$

17. Objective function $C = 200x - 20y$ with constraints $x \geq 0, y \geq 0, 3x + 2y \leq 24, x + y \leq 9, x + 2y \leq 16$

18. Objective function $C = 54x + 24y$ with constraints $x \geq 0, y \geq 0, 3x + 2y \leq 24, 3x - y \leq 15$, $3x - 4y \geq -12$

For Problems 19–26, solve each linear programming problem by graphing.
(a) Write a formula for the objective function.
(b) Write a system of inequalities for the constraints.
(c) Graph the set of feasible solutions.
(d) Find the optimum solution.

19. The math club is selling tickets for a show by a mathemagician. Student tickets will cost $1 and faculty tickets will cost $2. The ticket receipts must be at least $250 to cover the fee for the performer. An alumna promises to donate one calculator for each student ticket sold and three calculators for each faculty ticket sold. What is the minimum number of calculators that the alumna will donate?

20. The math department is having a book sale of unwanted textbooks to raise funds for $300 in scholarships. Paperback textbooks will be sold for $2 and the hardcover textbooks will be sold for $5. If paperback texts weigh 2 pounds each and hardcover books weigh 3 pounds each, find the minimum weight of textbooks the department must sell in order to raise its required funds.

21. Jeannette has 180 acres of farmland for growing wheat or soy. Each acre of wheat requires two hours of labor at harvest time, and each acre of soy needs one hour of labor. Jeannette will have 240 hours of labor available at harvest time. Find the maximum profit Jeannette can make from her two crops if she can get a profit of $36 per acre for wheat and $24 per acre for soy.

22. Vassilis has at most $10,000 to invest in two banks. Alpha Bank will pay 6% annual interest and Bank Beta pays 5% annual interest. Alpha Bank will only insure up to $6000, so Vassilis will invest no more than that with Alpha. What is the maximum amount of interest Vassilis can earn in 1 year?

23. Gary's pancake recipe includes corn meal and whole wheat flour. Corn meal has 2.4 grams of linoleic acid and 2.5 milligrams of niacin per cup. Whole wheat flour has 0.8 grams of linoleic acid and 5.2 milligrams of niacin per cup. These two dry ingredients do not exceed 3 cups total. They combine for at least 3.2 grams of linoleic acid and at least 10 milligrams of niacin. Minimize the number of calories possible in the recipe if corn meal has 433 calories per cup and whole wheat flour has 400 calories per cup.

24. Cho requires 1 hour of cutting and 2 hours of sewing to make a Batman costume. He requires 2 hours of cutting and 1 hour of sewing to make a Wonder Woman costume. At most 10 hours per day are available for cutting and at most 8 hours per day are available for sewing. At least one costume must be made each day to stay in business. Find Cho's maximum income from selling one day's costumes if a Batman costume costs $68 and a Wonder Woman costume costs $76.

For Problems 25–30, use a graphing calculator to find approximate values for the maximum and minimum of the objective function.

25. Objective function $C = 8.7x - 4.2y$ with constraints
$x \geq 0$, $y \geq 0$, $1.7x - 4.5y \geq -9$, $14.3x + 10.9y \leq 28.6$

26. Objective function $C = -142x + 83y$ with constraints
$x \geq 0$, $y \geq 0$, $21x - 49y \geq -147$, $19x + 21y \leq 171$

27. Objective function $C = 312x + 92y$ with constraints
$x \geq 0$, $y \geq 0$, $18x + 17y \leq 284$, $51x + 11y \leq 656$

28. Objective function $C = 5.3x + 4.2y$ with constraints
$x \geq 0$, $y \geq 0$, $2.5x + 1.7y \leq 20.1$,
$0.09x - 0.31y \geq -0.39$

29. Objective function $C = 202x + 220y$ with constraints
$x \geq 0$, $y \geq 0$, $38x + 24y \leq 294$, $35x + 34y \leq 310$,
$13x + 29y \leq 197$

30. Objective function $C = 54x + 24y$ with constraints
$x \geq 0$, $y \geq 0$, $43x + 32y \leq 333$, $23x - 9y \leq 152$,
$73x - 94y \geq -296$

CHAPTER 8 Summary and Review

KEY CONCEPTS

1. We can solve a 2×2 linear system by graphing. The solution is the intersection point of the two graphs.
2. A linear system may be **inconsistent** (has no solution), **dependent** (has infinitely many solutions), or **consistent and independent** (has one solution).
3.

Inconsistent and Dependent Systems

1. If an equation of the form

$$0x + 0y = k \qquad (k \neq 0)$$

is obtained as a linear combination of the equations in a system, the system is **inconsistent.**

2. If an equation of the form

$$0x + 0y = 0$$

is obtained as a linear combination of the equations in a system, the system is **dependent.**

4. We can use a system of equations to solve problems involving two unknown quantities.
5. In economics, the price at which the **supply** and **demand** are equal is called the **equilibrium price.**
6. The solution to a 3×3 linear system is an **ordered triple.**
7. A 3×3 system in **triangular form** can be solved by **back-substitution.**

8. **Gaussian reduction** is a generalized form of the elimination method that can be used to reduce a 3×3 linear system to triangular form.

9.

Steps for Solving a 3 × 3 Linear System

1. Clear each equation of fractions and put it in standard form.
2. Choose two of the equations and eliminate one of the variables by forming a linear combination.
3. Choose a different pair of equations and eliminate the *same* variable.
4. Form a 2×2 system with the equations found in steps (2) and (3). Eliminate one of the variables from this 2×2 system by using a linear combination.
5. Form a triangular system by choosing among the previous equations. Use back-substitution to solve the triangular system.

10. 3×3 linear systems may be **inconsistent** or **dependent.**
11. We can use a **matrix** to represent a system of linear equations. Each row of the matrix consists of the coefficients in one of the equations of the system.

12. We operate on a matrix by using the elementary row operations.

Elementary Row Operations

1. Multiply the entries of any row by a nonzero real number.

2. Add a constant multiple of one row to another row.

3. Interchange two rows.

13. We can solve a linear system by matrix reduction.

Solving a Linear System by Matrix Reduction

1. Write the augmented matrix for the system.

2. Using elementary row operations, transform the matrix into an equivalent one in upper triangular form.

3. Use back-substitution to find the solution to the system.

14.

Strategy for Matrix Reduction

1. If the first entry in the first row is zero, interchange that row with another. Obtain zeros in the *first* entries of the second and third rows by adding suitable multiples of the *first* row to the second and third rows.

2. If the second entry of the second row is zero, interchange the second and third rows. Obtain a zero in the *second* entry of the third row by adding a suitable multiple of the *second* row to the third row.

15. To reduce larger matrices, we start with the first row and work our way along the diagonal, using row operations to obtain nonzero entries on the diagonal and zeros below the diagonal entry.

16. The solutions of a linear inequality in two variables consist of a **half-plane** on one side of the line. The line itself is not included if the inequality is strict.

17. Once we have graphed the boundary line, we can decide which half-plane to shade by using a **test point.**

18.

To Graph an Inequality Using a Test Point:

1. Graph the corresponding equation to obtain the boundary line.

2. Choose a test point that does not lie on the boundary line.

3. Substitute the coordinates of the test point into the inequality.
   a. If the resulting statement is true, shade the half-plane that includes the test point.
   b. If the resulting statement is false, shade the half-plane that does not include the test point.

4. If the inequality is strict, make the boundary line a dashed line.

19. The solutions to a system of inequalities include all points that are solutions to each inequality in the system. The graph of the system is the intersection of the shaded regions for each inequality in the system.

20. To describe the solutions of a system of inequalities, it is useful to locate the **vertices,** or corner points, of the boundary.

21. **Linear programming** is a technique for finding the maximum or minimum value of an **objective function,** subject to a system of **constraints.**

22. The optimum solution occurs at one of the vertices of the set of **feasible solutions.**

23.

To Solve a Linear Programming Problem:

1. Represent the unknown quantities by variables. Write the objective function and the constraints in terms of the variables.

2. Graph the solutions to the constraint inequalities.

3. Find the coordinates of each vertex of the solution set.

4. Evaluate the objective function at each vertex.

5. The maximum and minimum values of the objective function occur at vertices of the set of feasible solutions.

REVIEW PROBLEMS

Solve each system by graphing. Use the *ZDecimal* window.

1. $y = -2.9x - 0.9$
$y = 1.4 - 0.6x$

2. $y = 0.6x - 1.94$
$y = -1.1x + 1.29$

Solve each system using substitution or elimination.

3. $x + 5y = 18$
$x - y = -3$

4. $x + 5y = 11$
$2x + 3y = 8$

5. $\frac{2}{3}x - 3y = 8$
$x + \frac{3}{4}y = 12$

6. $3x = 5y - 6$
$3y = 10 - 11x$

Decide whether each system is inconsistent, dependent, or consistent and independent.

7. $2x - 3y = 4$
$x + 2y = 7$

8. $2x - 3y = 4$
$6x - 9y = 4$

9. $2x - 3y = 4$
$6x - 9y = 12$

10. $x - y = 6$
$x + y = 6$

Solve each system using Gaussian reduction.

11. $x + 3y - z = 3$
$2x - y + 3z = 1$
$3x + 2y + z = 5$

12. $x + y + z = 2$
$3x - y + z = 4$
$2x + y + 2z = 3$

13. $x + z = 5$
$y - z = -8$
$2x + z = 7$

14. $x + 4y + 4z = 0$
$3x + 2y + z = -4$
$2x - 4y + z = -11$

15. $\frac{1}{2}x + y + z = 3$
$x - 2y - \frac{1}{3}z = -5$
$\frac{1}{2}x - 3y - \frac{2}{3}z = -6$

16. $\frac{3}{4}x - \frac{1}{2}y + 6z = 2$
$\frac{1}{2}x + y - \frac{3}{4}z = 0$
$\frac{1}{4}x + \frac{1}{2}y - \frac{1}{2}z = 0$

Use matrix reduction to solve each system.

17. $x - 2y = 5$
$2x + y = 5$

18. $4x - 3y = 16$
$2x + y = 8$

19. $2x - y = 7$
$3x + 2y = 14$

20. $2x - y + 3z = -6$
$x + 2y - z = 7$
$3x + y + z = 2$

21. $x + 2y - z = -3$
$2x - 3y + 2z = 2$
$x - y + 4z = 7$

22. $x + y + z = 1$
$2x - y - z = 2$
$2x - y + 3z = 2$

Solve the system by finding the reduced row echelon form of the augmented matrix.

23.
$$2a + 3b - 4c - 5d = 3$$
$$2a - 3b + 4c - 7d = -11$$
$$3a + b - 8c + d = 9$$
$$4a - 7b - 5c + 3d = -4$$

24.
$$-a - 2b + 5c + 2d = 15$$
$$-2a + 3b + 2c + d = 15$$
$$2a - 4b + 6c + 9d = 20$$
$$6a + 8b + 7c - 2d = 0$$

25.
$$2a - b - 3c + d + 5e = 7$$
$$4a + 6b - 3c - d + e = -6$$
$$5a + 2b - 9c - 4d + 7e = 3$$
$$6a - 2b + 7c + 2d - 8e = 11$$
$$7a + 8b + 2c + 6d + e = 2$$

26.
$$a - 4b + 2c + 3d - e = 7$$
$$a + 2b - 5c + 2d - 3e = -6$$
$$3a + 3b + 2c - 4d + 3e = 3$$
$$-2a - 3b + 5c + 4d - e = 11$$
$$-4a + 3b - c + d + 2e = 2$$

For Problems 27–32, solve by writing and solving a system of linear equations in two or three variables.

27. A math contest exam has 40 questions. A contestant scores 5 points for each correct answer but loses 2 points for each wrong answer. Lupe answered all the questions and her score was 102. How many questions did she answer correctly?

28. A game show contestant wins $25 for each correct answer he gives but loses $10 for each incorrect response. Roger answered 24 questions and won $355. How many answers did he get right?

29. Barbara wants to earn $500 a year by investing $5000 in two accounts, a savings plan that pays 8% annual interest and a high-risk option that pays 13.5% interest. How much should she invest in each account?

30. An investment broker promises his client a 12% return on her funds. If the broker invests $3000 in bonds paying 8% interest, how much must he invest in stocks paying 15% interest to keep his promise?

31. The perimeter of a triangle is 30 centimeters. The length of one side is 7 centimeters shorter than the second side, and the third side is 1 centimeter longer than the second side. Find the length of each side.

32. A company ships its product to three cities: Boston, Chicago, and Los Angeles. The cost of shipping is $10 per crate to Boston, $5 per crate to Chicago, and $12 per crate to Los Angeles. The company's shipping budget for April is $445. It has 55 crates to ship, and demand for its product is twice as high in Boston as in Los Angeles. How many crates should the company ship to each destination?

Graph each inequality.

33. $3x - 4y < 12$

34. $x > 3y - 6$

35. $y < -\dfrac{1}{2}$

36. $-4 \leq x < 2$

Graph the solutions to each system of inequalities.

37. $y > 3, \quad x \leq 2$

38. $y \geq x, \quad x > 2$

39. $3x - y < 6, \quad x + 2y > 6$

40. $x - 3y > 3, \quad y < x + 2$

For Problems 41–44,
(a) Graph the solutions to the system of inequalities.
(b) Find the coordinates of the vertices.

41. $3x - 4y \leq 12$
$x \geq 0, \; y \leq 0$

42. $x - 2y \leq 6$
$y \leq x$
$x \geq 0, \; y \geq 0$

43. $x + y \leq 5$
$y \geq x$
$y \geq 2, \; x \geq 0$

44. $x - y \leq -3$
$x + y \leq 6$
$x \leq 4$
$x \geq 0, \; y \geq 0$

45. Ruth wants to provide cookies for the customers at her video rental store. It takes 20 minutes to mix the ingredients for each batch of peanut butter cookies and 10 minutes to bake them. Each batch of granola cookies takes 8 minutes to mix and 10 minutes to bake. Ruth does not want to use the oven more than 2 hours a day or to spend more than 2 hours a day mixing ingredients. Write a system of inequalities for the number of batches of peanut butter cookies and of granola cookies that Ruth can make in one day and graph the solutions.

46. A vegetarian recipe calls for no more than 32 ounces of a combination of tofu and tempeh. Tofu provides 2 grams of protein per ounce and tempeh provides 1.6 grams of protein per ounce. Graham would like the dish to provide at least 56 grams of protein. Write a system of inequalities for the amount of tofu and the amount of tempeh for the recipe and graph the solutions.

For Problems 47–50,
(a) Graph the set of feasible solutions.
(b) Find the vertex that gives the minimum of the objective function, and find the minimum value.
(c) Find the vertex that gives the maximum of the objective function, and find the maximum value.

47. Objective function $C = 18x + 48y$ with constraints
$x \geq 0, y \geq 0, 3x + y \geq 3, 2x + y \leq 12, x + 5y \leq 15$

48. Objective function $C = 10x - 8y$ with constraints
$x \geq 0, y \geq 0, 5x - y \geq 2, x + 2y \leq 18, x - y \leq 3$

49. Ruth wants to provide cookies for the customers at her video rental store. It takes 20 minutes to mix the ingredients for each batch of peanut butter cookies and 10 minutes to bake them. Each batch of granola cookies takes 8 minutes to mix and 10 minutes to bake. Ruth does not want to use the oven more than 2 hours a day, or to spend more than 2 hours a day mixing ingredients.
   a. Write a system of inequalities for the number of batches of peanut butter cookies and granola cookies Ruth can make in one day and graph the solutions.
   b. Ruth decides to sell the cookies. If she charges 25¢ per peanut butter cookie and 20¢ per granola cookie, she will sell all the cookies she bakes. Each batch contains 50 cookies. How many batches of each type of cookie should she bake to maximize her income? What is the maximum income?

50. A vegetarian recipe calls for 32 ounces of a combination of tofu and tempeh. Tofu provides 2 grams of protein per ounce and tempeh provides 1.6 grams of protein per ounce. Graham would like the dish to provide at least 56 grams of protein.
   a. Write a system of inequalities for the amount of tofu and the amount of tempeh for the recipe and graph the solutions.
   b. Suppose that tofu costs 12¢ per ounce and tempeh costs 16¢ per ounce. What is the least expensive combination of tofu and tempeh Graham can use for the recipe? How much will it cost?

PROJECTS FOR CHAPTER 8

Projects 1–4 deal with the mean and the median.

1. According to the Bureau of the Census, the average age of the U.S. population is steadily rising. The table gives data for two types of average, the mean and the median, for the ages of women. (Consult the Glossary for the definitions of *mean* and *median*.)

Date	July 1990	July 1992	July 1994	July 1996	July 1998
Median age	34.0	34.6	35.2	35.8	36.4
Mean age	36.6	36.8	37.0	37.3	37.5

   a. Which is growing more rapidly, the mean age or the median age?
   b. Plot the data for median age versus the date, using July 1990 as $t = 0$. Draw a line through the data.
   c. What is the meaning of the slope of the line in part (b)?
   d. Plot the data for mean age versus date on the same axes. Draw a line that fits the data.
   e. Use your graph to estimate when the mean age and the median age will be the same.
   f. For the data given, the mean age of women in the United States is greater than the median age. What does this tell you about the U.S. population of women?

2. Repeat Project 1 for the mean and median ages of U.S. men, given in the table below.

Date	July 1990	July 1992	July 1994	July 1996	July 1998
Median age	31.6	32.2	32.9	33.5	34.0
Mean age	33.8	34.0	34.3	34.5	34.9

3. Refer to Project 1.
   a. Find an equation for the line you graphed in part (b), median age versus date.
   b. Find an equation for the line you graphed in part (d), mean age versus date.
   c. Solve the system of equations algebraically. How does your answer compare to the solution you found by graphing?

4. Refer to Project 2.
   a. Find an equation for the line you graphed in part (b), median age versus date.
   b. Find an equation for the line you graphed in part (d), mean age versus date.
   c. Solve the system of equations algebraically. How does your answer compare to the solution you found by graphing?

5. Water slowly contracts as it cools, causing it to become denser, until just before freezing, when the density decreases slightly. The salinity of water affects its freezing point and also the temperature at which it reaches its maximum density: The higher the salinity, the lower the temperature at which the maximum density occurs. Pure water is densest at $4°C$ and freezes at $0°C$. Water that is 15% saline is densest at $0.8°C$ and freezes at $-0.8°C$.
   a. Write an equation for the water's temperature of maximum density in terms of its salinity in percent.
   b. Write an equation for the water's freezing point in terms of its salinity in percent.
   c. Graph the system of equations.
   d. Solve the system to find the salinity of water that freezes when it is densest. What is the freezing point?

6. Charles's law says that the temperature of a gas is related to its volume by a linear equation. A gas that occupied one liter at $0°C$ expanded to 1.2 liters at a temperature of $54.6°C$. A second sample of gas that occupied 2 liters at $0°C$ expanded to 2.8 liters at $109.2°C$.
   a. Find a linear equation for temperature, T, in terms of volume, V, for the first sample of gas, and a second linear equation for the second sample of gas.
   b. Graph the system of equations.
   c. Solve the system. What is the significance of the intersection point?
   d. Show that both equations from part (a) can be written in the form $T = c\left(\frac{V}{V_0} - 1\right)$, where V_0 is the volume of the gas at $0°C$.

Grazing animals spend most of their time foraging for food. A small animal such as a ground squirrel must also be alert for predators. Which foraging strategy favors survival: Should the animal look for foods that satisfy its dietary requirements in minimum time, thus minimizing its exposure to predators and the elements, or should it try to maximize its intake of nutrients?

7. The Columbian ground squirrel eats forb, a type of flowering weed, and grass. One gram of forb provides 2.44 kilocalories, and one gram of grass provides 2.26 kilocalories. For survival, the squirrel needs at least 100 kilocalories per day. However, the squirrel can digest no more than 314 wet grams of food daily. Each dry gram of forb becomes 2.67 wet grams in the squirrel's stomach, and each dry gram of grass becomes 1.64 wet grams. In addition, the squirrel has at most 342 minutes available for grazing each day. It takes the squirrel 2.05 minutes to eat one gram of forb and 5.21 minutes to eat one gram of grass.
   a. Write a system of inequalities for the number of grams of grass and forb the squirrel can eat and survive.
   b. Graph the system.
   c. Give two possible combinations of grass and forb that satisfy the squirrel's feeding requirements.
   d. If the squirrel can find only grass on a given day, how much grass will it need to eat? If the squirrel can find only forb, how much forb will it need?

8. The Isle Royale moose eats deciduous leaves, which provide 3.01 kilocalories of energy per gram, and aquatic plants, which provide 3.82 kilocalories per gram. The moose needs at least 10,946 kilocalories of food per day, and its daily sodium requirement must be met by consuming at least 453 grams of aquatic plants. It takes the moose 0.05 minute to eat one gram of aquatic plants and 0.06 minute to eat each gram of leaves. To maintain its thermal balance, the moose cannot spend more than 150 minutes each day standing in water and eating aquatics plants, or more than 256 minutes eating leaves. Each gram of aquatic plants becomes 20 wet grams in the moose's stomach, and each gram of leaves becomes 4 wet grams, and the stomach can hold no more than 32,900 wet grams of food daily.
   a. Write a system of inequalities for the number of grams of aquatic plants and of leaves the moose can eat.
   b. Graph the system.
   c. Give two possible combinations of aquatic plants and leaves that satisfy the moose's feeding requirements.
   d. Can the moose survive by eating only aquatic plants?

9. Project 7 described the diet of the Colombian ground squirrel. Refer to the system of constraints and the set of feasible solutions you found in that problem.
   a. Write a formula for the squirrel's energy consumption. What is the optimum diet for the squirrel if it wants to maximize its energy consumption? Round your answers to one decimal place.
   b. Write a formula for the squirrel's foraging time. What is the optimum diet for the squirrel if it wants to minimize its time foraging? Round your answer to one decimal place.
   c. The actual observed diet of the squirrel consists of approximately 100 grams of forb and 25 grams of grass. Does the squirrel appear to be trying to maximize its caloric intake or to minimize its time eating? (Because 70% of the food available to the squirrel is grass, its diet is not determined by convenience.)

10. Project 8 described the diet of the Isle Royale moose. Refer to the system of constraints and the set of feasible solutions you found in that problem.
   a. Write a formula for the moose's energy consumption. What is the optimum diet for the moose if it wants to maximize its energy consumption? Round your answers to one decimal place.
   b. Write a formula for the moose's foraging time. What is the optimum diet for the moose if it wants to minimize its time foraging? Round your answer to one decimal place.
   c. The actual observed diet of the moose consists of approximately 868 grams of aquatic plants and 3437 grams of leaves. Does the moose appear to be trying to maximize its caloric intake or to minimize its time eating?

Algebra Skills Refresher

Numbers and Operations

Order of Operations

Numerical calculations often involve more than one operation. So that everyone agrees on how such expressions should be evaluated, we follow the **order of operations.**

Order of Operations

1. Simplify any expressions within grouping symbols (parentheses, brackets, square root bars, or fraction bars). Start with the innermost grouping symbols and work outward.
2. Evaluate all powers and roots.
3. Perform multiplications and divisions in order from left to right.
4. Perform additions and subtractions in order from left to right.

Parentheses and Fraction Bars

We can use parentheses to override the multiplication-first rule. Compare the two expressions below.

$$\text{The sum of 4 times 6 and 10} \qquad 4 \cdot 6 + 10$$
$$\text{4 times the sum of 6 and 10} \qquad 4(6 + 10)$$

In the first expression, we perform the multiplication 4×6 first, but in the second expression we perform the addition $6 + 10$ first, because it is enclosed in parentheses.

The location (or absence) of parentheses can drastically alter the meaning of an expression. In the following example, note how the location of the parentheses changes the value of the expression.

EXAMPLE 2

a. $5 - 3 \cdot 4^2$
$$= 5 - 3 \cdot 16$$
$$= 5 - 48 = -43$$

b. $5 - (3 \cdot 4)^2$
$$= 5 - 12^2$$
$$= 5 - 144 = -139$$

c. $(5 - 3 \cdot 4)^2$
$$= (5 - 12)^2$$
$$= (-7)^2 = 49$$

d. $(5 - 3) \cdot 4^2$
$$= 2 \cdot 16$$
$$= 32$$

CAUTION

In the expression $5 - 12^2$, which appears in Example 2b, the exponent 2 applies only to 12, not to -12. Thus, $5 - 12^2 \neq 5 + 144$.

The order of operations mentions other grouping devices besides parentheses: fraction bars and square root bars. Notice how the placement of the fraction bar affects the expressions in the next example.

EXAMPLE 3

a.
$$\frac{1+2}{3\cdot 4} = \frac{3}{12}$$
$$= \frac{1}{4}$$

b.
$$1 + \frac{2}{3\cdot 4} = 1 + \frac{2}{12}$$
$$= 1 + \frac{1}{6} = \frac{7}{6}$$

c.
$$\frac{1+2}{3}\cdot 4 = \frac{3}{3}\cdot 4$$
$$= 1\cdot 4 = 4$$

d.
$$1 + \frac{2}{3}\cdot 4 = 1 + \frac{8}{3}$$
$$= \frac{3}{3} + \frac{8}{3} = \frac{11}{3}$$

Radicals

You are already familiar with square roots. Every nonnegative number has two square roots, defined as follows:

$$s \quad \text{is a square root of} \quad n \quad \text{if} \quad s^2 = n$$

There are several other kinds of roots, one of which is called the **cube root,** denoted by $\sqrt[3]{n}$. We define the cube root as follows.

Cube Roots

$$b \quad \text{is the \textbf{cube root} of} \quad n \quad \text{if} \quad b \quad \text{cubed equals} \quad n.$$

In symbols, we write

$$b = \sqrt[3]{n} \quad \text{if} \quad b^3 = n$$

Although we cannot take the square root of a *negative* number, we can take the *cube root* of *any* real number. For example,

$$\sqrt[3]{64} = 4 \quad \text{because} \quad 4^3 = 64$$

and

$$\sqrt[3]{-27} = -3 \quad \text{because} \quad (-3)^3 = -27$$

In the order of operations, simplifying radicals and powers comes after parentheses but before products and quotients.

EXAMPLE 4

Simplify each expression.

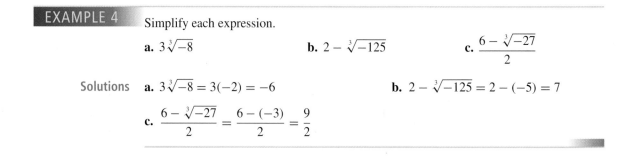

a. $3\sqrt[3]{-8}$

b. $2 - \sqrt[3]{-125}$

c. $\dfrac{6 - \sqrt[3]{-27}}{2}$

Solutions

a. $3\sqrt[3]{-8} = 3(-2) = -6$

b. $2 - \sqrt[3]{-125} = 2 - (-5) = 7$

c. $\dfrac{6 - \sqrt[3]{-27}}{2} = \dfrac{6 - (-3)}{2} = \dfrac{9}{2}$

Scientific Notation

Scientists and engineers regularly encounter very large numbers such as

$$5,980,000,000,000,000,000,000,000$$

(the mass of the Earth in kilograms) and very small numbers such as

$$0.000\,000\,000\,000\,000\,000\,000\,001\,67$$

(the mass of a hydrogen atom in grams). These numbers can be written in a more compact and useful form by using powers of 10.

In our base 10 number system, multiplying a number by a positive power of 10 has the effect of moving the decimal place k places to the right, where k is the exponent in the power of 10. For example,

$$3.529 \times 10^2 = 352.9 \qquad \text{and} \qquad 25 \times 10^4 = 250,000$$

Multiplying by a power of 10 with a negative exponent moves the decimal place to the left. For example,

$$1728 \times 10^{-3} = 1.728 \qquad \text{and} \qquad 4.6 \times 10^{-5} = 0.000046$$

Using this property, we can write any number as the product of a number between 1 and 10 (including 1) and a power of 10. For example, the mass of the Earth and the mass of a hydrogen atom can be expressed as

$$5.98 \times 10^{24} \text{ kilograms} \qquad \text{and} \qquad 1.67 \times 10^{-24} \text{ gram}$$

respectively. A number written in this form is said to be expressed in **scientific notation.**

To Write a Number in Scientific Notation:

1. Locate the decimal point so that there is exactly one nonzero digit to its left.
2. Count the number of places you moved the decimal point: This determines the power of 10.
   a. If the original number is greater than 10, the exponent is positive.
   b. If the original number is less than 1, the exponent is negative.

EXAMPLE 5

Write each number in scientific notation.

a. $478,000 = 4.78000 \times 10^5$
5 places

$$= 4.78 \times 10^5$$

b. $0.00032 = 00003.2 \times 10^{-4}$
4 places

$$= 3.2 \times 10^{-4}$$

EXAMPLE 6

The average American eats 110 kilograms of meat per year. It takes about 16 kilograms of grain to produce 1 kilogram of meat, and advanced farming techniques can produce about 6000 kilograms of grain on each hectare of arable land. (The hectare is 10,000 square meters, or just under $2\frac{1}{2}$ acres.) Now, the total land area of the Earth is about 13 billion hectares, but only about 11% of that land is arable. Is it possible for each of the 5.5 billion people on Earth to eat as much meat as Americans do?

Solution   First we will compute the amount of meat necessary to feed every person on Earth 110 kilograms per year. There are 5.5×10^9 people on Earth.

$$(5.5 \times 10^9 \text{ people}) \times (110 \text{ kg/person}) = 6.05 \times 10^{11} \text{ kg of meat}$$

Next we will compute the amount of grain needed to produce that much meat.

$$(16 \text{ kg of grain/kg of meat}) \times (6.05 \times 10^{11} \text{ kg of meat}) = 9.68 \times 10^{12} \text{ kg of grain}$$

Next we will see how many hectares of land are needed to produce that much grain.

$$(9.68 \times 10^{12} \text{ kg of grain}) \div (6000 \text{ kg/hectare}) = 1.61\overline{3} \times 10^9 \text{ hectares}$$

Finally, we will compute the amount of arable land available for grain production.

$$0.11 \times (13 \times 10^9 \text{ hectares}) = 1.43 \times 10^9 \text{ hectares}$$

Thus, even if we use every hectare of arable land to produce grain for livestock, we will not have enough to provide every person on Earth with 110 kilograms of meat per year.

SECTION A.1 SUMMARY

VOCABULARY   *Look up the definitions of new terms in the Glossary.*

Order of operations	Scientific notation	Radical	Cube root
Grouping symbol	Fraction bar	Square root bar	

SKILLS   *Practice each skill in the Exercises listed.*

1. Follow the order of operations: #1–26
2. Compute cube roots: #27–39
3. Use a calculator to simplify expressions: #31–42

4. Evaluate an expression: #43–50
5. Convert between standard and scientific notation: #51–54
6. Compute with scientific notation: #55–62

EXERCISES A.1

Simplify each expression according to the order of operations.

1. $\dfrac{3(6-8)}{-2} - \dfrac{6}{-2}$

2. $\dfrac{5(3-5)}{2} - \dfrac{18}{-3}$

3. $6[3-2(4+1)]$

4. $5[3+4(6-4)]$

5. $(4-3)[2+3(2-1)]$

6. $(8-6)[5+7(2-3)]$

7. $64 \div 8[4-2(3+1)]$

8. $27 \div 3[9-3(4-2)]$

9. $5[3+(8-1)] \div (-25)$

10. $-3[-2+(6-1)] \div 9$

11. $[-3(8-2)+3] \cdot [24 \div 6]$

12. $[-2+3(5-8)] \cdot [-15 \div 3]$

13. -5^2

14. $(-15)^2$

15. $(-3)^4$

16. -3^4

17. -4^3

18. $(-4)^3$

19. $(-2)^5$

20. -2^5

21. $\dfrac{4 \cdot 2^3}{16} + 3 \cdot 4^2$

22. $\dfrac{4 \cdot 3^2}{6} + (3 \cdot 4)^2$

23. $\dfrac{3^2 - 5}{6 - 2^2} - \dfrac{6^2}{3^2}$

24. $\dfrac{3 \cdot 2^2}{4 - 1} + \dfrac{(-3)(2)^3}{6}$

25. $\dfrac{(-5)^2 - 3^2}{4 - 6} + \dfrac{(-3)^2}{2 + 1}$

26. $\dfrac{7^2 - 6^2}{10 + 3} - \dfrac{8^2 \cdot (-2)}{(-4)^2}$

Compute each cube root. Round your answers to three decimal places if necessary. Verify your answers by cubing them.

27. a. $\sqrt[3]{512}$ **b.** $\sqrt[3]{-125}$
    c. $\sqrt[3]{-0.064}$ **d.** $\sqrt[3]{1.728}$

28. a. $\sqrt[3]{9}$ **b.** $\sqrt[3]{258}$
    c. $\sqrt[3]{-0.002}$ **d.** $\sqrt[3]{-3.1}$

Simplify each expression according to the order of operations.

29. a. $\dfrac{4 - 3\sqrt[3]{64}}{2}$

    b. $\dfrac{4 + \sqrt[3]{-216}}{8 - \sqrt[3]{8}}$

30. a. $\sqrt[3]{3^3 + 4^3 + 6^3}$

    b. $\sqrt[3]{9^3 + 10^3 - 1^3}$

Use a calculator to simplify each expression.

31. $\dfrac{-8398}{26 \cdot 17}$

32. $\dfrac{-415.112}{8.58 + 18.73}$

33. $\dfrac{112.78 + 2599.124}{27.56}$

34. $\dfrac{202{,}462 - 9510}{356}$

35. $\sqrt{24 \cdot 54}$

36. $\sqrt{\dfrac{1216}{19}}$

37. $\dfrac{116 - 35}{215 - 242}$

38. $\dfrac{842 - 987}{443 - 385}$

39. $\sqrt{27^2 + 36^2}$

40. $\sqrt{13^2 - 4 \cdot 21 \cdot 2}$

41. $\dfrac{-27 - \sqrt{27^2 - 4(4)(35)}}{2 \cdot 4}$

42. $\dfrac{13 + \sqrt{13^2 - 4(5)(-6)}}{2 \cdot 5}$

Evaluate each expression for the given values of the variable. Use your calculator where appropriate.

43. $\dfrac{5(F - 32)}{9}$; $F = 212$

44. $\dfrac{a - 4s}{1 - r}$; $r = 2, s = 12, \text{ and } a = 4$

45. $P + Prt$; $P = 1000, \quad r = 0.04, \quad \text{and} \quad t = 2$

46. $R(1 + at)$; $R = 2.5, \quad a = 0.05, \quad \text{and} \quad t = 20$

47. $\dfrac{1}{2}gt^2 - 12t$; $g = 32 \quad \text{and} \quad t = \dfrac{3}{4}$

48. $\dfrac{Mv^2}{g}$; $M = \dfrac{16}{3}, \quad v = \dfrac{3}{2}, \quad \text{and } g = 32$

49. $\dfrac{32(V - v)^2}{g}$; $V = 12.78, \quad v = 4.26, \quad \text{and} \quad g = 32$

50. $\dfrac{32(V - v)^2}{g}$; $V = 38.3, \quad v = -6.7, \quad \text{and} \quad g = 9.8$

Write each number in scientific notation.

51. a. 285   **b.** 8,372,000   **52. a.** 68,742   **b.** 481,000,000,000
      c. 0.024   **d.** 0.000523   **c.** 0.421   **d.** 0.000004

Write each number in standard notation.

53. a. 2.4×10^2 **b.** 6.87×10^{15} **54. a.** 4.8×10^3 **b.** 8.31×10^{12}
      c. 5.0×10^{-3} **d.** 2.02×10^{-4} **c.** 8.0×10^{-1} **d.** 4.31×10^{-5}

Compute with the aid of a calculator. Write your answers in standard notation.

55. a. $\dfrac{(2.4 \times 10^{-8})(6.5 \times 10^{32})}{5.2 \times 10^{18}}$ **56. a.** $\dfrac{(8.4 \times 10^{-22})(1.6 \times 10^{15})}{3.2 \times 10^{-11}}$

      b. $\dfrac{(7.5 \times 10^{-13})(3.6 \times 10^{-9})}{(1.5 \times 10^{-15})(1.6 \times 10^{-11})}$ **b.** $\dfrac{(9.4 \times 10^{24})(7.2 \times 10^{-18})}{(4.5 \times 10^{26})(6.4 \times 10^{-16})}$

57. In 2005, the public debt of the United States was over $8,062,156,000,000.
     a. Express this number in scientific notation.
     b. If the population of the United States in 2005 was 297,636,000, what was the per capita debt (the debt per person) in 2005?

58. A light-year is the number of miles traveled by light in 1 year (365 days). The speed of light is approximately 186,000 miles per second.
     a. Compute the number of miles in 1 light-year, and express your answer in scientific notation.
     b. The star nearest to the Sun is Proxima Centauri, at a distance of 4.3 light-years. How long would it take *Pioneer 10* (the first space vehicle to achieve escape velocity from the solar system), traveling at 32,114 miles per hour, to reach Proxima Centauri?

59. The diameter of the galactic disk is about 1.2×10^{18} kilometers, and our Sun lies about halfway from the center of the galaxy to the edge of the disk. The Sun orbits the galactic center once in 240 million years.
     a. What is the speed of the Sun in its orbit, in kilometers per year?
     b. What is its speed in meters per second?

60. Lake Superior has an area of 31,700 square miles and an average depth of 483 feet.
     a. Find the approximate volume of Lake Superior in cubic feet.
     b. If 1 cubic foot of water is equivalent to 7.48 gallons, how many gallons of water are in Lake Superior?

61. The average distance from the Earth to the Sun is 1.5×10^{11} meters. The distance from the Sun to Proxima Centauri, the next closest star, is 3.99×10^{16} meters. The most distant stars visible to the unaided eye are 2000 times as far away as Proxima Centauri.
     a. How many times farther is Proxima Centauri from the Sun than the Sun is from Earth?
     b. How far from the Sun are the most distant visible stars?

62. The radius of the Earth is 6.37×10^6 meters, and the radius of the Sun is 6.96×10^8 meters. The radii of the other stars range from 1% of the solar radius to 1000 times the solar radius.
     a. What fraction of the solar radius is Earth's radius?
     b. What is the range of stellar radii, in meters?

An **equation** is just a mathematical statement that two expressions are equal. Equations relating two variables are particularly useful. If we know the value of one of the variables, we can find the corresponding value of the other variable by solving the equation.

EXAMPLE 1

The equation $w = 6h$ gives Loren's wages, w, in terms of the number of hours she works, h. How many hours does Loren need to work next week if she wants to earn $225?

Solution

We know that $w = 225$, and we would like to know the value of h. We substitute the value for w into our equation and then solve for h.

$$w = 6h \qquad \text{Substitute 225 for } w.$$
$$225 = 6h \qquad \text{Divide both sides by 6.}$$
$$\frac{225}{6} = \frac{6h}{6} \qquad \text{Simplify.}$$
$$37.5 = h$$

Loren must work 37.5 hours in order to earn $225. In reality, Loren will probably have to work for 38 hours, because most employers do not pay for portions of an hour's work. Thus, Loren needs to work for 38 hours.

To solve an equation we can generate simpler equations that have the same solutions. Equations that have identical solutions are called **equivalent equations.** For example,

$$3x - 5 = x + 3$$

and

$$2x = 8$$

are equivalent equations because the solution of each equation is 4. Often we can find simpler equivalent equations by undoing in reverse order the operations performed on the variable.

Solving Linear Equations

Linear, or first-degree, equations can be written so that every term is either a constant or a constant times the variable. The equations above are examples of linear equations. Recall the following rules for solving linear equations.

To Generate Equivalent Equations

1. We can add or subtract the *same* number on *both* sides of an equation.
2. We can multiply or divide *both* sides of an equation by the *same* number (except zero).

Applying either of these rules produces a new equation equivalent to the old one and thus preserves the solution. We use the rules to isolate the variable on one side of the equation.

EXAMPLE 2   Solve the equation $3x - 5 = x + 3$.

Solution   We first collect all the variable terms on one side of the equation, and the constant terms on the other side.

$$3x - 5 - x = x + 3 - x \qquad \text{Subtract } x \text{ from both sides.}$$
$$2x - 5 = 3 \qquad \text{Simplify.}$$
$$2x - 5 + 5 = 5 + 5 \qquad \text{Add 5 to both sides.}$$
$$2x = 8 \qquad \text{Simplify.}$$
$$\frac{2x}{2} = \frac{8}{2} \qquad \text{Divide both sides by 2.}$$
$$x = 4 \qquad \text{Simplify.}$$

The solution is 4. (You can check the solution by substituting 4 into the original equation to show that a true statement results.)

The following steps should enable you to solve any linear equation. Of course, you may not need all the steps for a particular equation.

To Solve a Linear Equation:

1. Simplify each side of the equation separately.
   a. Apply the distributive law to remove parentheses.
   b. Collect like terms.
2. By adding or subtracting appropriate terms to both sides of the equation, get all the variable terms on one side and all the constant terms on the other.
3. Divide both sides of the equation by the coefficient of the variable.

EXAMPLE 3   Solve $3(2x - 5) - 4x = 2x - (6 - 3x)$.

Solution   We begin by simplifying each side of the equation.

$$3(2x - 5) - 4x = 2x - (6 - 3x) \qquad \text{Apply the distributive law.}$$
$$6x - 15 - 4x = 2x - 6 + 3x \qquad \text{Combine like terms on each side.}$$
$$2x - 15 = 5x - 6$$

Next, we collect all the variable terms on the left side of the equation, and all the constant terms on the right side.

$$2x - 15 - 5x + 15 = 5x - 6 - 5x + 15 \qquad \text{Add } -5x + 15 \text{ to both sides.}$$
$$-3x = 9$$

Finally, we divide both sides of the equation by the coefficient of the variable.

$$-3x = 9 \qquad \text{Divide both sides by } -3.$$
$$x = -3$$

The solution is -3.

Formulas

A **formula** is an equation that relates several variables. For example, the equation

$$P = 2l + 2w$$

gives the perimeter of a rectangle in terms of its length and width.

   Suppose we have some wire fence to enclose an exercise area for rabbits, and we would like to see what dimensions are possible for different rectangles with that perimeter. In this case, it would be more useful to have a formula for the length of the rectangle in terms of its perimeter and its width. We can find such a formula by solving the perimeter formula for l in terms of P and w.

$$2l + 2w = P \qquad \text{Subtact 2 } w \text{ from both sides.}$$
$$2l = P - 2w \qquad \text{Divide both sides by 2.}$$
$$l = \frac{P - 2w}{2}$$

The result is a new formula that gives the length of a rectangle in terms of its perimeter and its width.

EXAMPLE 4

The formula $5F = 9C + 160$ relates the temperature in degrees Fahrenheit, F, to the temperature in degrees Celsius, C. Solve the formula for C in terms of F.

Solution We begin by isolating the term that contains C.

$$5F = 9C + 160 \qquad \text{Subtract 160 from both sides.}$$
$$5F - 160 = 9C \qquad \text{Divide both sides by 9.}$$
$$\frac{5F - 160}{9} = C$$

We can also write the formula for C in terms of F as $C = \frac{5}{9}F - \frac{160}{9}$.

EXAMPLE 5

Solve $\quad 3x - 5y = 40 \quad$ for y in terms of x.

Solution We isolate y on one side of the equation.

$$3x - 5y = 40 \qquad \text{Subtract 3} x \text{ from both sides.}$$
$$-5y = 40 - 3x \qquad \text{Divide both sides by } -5.$$
$$\frac{-5y}{-5} = \frac{40 - 3x}{-5} \qquad \text{Simplify both sides.}$$
$$y = -8 + \frac{3}{5}x$$

Linear Inequalities

The symbol $>$ is called an **inequality symbol,** and the statement $a > b$ is called an **inequality.** There are four inequality symbols:

$>$	is greater than
$<$	is less than
$\geq$	is greater than or equal to
$\leq$	is less than or equal to

Inequalities that include the symbols $>$ or $<$ are called **strict inequalities;** those that include $\geq$ or $\leq$ are called **nonstrict.**

If we multiply or divide both sides of an inequality by a negative number, the direction of the inequality must be reversed. For example, if we multiply both sides of the inequality

$$2 < 5$$

by -3, we get

$$-3(2) > -3(5) \quad \text{Change inequality symbol from } < \text{ to } >.$$
$$-6 > -15.$$

Because of this property, the rules for solving linear equations must be revised slightly for solving linear inequalities.

To Solve a Linear Inequality:

1. We may add or subtract the same number to both sides of an inequality without changing its solutions.
2. We may multiply or divide both sides of an inequality by a *positive* number without changing its solutions.
3. If we multiply or divide both sides of an inequality by a *negative* number, we must *reverse the direction of the inequality symbol.*

EXAMPLE 6   Solve the inequality $4 - 3x \geq -17.$

Solution   Use the rules above to isolate x on one side of the inequality.

$$4 - 3x \geq -17 \quad \text{Subtract 4 from both sides.}$$
$$-3x \geq -21 \quad \text{Divide both sides by } -3.$$
$$x \leq 7$$

Notice that we reversed the direction of the inequality when we divided by -3. Any number less than or equal to 7 is a solution of the inequality.

A **compound inequality** involves two inequality symbols.

EXAMPLE 7   Solve $4 \le 3x + 10 \le 16$.

Solution   We isolate x by performing the same operations on all three sides of the inequality.

$$4 \le 3x + 10 \le 16 \qquad \text{Subtract 10.}$$
$$-6 \le 3x \le 6 \qquad \text{Divide by 3.}$$
$$-2 \le x \le 2$$

The solutions are all numbers between -2 and 2, inclusive.

Interval Notation

The solutions of the inequality in Example 7 form an interval. An **interval** is a set that consists of all the real numbers between two numbers a and b.

The set $-2 \le x \le 2$ includes its endpoints -2 and 2, so we call it a **closed interval,** and we denote it by $[-2, 2]$ (see Figure A.1a). The square brackets tell us that the endpoints are included in the interval. An interval that does not include its endpoints, such as $-2 < x < 2$, is called an **open interval,** and we denote it with round brackets, $(-2, 2)$ (see Figure A.1b).

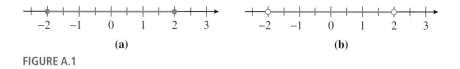

FIGURE A.1

CAUTION   Do not confuse the open interval $(-2, 2)$ with the point $(-2, 2)$! The notation is the same, so you must decide from the context whether an interval or a point is being discussed.

We can also discuss **infinite intervals,** such as $x < 3$ and $x \ge -1$, shown in Figure A.2. We denote the interval $x < 3$ by $(-\infty, 3)$, and the interval $x \ge -1$ by $[-1, \infty)$. The symbol ∞, for infinity, does not represent a specific real number; it indicates that the interval continues forever along the real line.

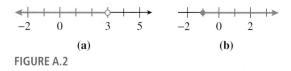

FIGURE A.2

Finally, we can combine two or more intervals into a larger set. For example, the set consisting of $x < -2$ or $x > 2$, shown in Figure A.3, is the **union** of two intervals and is denoted by $(-\infty, -2) \cup (2, \infty)$.

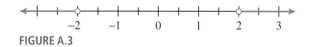

FIGURE A.3

Many solutions of inequalities are intervals or unions of intervals.

EXAMPLE 8 Write each of the solution sets with interval notation and graph the solution set on a number line.

a. $3 \le x < 6$ **b.** $x \ge -9$

c. $x \le 1$ or $x > 4$ **d.** $-8 < x \le -5$ or $-1 \le x < 3$

Solutions   **a.** $[3, 6)$. This is called a **half-open** or **half-closed** interval. (See Figure A.4.)

FIGURE A.4

b. $[-9, \infty)$. We always use round brackets next to the symbol ∞ because ∞ is not a specific number and is not included in the set. (See Figure A.5.)

FIGURE A.5

c. $(-\infty, 1] \cup (4, \infty)$. The word *or* describes the union of two sets. (See Figure A.6.)

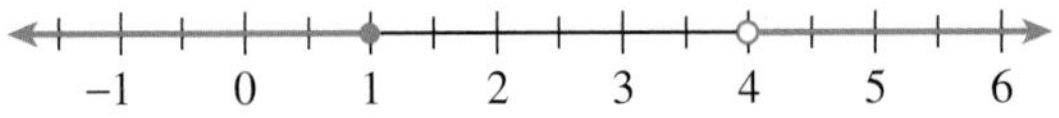

FIGURE A.6

d. $(-8, -5] \cup [-1, 3)$. (See Figure A.7.)

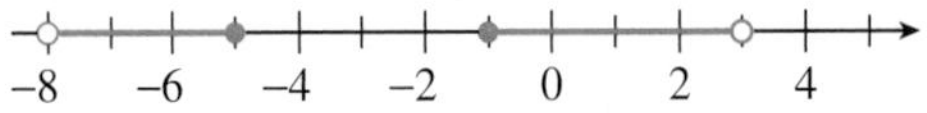

FIGURE A.7

SECTION A.2 SUMMARY

VOCABULARY   *Look up the definitions of new terms in the Glossary.*

Equation	Closed interval	Compound inequality	Inequality
Linear equation	Equivalent equation	Open interval	Interval
Strict inequality	Formula	Solve an equation	Union

SKILLS   *Practice each skill in the Exercises listed.*

1. Solve a linear equation: #1–10

2. Solve a formula for one variable in terms of the others: #11–28

3. Solve a linear inequality: #26–34

4. Solve a compound inequality: #35–41

5. Write solutions of inequalities in interval notation: #41–50

Solve each linear equation.

1. $3x + 5 = 26$

2. $2 + 5x = 37$

3. $3(z + 2) = 37$

4. $2(z - 3) = 15$

5. $3y - 2(y - 4) = 12 - 5y$

6. $5y - 3(y + 1) = 14 + 2y$

7. $0.8w - 2.6 = 1.4w + 0.3$

8. $4.8 - 1.3w = 0.7w + 2.1$

9. $0.25t + 0.10(t - 4) = 11.60$

10. $0.12t + 0.08(t + 10{,}000) = 12{,}000$

Solve for *y* in terms of *x*.

11. $4x + 3y = -2$

12. $x - 2y = -7$

13. $\dfrac{x}{8} - \dfrac{y}{2} = 1$

14. $\dfrac{x}{5} + \dfrac{y}{7} = 1$

15. $3x + \dfrac{2}{7}y = 1$

16. $\dfrac{5}{6}x + 8y = 1$

17. $-(x - 1) = 6(y - 3)$

18. $2y - 4 = 3(x + 5)$

19. $\dfrac{y + 8}{x - 1} = \dfrac{-7}{4}$

20. $\dfrac{2}{3} = \dfrac{y - 5}{x + 2}$

Solve each formula for the specified variable.

21. $v = k + gt$,   for t

22. $S = 3\pi d + \pi a$,   for d

23. $S = 2w(w + 2h)$,   for h

24. $A = P(1 + rt)$,   for r

25. $P = a + (n - 1)d$,   for n

26. $R = 2d + h(a + b)$,   for b

27. $A = \pi rh + \pi r^2$,   for h

28. $A = 2w^2 + 4lw$,   for l

Solve each inequality.

29. $3x - 2 > 1 + 2x$

30. $2x + 3 \leq x - 1$

31. $\dfrac{-2x - 6}{-3} > 2$

32. $\dfrac{-2x - 3}{2} \leq -5$

33. $\dfrac{2x - 3}{3} \leq \dfrac{3x}{-2}$

34. $\dfrac{3x - 4}{-2} > \dfrac{-2x}{5}$

35. $-6 < 4x + 10 < 20$

36. $3 < -2x - 5 < 15$

37. $-9 \leq -3x + 6 < 2$

38. $4 < 8x + 12 \leq 16$

39. $5 < \dfrac{8 - 2x}{4} \leq 7$

40. $-1 \leq \dfrac{4x - 6}{-3} \leq 0$

Write each set with interval notation, and graph the set on a number line.

41. $-5 < x \leq 3$

42. $0 \leq x < 4$

43. $0 \geq x \geq -4$

44. $8 > x > 5$

45. $x > -6$

46. $x \leq 1$

47. $x < -3$ or $x \geq -1$

48. $x \geq 3$ or $x \leq -3$

49. $-6 \leq x < -4$ or $-2 < x \leq 0$

50. $x < 2$ or $2 < x < 3$

You are familiar with the use of letters, or **variables,** to stand for unknown numbers in equations or formulas. Variables are also used to represent numerical quantities that change over time or in different situations. For example, p might stand for the atmospheric pressure at different heights above the Earth's surface. Or N might represent the number of people infected with cholera t days after the start of an epidemic.

An **algebraic expression** is any meaningful combination of numbers, variables, and symbols of operation. Algebraic expressions are used to express relationships between variable quantities.

EXAMPLE 1

Loren makes $6 an hour working at the campus bookstore.

a. Choose a variable for the number of hours Loren works per week.

b. Write an algebraic expression for the amount of Loren's weekly earnings.

Solutions

a. Let h stand for the number of hours Loren works per week.

b. The amount Loren earns is given by

$$6 \times \text{(number of hours Loren worked)}$$

or $6 \cdot h$. Loren's weekly earnings can be expressed as $6h$.

The algebraic expression $6h$ represents the amount of money Loren earns *in terms of* the number of hours she works. If we substitute a specific value for the variable in an expression, we find a numerical value for the expression. This is called **evaluating** the expression.

EXAMPLE 2

If Loren from Example 1 works for 16 hours in the bookstore this week, how much will she earn?

Solution

Evaluate the expression $6h$ for $h = 16$.

$$6h = 6(16) = 96$$

Loren will make $96.

EXAMPLE 3

April sells environmentally friendly cleaning products. Her income consists of $200 per week plus a commission of 9% of her sales.

a. Choose variables to represent the unknown quantities and write an algebraic expression for April's weekly income in terms of her sales.

b. Find April's income for a week in which she sells $350 worth of cleaning products.

Solutions

a. Let I represent April's total income for the week, and let S represent the total amount of her sales. We translate the information from the problem into mathematical language as follows:

> *Her income consists of $200 . . . plus . . . 9% of her sales*
> $I \quad = \quad 200 \quad + \quad 0.09 \quad S$

Thus, $I = 200 + 0.09S$.

b. We want to evaluate our expression from part (a) with $S = 350$. We substitute 350 for S to find

$$I = 200 + 0.09(350)$$

Following the order of operations, we perform the multiplication before the addition. Thus, we begin by computing 0.09(350).

$$
\begin{aligned}
I &= 200 + 0.09(350) && \text{Multiply 0.09 (350) first.}\\
&= 200 + 31.5 && \text{Add.}\\
&= 231.50
\end{aligned}
$$

April's income for the week is $231.50.

CALCULATOR TIP

On a scientific or a graphing calculator, we can enter the expression from Example 1b just as it is written:

$$200 \boxed{+} \, 0.09 \boxed{\times} \, 350 \boxed{\text{ENTER}}$$

The calculator will perform the operations in the correct order—multiplication first.

EXAMPLE 4

Economy Parcel Service charges $2.80 per pound to deliver a package from Pasadena to Cedar Rapids. Andrew wants to mail a painting that weighs 8.3 pounds, plus whatever packing material he uses.

a. Choose variables to represent the unknown quantities and write an expression for the cost of shipping Andrew's painting.

b. Find the shipping cost if Andrew uses 2.9 pounds of packing material.

Solutions

a. Let C stand for the shipping cost and let w stand for the weight of the packing material. Andrew must find the total weight of his package *first,* then multiply by the shipping charge. The total weight of the package is $8.3 + w$ pounds. We use parentheses around this expression to show that it should be computed first, and the sum should be multiplied by the shipping charge of $2.80 per pound. Thus,

$$C = 2.80(8.3 + w)$$

b. Evaluate the formula from part (a) with $w = 2.9$.

$$
\begin{aligned}
C &= 2.80(8.3 + 2.9) && \text{Add inside parentheses.}\\
&= 2.80(11.2) && \text{Multiply.}\\
&= 31.36
\end{aligned}
$$

The cost of shipping the painting is $31.36.

CALCULATOR TIP

On a calculator, we enter the expression for C in the order it appears, including the parentheses. (Experiment to see whether your calculator requires you to enter the $\boxed{\times}$ symbol after 2.80.) The keying sequence

$$2.80 \boxed{\times} \boxed{(} 8.3 \boxed{+} 2.9 \boxed{)} \boxed{\text{ENTER}}$$

gives the correct result, 31.36.

If we omit the parentheses, the calculator will perform the multiplication before the addition. Thus, the keying sequence

$$2.80 \;\boxed{\times}\; 8.3 \;\boxed{+}\; 2.9$$

gives an incorrect result for Example 2b. (The sequence

$$8.3 \;\boxed{+}\; 2.9 \;\boxed{\times}\; 2.80$$

does not work either!)

Problem Solving

Problem solving often involves translating a real-life problem into a computer programming language, or, in our case, into algebraic expressions. We can then use algebra to solve the mathematical problem and interpret the solution in the context of the original problem. Here are some guidelines for problem solving with algebraic equations.

> **Guidelines for Problem Solving**
>
> **Step 1** Identify the unknown quantity and assign a variable to represent it.
>
> **Step 2** Find some quantity that can be expressed in two different ways and write an equation.
>
> **Step 3** Solve the equation.
>
> **Step 4** Interpret your solution to answer the question in the problem.

In step 1, begin by writing an English phrase to describe the quantity you are looking for. Be as specific as possible—if you are going to write an equation about this quantity, you must understand its properties! Remember that your variable must represent a *numerical* quantity. For example, x can represent the *speed* of a train, but not just "the train."

Writing an equation is the hardest part of the problem. Note that the quantity mentioned in step 2 will probably *not* be the same unknown quantity you are looking for, but the algebraic expressions you write *will* involve your variable. For example, if your variable represents the *speed* of a train, your equation might be about the *distance* the train traveled.

Supply and Demand

The law of supply and demand is fundamental in economics. If you increase the price of a product, the supply increases because its manufacturers are willing to provide more of the product, but the demand decreases because consumers are not willing to buy as much at a higher price. The price at which the demand for a product equals the supply is called the *equilibrium price*.

EXAMPLE 5   The Coffee Connection finds that when it charges p dollars for a pound of coffee, it can sell $800 - 60p$ pounds per month. On the other hand, at a price of p dollars a pound, International Food and Beverage will supply the Connection with $175 + 40p$ pounds of coffee per month. What price should the Coffee Connection charge for a pound of coffee so that its monthly inventory will sell out?

Solution

Step 1   We are looking for the equilibrium price, p.

Step 2   The Coffee Connection would like the demand for its coffee to equal its supply. We equate the expressions for supply and for demand to obtain the equation

$$800 - 60p = 175 + 40p$$

Step 3   Solve the equation. To get all terms containing the variable, p, on one side of the equation, we add $60p$ to both sides and subtract 175 from both sides to obtain

$$800 - 60p + 60p - 175 = 175 + 40p + 60p - 175$$
$$625 = 100p \qquad \text{Divide both sides by 100.}$$
$$6.25 = p$$

Step 4   The Coffee Connection should charge $6.25 per pound for its coffee.

Percent Problems

Recall the basic formula for computing percents.

> **Percent Formula**
>
> $$P = rW$$
>
> the **P**art (or percent) $=$ the percentage **r**ate $\times$ the **W**hole Amount

A *percent increase* or *percent decrease* is calculated as a fraction of the *original* amount. For example, suppose you make $6.00 an hour now, but next month you are expecting a 5% raise. Your new salary should be

$$\underset{\text{Original salary}}{\$6.00} \quad + \quad \underset{\text{Increase}}{0.05(\$6.00)} \quad = \quad \underset{\text{New salary}}{\$6.30}$$

EXAMPLE 6   The price of housing in urban areas increased 4% over the past year. If a certain house costs $100,000 today, what was its price last year?

Solution

Step 1   Let c represent the cost of the house last year.

Step 2   Express the current price of the house in two different ways. During the past year, the price of the house increased by 4%, or $0.04c$. Its current price is thus

$$\underset{\text{Original cost}}{(1)c} \quad + \quad \underset{\text{Price increase}}{0.04c} \quad = \quad c(1 + 0.04) = 1.04c$$

This expression is equal to the value given for current price of the house:

$$1.04c = 100,000$$

Step 3   To solve this equation, we divide both sides by 1.04 to find

$$c = \frac{100,000}{1.04} = 96,153.846$$

Step 4   To the nearest cent, the cost of the house last year was $96,153.85.

In Example 3, it would be incorrect to calculate last year's price by subtracting 4% of $100,000 from $100,000 to get $96,000. (Do you see why?)

Weighted Averages

We find the *average,* or *mean,* of a set of values by adding up the values and dividing the sum by the number of values. Thus, the average, $\overline{x}$, of the numbers $x_1, x_2, \ldots, x_n$ is given by

$$\overline{x} = \frac{x_1 + x_2 + \cdots + x_n}{n}$$

In a **weighted average,** the numbers being averaged occur with different frequencies or are weighted differently in their contribution to the average value. For instance, suppose a biology class of 12 students takes a 10-point quiz. Of the 12 students, 2 receive 10s, three receive 9s, 5 receive 8s, and 2 receive scores of 6. The average score earned on the quiz is then

$$\overline{x} = \frac{2(10) + 3(9) + 5(8) + 2(6)}{12} = 8.25$$

The numbers in color are called the *weights*—in this example they represent the number of times each score was counted. Note that n, the total number of scores, is equal to the sum of the weights:

$$12 = 2 + 3 + 5 + 2$$

EXAMPLE 7   Kwan's grade in his accounting class will be computed as follows: Tests count for 50% of the grade, homework counts for 20%, and the final exam counts for 30%. If Kwan has an average of 84 on tests and 92 on homework, what score does he need on the final exam to earn a grade of 90?

Solution

Step 1   Let x represent the final exam score Kwan needs.

Step 2   Kwan's grade is the weighted average of his test, homework, and final exam scores.

$$\frac{0.50(84) + 0.20(92) + 0.30x}{1.00} = 90$$

(The sum of the weights is 1.00, or 100% of Kwan's grade.) Multiply both sides of the equation by 1.00 to get

$$0.50(84) + 0.20(92) + 0.30x = 1.00(90)$$

Step 3   Solve the equation. Simplify the left side first.

$$60.4 + 0.30x = 90 \qquad \text{Subtract 60.4 from both sides.}$$
$$0.30x = 29.6 \qquad \text{Divide both sides by 0.30.}$$
$$x = 98.7$$

Step 4   Kwan needs a score of 98.7 on the final exam to earn a grade of 90.

In step 2 of Example 5, we rewrote the formula for a weighted average in a simpler form.

This form is particularly useful for solving problems involving mixtures.

EXAMPLE 8 The vet advised Delbert to feed his dog Rollo with kibble that is no more than 8% fat. Rollo likes JuicyBits, which are 15% fat. LeanMeal is more expensive, but it is only 5% fat. How much LeanMeal should Delbert mix with 50 pounds of JuicyBits to make a mixture that is 8% fat?

Solution

Step 1 Let p represent the number of pounds of LeanMeal needed.

Step 2 In this problem, we want the weighted average of the fat contents in the two kibbles to be 8%. The weights are the number of pounds of each kibble we use. It is often useful to summarize the given information in a table.

	% fat	Total pounds	Pounds of fat
Juicy Bits	15%	50	0.15(50)
LeanMeal	5%	p	0.05p
Mixture	8%	$50 + p$	0.08(50 + p)

The amount of fat in the mixture must come from adding the amounts of fat in the two ingredients. This gives us an equation,

$$0.15(50) + 0.05p = 0.08(50 + p)$$

This equation is an example of the formula for weighted averages.

Step 3 Simplify each side of the equation, then solve.

$$7.5 + 0.05p = 4 + 0.08p$$
$$3.5 = 0.03p$$
$$p = 116.\overline{6}$$

Step 4 Delbert should mix $116\frac{2}{3}$ pounds of LeanMeal with 50 pounds of JuicyBits to make a mixture that is 8% fat.

SECTION A.3 SUMMARY

VOCABULARY *Look up the definitions of new terms in the Glossary.*

Variable	Weighted average	Demand	Equilibrium price
Supply	Algebraic expression	Evaluate an expression	

EXERCISES A.3

ThomsonNOW™ Test yourself on key content at **www.thomsonedu.com/login.**

Write algebraic expressions to describe each situation and then evaluate for the given values.

1. Jim was 27 years old when Ana was born.
   a. Write an expression for Jim's age in terms of Ana's age.
   b. Use your expression to find Jim's age when Ana is 22 years old.

2. Rani wants to replace the wheels on her in-line skates. New wheels cost $6.59 each.
   a. Write an expression for the total cost of new wheels in terms of the number of wheels Rani must replace.
   b. Use your expression to find the total cost if Rani must replace 8 wheels.

3. Helen decides to drive to visit her father. The trip is a distance of 1260 miles.
   a. Write an expression for the total number of hours Helen must drive in terms of her average driving speed.
   b. Use your expression to find how long Helen must drive if she averages 45 miles per hour.

4. Ben will inherit one million dollars on his twenty-first birthday.
   a. Write an expression for the number of years before Ben gets his inheritance in terms of his present age.
   b. Use your expression to calculate how many more years Ben must wait after he turns 13 years old.

5. The area of a circle is equal to π times the square of its radius.
   a. Write an expression for the area of a circle in terms of its radius.
   b. Find the area of a circle whose radius is 5 centimeters.

6. The volume of a sphere is equal to $\frac{4}{3}\pi$ times the cube of the radius.
   a. Write an expression for the volume of a sphere in terms of its radius.
   b. Find the volume of a sphere whose radius is 5 centimeters.

7. The sales tax in the city of Preston is 7.9%.
   a. Write an expression for the total bill for an item (price plus tax) in terms of the price of the item.
   b. Find the total bill for an item whose price is $490.

8. A savings account pays 6.4% annual interest on the amount deposited.
   a. Write an expression for the balance (initial deposit plus interest) in the account after one year in terms of the amount deposited.
   b. Find the total amount in the account after one year if $350 was deposited.

9. Your best friend moves to another state. To call her, a long-distance phone call costs $1.97 plus $0.39 for each minute.
   a. Write an expression for the cost of a long-distance phone call in terms of the number of minutes of the call.
   b. Find the cost of a 27-minute phone call.

10. Arenac Airlines charges 47 cents per pound on its flight from Omer to Pinconning, both for passengers and for luggage. Mr. Owsley wants to take the flight with 15 pounds of luggage.
    a. Write an expression for the cost of the flight in terms of Mr. Owsley's weight.
    b. Find the cost if Mr. Owsley weighs 162 pounds.

11. Juan buys a 50-pound bag of rice and consumes about 0.4 pound per week.
    a. Write an expression for the amount of rice Juan has consumed in terms of the number of weeks since he bought the bag.
    b. Write an expression for the amount of rice Juan has left in terms of the number of weeks since he bought the bag.
    c. Find the amount of rice Juan has left after 6 weeks.

Write and solve an equation to answer the question.

13. Celine's boutique carries a line of jewelry made by a local artists' co-op. If Celine charges p dollars for a pair of earrings, she finds that she can sell $200 - 5p$ pairs per month. On the other hand, the co-op will provide her with $56 + 3p$ pairs of earrings when she charges p dollars per pair. What price should Celine charge so that the demand for earrings will equal her supply?

15. Roger sets out on a bicycle trip at an average speed of 16 miles per hour. Six hours later, his wife finds his patch kit on the dining room table. If she heads after him in the car at 45 miles per hour, how long will it be before she catches him?
    a. What are we asked to find in this problem? Assign a variable to represent it.
    b. Write an expression in terms of your variable for the distance Roger's wife drives.
    c. Write an expression in terms of your variable for the distance Roger has cycled.
    d. Write an equation and solve it.

17. The reprographics department has a choice of 2 new copying machines. One sells for $20,000 and costs $0.02 per copy to operate. The other sells for $17,500, but its operating costs are $0.025 per copy. The repro department decides to buy the more expensive machine. How many copies must the repro department make before the higher price is justified?
    a. What are we asked to find in this problem? Assign a variable to represent it.
    b. Write expressions in terms of your variable for the total cost incurred by each machine.
    c. Write an equation and solve it.

19. The population of Midland has been growing at an annual rate of 8% over the past 5 years. Its present population is 135,000.
    a. Assuming the same rate of growth, what do you predict for the population of Midland next year?
    b. What was the population of Midland last year?

12. Trinh is bicycling down a mountain road that loses 500 feet in elevation for each 1 mile of road. She started at an elevation of 6300 feet.
    a. Write an expression for the elevation Trinh has lost in terms of the distance she has cycled.
    b. Write an expression for Trinh's elevation in terms of the number of miles she has cycled.
    c. Find Trinh's elevation after she has cycled 9 miles.

14. Curio Electronics sells garage door openers. If it charges p dollars per unit, it sells $120 - p$ openers per month. The manufacturer will supply $20 + 2p$ openers at a price of p dollars each. What price should Curio Electronics charge so that its monthly supply will meet its demand?

16. Kate and Julie set out in their sailboat on a straight course at 9 miles per hour. Two hours later, their mother becomes worried and sends their father after them in the speedboat. If their father travels at 24 miles per hour, how long will it be before he catches them?
    a. What are we asked to find in this problem? Assign a variable to represent it.
    b. Write an expression in terms of your variable for the distance Kate and Julie sailed.
    c. Write an expression in terms of your variable for the distance their father traveled.
    d. Write an equation and solve it.

18. Annie needs a new refrigerator and can choose between two models of the same size. One model sells for $525 and costs $0.08 per hour to run. A more energy-efficient model sells for $700 but runs for $0.05 per hour. If Annie buys the more expensive model, how long will it be before she starts saving money?
    a. What are we asked to find in this problem? Assign a variable to represent it.
    b. Write expressions in terms of your variable for the total cost incurred by each refrigerator.
    c. Write an equation and solve it.

20. For the past 3 years, the annual inflation rate has been 6%. This year, a steak dinner at Benny's costs $12.
    a. Assuming the same rate of inflation, what do you predict for the price of a steak dinner next year?
    b. What did a steak dinner cost last year?

21. Virginia took a 7% pay cut when she changed jobs last year. What percent pay increase must she receive this year in order to match her old salary of $24,000? (*Hint:* What was Virginia's salary after the pay cut?)

22. Clarence W. Networth took a 16% loss in the stock market last year. What percent gain must he realize this year in order to restore his original holdings of $85,000? (*Hint:* What was the value of Clarence's stock holdings after the loss?)

23. Delbert's test average in algebra is 77. If the final exam counts for 30% of the grade and the test average counts for 70%, what must Delbert score on the final exam to have a term average of 80?

24. Harold's batting average for the first 8 weeks of the baseball season is 0.385. What batting average must he maintain over the last 18 weeks so that his season average will be 0.350 (assuming he continues the same number of at-bats per week)?

25. A horticulturist needs a fertilizer that is 8% potash, but she can find only fertilizers that contain 6% and 15% potash. How much of each should she mix to obtain 10 pounds of 8% potash fertilizer?

Pounds of fertilizer	% potash	Pounds of potash

a. What are we asked to find in this problem? Assign a variable to represent it.
b. Write algebraic expressions in terms of your variable for the amounts of each fertilizer the horticulturist uses. Use the table.
c. Write expressions for the amount of potash in each batch of fertilizer.
d. Write two different expressions for the amount of potash in the mixture. Now write an equation and solve it.

26. A sculptor wants to cast a bronze statue from an alloy that is 60% copper. He has 30 pounds of a 45% alloy. How much 80% copper alloy should he mix with it to obtain the 60% copper alloy?

Pounds of alloy	% copper	Pounds of copper

a. What are we asked to find in this problem? Assign a variable to represent it.
b. Write algebraic expressions in terms of your variable for the amounts of each alloy the sculptor uses. Use the table.
c. Write expressions in terms of your variable for the amount of copper in each batch of alloy.
d. Write two different expressions for the amount of copper in the mixture. Now write an equation and solve it.

27. Lacy's Department Stores wants to keep the average salary of its employees under $19,000 per year. If the downtown store pays its 4 managers $28,000 per year and its 12 department heads $22,000 per year, how much can it pay its 30 clerks?
a. What are we asked to find in this problem? Assign a variable to represent it.
b. Write algebraic expressions for the total amounts Lacy's pays its managers, its department heads, and its clerks.
c. Write two different expressions for the total amount Lacy's pays in salaries each year.
d. Write an equation and solve it.

28. Federal regulations require that 60% of all vehicles manufactured next year comply with new emission standards. Major Motors can bring 85% of its small trucks in line with the standards, but only 40% of its automobiles. If Major Motors plans to manufacture 20,000 automobiles next year, how many trucks will it have to produce in order to comply with the federal regulations?
a. What are we asked to find in this problem? Assign a variable to represent it.
b. Write algebraic expressions for the number of trucks and the number of cars that will meet emission standards.
c. Write two different expressions for the total number of vehicles that will meet the standards.
d. Write an equation and solve it.

Graphs are useful tools for studying mathematical relationships. A graph provides an overview of a quantity of data, and it helps us identify trends or unexpected occurrences. Interpreting the graph can help us answer questions about the data.

For example, here are some data showing the atmospheric pressure at different altitudes. Altitude is given in feet, and atmospheric pressure is given in inches of mercury.

Altitude (ft)	0	5000	10,000	20,000	30,000	40,000	50,000
Pressure (in Hg)	29.7	24.8	20.5	14.6	10.6	8.5	7.3

We observe a generally decreasing trend in pressure as the altitude increases, but it is difficult to say anything more precise about this relationship. A clearer picture emerges if we plot the data. To do this, we use two perpendicular number lines called **axes.** We use the horizontal axis for the values of the first variable, altitude, and the vertical axis for the values of the second variable, pressure.

The entries in the table are called **ordered pairs,** in which the **first component** is the altitude and the **second component** is the atmospheric pressure measured at that altitude. For example, the first two entries can be represented by (0, 29.7) and (5000, 24.8). We plot the points whose **coordinates** are given by the ordered pairs, as shown in Figure A.8a.

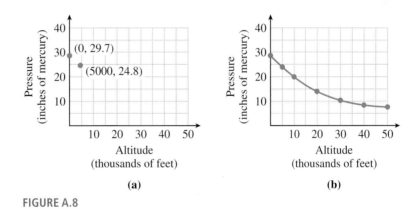

FIGURE A.8

We can connect the data points with a smooth curve as shown in Figure A.8b. In doing this, we are assuming that one variable changes smoothly with respect to the other, and in fact this is true for many physical situations. Thus, a smooth curve will thus serve as a good model.

Reading a Graph

Once we have constructed a graph, we can use it to estimate values of the variables between the known data points.

EXAMPLE 1

From the graph in Figure A.8b, estimate the following:

a. The atmospheric pressure measured at an altitude of 15,000 feet

b. The altitude at which the pressure is 12 inches of mercury

Solutions   **a.** The point with first coordinate 15,000 on the graph in Figure A.9 has second coordinate approximately 17.4. We estimate the pressure at 15,000 feet to be 17.4 inches of mercury.

b. The point on the graph with second coordinate 12 has first coordinate approximately 25,000, so an atmospheric pressure of 12 inches of mercury occurs at about 25,000 feet.

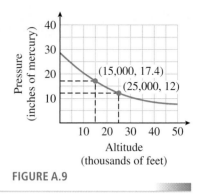

FIGURE A.9

We can also use the graph to obtain information about the relationship between altitude and pressure that would be difficult to see from the data alone.

EXAMPLE 2   **a.** For what altitudes is the pressure less than 18 inches of mercury?

b. How much does the pressure decrease as the altitude increases from 15,000 feet to 25,000 feet?

c. For which 10,000-foot increase in altitude does the pressure change most rapidly?

Solutions   **a.** From the graph in Figure A.8b, we see that the pressure has dropped to 18 inches of mercury at about 14,000 feet, and that it continues to decrease as the altitude increases. Therefore, the pressure is less than 18 inches of mercury for altitudes greater than 14,000 feet.

b. The pressure at 15,000 feet is approximately 17.4 inches of mercury, and at 25,000 feet it is 12 inches. This represents a decrease in pressure of 17.4 − 12, or 5.4, inches of mercury.

c. By studying the graph we see that the pressure decreases most rapidly at low altitudes, so we conclude that the greatest drop in pressure occurs between 0 and 10,000 feet.

Graphs of Equations

In Example 1, we used a graph to illustrate data given in a table. Graphs can also help us analyze models given by equations. Let us first review some facts about solutions of equations in two variables.

An equation in two variables, such as $y = 2x + 3$, is said to be *satisfied* if the variables are replaced by a pair of numbers that make the statement true. The pair of numbers is called a **solution** of the equation and is usually written as an ordered pair (x, y). (The first number in the pair is the value of x and the second number is the value of y.)

To find a solution of a given equation, we can assign a number to one of the variables and then solve for the second variable.

EXAMPLE 3   Find solutions to the equation $y = 2x + 3$.

Solution   We choose some values for x, say, -2, 0, and 1. Substitute these x-values into the equation to find a corresponding y-value for each.

$$\begin{array}{ll} \text{When } x = -2, & y = 2(-2) + 3 = -1 \\ \text{When } x = 0, & y = 2(0) + 3 = 3 \\ \text{When } x = 1, & y = 2(1) + 3 = 5 \end{array}$$

Thus, the ordered pairs $(-2, -1)$, $(0, 3)$, and $(1, 5)$ are three solutions of $y = 2x + 3$. We can also substitute values for y. For example, if we let $y = 10$, we have

$$10 = 2x + 3$$

Solving this equation for x, we find $7 = 2x$, or $x = 3.5$. This means that the ordered pair $(3.5, 10)$ is another solution of the equation $y = 2x + 3$.

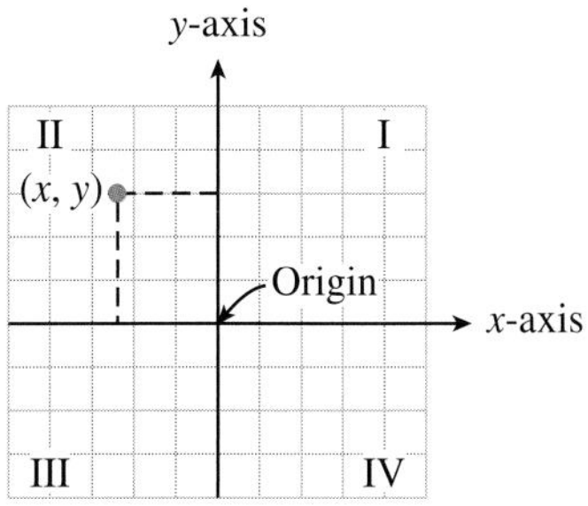

An equation in two variables may have infinitely many solutions, so we cannot list them all. However, we can display the solutions on a graph. For this we use a **Cartesian** (or **rectangular**) **coordinate system,** as shown in Figure A.10.

The graph of an equation is a picture of its solutions. A point is included in the graph if its coordinates satisfy the equation, and if the coordinates do not satisfy the equation, the point is not part of the graph. A graph of $y = 2x + 3$ is shown in Figure A.11.

This graph does not display *all* the solutions of the equation, but it shows important features such as the intercepts on the x- and y-axes. Because there is a solution corresponding to every real number x, the graph extends infinitely in either direction, as indicated by the arrows.

FIGURE A.10

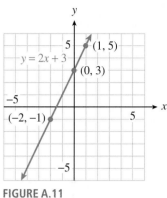

FIGURE A.11

EXAMPLE 4

Use the graph of $y = 0.5x^2 - 2$ in Figure A.12 to decide whether the given ordered pairs are solutions of the equation. Verify your answers algebraically.

a. $(-4, 6)$ **b.** $(3, 0)$

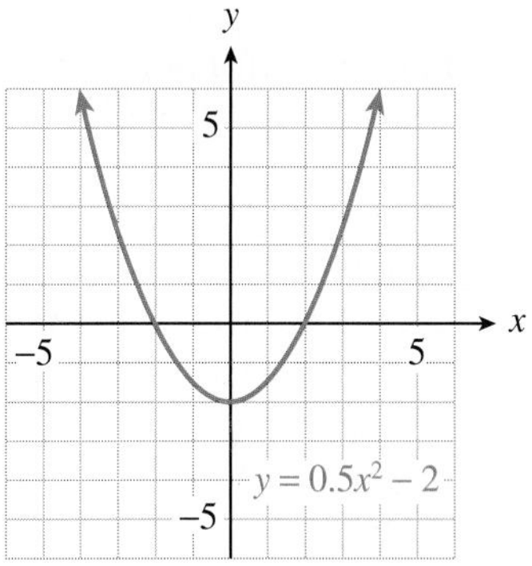

Solutions

a. Because the point $(-4, 6)$ does lie on the graph, the ordered pair $x = -4$, $y = 6$ is a solution of $y = 0.5x^2 - 2$. We can verify this by substituting -4 for x and 6 for y:

$$0.5(-4)^2 - 2 = 0.5(16) - 2$$
$$= 8 - 2 = 6$$

FIGURE A.12

b. Because the point $(3, 0)$ does not lie on the graph, the ordered pair $x = 3$, $y = 0$ is not a solution of $y = 0.5x^2 - 2$. We substitute 3 for x and 0 for y to verify this.

$$0.5(3)^2 - 2 = 0.5(9) - 2$$
$$= 4.5 - 2 = 2.5 \neq 0$$

SECTION A.4 SUMMARY

VOCABULARY   *Look up the definitions of new terms in the Glossary.*

Ordered pair	Component	Cartesian coordinate system	Solution
Equation in two variables	Satisfy an equation	Coordinate	Axis
Graph			

1. Read values from a graph: #1–4
2. Find solutions to an equation in two variables: #5–8
3. Make a table of values from an equation: #9–12
4. Make a table of values from a graph: #13–16

5. Estimate values from a graph: #17–20
6. Use the Trace feature on a calculator: #21–24
7. Find solutions to an equation in two variables from a graph: #25–32

EXERCISES A.4

Answer the questions about each graph.

1. The graph shows the temperatures recorded during a winter day in Billings, Montana.

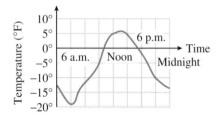

   a. What were the high and low temperatures recorded during the day?
   b. During what time intervals is the temperature above 5°F? Below −5°F?
   c. Estimate the temperatures at 7 a.m. and 2 p.m. At what time(s) is the temperature approximately 0°F? Approximately −12°F?
   d. How much did the temperature increase between 3 a.m. and 6 a.m.? Between 9 a.m. and noon? How much did the temperature decrease between 6 p.m. and 9 p.m.?
   e. During which 3-hour interval did the temperature increase most rapidly? Decrease most rapidly?

2. The graph shows the altitude of a commercial jetliner during its flight from Denver to Los Angeles.

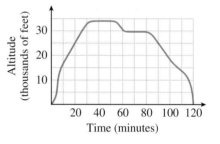

   a. What was the highest altitude the jet achieved? At what time(s) was this altitude recorded?
   b. During what time intervals was the altitude greater than 10,000 feet? Below 20,000 feet?
   c. Estimate the altitudes 15 minutes into the flight and 35 minutes into the flight. At what time(s) was the altitude approximately 16,000 feet? 32,000 feet?
   d. How many feet did the jet climb during the first 10 minutes of flight? Between 20 minutes and 30 minutes? How many feet did the jet descend between 100 minutes and 120 minutes?
   e. During which 10-minute interval did the jet ascend most rapidly? Descend most rapidly?

3. The graph shows the gas mileage achieved by an experimental model automobile at different speeds.

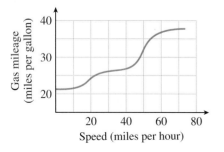

   a. Estimate the gas mileage achieved at 43 miles per hour.
   b. Estimate the speed at which a gas mileage of 34 miles per gallon is achieved.
   c. At what speed is the best gas mileage achieved? Do you think that the gas mileage will continue to improve as the speed increases? Why or why not?
   d. The data illustrated by the graph were collected under ideal test conditions. What factors might affect the gas mileage if the car were driven under more realistic conditions?

4. The graph shows the fish population of a popular fishing pond.

    a. During what months do the young fish hatch?

    b. During what months is fishing allowed?

    c. When does the park service restock the pond?

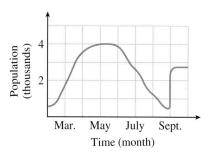

Find five solutions (ordered pairs) for each equation.

5. $y = 4 - \dfrac{x}{3}$

6. $\dfrac{x-5}{2} + 1 = y$

7. $3x^2 - 1 = y$

8. $y = 9 - (x - 2)^2$

Fill in the table of values for the given equation.

9. $3x + 2y = 1$

x	y
−3	
0	
	0
1	
	−2
	−4

10. $5y - 3x = 1$

x	y
−2	
	−1
	0
0	
1	
	3

11. $y = 1 - \dfrac{x}{4}$

x	y
−4	
	1
3	
	0
5	
	−1

12. $\dfrac{x+7}{3} = y$

x	y
−2	
0	
	3
5	
	5
	7

Fill in the table of values based on the graph.

13.

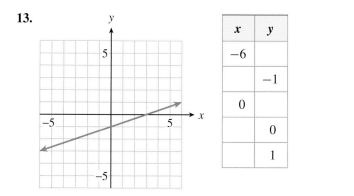

x	y
−6	
	−1
0	
	0
	1

14.

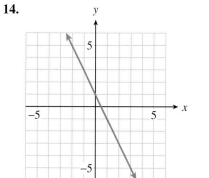

x	y
	5
−1	
0	
	−1
	−5

15.

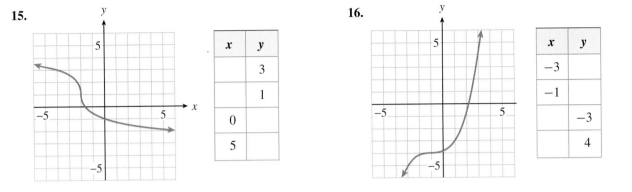

x	y
	3
	1
0	
5	

16.

x	y
-3	
-1	
	-3
	4

Estimate, based on the graph, any values of *x* with the given value of *y*.

17. $y = -3$

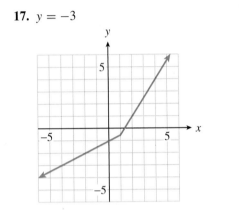

18. $y = -3$

19. $y = 0$

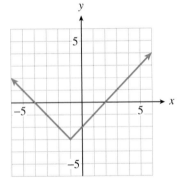

20. $y = -2$

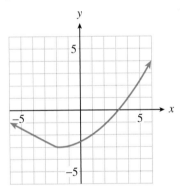

Graph the equation in the given friendly window. Use the calculator's *Trace* feature to make a table of values. (See Appendix B for help with entering expressions.) Round *y*-values to three decimal places.

21. $y = ||x + 2| - |x - 2||$,
  Xmin $= -4.7$, Xmax $= 4.7$;
  Ymin $= -6.2$, Ymax $= 6.2$

x	-3.2	-1.5	0.1	1.9		
y						

22. $y = |x^2 - x - 2|$,
  Xmin $= -4.7$, Xmax $= 4.7$;
  Ymin $= -9.3$, Ymax $= 9.3$

x	-3.1	-1.5	0.5	1.5		
y						

23. $y = \dfrac{x-2}{x+2}$,
Xmin = −4.7, Xmax = 4.7;
Ymin = −9.3, Ymax = 9.3

x	−3	−2.2	−2	4		
y						

24. $y = \sqrt{x^2 - 1.96}$,
Xmin = −4.7, Xmax = 4.7;
Ymin = −6.2, Ymax = 6.2

x	−3.0	−1.4	0.1	1.9		
y						

For Exercises 25–28,
(a) Use the graph to find the missing component in each solution of the equation.
(b) Verify your answers algebraically.

25. $s = 2t + 4$

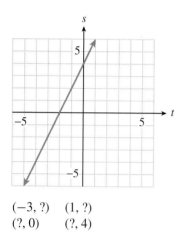

(−3, ?)   (1, ?)
(?, 0)   (?, 4)

26. $s = -2t + 4$

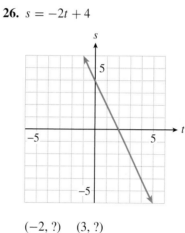

(−2, ?)   (3, ?)
(?, 0)   (?, 4)

27. $w = v^2 + 2$

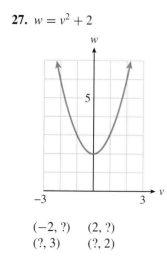

(−2, ?)   (2, ?)
(?, 3)   (?, 2)

28. $w = v^2 - 4$

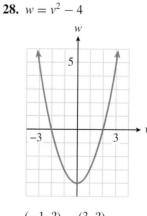

(−1, ?)   (3, ?)
(?, 0)   (?, −4)

29. $p = \dfrac{1}{m-1}$

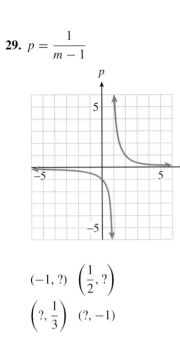

$(-1, ?)$ $\left(\dfrac{1}{2}, ?\right)$

$\left(?, \dfrac{1}{3}\right)$ $(?, -1)$

30. $p = \dfrac{1}{m+1}$

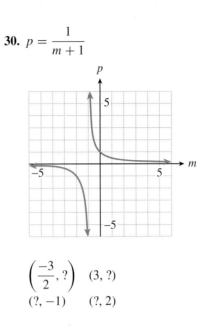

$\left(\dfrac{-3}{2}, ?\right)$ $(3, ?)$

$(?, -1)$ $(?, 2)$

31. $y = x^3$

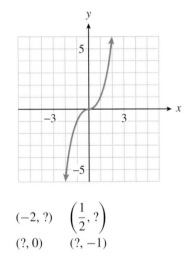

$(-2, ?)$ $\left(\dfrac{1}{2}, ?\right)$

$(?, 0)$ $(?, -1)$

32. $y = \sqrt{x+4}$

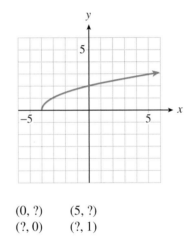

$(0, ?)$ $(5, ?)$

$(?, 0)$ $(?, 1)$

A.5   Linear Systems in Two Variables

A 2×2 **system** of equations is a set of 2 equations in the same 2 variables. A **solution** of a 2×2 system is an ordered pair that makes each equation in the system true. In this section, we review two algebraic methods for solving 2×2 linear systems: substitution and elimination.

Solving Systems by Substitution

The basic strategy for the **substitution** method can be described as follows.

> **Steps for Solving a 2 × 2 System by Substitution**
>
> 1. Solve one of the equations for one of the variables in terms of the other.
> 2. Substitute this expression into the second equation; doing so yields an equation in one variable.
> 3. Solve the new equation.
> 4. Use the result of step 1 to find the other variable.

EXAMPLE 1   Staci stocks two kinds of sleeping bags in her sporting goods store, a standard model and a down-filled model for colder temperatures. From past experience, she estimates that she will sell twice as many of the standard variety as of the down filled. She has room to stock 60 sleeping bags at a time. How many of each variety should Staci order?

Solution

Step 1   Number of standard sleeping bags: x

Number of down-filled sleeping bags: y

Step 2   Write two equations about the variables. Staci needs twice as many standard model as down filled, so

$$x = 2y \qquad (1)$$

Also, the total number of sleeping bags is 60, so

$$x + y = 60 \qquad (2)$$

These two equations give us a system.

Step 3   We will solve this system using substitution. Notice that Equation (1) is already solved for x in terms of y: $x = 2y$. Substitute $2y$ for x in Equation (2) to obtain

$$2y + y = 60$$
$$3y = 60$$

Solving for y, we find $y = 20$. Finally, substitute this value into Equation (1) to find

$$x = 2(20) = 40$$

The solution to the system is $x = 40$, $y = 20$.

Step 4   Staci should order 40 standard sleeping bags and 20 down-filled bags.

Solving Systems by Elimination

The method of substitution is convenient if one of the variables in the system has a coefficient of 1 or −1, because it is easy to solve for that variable. If none of the coefficients is 1 or −1, then a second method, called **elimination,** is usually more efficient.

The method of elimination is based on the following properties of linear equations.

Properties of Linear Systems

1. Multiplying a linear equation by a (nonzero) constant does not change its solutions. That is, any solution of the equation

$$ax + by = c$$

   is also a solution of the equation

$$kax + kby = kc$$

2. Adding (or subtracting) two linear equations does not change their common solutions. That is, any solution of the system

$$a_1x + b_1y = c_1$$
$$a_2x + b_2y = c_2$$

   is also a solution of the equation

$$(a_1 + a_2)\,x + (b_1 + b_2)\,y = c_1 + c_2$$

EXAMPLE 2 Solve the system by the method of elimination.

$$2x + 3y = 8 \qquad (1)$$
$$3x - 4y = -5 \qquad (2)$$

Solution We first decide which variable to eliminate, x or y. We can choose whichever looks easiest. In this problem, we choose to eliminate x. We next look for the smallest number that both coefficients, 2 and 3, divide into evenly. This number is 6. We want the coefficients of x to become 6 and -6, so we will multiply Equation (1) by 3 and Equation (2) by -2 to obtain

$$6x + 9y = 24 \qquad (1a)$$
$$-6x + 8y = 10 \qquad (2a)$$

Now add the corresponding terms of (1a) and (2a). The x-terms are eliminated, yielding an equation in one variable.

$$\begin{array}{r} 6x + 9y = 24 \\ -6x + 8y = 10 \\ \hline 17y = 34 \end{array} \qquad (3)$$

Solve this equation for y to find $y = 2$. We can substitute this value of y into any of our equations involving both x and y. If we choose Equation (1), then

$$2x + 3\,(2) = 8$$

and solving this equation yields $x = 1$. The ordered pair $(1, 2)$ is a solution to the system. You should verify that these values satisfy both original equations.

We summarize the strategy for solving a linear system by elimination.

> ### Steps for Solving a 2 × 2 Linear System by Elimination
>
> 1. Choose one of the variables to eliminate. Multiply each equation by a suitable factor so that the coefficients of that variable are opposites.
> 2. Add the two new equations termwise.
> 3. Solve the resulting equation for the remaining variable.
> 4. Substitute the value found in step 3 into either of the original equations and solve for the other variable.

In Example 2, we added 3 times the first equation to -2 times the second equation. The result from adding a constant multiple of one equation to a constant multiple of another equation is called a **linear combination** of the two equations. The method of elimination is also called the method of linear combinations.

If either equation in a system has fractional coefficients, it is helpful to clear the fractions before applying the method of linear combinations.

EXAMPLE 3   Solve the system by linear combinations.

$$\frac{2}{3}x - y = 2 \qquad (1)$$

$$x + \frac{1}{2}y = 7 \qquad (2)$$

Solution   Multiply each side of Equation (1) by 3 and each side of Equation (2) by 2 to clear the fractions:

$$2x - 3y = 6 \qquad (1a)$$
$$2x + \ y = 14 \qquad (2a)$$

To eliminate the variable x, multiply Equation (2a) by -1 and add the result to Equation (1a) to get

$$-4y = -8 \quad \text{Divide both sides by } -4.$$
$$y = 2$$

Substitute 2 for y in one of the original equations and solve for x. We use Equation (2).

$$x + \frac{1}{2}(2) = 7 \quad \text{Subtract 1 from both sides.}$$
$$x = 6$$

Verify that $x = 6$ and $y = 2$ satisfy both Equations (1) and (2). The solution to the system is the ordered pair $(6, 2)$.

VOCABULARY   *Look up the definitions of new terms in the Glossary.*

System of equations   Dependent   Linear combination   Inconsistent
Elimination   Solution of a system   Substitution

SKILLS   *Practice each skill in the Exercises listed.*

1. Solve a system by substitution: #1–4

2. Solve a system by elimination: #5–8

3. Choose a method and solve the system: #9–18, 29–32

4. Solve problems by writing and solving a system: #19–28

EXERCISES A.5

ThomsonNOW™ Test yourself on key content at **www.thomsonedu.com/login.**

Solve the system by substitution.

1. $a + 2b = -6$
$\quad 2a - 3b = 16$

2. $7r - 4s = 1$
$\quad 3r + s = 14$

3. $2x - 3y = 6$
$\quad x + 3y = 3$

4. $2r = s + 7$
$\quad 2s = 14 - 3r$

Solve the system by elimination.

5. $5x - 2y = -4$
$\quad -6x + 3y = 5$

6. $2p + 3q = 38$
$\quad 6p - 5q = 2$

7. $3x - 4y = -11$
$\quad 2x + 6y = -3$

8. $2u - 3v = -4$
$\quad 5u + 2v = 9$

In Problems 9–12, solve each system by substitution or by linear combinations.

9. $3m + n = 7$
$\quad 2m = 5n - 1$

10. $3x + 5y = 1$
$\quad 2x - 3y = 7$

11. $3y = 2x - 8$
$\quad 4y + 11 = 3x$

12. $4L - 3 = 3W$
$\quad 25 + 5L = -2W$

In Problems 13–18, clear the fractions in each equation first, then solve the system by substitution or by linear combinations.

13. $\dfrac{2}{3}A - B = 4$
$\quad A - \dfrac{3}{4}B = 6$

14. $\dfrac{1}{8}w - \dfrac{3}{8}z = 1$
$\quad \dfrac{1}{2}w - \dfrac{1}{4}z = -1$

15. $\dfrac{M}{4} = \dfrac{N}{3} - \dfrac{5}{12}$
$\quad \dfrac{N}{5} = \dfrac{1}{2} - \dfrac{M}{10}$

16. $\dfrac{R}{3} = \dfrac{S}{3} + 2$
$\quad \dfrac{S}{3} = \dfrac{R}{6} - 1$

17. $\dfrac{s}{2} = \dfrac{7}{6} - \dfrac{t}{3}$
$\quad \dfrac{s}{4} = \dfrac{3}{4} - \dfrac{t}{4}$

18. $\dfrac{2p}{3} + \dfrac{8q}{9} = \dfrac{4}{3}$
$\quad \dfrac{p}{3} = 2 + \dfrac{q}{2}$

In Problems 19–28, write a system of equations for each problem and solve algebraically.

19. Francine has $2000, part of it invested in bonds paying 10%, and the rest in a certificate account at 8%. Her annual income from the two investments is $184. How much did Francine invest at each rate?
   a. Choose variables for the unknown quantities, and fill in the table.

	Principal	Interest rate	Interest
Bonds			
Certificate			
Total		——	

   b. Write one equation about the amount Francine invested.
   c. Write a second equation about Francine's annual interest.
   d. Solve the system and answer the question in the problem.

20. Carmella has $1200 invested in two stocks; one returns 8% per year, and the other returns 12% per year. The income from the 8% stock is $3 more than the income from the 12% stock. How much did Carmella invest in each stock?
   a. Choose variables for the unknown quantities, and fill in the table.

	Principal	Interest rate	Interest
First stock			
Second stock			
Total		——	——

   b. Write one equation about the amount Carmella invested.
   c. Write a second equation about Carmella's annual interest.
   d. Solve the system and answer the question in the problem.

21. Paul needs 40 pounds of 48% silver alloy to finish a collection of jewelry. How many pounds of 45% silver alloy should he melt with 60% silver alloy to obtain the alloy he needs?
   a. Choose variables for the unknown quantities, and fill in the table.

	Pounds	% silver	Amount of silver
First alloy			
Second alloy			
Mixture			

   b. Write one equation about the amount of alloy Paul needs.
   c. Write a second equation about the amount of silver in the alloys.
   d. Solve the system and answer the question in the problem.

22. Amal plans to make 10 liters of a 17% acid solution by mixing a 20% acid solution with a 15% acid solution. How much of each should she use?
   a. Choose variables for the unknown quantities, and fill in the table.

	Liters	% acid	Amount of acid
First solution			
Second solution			
Mixture			

   b. Write one equation about the amount of solution Amal needs.
   c. Write a second equation about the amount of acid in the solutions.
   d. Solve the system and answer the question in the problem.

23. Delbert answered 13 true-false and 9 fill-in questions correctly on his last test and got a score of 71. If he had answered 9 true-false and 13 fill-ins correctly, he would have made an 83. How many points was each type of problem worth?

24. In a recent election, 7179 votes were cast for the two candidates. If 6 votes had been switched from the winner to the loser, the loser would have won by 1 vote. How many votes were cast for each candidate?

25. Because of prevailing winds, a flight from Detroit to Denver, a distance of 1120 miles, takes 4 hours on Econoflite, while the return trip takes 3.5 hours. What were the speed of the airplane and the speed of the wind?

    a. Choose variables for the unknown quantities, and fill in the table.

	Rate	Time	Distance
Detroit to Denver			
Denver to Detroit			

    b. Write one equation about the trip from Detroit to Denver.

    c. Write a second equation about the return trip.

    d. Solve the system and answer the question in the problem.

27. A cup of rolled oats provides 310 calories. A cup of rolled wheat flakes provides 290 calories. A new breakfast cereal combines wheat and oats to provide 302 calories per cup. How much of each grain does 1 cup of the cereal include?

    a. Choose variables for the unknown quantities and fill in the table.

	Cups	Calories per cup	Calories
Oat flakes			
Wheat flakes			
Mixture		—	

    b. Write one equation about the amounts of each grain.

    c. Write a second equation about the number of calories.

    d. Solve the system and answer the question in the problem.

26. On a breezy day, Bonnie propelled her human-powered aircraft 100 meters in 15 seconds going into the wind and made the return trip in 10 seconds with the wind. What were the speed of the wind and Bonnie's speed in still air?

    a. Choose variables for the unknown quantities, and fill in the table.

	Rate	Time	Distance
Against the wind			
With the wind			

    b. Write one equation about Bonnie's initial flight.

    c. Write a second equation Bonnie's return trip.

    d. Solve the system and answer the question in the problem.

28. Acme Motor Company is opening a new plant to produce chassis for two of its models, a sports coupe and a wagon. Each sports coupe requires a riveter for 3 hours and a welder for 4 hours; each wagon requires a riveter for 4 hours and a welder for 5 hours. The plant has available 120 hours of riveting and 155 hours of welding per day. How many of each model of chassis can it produce in a day?

    a. Choose variables for the unknown quantities and fill in the table.

	Sports coupes	Wagons	Total
Hours of riveting			
Hours of welding			

    b. Write one equation about the hours of riveting.

    c. Write a second equation about the hours of welding.

    d. Solve the system and answer the question in the problem.

Use a calculator to solve each system.

29. $4.8x - 3.5y = 5.44$
$2.7x + 1.3y = 8.29$

30. $6.4x + 2.3y = -14.09$
$-5.2x - 3.7y = -25.37$

31. $0.9x = 25.78 + 1.03y$
$0.25x + 0.3y = 85.7$

32. $0.02x = 0.6y - 78.72$
$1.1y = -0.4x + 108.3$

In this section, we review the rules for performing operations on powers.

Product of Powers

Consider a product of two powers with the same base.

$$(a^3)(a^2) = a\,a\,a \cdot a\,a = a^5$$

because a occurs as a factor five times. The number of a's in the product is the *sum* of the number of a's in each factor.

First Law of Exponents: Product of Powers

To multiply two powers with the same base, add the exponents and leave the base unchanged.

$$a^m \cdot a^n = a^{m+n}$$

EXAMPLE 1

                                      Add exponents.

a. $5^3 \cdot 5^4 = 5^{3+4} = 5^7$

Same base

                                        Add exponents.

b. $x^4 \cdot x^2 = x^{4+2} = x^6$

Same base

Here are some mistakes to avoid.

CAUTION

1. Note that we do not *multiply* the exponents when simplifying a product. For example,

$$b^4 \cdot b^2 \neq b^8$$

You can check this with your calculator by choosing a value for b, for instance, $b = 3$:

$$3^4 \cdot 3^2 \neq 3^8$$

2. In order to apply the first law of exponents, the bases must be the same. For example,

$$2^3 \cdot 3^5 \neq 6^8$$

(Check this on your calculator.)

3. We do not multiply the bases when simplifying a product. In Example 1a, note that

$$5^3 \cdot 5^4 \neq 25^7$$

4. Although we can simplify the product $x^2 x^3$ as x^5, we cannot simplify the *sum* $x^2 + x^3$, because x^2 and x^3 are not like terms.

EXAMPLE 2   Multiply $(-3x^4z^2)(5x^3y)$.

Solution   Rearrange the factors to group the numerical coefficients and the powers of each base. Apply the first law of exponents.

$$(-3x^4z^2)(5x^3z) = (-3)(5)x^4x^3z^2z$$
$$= -15x^7z^3$$

Quotients of Powers

To reduce a fraction, we divide both numerator and denominator by any common factors.

$$\frac{x^7}{x^4} = \frac{xxx\cancel{x}\cancel{x}\cancel{x}\cancel{x}}{\cancel{x}\cancel{x}\cancel{x}\cancel{x}} = \frac{x^3}{1} = x^3$$

We can obtain the same result more quickly by *subtracting* the exponent of the denominator from the exponent of the numerator.

$$\frac{x^7}{x^4} = x^{7-4} = x^3$$

What if the larger power occurs in the denominator of the fraction?

$$\frac{x^4}{x^7} = \frac{\cancel{x}\cancel{x}\cancel{x}\cancel{x}}{\cancel{x}\cancel{x}\cancel{x}\cancel{x}xxx} = \frac{1}{x^3}$$

In this case, we subtract the exponent of the numerator from the exponent of the denominator.

$$\frac{x^4}{x^7} = \frac{1}{x^{7-4}} = \frac{1}{x^3}$$

These examples suggest the following law.

Second Law of Exponents: Quotient of Powers

To divide two powers with the same base, subtract the smaller exponent from the larger one, keeping the same base.

a. If the larger exponent occurs in the numerator, put the power in the numerator.

$$\text{If } m > n, \quad \text{then} \quad \frac{a^m}{a^n} = a^{m-n} \quad (a \neq 0)$$

b. If the larger exponent occurs in the denominator, put the power in the denominator.

$$\text{If } m < n, \quad \text{then} \quad \frac{a^m}{a^n} = \frac{1}{a^{n-m}} \quad (a \neq 0)$$

EXAMPLE 3

a. $\dfrac{3^8}{3^2} = 3^{8-2} = 3^6$ Subtract exponents: $8 > 2$.

b. $\dfrac{w^3}{w^6} = \dfrac{1}{w^{6-3}} = \dfrac{1}{w^3}$ Subtract exponents: $3 < 6$.

744   Appendix A ■ Algebra Skills Refresher

EXAMPLE 4   Divide $\dfrac{3x^2y^4}{6x^3y}$.

Solution   Consider the numerical coefficients and the powers of each variable separately. Use the second law of exponents to simplify each quotient of powers.

$$\frac{3x^2y^4}{6x^3y} = \frac{3}{6} \cdot \frac{x^2}{x^3} \cdot \frac{y^4}{y}$$

$$= \frac{1}{2} \cdot \frac{1}{x^{3-2}} \cdot y^{4-1} \qquad \text{Subtract exponents.}$$

$$= \frac{1}{2} \cdot \frac{1}{x} \cdot \frac{y^3}{1} = \frac{y^3}{2x}$$

Power of a Power

Consider the expression $(a^4)^3$, the third power of a^4.

Add exponents.

$$(a^4)^3 = (a^4)(a^4)(a^4) = a^{4+4+4} = a^{12}$$

We can obtain the same result by multiplying the exponents together.

$$(a^4)^3 = a^{4 \cdot 3} = a^{12}$$

> **Third Law of Exponents: Power of a Power**
>
> To raise a power to a power, keep the same base and multiply the exponents.
>
> $$(a^m)^n = a^{mn}$$

EXAMPLE 5

Multiply exponents.

a. $(4^3)^5 = 4^{3 \cdot 5} = 4^{15}$

Multiply exponents.

b. $(y^5)^2 = y^{5 \cdot 2} = y^{10}$

CAUTION   Notice the difference between the expressions

$$(x^3)(x^4) = x^{3+4} = x^7$$

and

$$(x^3)^4 = x^{3 \cdot 4} = x^{12}$$

The first expression is a product, so we add the exponents. The second expression raises a power to a power, so we multiply the exponents.

Power of a Product

To simplify the expression $(5a)^3$, we use the associative and commutative laws to regroup the factors as follows.

$$(5a)^3 = (5a)(5a)(5a)$$
$$= 5 \cdot 5 \cdot 5 \cdot a \cdot a \cdot a$$
$$= 5^3 a^3$$

Thus, to raise a product to a power, we can simply raise each factor to the power.

Fourth Law of Exponents: Power of a Product

A power of a product is equal to the product of the powers of each of its factors.

$$(ab)^n = a^n b^n$$

EXAMPLE 6

a. $(5a)^3 = 5^3 a^3 = 125a^3$ Cube each factor.

b. $(-xy^2)^4 = (-x)^4 (y^2)^4$ Raise each factor to the fourth power.

         $= x^4 y^8$ Apply the third law of exponents.

CAUTION

1. Compare the two expressions $3a^2$ and $(3a)^2$; they are not the same. In the expression $3a^2$, only the factor a is squared. But in $(3a)^2$, both 3 and a are squared. Thus,

$$3a^2 \text{ cannot be simplified}$$

but

$$(3a)^2 = 3^2 a^2 = 9a^2$$

2. Compare the two expressions $(3a)^2$ and $(3+a)^2$. The fourth law of exponents applies to the *product* $3a$, but not to the *sum* $3 + a$. Thus,

$$(3+a)^2 \neq 3^2 + a^2$$

In order to simplify $(3+a)^2$, we must expand the binomial product:

$$(3+a)^2 = (3+a)(3+a) = 9 + 6a + a^2$$

Power of a Quotient

To simplify the expression $\left(\dfrac{x}{3}\right)^4$, we multiply together 4 copies of the fraction $\dfrac{x}{3}$.

$$\left(\frac{x}{3}\right)^4 = \frac{x}{3} \cdot \frac{x}{3} \cdot \frac{x}{3} \cdot \frac{x}{3} = \frac{x \cdot x \cdot x \cdot x}{3 \cdot 3 \cdot 3 \cdot 3}$$
$$= \frac{x^4}{3^4} = \frac{x^4}{81}$$

In general, we have the following rule.

Fifth Law of Exponents: Power of a Quotient

To raise a quotient to a power, raise both the numerator and denominator to the power.

$$\left(\frac{a}{b}\right)^n = \frac{a^n}{b^n} \qquad (b \neq 0)$$

For reference, we state all of the laws of exponents together. All the laws are valid when a and b are not equal to zero and when the exponents m and n are whole numbers.

Laws of Exponents

I. $a^m \cdot a^n = a^{m+n}$ **III.** $(a^m)^n = a^{mn}$

II. a. $\left(\dfrac{a^m}{a^n}\right) = a^{m-n} \quad (m > n)$ **IV.** $(ab)^n = a^m b^n$

b. $\left(\dfrac{a^m}{a^n}\right) = \dfrac{1}{a^{n-m}} \quad (m < n)$ **V.** $\left(\dfrac{a}{b}\right)^n = \dfrac{a^n}{b^n}$

EXAMPLE 7   Simplify $5x^2 y^3 (2xy^2)^4$.

Solution   According to the order of operations, we should perform any powers before multiplications. Thus, we begin by simplifying $(2xy^2)^4$. We apply the fourth law.

$$5x^2 y^3 (2xy^2)^4 = 5x^2 y^3 \cdot 2^4 x^4 (y^2)^4 \quad \text{Apply the fourth law.}$$
$$= 5x^2 y^3 \cdot 2^4 x^4 y^8$$

Finally, multiply powers with the same base. Apply the first law.

$$5x^2 y^3 \cdot 2^4 x^4 y^8 = 5 \cdot 2^4 x^2 x^4 y^3 y^8 = 80 x^6 y^{11}$$

EXAMPLE 8   Simplify $\left(\dfrac{2x}{z^2}\right)^3$.

Solution   Begin by applying the fifth law.

$$\left(\frac{2x}{z^2}\right)^3 = \frac{(2x)^3}{\left(z^2\right)^3} \quad \begin{array}{l}\text{Apply the fourth law to the numerator}\\ \text{and the third law to the denominato.}\end{array}$$

$$= \frac{2^3 x^3}{z^6} = \frac{8x^3}{z^6}$$

SECTION A.6 SUMMARY

VOCABULARY   *Look up the definitions of new terms in the Glossary.*

Exponent   Power

1. Apply the laws of exponents: #1–8
2. Simplify expressions: #9–16, 25–32

3. Multiply and divide powers: #17–24

EXERCISES A.6

ThomsonNOW™ Test yourself on key content at www.thomsonedu.com/login.

Simplify by applying the appropriate law of exponents.

1. a. $b^4 \cdot b^5$ **b.** $b^2 \cdot b^8$
   c. $(q^3)(q)(q^5)$ **d.** $(p^2)(p^4)(p^4)$

2. a. $\dfrac{w^6}{w^3}$ **b.** $\dfrac{c^{12}}{c^4}$
   c. $\dfrac{z^6}{z^9}$ **d.** $\dfrac{b^4}{b^8}$

3. a. $2^7 \cdot 2^2$ **b.** $6^5 \cdot 6^3$
   c. $\dfrac{2^9}{2^4}$ **d.** $\dfrac{8^6}{8^2}$

4. a. $(d^3)^5$ **b.** $(d^4)^2$
   c. $(5^4)^3$ **d.** $(4^3)^3$

5. a. $(6x)^3$ **b.** $(3y)^4$
   c. $(2t^3)^5$ **d.** $(6s^2)^2$

6. a. $\left(\dfrac{w}{2}\right)^6$ **b.** $\left(\dfrac{5}{u}\right)^4$
   c. $\left(\dfrac{-4}{p^5}\right)^3$ **d.** $\left(\dfrac{-3}{q^4}\right)^5$

7. a. $\left(\dfrac{h^2}{m^3}\right)^4$ **b.** $\left(\dfrac{n^3}{k^4}\right)^8$
   c. $(-4a^2b^4)^4$ **d.** $(-5ab^8)^3$

8. a. $\dfrac{ab^2}{(ab)^2}$ **b.** $\dfrac{(x^2y)^2}{x^2y^2}$
   c. $\dfrac{(2mp)^3}{2m^3p}$ **d.** $\dfrac{4^2rt^4}{2^4r^4t}$

Simplify if possible.

9. a. $w + w$
  b. $w(w)$

10. a. $m^2 - m^2$
   b. $m^2(-m^2)$

11. a. $4z^2 - 6z^2$
   b. $4z^2(-6z^2)$

12. a. $t^3 + 3t^3$
   b. $t^3(3t^3)$

13. a. $4p^2 + 3p^3$
   b. $4p^2(3p^3)$

14. a. $2w^2 - 5w^4$
   b. $(2w^2)(-5w^4)$

15. a. $3^9 \cdot 3^8$
   b. $3^9 + 3^8$

16. a. $(-2)^7(-2)^5$
   b. $-2^7 - 2^5$

Multiply.

17. a. $(4y)(-6y)$
   b. $(-4z)(-8z)$

18. a. $(2wz^3)(-8z)$
   b. $(4wz)(-9w^2z^2)$

19. a. $-4x(3xy)(xy^3)$
   b. $-5x^2(2xy)(5x^2)$

20. a. $-7ab^2(-3ab^3)$
   b. $-4a^2b(-3a^3b^2)$

Divide.

21 a. $\dfrac{2a^3b}{8a^4b^5}$ **b.** $\dfrac{8a^2b}{12a^5b^3}$

22 a. $\dfrac{-12qw^4}{8qw^2}$ **b.** $\dfrac{-12rz^6}{20rz}$

Multiply or divide.

23. a. $\dfrac{-15bc(b^2c)}{-3b^3c^4}$ **b.** $\dfrac{-25c(c^2d^2)}{-5c^8d^2}$

24. a. $-2x^3(x^2y)(-4y^2)$ **b.** $3xy^3(-x^4)(-2y^2)$

Simplify by applying the laws of exponents.

25. a. $b^3(b^2)^5$
 b. $b(b^4)^6$

26. a. $(p^2q)^3\,(pq^3)$
 b. $(p^3)^4\,(p^3q^4)$

27. a. $(2x^3y)^2\,(xy^3)^4$
 b. $(3xy^2)^3\,(2x^2y^2)^2$

28. a. $-a^2(-a)^2$
 b. $-a^3(-a)^3$

Simplify by applying the laws of exponents.

29. a. $\left(\dfrac{-2x}{3y^2}\right)^3$
 b. $\left(\dfrac{-x^2}{2y}\right)^4$

30. a. $\dfrac{(4x)^3}{(-2x^2)^2}$
 b. $\dfrac{(5x)^2}{(-3x^2)^3}$

31. a. $\dfrac{(xy)^2(-x^2y)^3}{(x^2y^2)^2}$
 b. $\dfrac{(-x)^2(-x^2)^4}{(x^2)^3}$

32. a. $\left(\dfrac{-2x}{y^2}\right)\left(\dfrac{y^2}{3x}\right)^2$
 b. $\left(\dfrac{x^2z}{2}\right)\left(\dfrac{-2}{x^2z}\right)^3$

A.7 Polynomials and Factoring

In Section A.6, we used the first law of exponents to multiply two or more monomials. In this section, we review techniques for multiplying and factoring polynomials of several terms.

Polynomials

A **polynomial** is a sum of terms in which all the exponents on the variables are whole numbers and no variables appear in the denominator or under a radical. The expressions

$$0.1\,R^4, \quad d^2 + 32\,d - 21, \quad \text{and} \quad 128x^3 - 960x^2 + 8000$$

are all examples of polynomials in one variable.

An algebraic expression consisting of one term of the form cx^n, where c is a constant and n is a whole number, is called a **monomial.** For example,

$$y^3, \quad -3x^8, \quad \text{and} \quad 0.1R^4$$

are monomials. A polynomial is just a sum of one or more monomials.

A polynomial with exactly two terms, such as $\frac{1}{2}n^2 + \frac{1}{2}n$, is called a **binomial.** A polynomial with exactly three terms, such as $d^2 + 32\,d - 21$ or $128x^3 - 960x^2 + 8000$, is called a **trinomial.** We have no special names for polynomials with more than three terms.

EXAMPLE 1 Which of the following expressions are polynomials?

a. πr^2

b. $23.4s^6 - 47.9s^4$

c. $\dfrac{2}{3}w^3 - \dfrac{7}{3}w^2 + \dfrac{1}{3}w$

d. $7 + m^{-2}$

e. $\dfrac{x-2}{x+2}$

f. $\sqrt[3]{4y}$

Solutions The first three are all polynomials. In fact, (a) is a monomial, (b) is a binomial, and (c) is a trinomial. The last three are not polynomials. The variable in (d) has a negative exponent, the variable in (e) occurs in the denominator, and the variable in (f) occurs under a radical.

In a polynomial containing only one variable, the greatest exponent that appears on the variable is called the **degree** of the polynomial. If there is no variable at all, then the polynomial is called a **constant,** and the degree of a constant is zero.

EXAMPLE 2 Give the degree of each polynomial.

a. $b^3 - 3b^2 + 3b - 1$ **b.** 10^{10}

c. $-4w^3$ **d.** $s^2 - s^6$

Solutions **a.** This is a polynomial in the variable b, and because the greatest exponent on b is 3, the degree of this polynomial is 3.

b. This is a constant polynomial, so its degree is 0. (The exponent on a *constant* does not affect the degree.)

c. This monomial has degree 3.

d. This is a binomial of degree 6.

We can evaluate a polynomial just as we evaluate any other algebraic expression: We replace the variable with a number and simplify the result.

EXAMPLE 3 Let $p(x) = -2x^2 + 3x - 1$. Evaluate each of the following.

a. $p(2)$ **b.** $p(-1)$ **c.** $p(t)$ **d.** $p(t + 3)$

Solutions In each case, we replace x by the given value.

a. $p(2) = -2(2)^2 + 3(2) - 1 = -8 + 6 - 1 = -3$

b. $p(-1) = -2(-1)^2 + 3(-1) - 1 = -2 + (-3) - 1 = -6$

c. $p(t) = -2(t)^2 + 3(t) - 1 = -2t^2 + 3t - 1$

d. $p(t + 3) = -2(t + 3)^2 + 3(t + 3) - 1$
$$= -2(t^2 + 6t + 9) + 3(t + 3) - 1$$
$$= -2t^2 - 9t - 10$$

Products of Polynomials

To multiply polynomials, we use a generalized form of the distributive property:

$$a(b + c + d + \cdots) = ab + ac + ad + \cdots$$

To multiply a polynomial by a monomial, we multiply each term of the polynomial by the monomial.

EXAMPLE 4 **a.** $3x(x + y + z) = 3x(x) + 3x(y) + 3x(z)$
$$= 3x^2 + 3xy + 3xz$$

b. $-2ab^2(3a^2 - ab + 2b^2) = -2ab^2(3a^2) - 2ab^2(-ab) - 2ab^2(2b^2)$
$$= -6a^3b^2 + 2a^2b^3 - 4ab^4$$

Products of Binomials

Products of binomials occur so frequently that it is worthwhile to learn a shortcut for this type of multiplication. We can use the following scheme to perform the multiplication mentally. (See Figure A.13.)

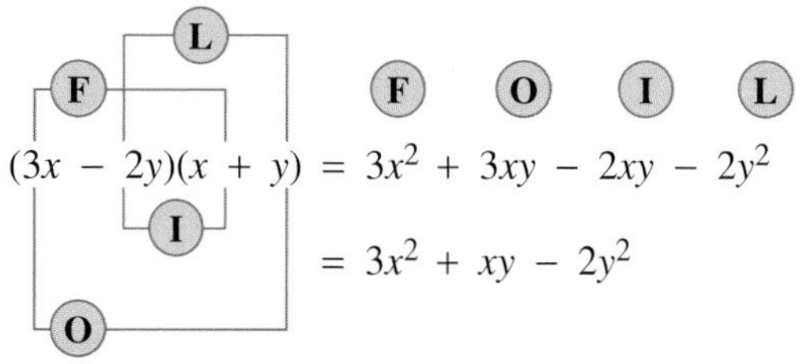

$$(3x - 2y)(x + y) = 3x^2 + 3xy - 2xy - 2y^2$$
$$= 3x^2 + xy - 2y^2$$

FIGURE A.13

This process is sometimes called the **FOIL** method, where FOIL represents

the product of the **F**irst terms

the product of the **O**uter terms

the product of the **I**nner terms

the product of the **L**ast terms

EXAMPLE 5

$$(2x - 1)(x + 3) = 2x^2 + 6x - x - 3$$
$$= 2x^2 + 5x - 3$$

Factoring

We sometimes find it useful to write a polynomial as a single *term* composed of two or more *factors*. This process is the reverse of multiplication and is called **factoring.** For example, observe that

$$3x^2 + 6x = 3x(x + 2)$$

We will only consider factorization in which the factors have integer coefficients.

Common Factors

We can factor a common factor from a polynomial by using the distributive property in the form

$$ab + ac = a(b + c)$$

We first identify the common factor. For example, each term of the polynomial

$$6x^3 + 9x^2 - 3x$$

contains the monomial $3x$ as a factor; therefore,

$$6x^3 + 9x^2 - 3x = 3x(\underline{\qquad})$$

Next, we insert the proper polynomial factor within the parentheses. This factor can be determined by inspection. We ask ourselves for monomials that, when multiplied by $3x$, yield $6x^3$, $9x^2$, and $-3x$, respectively, and obtain

$$6x^3 + 9x^2 - 3x = 3x(2x^2 + 3x - 1)$$

We can check the result of factoring an expression by multiplying the factors. In the example above,

$$3x(2x^2 + 3x - 1) = 6x^3 + 9x^2 - 3x$$

EXAMPLE 6 **a.** $18x^2y - 24xy^2 = 6xy(?-?)$

$$= 6xy(3x - 4y)$$

because

$$6xy(3x - 4y) = 18x^2y - 24xy^2$$

b. $y(x - 2) + z(x - 2) = (x - 2)(?-?)$

$$= (x - 2)(y + z)$$

because

$$(x - 2)(y + z) = y(x - 2) + z(x - 2)$$

Opposite of a Binomial

It is often useful to factor -1 from the terms of a binomial.

$$a - b = (-1)(-a + b)$$
$$= (-1)(b - a) = -(b - a)$$

Hence, we have the following important relationship.

Opposite of a Binomial

$$a + b = -(b - a)$$

That is, $a - b$ and $b - a$ are opposites or negatives of each other.

EXAMPLE 7 **a.** $3x - y = -(y - 3x)$

b. $a - 2b = -(2b - a)$

Polynomial Division

We can divide one polynomial by a polynomial of lesser degree. The quotient will be the sum of a polynomial and a simpler algebraic fraction.

If the divisor is a monomial, we can simply divide the monomial into each term of the numerator.

EXAMPLE 8 Divide $\dfrac{9x^3 - 6x^2 + 4}{3x}$.

Solution Divide $3x$ into each term of the numerator.

$$\frac{9x^3 - 6x^2 + 4}{3x} = \frac{9x^3}{3x} - \frac{6x^2}{3x} + \frac{4}{3x}$$

$$= 3x^2 - 2x + \frac{4}{3x}$$

The quotient is the sum of a polynomial, $3x^2 - 2x$, and an algebraic fraction, $\frac{4}{3x}$.

If the denominator is not a monomial, we can use a method similar to the long division algorithm used in arithmetic.

EXAMPLE 9 Divide $\dfrac{2x^2 + x - 7}{x + 3}$.

Solution First write

$$x + 3 \overline{)\, 2x^2 + x - 7}$$

and divide $2x^2$ (the first term of the numerator) by x (the first term of the denominator) to obtain $2x$. (It may be helpful to write down the division: $\frac{2x^2}{x} = 2x$.) Write $2x$ above the quotient bar as the first term of the quotient, as shown below.

Next, multiply $x + 3$ by $2x$ to obtain $2x^2 + 6x$, and subtract this product from $2x^2 + x - 7$:

$$
\begin{array}{r}
2x \\
x + 3 \overline{)\, 2x^2 + x - 7} \\
\underline{-(2x^2 + 6x)} \\
-5x - 7
\end{array}
$$

Repeating the process, divide $-5x$ by x to obtain -5. Write -5 as the second term of the quotient. Then multiply $x + 3$ by -5 to obtain $-5x - 15$, and subtract:

$$
\begin{array}{r}
2x - 5 \\
x + 3 \overline{)\, 2x^2 + x - 7} \\
\underline{-(2x^2 + 6x)} \\
-5x - 7 \\
\underline{-(-5x - 15)} \\
8
\end{array}
$$

Because the degree of 8 is less than the degree of $x + 3$, the division is finished. The quotient is $2x - 5$, with a remainder of 8. We write the remainder as a fraction to obtain

$$\frac{2x^2 + x - 7}{x + 3} = 2x - 5 + \frac{8}{x + 3}$$

When using polynomial division, it helps to write the polynomials in descending powers of the variable. If the numerator is missing any terms, we can insert terms with zero coefficients so that like powers will be aligned. For example, to perform the division

$$\frac{3x - 1 + 4x^3}{2x - 1}$$

we first write the numerator in descending powers as $4x^3 + 3x - 1$. We then insert $0x^2$ between $4x^3$ and $3x$ and set up the quotient as

$$2x - 1 \overline{)\, 4x^3 + 0x^2 + 3x - 1}$$

We then proceed as in Example 7. You can check that the quotient is

$$2x^2 + x + 2 + \frac{1}{2x - 1}$$

VOCABULARY *Look up the definitions of new terms in the Glossary.*

Polynomial	Common factor	Degree	Constant
Trinomial	Monomial	Binomial	

SKILLS *Practice each skill in the Exercises listed.*

1. Identify polynomials: #1–12
2. Evaluate polynomials: #13–20
3. Multiply polynnomials: #21–42

4. Factor out a common factor: #43–68
5. Divide polynomials: #69–80

EXERCISES A.7

ThomsonNOW Test yourself on key content at www.thomsonedu.com/login.

Identify each polynomial as a monomial, a binomial, or a trinomial. Give the degree of each polynomial.

1. $2x^3 - x^2$

2. $x^2 - 2x + 1$

3. $5n^4$

4. $3n + 1$

5. $3r^2 - 4r + 2$

6. r^3

7. $y^3 - 2y^2 - y$

8. $3y^2 + 1$

Which are not polynomials?

9. **a.** $1 - 0.04t^2$

   b. $3x^2 - 4x + \dfrac{2}{x}$

   c. $2\sqrt{z} - 7z^3 + 2$

   d. $\sqrt{2}\,w^3 + \dfrac{3}{4}w^2 - w$

10. **a.** $\sqrt{3}\,p^2 - 7p + 2$

    b. $2h^{4/3} + 6h^{1/3} - 2$

    c. $\dfrac{2}{x^2 - 6x + 5}$

    d. $\dfrac{1}{4}y^{-2} + 3y^{-1} + 4$

11. **a.** $\dfrac{1}{m^2 + 3}$

    b. $v^2 - 16 + 2^v$

    c. $\sqrt{x^3 - 4x}$

    d. $\dfrac{m^4}{12}$

12. **a.** $3^t - 5t^3 + 2$

    b. $\dfrac{q + 3}{q - 1}$

    c. $c^{1/2} - c$

    d. $\sqrt[3]{d + 1}$

Evaluate each polynomial function for the given values of the variable.

13. $P(x) = x^3 - 3x^2 + x + 1$

    a. $x = 2$

    b. $x = -2$

    c. $x = 2b$

14. $P(x) = 2x^3 + x^2 - 3x + 4$

    a. $x = 3$

    b. $x = -3$

    c. $x = -a$

15. $Q(t) = t^2 + 3t + 1$

    a. $t = \dfrac{1}{2}$

    b. $t = -\dfrac{1}{3}$

    c. $t = -w$

16. $Q(t) = 2t^2 - t + 1$

    a. $t = \dfrac{1}{4}$

    b. $t = -\dfrac{1}{2}$

    c. $t = 3v$

17. $R(z) = 3z^4 - 2z^2 + 3$

    a. $z = 1.8$

    b. $z = -2.6$

    c. $z = k - 1$

18. $R(z) = z^4 + 4z - 2$

    a. $z = 2.1$

    b. $z = -3.1$

    c. $z = h + 2$

19. $N(a) = a^6 - a^5$

    a. $a = -1$

    b. $a = -2$

    c. $a = \dfrac{m}{3}$

20. $N(a) = a^5 - a^4$

    a. $a = -1$

    b. $a = -2$

    c. $a = \dfrac{q}{2}$

Write each product as a polynomial and simplify.

21. $4y(x - 2y)$

22. $3x(2x + y)$

23. $-6x(2x^2 - x + 1)$

24. $-2y(y^2 - 3y + 2)$

25. $a^2b(3a^2 - 2ab - b)$

26. $ab^3(-a^2b^2 + 4ab - 3)$

27. $2x^2y(4xy^4 - 2xy - 3x^3y^2)$

28. $5x^2y^2(3x^4y^2 + 3x^2y - xy^6)$

29. $(n + 2)(n + 8)$

30. $(r - 1)(r - 6)$

31. $(r + 5)(r - 2)$

32. $(z - 3)(z + 5)$

33. $(2z + 1)(z - 3)$

34. $(3t - 1)(2t + 1)$

35. $(4r + 3s)(2r - s)$

36. $(2z - w)(3z + 5w)$

37. $(2x - 3y)(3x - 2y)$

38. $(3a + 5b)(3a + 4b)$

39. $(3t - 4s)(3t + 4s)$

40. $(2x - 3z)(2x + 3z)$

41. $(2a^2 + b^2)(a^2 - 3b^2)$

42. $(s^2 - 5t^2)(3s^2 + 2t^2)$

Factor completely. Check your answers by multiplying factors.

43. $4x^2z + 8xz$

44. $3x^2y + 6xy$

45. $3n^4 - 6n^3 + 12n^2$

46. $2x^4 - 4x^2 + 6x$

47. $15r^2s + 18rs^2 - 3r$

48. $2x^2y^2 - 3xy + 5x^2$

49. $3m^2n^4 - 6m^3n^3 + 14m^3n^2$

50. $6x^3y - 6xy^3 + 12x^2y^2$

51. $15a^4b^3c^4 - 12a^2b^2c^5 + 6a^2b^3c^4$

52. $14xy^4z^3 + 21x^2y^3z^2 - 28x^3y^2z^5$

53. $a(a + 3) + b(a + 3)$

54. $b(a - 2) + a(a - 2)$

55. $y(y - 2) - 3x(y - 2)$

56. $2x(x + 3) - y(x + 3)$

57. $4(x - 2)^2 - 8x(x - 2)^3$

58. $6(x + 1) - 3x(x + 1)^2$

59. $x(x - 5)^2 - x^2(x - 5)^3$

60. $x^2(x + 3)^3 - x(x + 3)^2$

Supply the missing factors or terms.

61. $3m - 2n = -(\quad ? \quad)$

62. $2a - b = -(\quad ? \quad)$

63. $-2x + 2 = -2(\quad ? \quad)$

64. $-6x - 9 = -3(\quad ? \quad)$

65. $-ab - ac = ?(b + c)$

66. $-a^2 + ab = ? \ (a - b)$

67. $2x - y + 3z = -(\quad ? \quad)$

68. $3x + 3y - 2z = -(\quad ? \quad)$

Divide.

69. $\dfrac{18r^2s^2 - 15rs + 6}{3rs}$

70. $\dfrac{8a^2x^2 - 4ax^2 + ax}{2ax}$

71. $\dfrac{15s^{10} - 21s^5 + 6}{-3s^2}$

72. $\dfrac{25m^6 - 15m^4 + 7}{-5m^3}$

73. $\dfrac{4y^2 + 12y + 7}{2y + 1}$

74. $\dfrac{4t^2 - 4t - 5}{2t - 1}$

75. $\dfrac{x^3 + 2x^2 + x + 1}{x - 2}$

76. $\dfrac{2x^3 - 3x^2 - 2x + 4}{x + 1}$

77. $\dfrac{4z^2 + 5z + 8z^4 + 3}{2z + 1}$

78. $\dfrac{7 - 3t^3 - 23t^2 + 10t^4}{2t + 3}$

79. $\dfrac{x^4 - 1}{x - 2}$

80. $\dfrac{y^5 + 1}{y - 1}$

Consider the trinomial

$$x^2 + 10x + 16$$

Can we find two binomial factors,

$$(x + a)(x + b)$$

whose product is the given trinomial? The product of the binomials is

$$(x + a)(x + b) = x^2 + (a + b)x + ab$$

Thus, we are looking for two numbers, a and b, that satisfy

$$(x + a)(x + b) = x^2 + (a + b)x + ab = x^2 + 10x + 16$$

By comparing the coefficients of the terms in the two trinomials, we see that $a + b = 10$ and $ab = 16$. That is, the sum of the two numbers is the coefficient of the linear term, 10, and their product is the constant term, 16.

To find the numbers, we list all the possible integer factorizations of 16:

$$1 \cdot 16, \quad 2 \cdot 8, \quad \text{and} \quad 4 \cdot 4$$

We see that only one combination gives the correct linear term: 8 and 2. These are the numbers a and b, so

$$x^2 + 10x + 16 = (x + 8)(x + 2)$$

In Example, 1 we factor quadratic trinomials in which one or more of the coefficients is negative.

EXAMPLE 1

Factor.

a. $x^2 - 7x + 12$ **b.** $x^2 - x - 12$

Solutions

a. Find two numbers whose product is 12 and whose sum is -7. Because the product is positive and the sum is negative, the two numbers must both be negative. The possible factors of 12 are -1 and -12, -2 and -6, or -3 and -4. Only -4 and -3 have the correct sum, -7. Hence,

$$x^2 - 7x + 12 = (x - 4)(x - 3)$$

b. Find two numbers whose product is -12 and whose sum is -1. Because the product is negative, the two numbers must be of opposite sign and their sum must be -1. By listing the possible factors of -12, we find that the two numbers are -4 and 3. Hence,

$$x^2 - x - 12 = (x - 4)(x + 3)$$

If the coefficient of the quadratic term is not 1, we must also consider its factors.

EXAMPLE 2 Factor $8x^2 - 9 - 21x$.

Solutions

Step 1 Write the trinomial in decreasing powers of x.

$$8x^2 - 21x - 9$$

Step 2 List the possible factors for the quadratic term.

$$(8x \quad)\,(x \quad)$$
$$(4x \quad)\,(2x \quad)$$

Step 3 Consider possible factors for the constant term: 9 may be factored as $9 \cdot 1$ or as $3 \cdot 3$. Form all possible pairs of binomial factor using these factorizations. (See Figure A.14.)

```
          ┌─ 1 ─┐
          │  ┌ 2 ┐ │
(8x   9)(x   1)
(8x   1)(x   9)
(8x   3)(x   3)
(4x   9)(2x   1)
(4x   1)(2x   9)
(4x   3)(2x   3)
```
FIGURE A.14

Step 4 Select the combinations of the products ❶ and ❷ whose sum or difference could be the linear term, $-21x$.

$$(8x \quad 3)\,(x \quad 3)$$

Step 5 Insert the proper signs:

$$(8x + 3)\,(x - 3)$$

With practice, you can usually factor trinomials of the form $Ax^2 + Bx + C$ mentally. The following observations may help.

1. If A, B and C are all positive, both signs in the factored form are positive. For example, as a first step in factoring $6x^2 + 11x + 4$, we could write

$$(\quad + \quad)(\quad + \quad)$$

2. If A and C are positive and B is negative, both signs in the factored form are negative. Thus as a first step in factoring $6x^2 - 11x + 4$, we could write

$$(\quad - \quad)(\quad - \quad)$$

3. If C is negative, the signs in the factored form are opposite. Thus as a first step in factoring $6x^2 - 5x - 4$, we could write

$$(\quad + \quad)(\quad - \quad) \quad \text{or} \quad (\quad - \quad)(\quad + \quad)$$

EXAMPLE 3
  a. $6x^2 + 5x + 1 = (\quad + \quad)(\quad + \quad)$
$$= (3x + 1)\,(2x + 1)$$

  b. $6x^2 - 5x + 1 = (\quad - \quad)(\quad - \quad)$
$$= (3x - 1)\,(2x - 1)$$

  c. $6x^2 - x - 1 = (\quad + \quad)(\quad - \quad)$
$$= (3x + 1)\,(2x - 1)$$

  d. $6x^2 - xy - y^2 = (\quad + \quad)(\quad - \quad)$
$$= (3x + y)\,(2x - y)$$

Special Products and Factors

The products below are special cases of the multiplication of binomials. They occur so often that you should learn to recognize them on sight.

> ### Special Products
>
> I. $(a+b)^2 = (a+b)(a+b) = a^2 + 2ab + b^2$
>
> II. $(a-b)^2 = (a-b)(a-b) = a^2 - 2ab + b^2$
>
> III. $(a+b)(a-b) = a^2 - b^2$

CAUTION   Notice that in (I) $(a+b)^2 \neq a^2 + b^2$, and that in (II) $(a-b)^2 \neq a^2 - b^2$. For example,

$$(x+4)^2 \neq x^2 + 16, \quad \text{instead} \quad (x+4)^2 = x^2 + 8x + 16$$

$$(t-5)^2 \neq t^2 - 25, \quad \text{instead} \quad (t-5)^2 = t^2 - 10t + 25$$

EXAMPLE 4   **a.** $3(x+4)^2 = 3(x^2 + 2 \cdot 4x + 4^2)$

$$= 3x^2 + 24x + 48$$

b. $(y+5)(y-5) = y^2 - 5^2$

$$= y^2 - 25$$

c. $(3x - 2y)^2 = (3x)^2 - 2(3x)(2y) + (2y)^2$

$$= 9x^2 - 12xy + 4y^2$$

Each of the formulas for special products, when viewed from right to left, also represents a special case of factoring quadratic polynomials.

> ### Special Factorizations
>
> I. $a^2 + 2ab + b^2 = (a+b)^2$ II. $a^2 - 2ab + b^2 = (a-b)^2$
>
> III. $a^2 - b^2 = (a+b)(a-b)$ IV. $a^2 + b^2$ cannot be factored

The trinomials in (I) and (II) are sometimes called **perfect-square trinomials** because they are squares of binomials. Note that the sum of two squares, $a^2 + b^2$, cannot be factored.

EXAMPLE 5   Factor.

a. $x^2 + 8x + 16$ **b.** $y^2 - 10y + 25$

c. $4a^2 - 12ab + 9b^2$ **d.** $25m^2n^2 + 20mn + 4$

Solutions   **a.** Because 16 is equal to 4^2 and 8 is equal to $2 \cdot 4$,

$$x^2 + 8x + 16 = x^2 - 2 \cdot 4x + 4^2$$

$$= (x+4)^2$$

b. Because $25 = 5^2$ and $10 = 2 \cdot 5$,

$$y^2 - 10y + 25 = y^2 - 2 \cdot 5y + 5^2$$
$$= (y - 5)^2$$

c. Because $4a^2 = (2a)^2, 9b^2 = (3b)^2,$ and $12ab = 2(2a)(3b),$

$$4a^2 - 12ab + 9b^2 = (2a)^2 - 2(2a)(3b) + (3b)^2$$
$$= (2a - 3b)^2$$

d. Because $25m^2n^2 = (5mn)^2, 4 = 2^2$ and $20mn = 2(5mn)(2),$

$$25m^2n^2 + 20mn + 4 = (5mn)^2 + 2(5mn)(2) + 2^2$$
$$= (5mn + 2)^2$$

Binomials of the form $a^2 - b^2$ are often called the **difference of two squares.**

EXAMPLE 6 Factor if possible.

a. $x^2 - 81$ **b.** $4x^2 - 9y^2$ **c.** $x^2 + 81$

Solutions **a.** The expression $x^2 - 81$ is the difference of two squares, $x^2 - 9^2$, and thus can be factored according to (III) on the preceding page.

$$x^2 - 81 = x^2 - 9^2$$
$$= (x + 9)(x - 9)$$

b. Because $4x^2 - 9y^2$ can be written as $(2x)^2 - (3y)^2,$

$$4x^2 - 9y^2 = (2x)^2 - (3y)^2$$
$$= (2x + 3y)(2x - 3y)$$

c. The expression $x^2 + 81$, or $x^2 + 0x + 81$, is *not* factorable, because no two real numbers have a product of 81 and a sum of 0.

CAUTION $x^2 + 81 \neq (x + 9)(x + 9)$, which you can verify by multiplying

$$(x + 9)(x + 9) = x^2 + 18x + 8$$

The factors $x + 9$ and $x - 9$ in Example 6a are called **conjugates** of each other. In general, any binomials of the form $a + b$ and $a - b$ are called a **conjugate pair.**

SECTION A.8 SUMMARY

VOCABULARY *Look up the definitions of new terms in the Glossary.*

Perfect-square trinomial Difference of squares Conjugate

SKILLS *Practice each skill in the Exercises listed.*

1. Factor quadratic trinomials: #1–36
2. Expand special products: #37–48

3. Factor special quadratic expressions: #49–68

Thomson NOW™ Test yourself on key content at www.thomsonedu.com/login.

Factor completely.

1. $x^2 + 5x + 6$ **2.** $x^2 + 5x + 4$ **3.** $y^2 - 7y + 12$ **4.** $y^2 - 7y + 10$

5. $x^2 - 6x - x$ **6.** $x^2 - 15 - 2x$ **7.** $2x^2 + 3x - 2$ **8.** $3x^2 - 7x + 2$

9. $7x + 4x^2 - 2$ **10.** $1 - 5x + 6x^2$ **11.** $9y^2 - 21y - 8$ **12.** $10y^2 - 3y - 18$

13. $10u^2 - 3 - u$ **14.** $8u^2 - 3 + 5u$ **15.** $21x^2 - 43x - 14$ **16.** $24x^2 - 29x + 5$

17. $5a + 72a^2 - 25$ **18.** $-30a + 72a^2 - 25$ **19.** $12 - 53x + 30x^2$ **20.** $39x + 80x^2 - 20$

21. $-30t - 44 + 54t^2$ **22.** $48t^2 - 122t + 39$ **23.** $3x^2 - 7ax + 2a^2$ **24.** $9x^2 + 9ax - 10a^2$

25. $15x^2 - 4xy - 4y^2$ **26.** $12x^2 + 7xy - 12y^2$ **27.** $18u^2 + 20v^2 - 39uv$ **28.** $24u^2 - 20v^2 + 17uv$

29. $12a^2 - 14b^2 - 13ab$ **30.** $24a^2 - 15b^2 - 2ab$ **31.** $10a^2b^2 - 19ab + 6$ **32.** $12a^2b^2 - ab - 20$

33. $56x^2y^2 - 2xy - 4$ **34.** $54x^2y^2 + 3xy - 2$ **35.** $22a^2z^2 - 21 - 19az$ **36.** $26a^2z^2 - 24 + 23az$

Write each expression as a polynomial and simplify.

37. $(x + 3)^2$ **38.** $(y - 4)^2$ **39.** $(2y - 5)^2$ **40.** $(3x + 2)^2$

41. $(x + 3)(x - 3)$ **42.** $(x - 7)(x + 7)$ **43.** $(3t - 4s)(3t + 4s)$ **44.** $(2x + a)(2x - a)$

45. $(5a - 2)(5a - 2)$ **46.** $(4u + 5v)(4u + 5v)$ **47.** $(8xz + 3)(8xz + 3)$ **48.** $(7yz - 2)(7yz - 2)$

Factor completely.

49. $x^2 - 25$ **50.** $x^2 - 36$ **51.** $x^2 - 24x + 144$ **52.** $x^2 + 26x + 169$

53. $x^2 - 4y^2$ **54.** $9x^2 - y^2$ **55.** $4x^2 + 12x + 9$ **56.** $4y^2 + 4y + 1$

57. $9u^2 - 30uv + 25v^2$ **58.** $16s^2 - 56st + 49t^2$ **59.** $4a^2 - 25b^2$ **60.** $16a^2 - 9b^2$

61. $x^2y^2 - 81$ **62.** $x^2y^2 - 64$ **63.** $9x^2y^2 + 6xy + 1$ **64.** $4x^2y^2 + 12xy + 9$

65. $16x^2y^2 - 1$ **66.** $64x^2y^2 - 1$ **67.** $(x + 2)^2 - y^2$ **68.** $x^2 - (y - 3)^2$

A quotient of two polynomials is called a **rational expression** or an **algebraic fraction.** Operations on algebraic fractions follow the same rules as operations on common fractions.

Reducing Fractions

When we reduce an ordinary fraction such as $\frac{24}{36}$, we are using the fundamental principle of fractions.

Fundamental Principle of Fractions

If we multiply or divide the numerator and denominator of a fraction by the same (nonzero) number, the new fraction is equivalent to the old one. In symbols,
$$\frac{ac}{bc} = \frac{a}{b}, \qquad (b, c \neq 0)$$

Thus, for example,
$$\frac{24}{36} = \frac{2 \cdot 12}{3 \cdot 12} = \frac{2}{3}$$

We use the same procedure to reduce algebraic fractions: We look for common factors in the numerator and denominator and then apply the fundamental principle.

EXAMPLE 1 Reduce each algebraic fraction.

a. $\dfrac{8x^3 y}{6x^2 y^3}$ **b.** $\dfrac{6x - 3}{3}$

Solutions Factor out any common factors from the numerator and denominator. Then divide numerator and denominator by the common factors.

a. $\dfrac{8x^3 y}{6x^2 y^3} = \dfrac{4x \cdot 2x^2 y}{3y^2 \cdot 2x^2 y}$ **b.** $\dfrac{6x + 3}{3} = \dfrac{3(2x + 1)}{3}$

$\qquad\quad = \dfrac{4x}{3y^2}$ $\qquad\qquad = 2x + 1$

If the numerator or denominator of the fraction contains more than one term, it is especially important to *factor* before attempting to apply the fundamental principle. We can divide out common *factors* from the numerator and denominator of a fraction, but the fundamental principle does *not* apply to common *terms.*

CAUTION We can reduce
$$\frac{2xy}{3y} = \frac{2x}{3}$$

because y is a common factor in the numerator and denominator. However,
$$\frac{2x + y}{3 + y} \neq \frac{2x}{3}$$

because y is a common term but is *not a common factor* of the numerator and denominator. Furthermore,

$$\frac{5x + 3}{5y} \neq \frac{x + 3}{y}$$

because 5 is not a factor of the *entire* numerator.

EXAMPLE 2 Reduce each fraction.

    a. $\dfrac{4x + 2}{4}$ **b.** $\dfrac{9x^2 + 3}{6x + 3}$

Solutions Factor the numerator and denominator. Then divide numerator and denominator by the common factors.

    a. $\dfrac{4x + 2}{4} = \dfrac{\cancel{2}(2x + 1)}{\cancel{2}(2)}$ **b.** $\dfrac{9x^2 + 3}{6x + 3} = \dfrac{\cancel{3}(3x^2 + 1)}{\cancel{3}(2x + 1)}$

               $= \dfrac{2x + 1}{2}$ $= \dfrac{3x^2 + 1}{2x + 1}$

CAUTION Note that in Example 2a above,

$$\frac{4x + 2}{4} \neq x + 2$$

and in Example 2b,

$$\frac{9x^2 + 3}{6x + 3} \neq \frac{9x^2}{6x}$$

We summarize the procedure for reducing algebraic fractions as follows.

> **To Reduce an Algebraic Fraction:**
>
> **1.** Factor the numerator and denominator.
> **2.** Divide the numerator and denominator by any common factors.

EXAMPLE 3 Reduce each fraction.

    a. $\dfrac{x^2 - 7x + 6}{36 - x^2}$ **b.** $\dfrac{27x^3 - 1}{9x^2 - 1}$

Solutions **a.** Factor numerator and denominator to obtain

$$\frac{(x - 6)(x - 1)}{(6 - x)(6 + x)}$$

The factor $x - 6$ in the numerator is the opposite of the factor $6 - x$ in the denominator. That is, $x - 6 = -1(6 - x)$. Thus,

$$\frac{-1(\cancel{6 - x})(x - 1)}{(\cancel{6 - x})(6 + x)} = \frac{-1(x - 1)}{6 + x} = \frac{1 - x}{6 + x}$$

b. The numerator of the fraction is a difference of two cubes, and the denominator is a difference of two squares. Factor each to obtain

$$\frac{(3x-1)(9x^2+3x+1)}{(3x-1)(3x+1)} = \frac{9x^2+3x+1}{3x+1}$$

Products of Fractions

To multiply two or more common fractions together, we multiply their numerators together and multiply their denominators together. The same is true for a product of algebraic fractions. For example,

$$\frac{6x^2}{y} \cdot \frac{xy}{2} = \frac{6x^2 \cdot xy}{y \cdot 2} = \frac{6x^3 y}{2y} \quad \text{Reduce.}$$
$$= \frac{3x^3(2y)}{2y} = 3x^3$$

We can simplify the process by first factoring each numerator and denominator and dividing out any common factors.

$$\frac{6x^2}{y} \cdot \frac{xy}{2} = \frac{2 \cdot 3x^2}{\cancel{y}} \cdot \frac{x\cancel{y}}{\cancel{2}} = 3x^3$$

In general, we have the following procedure for finding the product of algebraic fractions.

> **To Multiply Algebraic Fractions:**
>
> **1.** Factor each numerator and denominator.
> **2.** Divide out any factors that appear in both a numerator and a denominator.
> **3.** Multiply together the numerators; multiply together the denominators.

EXAMPLE 4 Find each product.

a. $\dfrac{5}{x^2-1} \cdot \dfrac{x+2}{x}$

b. $\dfrac{4y^2-1}{4-y^2} \cdot \dfrac{y^2-2y}{4y+2}$

Solutions **a.** The denominator of the first fraction factors into $(x+1)(x-1)$. There are no common factors to divide out, so we multiply the numerators together and multiply the denominators together.

$$\frac{5}{x^2-1} \cdot \frac{x+2}{x} = \frac{5(x+2)}{x(x^2-1)} = \frac{5x+10}{x^3-x}$$

b. Factor each numerator and each denominator. Look for common factors.

$$\frac{4y^2-1}{4-y^2} \cdot \frac{y^2-2y}{4y+2} = \frac{(2y-1)\cancel{(2y+1)}}{\cancel{(2-y)}(2+y)} \cdot \frac{y\,\overset{-1}{\cancel{(y-2)}}}{2\cancel{(2y+1)}} \quad \text{Divide out common factors.}$$
$$= \frac{-y(2y-1)}{2(y+2)} \quad \text{Note: } y-2 = -(2-y).$$

Quotients of Fractions

To divide two algebraic fractions we multiply the first fraction by the reciprocal of the second fraction. For example,

$$\frac{2x^3}{3y} \div \frac{4x}{5y^2} = \frac{2x^3}{3y} \cdot \frac{5y^2}{4x}$$

$$= \frac{2x \cdot x^2}{3y} \cdot \frac{5y \cdot y}{2 \cdot 2x} = \frac{5x^2 y}{6}$$

If the fractions involve polynomials of more than one term, we may need to factor each numerator and denominator in order to recognize any common factors. This suggests the following procedure for dividing algebraic fractions.

To Divide Algebraic Fractions:

1. Multiply the first fraction by the reciprocal of the second fraction.
2. Factor each numerator and denominator.
3. Divide out any factors that appear in both a numerator and a denominator.
4. Multiply together the numerators; multiply together the denominators.

EXAMPLE 5   Find each quotient.

a. $\dfrac{x^2 - 1}{x + 3} \div \dfrac{x^2 - x - 2}{x^2 + 5x + 6}$

b. $\dfrac{6ab}{2a + b} \div (4a^2 b)$

Solutions   **a.** Multiply the first fraction by the reciprocal of the second fraction.

$$\frac{x^2 - 1}{x + 3} \div \frac{x^2 - x - 2}{x^2 + 5x + 6} = \frac{x^2 - 1}{x + 3} \cdot \frac{x^2 + 5x + 6}{x^2 - x - 2} \qquad \text{Factor.}$$

$$= \frac{(x - 1)(x + 1)}{x + 3} \cdot \frac{(x + 3)(x + 2)}{(x + 1)(x - 2)}$$

$$= \frac{(x - 1)(x + 2)}{x - 2}$$

b. Multiply the first fraction by the reciprocal of the second fraction.

$$\frac{6ab}{2a + b} \div 4a^2 b = \frac{\overset{3}{\cancel{6ab}}}{2a + b} \cdot \frac{1}{\underset{2}{\cancel{4a^2 b}}} \qquad \text{Divide out common factors.}$$

$$= \frac{3}{2a(2a + b)} = \frac{3}{4a^2 + 2ab}$$

Sums and Differences of Like Fractions

Algebraic fractions with the same denominator are called **like fractions.** To add or subtract like fractions, we combine their numerators and keep the same denominator for the sum or difference. This method is an application of the distributive law.

EXAMPLE 6 Find each sum or difference.

a. $\dfrac{2x}{9z^2} + \dfrac{5x}{9z^2}$

b. $\dfrac{2x - 1}{x + 3} - \dfrac{5x - 3}{x + 3}$

Solutions **a.** Because these are like fractions, we add their numerators and keep the same denominator.

$$\frac{2x}{9z^2} + \frac{5x}{9z^2} = \frac{2x + 5x}{9z^2} = \frac{7x}{9z^2}$$

b. Be careful to subtract the *entire* numerator of the second fraction: Use parentheses to show that the subtraction applies to both terms of $5x - 3$.

$$\frac{2x - 1}{x + 3} - \frac{5x - 3}{x + 3} = \frac{2x - 1 - (5x - 3)}{x + 3}$$

$$= \frac{2x - 1 - 5x + 3}{x + 3} = \frac{-3x + 2}{x + 3}$$

Lowest Common Denominator

To add or subtract fractions with different denominators, we must first find a *common denominator*. For arithmetic fractions, we use the smallest natural number that is exactly divisible by each of the given denominators. For example, to add the fractions $\frac{1}{6}$ and $\frac{3}{8}$, we use 24 as the common denominator because 24 is the smallest natural number that both 6 and 8 divide into evenly.

We define the **lowest common denominator (LCD)** of two or more algebraic fractions as the polynomial of least degree that is exactly divisible by each of the given denominators.

EXAMPLE 7 Find the LCD for the fractions $\dfrac{3x}{x + 2}$ and $\dfrac{2x}{x - 3}$.

Solution The LCD is a polynomial that has as factors both $x + 2$ and $x - 3$. The simplest such polynomial is $(x + 2)(x - 3)$, or $x^2 - x + 6$. For our purposes, it will be more convenient to leave the LCD in factored form, so the LCD is $(x + 2)(x - 3)$.

The LCD in Example 9 was easy to find because each original denominator consisted of a single factor; that is, neither denominator could be factored. In that case, the LCD is just the product of the original denominators. We can always find a common denominator by multiplying together all the denominators in the given fractions, but this may not give us the *simplest* or *lowest* common denominator. Using anything other than the simplest possible common denominator will complicate our work needlessly.

If any of the denominators in the given fractions can be factored, we factor them before looking for the LCD.

> **To Find the LCD of Algebraic Fractions:**
>
> **1.** Factor each denominator completely.
>
> **2.** Include each different factor in the LCD as many times as it occurs in any *one* of the given denominators.

EXAMPLE 8 Find the LCD for the fractions $\dfrac{2x}{x^2-1}$ and $\dfrac{x+3}{x^2+x}$.

Solution Factor the denominators of each of the given fractions.

$$x^2 - 1 = (x-1)(x+1) \qquad \text{and} \qquad x^2 + x = x(x+1)$$

The factor $(x-1)$ occurs once in the first denominator, the factor x occurs once in the second denominator, and the factor $(x+1)$ occurs once in each denominator. Therefore, we include in our LCD one copy of each of these factors. The LCD is $x(x+1)(x-1)$.

CAUTION In Example 8, we do not include two factors of $(x+1)$ in the LCD. We need only one factor of $(x+1)$ because $(x+1)$ occurs only once in either denominator. You should check that each original denominator divides evenly into our LCD, $x(x+1)(x-1)$.

Building Fractions

After finding the LCD, we *build* each fraction to an equivalent one with the LCD as its denominator. The new fractions will be like fractions, and we can combine them as explained above.

Building a fraction is the opposite of reducing a fraction, in the sense that we multiply, rather than divide, the numerator and denominator by an appropriate factor. To find the **building factor,** we compare the factors of the original denominator with those of the desired common denominator.

EXAMPLE 9 Build each of the fractions $\dfrac{3x}{x+2}$ and $\dfrac{2x}{x-3}$ to equivalent fractions with the LCD $(x+2)$ $(x-3)$ as denominator.

Solution Compare the denominator of the given fraction to the LCD. We see that the fraction $\dfrac{3x}{x+2}$ needs a factor of $(x-3)$ in its denominator, so $(x-3)$ is the building factor for the first fraction. We multiply the numerator and denominator of the first fraction by $(x-3)$ to obtain an equivalent fraction:

$$\frac{3x}{x+2} = \frac{3x(x-3)}{(x+2)(x-3)} = \frac{3x^2-9x}{x^2-x-6}$$

The fraction $\dfrac{2x}{x-3}$ needs a factor of $(x+2)$ in the denominator, so we multiply numerator and denominator by $(x+2)$:

$$\frac{2x}{x-3} = \frac{2x(x+2)}{(x-3)(x+2)} = \frac{2x^2+4x}{x^2-x-6}$$

The two new fractions we obtained in Example 9 are like fractions; they have the same denominator.

Sums and Differences of Unlike Fractions

We are now ready to add or subtract algebraic fractions with unlike denominators. We will do this in four steps.

To Add or Subtract Fractions with Unlike Denominators:

1. Find the LCD for the given fractions.

2. Build each fraction to an equivalent fraction with the LCD as its denominator.

3. Add or subtract the numerators of the resulting like fractions. Use the LCD as the denominator of the sum or difference.

4. Reduce the sum or difference, if possible.

EXAMPLE 10   Subtract $\dfrac{3x}{x+2} - \dfrac{2x}{x-3}$.

Solution

Step 1   The LCD for these fractions is $(x+2)(x-3)$.

Step 2   We build each fraction to an equivalent one with the LCD, as we did in Example 9.

$$\frac{3x}{x+2} = \frac{3x^2 - 9x}{x^2 - x - 6} \quad \text{and} \quad \frac{2x}{x-3} = \frac{2x^2 + 4x}{x^2 - x - 6}$$

Step 3   Combine the numerators over the same denominator.

$$\frac{3x}{x+2} - \frac{2x}{x-3} = \frac{3x^2 - 9x}{x^2 - x - 6} - \frac{2x^2 + 4x}{x^2 - x - 6} \qquad \text{Subtract the numerators.}$$

$$= \frac{(3x^2 - 9x) - (2x^2 + 4x)}{x^2 - x - 6} = \frac{x^2 - 13x}{x^2 - x - 6}$$

Step 4   Reduce the result, if possible. If we factor both numerator and denominator, we find

$$\frac{x(x-13)}{(x-3)(x+2)}$$

The fraction cannot be reduced.

EXAMPLE 11   Write as a single fraction: $1 + \dfrac{2}{a} - \dfrac{a^2 + 2}{a^2 + a}$.

Solution

Step 1   To find the LCD, factor each denominator:

$$a = a$$
$$a^2 + a = a(a+1)$$

The LCD is $a(a+1)$.

Step 2   Build each term to an equivalent fraction with the LCD as denominator. (The building factors for each fraction are shown in color.) The third fraction already has the LCD for its denominator.

$$1 = \frac{1 \cdot a(a+1)}{1 \cdot a(a+1)} = \frac{a^2 + a}{a(a+1)}$$

$$\frac{2}{a} = \frac{2 \cdot (a+1)}{a \cdot (a+1)} = \frac{2a + 2}{a(a+1)}$$

$$\frac{a^2 + 2}{a^2 + a} = \frac{a^2 + 2}{a(a+1)}$$

Step 3 Combine the numerators over the LCD.

$$1 + \frac{2}{a} - \frac{a^2 + 2}{a^2 + a} = \frac{a^2 + a}{a(a+1)} + \frac{2a + 2}{a(a+1)} - \frac{a^2 + 2}{a(a+1)}$$

$$= \frac{a^2 + a + (2a + 2) - (a^2 + 2)}{a(a+1)}$$

$$= \frac{3a}{a(a+1)}$$

Step 4 Reduce the fraction to find

$$\frac{3\cancel{a}}{\cancel{a}(a+1)} = \frac{3}{a+1}$$

Complex Fractions

A fraction that contains one or more fractions in either its numerator or its denominator or both is called a **complex fraction.** For example,

$$\frac{\frac{2}{3}}{\frac{5}{6}} \qquad \text{and} \qquad \frac{x + \frac{3}{4}}{x - \frac{1}{2}}$$

are complex fractions. Like simple fractions, complex fractions represent quotients. For the examples above,

$$\frac{\frac{2}{3}}{\frac{5}{6}} = \frac{2}{3} \div \frac{5}{6} \qquad \text{and} \qquad \frac{x + \frac{3}{4}}{x - \frac{1}{2}} = \left(x + \frac{3}{4} \right) \div \left(x - \frac{1}{2} \right)$$

We can always simplify a complex fraction into a standard algebraic fraction. If the denominator of the complex fraction is a single term, we can treat the fraction as a division problem and multiply the numerator by the reciprocal of the denominator. Thus,

$$\frac{\frac{2}{3}}{\frac{5}{6}} = \frac{2}{3} \div \frac{5}{6} = \frac{2}{3} \cdot \frac{6}{5} = \frac{4}{5}$$

If the numerator or denominator of the complex fraction contains more than one term, it is easier to use the fundamental principle of fractions to simplify the expression.

EXAMPLE 12

Simplify $\dfrac{x + \dfrac{3}{4}}{x - \dfrac{1}{2}}$.

Solution Consider all of the simple fractions that appear in the complex fraction; in this example $\frac{1}{2}$ and $\frac{3}{4}$. The LCD of these fractions is 4. If we multiply the numerator and denominator of the complex fraction by 4, we will eliminate the fractions within the fraction. Be sure to multiply *each* term of the numerator and *each* term of the denominator by 4.

$$\frac{4\left(x+\dfrac{3}{4}\right)}{4\left(x-\dfrac{1}{2}\right)} = \frac{4(x)+4\left(\dfrac{3}{4}\right)}{4(x)-4\left(\dfrac{1}{2}\right)} = \frac{4x+3}{4x-2}$$

Thus, the original complex fraction is equivalent to the simple fraction $\dfrac{4x+3}{4x-2}$.

We summarize the method for simplifying complex fractions as follows.

To Simplify a Complex Fraction:

1. Find the LCD of all the fraction contained in the complex fraction.
2. Multiply the numerator and the denominator of the complex fraction by the LCD.
3. Reduce the resulting simple fraction, if possible.

Negative Exponents

Algebraic fractions are sometimes written using negative exponents. (You can review negative exponents in Section 3.1.)

EXAMPLE 13   Write each expression as a single algebraic fraction.

a. $x^{-1} - y^{-1}$ **b.** $(x^{-2} + y^{-2})^{-1}$

Solutions   **a.** $x^{-1} - y^{-1} = \dfrac{1}{x} - \dfrac{1}{y}$,   or $\dfrac{y-x}{xy}$

b. $(x^{-2} + y^{-2})^{-1} = \left(\dfrac{1}{x^2} + \dfrac{1}{y^2}\right)^{-1} = \left(\dfrac{y^2+x^2}{x^2 y^2}\right)^{-1} = \left(\dfrac{x^2 y^2}{y^2+x^2}\right)$

When working with fractions and exponents, it is important to avoid some tempting but *incorrect* algebraic operations.

CAUTION

1. In Example 13a, note that

$$\frac{1}{x} - \frac{1}{y} \neq \frac{1}{x-y}$$

For example, you can check that for $x = 2$ and $y = 3$,

$$\frac{1}{2} - \frac{1}{3} \neq \frac{1}{2-3} = -1$$

2. In Example 13b, note that

$$(x^{-2} + y^{-2})^{-1} \neq x^2 + y^2 \qquad \frac{1}{x^{-2}+y^{-2}}$$

In general, the fourth law of exponents does *not* apply to sums and differences; that is,

$$(a+b)^n \neq a^n + b^n$$

VOCABULARY *Look up the definitions of new terms in the Glossary.*

Rational expression	Building factor	Like fractions	Reciprocal
Common factor	Algebraic fraction	Complex fraction	Common denominator
Numerator	Common term	Reduce	
Polynomial division	Denominator	Opposite	

SKILLS *Practice each skill in the Exercises listed.*

1. Reduce fractions: #1–24

2. Multiply fractions: #25–36

3. Divide fractions: #37–48

4. Add like fractions: #49–56

5. Find the LCD: #57–62

6. Add unlike fractions: #63–82

7. Simplify complex fractions: #83–106

EXERCISES A.9

ThomsonNOW™ Test yourself on key content at **www.thomsonedu.com/login.**

Reduce each algebraic fraction.

1. $\dfrac{14c^2d}{-7c^2d^3}$ **2.** $\dfrac{-12r^2st}{-6rst^2}$ **3.** $\dfrac{4x+6}{6}$ **4.** $\dfrac{2y-8}{8}$

5. $\dfrac{6a^3-4a^2}{4a}$ **6.** $\dfrac{3x^3-6x^2}{6x^2}$ **7.** $\dfrac{6-6t^2}{(t-1)^2}$ **8.** $\dfrac{4-4x^2}{(x+1)^2}$

9. $\dfrac{2y^2-8}{2y+4}$ **10.** $\dfrac{5y^2-20}{2y-4}$ **11.** $\dfrac{6-2v}{v^3-27}$ **12.** $\dfrac{4-2u}{u^3-8}$

13. $\dfrac{4x^3-36x}{6x^2+18x}$ **14.** $\dfrac{5x^2+10x}{5x^3+20x}$ **15.** $\dfrac{y^2-9x^2}{(3x-y)^2}$ **16.** $\dfrac{(2x-y)^2}{y^2-4x^2}$

17. $\dfrac{2x^2+x-6}{x^2+x-2}$ **18.** $\dfrac{6x^2-x-1}{2x^2+9x-5}$ **19.** $\dfrac{8z^3-27}{4z^2-9}$ **20.** $\dfrac{8z^3-1}{4z^2-1}$

21. Which of the following fractions are equivalent to $2a$ (on their common domain)?

a. $\dfrac{2a+4}{4}$ **b.** $\dfrac{4a^2-2a}{2a-1}$ **c.** $\dfrac{4a^2-2a}{2a}$ **d.** $\dfrac{a+3}{2a^2+6a}$

22. Which of the following fractions are equivalent to $3b$ (on their common domain)?

a. $\dfrac{9b^2-3b}{3b}$ **b.** $\dfrac{b+2}{3b^2+6b}$ **c.** $\dfrac{3b-9}{9}$ **d.** $\dfrac{9b^2-3b}{3b-1}$

23. Which of the following fractions are equivalent to -1 (where they are defined)?

a. $\dfrac{2a+b}{2a-b}$ **b.** $\dfrac{-(a+b)}{b-a}$ **c.** $\dfrac{2a^2-1}{2a^2}$ **d.** $\dfrac{-a^2+3}{a^2+3}$

24. Which of the following fractions are equivalent to -1 (where they are defined)?

a. $\dfrac{2a - b}{b - 2a}$

b. $\dfrac{-b^2 - 2}{b^2 + 2}$

c. $\dfrac{3b^2 - 1}{3b^2 + 1}$

d. $\dfrac{b - 1}{b}$

Write each product as a single fraction in lowest terms.

25. $\dfrac{-4}{3np} \cdot \dfrac{6n^2 p^3}{16}$

26. $\dfrac{14a^3 b}{3b} \cdot \dfrac{-6}{7a^2}$

27. $5a^2 b^2 \cdot \dfrac{1}{a^3 b^3}$

28. $15x^2 y \cdot \dfrac{3}{35xy^2}$

29. $\dfrac{5x + 25}{2x} \cdot \dfrac{4x}{2x + 10}$

30. $\dfrac{3y}{4xy - 6y^2} \cdot \dfrac{2x - 3y}{12x}$

31. $\dfrac{4a^2 - 1}{a^2 - 16} \cdot \dfrac{a^2 - 4a}{2a + 1}$

32. $\dfrac{9x^2 - 25}{2x - 2} \cdot \dfrac{x^2 - 1}{6x - 10}$

33. $\dfrac{2x^2 - x - 6}{3x^2 + 4x + 1} \cdot \dfrac{3x^2 + 7x + 2}{2x^2 + 7x + 6}$

34. $\dfrac{3x^2 - 7x - 6}{2x^2 - x - 1} \cdot \dfrac{2x^2 - 9x - 5}{3x^2 - 13x - 10}$

35. $\dfrac{3x^4 - 48}{x^4 - 4x^2 - 32} \cdot \dfrac{4x^4 - 8x^3 + 4x^2}{2x^4 + 16x}$

36. $\dfrac{x^4 - 3x^3}{x^4 + 6x^2 - 27} \cdot \dfrac{x^4 - 81}{3x^4 - 81x}$

Write each quotient as a single fraction in lowest terms.

37. $\dfrac{4x - 8}{3y} \div \dfrac{6x - 12}{y}$

38. $\dfrac{6y - 27}{5x} \div \dfrac{4y - 18}{x}$

39. $\dfrac{a^2 - a - 6}{a^2 + 2a - 15} \div \dfrac{a^2 - 4}{a^2 + 6a + 5}$

40. $\dfrac{a^2 + 2a - 15}{a^2 + 3a - 10} \div \dfrac{a^2 - 9}{a^2 - 9a + 14}$

41. $\dfrac{x^3 + y^3}{x} \div \dfrac{x + y}{3x}$

42. $\dfrac{8x^3 - y^3}{x + y} \div \dfrac{2x - y}{x^2 - y^2}$

43. $1 \div \dfrac{x^2 - 1}{x + 2}$

44. $1 \div \dfrac{x^2 + 3x + 1}{x - 2}$

45. $(x^2 - 5x + 4) \div \dfrac{x^2 - 1}{x^2}$

46. $(x^2 - 9) \div \dfrac{x^2 - 6x + 9}{3x}$

47. $\dfrac{x^2 + 3x}{2y} \div 3x$

48. $\dfrac{2y^2 + y}{3x} \div 2y$

Write each sum or difference as a single fraction in lowest terms.

49. $\dfrac{x}{2} - \dfrac{3}{2}$

50. $\dfrac{y}{7} - \dfrac{5}{7}$

51. $\dfrac{1}{6}a + \dfrac{1}{6}b - \dfrac{5}{6}c$

52. $\dfrac{1}{3}x - \dfrac{2}{3}y + \dfrac{1}{3}z$

53. $\dfrac{x - 1}{2y} + \dfrac{x}{2y}$

54. $\dfrac{y + 1}{b} - \dfrac{y - 1}{b}$

55. $\dfrac{3}{x + 2y} - \dfrac{x - 3}{x + 2y} - \dfrac{x - 1}{x + 2y}$

56. $\dfrac{2}{a - 3b} - \dfrac{b - 2}{a - 3b} + \dfrac{b}{a - 3b}$

Find the LCD for each pair of fractions.

57. $\dfrac{5}{6(x+y)^2}$, $\dfrac{3}{4xy^2}$

58. $\dfrac{1}{8(a-b)^2}$, $\dfrac{5}{12a^2b^2}$

59. $\dfrac{2a}{a^2+5a+4}$, $\dfrac{2}{(a+1)^2}$

60. $\dfrac{3x}{x^2-3x+2}$, $\dfrac{3}{(x-1)^2}$

61. $\dfrac{x+2}{x^2-x}$, $\dfrac{x+1}{(x-1)^3}$

62. $\dfrac{y-1}{y^2+2y}$, $\dfrac{y-3}{(y+2)^2}$

Write each sum or difference as a single fraction in lowest terms.

63. $\dfrac{x}{2}+\dfrac{2x}{3}$

64. $\dfrac{3y}{4}+\dfrac{y}{3}$

65. $\dfrac{5}{6}y-\dfrac{3}{4}y$

66. $\dfrac{3}{4}x-\dfrac{1}{6}x$

67. $\dfrac{x+1}{2x}+\dfrac{2x-1}{3x}$

68. $\dfrac{y-2}{4y}+\dfrac{2y-3}{3y}$

69. $\dfrac{5}{x}+\dfrac{3}{x-1}$

70. $\dfrac{2}{y+2}+\dfrac{3}{y}$

71. $\dfrac{y}{2y-1}-\dfrac{2y}{y+1}$

72. $\dfrac{2x}{3x+1}-\dfrac{x}{x-2}$

73. $\dfrac{y-1}{y+1}-\dfrac{y-2}{2y-3}$

74. $\dfrac{x-2}{2x+1}-\dfrac{x+1}{x-1}$

75. $\dfrac{7}{5x-10}-\dfrac{5}{3x-6}$

76. $\dfrac{2}{3y+6}-\dfrac{3}{2y+4}$

77. $\dfrac{y-1}{y^2-3y}-\dfrac{y+1}{y^2+2y}$

78. $\dfrac{x+1}{x^2+2x}-\dfrac{x-1}{x^2-3x}$

79. $x-\dfrac{1}{x}$

80. $1+\dfrac{1}{y}$

81. $x+\dfrac{1}{x-1}-\dfrac{1}{(x-1)^2}$

82. $y-\dfrac{2}{y^2-1}+\dfrac{3}{y+1}$

Write each complex fraction as a simple fraction in lowest terms.

83. $\dfrac{\dfrac{2}{a}+\dfrac{3}{2a}}{5+\dfrac{1}{a}}$

84. $\dfrac{\dfrac{2}{y}+\dfrac{1}{2y}}{y+\dfrac{y}{2}}$

85. $\dfrac{1+\dfrac{2}{a}}{1-\dfrac{4}{a^2}}$

86. $\dfrac{9-\dfrac{1}{x^2}}{3-\dfrac{1}{x}}$

87. $\dfrac{h+\dfrac{h}{m}}{1+\dfrac{1}{m}}$

88. $\dfrac{1+\dfrac{1}{p}}{1-\dfrac{1}{p}}$

89. $\dfrac{1}{1-\dfrac{1}{q}}$

90. $\dfrac{4}{\dfrac{2}{v}+2}$

91. $\dfrac{L+C}{\dfrac{1}{L}+\dfrac{1}{C}}$

92. $\dfrac{H-T}{\dfrac{H}{T}-\dfrac{T}{H}}$

93. $\dfrac{\dfrac{4}{x^2}-\dfrac{4}{z^2}}{\dfrac{2}{z}-\dfrac{2}{x}}$

94. $\dfrac{\dfrac{6}{b}-\dfrac{6}{a}}{\dfrac{3}{a^2}-\dfrac{3}{b^2}}$

Write each expression as a single algebraic fraction.

95. $x^{-2}+y^{-2}$

96. $x^{-2}-y^{-2}$

97. $2w^{-1}-(2w)^{-2}$

98. $3w^{-3}+(3w)^{-1}$

99. $a^{-1}b-ab^{-1}$

100. $a-b^{-1}a-b^{-1}$

101. $(x^{-1}+y^{-1})^{-1}$

102. $(1-xy^{-1})^{-1}$,

103. $\dfrac{x+x^{-2}}{x}$

104. $\dfrac{x^{-1}-y}{x^{-1}}$

105. $\dfrac{a^{-1}+b^{-1}}{(ab)^{-1}}$

106. $\dfrac{x}{x^{-2}-y^{-2}}$

In some situations, radical notation is more convenient to use than exponents. In these cases, we usually simplify radical expressions algebraically as much as possible before using a calculator to obtain decimal approximations.

Properties of Radicals

Because $\sqrt[n]{a} = a^{1/n}$, we can use the laws of exponents to derive two important properties that are useful in simplifying radicals.

Properties of Radicals

1. $\sqrt[n]{ab} = \sqrt[n]{a}\ \sqrt[n]{b}$ for $a,\ b \geq 0$

2. $\sqrt[n]{\dfrac{a}{b}} = \dfrac{\sqrt[n]{a}}{\sqrt[n]{b}}$ for $a \geq 0,\ \ b > 0$

As examples, you can verify that

$$\sqrt{36} = \sqrt{4}\sqrt{9} \qquad \text{and} \qquad \sqrt[3]{\frac{1}{8}} = \frac{\sqrt[3]{1}}{\sqrt[3]{8}}$$

EXAMPLE 1 Which of the following are true?

a. Is $\sqrt{36 + 64} = \sqrt{36} + \sqrt{64}$? **b.** Is $\sqrt[3]{8(64)} = \sqrt[3]{8}\ \sqrt[3]{64}$?

c. Is $\sqrt{x^2 + 4} = x + 2$? **d.** Is $\sqrt[3]{8x^3} = 2x$?

Solution The statements in (b) and (d) are true, and both are examples of the first property of radicals. Statements (a) and (c) are false. In general, $\sqrt[n]{a + b}$ is not equal to $\sqrt[n]{a} + \sqrt[n]{b}$, and $\sqrt[n]{a - b}$ is not equal to $\sqrt[n]{a} - \sqrt[n]{b}$.

Simplifying Radicals

We use Property (1) to simplify radical expressions by factoring the radicand. For example, to simplify $\sqrt[3]{108}$, we look for perfect cubes that divide evenly into 108. The easiest way to do this is to try the perfect cubes in order: 1, 8, 27, 64, 125, . . . and so on, until we find one that is a factor. For this example, we find that $108 = 27 \cdot 4$. Using Property (1), we write

$$\sqrt[3]{108} = \sqrt[3]{27}\ \sqrt[3]{4}$$

Simplify the first factor to find

$$= \sqrt[3]{108} = 3\sqrt[3]{4}$$

This expression is considered simpler than the original radical because the new radicand, 4, is smaller than the original, 108.

We can also simplify radicals containing variables. If the exponent on the variable is a multiple of the index, we can extract the variable from the radical. For instance,

$$\sqrt[3]{x^{12}} = x^{12/3} = x^4$$

(You can verify this by noting that $(x^4)^3 = x^{12}$.) If the exponent on the variable is not a multiple of the index, we factor out the highest power that is a multiple. For example,

$$\sqrt[3]{x^{11}} = \sqrt[3]{x^9 \cdot x^2} \qquad \text{Apply Property (1).}$$
$$= \sqrt[3]{x^9} \cdot \sqrt[3]{x^2} \quad \text{Simplify } \sqrt[3]{x^9} = x^{9/3}.$$
$$= x^3 \sqrt[3]{x^2}$$

EXAMPLE 2   Simplify each radical.

a. $\sqrt{18x^5}$ **b.** $\sqrt[3]{24x^6 y^8}$

Solutions   **a.** The index of the radical is 2, so we look for perfect square factors of $18x^5$. The factor 9 is a perfect square, and x^4 has an exponent divisible by 2. Thus,

$$\sqrt{18x^5} = \sqrt{9x^4 \cdot 2x} \qquad \text{Apply Property (1).}$$
$$= \sqrt{9x^4}\sqrt{2x} \qquad \text{Take square roots.}$$
$$= 3x^2\sqrt{2x}$$

b. The index of the radical is 3, so we look for perfect cube factors of $24x^6 y^8$. The factor 8 is a perfect cube, and x^6 and y^6 have exponents divisible by 3. Thus,

$$\sqrt[3]{24x^6 y^8} = \sqrt[3]{8x^6 y^6 \cdot 3y^2} \qquad \text{Apply Property (1).}$$
$$= \sqrt[3]{8x^6 y^6} \sqrt[3]{3y^2} \qquad \text{Take cube roots.}$$
$$= 2x^2 y^2 \sqrt[3]{3y^2}$$

CAUTION   Property (1) applies only to *products* under the radical, not to sums or differences. Thus, for example,

$$\sqrt{4 \cdot 9} = \sqrt{4}\sqrt{9} = 2 \cdot 3, \qquad \text{but} \qquad \sqrt{4+9} \neq \sqrt{4} + \sqrt{9}$$

and

$$\sqrt[3]{x^3 y^6} = \sqrt[3]{x^3}\sqrt[3]{y^6} = xy^2, \qquad \text{but} \qquad \sqrt[3]{x^3 - y^6} \neq \sqrt[3]{x^3} - \sqrt[3]{y^6}$$

To simplify roots of fractions, we use Property (2), which allows us to write the expression as a quotient of two radicals.

EXAMPLE 3   **a.** $\sqrt{\dfrac{3}{4}} = \dfrac{\sqrt{3}}{\sqrt{4}} = \dfrac{\sqrt{3}}{2}$ **b.** $\sqrt[3]{\dfrac{5}{8}} = \dfrac{\sqrt[3]{5}}{\sqrt[3]{8}} = \dfrac{\sqrt[3]{5}}{2}$

We can also use Properties (1) and (2) to simplify products and quotients of radicals.

EXAMPLE 4   Simplify.

a. $\sqrt[4]{6x^2} \sqrt[4]{8x^3}$ **b.** $\dfrac{\sqrt[3]{16y^5}}{\sqrt[3]{y^2}}$

Solutions **a.** First apply Property (1) to write the product as a single radical, then simplify.

$$\sqrt[4]{6x^2}\ \sqrt[4]{8x^3} = \sqrt[4]{48x^5} \qquad \text{Factor out perfect fourth powers.}$$

$$= \sqrt[4]{16x^4}\ \sqrt[4]{3x} \quad \text{Simplify.}$$

$$= 2x\ \sqrt[4]{3x}$$

b. Apply Property (2) to write the quotient as a single radical.

$$\frac{\sqrt[3]{16y^5}}{\sqrt[3]{y^2}} = \sqrt[3]{\frac{16y^5}{y^2}} \qquad \text{Reduce.}$$

$$= \sqrt[3]{16y^3} \qquad \text{Simplify: factor out perfect cubes.}$$

$$= \sqrt[3]{8y^3}\ \sqrt[3]{2}$$

$$= 2y\ \sqrt[3]{2}$$

Sums and Differences of Radicals

You know that sums or differences of like terms can be combined by adding or subtracting their coefficients:

$$3xy + 5xy = (3+5)xy = 8xy$$

Like radicals, that is, radicals of the same index and radicand, can be combined in the same way.

EXAMPLE 5

a. $3\sqrt{3} + 4\sqrt{3} = (3+4)\sqrt{3}$
$= 7\sqrt{3}$

b. $4\sqrt[3]{y} - 6\sqrt[3]{y} = (4-6)\sqrt[3]{y}$
$= -2\sqrt[3]{y}$

CAUTION

1. In Example 5a, $3\sqrt{3} + 4\sqrt{3} \neq 7\sqrt{6}$. Only the coefficients are added; the radicand does not change.

2. Sums of radicals with different radicands or different indices *cannot* be combined. Thus,

$$\sqrt{11} + \sqrt{5} \neq \sqrt{16}$$
$$\sqrt[3]{10x} - \sqrt[3]{2x} \neq \sqrt[3]{8x}$$

and

$$\sqrt[3]{7} + \sqrt{7} \neq \sqrt[5]{7}$$

None of the expressions above can be simplified.

Products of Radicals

According to Property (1), radicals of the same index can be multiplied together.

Product of Radicals

$$\sqrt[n]{a}\ \sqrt[n]{b} = \sqrt[n]{ab} \qquad (a, b \geq 0)$$

Thus, for example,

$$\sqrt{2}\sqrt{18} = \sqrt{36} = 6 \qquad \text{and} \qquad \sqrt[3]{2x}\,\sqrt[3]{4x^2} = \sqrt[3]{8x^3} = 2x$$

For products involving binomials, we can apply the distributive law.

EXAMPLE 6

a. $\sqrt{3}\left(\sqrt{2x} + \sqrt{6}\right) = \sqrt{3 \cdot 2x} + \sqrt{3 \cdot 6}$
$$= \sqrt{6x} + \sqrt{18} = \sqrt{6x} + 3\sqrt{2}$$

b. $\left(\sqrt{x} - \sqrt{y}\right)\left(\sqrt{x} + \sqrt{y}\right) = \sqrt{x^2} - \sqrt{xy} + \sqrt{xy} - \sqrt{y^2}$
$$= x - y$$

Rationalizing the Denominator

It is easier to work with radicals if there are no roots in the denominators of fractions. We can use the fundamental principle of fractions to remove radicals from the denominator. This process is called **rationalizing the denominator.** For square roots, we multiply the numerator and denominator of the fraction by the radical in the denominator.

EXAMPLE 7

Rationalize the denominator of each fraction.

a. $\sqrt{\dfrac{1}{3}}$

b. $\dfrac{\sqrt{2}}{\sqrt{50x}}$

Solutions

a. Apply Property (2) to write the radical as a quotient.

$$\sqrt{\dfrac{1}{3}} = \dfrac{\sqrt{1}}{\sqrt{3}} = \dfrac{1}{\sqrt{3}} \qquad \text{Multiply numerator and denominator by } \sqrt{3}.$$
$$= \dfrac{1 \cdot \sqrt{3}}{\sqrt{3} \cdot \sqrt{3}} = \dfrac{\sqrt{3}}{3}$$

b. It is always best to simplify the denominator before rationalizing.

$$\dfrac{\sqrt{2}}{\sqrt{50x}} = \dfrac{\sqrt{2}}{5\sqrt{2x}} \qquad \text{Multiply numerator and denominator by } \sqrt{2x}.$$
$$= \dfrac{\sqrt{2} \cdot \sqrt{2x}}{5\sqrt{2x} \cdot \sqrt{2x}} \qquad \text{Simplify.}$$
$$= \dfrac{\sqrt{4x}}{5\,(2x)} = \dfrac{2\sqrt{x}}{10x}$$

If the denominator of a fraction is a *binomial* in which one or both terms is a radical, we can use a special building factor to rationalize it. First, recall that

$$(p - q)(p + q) = p^2 - q^2$$

where the product consists of perfect squares only. Each of the two factors $p - q$ and $p + q$ is said to be the **conjugate** of the other.

Now consider a fraction of the form

$$\dfrac{a}{b + \sqrt{c}}$$

If we multiply the numerator and denominator of this fraction by the conjugate of the denominator, we get

$$\frac{a(b - \sqrt{c}\,)}{(b + \sqrt{c}\,)(b - \sqrt{c}\,)} = \frac{ab - a\sqrt{c}}{b^2 - (\sqrt{c}\,)^2} = \frac{ab - a\sqrt{c}}{b^2 - c}$$

The denominator of the fraction no longer contains any radicals—it has been rationalized.

Multiplying numerator and denominator by the conjugate of the denominator also works on fractions of the form

$$\frac{a}{\sqrt{b} + c} \qquad \text{and} \qquad \frac{a}{\sqrt{b} + \sqrt{c}}$$

We leave the verification of these cases as exercises.

EXAMPLE 8

Rationalize the denominator: $\dfrac{x}{\sqrt{2} + \sqrt{x}}$.

Solution

Multiply numerator and denominator by the conjugate of the denominator, $\sqrt{2} - \sqrt{x}$.

$$\frac{x\,(\sqrt{2} - \sqrt{x}\,)}{(\sqrt{2} + \sqrt{x}\,)(\sqrt{2} - \sqrt{x}\,)} = \frac{x\,(\sqrt{2} - \sqrt{x}\,)}{2 - x}$$

Simplifying $\sqrt[n]{a^n}$

Raising to a power is the inverse operation for extracting roots; that is,

$$\boxed{(\sqrt[n]{a})^n = a}$$

as long as $\sqrt[n]{a}$ is a real number. For example,

$$(\sqrt[4]{16})^4 = 2^4 = 16, \quad \text{and} \quad (\sqrt[3]{-125})^3 = (-5)^3 = -125$$

Now consider the power and root operations in the opposite order; is it true that $\sqrt[n]{a^n} = a$? If the index n is an odd number, then the statement is always true. For example,

$$\sqrt[3]{2^3} = \sqrt[3]{8} = 2 \qquad \text{and} \qquad \sqrt[3]{(-2)^3} = \sqrt[3]{-8} = -2$$

However, if n is even, we must be careful. Recall that the principal root $\sqrt[n]{x}$ is always positive, so if a is a negative number, it cannot be true that $\sqrt[n]{a^n} = a$. For example, if $a = -3$, then

$$\sqrt{(-3)^2} = \sqrt{9} = 3$$

Instead, we see that, for even roots, $\sqrt[n]{a^n} = |a|$.

We summarize our results in below.

Roots of Powers

1. If n is odd, $\qquad \sqrt[n]{a^n} = a$
2. If n is even, $\qquad \sqrt[n]{a^n} = |a|$
   In particular, $\qquad \sqrt{a^2} = |a|$

EXAMPLE 9

a. $\sqrt{16x^2} = 4|x|$ $\qquad\qquad\qquad$ **b.** $\sqrt{(x - 1)^2} = |x - 1|$

Extraneous Solutions to Radical Equations

It is important to check the solution to a radical equation, because it is possible to introduce false, or **extraneous,** solutions when we square both sides of the equation. For example, the equation

$$\sqrt{x} = -5$$

has no solution, because $\sqrt{x}$ is never a negative number. However, if we try to solve the equation by squaring both sides, we find

$$(\sqrt{x})^2 = (-5)^2$$
$$x = 25$$

You can check that 25 is *not* a solution to the original equation, $\sqrt{x} = -5$, because $\sqrt{25}$ does not equal -5.

If each side of an equation is raised to an odd power, extraneous solutions will not be introduced. However, if we raise both sides to an even power, we should check each solution in the original equation.

EXAMPLE 10 Solve the equation $\sqrt{x+2} + 4 = x$.

Solution First, isolate the radical expression on one side of the equation. (This will make it easier to square both sides.)

$$\sqrt{x+2} = x - 4 \qquad \text{Square both sides of the equation.}$$
$$(\sqrt{x+2})^2 = (x-4)^2$$
$$x + 2 = x^2 - 8x + 16$$
$$x^2 - 9x + 14 = 0 \qquad \text{Subtract } x + 2 \text{ from both sides.}$$
$$(x-2)(x-7) = 0 \qquad \text{Factor the left side.}$$
$$x = 2 \quad \text{or} \quad x = 7 \qquad \text{Set each factor equal to zero.}$$

Check Does $\sqrt{2+2} + 4 = 2$? No; 2 is not a solution.

Does $\sqrt{7+2} + 4 = 7$? Yes; 7 is a solution.

The apparent solution 2 is extraneous. The only solution to the original equation is 7. We can verify the solution by graphing the equations

$$y_1 = \sqrt{x+2} \quad \text{and} \quad y_2 = x - 4$$

as shown in Figure A.15. The graphs intersect in only one point, $(7, 3)$, so there is only one solution, $x = 7$.

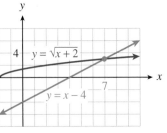

FIGURE A.15

CAUTION When we square both sides of an equation, it is *not* correct to square each term of the equation separately. Thus, in Example 9, the original equation is *not* equivalent to

$$(\sqrt{x+2})^2 + 4^2 = x^2$$

This is because $(a + b)^2 \neq a^2 + b^2$. Instead, we must square the *entire* left side of the equation as a binomial, like this,

$$(\sqrt{x+2} + 4)^2 = x^2$$

or we may proceed as shown in Example 10.

Equations with More than One Radical

Sometimes it is necessary to square both sides of an equation more than once in order to eliminate all the radicals.

EXAMPLE 11   Solve $\sqrt{x-7} + \sqrt{x} = 7$.

Solution   First, isolate the more complicated radical on one side of the equation. (This will make it easier to square both sides.) We will subtract $\sqrt{x}$ from both sides.

$$\sqrt{x-7} = 7 - \sqrt{x}$$

Now square each side to remove one radical. Be careful when squaring the binomial $7 - \sqrt{x}$.

$$(\sqrt{x-7})^2 = (7 - \sqrt{x})^2$$
$$x - 7 = 49 - 14\sqrt{x} + x$$

Collect like terms, and isolate the radical on one side of the equation.

$$-56 = -14\sqrt{x} \quad \text{Divide both sides by } -14.$$
$$4 = \sqrt{x}$$

Now square again to obtain

$$(4)^2 = (\sqrt{x})^2$$
$$16 = x$$

Check   Does $\sqrt{16-7} + \sqrt{16} = 7$?   Yes.   The solution is 16.

CAUTION   Recall that we cannot solve a radical equation by squaring each term separately. In other words, it is *incorrect* to begin Example 4 by writing

$$(\sqrt{x-7})^2 + (\sqrt{x})^2 = 7^2$$

We must square the *entire expression* on each side of the equal sign as one piece.

SECTION A.10 SUMMARY

VOCABULARY   *Look up the definitions of new terms in the Glossary.*

Radical	Extraneous	Rationalize	Conjugate
Like radicals	Radicand	Index	

SKILLS   *Practice each skill in the Exercises listed.*

1. Simplify radicals: #1–6
2. Simplify products and quotients of radcials: #7–10
3. Combine like radicals: #11–18
4. Multiply radical expressions: #19–36

5. Rationalize the denominator: #37–50
6. Simplify $\sqrt[n]{a^n}$: #51–54
7. Solve radical equations: #55–80

ThomsonNOW™ Test yourself on key content at **www.thomsonedu.com/login.**

Simplify. Assume that all variables represent positive numbers.

1. a. $\sqrt{18}$ **b.** $\sqrt[3]{24}$ **c.** $-\sqrt[4]{64}$ **2. a.** $\sqrt{50}$ **b.** $\sqrt[3]{54}$ **c.** $-\sqrt[4]{162}$

3. a. $\sqrt{60,000}$ **b.** $\sqrt[3]{900,000}$ **c.** $\sqrt[3]{\dfrac{-40}{27}}$ **4. a.** $\sqrt{800,000}$ **b.** $\sqrt[3]{24,000}$ **c.** $\sqrt[4]{\dfrac{80}{625}}$

5. a. $\sqrt[3]{x^{10}}$ **b.** $\sqrt{27z^3}$ **c.** $\sqrt[4]{48a^9}$ **6. a.** $\sqrt[3]{y^{16}}$ **b.** $\sqrt{12t^5}$ **c.** $\sqrt[3]{81b^8}$

Simplify.

7. a. $-\sqrt{18s}\sqrt{2s^3}$ **b.** $\sqrt[3]{7h^2}\,\sqrt[3]{-49h}$ **c.** $\sqrt{16 - 4x^2}$

8. a. $\sqrt{3w^3}\sqrt{27w^3}$ **b.** $-\sqrt[4]{2m^3}\,\sqrt[4]{8m}$ **c.** $\sqrt{9Y^2 + 18}$

9. a. $\sqrt[3]{8A^3 + A^6}$ **b.** $\dfrac{\sqrt{45x^3 y^3}}{\sqrt{5y}}$ **c.** $\dfrac{\sqrt[3]{8b^7}}{\sqrt[3]{a^6 b^2}}$

10. a. $\sqrt[3]{b^9 - 27b^3}$ **b.** $\dfrac{\sqrt{98x^2 y^3}}{\sqrt{xy}}$ **c.** $\dfrac{\sqrt[3]{16r^4}}{\sqrt[3]{4t^3}}$

Simplify and combine like terms.

11. $3\sqrt{7} + 2\sqrt{7}$ **12.** $5\sqrt{2} - 3\sqrt{2}$ **13.** $4\sqrt{3} - \sqrt{27}$

14. $\sqrt{75} + 2\sqrt{3}$ **15.** $\sqrt{50x} + \sqrt{32x}$ **16.** $\sqrt{8y} - \sqrt{18y}$

17. $3\sqrt[3]{16} - \sqrt[3]{2} - 2\sqrt[3]{54}$ **18.** $\sqrt[3]{81} + 2\sqrt[3]{24} - 3\sqrt[3]{3}$

Multiply.

19. $2(3 - \sqrt{5})$ **20.** $5(2 - \sqrt{7})$ **21.** $\sqrt{2}(\sqrt{6} + \sqrt{10})$

22. $\sqrt{3}(\sqrt{12} - \sqrt{15})$ **23.** $\sqrt[3]{2}(\sqrt[3]{20} - 2\sqrt[3]{12})$ **24.** $\sqrt[3]{3}(2\sqrt[3]{18} + \sqrt[3]{36})$

25. $(\sqrt{x} - 3)(\sqrt{x} + 3)$ **26.** $(2 + \sqrt{x})(2 - \sqrt{x})$ **27.** $(\sqrt{2} - \sqrt{3})(\sqrt{2} + 2\sqrt{3})$

28. $(\sqrt{3} - \sqrt{5})(2\sqrt{3} + \sqrt{5})$ **29.** $(\sqrt{5} - \sqrt{2})^2$ **30.** $(\sqrt{2} - 2\sqrt{3})^2$

31. $(\sqrt{a} - 2\sqrt{b})^2$ **32.** $(\sqrt{2a} - 2\sqrt{b})(\sqrt{2a} + 2\sqrt{b})$

Verify by substitution that the number is a solution of the quadratic equation.

33. $x^2 - 2x - 2 = 0;\ \ 1 + \sqrt{3}$ **34.** $x^2 + 4x - 1 = 0;\ \ -2 + \sqrt{5}$

35. $x^2 + 6x - 9 = 0;\ \ -3 + 3\sqrt{2}$ **36.** $4x^2 - 20x + 22 = 0;\ \ \dfrac{5 - \sqrt{3}}{2}$

Rationalize the denominator.

37. $\dfrac{6}{\sqrt{3}}$

38. $\dfrac{10}{\sqrt{5}}$

39. $\sqrt{\dfrac{7x}{18}}$

40. $\sqrt{\dfrac{27x}{20}}$

41. $\sqrt{\dfrac{2a}{b}}$

42. $\sqrt{\dfrac{5p}{q}}$

43. $\dfrac{2\sqrt{3}}{\sqrt{2k}}$

44. $\dfrac{6\sqrt{2}}{\sqrt{3v}}$

45. $\dfrac{4}{1+\sqrt{3}}$

46. $\dfrac{3}{7-\sqrt{2}}$

47. $\dfrac{x}{x-\sqrt{3}}$

48. $\dfrac{y}{\sqrt{5}-y}$

49. $\dfrac{\sqrt{6}-3}{2-\sqrt{6}}$

50. $\dfrac{\sqrt{x}+\sqrt{y}}{\sqrt{x}-\sqrt{y}}$

51. Use your calculator to graph each function, and explain the result.

    a. $y = \sqrt{x^2}$ **b.** $y = \sqrt[3]{x^3}$

52. Use your calculator to graph each function, and explain the result.

    a. $y = (x^4)^{1/4}$ **b.** $y = (x^5)^{1/5}$

For these problems, do not assume that variables represent positive numbers. Use absolute value bars as necessary to simplify the radicals.

53. a. $\sqrt{4x^2}$ **b.** $\sqrt{(x-5)^2}$ **c.** $\sqrt{x^2-6x+9}$

54. a. $\sqrt{9x^2y^4}$ **b.** $\sqrt{(2x-1)^2}$ **c.** $\sqrt{9x^2-6x+1}$

Solve.

55. $\sqrt{x}-5=3$

56. $\sqrt{x}-4=1$

57. $\sqrt{y+6}=2$

58. $\sqrt{y-3}=5$

59. $4\sqrt{z}-8=-2$

60. $-3\sqrt{z}+14=8$

61. $5+2\sqrt{6-2w}=13$

62. $8-3\sqrt{9+2w}=-7$

63. $3z+4=\sqrt{3z+10}$

64. $2x-3=\sqrt{7x-3}$

65. $2x+1=\sqrt{10x+5}$

66. $4x+5=\sqrt{3x+4}$

67. $\sqrt{y+4}=y-8$

68. $4\sqrt{x-4}=x$

69. $\sqrt{2y-1}=\sqrt{3y-6}$

70. $\sqrt{4y+1}=\sqrt{6y-3}$

71. $\sqrt{x-3}\sqrt{x}=2$

72. $\sqrt{x}\sqrt{x-5}=6$

73. $\sqrt{y+4}=\sqrt{y+20}-2$

74. $4\sqrt{y}+\sqrt{1+16y}=5$

75. $\sqrt{x}+\sqrt{2}=\sqrt{x+2}$

76. $\sqrt{4x+17}=4-\sqrt{x+1}$

77. $\sqrt{5+x}+\sqrt{x}=5$

78. $\sqrt{y+7}+\sqrt{y+4}=3$

79. Explain why the following first step for solving the radical equation is incorrect:

$$\sqrt{x-5}+\sqrt{2x-1}=8$$
$$(x-5)+(2x-1)=64$$

80. Explain why the following first step for solving the radical equation is incorrect:

$$\sqrt{x+2}+1=\sqrt{2x-3}$$
$$(x+2)+1=2x-3$$

Write each complex fraction as a simple fraction in lowest terms, and rationalize the denominator.

81. $\dfrac{\dfrac{2}{\sqrt{7}}}{1-\dfrac{\sqrt{3}}{\sqrt{7}}}$

82. $\dfrac{\dfrac{1}{4}}{\dfrac{\sqrt{5}}{2\sqrt{2}}-\dfrac{\sqrt{3}}{2\sqrt{2}}}$

83. $\dfrac{\dfrac{\sqrt{3}}{2}+\dfrac{1}{\sqrt{2}}}{1-\dfrac{\sqrt{3}}{2}\cdot\dfrac{1}{\sqrt{2}}}$

84. $\dfrac{\dfrac{1}{\sqrt{3}}-\dfrac{\sqrt{5}}{3}}{1+\dfrac{1}{\sqrt{3}}\cdot\dfrac{\sqrt{5}}{3}}$

In this section, we review some information you will need from geometry. You are already familiar with the formulas for the area and perimeter of common geometric figures; you can find these formulas in the reference section at the front of the book.

Right Triangles and the Pythagorean Theorem

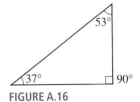

FIGURE A.16

A **right triangle** is a triangle in which one of the angles is a right angle, or 90°. Because the sum of the three angles in *any* triangle is 180°, this means that the other two angles in a right triangle must have a sum of 180° − 90°, or 90°. For instance, if we know that one of the angles in a right triangle is 37°, then the remaining angle must be 90° − 37°, or 53°, as shown in Figure A.16.

EXAMPLE 1   In a right triangle, the medium-sized angle is 15° less than twice the smallest angle. Find the sizes of the three angles in Figure A.17.

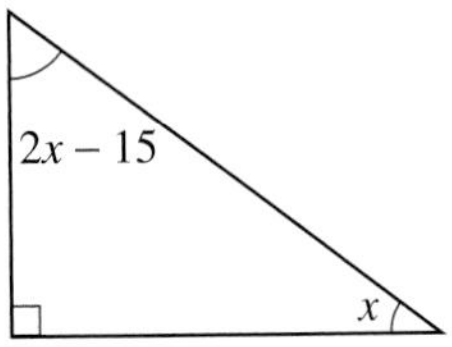

Solution

Step 1   Let x stand for the size of the smallest angle. Then the medium-sized angle must be $2x − 15$.

Step 2   Because the right angle is the largest angle, the sum of the smallest and medium-sized angles must be the remaining 90°. Thus,

$$x + (2x − 15) = 90$$

FIGURE A.17

Step 3   Solve the equation. Begin by simplifying the left side.

$$3x − 15 = 90 \quad \text{Add 15 to both sides.}$$
$$3x = 105 \quad \text{Divide both sides by 3.}$$
$$x = 35$$

Step 4   The smallest angle is 35°, and the medium-sized angle is $2(35°) − 15°$, or 55°.

In a right triangle (see Figure A.18), the longest side is opposite the right angle and is called the **hypotenuse.** Ordinarily, even if we know the lengths of two sides of a triangle, it is not easy to find the length of the third side (to solve this problem we need trigonometry), but for the special case of a right triangle, there is an equation that relates the lengths of the three sides. This property of right triangles was known to many ancient cultures, and we know it today by the name of a Greek mathematician, Pythagoras, who provided a proof of the result.

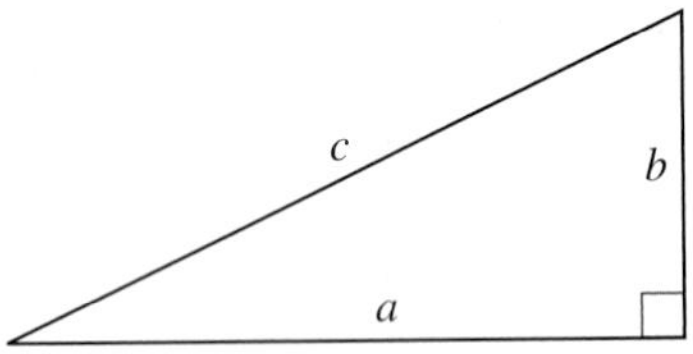

FIGURE A.18

Pythagorean Theorem

In a right triangle, if c stands for the length of the hypotenuse and a and b stand for the lengths of the two sides, then

$$a^2 + b^2 = c^2$$

EXAMPLE 2

The hypotenuse of a right triangle is 15 feet long. The third side is twice the length of the shortest side. Find the lengths of the other 2 sides.

Solution

Step 1   Let x represent the length of the shortest side, so that the third side has length $2x$. (See Figure A.19.)

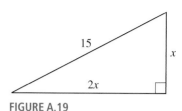

FIGURE A.19

Step 2   Substituting these expressions into the Pythagorean theorem, we find

$$x^2 + (2x)^2 = 15^2$$

Step 3   This is a quadratic equation with no linear term, so we simplify and then isolate x^2.

$$x^2 + 4x^2 = 225 \quad \text{Combine like terms.}$$
$$5x^2 = 225 \quad \text{Divide both sides by 5.}$$
$$x^2 = 45$$

Taking square roots of both sides yields

$$x = \pm\sqrt{45} \simeq \pm 6.708203932$$

Step 4   Because a length must be a positive number, the shortest side has length approximately 6.71 feet, and the third side has length 2 (6.71), or approximately 13.42 feet.

Isosceles and Equilateral Triangles

Recall also that an **isosceles** triangle is one that has at least two sides of equal length. In an isosceles triangle, the angles opposite the equal sides, called the **base** angles, are equal in measure, as shown in Figure A.20a. In an **equilateral** triangle (Figure A.20b), all three sides have equal length, and all three angles have equal measure.

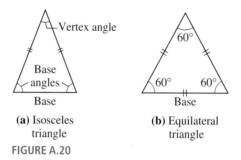

(a) Isosceles triangle   **(b)** Equilateral triangle

FIGURE A.20

The Triangle Inequality

The longest side in a triangle is always opposite the largest angle, and the shortest side is opposite the smallest angle. It is also true that the sum of the lengths of any two sides of a triangle must be greater than the third side, or else the two sides will not meet to form a triangle! This fact is called the **triangle inequality.**

In Figure A.21, we must have that

$$p + q > r$$

where p, q, and r are the lengths of the sides of the triangle.

Now we can use the triangle inequality to discover information about the sides of a triangle.

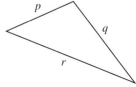

FIGURE A.21

EXAMPLE 4

Two sides of a triangle have lengths 7 inches and 10 inches. What can you say about the length of the third side (see Figure A.22)?

Solution   Let x represent the length of the third side of the triangle. By the triangle inequality, we must have that

$$x < 7 + 10, \quad \text{or} \quad x < 17$$

Looking at another pair of sides, we must also have that

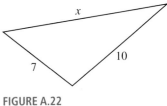

FIGURE A.22

$$10 < x + 7, \quad \text{or} \quad x > 3$$

Thus the third side must be greater than 3 inches but less than 17 inches long.

Similar Triangles

Two triangles are said to be **similar** if their corresponding angles are equal. This means that the two triangles will have the same shape but not necessarily the same size. One of the triangles will be an enlargement or a reduction of the other; so their corresponding sides are proportional. In other words, for similar triangles, the ratios of the corresponding sides are equal (see Figure A.23).

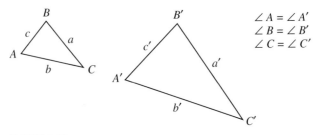

FIGURE A.23

If any two pairs of corresponding angles of two triangles are equal, then the third pair must also be equal, because in both triangles the sum of the angles is $180°$. Thus, to show that two triangles are similar, we need only show that two pairs of angles are equal.

EXAMPLE 5

The roof of an A-frame ski chalet forms an isosceles triangle with the floor as the base (see Figure A.24). The floor of the chalet is 24 feet wide, and the ceiling is 20 feet tall at the center. If a loft is built at a height of 8 feet from the floor, how wide will the loft be?

Solution   From Figure A.25, we can show that $\triangle ABC$ is similar to $\triangle ADE$. Both triangles include $\angle A$, and because $\overline{DE}$ is parallel to $\overline{BC}$, $\angle ADE$ is equal to $\angle ABC$. Thus, the triangles have two pairs of equal angles and are therefore similar triangles.

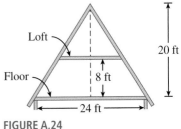

FIGURE A.24

Step 1   Let w stand for the width of the loft.

Step 2   First note that if $FG = 8$, then $AF = 12$. Because $\triangle ABC$ is similar to $\triangle ADE$, the ratios of their corresponding sides (or corresponding altitudes) are equal. In particular,

$$\frac{w}{24} = \frac{12}{20}$$

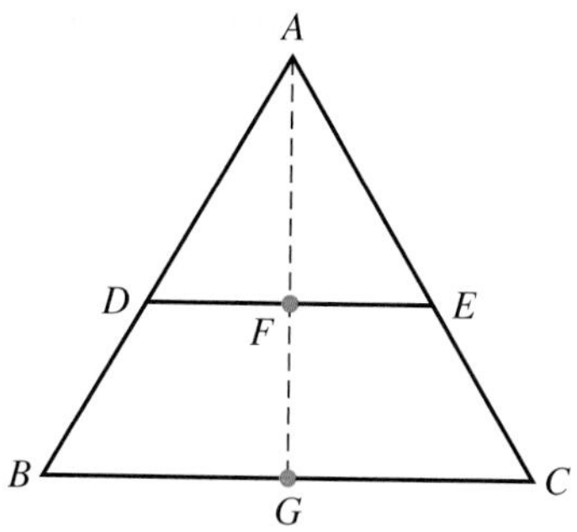

Step 3   Solve the proportion for w. Begin by cross-multiplying.

$$20w = (12)(24) \qquad \text{Apply the fundamental principle}$$

$$w = \frac{288}{20} = 14.4 \quad \text{Divide by 20.}$$

Step 4   The floor of the loft will be 14.4 feet wide.

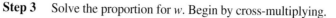

FIGURE A.25

Volume and Surface Area

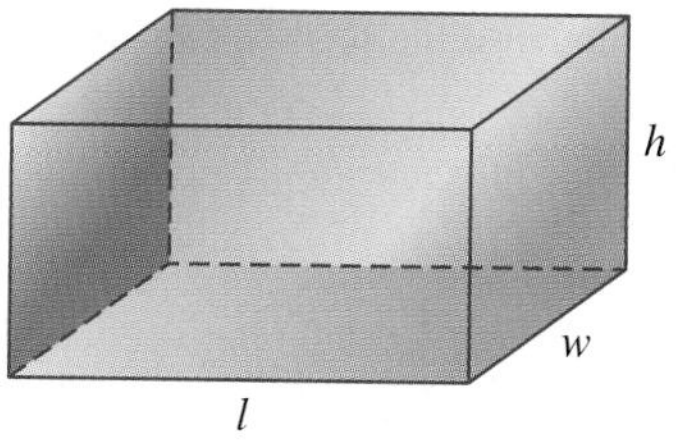

FIGURE A.26

The **volume** of a three-dimensional object measures its capacity, or how much space it encloses. Volume is measured in cubic units, such as cubic inches or cubic meters. The volume of a rectangular prism, or box, is given by the product of its length, width, and height. For example, the volume of the box of length 4 inches, width 3 inches, and height 2 inches shown in Figure A.26 is

$$V = lwh = 4(3)(2) = 24 \text{ cubic inches}$$

Formulas for the volumes of other common objects can be found inside the front cover of the book.

EXAMPLE 6   A cylindrical can must have a height of 6 inches, but it can have any reasonable radius.

a. Write an algebraic expression for the volume of the can in terms of its radius.

b. If the volume of the can should be approximately 170 cubic inches, what should its radius be?

Solutions   **a.** The formula for the volume of a right circular cylinder is $V = \pi r^2 h$. If the height of the cylinder is 6 inches, then $V = \pi r^2$ (6), or $V = 6\pi r^2$. (See Figure A.27.)

b. Substitute 170 for V and solve for r.

$$170 = 6\pi r^2 \qquad \text{Divide both sides by } 6\pi.$$

$$r^2 = \frac{170}{6\pi} \qquad \text{Take square roots.}$$

$$r = \sqrt{\frac{170}{6\pi}} \approx 3.00312$$

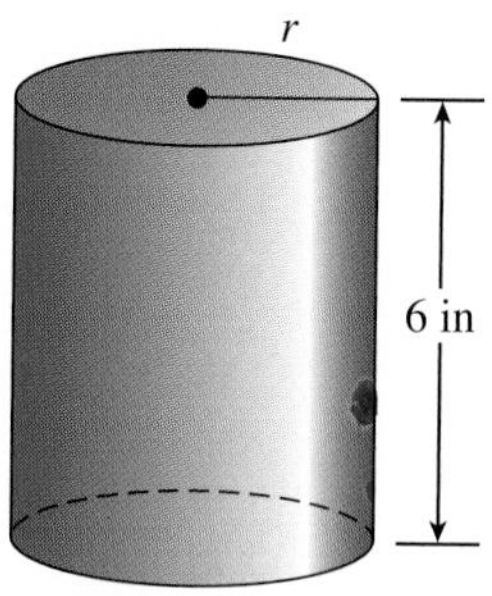

Thus, the radius of the can should be approximately 3 inches. A calculator keying sequence for the expression above is

FIGURE A.27

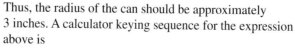

The **surface area** of a solid object is the sum of the areas of all the exterior faces of the object. It measures the amount of paper that would be needed to cover the object entirely. Since it is an area, it is measured in square units.

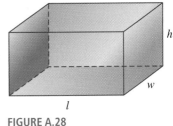

FIGURE A.28

EXAMPLE 7

Write a formula for the surface area of a closed box in terms of its length, width, and height. (See Figure A.28.)

Solution   The box has six sides; we must find the area of each side and add them. The top and bottom of the box each have area lw, so together they contribute $2lw$ to the surface area. The back and front of the box each have area lh, so they contribute $2lh$ to the surface area. Finally, the left and right sides of the box each have area wh, so they add $2wh$ to the surface area. Thus, the total surface area is

$$S = 2lw + 2lh + 2wh$$

Formulas for the surface areas of other common solids can be found on the insert in the front of the book.

The Distance Formula

By using the Pythagorean theorem, we can derive a formula for the distance between two points, P_1 and P_2, in terms of their coordinates. We first label a right triangle, as we did in the example above. Draw a horizontal line through P_1 and a vertical line through P_2. These lines meet at a point P_3, as shown in Figure A.29. The x-coordinate of P_3 is the same as the x-coordinate of P_2, and the y-coordinate of P_3 is the same as the y-coordinate of P_1. Thus, the coordinates of P_3 are (x_2, y_1).

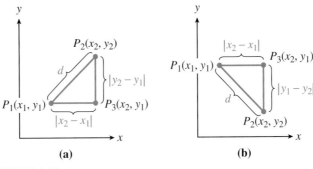

(a)   **(b)**

FIGURE A.29

The distance between P_1 and P_3 is $|x_2 - x_1|$, and the distance between P_2 and P_3 is $|y_2 - y_1|$. (See Section 2.5 to review distance and absolute value.) These two numbers are the lengths of the legs of the right triangle. The length of the hypotenuse is the distance between p_1 and P_2, which we will call d. By the Pythagorean theorem,

$$d^2 = (x_2 - x_1)^2 + (y_2 - y_1)^2$$

Taking the (positive) square root of each side of this equation gives us the **distance formula.**

Distance Formula

The **distance** d between points $P_1(x_1, y_1)$ and $P_2(x_2, y_2)$ is

$$d = \sqrt{(x_2 - x_1)^2 + (y_2 - y_1)^2}$$

EXAMPLE 8   Find the distance between $(2, -1)$ and $(4, 3)$.

Solution   Substitute $(2, -1)$ for (x_1, y_1) and substitute $(4, 3)$ for (x_2, y_2) in the distance formula to obtain

$$
\begin{aligned}
d &= \sqrt{(x_2 - x_1)^2 + (y_2 - y_1)^2} \\
&= \sqrt{(4 - 2)^2 + [3 - (-1)]^2} \\
&= \sqrt{4 + 16} = \sqrt{20} \approx 4.47
\end{aligned}
$$

In Example 8, we obtain the same answer if we use $(4, 3)$ for P_1 and use $(2, -1)$ for P_2:

$$
\begin{aligned}
d &= \sqrt{(2 - 4)^2 + [(-1) - 3]^2} \\
&= \sqrt{4 + 16} = \sqrt{20}
\end{aligned}
$$

The Midpoint Formula

If we know the coordinates of two points, we can calculate the coordinates of the point halfway between them using the **midpoint** formula. Each coordinate of the midpoint is the average of the corresponding coordinates of the two points.

Midpoint Formula

The **midpoint** of the line segment joining the points $P_1(x_1, y_1)$ and $P_2(x_2, y_2)$ is the point $M(\bar{x}, \bar{y})$, where

$$\bar{x} = \frac{x_1 + x_2}{2} \quad \text{and} \quad \bar{y} = \frac{y_1 + y_2}{2}$$

The proof of the formula uses similar triangles and is left as an exercise.

EXAMPLE 9   Find the midpoint of the line segment joining the points $(-2, 1)$ and $(4, 3)$.

Solution   Substitute $(-2, 1)$ for (x_1, y_1) and $(4, 3)$ for (x_2, y_2) in the midpoint formula to obtain

$$\bar{x} = \frac{x_1 + x_2}{2} = \frac{-2 + 4}{2} = 1 \text{ and}$$

$$\bar{y} = \frac{y_1 + y_2}{2} = \frac{1 + 3}{2} = 2$$

The midpoint of the segment is the point $(\bar{x}, \bar{y}) = (1, 2)$.

Circles

A **circle** is the set of all points in a plane that lie at a given distance, called the **radius,** from a fixed point called the **center.** We can use the distance formula to find an equation for a circle. First consider the circle in Figure A.30a, whose center is the origin, $(0, 0)$.

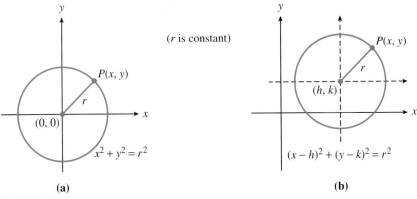

FIGURE A.30

The distance from the origin to any point $P(x, y)$ on the circle is r. Therefore,

$$\sqrt{(x - 0)^2 + (y - 0)^2} = r$$

Or, squaring both sides,

$$(x - 0)^2 + (y - 0)^2 = r^2$$

Thus, the equation for a circle of radius r centered at the origin is

$$x^2 + y^2 = r^2$$

Now consider the circle in Figure A.30b, whose center is the point (h, k). Every point $P(x, y)$ on the circle lies a distance r from (h, k), so the equation of the circle is given by the following formula.

> **Standard Form for a Circle**
>
> The equation for a **circle** of **radius** r centered at the point (h, k) is
> $$(x - h)^2 + (y - k)^2 = r^2$$

This equation is the *standard form* for a circle of radius r with center at (h, k). It is easy to graph a circle if its equation is given in standard form.

EXAMPLE 10   Graph the circles.

a. $(x - 2)^2 + (y + 3)^2 = 16$ **b.** $x^2 + (y - 4)^2 = 7$

Solutions   **a.** The graph of $(x - 2)^2 + (y + 3)^2 = 16$ is a circle with radius 4 and center at $(2, -3)$. To sketch the graph, first locate the center of the circle. (The center is not part of the graph of the circle.) From the center, move a distance of 4 units (the radius of the circle) in each of four directions: up, down, left, and right. This locates four points that lie on the circle: $(2, 1)$, $(2, -7)$, $(-2, -3)$, and $(6, -3)$. Sketch the circle through these four points. (See Figure A.31a.)

b. The graph of $x^2 + (y - 4)^2 = 7$ is a circle with radius $\sqrt{7}$ and center at $(0, 4)$. From the center, move approximately $\sqrt{7}$, or 2.6, units in each of the four coordinate directions to obtain the points $(0, 6.6)$, $(0, 1.4)$, $(-2.6, 4)$, and $(2.6, 4)$. Sketch the circle through these four points. (See Figure A.31b.)

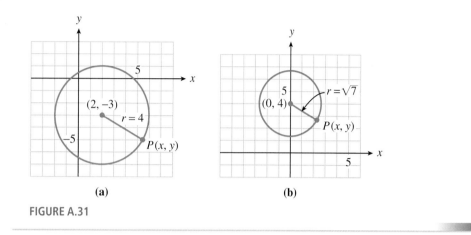

(a)   (b)

FIGURE A.31

We can write an equation for any circle if we can find its center and radius.

EXAMPLE 11   Find an equation for the circle whose diameter has endpoints $(7, 5)$ and $(1, -1)$.

Solution   The center of the circle is the midpoint of its diameter (see Figure A.32). Use the midpoint formula to find the center:

$$h = \bar{x} = \frac{7 + 1}{2} = 4$$

$$k = \bar{y} = \frac{5 - 1}{2} = 2$$

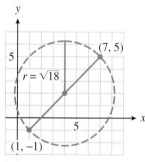

FIGURE A.32

Thus, the center is the point $(h, k) = (4, 2)$. The radius is the distance from the center to either of the endpoints of the diameter, say the point $(7, 5)$. Use the distance formula with the points $(7, 5)$ and $(4, 2)$ to find the radius.

$$r = \sqrt{(7 - 4)^2 + (5 - 2)^2}$$

$$= \sqrt{3^2 + 3^2} = \sqrt{18}$$

Finally, substitute 4 for h and 2 for k (the coordinates of the center) and $\sqrt{18}$ for r (the radius) into the standard form to obtain

$$(x - h)^2 + (y - k)^2 = r^2$$

$$(x - 4)^2 + (y - 2)^2 = 18$$

VOCABULARY   *Look up the definitions of new terms in the Glossary.*

Right triangle   Circle   Surface area   Similar triangles
Isosceles   Hypotenuse   Center   Midpoint
Volume   Triangle inequality Equilateral   Radius

SKILLS   *Practice each skill in the Exercises listed.*

1. Use properties of triangles: #1–10
2. Use similar triangles to solve problems: #11–16
3. Calculate volumes and surface areas: #17–20

4. Use the distance and midpoint formulas: #21–32
5. Sketch a circle: #33–40
6. Find the equation for a circle: #41–46

EXERCISES A.11

ThomsonNOW™ Test yourself on key content at www.thomsonedu.com/login.

Use properties of triangles to answer the questions.

1. One angle of a triangle is 10° larger than another, and the third angle is 29° larger than the smallest. How large is each angle?

2. One angle of a triangle is twice as large as the second angle, and the third angle is 10° less than the larger of the other two. How large is each angle?

3. One acute angle of a right triangle is twice the other acute angle. How large is each acute angle?

4. One acute angle of a right triangle is 10° less than three times the other acute angle. How large is each acute angle?

5. The vertex angle of an isosceles triangle is 20° less than the sum of the equal angles. How large is each angle?

6. The vertex angle of an isosceles triangle is 30° less than one of the equal angles. How large is each angle?

7. The perimeter of an isosceles triangle is 42 centimeters and its base is 12 centimeters long. How long are the equal sides?

8. The altitude of an equilateral triangle is $\frac{\sqrt{3}}{2}$ times its base. If the perimeter of an equilateral triangle is 18 inches, what is its area?

9. If two sides of a triangle are 6 feet and 10 feet long, what can you say about the length of the third side?

10. If one of the equal sides of an isosceles triangle is 8 millimeters long, what can you say about the length of the base?

Use the properties of similar triangles to answer the questions.

11. A 6-foot man stands 12 feet from a lamppost. His shadow is 9 feet long. How tall is the lamppost?

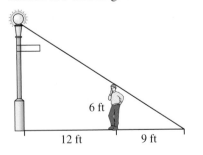

6 ft

12 ft   9 ft

12. A rock climber estimates the height of a cliff she plans to scale as follows: She places a mirror on the ground so that she can just see the top of the cliff in the mirror while she stands straight. (The angles 1 and 2 formed by the light rays are equal.) She then measures the distance to the mirror (2 feet) and the distance from the mirror to the base of the cliff. If she is 5 feet 6 inches tall, how high is the cliff?

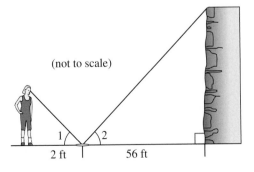

(not to scale)

1   2

2 ft   56 ft

13. A conical tank is 12 feet deep and the diameter of the top is 8 feet. If the tank is filled with water to a depth of 7 feet, what is the area of the exposed surface of the water?

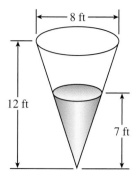

14. A florist fits a cylindrical piece of foam into a conical vase that is 10 inches high and measures 8 inches across the top. If the radius of the foam cylinder is $2\frac{1}{2}$ inches, how tall should it be just to reach the top of the vase?

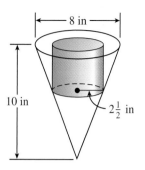

15. To measure the distance across a river, stand at point A and sight across the river to a convenient landmark at B. Then measure the distances AC, CD, and DE. If $AC = 20$ feet, $CD = 13$ feet, and $DE = 58$ feet, how wide is the river?

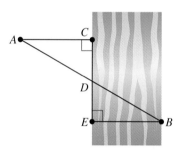

16. To measure the distance EC across a lake, stand at point A and sight point C across the lake, then mark point B. Then sight to point E and mark point D so that DB is parallel to CE. If $AD = 25$ yards, $AE = 60$ yards, and $BD = 30$ yards, how wide is the lake?

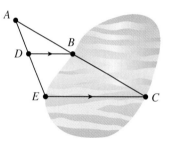

Use the Geometry formulas in the reference card in the front of this book to find volumes and surface areas.

17. a. How much helium (in cubic meters) is needed to inflate a spherical balloon to a radius of 1.2 meters?
   b. How much gelatin (in square centimeters) is needed to coat a spherical pill whose radius is 0.7 centimeter?

18. a. How much storage space is there in a rectangular box whose length is 12.3 inches, whose width is 4 inches, and whose height is 7.3 inches?
   b. How much marine sealer will be needed to paint a rectangular wooden storage locker with length 6.2 feet, width 5.8 feet, and height 2.6 feet?

19. a. How much grain can be stored in a cylindrical silo whose radius is 6 meters and whose height is 23.2 meters?
   b. How much paint is needed to cover a cylindrical storage drum whose radius is 15.3 inches and whose height is 4.5 inches?

20. a. A conical pile of sand is 8.1 feet high and has a radius of 4.6 feet. How much sand is in the pile?
   b. How much plastic is needed to line a conical funnel with a radius of 16 centimeters and a slant height of 42 centimeters?

Find the distance between each of the given pairs of points, and find the midpoint of the segment joining them.

21. $(1, 1)$, $(4, 5)$

22. $(-1, 1)$, $(5, 9)$

23. $(2, -3)$, $(-2, -1)$

24. $(5, -4)$, $(-1, 1)$

25. $(3, 5)$, $(-2, 5)$

26. $(-2, -5)$, $(-2, 3)$

27. Leanne is sailing 3 miles west and 5 miles south of the harbor. She heads directly toward an island that is 8 miles west and 7 miles north of the harbor.
   a. How far is Leanne from the island?
   b. How far will Leanne be from the harbor when she is halfway to the island?

28. Dominic is 100 meters east and 250 meters north of Kristy. He is walking directly toward a tree that is 220 meters east and 90 meters north of Kristy.
   a. How far is Dominic from the tree?
   b. How far will Dominic be from the Kristy when he is halfway to the tree?

Sketch a diagram for each problem on graph paper, then solve the problem.

29. Find the perimeter of the triangle with vertices $(10, 1)$, $(3, 1)$, $(5, 9)$.

30. Find the perimeter of the triangle with vertices $(-1, 5)$, $(8, -7)$, $(4, 1)$.

31. Show that the point $C(\sqrt{5}, 2 + \sqrt{5})$ is the same distance from $A(2, 0)$ and $B(-2, 4)$.

32. Show that the points $(-2, 1)$, $(0, -1)$, and $(\sqrt{3} - 1, \sqrt{3})$ are the vertices of an equilateral triangle.

Graph each equation.

33. $x^2 + y^2 = 25$

34. $x^2 + y^2 = 16$

35. $4x^2 + 4y^2 = 16$

36. $2x^2 + 2y^2 = 18$

37. $(x - 4)^2 + (y + 2)^2 = 9$

38. $(x - 1)^2 + (y - 3)^2 = 16$

39. $(x + 3)^2 + y^2 = 10$

40. $x^2 + (y + 4)^2 = 12$

Write an equation for the circle with the given properties.

41. Center at $(-2, 5)$, radius $2\sqrt{3}$.

42. Center at $(4, -3)$, radius $2\sqrt{6}$.

43. Center at $\left(\frac{3}{2}, -4\right)$, one point on the circle $(4, -3)$.

44. Center at $\left(\frac{-3}{2}, \frac{-1}{2}\right)$, one point on the circle $(-4, -2)$.

45. Endpoints of a diameter at $(1, 5)$ and $(3, -1)$.

46. Endpoints of a diameter at $(3, 6)$ and $(-5, 2)$.

A.12 The Real Number System

Subsets of the Real Numbers

The numbers associated with points on a number line are called the **real numbers.** The set of real numbers is denoted by $\mathbb{R}$. You are already familiar with several types, or subsets, of real numbers:

- The set $\mathbb{N}$ of **natural,** or **counting numbers,** as its name suggests, consists of the numbers $1, 2, 3, 4, \ldots$, where "$\ldots$" indicates that the list continues without end.

- The set $\mathbb{W}$ of **whole numbers** consists of the natural numbers and zero: $0, 1, 2, 3 \ldots$.

- The set $\mathbb{Z}$ of **integers** consists of the natural numbers, their negatives, and zero: $\ldots, -3, -2, -1, 0, 1, 2, 3, \ldots$.

All of these numbers are subsets of the rational numbers.

Rational Numbers

A number that can be expressed as the quotient of two integers $\frac{a}{b}$, where $b \neq 0$, is called a **rational number.** The integers are rational numbers, and so are common fractions. Some examples of rational numbers are 5, -2, 0, $\frac{2}{9}$, $\sqrt{16}$, and $\frac{-4}{17}$. The set of rational numbers is denoted by $\mathbb{Q}$.

Every rational number has a decimal form that either terminates or repeats a pattern of digits. For example,

$$\frac{3}{4} = 3 \div 4 = 0.75, \quad \text{a \textbf{terminating decimal}}$$

and

$$\frac{9}{37} = 9 \div 37 = 0.243243243\ldots$$

where the pattern of digits 243 is repeated endlessly. We use the **repeater bar** notation to write a repeating decimal fraction:

$$\frac{9}{37} = 0.\overline{243}$$

Irrational Numbers

Some real numbers *cannot* be written in the form $\frac{a}{b}$, where a and b are integers. For example, the number $\sqrt{2}$ is not equal to any common fraction. Such numbers are called **irrational numbers.** Examples of irrational numbers are $\sqrt{15}$, π, and $-\sqrt[3]{7}$.

The decimal form of an irrational number never terminates, and its digits do not follow a repeating pattern, so it is impossible to write down an exact decimal equivalent for an irrational number. However, we can obtain decimal *approximations* correct to any desired degree of accuracy by rounding off. A graphing calculator gives the decimal representation of π as 3.141592654. This is not the *exact* value of π, but for most calculations it is quite adequate.

Some nth roots are rational numbers and some are irrational numbers. For example,

$$\sqrt{49}, \quad \sqrt[3]{\frac{27}{8}}, \quad \text{and} \quad 81^{1/4}$$

are rational numbers because they are equal to 7, $\frac{3}{2}$, and 3, respectively. On the other hand,

$$\sqrt{5}, \quad \sqrt[3]{56}, \quad \text{and} \quad 7^{1/5}$$

are irrational numbers. We can use a calculator to obtain decimal approximations for each of these numbers:

$$\sqrt{5} \approx 2.236, \quad \sqrt[3]{56} \approx 3.826, \quad \text{and} \quad 7^{1/5} \approx 1.476$$

The subsets of the real numbers are related as shown in Figure A.33. Every natural number is also a whole number, every whole number is an integer, every integer is a rational number, and every rational number is real. Also, every real number is either rational or irrational.

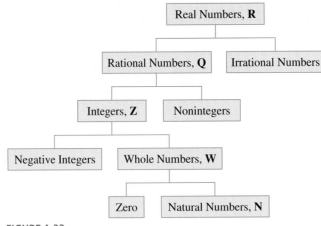

FIGURE A.33

 EXAMPLE 1

a. 2 is a natural number, a whole number, an integer, a rational number, and a real number.

b. $\sqrt{15}$ is an irrational number and a real number.

c. The number π, whose decimal representation begins 3.14159... is irrational and real.

d. 3.14159 is a rational and real number (which is close but not exactly equal to π).

Properties of the Real Numbers

The real numbers have several useful properties governing the operations of addition and multiplication. If a, b, and c represent real numbers, then each of the following equations is true:

- $a + b = b + a$ Commutative properties
  $ab = ba$

- $(a + b) + c = a + (b + c)$ Associative properties
  $(ab)c = a(bc)$

- $a(b + c) = ab + ac$ Distributive property

- $a + 0 = a$ Identity properties
  $a \cdot 1 = a$

These properties do not mention subtraction or division. But we can define *subtraction* and *division* in terms of addition and multiplication. For example, we can define the difference $a - b$ as follows:

$$a - b = a + (-b)$$

where $-b$, the **additive inverse** (or **opposite**) of b, is the number that satisfies

$$b + (-b) = 0$$

Similarly, we can define the quotient $\frac{a}{b}$:

$$\frac{a}{b} = a\left(\frac{1}{b}\right) \qquad (b \neq 0)$$

where $\frac{1}{b}$, the **multiplicative inverse** (or **reciprocal**) of b, is the number that satisfies

$$b \cdot \frac{1}{b} = 1 \quad (b \neq 0)$$

Division by zero is not defined.

EXAMPLE 2   Use the commutative and associative laws to simplify the computations.

a. $24 + 18 + 6$ **b.** $4 \cdot 27 \cdot 25$

Solutions   **a.** Apply the commutative law of addition.

$$24 + 18 + 6 = (24 + 6) + 18$$
$$= 30 + 18 = 48$$

b. Apply the commutative law of multiplication.

$$4 \cdot 27 \cdot 25 = (4 \cdot 25) \cdot 27$$
$$= 100 \cdot 27 = 2700$$

Order Properties of the Real Numbers

Real numbers obey properties about order, that is, properties about inequalities. The familiar inequality symbols, $<$ and $>$, have the following properties:

- If a and b are any real numbers, then one of three things is true:

$$a < b, \quad \text{or} \quad a > b, \quad \text{or} \quad a = b$$

- (Transitive property) For real numbers a, b, and c,

$$\text{if} \quad a < b \quad \text{and} \quad b < c, \quad \text{then} \quad a < c$$

We also have three properties that are useful for solving inequalities:

- If $a < b$, then $a + c < b + c$.
- If $a < b$ and $c > 0$, then $ac < bc$.
- If $a < b$ and $c < 0$, then $ac > bc$.

EXAMPLE 3   **a.** If $x < y$ and $y < -2$, then $x < -2$.   **b.** $\pi < 3.1416$, so $10\pi < 31.416$.
c. $\frac{1}{3} > 0.33$, so $-\frac{1}{3} < -0.33$.

SECTION A.12 SUMMARY

VOCABULARY   *Look up the definitions of new terms in the Glossary.*

Real number	Multiplicative inverse	Additive inverse	Distributive property
Whole number	Natural number	Reciprocal	Opposite
Irrational number	Integers	Counting number	Transitive property
Commutative property	Terminating decimal	Rational number	
Identity property	Associative property	Repeater bar	

SKILLS   *Practice each skill in the Exercises listed.*

1. Identify types of numbers: #1–12
2. Write the decimal form of a fraction: #13–20

3. Use the properties governing arithmetic operations: #21–40
4. Use the properties of order: #41–46

ThomsonNOW™ Test yourself on key content at www.thomsonedu.com/login.

Name the subsets of the real numbers to which each of the numbers in Exercises 1–12 belongs.

1. $-\dfrac{5}{8}$ **2.** 137   **3.** $\sqrt{8}$ **4.** 2.71828...   **5.** -36 **6.** $\sqrt{49}$

7. 0   **8.** $0.\overline{357}$ **9.** $13.\overline{289}$ **10.** $\sqrt{\dfrac{4}{9}}$ **11.** 2π **12.** $\dfrac{13}{7}$

Write each rational number in decimal form. Does the decimal terminate or does it repeat a pattern?

13. $\dfrac{3}{8}$ **14.** $\dfrac{5}{6}$ **15.** $\dfrac{2}{7}$ **16.** $\dfrac{43}{11}$ **17.** $\dfrac{7}{16}$ **18.** $\dfrac{5}{12}$ **19.** $\dfrac{11}{13}$ **20.** $\dfrac{25}{6}$

Fill in the blank according to the indicated property.

21. Commutative property
$7 + 10 = 10 +$ _____

22. Associative property
$(6 \cdot 4) \cdot 3 = 6 \cdot (4 \cdot$ _____$)$

23. Associative property
$(3 + 6) + 9 =$ _____$+ (6 + 9)$

24. Commutative property
$8 \cdot 12 =$ _____ $\cdot 8$

25. Commutative property
$36 \cdot 147 =$ _____ $\cdot 36$

26. Commutative property
$13 + 87 = 87+$ _____

27. Associative property
$(17 \cdot 2) \cdot 5 = 17 \cdot ($ _____ $\cdot$ _____ $)$

28. Associative property
$(44 + 12) + 8 = 44 + ($ _____ $+$ _____ $)$

29. Commutative property
$(5 + 9) + 4 = (9 +$ _____ $) + 4$

30. Commutative property
$(8 \cdot 9) \cdot 3 = (9 \cdot$ _____ $) \cdot 3$

Use the commutative and associative properties to compute mentally.

31. $47 + 28 + 3$ **32.** $12 + 147 + 8$ **33.** $26 + 37 + 3 + 4$ **34.** $55 + 32 + 5 + 8$

35. $2 \cdot 7 \cdot 5$ **36.** $15 \cdot 6 \cdot 2$ **37.** $50 \cdot 13 \cdot 2$ **38.** $4 \cdot 26 \cdot 25$

39. $4 \cdot 6 \cdot 5 \cdot 5$ **40.** $8 \cdot 8 \cdot 5 \cdot 5$

Fill in the blank with the correct symbol: <, >, or =.

41. -0.667 _____ $-\frac{2}{3}$

42. $\sqrt{2}$ _____ 1.4

43. If $x > 8$, then $x - 7$ _____ 1.

44. If $x < -6$, then $x - 6$ _____ -12.

45. If $x > -2$, then $-9x$ _____ 18.

46. If $x < -4$, then $3x$ _____ -12.

Using a Graphing Calculator

This appendix provides instructions for TI-84 or TI-83 calculators from Texas Instruments, but most other calculators work similarly. We describe only the basic operations and features of the graphing calculator used in your textbook.

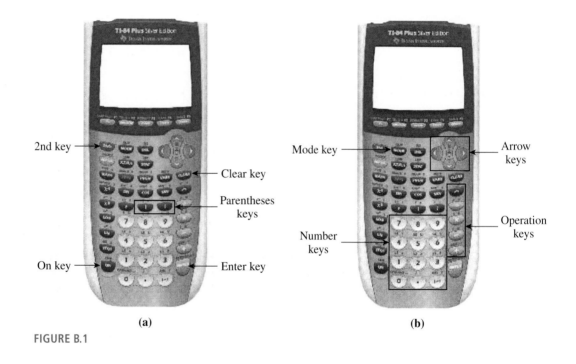

2nd key

Clear key

Parentheses keys

On key

Enter key

(a)

Mode key

Arrow keys

Operation keys

Number keys

(b)

FIGURE B.1

Getting Started

A. On and Off

Press $\boxed{\text{ON}}$ to turn **on** the calculator (see Figure B.1a). You will see a cursor blinking in the upper left corner of the Home screen. Press $\boxed{\text{2nd}}$ $\boxed{\text{ON}}$ to turn **off** the calculator.

B. Numbers and Operations

The parentheses keys, the Clear key, and the Enter key are shown in Figure B.1a. Locate the number keys, operation keys, and arrow keys on your calculator, as shown in Figure B.1b.

1. We use the $\boxed{-}$ key for subtraction, but we use the $\boxed{(-)}$ key (located next to $\boxed{\text{ENTER}}$) for negative numbers.

EXAMPLE Compute $5 - 8$. Press

$$5 \boxed{-} 8 \boxed{\textbf{ENTER}} \qquad \textit{Ans.} \quad -3$$

EXAMPLE Compute $-5 + 8$. Press

$$\boxed{(-)} 5 \boxed{+} 8 \boxed{\textbf{ENTER}} \qquad \textit{Ans.} \quad 3$$

We press $\boxed{\textbf{ENTER}}$ to tell the calculator to compute.

2. The calculator has a key for the value of π.

EXAMPLE Compute 2π. Press

$$2 \boxed{\times} \boxed{\textbf{2nd}} \boxed{\wedge} \boxed{\textbf{ENTER}} \quad \text{or} \quad 2 \boxed{\textbf{2nd}} \boxed{\wedge} \boxed{\textbf{ENTER}} \qquad \textit{Ans.} \quad 6.283185307$$

C. Clear and Delete

Press $\boxed{\textbf{DEL}}$ to delete the character under the cursor.

Press $\boxed{\textbf{CLEAR}}$ to clear the contents of the current input line.

In the Home screen, press $\boxed{\textbf{CLEAR}}$ $\boxed{\textbf{CLEAR}}$ to clear the entire screen.

Troubleshooting

1. If your screen is too light, press $\boxed{\textbf{2nd}}$ $\boxed{\Delta}$ several times to make it darker. If it is too dark, press $\boxed{\textbf{2nd}}$ $\boxed{\nabla}$.

2. For the features we use in this book, the **MODE** and **FORMAT** should be in their default settings. Press $\boxed{\textbf{MODE}}$ to see the menu in Figure B.2a, and $\boxed{\textbf{2nd}}$ $\boxed{\textbf{ZOOM}}$ to see the format menu in Figure B2.b. Use the Arrow keys and $\boxed{\textbf{ENTER}}$ to alter the menus to the default settings if necessary. *Note:* The Set Clock function does not appear on the T1-83.

        (a)   **(b)**

FIGURE B.2

B.2   Entering Expressions

A. Parentheses

1. Order of Operations The calculator follows the standard order of operations.

EXAMPLE Compute $2 + 3 \cdot 4$.

$$2 \boxed{+} 3 \boxed{\times} 4 \boxed{\textbf{ENTER}} \qquad \textit{Ans.} \quad 14$$

Compute $(2 + 3) \cdot 4$.

$(\; 2 \; + \; 3 \;) \; \times \; 4 \; \boxed{\text{ENTER}}$ *Ans.* 20

---

2. Fractions If the numerator or denominator of a fraction includes an operation, we must enclose the entire numerator or denominator within parentheses.

Evaluate $\dfrac{1}{2 \cdot 3}$.

$1 \; \div \; (\; 2 \; \times \; 3 \;) \; \boxed{\text{ENTER}}$ *Ans.* 0.1666666667

Evaluate $\dfrac{1 + 3}{2}$.

$(\; 1 \; + \; 3 \;) \; \div \; 2 \; \boxed{\text{ENTER}}$ *Ans.* 2

B. Exponents and Powers

1. Exponents We use the caret key, $\boxed{\land}$ to enter exponents or powers.

Evaluate 2^{10}.

$2 \; \boxed{\land} \; 10 \; \boxed{\text{ENTER}}$ *Ans.* 1024

---

2. Squaring There is a short-cut key for squaring, $\boxed{x^2}$.

Evaluate 57^2.

$57 \; \boxed{x^2} \; \boxed{\text{ENTER}}$ *Ans.* 3249

---

3. Fractional Exponents Fractional exponents must be enclosed in parentheses!

Evaluate $8^{2/3}$.

$8 \; \boxed{\land} \; (\; 2 \; \div \; 3 \;) \; \boxed{\text{ENTER}}$ *Ans.* 4

C. Roots

1. Square Roots We access the square root by pressing $\boxed{\text{2nd}}$ $\boxed{x^2}$, and the display shows $\sqrt{\;}($. The calculator automatically gives an open parenthesis for the square root, but not a close parenthesis.

Evaluate $\sqrt{2}$.

$\boxed{\text{2nd}} \; \boxed{x^2} \; 2 \;) \; \boxed{\text{ENTER}}$ *Ans.* 1.414213562

EXAMPLE   Evaluate $\sqrt{9+16}$.

$$\boxed{\text{2nd}}\ \boxed{x^2}\ \textbf{9}\ \boxed{+}\ \textbf{16}\ \boxed{)}\ \boxed{\text{ENTER}}\qquad Ans.\ 5$$

In the next example, note that we must enter $\boxed{)}$ *at the end of the radicand* to tell the calculator where the radical ends.

EXAMPLE   Evaluate $\sqrt{9}+16$.

$$\boxed{\text{2nd}}\ \boxed{x^2}\ \textbf{9}\ \boxed{)}\ \boxed{+}\ \textbf{16}\ \boxed{\text{ENTER}}\qquad Ans.\ 19$$

2. Cube Roots For cube roots, we press $\boxed{\text{MATH}}$ to open the **Math** menu and press 4 (see Figure B.3).

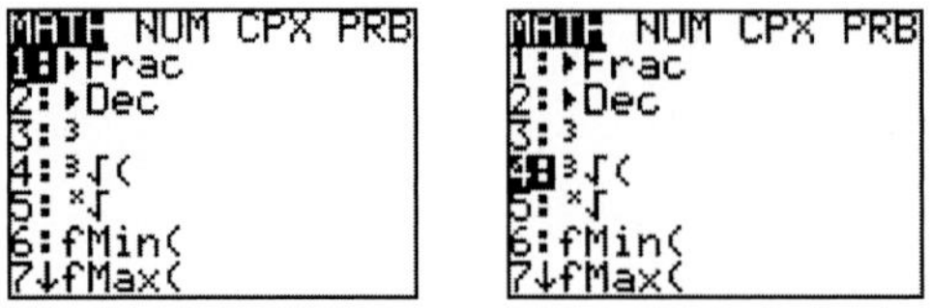

(a)   (b)

FIGURE B.3

EXAMPLE   Compute $\sqrt[3]{1728}$.

$$\boxed{\text{MATH}}\ \boxed{4}\ \textbf{1728}\ \boxed{)}\ \boxed{\text{ENTER}}\qquad Ans.\ 12$$

For evaluating cube roots and square roots, $\boxed{)}$ can be omitted if there are no operations following the radical.

3. Other Roots For nth roots, we press $\boxed{\text{MATH}}$ to open the **Math** menu and press 5 (see Figure B.3a). The calculator symbol for nth roots, $\sqrt[x]{\ }$, does not include an open parenthesis, $\boxed{(}$. If the radicand includes an operation, we must enclose it in parentheses.

EXAMPLE   Compute $\sqrt[10]{2.512}$

$$\textbf{10}\ \boxed{\text{MATH}}\ \boxed{5}\ \boxed{(}\ \textbf{2}\ \boxed{\times}\ \textbf{512}\ \boxed{)}\ \boxed{\text{ENTER}}\qquad Ans.\ 2$$

Notice that we enter the index 10 *before* the radical symbol.

FIGURE B.4

D. Absolute Value

TI calculators use **abs (x)** instead of |x| to denote the absolute value of x. The absolute value function is the first entry in the **MATH NUM** menu (see Figure B.4). The calculator gives $\boxed{(}$ for the absolute value function, but not $\boxed{)}$.

EXAMPLE   Evaluate $\dfrac{|21 \cdot 54 - 81|}{-9}$.

$$\boxed{\text{MATH}}\ \boxed{\triangleright}\ \boxed{\text{ENTER}}\ \textbf{21}\ \boxed{\times}\ \textbf{54}\ \boxed{-}\ \textbf{81}\ \boxed{)}\ \boxed{\div}\ \boxed{(-)}\ \textbf{9}\ \boxed{\text{ENTER}}\qquad Ans.\ -117$$

E. Scientific Notation

The TI calculators display numbers in scientific notation if they are too large or too small to fit in the display screen.

EXAMPLE   Compute $123,456,789^2$. Enter

$$\textbf{123456789} \boxed{x^2} \qquad \textit{Ans.} \quad 1.524157875 \text{ E } 16$$

This is how the calculator displays the number $1.524157875 \times 10^{16}$. Notice that the power 10^{16} is displayed as **E** 16.

To enter a number in scientific form, we use the key labeled **EE**, or $\boxed{\text{2nd}}$ $\boxed{,}$.

EXAMPLE   To enter 3.26×10^{-18}, use the keying sequence

$$\textbf{3.26} \boxed{\text{2nd}} \boxed{,} \boxed{(-)} \textbf{18} \qquad \textit{Ans.} \quad 3.26 \text{ E--18}$$

Troubleshooting

If your calculator gives you an error message like this, you may have made one of the following common mistakes:

```
ERR:SYNTAX
1:Quit
2:Goto
```

1. Using the negative key, $\boxed{(-)}$, when you wanted the subtraction key, $\boxed{-}$, or vice versa
2. Omitting a $\boxed{(}$ or $\boxed{)}$. Each $\boxed{(}$ should have a matching $\boxed{)}$

Press $\boxed{2}$ to **Go to** the error, and see Editing Expressions below.

B.3   Editing Expressions

A. Overwriting

We can edit an expression without starting again. If we place the cursor over a symbol and press a new key, the new symbol replaces (overwrites) the old one.

EXAMPLE   Correct the error in the following keystrokes for $120 - 36$:

$$\textbf{120} \boxed{(-)} \textbf{36} \boxed{\text{ENTER}}$$

The calculator gives an error message. Select option 2, and a blinking cursor appears over the error. Press $\boxed{-}$ to replace the negative symbol by the subtraction symbol.

B. Recalling an Entry

We can recall a previous entry by pressing $\boxed{\text{2nd}}$ $\boxed{\text{ENTER}}$.

EXAMPLE   Evaluate $\sqrt{3} + \sqrt{2}$ and $\sqrt{3} - \sqrt{2}$. First evaluate $\sqrt{3} + \sqrt{2}$:

$$\boxed{\text{2nd}} \boxed{x^2} \textbf{3} \boxed{)} \boxed{+} \boxed{\text{2nd}} \boxed{x^2} \textbf{2} \boxed{\text{ENTER}} \qquad \textit{Ans.} \quad 3.14626437$$

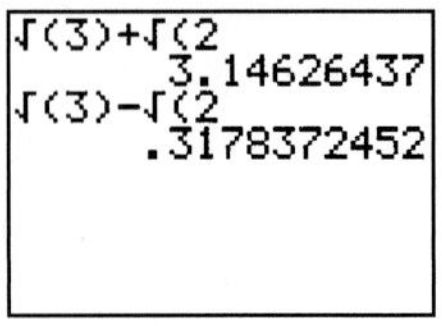

FIGURE B.5

Now press `2nd` `ENTER` to recall the last entry, and use the left arrow key `◄` to position the cursor over + . Press `-` to change to −, then press `ENTER` . Your screen should look like Figure B.5.

C. Inserting a Character

To insert a new character *before* a symbol, position the cursor over that symbol and press `2nd` `DEL` to get the **INS** (insert) command.

EXAMPLE   Evaluate $\sqrt{3} - 5\sqrt{2}$ by editing the example above.

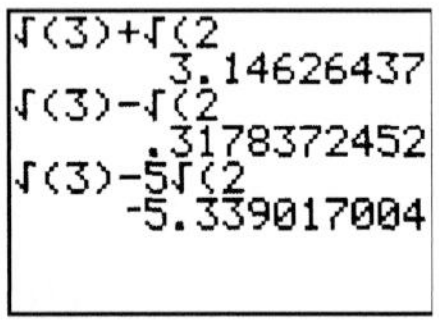

FIGURE B.6

Press `2nd` `ENTER` to recall the last entry, and use the left arrow key `◄` to position the cursor over the second $\sqrt{\ }$ from left to right. Press `2nd` `DEL` **5** to insert 5 *before* the $\sqrt{\ }$ symbol, then press `ENTER` . Your screen should look like Figure B.6.

D. Recalling an Answer

We often want to use the result from a previous calculation in a new calculation, without having to type in the number. We use the **ANS** key, `2nd` `(−)` , to recall the answer to the last calculation.

EXAMPLE   **a.** Evaluate $\sqrt{5} - 1$.

$$`2nd` \; `x^2` \; \textbf{5} \; `)` \; `-` \; \textbf{1} \; `ENTER` \qquad Ans. \; 1.236067977$$

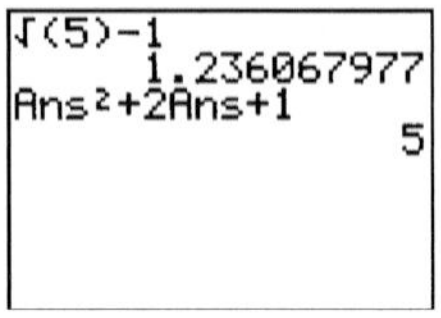

FIGURE B.7

b. Evaluate $x^2 + 2x + 1$ for $x = \sqrt{5} - 1$. Because x is the last answer the calculator computed, we enter

$$`2nd` \; `(−)` \; `x^2` \; `+` \; \textbf{2} \; `2nd` \; `(−)` \; `+` \; \textbf{1} \; `ENTER` \qquad Ans. \; 5$$

Your screen should look like Figure B.7.

B.4   Graphing an Equation

We can graph equations written in the form $y = $ (expression in x). The graphing keys are located on the top row of the keypad. There are two steps to graphing an equation:

1. Entering the equation
2. Setting the graphing window

A. Standard Window

The standard window displays values from -10 to 10 on both axes.

EXAMPLE   Graph $y = 2x - 5$ in the standard window.

1. Press `Y=` and enter $2X - 5$ after $Y_1 = $ by keying in

$$\textbf{2} \; `X, T, \theta, n` \; `-` \; \textbf{5} \qquad \text{Use the } `X, T, \theta, n` \text{ key to enter } X.$$

2. Press the `ZOOM` `6` to set the standard window, and the graph will appear (see Figure B.8).

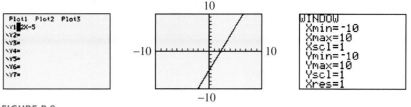

FIGURE B.8

You can press **WINDOW** to see the settings for the standard window. Xscl = 1 means that the tick marks on the *x*-axis are spaced 1 unit apart.

Press **2nd** **MODE** to Quit the graph and return to the Home screen, where we enter computations. From the Home screen, press **GRAPH** to return to the graph.

B. Tracing

The calculator can display the coordinates of selected points on the graph. Press the **TRACE** key to see a "bug" blinking on the graph. The coordinates of the bug are displayed at the bottom of the screen. Use the left and right arrow keys to move the bug along the graph, as shown in Figure B.9. Note that the **Trace** feature does not show *every* point on the graph!

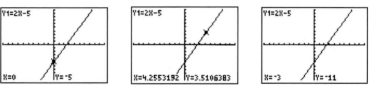

FIGURE B.9

EXAMPLE   Use the **Trace** to find the point on the graph with $x = -3$. Press

TRACE **(−)** 3 **ENTER**

The bug is off the bottom of the screen, but the coordinates are still shown.

C. Multiple Graphs

You can enter more than one graph at a time. Press **▽** to enter a second equation at $Y_2 =$, at $Y_3 =$, and so on. When Tracing, press the **▽** and **△** keys to move from one graph to another.

To turn off a graph without deleting its equation, press **Y=** and move the cursor over the = sign in the equation. Press **ENTER** to deactivate that equation. (When you move the cursor away, the = sign is no longer highlighted.) To reactivate the equation, move the cursor back over the = sign and press **ENTER** again.

D. Setting the Window

Of course, the standard window is not suitable for every graph.

Graph $y = 0.01x^2 - 50$ in the window

$$\text{Xmin} = -100 \qquad \text{Xmax} = 100$$
$$\text{Ymin} = -60 \qquad \text{Ymax} = 50$$

1. Press $\boxed{\text{Y=}}$ and enter $0.01X^2 - 50$ after $Y_1 =$ by keying in

   0.01 $\boxed{\text{X, T, }\theta\text{, }n}$ $\boxed{x^2}$ $\boxed{-}$ **50**   Use the $\boxed{\text{X, T, }\theta\text{, }n}$ key to enter X.

2. Press $\boxed{\text{WINDOW}}$ and enter the settings as shown in Figure B.10. Use the up and down arrow keys to move from line to line. Then press $\boxed{\text{GRAPH}}$.

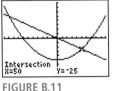

```
WINDOW
 Xmin=-100
 Xmax=100
 Xscl=10
 Ymin=-60
 Ymax=50
 Yscl=10
 Xres=1
```

FIGURE B.10

E. Intersect Feature

We can use the calculator to find the intersection point of two graphs:

1. Enter the equations for the two graphs in the $\boxed{\text{Y=}}$ menu.
2. Choose window settings so that the intersection point is visible in the window.
3. Press $\boxed{\text{2nd}}$ $\boxed{\text{TRACE}}$ $\boxed{5}$ to activate the **intersect** feacture.
4. Use the left and right arrow keys to position the bug near the intersection point.
5. Respond to each of the calculator's questions, **First curve?, Second curve?,** and **Guess?** by pressing $\boxed{\text{ENTER}}$. The coordinates of the intersection point are then displayed at the bottom of the screen.

FIGURE B.11

```
Intersection
X=50        Y=-25
```

Figure B.11 shows one of the intersection points of $y = 0.01x^2 - 50$ and $y = -0.5x$.

F. Other Windows

1. The **ZDecimal** (Zoom Decimal) window, accessed by pressing $\boxed{\text{ZOOM}}$ $\boxed{4}$, shows x-values from -4.7 to 4.7 only, but the **Trace** feature shows "nice" x-values in increments of 0.1.

2. The **ZInteger** (Zoom Integer) window shows nice x-values in increments of 1 unit. Access the **ZInteger** window as follows: Press $\boxed{\text{ZOOM}}$ $\boxed{8}$, move the bug with the arrow keys to the center of your new window, and press $\boxed{\text{ENTER}}$.

3. The **ZSquare** window, accessed by pressing $\boxed{\text{ZOOM}}$ $\boxed{5}$, makes the tick marks on both axes have the same size. In this window, squares look like squares, circles look like circles, and all angles appear true.

4. **"Friendly" Windows:** If the difference between Xmin and Xmax is a multiple of 94, the **Trace** feature gives nice values for x. A useful example of a friendly window is Xmin $= -9.4$, Xmax $= 9.4$.

Troubleshooting

1. If the graph is not visible, you may need to adjust your window. Or, the equation may not be activated. Press $\boxed{\text{Y=}}$ and check to see if the $=$ sign is highlighted.

2. If you get a range error, **ERR: WINDOW RANGE,** quit the message and press $\boxed{\text{WINDOW}}$. Alter the window settings so that Xmin is smaller than Xmax and so that Ymin is smaller than Ymax.

3. If you press $\boxed{\text{Y=}}$ and get an unfamiliar window, or if the axes are not visible in the ZStandard window, you may need to return the **Mode** or **Format** menus to their default settings. See Troubleshooting for Part I.

4. If you get a dimension error, **ERR: INVALID DIM,** you may have a StatPlot turned on. Press 2nd Y= 4 ENTER to turn off the StatPlots.

5. If the bug does not move along the curve, TRACE may not be activated. Press TRACE and then the left or right arrow key.

6. If you get the error, **ERR: INVALID,** you have probably entered a value of x that is outside the window. Adjust the window settings accordingly.

7. If the x-axis or y-axis is too thick, the tick marks are too close together. Press WINDOW and make Xscl or Yscl larger. Set Xscl $= 0$ or Yscl $= 0$ to remove the tick marks.

8. If you get **ERR: NO SIGN CHNG** when using the **intersect** feature, the calculator did not find any intersection point within the current window. Alter the window settings so that the two curves meet within the window. If the two curves are tangent, the calculator may simply fail to find the point of intersection.

B.5   Making a Table

The table feature gives us a convenient tool for evaluating expressions. There are two steps to making a table:

1. Entering the equation
2. Setting the Table features

A. Using the Auto Option

If we want a table with evenly spaced x-values, we use the Automatic setting.

1. Press Y= , clear any previous entries, and enter the expression or expressions you want to evaluate.

2. Press 2nd WINDOW (TBLSET) to access the Table Setup menu. Enter the first x-value after **TblStart =** , and the x-increment after Δ**Tbl =**. Highlight **Auto** for both the Independent and Dependent variables.

3. Finally, press 2nd GRAPH (TABLE) to see the table. You can use the arrows to scroll up and down the table. Figure B.12 shows the Table Setup and the resulting table for $y = 5 - x^3$.

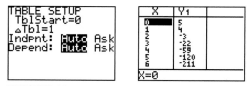

FIGURE B.12

B. Using the Ask Option

If we want to choose specific x-values for the table, we proceed as above but choose **Ask** instead of **Auto** for the Independent variable. Press 2nd GRAPH to see the table, then enter the x-values, pressing ENTER after each one.

Troubleshooting

1. If the table shows **ERROR** for a *y*-value, the expression you entered is undefined at the corresponding *x*-value.

2. If the table shows only *x*-values, no equation is activated. Press $\boxed{\textbf{Y=}}$ to enter or activate an equation.

B.6 | Regression

We use the statistics features to plot data and calculate regression equations. Press $\boxed{\textbf{STAT}}$ to see the Statistics menus. We will use only the first two, **EDIT** and **CALC**.

A. Making a Scatterplot

1. Press $\boxed{\textbf{STAT}}$ $\boxed{\textbf{ENTER}}$ to access the list editor.

2. Enter the *x*-coordinates of the data points under **L1** and the *y*-coordinates in **L2**. An example is shown in Figure B.13a. (If there is a previous list under **L1** or **L2**, move the cursor up to **L1** or **L2** and press $\boxed{\textbf{CLEAR}}$ $\boxed{\textbf{ENTER}}$.)

3. Access the **STAT PLOT** menu by pressing $\boxed{\textbf{2nd}}$ $\boxed{\textbf{Y=}}$, select and turn on **Plot1** by pressing $\boxed{\textbf{ENTER}}$ $\boxed{\textbf{ENTER}}$, and set the menu options as shown in Figure B.13b.

4. Clear out any old equations from the $\boxed{\textbf{Y=}}$ menu, then press $\boxed{\textbf{ZOOM}}$ $\boxed{\textbf{9}}$ (ZoomStat) to see the scatterplot.

    (a)   (b)   (c)

FIGURE B.13

Note: You can use any of the lists **L1–L6** to store the data. Change **Xlist** and **Ylist** to reflect the appropriate lists.

CAUTION

When you are through with the scatterplot, press $\boxed{\textbf{Y=}}$ $\boxed{\triangle}$ $\boxed{\textbf{ENTER}}$ to turn off **Plot1** (or press $\boxed{\textbf{2nd}}$ $\boxed{\textbf{Y=}}$ $\boxed{\textbf{4}}$ to turn off all the StatPlots). If you neglect to do this, you will continue to see the scatterplot even after you graph a new equation.

B. Finding a Regression Equation

1. Enter the data as in steps 1 and 2 of Making a Scatterplot.

2. Press $\boxed{\textbf{STAT}}$ $\boxed{\triangleright}$ to open the **Calculate** menu, and select the type of regression equation you want.

3. Press $\boxed{\textbf{ENTER}}$ and the calculator will display the parameters of the regression equation. See the example in Figure B.14. (You may also see information about r and r^2.)

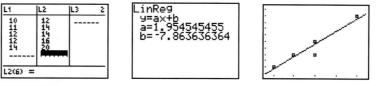

FIGURE B.14

Note: If you do not use **L1** and **L2** to store the data, enter the appropriate lists, separated by a comma, after the regression command.

C. Graphing the Regression Equation

If you would like to graph the regression equation on top of the scatterplot, first follow the steps in A and B on the previous page.

1. Press $\boxed{\textbf{Y=}}$ and clear out any old equations.
2. Position the cursor after $Y_1 =$.
3. Press $\boxed{\textbf{VARS}}$ $\boxed{\textbf{5}}$ $\boxed{\triangleright}$ $\boxed{\triangleright}$ $\boxed{\textbf{ENTER}}$ to copy the regression equation.
4. Press $\boxed{\textbf{GRAPH}}$ to see graph of the regression equation and the scatterplot.

Troubleshooting

1. If you get a dimension error, **ERR: DIM MISMATCH,** you have more data entries in one list than in the other (or no data at all). Press $\boxed{\textbf{STAT}}$ $\boxed{\textbf{ENTER}}$ and check for missing data.
2. If no scatterplot appears, then the **StatPlot** is not activated or the window in inappropriate. Press $\boxed{\textbf{Y=}}$ and make sure that **Plot1** is highlighted, then press $\boxed{\textbf{ZOOM}}$ $\boxed{\textbf{9}}$.

B.7 Function Notation and Transformations of Graphs

A. Function Notation

The calculator uses $Y_1(X)$, $Y_2(X)$, and so on, instead of $f(x)$, $g(x)$, and so on, for function notation.

EXAMPLE   Evaluate $f(x) = x^2 + 6x + 9$ for $x = 3$.

1. Set $Y_1 = X^2 + 6X + 9$, and quit ($\boxed{\textbf{2nd}}$ $\boxed{\textbf{MODE}}$) to the Home screen.
2. To evaluate this function for $X = 3$, press

$$\boxed{\textbf{VARS}} \; \boxed{\triangleright} \; \boxed{\textbf{ENTER}} \; \boxed{\textbf{ENTER}} \; \boxed{\textbf{(}} \; \boxed{\textbf{3}} \; \boxed{\textbf{)}} \; \boxed{\textbf{ENTER}}$$

See Figure B.15.

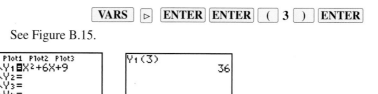

FIGURE B.15

B. Transformation of Graphs

We can use function notation to facilitate graphing transformations. In the examples below, we use $f(x) = x^2$.

1. Translations

EXAMPLE

Compare the graphs of $y = f(x) - 8$ and $y = f(x - 8)$ with that of $y = f(x)$.

Define $Y_1 = X^2$ and $Y_2 = Y_1(X) - 8$. Press **ZOOM** **6** to see the graphs (Figure B.16).

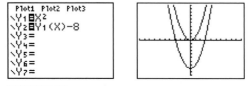

FIGURE B.16

FIGURE B.17

Define $Y_1 = X^2$ and $Y_2 = Y_1(X - 8)$. Press **ZOOM** **6** to see the graphs (Figure B.17).

2. Vertical Scalings and Reflections

EXAMPLE

Compare the graph of $y = \frac{-1}{2} f(x)$ with that of $y = f(x)$.

Define $Y_1 = X^2$ and $Y_2 = -1/2*Y_1(X)$. Press **ZOOM** **6** to see the graphs (Figure B.18).

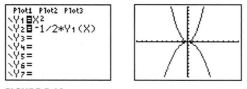

FIGURE B.18

GLOSSARY

Abbreviations used in this glossary: *n* (noun), *v* (verb), *adj* (adjective)

absolute value, *n*, the distance on the number line from a number to 0. For example, the absolute value of -7 is 7. This fact is expressed by the equation $|-7| = 7$.

absolute value equation, *n*, an equation in which the variable occurs between the absolute value bars.

absolute value inequality, *n*, an inequality in which the variable occurs between the absolute value bars.

algebraic expression, *n*, a meaningful combination of numbers, variables, and operation symbols. Also called an *expression*.

algebraic fraction, *n*, a fraction whose numerator and denominator are polynomials. Also called a *rational expression*.

algebraic solution, *n*, a method for solving equations (or inequalities) by manipulating the equations (or inequalities). Compare with *graphical solution* and *numerical solution*.

allometric equation, *n*, an equation showing the (approximate) relationship between a living organism's body mass and another of the organisms properties or processes, usually given in the form $y = k \, (\text{mass})^p$.

altitude, *n*, (i) the distance above the ground or above sea level; (ii) the vertical distance between the base and the opposite vertex of a triangle, pyramid, or cone; (iii) the distance between parallel sides of a parallelogram, trapezoid, or rectangle. Also called *height*.

amortization, *n*, the payment of a debt through regular installments over a period of time.

amount (in an interest-bearing account), *n*, the sum of the principal that was invested and all the interest earned.

amplitude, *n*, the vertical distance between the midline and the maximum value of a sinusoidal function.

annuity, *n*, sequence of equal payments or deposits made at equal time intervals.

approximation, *n*, an inexact result.

area, *n*, a measure of the two-dimensional space enclosed by a polygon or curve, typically expressed in terms of square units, such as square meters or square feet, etc.

ascending powers, *n*, an ordering of the terms of a polynomial so that the exponents on the variable are increasing, such as in the polynomial $1 + x + x^2$.

associative law of addition, *n*, the property that when adding three or more terms, the grouping of terms does not affect the sum. We express this formally by saying that if a, b, and c are any numbers, then $(a + b) + c = a + (b + c)$.

associative law of multiplication, *n*, the property that when multiplying three or more factors, the grouping of factors does not affect the product. We express this formally by saying that if a, b, and c are any numbers, then $(a \cdot b) \cdot c = a \cdot (b \cdot c)$.

asymptote, *n*, a reference line (or curve) towards which the graph of an equation tends as the value of x and/or y grows or diminishes without bound.

augmented matrix (for a linear system with n variables in standard form), *n*, the matrix obtained by making each row of the matrix correspond to an equation of the system, with the coefficients of the variables filling the first n columns, and the last (that is, the $n + 1$) column having the constants.

axis, *n*, (*plural* **axes**), a line used as a reference for position and/or orientation (*plural* **axes**).

axis of symmetry, *n*, a line that cuts a plane figure into two parts, each a mirror image of the other.

back substitution, *n*, a technique for solving a triangular system of linear equations.

bar graph, *n*, a picture of numerical information in which the lengths or heights of bars are used to represent the values of variables.

base, *n*, (i) a number or algebraic expression that is used as a repeated factor, where an exponent indicates how many times the base is used as a factor. For example, when we write 3^5, the base is 3. (ii) The bottom side of a polygon. (iii) The bottom face of a solid.

base angles, *n*, the angles opposite the equal sides in an isosceles triangle.

binomial, *n*, a polynomial with exactly two terms.

binomial expression, *n*, a sum of two unlike terms, such as $\sqrt{3} + \sqrt{2}$.

build (a fraction), *v*, to find an equivalent fraction by multiplying numerator and denominator by the same nonzero expression.

building factor, *n*, an expression by which both numerator and denominator of a given fraction are multiplied (in order to *build the fraction*).

cartesian coordinate system, *n*, the grid that associates points in the coordinate plane to ordered pairs of numbers.

cartesian plane, *n*, a plane with a pair of coordinate axes. Also called a *coordinate plane*.

change in (a variable), *n*, the final value (of the variable) minus the starting value.

change of variables, *n*, (i) a *transformation* of data, (ii) substitution of a new variable for a variable expression, for example, replacing t^2 with x so that the equation $y = at^2 + b$ becomes $y = ax + b$.

circle, *n*, the set of all points in a plane at a fixed distance (the *radius*) from the center.

circumference, *n*, the distance around a circle.

closed interval, *n*, a set of numbers, denoted by [*a*, *b*], which includes all the numbers between *a* and *b* as well as the numbers *a* and *b* themselves, where *a* and *b* are real numbers and *a* < *b*. Or the set of numbers denoted by (−∞, *b*], which includes the real number *b* and all numbers less than *b*, or the set of numbers denoted by [*a*, ∞), which includes the real number *a* and all numbers greater than *a*.

coefficient, *n*, the numerical factor in a term. For example, in the expression $32a + 7b$, the coefficient of *a* is 32 and the coefficient of *b* is 7.

coefficient matrix (for a linear system with *n* variables in standard form), *n*, the matrix of *n* columns obtained by making each row of the matrix correspond to an equation of the system, with the coefficients of the variables filling the *n* columns (and the constants are not represented in the matrix).

common factor (of two or more expressions), *n*, a quantity that divides evenly into each of the given expressions.

common log or **common logarithm** (of a given positive number *x*), *n*, the exponent, denoted by log *x* (or by $\log_{10} x$) for the number 10 to obtain the value *x*, that is, $10^{\log x} = x$.

commutative law of addition, *n*, the property that when adding terms, the order of the terms does not affect the sum. We express this formally by saying that if *a* and *b* are any numbers, then $a + b = b + a$.

commutative law of multiplication, *n*, the property that when multiplying factors, the order of the factors does not affect the product. We express this formally by saying that if *a* and *b* are any numbers, then $a \cdot b = b \cdot a$.

complementary angles, *n*, two angles whose measures add up to 90°.

complete the square, *v*, to determine the appropriate constant to add to a binomial of the form $ax^2 + bx$ so that the result can be written in the form $a(x + k)^2$.

complex conjugate (of a complex number), *n*, the complex number with the same real part and opposite imaginary part; for example, the complex conjugate of $1 + i$ is $1 - i$.

complex fraction, *n*, a fraction that contains one or more fractions in its numerator and/or in its denominator.

complex plane, *n*, a coordinate plane representing complex numbers, with the real parts corresponding to the values on the horizontal axis and imaginary parts corresponding to values on the vertical axis.

complex number, *n*, a number that can be written in the form $a + bi$, where *a* and *b* are real numbers and $i^2 = -1$.

component, *n*, one of the values of an ordered pair or ordered triple.

compound inequality, *n*, a mathematical statement involving two order symbols. For example, the compound inequality $1 < x < 2$ says that "1 is less than *x*, and *x* is less than 2."

compound interest (or **compounded interest**), *n*, an interest earning agreement in which the interest payment at a given time is computed based on the sum of the original principal and any interest money already accrued.

compounding period, *n*, the time interval between consecutive interest payments to an account that earns interest.

concave down (of a graph), *adj*, curving so that the ends of a flexible rod would need to be bent downward (compared with a straight rod) to lie along the graph. Or equivalently, curving so that a line segment tangent to the curve will lie above the curve.

concave up (of a graph), *adj*, curving so that the ends of a flexible rod would need to be bent upward (compared with a straight rod) to lie along the graph. Or equivalently, curving so that a line segment tangent to the curve will lie below the curve.

concavity, *n*, a description of a curve as either concave up or concave down.

concentric (of circles or spheres), *adj*, having the same center.

conditional equation, *n*, an equation that is true for some (but not all) values of the variable(s).

cone, *n*, a three-dimensional object whose base is a circle and whose vertex is a point above the circle. The points on the segments joining the circle to the vertex make up the cone.

congruent, *adj*, having all measure(s) matching exactly. For example, two line segments are congruent when they have the same length; two triangles are congruent if all three sides and all three angles of one match exactly with the corresponding parts of the other triangle.

conjugate, *n*, (i) (of a complex number) the complex number with the same real part and opposite imaginary part; (ii) (of a binomial expression) the binomial expression with the same first term and opposite second term.

conjugate pair, *n*, (i) (of a complex number) a complex number and its conjugate; (ii) (of a binomial expression) the binomial expression and its conjugate.

consistent (of a system of equations), *adj*, having at least one solution.

consistent and independent (of a system of linear equations), *adj*, having exactly one solution.

constant, *adj*, unchanging, not variable. For example, we say that the product of two variables is *constant* if the product is always the same number, for any values of the variables.

constant, *n*, a number (as opposed to a *variable*).

constant of proportionality, *n*, the quotient of two directly proportional variables, or the product of two inversely proportional variables. Also called the *constant of variation*.

constant of variation, *see* constant of proportionality.

constraint, *n*, an equation or inequality involving one or more variables, typically specifying a condition that must be true in the given context.

continuous, *adj*, without holes or gaps. For example, a curve is continuous if it can be drawn without lifting the pencil from the page, and a function is continuous if its graph can be drawn without lifting the pencil from the page.

continuous compounding, *n*, an interest earning agreement in which the *amount* in the account is Pe^{rt}, where *P* is the initial principal, *r* is the annual interest rate, and $e \approx 2.71828$ is the base of the natural logarithm.

conversion factor, *n*, a ratio used to convert from one unit of measure to another.

coordinate, *n*, a number used with a number line or an axis to designate position.

coordinate axis, *n*, one of the two perpendicular number lines used to define the coordinates of points in the plane.

coordinate plane, *n*, a plane with a pair of coordinate axes. Also called the *Cartesian plane or xy-plane*.

corollary, *n*, a mathematical fact that is a consequence of a previously known fact.

costs, *n*, money that an individual or group must pay out. For example, the costs of a company might include payments for wages, supplies, and rent.

counting number, *n*, one of the numbers 1, 2, 3, 4, ...

cube, *n*, (i) a three-dimensional box whose six faces all consist of squares; (ii) an expression raised to the power 3.

cube, *v*, to raise an expression to the power 3. For example, to cube 2 means to form the product of three 2s: $2^3 = 2 \times 2 \times 2 = 8$.

cube root, *n*, a number that when raised to the power 3 gives a desired value. For example, 2 is the cube root of 8 because $2^3 = 8$.

cubic, *adj*, having to do with the third degree of a variable or with a polynomial of degree 3.

cylinder, *n*, a three-dimensional figure in the shape of a soft drink can. The top and base are circles of identical size, and the line segments joining the two circles are perpendicular to the planes containing the two circles.

decay factor, *n*, the factor by which an initial value of a diminishing quantity is multiplied to obtain the final value.

decimal, *adj*, having to do with a base-10 numeration system.

decimal place, *n*, the position of a digit relative to the decimal point. For example, in the number 3.14159, the digit 4 is in the second decimal place, or hundredths place.

decimal point, *n*, the mark "." that is written between the whole number part and the fractional part of a decimal number. For example, the decimal form of $1\frac{3}{10}$ is 1.3.

decreasing, *adj*, (i) (of numbers) moving to the left on a number line: Positive numbers are decreasing when getting closer to zero, and negative numbers are decreasing when they move farther from 0; (ii) (of a graph) having decreasing values of *y* when moving along the graph from left to right; (iii) (of a function) having a decreasing graph.

degree, *n*, a measure of angle equal to $\frac{1}{360}$ of a complete revolution.

degree, *n*, (i) (of a monomial) the exponent on the variable, or if there are more than one variable, the sum of the exponents of all the variables; (ii) (of a polynomial) the largest degree of the monomials in the polynomial.

demand equation, *n*, an equation that gives the quantity of some product that consumers are willing to purchase in terms of the price of that product.

denominator, *n*, the expression below the fraction bar in a fraction.

dependent, *adj*, (of a system of equations) having infinitely many solutions.

dependent variable, *n*, a variable whose value is determined by specifying the value of the *independent variable*.

descending powers, *n*, expressed with the term with the highest degree written first, then the term with the second highest degree, etc.

diagonal, *n*, (i) a line segment joining one vertex of a quadrilateral to the opposite vertex; (ii) a line segment joining opposite corners of a box; (iii) the entries of a matrix whose row number match the column number, that is, the (1, 1), (2, 2), . . . , (*n*, *n*) entries

diameter, *n*, (i) a line segment passing through the center of a circle (or sphere) with endpoints on the circle (sphere); (ii) the length of that line segment.

difference, *n*, the result of a subtraction. For example, the expression $a - b$ represents the difference between *a* and *b*.

difference of squares, *n*, an expression of the form $a^2 - b^2$.

dimension, *n*, (i) (of a matrix) the numbers of rows and columns respectively of the matrix, also called the *order* of the matrix. For example, a matrix with dimension 2 by 3 (or 2×3) has two rows and three columns; (ii) a measurement defining a geometric figure, for example, the length and width are dimensions of a rectangle.

direct variation, *n*, a relation between two variables in which one is a constant multiple of the other (so that the ratio between the two variables is the constant), or in which one is a constant multiple of a positive exponent power of the other variable.

directed distance, *n*, the difference between the ending coordinate and the starting coordinate of points on a number line; the directed distance is negative if the ending value is smaller than the starting value. For example, the directed distance from 5 to 2 is $2 - 5 = -3$.

directly proportional, *adj*, describing variables related by direct variation.

discriminant, *n*, (for the quadratic polynomial $ax^2 + bx + c$) the quantity $b^2 - 4ac$.

distributive law, *n*, the property that for any numbers *a*, *b*, and *c*, $a(b + c) = ab + ac$.

divisor, *n*, a quantity that is divided into another quantity. For example, in the expression $a \div b$, the divisor is *b*.

domain, *n*, the set of all acceptable inputs for a function or equation.

doubling time, *n*, (of exponential growth) the time required for a quantity to double in size.

elementary row operation, *n*, one of the three following operations: (1) an exchange of two rows, (2) multiplying all entries of a row by a nonzero constant, (3) adding a multiple of any row to another row.

elimination, *n*, a method for solving a system of equations that involves adding together the equations of the system or multiples of the equations of the system.

empirical model, *n*, an equation whose graph (approximately) fits a given set of data (but gives no information about the physical processes involved).

entry, *n*, a value in a matrix, often identified by specifying location by row and column.

equation, *n*, a mathematical statement that two expressions are equal, for example, $1 + 1 = 2$.

equation in two variables, *n*, an equation that involves two variables.

equilateral, *adj*, (of a polygon) having all sides of equal length.

equilibrium point, *n*, the point where the graphs of the supply and demand equations intersect

equivalent, *adj*, representing the same value.

equivalent equations, *n*, equations that have the same solutions.

equivalent expressions, *n*, expressions that have the same value for all permissible values of their variables.

error tolerance, *n*, the allowable difference between an estimate and the actual value.

evaluate, *v*, to determine the value of an expression when the variable in the expression is replaced by a number.

exact, *adj*, not simply close, but with absolutely no deviation from an intended value.

exact solution, *n*, the exact value of a solution, i.e., not an approximation.

exponent, *n*, the expression that indicates how many times the *base* is used as a factor. For example, when we write 3^5, the exponent is 5, and $3^5 = 3 \times 3 \times 3 \times 3 \times 3$.

exponential decay, *n*, a manner of decreasing characterized by a constant decay factor for any fixed specified interval of time, or equivalently, modeled by a function f with the form $f(t) = ab^t$, where a and b are positive constants and $0 < b < 1$.

exponential equation, *n*, an equation containing a variable expression as an exponent.

exponential function, *n*, a function f which can be put in the form $f(x) = ab^x$, where a is a nonzero constant and $b \neq 1$ is a positive constant.

exponential growth, *n*, growth characterized by a constant growth factor for any fixed specified interval of time, or equivalently, modeled by a function f with the form $f(t) = ab^t$, where a and b are positive constants and $b > 1$.

exponential notation, *n*, a way of writing an expression that involves radicals and/or reciprocals in terms of powers that have fractional and/or negative exponents. For example, the exponential notation for $\sqrt{3}$ is $3^{1/2}$.

expression, *see* algebraic expression.

extraction of roots, *n*, a method used to solve (quadratic) equations.

extraneous solution, *n*, a value that is not a solution to a given equation but is a solution to an equation derived from the original.

extrapolate, *v*, to estimate the value of a dependent variable for a value of the independent variable that is outside the range of the data.

factor, *n*, an expression that divides evenly into another expression. For example, 2 is a factor of 6.

factor, *v*, to write as a product. For example, to factor 6 we write $6 = 2 \times 3$.

factored form, *n*, (i) (of a polynomial or algebraic expression) an expression written as a product of two or more factors, where the algebraic factors cannot be further factored; (ii) (of an equation of a parabola) the form $y = a(x - r_1)(x - r_2)$.

feasible solution, *n*, an ordered pair which satisfies the constraints of a linear programming problem.

FOIL, *n*, an acronym for a method for computing the product of two binomials: F stands for First terms, O for Outer terms, I for Inner terms, and L stands for Last terms.

formula, *n*, an equation involving two or more variables.

fraction bar, *n*, the line segment separating the numerator and denominator of a fraction. In the fraction $\frac{1}{2}$, the fraction bar is the short segment between the 1 and the 2.

function, *n*, a relationship between two variables in which each value of the input variable determines a unique value of the output variable.

function of two variables, *n*, a relationship between an output variable and an ordered pair of input variables in which each ordered pair of the input variables determines a unique value of the output variable.

function value, *n*, an output value of a function.

fundamental principle of fractions, *n*, the property that the value of a fraction is unchanged when both its numerator and denominator are multiplied by the same nonzero value. We express this formally by saying if a is any number, and b and c are nonzero numbers, then $\dfrac{a \cdot c}{b \cdot c} = \dfrac{a}{b}$.

Gaussian reduction, *n*, the process of performing elementary row operations on a matrix to obtain a matrix in echelon form.

geometrically similar, *adj*, having the same shape (but possibly different size).

graph, *n*, a visual representation of the values of a variable or variables, typically drawn on a number line or on the Cartesian plane.

graph, *v*, to draw a graph.

graph of an equation (or **inequality**), *n*, a picture of the solutions of an equation (or inequality) using a number line or coordinate plane.

graphical solution, *n*, a method for solving equations (or inequalities) by reading values off an appropriate graph. Compare with *algebraic solution* and *numerical solution*.

greatest common factor (GCF) of two or more expressions, *n*, the largest factor that divides evenly into each expression.

growth factor, *n*, the factor by which an initial value of a growing quantity is multiplied to obtain the final value.

guidepoints, *n*, individual points that are plotted to help draw a graph (by hand).

half-life, *n*, (of exponential decay) the time required for a quantity to diminish to half its original size.

half-plane, *n*, either of the two regions of a plane that has been divided into two regions by a straight line

height, *see* altitude.

hemisphere, *n*, half a sphere (on one side or the other of a plane passing through the center).

horizontal asymptote, *n*, a line parallel to the *x*-axis toward which the graph of an equation tends as the value of *x* grows or diminishes without bound.

horizontal axis, *n*, the horizontal coordinate axis. Often called the *x-axis*.

horizontal intercept, *n*, where the graph meets the horizontal axis. Also called *x-intercept*.

horizontal line test, *n*, a test to determine if a function has an inverse function: If no horizontal line intersects the graph of a function more than once, then the inverse is also a function.

horizontal translation (of a graph), *n*, the result of moving all points of the graph straight left (or all straight right) by the same distance.

hypotenuse, *n*, the longest side of a right triangle. (It is always the side opposite the right angle.)

identity, *n*, an equation that is true for all permissible values of the variable(s).

imaginary axis, *n,* the vertical axis in the complex plane.

imaginary number, *n,* a complex number of the form *bi,* where *b* is a real number and $i^2 = -1$.

imaginary part, *n,* (of a complex number) the coefficient of *i* when the complex number is written in the form $a + bi$, where *a* and *b* are real numbers. For example, the imaginary part of $4 - 7i$ is -7.

imaginary unit, *n,* a nonreal number denoted by *i* and which satisfies $i^2 = -1$, that is, *i* is defined to be a square root of -1.

inconsistent, *adj,* (of a system of equations) having no solution.

increasing, *adj,* (i) (of numbers) moving to the right on a number line: Positive numbers are increasing when moving farther from zero, and negative numbers are increasing when they move closer to 0; (ii) (of a graph) having increasing values of *y* when moving along the graph from left to right; (iii) (of a function) having an increasing graph.

independent, *adj,* (i) (of a system of 2 linear equations in 2 variables) having different graphs for the two equations; (ii) (of a system of *n* linear equations in *n* variables) having no one equation equal to a linear combination of the others.

independent variable, *n,* a variable whose value determines the value of the *dependent variable.*

index, *n,* (of a radical) the number at the left of the radical symbol that indicates the type of root involved; for example, the index of 3 in the expression $\sqrt[3]{x}$ indicates a cube root.

inequality, *n,* a mathematical statement of the form $a < b$, $a \leq b$, $a > b$, $a \geq b$, or $a \neq b$.

inflation, *n,* a persistent increase over time of consumer prices.

inflection point, *n,* a point where a graph changes concavity.

initial value, *n,* the starting value of a variable, often when $t = 0$.

input, *n,* value of the independent variable.

integer, *n,* a whole number or the negative of a whole number.

intercept, *n,* a point where a graph meets a coordinate axis.

intercept method, *n,* a method for graphing a line by finding its horizontal and vertical intercepts.

interest, *n,* money paid for the use of money. For example, after borrowing money, the borrower must pay the lender not only the original amount of money borrowed (known as the *principal*) but also the interest on the principal.

interest rate, *n,* the fraction of the principal that is paid as interest for one year. For example, an interest rate of 10% means that the interest for one year will be 10% of the principal.

interpolate, *v,* to estimate the value of a dependent variable based on data that include both larger and smaller values of the independent variable.

intersection point, *n,* a point in common to two graphs.

interval, *n,* a set of numbers that includes all the numbers between *a* and *b* (possibly but not necessarily including *a* and/or *b*), where *a* and *b* are real numbers. Or the set of all numbers less than *b* (and possibly including *b*), or the set all numbers greater than *a* (and possibly including *a*).

interval notation, *n,* notation used to designate an interval. For example, [2, 3] is the interval notation to designate all the real numbers from 2 to 3, including both 2 and 3.

inverse function, *n,* a function whose inputs are outputs of a given function *f,* and whose outputs are the corresponding inputs of *f.*

inverse square law, *n,* a physical law that states that the magnitude of some quantity is inversely proportional to the square of the distance to the source of that quantity.

inverse variation, *n,* a relation between two variables in which one is a constant divided by the other (so that the product of the two variables is the constant), or in which one is a constant divided by a positive exponent power of the other.

inversely proportional, *adj,* describing variables related by inverse variation.

irrational number, *n,* a number that is not rational but does correspond to a point on the number line.

isolate, *v,* (a variable or expression) to create an equivalent equation (or inequality) in which the variable or expression is by itself on one side of the equation (or inequality).

isosceles triangle, *n,* a triangle with two sides of equal length.

joint variation, *n,* a relationship among three or more variables in which whenever all but two variables are held constant, those remaining two variables vary directly or inversely with each other.

law of exponents, a basic property about powers and exponents.

lead coefficient, *n,* (of a polynomial) the coefficient of term with highest degree.

leading entry, *n,* (of a row in a matrix) the first nonzero entry of the row, when read from left to right.

leg, *n,* one of the two shorter sides of a right triangle, or the length of that side.

like fractions, *n,* fractions with equivalent denominators.

like terms, *n,* terms with equivalent variable parts.

line segment, *n,* the points on a single line that join two specified points (the *endpoints*) on that line.

linear combination, *n,* (i) the sum of a nonzero constant multiple of one equation and a nonzero constant multiple of a second equation; (ii) the sum of constant multiples of quantities.

linear combinations, *n,* a procedure for solving a linear system of equations which requires taking one or more linear combination of equations.

linear equation, *n,* an equation such as $2x + 3y = 4$ or $x - 3y = 7$ in which each term has degree 0 or 1.

linear programming, *n,* the study of optimizing functions with constraint equations and/or constraint inequalities.

linear regression, *n,* the process of using a line to predict values of a (dependent) variable.

linear system, *n,* a set of linear equations.

linear term, *n,* a term that consists of a constant times a variable.

log, *see* logarithm.

log scale, *n,* a scale of measurement that uses the logarithm of a physical quantity rather than the quantity itself.

log-log paper, *n,* a type of graph paper in which both horizontal and vertical axes use log scales.

logarithm, *n,* (i) an exponent; (ii) a function whose outputs are exponents associated with a given base.

logarithmic equation, *n,* an equation involving the logarithm of a variable expression.

logarithmic function, *n,* a function of the form $f(x) = \log_b(x)$, where *b* is a positive constant different from 1.

lowest common denominator (LCD), *n,* (of two or more fractions) the smallest denominator that is a multiple of the denominators in the given fractions.

lowest common multiple (LCM), *n,* (of two or more counting numbers) the smallest counting number that the given numbers divide into evenly.

mathematical model, *n,* a representation of relationships among quantities using equations, tables, and/or graphs.

matrix, *n,* a rectangular array of numbers.

maximum, *adj,* largest or greatest.

maximum, *n,* largest value.

maximum value, *n,* (of a variable expression) the largest value that the expression can equal when the variable is allowed to assume all possible values.

mean, *n,* the average of a set of numbers, computed by adding the numbers and dividing by how many are in the set. For example, the mean of 5, 2, and 11 is $\frac{5+2+11}{3} = 6$.

mechanistic model, *n,* an equation whose graph (approximately) fits a given set of data and whose parameters are estimates about the physical properties involved.

median, *n,* the middle number in a set of numbers when written in increasing order. For example, the median of 5, 2, and 11 is 5. If the set has two numbers in the middle when written in order, then the median of the set is the mean of those middle numbers. For example, the median of 6, 1, 9, and 27 is $\frac{6+9}{2} = 7.5$.

minimum, *adj,* least or smallest.

minimum, *n,* smallest value.

minimum value, *n,* (of a variable expression) the smallest value that the expression can equal when the variable is allowed to assume all possible values.

mode, *n,* the number that occurs most frequently in a set of numbers. For example, the mode of 1, 1, 2, and 3 is 1.

model, *n,* a mathematical equation or graph or table used to represent a situation in the world or a situation described in words. For example, the equation $P = R - C$ is a model for the relationship among the variables of profit, revenue, and cost.

model, *v,* to create a model.

monomial, *n,* an algebraic expression with only one term.

monotonic, *adj,* (of a function or graph) either never increasing or never decreasing.

multiplicative property (of absolute values), *n,* the property that $|a \cdot b| = |a| \cdot |b|$ for any real numbers a and b.

multiplicity, *n,* (i) (of a zero of a polynomial) the number of times the corresponding linear factor appears as a factor of the polynomial. For example, -9 is a zero of multiplicity one and 7 is a zero of multiplicity two for the polynomial $p(x) = x^3 - 5x^2 - 77x + 441$ because $p(x)$ factors as $p(x) = (x + 9)(x - 7)^2$; (ii) (of a solution to a polynomial equation) the multiplicity of the zero of the corresponding polynomial. For example, -9 is a solution of multiplicity one and 7 is a solution of multiplicity two for the polynomial equation $x^3 = 5x^2 + 77x - 441$ because the equation can be written in the standard form $p(x) = 0$, where $p(x)$ factors as $p(x) = (x + 9)(x - 7)^2$.

natural base, *n,* the irrational number $e \approx 2.71828182846$, which is useful in calculus, statistics, and other mathematical topics.

natural exponential function, *n,* the function $f(x) = e^x$, where e is the natural base.

natural log or **natural logarithm,** *n,* the logarithm with base e, where e is the natural base.

natural number, *n,* a counting number.

negative number, *n,* a number that is less than zero.

negative of, *n,* the opposite of.

net change, *n,* the final value of a variable minus the initial value. For example, if an object's weight decreases from 15 pounds to 13 pounds, the net change in weight is -2 pounds.

nonstrict inequality, *n,* a mathematical statement of the form $a \leq b$ or $a \geq b$.

normal, *adj,* perpendicular.

nth root, *n,* a number which when raised to the power n gives a desired value. When $b^n = a$, then b is an nth root of a.

number line, *n,* a line with coordinates marked on it representing the real numbers.

numerator, *n,* the expression in a fraction that is above the fraction bar.

numerical solution, *n,* a method for solving equations by reading values from an appropriate table of values. Compare with *algebraic solution* and *graphical solution*.

objective function, *n,* (in linear programming) the function that is to be optimized.

one-to-one, *adj,* (pertaining to a function) having the property that every output comes from one and only one input.

open interval, *n,* a set of numbers denoted by (a, b), which includes all the numbers between a and b but not the numbers a and b themselves, where a and b are real numbers and $a < b$. Or the set of numbers denoted by $(-\infty, b)$, which includes all numbers less than b, or the set of numbers denoted by (a, ∞), which includes all numbers greater than a.

operation, *n,* addition, subtraction, multiplication, or division (or raising to a power or taking a root).

opposite, *n,* the number on the number line that is on the other side of 0 and at the same distance. For example, 5 and -5 are opposites.

order, *n,* (of a matrix) the numbers of rows and columns respectively of the matrix, also called the *dimension* of the matrix. For example, a matrix with order 2 by 3 (or 2×3) has two rows and three columns.

order of operations, *n,* rules that prescribe the order in which to carry out the operations in an expression.

order symbol, *n,* one of the four symbols $<$, or $\leq$, or $>$, or $\geq$.

ordered pair, *n,* a pair of numbers enclosed in parentheses, like this: (x, y). Often used to specify a point or a location on the coordinate plane.

ordered triple, *n,* three numbers enclosed in parentheses, like this: (x, y, z). Often used to specify a solution to a system of equations in three variables or a point in three-dimensional space.

origin, *n*, the point where the coordinate axes meet. It has coordinates $(0, 0)$.

output, *n*, value of the dependent variable.

parabola, *n*, a curve with the shape of the graph of $y = ax^2$, where $a \neq 0$.

parallel lines, *n*, lines that lie in the same plane but do not intersect, even if extended indefinitely.

parameter, *n*, a constant in an equation that varies in other equations of the same form. For example, in the slope-intercept formula $y = b + mx$, the constants b and m are parameters.

percent, *n*, a fraction with (an understood) denominator of 100. For example, to express the fraction $\frac{51}{100}$ as a percent, we write 51% or say "51 percent."

percent increase, *n*, the change in some quantity, expressed as a percentage of the starting amount.

perfect square, *n*, the square of an integer. For example, 9 is a perfect square because $9 = 3^2$.

perimeter, *n*, the distance around the edge or boundary of a two-dimensional figure.

perpendicular lines, *n*, lines that meet and form right angles with each other.

piecewise defined function, *n*, a function defined by multiple expressions, one expression for each specified interval of the independent variable.

point-slope form, *n*, one way of writing the equation for a line: $y - y_1 = m(x - x_1)$ or $\frac{y - y_1}{x - x_1} = m$.

polygon, *n*, a simple closed geometric figure in the plane consisting of line segments (called sides) that meet only at their endpoints. For example, triangles are polygons with three sides.

polynomial, *n*, a sum of terms, where each term is either a constant or a constant times a power of a variable, and the exponent is a positive integer.

polynomial function, *n*, a function that can be written in the form $f(x) = a_n x^n + a_{n-1} x^{n-1} + a_{n-2} x^{n-2} + \cdots + a_2 x^2 + a_1 x + a_0$ where $a_0, a_1, a_2, \ldots a_n$ are constants.

positive number, *n*, a number greater than zero.

power, *n*, an expression that consists of a base and an exponent.

power function, *n*, a function of the form $f(x) = ax^p$, where a and p are constants.

prime (or **prime number**), *n*, an integer greater than 1 whose only whole number factors are itself and 1.

principal, *n*, the original amount of money deposited in an account or borrowed from a lender. (Compare with *interest*.)

principal root, *see* principal square root.

principal square root, *n*, the nonnegative square root.

product, *n*, the result of a multiplication. For example, the expression $a \cdot b$ represents the product of a and b.

profit, *n*, the money left after counting all the revenue that came in and subtracting the costs that had to be paid out.

proportion, *n*, an equation in which each side is a ratio.

proportional, *see* directly proportional, inversely proportional.

pyramid, *n*, a three-dimensional object like a cone except that the base is a polygon instead of a circle.

Pythagorean theorem: If the legs of a right triangle are a and b and the hypotenuse is c, then $a^2 + b^2 = c^2$.

quadrant, *n*, any of the four regions into which the coordinate axes divide the plane. The **first quadrant** consists of the points where both coordinates are positive; the **second quadrant** where the first coordinate is negative and the second coordinate positive; the **third quadrant** consists of points where both coordinates are negative; and the **fourth quadrant** contains the points where the first coordinate is positive and the second coordinate is negative.

quadratic, *adj*, relating to the square of a variable (or of an expression).

quadratic equation, *n*, an equation that equates zero to a polynomial of degree 2 (or an equivalent equation).

quadratic formula, *n*, the formula that gives the solutions of the quadratic equation $ax^2 + bx + c = 0$, namely $x = \frac{-b \pm \sqrt{b^2 - 4ac}}{2a}$.

quadratic function, *n*, a function of the form $f(x) = ax^2 + bx + c$.

quadratic polynomial, *n*, a polynomial whose degree is 2.

quadratic regression, *n*, the process of using a quadratic function to predict values of a (dependent) variable.

quadratic term, *n*, a term whose degree is 2.

quadratic trinomial, *n*, a polynomial of degree 2 and having exactly 3 terms.

quadrilateral, *n*, a polygon with exactly 4 sides.

quartic, *adj*, (pertaining to a polynomial) having degree 4.

quotient, *n*, the result of a division. For example, the expression $a \div b$ represents the quotient of a and b.

radical, *n*, a root of a number, such as a square root or a cube root.

radical expression, *n*, a square root, a cube root, or an *n*th root.

radical equation, *n*, an equation in which the variable appears under a radical sign.

radical notation, *n*, notation using the radical sign to indicate a root.

radical sign, *n*, the symbol $\sqrt{}$, which is used to indicate the principal square root, or the symbol $\sqrt[3]{}$, which is used to indicate cube root, or the symbol $\sqrt[n]{}$, which is used to indicate *n*th root, where *n* is a counting number greater than 2.

radicand, *n*, the expression under a radical sign.

radius, *n*, (i) a line segment from the center of a circle (or sphere) to a point on the circle (sphere), (ii) the length of that line segment.

raise to a power, *v*, use as a repeated factor, for example, to raise x to the power 2 is the same as multiplying $x \cdot x$.

range, *n*, (i) the set of all output values for a function; (ii) the difference between the largest and smallest values in a set of data.

rate, *n*, a ratio that compares two quantities (typically) with different units.

rate of change, *n*, the ratio of change in the dependent variable to the corresponding change in the independent variable, measuring the change in the dependent variable per unit change in the independent variable.

ratio, *n*, (i) a way to compare two quantities by division, (ii) a fraction. For example, "the ratio of 1 to 2" can be written as $\frac{1}{2}$.

rational, *adj*, having to do with ratios.

rational exponent, *n*, an exponent that is a rational number. For example, the expression $x^{1/3}$ has a rational exponent of 1/3, and $x^{1/3} = \sqrt[3]{x}$.

rational expression, *n,* a ratio of two polynomials. Also called an *algebraic fraction.*

rational function, *n,* a function of the form $f(x) = \frac{p(x)}{q(x)}$, where p and q are polynomial functions.

rational number, *n,* a number that can be expressed as the ratio of two integers.

rationalize the denominator, *v,* to find an equivalent fraction that contains no radical in the denominator. For example, when we rationalize $\frac{1}{\sqrt{2}}$, we obtain $\frac{\sqrt{2}}{2}$.

real axis, *n,* the horizontal axis in the complex plane.

real line, *see* number line.

real part, *n,* (of a complex number) the term which does not include i when the complex number is written in the form $a + bi$, where a and b are real numbers. For example, the real part of $4 - 7i$ is 4.

real number, *n,* a number that corresponds to a point on a number line.

reciprocal (of a number), *n,* the result of dividing 1 by the given number. For example, the reciprocal of 2 is $\frac{1}{2}$. Two numbers are reciprocals of each other when their product is 1.

rectangle, *n,* a four-sided figure (in the plane) with four right angles. The opposite sides are equal in length and parallel.

reduce a fraction, *v,* to find an equivalent fraction whose numerator and denominator share no common factors (other than 1 and -1).

reduced row echelon form, *n,* (of a matrix) a row echelon form matrix that also satisfies (1) the leading entry in each nonzero row is a 1, (2) each leading 1 is the only nonzero entry in its column.

reflection (of a point or graph across a line), *n,* the transformation that replaces each point of a graph with its mirror image on the other side of the line.

regression line, *n,* the line used for linear regression.

regular polygon, *n,* a polygon all of whose sides have equal length and all of whose angles are congruent.

restricted domain, *n,* a domain of a function that does not include all real numbers.

revenue, *n,* money that an individual or group receives. For example, a person might have revenues from both a salary and from earnings on investments.

right angle, *n,* an angle of 90°.

right triangle, *n,* a triangle that includes one right angle.

root, *n,* the solution to an equation. *See also* cube root; *n*th root; principal square root; square root.

round, *v,* to give an approximate value of a number by choosing the nearest number of a specified form. For example, to round 3.14159 to two decimal places, we use 3.14.

row echelon form, *n,* (of a matrix) a matrix in which (1) only zeros occur below each nonzero leading entry, (2) the leading entry in any row is to the right of any leading entry above it, and (3) any row consisting entirely of zeros is below all rows with any nonzero entry.

satisfy, *v,* to make an equation true (when substituted for the variable or variables). For example, the number 5 satisfies the equation $x - 2 = 3$.

scale, *n,* marked values on a number line or axes to establish how wide an interval of numbers is represented by a physical distance on the number line.

scale, *v,* (i) to determine the scale on an axis or axes; (ii) to multiply (measurements) by a fixed number (the *scale factor*).

scale factor, *n,* a fixed number by which measurements or values are multiplied.

scaling exponent, *n,* the exponent defining direct variation or a power function. For example, if $y = 3x^4$, then the scaling exponent is 4.

scatterplot, *n,* a type of graph used to represent pairs of data values. Each pair of data values provides the coordinates for one point on the scatterplot. Also called a *scatter diagram.*

scientific notation, *n,* a standard method for writing very large or very small numbers that uses powers of 10. For example, the scientific notation for 12,000 is 1.2×10^4.

semicircle, *n,* half a circle (on one side or the other of a diameter).

signed number, *n,* a positive or negative number.

significant digit, *n,* (in the decimal form of a number) a digit warranted by the accuracy of the measuring device. When the decimal point is present, the significant digits are all those from the leftmost nonzero digit to the rightmost digit after the decimal point. When there is no decimal point, the significant digits are all those from the leftmost nonzero digit to the rightmost nonzero digit. For example, 123.40 has five significant digits, but 12,340 has only four significant digits. Also called *significant figure.*

significant figure, *n, see* significant digit.

similar, *adj, see* geometrically similar.

simplify, *v,* to write in an equivalent but simpler or more convenient form. For example, we can simplify the expression $\sqrt{16}$ to 4.

sinusoidal, *adj,* having the shape of a sine or cosine graph.

slope, *n,* a measure of the steepness of a line or of the rate of change of one variable with respect to another.

slope-intercept form, *n,* a standard form for the equation of a nonvertical line: $y = b + mx$.

slope-intercept method, *n,* a method for graphing a line that uses the slope and the *y*-intercept.

solution, *n,* a value for the variable that makes an equation or an inequality true. A solution to an equation in two variables is an ordered pair that satisfies the equation. A solution to a system is an ordered pair that satisfies each equation of the system.

solve, *v,* (i) (an equation) to find any and all solutions to an equation, inequality, or system; (ii) (a formula) to write an equation for one variable in terms of any other variables, for example, when we solve $5x + y = 3$ for y to get $y = -5x + 3$; (iii) (a triangle) to find the measures of all three sides and of all three angles.

sphere, *n,* a three-dimensional object in the shape of a ball. A sphere consists of all the points in space at a fixed distance (the radius) from the center of the sphere.

square, *n,* (i) any expression times itself; (ii) a rectangle whose sides are all the same length.

square, *v,* to multiply by itself, that is, to raise to the power 2.

square matrix, *n,* a matrix with the same number of rows as columns.

square root, *n,* a number that when squared gives a desired value. For example, 7 is a square root of 49 because $7^2 = 49$.

standard form, *n*, (i) (of a linear, quadratic, or other polynomial equation) an equation in which the right side is 0, so the equation has the form $p(x) = 0$; (ii) (of a system of equations) a system in which the right side of each equation is 0.

strict inequality, *n*, a mathematical statement of the form $a < b$ or $a > b$.

subscript, *n*, a small number written below and to the right of a variable. For example, in the equation $x_1 = 3$, the variable x has the subscript 1.

substitution method, *n*, a method for solving a system of equations that begins by expressing one variable in terms of the other.

sum, *n*, the result of an addition. For example, the expression $a + b$ represents the sum of a and b.

surface area, *n*, the total area of the faces or surfaces of a three-dimensional object.

supplementary angles, *n*, two angles whose measures add up to $180°$.

supply equation, *n*, an equation that gives the quantity of some product that producers are willing to produce in terms of the price of that product.

symmetry, *n*, a geometric property of having sameness on opposite sides of a line (or plane) or about a point.

system of equations, *n*, two or more equations involving the same variables.

term, *n*, (i) (in a sum) a quantity that is added to another. For example, in the expression $x + y - 4$, *x*, *y*, and -4 are the terms; (ii) an algebraic expression that is not a sum or difference, for example, $4x$ is one term.

test point, *n*, (for an inequality) a point in the plane (or on a number line) used to determine which side of the plane (or number line) is included in the solution.

transform, *v*, to apply a *transformation*.

transformation, *n*, (i) (of data) applying a function to one or both components in a set of data, typically so that the resulting data becomes approximately linear; (ii) (of a graph) a change that occurs in the graph of an equation when one or more of the parameters defining that equation are altered.

translation, *n*, (of a graph or geometric figure) sliding horizontally and/or vertically without rotating or changing any shapes.

trapezoid, *n*, a four-sided figure in the plane with one pair of parallel sides.

triangle, *n*, a three-sided figure in the plane.

triangle inequality, *n*, the inequality $|a + b| \le |a| + |b|$, which is true for any two real numbers a and b.

triangular form, *n*, a system of linear equations in which the first variable does not occur in the second equation, the first two variables do not occur in the third equation (and the first three variables do not occur in the fourth equation if there are more than three variables, and so on).

trinomial, *n*, a polynomial with exactly three terms.

turning point, *n*, (of a graph) where the graph either changes from increasing to decreasing or vice versa.

union, *n*, the set obtained by collecting all the elements of one set along with all the elements of another set.

unit circle, *n*, a circle of radius 1 unit (usually centered at the origin).

unlike fractions, *n*, fractions whose denominators are not equivalent.

unlike terms, *n*, terms with variable parts that are not equivalent.

upper triangular form, *n*, (of a matrix) a matrix with all zeros in the lower left corner. More precisely, the entry in the *i*th row and *j*th column is 0 whenever $i > j$.

variable, *adj*, not constant, subject to change.

variable, *n*, a numerical quantity that changes over time or in different situations.

variation, *see* direct variation; inverse variation.

verify, *v*, to prove the truth or validity of an assertion.

vertex, *n*, (*plural* **vertices**), (i) a point where two sides of a polygon meet; (ii) a corner or extreme point of a geometric object; (iii) the highest or lowest point on a parabola.

vertex angle, *n*, the angle between the equal sides in an isosceles triangle.

vertex form, *n*, one way of writing a quadratic equation, $y = a(x - x_v)^2 + y_v$, which displays the vertex, (x_v, y_v).

vertical asymptote, *n*, a line $x = a$ parallel to the *y*-axis toward which the graph of an equation tends as the value of *x* approaches *a*.

vertical axis, *n*, the vertical coordinate axis. Often called the *y-axis*.

vertical compression, *n*, (of a graph) the result of replacing each point of the graph with the point obtained by scaling the *y*-coordinate by a fixed factor (when that factor is between 0 and 1).

vertical intercept, *n*, where the graph meets the vertical axis. Also called the *y-intercept*.

vertical line test, *n*, a test to decide whether a graph defines a function: A graph represents a function if and only if every vertical line intersects the graph in at most one point.

vertical stretch, *n*, (of a graph) the result of replacing each point of the graph with the point obtained by scaling the *y*-coordinate by a fixed factor (when that factor is greater than 1).

vertical translation, *n*, (of a graph) the result of moving all points of the graph straight up (or all straight down) by the same distance.

vertices, *n*, the plural of *vertex*.

volume, *n*, a measure of the three-dimensional space enclosed by a three-dimensional object, typically expressed in terms of cubic units, such as cubic meters or cubic feet.

whole number, *n*, one of the numbers 0, 1, 2, 3, ...

x-axis, *see* horizontal axis.

x-intercept, *see* horizontal intercept.

xy-plane, *see* coordinate plane.

y-axis, *see* vertical axis.

y-intercept, *see* vertical intercept.

zero, *n*, (i) the number 0, with the property that when it is added to any other number, the resulting sum is equal to that second number; (ii) an input to a function which yields an output of 0.

zero-factor principle, *see* Properties of Numbers on the next page.

Properties of Numbers

Associative Laws

Addition: If a, b, and c are any numbers, then $(a + b) + c = a + (b + c)$.

Multiplication: If a, b, and c are any numbers, then $(a \cdot b) \cdot c = a \cdot (b \cdot c)$.

Commutative Laws

Addition: If a and b are any numbers, then $a + b = b + a$.

Multiplication: If a and b are any numbers, then $a \cdot b = b \cdot a$.

Distributive Law: $a(b + c) = ab + ac$ for any numbers a, b, and c.

Properties of Equality

Addition: If $a = b$ and c is any number, then $a + c = b + c$.

Subtraction: If $a = b$ and c is any number, then $a - c = b - c$.

Multiplication: If $a = b$ and c is any number, then $a \cdot c = b \cdot c$.

Division: If $a = b$ and c is any nonzero number, then $\frac{a}{c} = \frac{b}{c}$.

Fundamental Principle of Fractions

If a is any number, and b and c are nonzero numbers, then $\frac{a \cdot c}{b \cdot c} = \frac{a}{b}$.

Laws of Exponents:

(1) $a^m \cdot a^n = a^{m+n}$

(2) $\frac{a^m}{a^n} = a^{m-n}$ $(n < m)$

    $\frac{a^m}{a^n} = \frac{1}{a^{n-m}}$ $(n > m)$

(3) $(a^m)^n = a^{mn}$

(4) $(ab)^n = a^n b^n$

(5) $\left(\frac{a}{b}\right)^n = \frac{a^n}{b^n}$

Product Rule for Radicals

If a and b are both nonnegative, then $\sqrt{ab} = \sqrt{a}\sqrt{b}$.

Quotient Rule for Radicals

If $a \geq 0$ and $b > 0$, then $\sqrt{\frac{a}{b}} = \frac{\sqrt{a}}{\sqrt{b}}$.

Zero Factor Principle

If $ab = 0$, then either $a = 0$ or $b = 0$.

Properties of Absolute Value

$|a + b| \leq |a| + |b|$ Triangle Inequality

$|ab| = |a| \cdot |b|$ Multiplicative Property

BIBLIOGRAPHY

Ahrens, C. Donald. *Essentials of Meteorology*. Belmont, CA: Wadsworth, 1998.

Alexander, R. M. *The Human Machine*. Natural History Museum Publications, 1992.

Altenburg, Edgar. *Genetics*. New York: Holt Rinehart and Winston, 1956.

Atkins, P. W. and Beran, J. A. *General Chemistry*. Scientific American Books, 1989.

Baddeley, Alan D. *Essentials of Human Memory*. Psychology Press Ltd., 1999.

Belovsky, G. E. "Diet Optimization in a Generalist Herbivore: The Moose." *Theoretical Population Biology*, 14 (1978), 105–134.

Belovsky, G. E. "Generalist Herbivore Foraging and Its Role in Competitive Interactions." *American Zoologist*, 26 (1986), 51–69.

Bender, Edward. *An Introduction to Mathematical Modeling*. Mineola, NY: Dover Publications, 1978.

Berg, Richard and Stork, David. *The Physics of Sound*. Englewood Cliffs, NJ: Prentice-Hall, 1982.

Boleman, Jay. *Physics, a Window on Our World*. Englewood Cliffs, NJ: Prentice Hall, 1982.

Bolton, William. *Patterns in Physics*. McGraw Hill, 1974.

Brandt, John C. and Maran, Stephen P. *New Horizons in Astronomy*. San Francisco: W. H. Freeman, 1972.

Briggs, David and Courtney, Frank. *Agriculture and Environment*. Singapore: Longman Scientific and Technical, 1985.

Burton, R. F. *Biology by Numbers*. Cambridge: Cambridge University Press, 1998.

Carr, Bernard, and Giddings. Steven. "Quantum Black Holes." *Scientific American*, May 2005.

Chapman J. L. and Reiss M. J. *Ecology: Principles and Applications*. Cambridge: Cambridge University Press, 1992.

Davis, Kimmet, and Autry. *Physical Education: Theory and Practice*. Macmillan Education Australia PTY Ltd., 1986.

Deffeyes, Kenneth. *Hubbert's Peak*. Princeton: Princeton University Press, 2001.

The Earth and Its Place in the Universe. Open University, Milton Keynes, 1998.

Fowler, Murray and Miller, Eric. *Zoo and Wild Animal Medicine*. Saunders, 1999.

Gillner, Thomas C. *Modern Ship Design*. U.S. Naval Institute Press, 1972.

The Guiness Book of Records, 1998. Guinness Publishing Ltd., 1997.

Hayward, Geoff. *Applied Ecology*. Thomas Nelson and Sons, Ltd., 1992.

Holme, David J., and Peck, Hazel. *Analytical Biochemistry*. Longman Scientific and Technical, 1993.

Hunt, J. A. and Sykes, A. *Chemistry*. Longman Group Ltd., 1984.

Hutson, A. H. *The Pocket Guide to Mammals of Britain and Europe*. Dragon's World Ltd., 1995.

Ingham, Neil. *Astrophysics*. International Thomson Publishing, 1997.

Karttunen, H. et al. *Fundamental Astronomy*. Springer-Verlag, 1987.

Krebs J. R. and Davies N. B. *An Introduction to Behavioural Ecology*. Oxford: Blackwell Scientific Publications, 1993.

Leopold, Luna B., Wolman, M. Gordon, and Miller, John P. *Fluvial Processes in Geomorphology*, New York: Dover Publications, Inc., 1992.

Mannering, Fred, and Kilareski, Walter. *Principles of Highway Engineering and Traffic Analysis*. New York: John Wiley and Sons, Inc., 1998.

Meadows, Donella, Randers, Jorgen, and Meadows, Dennis. *Limits to Growth, the 30-Year Update*. White River Junction, VT: Chelsea Green Publishing Co., 2004.

Mee, Frederick. *Sound*. Heinemann Educational Books, 1967.

Oglesby, Clarkson and Hicks, R. Gary. *Highway Engineering*. New York: John Wiley and Sons, Inc., 1982.

Oliver, C. P. "The Effect of Varying the Duration of X-Ray Treatment upon the Frequency of Mutation," *Science,* 71 (1930), 44–46.

Perrins, C. M. *British Tits*. Collins, London, 1979.

Perrins, C. M. and Moss, D. "Reproductive Rates in the Great Tit." *Journal of Animal Ecology*, 44 (1975), 695–706.

Plummer, David. *Biochemistry, the Chemistry of Life*. London: McGraw-Hill Book Co., 1989.

Pope, Jean A. *Medical Physics*. Heinemann Educational, 1989.

P. E. di Prampero et al. *Journal of Applied Physiology,* 37 (1974), 1–5.

Pratt, Paul W. *Principles and Practices of Veterinary Technology*. Mosby, 1998.

Pugh, L. G. C. E. "Relation of Oxygen Intake and Speed in Competition Cycling and Comparative Observations on the Bicycle Ergometer." *Journal of Physiology*, 241 (1974), 795–808

Scarf, Philip. "An Empirical Basis for Naismith's Rule." *Mathematics Today*, vol. 34 no. 5 (1998), 149–151

Schad, Jerry. *Afoot and Afield in Los Angeles County*. Berkeley: Wilderness Press, 1991.

Schmidt-Nielsen, Knut. *How Animals Work*. Cambridge: Cambridge University Press, 1972.

Schmidt-Nielsen, Knut. *Scaling: Why Is Animal Size so Important?* Cambridge: Cambridge University Press, 1984.

Sears, Francis. *Mechanics, Heat, and Sound*.

Sternberg, Robert J. *In Search of the Human Mind*. Harcourt Brace College Publishing, 1995.

Storch, Hammon, and Bunch. *Ship Production*. Cornell Maritime Press, 1988.

Strickberger, Monroe W. *Genetics*. Macmillan, 1976.

Underwood, Benton J. "Forgetting." *Scientific American,* vol. 210, no. 3, 91–99.

Weisberg, Joseph and Parish, Howard. *Introductory Oceanography*. McGraw-Hill, 1974.

Wilkinson, G. S. "Reciprocal Food Sharing in the Vampire Bat. *Nature*, 308 (1984), 181–184.

Wood, Alexander. *The Physics of Music*. Chapman and Hall, 1975.

Wright, Paul and Paquette, Radnor. *Highway Engineering*. New York: John Wiley and Sons, Inc., 1987.

ANSWERS TO ODD-NUMBERED PROBLEMS

HOMEWORK 1.1

1.

h	0	3	6	9	10
T	65	80	95	110	115

a.

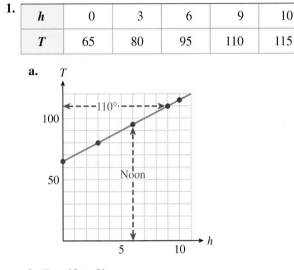

b. $T = 65 + 5h$

c. $95°$

d. 3 p.m.

3.

w	0	4	8	12	16
A	250	190	130	70	10

a.

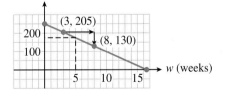

b. $A = 250 - 15w$

c. 75 gallons

d. Until the fifth week

5. a. $P = -800 + 40t$

b. $(0, -800), (20, 0)$

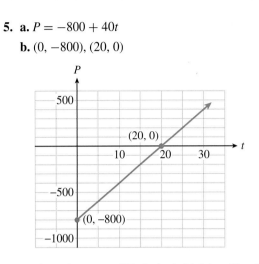

c. The P-intercept, 800, is the initial ($t = 0$) value of the profit. Phil and Ernie start out $800 in debt. The t-intercept, 20, is the number of hours required for Phil and Ernie to break even.

7. a. $C = 5000 + 0.125d$

b.

Miles driven	4000	8000	12,000	16,000	20,000
Cost ($)	5500	6000	6500	7000	7500

c.

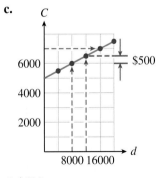

d. $500

e. More than 16,000 miles

9.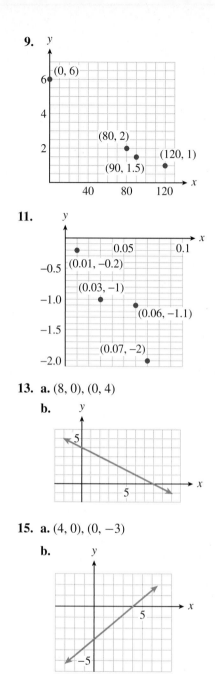

11.

13. a. $(8, 0), (0, 4)$

b.

15. a. $(4, 0), (0, -3)$

b.

17. a. $(9, 0), (0, -4)$

b.

19. a. $(-2250, 0), (0, 1500)$

b.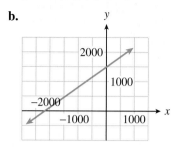

21. a. $(12, 0), (0, 4)$

b.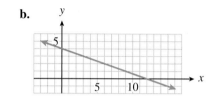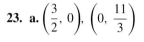

23. a. $\left(\dfrac{3}{2}, 0\right), \left(0, \dfrac{11}{3}\right)$

b.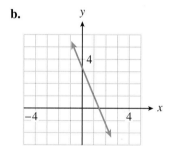

25. a. $\$2.40x, \$3.20y$

b. $2.40x + 3.20y = 19{,}200$

c. $(8000, 0), (0, 6000)$

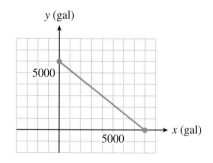

d. The y-intercept, 6000 gallons, is the amount of premium that the gas station owner can buy if he buys no regular. The x-intercept, 8000 gallons, is the amount of regular he can buy if he buys no premium.

27. a. $9x$ mg, $4y$ mg

b. $9x + 4y = 1800$

c. (200, 0), (0, 450)

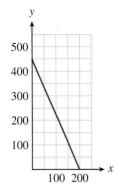

d. The x-intercept, 200 grams, tells how much fig Delbert should eat if he has no bananas, and the y-intercept, 450 grams, tells how much banana he should eat if he has no figs.

29. a. (3, 0), (0, 5) **b.** $\left(\dfrac{1}{2}, 0\right), \left(0, \dfrac{-1}{4}\right)$

c. $\left(\dfrac{5}{2}, 0\right), \left(0, \dfrac{-3}{2}\right)$ **d.** $(p, 0), (0, q)$

e. The value of a is the x-intercept, and the value of b is the y-intercept.

31. a. (0, b) **b.** $\left(\dfrac{-b}{m}, 0\right)$ if $m \neq 0$

33. $-2x + 3y = 2400$ **35.** $3x + 400y = 240$

37. a. $y = 6 - 2x$

c.

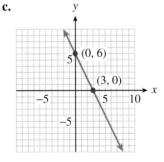

39. a. $y = \dfrac{3}{4}x - 300$

c.

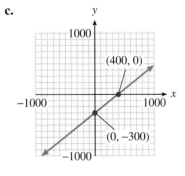

41. a. $y = 0.02 - 0.04x$

c.

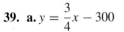

43. a. $y = 210 - 35x$

c.

45. a. (100, 0), (0, 25) **b.** $y = 25 - \dfrac{1}{4}x$

c.

47. a. (0.04, 0), (0, −0.05) **b.** $y = 1.25x - 0.05$

c.

49. a. (−60, 0), (0, 12) **b.** $y = 12 + \dfrac{1}{5}x$

c.

51. a. (−42, 0), (0, −28) **b.** $y = \dfrac{-2}{3}x - 28$

c.

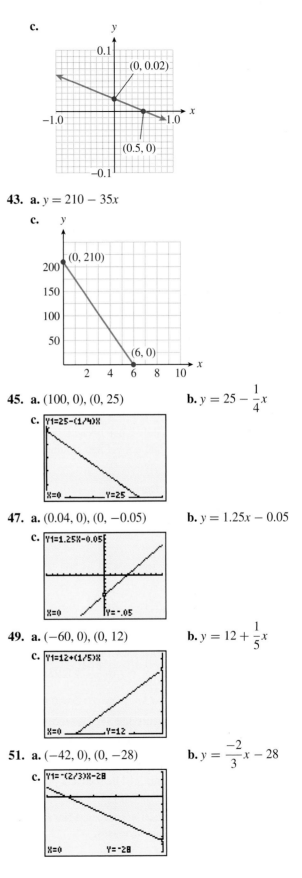

HOMEWORK 1.2

1. Function; the tax is determined by the price of the item.

3. Not a function; incomes may differ for same number of years of education.

5. Function; weight is determined by volume.

7. Input: items purchased; output: price of item. Yes, a function because each item has only one price.

9. Input: topics; output: page or pages on which topic occurs. No, not a function because the same topic may appear in more than one page.

11. Input: students' names; output: students' scores on quizzes, tests, etc. No, not a function because the same student can have different grades on different tests.

13. Input: person stepping on scales; output: person's weight. Yes, a function because a person cannot have two different weights at the same time.

15. No   17. Yes   19. Yes

21. Yes   23. No   25. Yes

27. **a.** 60   **b.** 37.5   **c.** 30

29. **a.** 15%   **b.** 14%   **c.** \$7010–\$9169

31. **a.** 67.7: In 1985, 67.7% of 20–24 year old women had not yet had children.

   b. 1987: 1987 is approximately when 68% of 20–24 year old women had not yet had children.

   c. $f(1997) = 64.9$

33. **a.** No   **b.** 60; no   **c.** Decreasing

35. **a.** 1991   **b.** 1 yr   **c.** 1 yr   **d.** About 7300

37. **a.** Approximately \$1920   **b.** \$5 or \$15

   c. $7.50 < d < 12.50$

39. **a.** 1968, about \$8.70   **b.** 1989, about \$5.10

   c. 1967, $\approx$ 1970

41. **a.** 0   **b.** 10   **c.** -19.4 **d.** $\dfrac{14}{3}$

43. **a.** 1   **b.** 6   **c.** $\dfrac{3}{8}$ **d.** 96.48

45. **a.** $\dfrac{5}{6}$ **b.** 9   **c.** $\dfrac{-1}{10}$ **d.** $\dfrac{12}{13} \approx 0.923$

47. **a.** $\sqrt{12}$ **b.** 0   **c.** $\sqrt{3}$ **d.** $\sqrt{0.2} \approx 0.447$

49. **a.** $V(12) = 1120$: After 12 years, the SUV is worth \$1120.

   b. $t = 12.5$: The SUV has zero value after $12\frac{1}{2}$ years.

   c. The value 2 years later

51. **a.** $N(6000) = 2000$: 2000 cars will be sold at a price of \$6000.

   b. $N(p)$ decreases with increasing p because fewer cars will be sold when the price increases.

   c. $2N(D)$ represents twice the number of cars that can be sold at the current price.

53. **a.** $v(250) = 54.8$ is the speed of a car that left 250-foot skid marks.

   b. $833\dfrac{1}{3}$ feet   **c.** $v\left(833\dfrac{1}{3}\right) = 100$

55. **a.** June 21–24, June 29–July 3, July 8–14

   b. June 17–21, June 25–29, July 4–7

   c. June 22–24, June 27, June 29–July 4, July 8–14

   d. Avalanches occur when temperatures rise above freezing immediately after snowfall.

57. **a.** No   **b.** Yes

   c. Moving downwards on the graph corresponds to moving downwards in the ocean.

59. **a.** $27a^2 - 18a$ **b.** $3a^2 + 6a$

   c. $3a^2 - 6a + 2$ **d.** $3a^2 + 6a$

61. **a.** 8   **b.** 8   **c.** 8   **d.** 8

63. **a.** $8x^3 - 1$ **b.** $2x^3 - 2$

   c. $x^6 - 1$ **d.** $x^6 - 2x^3 + 1$

65. **a.** a^6 **b.** a^{12} **c.** a^3b^3 **d.** $a^3 + 3a^2b + 3ab^2 + b^3$

67. **a.** $96a^5$ **b.** $6a^5$ **c.** $3a^{10}$ **d.** $9a^{10}$

69. **a.** 11   **b.** 13   **c.** $3a + 3b - 4$ **d.** $3a + 3b - 2$
   This function does NOT satisfy $f(a + b) = f(a) + f(b)$.

71. **a.** 19   **b.** 28
   c. $a^2 + b^2 + 6$ **d.** $a^2 + 2ab + b^2 + 3$
   This function does NOT satisfy $f(a + b) = f(a) + f(b)$.

73. **a.** $\sqrt{3} + 2$ **b.** $\sqrt{6}$
   c. $\sqrt{a + 1} + \sqrt{b + 1}$ **d.** $\sqrt{a + b + 1}$
   This function does NOT satisfy $f(a + b) = f(a) + f(b)$.

75. **a.** $\dfrac{-5}{3}$ **b.** $\dfrac{-2}{5}$ **c.** $\dfrac{-2}{a} - \dfrac{2}{b}$ **d.** $\dfrac{-2}{a + b}$
   This function does NOT satisfy $f(a + b) = f(a) + f(b)$.

77. **a.** $x = 50$ and $x = 60$

x	0	10	20	30	40	50
$f(x)$	800	840	840	800	720	600

x	60	70	80	90	100
$f(x)$	440	240	0	-280	-600

b. $x = 56$ and $x = 57$

x	50	51	52	53	54	55
$f(x)$	600	585.8	571.2	556.2	540.8	525

x	56	57	58
$f(x)$	508.8	492.2	475.2

c. $x = 56.5$ and $x = 56.6$

x	56	56.1	56.2	56.3
$f(x)$	508.8	507.158	505.512	503.862

x	56.4	56.5	56.6
$f(x)$	502.208	500.55	498.888

d. $s = 56.55$ **e.** $f(56.55) = 499.7195$

79. 94.85

HOMEWORK 1.3

1. a. $-2, 0, 5$ **b.** 2   **c.** $h(-2) = 0, h(1) = 0, h(0) = -2$

   d. 5   **e.** 3

   f. Increasing: $(-4, -2)$ and $(0, 3)$; Decreasing: $(-2, 0)$

3. a. $-1, 2$ **b.** $3, -1.3$

   c. $R(-2) = 0, R(2) = 0, R(4) = 0, R(0) = 4$

   d. Max: 4; Min: -5 **e.** Max at $p = 0$; Min at $p = 5$

   f. Increasing: $(-3, 0)$ and $(1, 3)$; Decreasing: $(0, 1)$ and
   $(3, 5)$

5. a. $0, \dfrac{1}{2}, 0$ **b.** 0.9   **c.** $\dfrac{-5}{6}, \dfrac{-1}{6}, \dfrac{7}{6}, \dfrac{11}{6}$

   d. Max: 1; Min: -1

   e. Max at $x = -1.5, 0.5$; Min at $x = -0.5, 1.5$

7. a. $2, 2, 1$ **b.** $-6 \le s < -4$ or $0 \le s < 2$

   c. Max: 2; Min: -1

   d. Max for $-3 \le s < -1$ or $3 \le s < 5$;
   Min for $-6 \le s < -4$ or $0 \le s < 2$

9. (a) and **(d)**

11.   **13.**

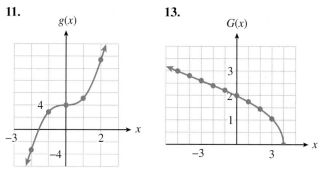

15.

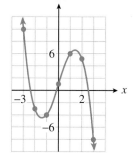

17. a. $f(1000) = 1495$: The speed of sound at a depth of 1000 meters is approximately 1495 meters per second.

   b. $d = 570$ or $d = 1070$: The speed of sound is 1500 meters per second at both a depth of 570 meters and a depth of 1070 meters.

   c. The slowest speed occurs at a depth of about 810 meters and the speed is about 1487 meters per second, so $f(810) = 1487$.

   d. f increases from about 1533 to 1541 in the first 110 meters of depth, then drops to about 1487 at 810 meters, then rises again, passing 1553 at a depth of about 1600 meters.

19. a. $f(1985) = 41$: The federal debt in 1985 was about 41% of the gross domestic product.

   b. $t = 1942$ or $t = 1955$: The federal debt was 70% of the gross domestic product in 1942 and 1955.

   c. In about 1997, the debt was about 67% of the gross domestic products of $f(1997) \approx 67.3$.

   d. The percentage basically dropped from 1946 to 1973, but there were small rises around 1950, 1954, 1958, and 1968, so the longest time interval was from 1958 to 1967.

21. a. i) $x = -3$ ii) $x < -3$ iii) $x > -3$

   b. i) $x = -3$ ii) $x < -3$ iii) $x > -3$

   c. On the graph of $y = -2x + 6$, a value of y is the same as a value of $-2x + 6$, so parts (a) and (b) are asking for the same x's.

23. a. $x = 0.6$ **b.** $x = -0.4$ **c.** $x > 0.6$ **d.** $x < -0.4$

25. a. $x = -1$ or $x = 1$

   b. Approximately $-3 \le x \le -2$ or $2 \le x \le 3$

27. a. 3.5   **b.** $-2.2, -1.2, 3.4$

   c. $p < -3.1$ or $0.3 < p < 2.8$ **d.** $0.5 < B < 5.5$

   e. $p < -1.7$ or $p > 1.7$

29. a. $-2, 2$ **b.** $-2.8, 0, 2.8$

   c. $-2.5 < q < -1.25$ or $1.25 < q < 2.5$

   d. $-2 < q < 0$ or $q > 2$

31. a. He has an error: $f(-3)$ cannot have both the value 5 and also the value -2 and $f(-1)$ cannot have both the value of 2 and -4.

    b. Her readings are possible for a function: each input has only one output.

33. a.

x	-4	-2	3	5
$g(x)$	4.5	5.7	5.2	3.3

    b. $-4.8,\ 4.8$

35. a. $(-1.6, 4.352),\ (1.6, -4.352)$

    b. $F(-1.6) = 4.352;\ F(1.6) = -4.352$

37. a.

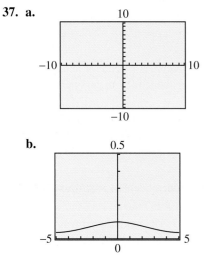

    b.

The curve cannot be distinguished from the x-axis in the standard window because the values of y are closer to zero than the resolution of the calculator can display. The second window provides sufficient resolution to see the curve.

39. a.

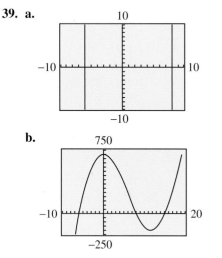

    b.

The curve looks like two vertical lines in the standard window because that window covers too small a region of the plane. The second window allows us to see the turning points of the curve.

41. a. $x = 4$ **b.** $x = -5$ **c.** $x > 1$ **d.** $x < 14$

43. a. $x = 11$ **b.** $x = -10$ **c.** $x \geq -5$ **d.** $x \leq 8$

45. a. $x = 4$ **b.** $x < 22$

47. a. $x = 20$ **b.** $x \leq 7$

49. a. $-15, 5, 20$ **b.** $-13, 2, 22$

HOMEWORK 1.4

1. Anthony   **3.** Bob's driveway

5. -1 **7.** $\dfrac{-2}{3}$

9. a.

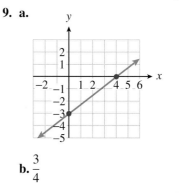

    b. $\dfrac{3}{4}$

11. a.   **b.** -3

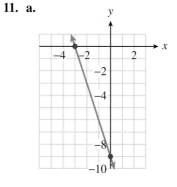

13. a.   **b.** $\dfrac{8}{5}$

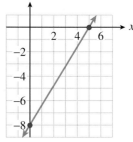

15. a.

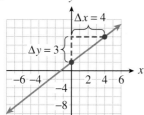

b.

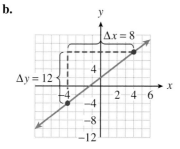

c.

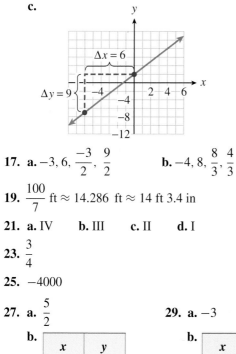

17. a. $-3, 6, \dfrac{-3}{2}, \dfrac{9}{2}$ **b.** $-4, 8, \dfrac{8}{3}, \dfrac{4}{3}$

19. $\dfrac{100}{7}$ ft ≈ 14.286 ft ≈ 14 ft 3.4 in

21. a. IV **b.** III **c.** II **d.** I

23. $\dfrac{3}{4}$

25. -4000

27. a. $\dfrac{5}{2}$

b.

x	y
−4	−14
−2	−9
2	1
3	$\frac{7}{2}$
6	11

29. a. -3

b.

x	y
−3	36
−1	30
5	12
6	9
10	−3

31. a.

t	4	8	20	40
S	32	64	160	320

b.

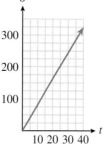

c. 8 dollars/hour

d. The typist is paid $8 per hour.

33. a. 1250 barrels/day

b. The slope indicates that oil is pumped at a rate of 1250 barrels per day.

35. a. −6 liters/day

b. The slope indicates that the water is diminishing at a rate of 6 liters per day.

37. a. 12 inches/foot

b. The slope gives the conversion rate of 12 inches per foot.

39. a. 4 dollars/kilogram

b. The slope gives the unit price of $4 per kilogram.

41. (a)

43. a. Yes, the slope between any two points is $\frac{1}{2}$.

b. 0.5 grams of salt per degree Celsius

45. a. Yes **b.** 2π

47. a. $\dfrac{1500 \text{ meters}}{1 \text{ second}}$ **b.** 3375 meters

49. a. The distances are known. **b.** 5.7 km per second

51. a. About 18°C **b.** 0.3 km to 0.4 km

c. About −28°C per kilometer

53. a. −3

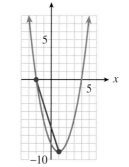

b. 2

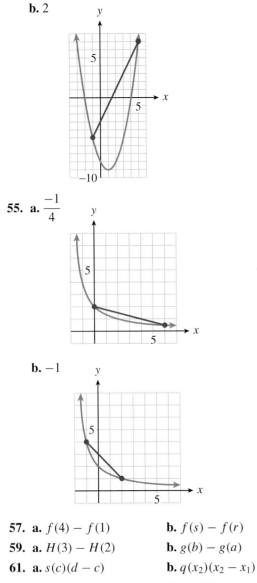

55. a. $\dfrac{-1}{4}$

b. -1

57. a. $f(4) - f(1)$ **b.** $f(s) - f(r)$

59. a. $H(3) - H(2)$ **b.** $g(b) - g(a)$

61. a. $s(c)(d - c)$ **b.** $q(x_2)(x_2 - x_1)$

63. a. $(1, f(1)),\ (5, f(5));\ \dfrac{f(5) - f(1)}{4}$

     b. $(-1, f(-1)),\ (2, f(2));\ \dfrac{f(2) - f(-1)}{3}$

65. a. $(a, f(a)),\ (b, f(b)),\ \dfrac{f(b) - f(a)}{b - a}$

     b. $(a, f(a)),\ (a + \Delta x, f(a + \Delta x)),\ \dfrac{f(a + \Delta x) - f(a)}{\Delta x}$

HOMEWORK 1.5

1. a. $y = \dfrac{1}{2} - \dfrac{3}{2}x$ **b.** Slope $\dfrac{-3}{2}$, y-intercept $\dfrac{1}{2}$

3. a. $y = \dfrac{1}{9} - \dfrac{1}{6}x$ **b.** Slope $\dfrac{-1}{6}$, y-intercept $\dfrac{1}{9}$

5. a. $y = -22 + 14x$ **b.** Slope 14, y-intercept -22

7. a. $y = -29$ **b.** Slope 0, y-intercept -29

9. a. $y = \dfrac{49}{3} - \dfrac{5}{3}x$ **b.** Slope $\dfrac{-5}{3}$, y-intercept $\dfrac{49}{3}$

11. a.

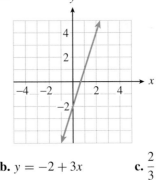

     b. $y = -2 + 3x$ **c.** $\dfrac{2}{3}$

13. a.

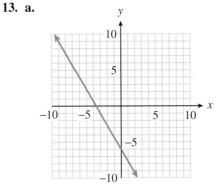

     b. $y = -6 - \dfrac{5}{3}x$ **c.** $\dfrac{-18}{5}$

15. 5   **17.** $\dfrac{-1}{4}$

19. $m = \dfrac{-A}{B}$, x-intercept $\left(\dfrac{C}{A}, 0\right)$, y-intercept $\left(0, \dfrac{C}{B}\right)$

21. a. $a = 100 + 150t$

     b. The slope tells us that the skier's altitude is increasing at a rate of 150 feet per minute, the vertical intercept that the skier began at an altitude of 200 feet.

23. a. $G = 25 + 12.5t$

     b. The slope tells us that the garbage is increasing at a rate of 12.5 tons per year, the vertical intercept that the dump already had 25 tons (when the new regulations went into effect).

25. a. $M = 7000 - 400w$

     b. The slope tells us that Tammy's bank account is diminishing at a rate of $400 per week, the vertical intercept that she had $7000 (when she lost all sources of income).

27. a. 50°F   **b.** −20°C

c.

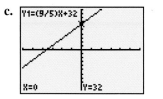

d. The slope, $\frac{9}{5} = 1.8$, tells us that Fahrenheit temperatures increase by 1.8° for each increase of 1° Celsius.

e. C-intercept $(-17\frac{7}{9}, 0)$: $-17\frac{7}{9}$ °C is the same as 0°F; F-intercept $(0, 32)$: 0°C is the same as 32°F.

29. a.

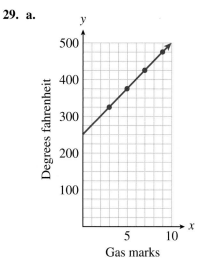

b. $m = 25,\ b = 250$ **c.** $y = 250 + 25x$

31. a.

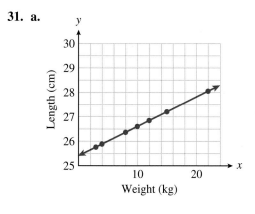

b. $y = 0.12x + 25.4$ **c.** 18 kg

33. a.

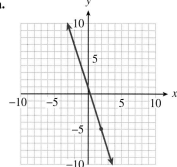

b. $y + 5 = -3(x - 2)$ **c.** $y = 1 - 3x$

35. a.

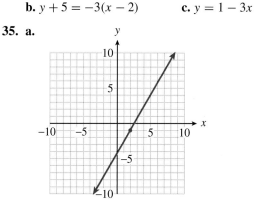

b. $y + 1 = \frac{5}{3}(x - 2)$ **c.** $y = \frac{-13}{3} + \frac{5}{3}x$

37. a. $y + 3.5 = -0.25(x + 6.4)$

b. $y = -5.1 - 0.25x$

c.

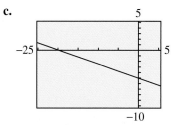

39. a. $y + 250 = 2.4(x - 80)$ **b.** $y = -442 + 2.4x$

c.

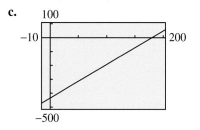

41. a. $m = \dfrac{2}{3}$ **b.** $y = \dfrac{-1}{3} + \dfrac{2}{3}x$

43. a. $(-4, 4)$: neither; $(0, 3)$: $y = px + q$; $(3, 2)$: both; $(2, 1)$: neither; $(1, -2)$: $y = tx + v$

  b. $p = \dfrac{-1}{3}$, $q = 3$, $t = 2$, $v = -4$

45. a. $m = 4$, $b = 40$ **b.** $y = 40 + 4x$

47. a. $m = -80$, $b = -2000$ **b.** $P = -2000 - 80t$

49. a. $m = \dfrac{1}{4}$, $b = 0$ **b.** $V = \dfrac{1}{4}d$

51. a. $y = \dfrac{3}{4}x$, $y = 1 + \dfrac{3}{4}x$, $y = -2.7 + \dfrac{3}{4}x$

  b.

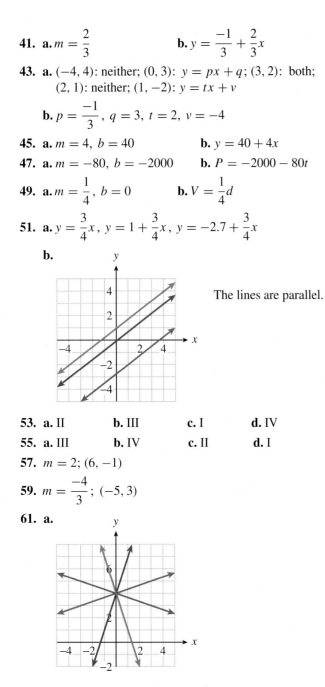

The lines are parallel.

53. a. II **b.** III **c.** I **d.** IV

55. a. III **b.** IV **c.** II **d.** I

57. $m = 2$; $(6, -1)$

59. $m = \dfrac{-4}{3}$; $(-5, 3)$

61. a.

b. The lines with slope 3 and $\dfrac{-1}{3}$ are perpendicular to each other, and the lines with slope -3 and $\dfrac{1}{3}$ are perpendicular to each other.

63. $m = -0.0018$ degree/foot, so the boiling point drops with altitude at a rate of 0.0018 degree per foot.
$b = 212$, so the boiling point is $212°$ at sea level (where the elevation $h = 0$).

HOMEWORK 1.6

1. a.

x	50	125
C	9000	15,000

b. $C = 5000 + 80x$

c. $m = 80$ dollars/bike, so it costs the company \$80 per bike it manufactures.

3. a.

g	12	5
d	312	130

b. $d = 26g$

c. $m = 26$ miles/gallon, so the Porche's fuel efficiency is 26 miles per gallon.

5. a.

C	15	-5
F	59	23

b. $F = 32 + \dfrac{9}{5}C$

c. $m = \frac{9}{5}$, so an increase of $1°$C is equivalent to an increase of $\frac{9}{5}°$F.

7.

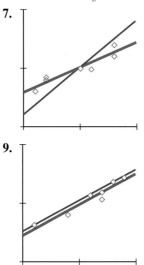

9.

11. a. 12 seconds **b.** 39

  c.

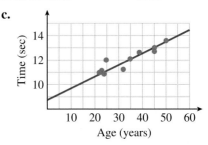

d. 11.6 seconds   **e.** $y = 8.5 + 0.1x$

f. 12.7 seconds; 10.18 seconds; The prediction for the 40-year-old is reasonable, but not the prediction for the 12-year-old. Yes, if you can find a gifted 12-year old sergeant!

13. a.

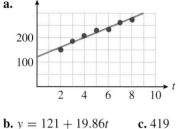

b. $y = 121 + 19.86t$ **c.** 419

15. a.

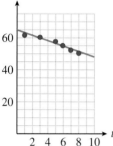

b. $y = 64.2 - 1.63t$ **c.** 58 births per 1000 women

d. 32 births per 1000 women

17. a.

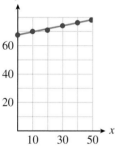

b. $y = 0.18x + 68.2$ **c.** 74.9 years

d. 79 years

19. a.

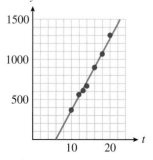

b. $y = 90.49t - 543.7$

c. 90.49 dollars/year: Each additional year of education corresponds to an additional \$90.49 in weekly earnings.

d. No: The degree or diploma attained is more significant than the number of years. So, for example, interpolation for the years of education between a bachelor's and master's degree may be inaccurate because earnings with just the bachelor's degree will not change until the master's degree is attained. And the years after the professional degree will not add significantly to earnings, so extrapolation is inappropriate.

21. a.

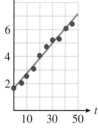

b. $y = 1.6 + 0.11t$ **c.** 6.2 billion tons

23. a. 0.34 meters per year

b. $y = 0.34x$ ($b = 0$ because the plant has zero size until it begins.)

c. Over 1300 years

25. a, b. Yes.

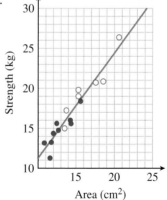

b. $y = 1.29x - 1.62$

c. The slope, 1.29 kg/sq cm, tells us that strength increases by 1.29 kg when the muscle cross-sectional area increases by 1 sq cm.

27. a. E   **b.**

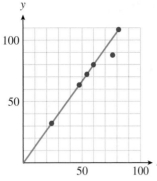

c. $y = 1.33x$; There should be no loss in mass when no gas evaporates.

d. 1333 mg   **e.** Oxygen

29. a. 75°

b. The slope of -2 degrees/hour says that temperatures are dropping at a rate of 2° per hour.

31. a. 20 mph

b. The slope of 10 mph/second says the car accelerates at a rate of 10 mph per second.

33. a. 2 min: 21°C; 2 hr: 729°C; The estimate at 2 minutes is reasonable; the estimate at 2 hours is not reasonable.

35. 128 lb.

37. a. $y \approx -0.54x + 58.7$

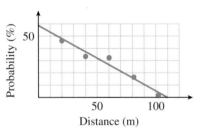

b. 31.7%   **c.** 90 meters

d. The regression line gives a negative probability, which is not reasonable.

39. a.

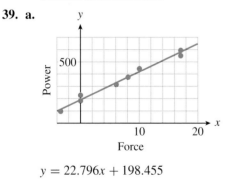

$y = 22.796x + 198.455$

b. $\approx$540 watts   **c.** 198.5 watts   **d.** ≈ -8.7 newtons

e. 3.48 watts   **f.** about 0.018 or 1.8%

CHAPTER 1 REVIEW PROBLEMS

1. a.

n	100	500	800	1200	1500
C	4000	12,000	18,000	26,000	32,000

b. $C = 20n + 2000$

c.

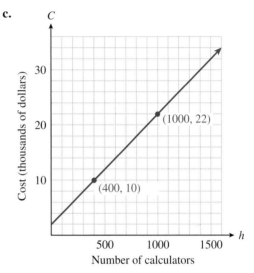

d. $22,000   **e.** 400

3. a. $R = 2100 - 28t$ **b.** (75, 0), (0, 2100)

c. t-intercept: The oil reserves will be gone in 2080; R-intercept: There were 2100 billion barrels of oil reserves in 2005.

5. a. $2C + 5A = 1000$

b. (500, 0), (0, 200)

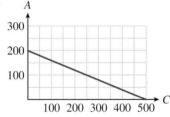

c. 200

d. C-intercept: If no adult tickets are sold, he must sell 500 children's tickets. A-intercept: If no children's tickets are sold, he must sell 200 adult tickets.

7.

9.

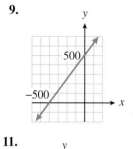

11.

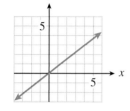

13.

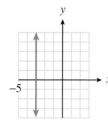

15. A function: Each x has exactly one associated y-value.

17. Not a function: The IQ of 98 has two possible SAT scores.

19. $N(10) = 7000$: Ten days after the new well is opened, the company has pumped a total of 7000 barrels of oil.

21. Function

23. Not a function

25. $F(0) = 1$, $F(-3) = \sqrt{37}$

27. $h(8) = -6$, $h(-8) = -14$

29. a. $f(-2) = 3$, $f(2) = 5$ **b.** $t = 1$, $t = 3$

   c. t-intercepts $(-3, 0)$, $(4, 0)$; $f(t)$-intercept: $(0, 2)$

   d. Maximum value of 5 occurs at $t = 2$

31. a. $x = \dfrac{1}{2} = 0.5$ **b.** $x = \dfrac{27}{8} \approx 3.4$

   c. $x > 4.9$ **d.** $x \le 2.0$

33. a. $x \approx \pm 5.8$ **b.** $x = \pm 0.4$

   c. $-2.5 < x < 0$ or $0 < x < 2.5$

   d. $x \le -0.5$ or $x \ge 0.5$

35. $H(2a) = 4a^2 + 4a$; $H(a + 1) = a^2 + 4a + 3$

37. $f(a) + f(b) = 2a^2 + 2b^2 - 8$;
   $f(a + b) = 2a^2 + 4ab + 2b^2 - 4$

39. The volleyball

41. Highway 33

43. a. $B = 800 - 5t$

   b.

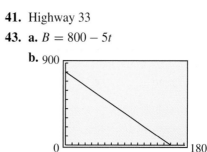

   c. $m = -5$ thousand barrels/minute: The amount of oil in the tanker is decreasing by 5000 barrels per minute.

45. a. $F = 500 + 0.10C$

   b.

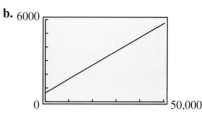

   c. $m = 0.10$: The fee increases by \$0.10 for each dollar increase in the remodeling job.

47. $\dfrac{-3}{2}$ **49.** $\dfrac{-34}{83} \approx -0.4$ **51.** 80 feet

53. a. $h(x_2) - h(x_1)$ **b.** $\dfrac{h(x_2) - h(x_1)}{x_2 - x_1}$

55. Neither

57.

d	V
-5	-4.8
-2	-3
1	-1.2
6	1.8
10	4.2

59. $m = \dfrac{1}{2}$, $b = \dfrac{-5}{4}$ **61.** $m = -4$, $b = 3$

63. a.

   b. $y = \dfrac{10}{3} - \dfrac{2}{3}x$

65. a. $m = -2$, $b = 3$ **b.** $y = 3 - 2x$

67. $\dfrac{3}{5}$

69. a. $\dfrac{3}{2}$ **b.** $(4, 2)$, no   **c.** $(6, 5)$

71. $(3, -14)$, $(-7, 2)$

73. a. $T = 62 - 0.0036h$ **b.** $-46°F$; $108°F$ **c.** $-71°F$

75. a. $y = \dfrac{2}{5} - \dfrac{9}{5}x$

77. a.

t	0	15
P	4800	6780

b. $P = 4800 + 132t$

c. $m = 132$ people/year: the population grew at a rate of 132 people per year.

79. 6

81. a.

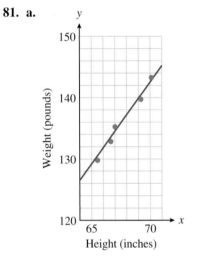

b. 129 lb, 145 lb   **c.** $y = 2.\overline{6}x - 44.\overline{3}$

d. 137 lb   **e.** $y = 2.84x - 55.74$

f. 137.33 lb

83. a.

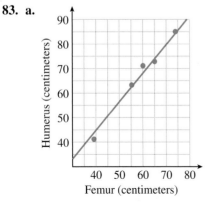

b. 45 cm   **c.** 87 cm   **d.** $y = 1.2x - 3$

e. 69 cm   **f.** $y = 1.197x - 3.660$; 68.16 cm

HOMEWORK 2.1

1. $\pm\dfrac{5}{3}$ **3.** $\pm\sqrt{6}$ **5.** $\pm\sqrt{6}$ **7.** ± 2.65 **9.** ± 5.72

11. ± 5.73 **13.** $\pm\sqrt{\dfrac{Fr}{m}}$ **15.** $\pm\sqrt{\dfrac{2s}{g}}$

17. a. $V = 2.8\pi r^2 \approx 8.8r^2$

b.

r	1	2	3	4	5	6	7	8
V	8.8	35.2	79.2	140.8	220.0	316.8	431.2	563.2

The volume increases by a factor of 4.

c. 5.86 cm   **d.**

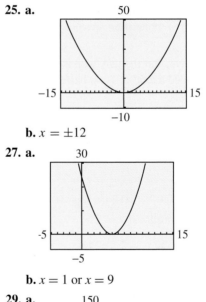

19. 21 in.   **21.** $\sqrt{1800} \approx 42.4$ m

23. $\sqrt{128}$ in. by $\sqrt{128}$ in. ≈ 11.3 in. $\times 11.3$ in.

25. a.

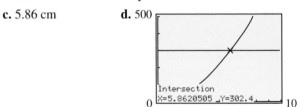

b. $x = \pm 12$

27. a.

b. $x = 1$ or $x = 9$

29. a.

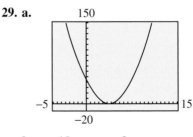

b. $x = 10$ or $x = -2$

31. $5, -1$ **33.** $\dfrac{5}{2}, \dfrac{-3}{2}$ **35.** $-2 \pm \sqrt{3}$ **37.** $\dfrac{1}{2} \pm \dfrac{\sqrt{3}}{2}$

39. $\dfrac{-2}{9}, \dfrac{-4}{9}$ **41.** $\dfrac{7}{8} \pm \dfrac{\sqrt{8}}{8}$ **43.** 6 **45.** 64 **47.** $\dfrac{13}{6}$

49. 8 **51.** 9 **53.** $\dfrac{33}{64}$

55. a. $B = 5000\,(1 + r)^2$ **b.** 11.8%

c. 10,000

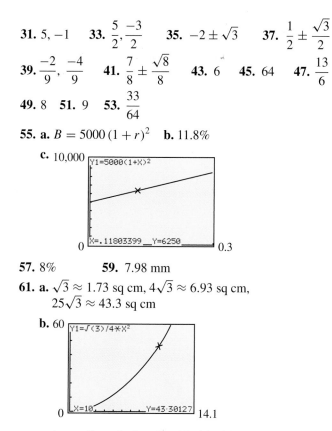

0.3

57. 8% **59.** 7.98 mm

61. a. $\sqrt{3} \approx 1.73$ sq cm, $4\sqrt{3} \approx 6.93$ sq cm,
$25\sqrt{3} \approx 43.3$ sq cm

b. 60

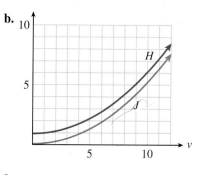

14.1

c. An equilateral triangle with side 5.1 cm has an area of 11.263 cm².

d. side ≈ 6.8

e. equation: $\dfrac{\sqrt{3}}{4} s^2 = 20; s \approx 6.79 \approx 6.8$

f. Each side is ≈ 20 cm long.

63. $\pm\sqrt{\dfrac{bc}{a}}$ **65.** $a \pm 4$ **67.** $\dfrac{-b \pm 3}{a}$

69. a.

Height	Base	Area
1	34	34
2	32	64
3	30	90
4	28	112
5	26	130
6	24	144
7	22	154
8	20	160
9	18	162

Height	Base	Area
10	16	160
11	14	154
12	12	144
13	10	130
14	8	112
15	6	90
16	4	64
17	2	34
18	0	0

b.

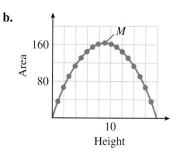

c. 162 sq ft, with base 18 ft, height 9 ft

d. Base: $36 - 2x$; Area: $x\,(36 - 2x)$ **f.** 6.5 ft or 11.5 ft

71. a.

v	0	1	2	3	4	5	6	7	8	9	10	11
J	0	.05	.2	.46	.82	1.28	1.84	2.5	3.27	4.13	5.1	6.17

b.

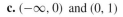

c.

v	0	1	2	3	4	5	6	7	8	9	10	11
H	.9	.95	1.1	1.36	1.72	2.18	2.74	3.4	4.17	5.03	6.0	7.07

d. 5.5 meters   **e.** 10.14 meters per second

HOMEWORK 2.2

1. a. -9 **b.** 9   **3. a.** -4 **b.** 20   **5.** -50 **7. a.** 144

9. 1   **11. a.** 2.7   **b.** -2.7 **c.** 1.8   **d.** 2.9

13. a. 0.3   **b.** -0.4 **c.** 0.2

15. a.   **b.** $x = 0, x = 1$

c. $(-\infty, 0)$ and $(0, 1)$

17. a.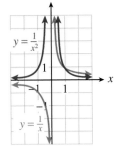

b. $x = 1$ **c.** $(1, +\infty)$

19. Graph (b) is the basic graph shifted 2 units down; graph (c) is the basic graph shifted 1 unit up.

21. Graph (b) is the basic graph shifted 1.5 units left; graph (c) is the basic graph shifted 1 unit right.

23. Graph (b) is the basic graph reflected about the x-axis; graph (c) is the basic graph reflected about the y-axis.

25. a. $\sqrt{x}$ **b.** $\sqrt[3]{x}$ **c.** $|x|$ **d.** $\dfrac{1}{x}$ **e.** x^3 **f.** $\dfrac{1}{x^2}$

27. a. $x \approx 12$ **b.** $x \approx 18$ **c.** $x < 9$ **d.** $x > 3$

29. a. $t \approx -3.1$ **b.** $t \approx 1.5$ **c.** $t < 0.8$ **d.** $-2.4 < t < 0.4$

31. a. $x = 41$ **b.** $29 < x < 61$

33. a. $x = -5$ or $x = 17$ **b.** $-1 < x < 13$

35. a.

x	4	1	$\frac{1}{4}$	0	$\frac{1}{4}$	1	4
y	-2	-1	$-\frac{1}{2}$	0	$\frac{1}{2}$	1	2

b. no

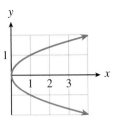

37. a.

x	2	1	$\frac{1}{2}$	0	$\frac{1}{2}$	1	2
y	-2	-1	$-\frac{1}{2}$	0	$\frac{1}{2}$	1	2

b. no

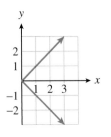

39. a.

x	$-\frac{1}{2}$	-1	-2	undefined	2	1	$\frac{1}{2}$
y	-2	-1	$-\frac{1}{2}$	0	$\frac{1}{2}$	1	2

b. Yes

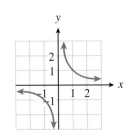

41.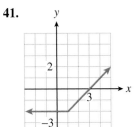

43.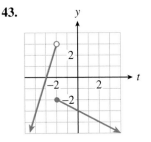

45.

47.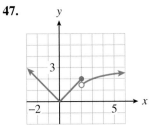

49.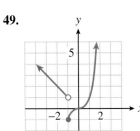

51.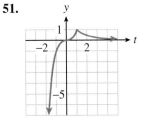

53. $f(x) = \begin{cases} 8 - 2x & x < 4 \\ 2x - 8 & x \geq 4 \end{cases}$

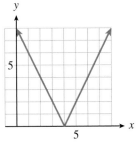

55. $g(t) = \begin{cases} -1 - \dfrac{t}{3} & t < -3 \\ 1 + \dfrac{t}{3} & t \geq -3 \end{cases}$

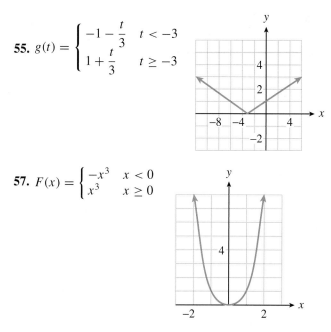

57. $F(x) = \begin{cases} -x^3 & x < 0 \\ x^3 & x \geq 0 \end{cases}$

59. a. Not always true: $f(1+2) \neq f(1) + f(2)$ because $9 \neq 5$.

    b. True: $(ab)^2 = a^2 b^2$

61. a. Not always true: $f(1+2) \neq f(1) + f(2)$ because $\frac{1}{3} \neq \frac{3}{2}$.

    b. True: $\dfrac{1}{ab} = \dfrac{1}{a} \cdot \dfrac{1}{b}$

63. a. Not always true (unless $b = 0$):
$f(1+2) \neq f(1) + f(2)$ because $3m + b \neq 3m + 2b$.

    b. Not always true: $f(1 \cdot 2) \neq f(1) \cdot f(2)$ because
$2m + b \neq 2m^2 + 3mb + b^2$.

65. a.

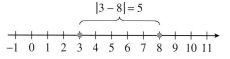

    b.

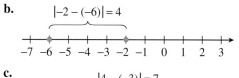

    c.

    d.

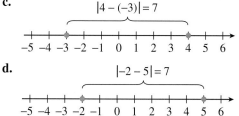

67. The distributive law shows a relationship between multiplication and addition that always holds. The equation $f(a + b) = f(a) + f(b)$ is not about multiplication and may or may not be true.

HOMEWORK 2.3

1. $y = \sqrt{x + 2}$ **3.** $y = x^3 - 1$ **5.** $y = \dfrac{1}{x - 4}$

7. a. Translate $y = |x|$ by 2 units down.

    b.

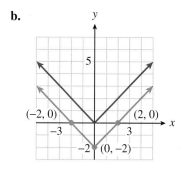

9. a. Translate $y = \sqrt[3]{s}$ by 4 units right.

    b.

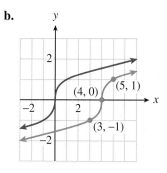

11. a. Translate $y = \dfrac{1}{t^2}$ by 1 unit up.

    b.

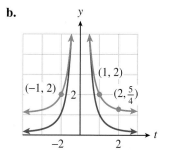

13. a. Translate $y = r^3$ by 2 units left.

    b.

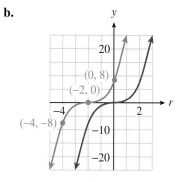

15. a. Translate $y = \sqrt{d}$ by 3 units down.

b.

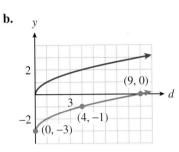

17. a. Translate $y = \frac{1}{v}$ by 6 units left.

b.

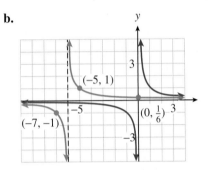

19. A vertical stretch by a factor of 3: $y = \frac{3}{x}$

21. A vertical compression, the scale factor is $\frac{1}{2}$: $y = \frac{1}{2}x^3$

23. a. Scale factor $\frac{1}{3}$; $y = |x|$ is compressed vertically by the scale factor.

b.

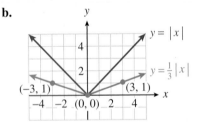

25. a. Scale factor -2: $y = \frac{1}{z^2}$ is reflected over the x-axis and stretched vertically by a factor of 2.

b.

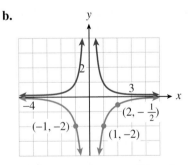

27. a. Scale factor -3: $y = \sqrt{v}$ is reflected over the x-axis and stretched vertically by a factor of 3.

b.

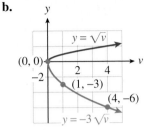

29. a. Scale factor $\frac{-1}{2}$: $y = s^3$ is reflected over the x-axis and compressed vertically by a factor of $\frac{1}{2}$.

b.

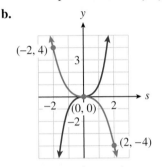

31. a. Scale factor $\frac{1}{3}$; $y = \frac{1}{x}$ is compressed vertically by the scale factor.

b.

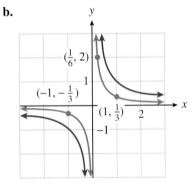

33. a. vi   **b.** ii   **c.** iv   **d.** i   **e.** v   **f.** iii

35. a. Vertically stretched by a factor of 3: $y = 3f(x)$

    b. Reflected about the x-axis: $y = -f(x)$

    c. Translated 1 unit right: $y = f(x - 1)$

    d. Translated 4 units up: $y = f(x) + 4$

37. a. Reflected about the x-axis and stretched vertically by a factor of 2: $T = -2h(v)$

    b. Stretched vertically by a factor of 3: $T = 3h(v)$

    c. Translated 3 units up: $T = h(v) + 3$

    d. Translated 3 units left: $T = h(v + 3)$

39. a. Translated 2 units up: $y = f(x) + 2$

    b. Translated 4 units down: $y = f(x) - 4$

    c. Vertically compressed with a factor of $\frac{1}{2}$: $y = \frac{1}{2}f(x)$

    d. Translated 1 unit right: $y = f(x - 1)$

41. a. Translated 1 unit right: $y = f(x - 1)$

 b. Part (a) is translated 30 units up: $y = f(x - 1) + 30$

 c. f is reflected about the x-axis and stretched vertically by a factor of 2: $y = -2f(x)$

 d. Part (c) is translated 10 units down: $y = -2f(x) - 10$

43. $y = \frac{1}{2} \cdot \frac{1}{x^2}$ is a vertical compression with factor $\frac{1}{2}$ of $y = \frac{1}{x^2}$.

45. $y = 2\sqrt[3]{x}$ is a vertical stretch with factor 2 of $y = \sqrt[3]{x}$.

47. $y = 3|x|$ is a vertical stretch with factor 3 of $y = |x|$.

49. $y = \frac{1}{8}x^3$ is a vertical compression with factor $\frac{1}{8}$ of $y = x^3$.

51. a. Translation by 2 units up and 3 units right.

 b.

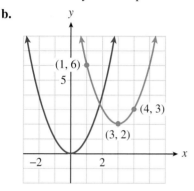

53. a. Translation by 2 units left and 3 units down.

 b.

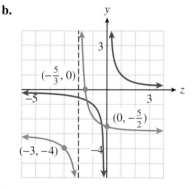

55. a. Reflection across the x-axis, vertical stretch by a factor of 3, translation by 4 units left and 4 units up

 b.

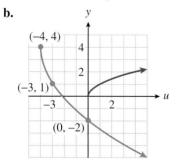

57. a. Vertical stretch by a factor of 2, translation by 5 units right and 1 down

 b.

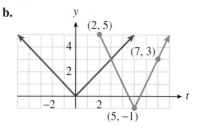

59. a. Reflection across the x-axis, vertical stretch by a factor of 2, translation by 6 units up and 1 unit right

 b.

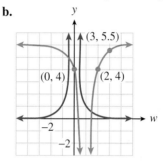

61. Translation by 8 units right and 1 unit down

 b.

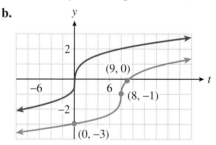

63. a. Translation by 4 units up and 1 unit right: $y = f(x - 1) + 4$

 b. Vertical stretch by a factor of 2 and a translation by 4 units up: $y = 2f(x) + 4$

65. a. $y = |x|$ translated by 1 unit left and 2 units down

 b. $y = |x + 1| - 2$

67. $y = \sqrt{x}$ reflected about the x-axis and shifted 3 units up

 b. $y = -\sqrt{x} + 3$

69. a. $y = x^3$ translated by 3 units right and 1 unit up

 b. $y = (x - 3)^3 + 1$

71. a. $y = f(x - 20)$: Students scored 20 points higher than Professor Hilbert's class.

 b. $y = 1.5f(x)$: The class is about 50% larger than Hilbert's, but the classes scored the same.

73. a. $y = f(x - 5000)$: Taxpayers earn $5000 more than Californians in each tax rate.

b. $y = f(x) - \frac{1}{2}$: Taxpayers pay 2% less tax than Californians on the same income.

75. a. $y = f(t + 2)$: This population has its maximum and minimum two months before the marmots.

b. $y = f(t) - 20$: This population remains 20 fewer that of the marmots.

HOMEWORK 2.4

1. (b)   **3.** (a)

5. d

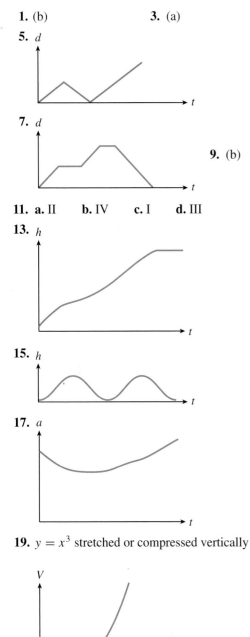

7. d

9. (b)

11. a. II   **b.** IV   **c.** I   **d.** III

13. h

15. h

17. a

19. $y = x^3$ stretched or compressed vertically

V

21. $y = \frac{1}{x}$ stretched or compressed vertically

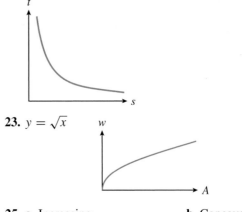

23. $y = \sqrt{x}$

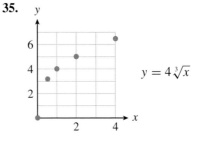

25. a. Increasing   **b.** Concave up

27. a. Increasing   **b.** Concave down

29. a. Increasing, linear (neither concave up nor down)

   b. C

31. a. Increasing, concave down   **b.** F

33. a. Decreasing, linear (neither concave up nor down)

   b. D

35.

$y = 4\sqrt[3]{x}$

37.

$y = 3 \cdot \dfrac{1}{x^2}$

39.

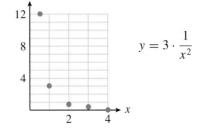

$y = 0.5x^2$

41. a. Table (4), Graph (c)   **b.** Table (3), Graph (b)

   c. Table (1), Graph (d)   **d.** Table (2), Graph (a)

43. a. III   **b.** III

45. a. $S(x) = \begin{cases} 5.95 & x \le 25 \\ 7.95 & 25 < x \le 50 \\ 9.95 & 50 < x \le 75 \\ 10.95 & 75 < x \le 100 \end{cases}$

b.

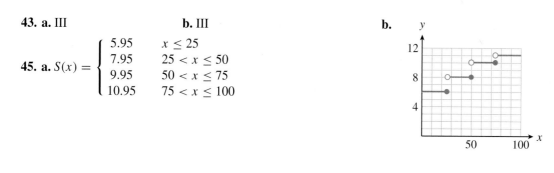

47. a.

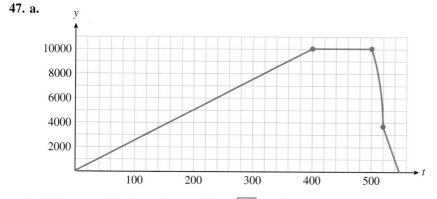

b. 240 seconds (4 minutes) and $500 + \sqrt{250} \approx 515.8$

49. a.

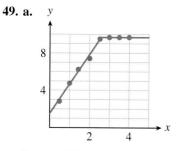

b. $m \approx 3.2$ mm/cc: The height of precipitate increases by 1 mm for each additional cc of lead nitrate.

c. $f(x) = \begin{cases} 1.34 + 3.2x & x < 2.6 \\ 9.6 & x \ge 2.6 \end{cases}$

d. The increasing portion of the graph corresponds to the period when the reaction was occuring, and the horizontal section corresponds to when the potassium iodide is used up.

51. a. II   **b.** IV   **c.** I   **d.** III

HOMEWORK 2.5

1. a. $|x| = 6$

b.

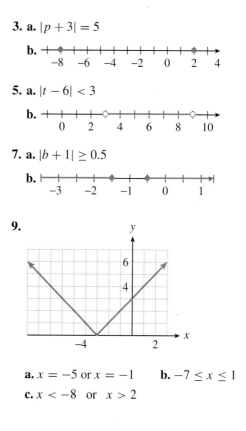

3. a. $|p + 3| = 5$

b.

5. a. $|t - 6| < 3$

b.

7. a. $|b + 1| \ge 0.5$

b.

9.

a. $x = -5$ or $x = -1$ **b.** $-7 \le x \le 1$
c. $x < -8$ or $x > 2$

11.

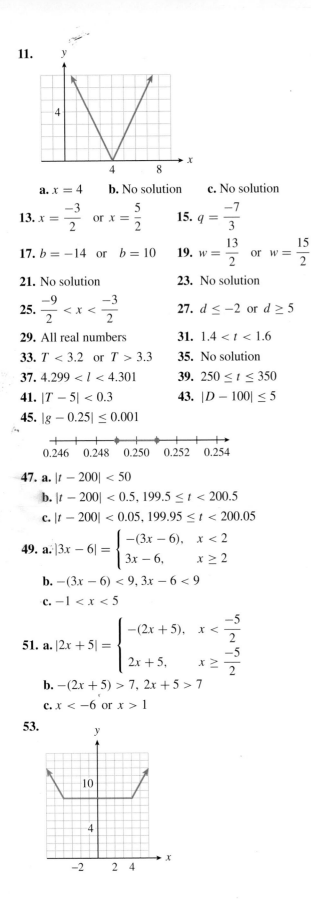

a. $x = 4$ **b.** No solution   **c.** No solution

13. $x = \dfrac{-3}{2}$ or $x = \dfrac{5}{2}$ **15.** $q = \dfrac{-7}{3}$

17. $b = -14$ or $b = 10$ **19.** $w = \dfrac{13}{2}$ or $w = \dfrac{15}{2}$

21. No solution   **23.** No solution

25. $\dfrac{-9}{2} < x < \dfrac{-3}{2}$ **27.** $d \le -2$ or $d \ge 5$

29. All real numbers   **31.** $1.4 < t < 1.6$

33. $T < 3.2$ or $T > 3.3$ **35.** No solution

37. $4.299 < l < 4.301$ **39.** $250 \le t \le 350$

41. $|T - 5| < 0.3$ **43.** $|D - 100| \le 5$

45. $|g - 0.25| \le 0.001$

47. a. $|t - 200| < 50$

   b. $|t - 200| < 0.5,\ 199.5 \le t < 200.5$

   c. $|t - 200| < 0.05,\ 199.95 \le t < 200.05$

49. a. $|3x - 6| = \begin{cases} -(3x - 6), & x < 2 \\ 3x - 6, & x \ge 2 \end{cases}$

   b. $-(3x - 6) < 9,\ 3x - 6 < 9$

   c. $-1 < x < 5$

51. a. $|2x + 5| = \begin{cases} -(2x + 5), & x < \dfrac{-5}{2} \\ 2x + 5, & x \ge \dfrac{-5}{2} \end{cases}$

   b. $-(2x + 5) > 7,\ 2x + 5 > 7$

   c. $x < -6$ or $x > 1$

53.

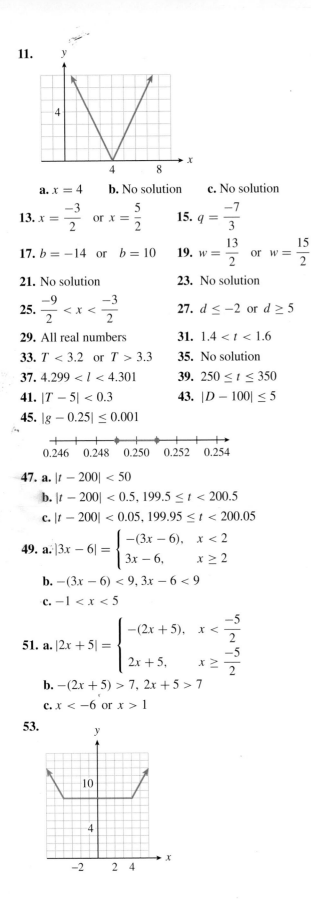

a. $f(x) = \begin{cases} -2x, & x < -4 \\ 8, & -4 \le x \le 4 \\ 2x, & x > 4 \end{cases}$

b. The graphs looks like like a trough. The middle horizontal section is $y = p + q$ for $-p < x < q$, the left side, $x < -p$, has slope -2 and the right side, $x > q$, has slope 2.

55.

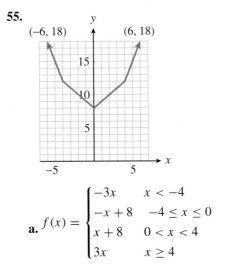

a. $f(x) = \begin{cases} -3x & x < -4 \\ -x + 8 & -4 \le x \le 0 \\ x + 8 & 0 < x < 4 \\ 3x & x \ge 4 \end{cases}$

b. 8   **c.** $p + q$

57. a. $|x + 12|,\ |x + 4|,\ |x - 24|$

   b. $f(x) = |x + 12| + |x + 4| + |x - 24|$

   c.

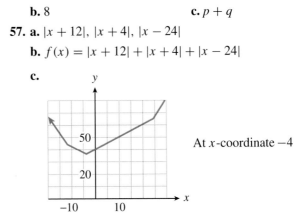

At x-coordinate -4

59. 2 miles east of the river

HOMEWORK 2.6

1. Domain: $[-5, 3]$; Range: $[-3, 7]$

3. Domain: $[4, 5]$; Range: $[1, 1) \cup [3, 6]$

5. Domain: $[-2, 2]$; Range: $[-1, 1]$

7. Domain: $(-5, 5]$; Range: $\{-1, 0, 2, 3\}$

9. a. Domain: all real numbers; Range: all real numbers

   b. Domain: all real numbers; Range: $[0, \infty)$

11. a. Domain: all real numbers except zero; Range: $(0, \infty)$

b. Domain: all real numbers except zero; Range: all real numbers except zero

13. Domain: [0, 26.2]; Range: [90, 300]

15. Domain: [0, 600]; Range: [−90, 700]

17. a. $V(t) = 6000 - 550t$

 b. Domain: [0, 10]; Range: [500, 6000]

19. a.

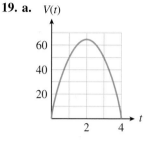

 b. Domain: [0, 4]; Range: [0, 64]

21. a.

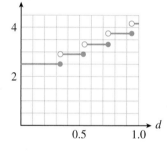

 b. Range: 2.50, 2.90, 3.30, 3.70, 4.10

 c. $13.30

23. Domain: nonnegative integers; The range includes all whole number multiples of 2.50 up to 20 × 2.50 = 50, all integer multiples of 2.25 from 21 × 2.25 = 47.25 to 50 × 2.25 = 112.50, and all integer multiples of 2.10 from 51 × 2.10 = 107.10 onwards: 0, 2.50, 5.00, 7.50, . . . , 50, 47.25, 49.50, 51.75, . . . , 112.50, 107.10, 109.20, 111.30, . . .

25. a. $f(x)$ domain: $x \neq 4$; Range: $(0, \infty)$

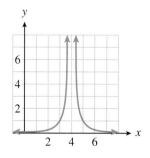

b. $h(x)$ domain: $x \neq 0$; Range: $(-4, \infty)$

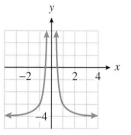

27. a. $G(v)$ domain: all real numbers; Range: all real numbers

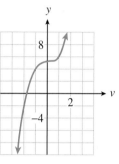

b. $H(v)$ domain: all real numbers; Range: all real numbers

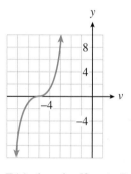

29. a. $T(z)$ domain: $[2, \infty)$; Range: $[0, \infty)$

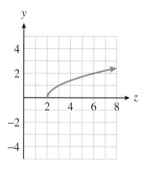

b. $S(z)$ domain: $[0, \infty)$; Range: $[-2, \infty)$

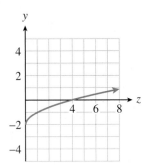

31. a. Not in range   **b.** $x = -6$ or $x = 2$

33. a. $t = -64$ **b.** $t = -8$

35. a. $w = \dfrac{1}{2}$ **b.** Not in range

37. a. Not in range   **b.** $h = 4$

39. Domain: $[-2, 5]$; Range: $[-4, 12]$

41. Domain: $[-5, 3]$; Range: $[-15, 1]$

43. Domain: $[-2, 2]$; Range: $[-9, 7]$

45. Domain: $[-1, 8]$; Range: $[0, 3]$

47. Domain: $[-1.25, 2.75]$; Range: $\left[\dfrac{4}{17}, 4\right]$

49. Domain $(3, 6]$; Range: $\left(-\infty, \dfrac{-1}{3}\right]$

51. a. Squaring both sides of the equation gives the equation of the circle centered on the origin with radius 4, but the points in the third and fourth quadrants are extraneous solutions introduced by squaring. (The original equation allowed only $y \geq 0$.)

    b. Domain: $[-4, 4]$; Range: $[0, 4]$

    c.
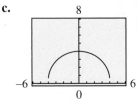
The calculator does not show the graph extending down to the x-axis.

53. a. Domain: $x \neq 2$; Range: $(0, \infty)$

    b. Domain: $x \neq 0$; Range: $(-2, \infty)$

    c. Domain: $x \neq 3$; Range: $(-5, \infty)$

55. a. Domain: all real numbers; Range: $(-\infty, 0]$

    b. Domain: all real numbers; Range: $(-\infty, 6]$

    c. Domain: all real numbers; Range: $(-\infty, 6]$

57. a. Domain: $[3, 13]$; Range: $[-2, 2]$

    b. Domain: $[0, 10]$; Range: $[-6, 6]$

    c. Domain: $[5, 15]$; Range: $[-4, 4]$

59. a. Domain: $(0, \infty)$; Range: $(0, 5)$

    b. Domain: $(-2, \infty)$; Range: $(0, 3)$

    c. Domain: $(3, \infty)$; Range: $(2, 4)$

61. a. $f(x)$ **b.** $g(x)$

63. a. $g(x)$ **b.** $f(x)$

65. Domain: $[0°, 90°]$; Range: $[12, 24]$

CHAPTER 2 REVIEW PROBLEMS

1. $x = 1$ or $x = 4$ **3.** $w = -2$ or $w = 4$

5. $r = -1 \pm \sqrt{\dfrac{A}{P}}$ **7.** 11%

9. $P = 1.001$ **11.** $m = 29$

13. $r = 3$ **15.** $\sqrt{12132} \approx 110$ cm

17. -2 **19.** 24

21. a. $x = -2$ or $x = 6$ **b.** $(-2, 6)$

    c. $(-\infty, -2] \cup [6, +\infty)$

23. a. $x = 0$ or $x = 3$ **b.** $(0, 3)$

    c. $(-\infty, 0) \cup (3, \infty)$

25.
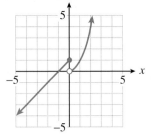

27.
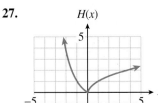

29.
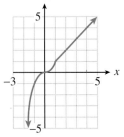

31. a. $y = |x|$ shifted up 2 units

b.

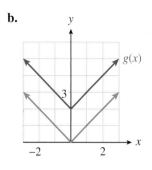

33. a. $y = \sqrt{x}$ shifted up 3 units

b.

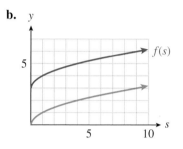

35. a. $y = |x|$ shifted left 2 units and down 3 units

b.

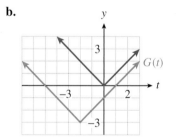

37. a. $y = \sqrt{x}$ reflected across the horizontal axis and stretched vertically by a factor of 2

b.

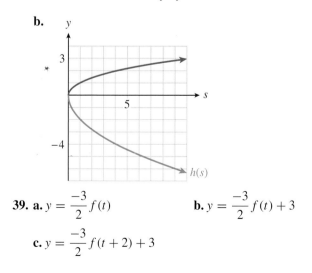

39. a. $y = \dfrac{-3}{2} f(t)$ **b.** $y = \dfrac{-3}{2} f(t) + 3$

    c. $y = \dfrac{-3}{2} f(t + 2) + 3$

41. a. $y = f(t - 1)$ **b.** $y = -f(t - 1)$

    c. $y = -f(t - 1) + 300$

43. $y = (x - 2)^2 - 4$

45.

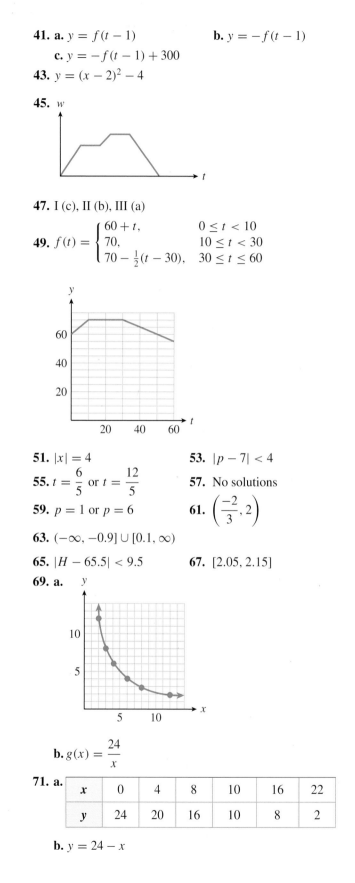

47. I (c), II (b), III (a)

49. $f(t) = \begin{cases} 60 + t, & 0 \le t < 10 \\ 70, & 10 \le t < 30 \\ 70 - \frac{1}{2}(t - 30), & 30 \le t \le 60 \end{cases}$

51. $|x| = 4$ **53.** $|p - 7| < 4$

55. $t = \dfrac{6}{5}$ or $t = \dfrac{12}{5}$ **57.** No solutions

59. $p = 1$ or $p = 6$ **61.** $\left(\dfrac{-2}{3}, 2 \right)$

63. $(-\infty, -0.9] \cup [0.1, \infty)$

65. $|H - 65.5| < 9.5$ **67.** $[2.05, 2.15]$

69. a.

b. $g(x) = \dfrac{24}{x}$

71. a.

x	0	4	8	10	16	22
y	24	20	16	10	8	2

b. $y = 24 - x$

73. a.

x	0	1	4	9	16	25
y	0	1	2	3	4	5

b. $y = \sqrt{x}$

75. a.

x	-3	-2	-2	0	1	2
y	5	0	-3	-4	-3	0

b. $y = x^2 - 4$

77. Domain: $[-2, 4]$; Range: $[-10, -4]$

79. Domain: $(-2, 4]$; Range: $\left[\dfrac{1}{6}, \infty\right)$

HOMEWORK 3.1

1. a. Yes; 6.5% **b.** $T = 0.065p$

c.

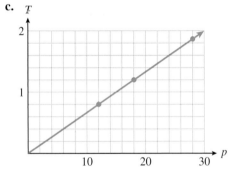

3. a. 24 square feet

b. $L = \dfrac{24}{w}$

c.

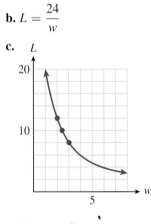

5. a. The ratio $\frac{y}{x}$ is a constant.

b. The product xy is a constant.

7. a.

Length	Width	Perimeter	Area
10	8	36	80
12	8	40	96
15	8	46	120
20	8	56	160

b. No **c.** $P = 16 + 2l$ **d.** Yes **e.** $A = 8l$

9. (b) **11.** (c)

13. a. $m = 0.165w$

w	50	100	200	400
m	8.25	16.5	33	66

b. 19.8 lb **c.** 303.03 lb **d.** It will double.

15. a. $L = 0.8125T^2$

T	1	5	10	20
L	0.8125	20.3	81.25	325

b. 234.8125 ft **c.** 0.96 sec

d. It must be four times as long.

17. a. $b = \dfrac{88}{d}$

d	1	2	12	24
B	88	44	7.3	3.7

b. 8.8 milligauss **c.** More than 20.47 in

d. It is one half as strong.

19. a. $P = \dfrac{1825}{8192}w^3 \approx 0.2228w^3$

w	10	20	40	80
p	223	1782	14,259	114,074

b. 752 kilowatts **c.** 35.54 mph

d. It is multiplied by 8.

21. a. $y = 0.3x$

b.

x	2	5	8	12	15
y	0.6	1.5	2.4	3.6	4.5

c. y doubles.

23. a. $y = \dfrac{2}{3}x^2$

b.

x	3	6	9	12	15
y	6	24	54	96	150

c. y is quadrupled.

25. a. $y = \dfrac{120}{x}$

b.

x	4	8	20	30	40
y	30	15	6	4	3

c. y is halved.

27. (b) $y = 0.5x^2$

29. (c) $\dfrac{y}{x^p}$ is not constant for any exponent p.

31. (b) $y = \dfrac{72}{x^2}$

33. (c) $x^p y$ is not constant for any exponent p.

35. a. $d = 0.005v^2$ **b.** 50 m

37. a. $m = \dfrac{8}{p}$ **b.** 0.8 ton

39. a. $T = \dfrac{6}{d}$ **b.** 1°C

41. a. $W = 600d^2$ **b.** 864 newtons

43. a. Wind resistance quadruples.

     b. It is one-ninth as great.

     c. It is decreased by 19% because it is 81% of the original.

45. a. It is one-fourth the original illumination.

     b. It is one-ninth the illumination.

     c. It is 64% of the illumination.

47. a. $D = (2.5 \times 10^{-52})m^{\frac{2}{?}}$ **b.** 2×10^{74} kg

     c. 1.7×10^{-8} m

49. a. $L = (1.25 \times 10^{-27})m^3$ **b.** 2×10^{12} kg

51. $xy = k$ implies that $(cx)\left(\dfrac{y}{c}\right) = k$.

53. If $y = \dfrac{k}{x^2}$, then multiplying both sides of the equation by x^2 gives $yx^2 = k$.

55. Yes

HOMEWORK 3.2

1.

n	-5	-4	-3	-2	-1	0	1	2	3	4	5
3^n	$\frac{1}{243}$	$\frac{1}{81}$	$\frac{1}{27}$	$\frac{1}{9}$	$\frac{1}{3}$	1	3	9	27	81	243

Each time n dicreases by 1, we divide the power in the bottom row by 3.

3. a. 8   **b.** -8 **c.** $\dfrac{1}{8}$ **d.** $\dfrac{-1}{8}$

5. a. $\dfrac{1}{8}$ **b.** $\dfrac{-1}{8}$ **c.** 8   **d.** -8

7. a. $\dfrac{1}{2^1} = \dfrac{1}{2}$ **b.** $\dfrac{1}{(-5)^2} = \dfrac{1}{25}$ **c.** $3^3 = 27$

     d. $(-2)^4 = 16$

9. a. $5 \cdot 4^3 = 320$ **b.** $\dfrac{1}{(2q)^5} = \dfrac{1}{32q^5}$ **c.** $\dfrac{-4}{x^2}$ **d.** $8b^3$

11. a. $\dfrac{1}{(m-n)^2}$ **b.** $\dfrac{1}{y^2} + \dfrac{1}{y^3}$ **c.** $\dfrac{2p}{q^4}$ **d.** $\dfrac{-5x^5}{y^2}$

13. a.

x	1	2	4	8	16
x^{-2}	1	0.25	0.06	0.02	0.00

     b. The values of $f(x)$ decrease, because x^{-2} is the reciprocal of x^2.

     c.

x	1	0.5	0.25	0.125	0.0625
x^{-2}	1	4	16	64	256

     d. The values of $f(x)$ increase toward infinity, because x^{-2} is the reciprocal of x^2.

15. (ii), (iii), and (iv) have the same graph, because they represent the same function.

17. a. $F(r) = 3r^{-4}$ **b.** $G(w) = \dfrac{2}{5}w^{-3}$

     c. $H(z) = \dfrac{1}{9}z^{-2}$

19. $x = -1.25$ or $x = 1.25$ **21.** $t = \dfrac{1}{16}$

23. $v = \dfrac{1}{5}$ or $v = \dfrac{-1}{5}$

25. a. $P = 0.355v^3$ **b.** $v \approx 52.03$ mph   **c.** 3.375

27. a. $D = \dfrac{70}{i}$ **b.** It decreases by about 2.3 years.

29. a. $L = (4\pi s R^2)T^4 \approx 7.2 \times 10^{-7} R^2 T^4$ **b.** 4840 K

31. a. 500 picowatts   **b.** $P = 8000d^{-4}$

     c.

D (nautical miles)	4	5	7	10
P (picowatts)	31.3	12.8	3.3	0.8

     d. 16.8 nautical miles

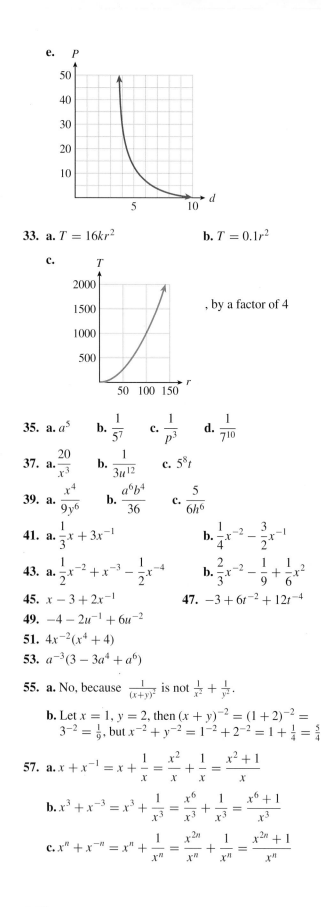

e.

33. a. $T = 16kr^2$ **b.** $T = 0.1r^2$

c.

, by a factor of 4

35. a. a^5 **b.** $\dfrac{1}{5^7}$ **c.** $\dfrac{1}{p^3}$ **d.** $\dfrac{1}{7^{10}}$

37. a. $\dfrac{20}{x^3}$ **b.** $\dfrac{1}{3u^{12}}$ **c.** $5^8 t$

39. a. $\dfrac{x^4}{9y^6}$ **b.** $\dfrac{a^6 b^4}{36}$ **c.** $\dfrac{5}{6h^6}$

41. a. $\dfrac{1}{3}x + 3x^{-1}$ **b.** $\dfrac{1}{4}x^{-2} - \dfrac{3}{2}x^{-1}$

43. a. $\dfrac{1}{2}x^{-2} + x^{-3} - \dfrac{1}{2}x^{-4}$ **b.** $\dfrac{2}{3}x^{-2} - \dfrac{1}{9} + \dfrac{1}{6}x^2$

45. $x - 3 + 2x^{-1}$ **47.** $-3 + 6t^{-2} + 12t^{-4}$

49. $-4 - 2u^{-1} + 6u^{-2}$

51. $4x^{-2}(x^4 + 4)$

53. $a^{-3}(3 - 3a^4 + a^6)$

55. a. No, because $\dfrac{1}{(x+y)^2}$ is not $\dfrac{1}{x^2} + \dfrac{1}{y^2}$.

  b. Let $x = 1$, $y = 2$, then $(x + y)^{-2} = (1 + 2)^{-2} = 3^{-2} = \frac{1}{9}$, but $x^{-2} + y^{-2} = 1^{-2} + 2^{-2} = 1 + \frac{1}{4} = \frac{5}{4}$

57. a. $x + x^{-1} = x + \dfrac{1}{x} = \dfrac{x^2}{x} + \dfrac{1}{x} = \dfrac{x^2 + 1}{x}$

  b. $x^3 + x^{-3} = x^3 + \dfrac{1}{x^3} = \dfrac{x^6}{x^3} + \dfrac{1}{x^3} = \dfrac{x^6 + 1}{x^3}$

  c. $x^n + x^{-n} = x^n + \dfrac{1}{x^n} = \dfrac{x^{2n}}{x^n} + \dfrac{1}{x^n} = \dfrac{x^{2n} + 1}{x^n}$

59. $a^{-2}a^{-3} = \dfrac{1}{a^2} \cdot \dfrac{1}{a^3} = \dfrac{1}{a^2 \cdot a^3}$

  $= \dfrac{1}{a^{2+3}}$ By the first law of exponents.

  $= \dfrac{1}{a^5} = a^{-5}$

61. $\dfrac{a^{-2}}{a^{-6}} = a^{-2} \div a^{-6} = \dfrac{1}{a^2} \div \dfrac{1}{a^6}$

  $= \dfrac{1}{a^2} \cdot \dfrac{a^6}{1} = \dfrac{a^6}{a^2}$ By the second law of exponents.

  $= a^{6-2}$

  $= a^4$

HOMEWORK 3.3

1. a. 11 **b.** 3 **c.** 5

3. a. 2 **b.** 2 **c.** 9

5. a. 3 **b.** 3 **c.** 2

7. a. 2 **b.** $\dfrac{1}{2}$ **c.** $\dfrac{1}{8}$

9. a. $\sqrt{3}$ **b.** $4\sqrt[3]{x}$ **c.** $\sqrt[5]{4x}$

11. a. $\dfrac{1}{\sqrt[3]{6}}$ **b.** $\dfrac{3}{\sqrt[8]{xy}}$ **c.** $\sqrt[4]{x - 2}$

13. a. $7^{1/2}$ **b.** $(2x)^{1/3}$ **c.** $2z^{1/5}$

15. a. $-3 \cdot 6^{-1/4}$ **b.** $(x - 3y)^{1/4}$ **c.** $-(1 + 3b)^{-1/5}$

17. a. 125 **b.** 2 **c.** 63 **d.** $-2x^7$

19. a. 1.414 **b.** 4.217 **c.** 1.125 **d.** 0.140 **e.** 2.782

21. a. $g(x) = 3.7x^{1/3}$ **b.** $H(x) = 85^{1/4}x^{1/4}$

  c. $F(t) = 25t^{-1/5}$

23. $x = 91.125$ **25.** $x = 241$ **27.** $x = \dfrac{19}{2}$

29. $\pm\sqrt{30}$ **31.** $L = \dfrac{gT^2}{4\pi^2}$ **33.** $s = \pm\sqrt{t^2 - r^2}$

35. $V = \dfrac{4}{3}\pi r^3$ **37.** $p = \dfrac{8Lvf}{\pi R^4}$

39. a. $T = \dfrac{2\pi}{\sqrt{32}}L^{1/2}$ **b.** 90 feet

  c.

41. a. 3 meters per second **b.** ≈2.2 meters per second

c.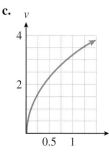

d. 1.9 meters **e.** 1.2 meters per second

43. a. 6.5×10^{-13} cm; 1.17×10^{-36} cm^3

b. 1.8×10^{14} g/cm^3

c.

Element	Carbon	Potassium	Cobalt
Mass number, A	14	40	60
Radius, r (10^{-13} cm)	3.1	4.4	5.1

Element	Technetium	Radium
Mass number, A	99	226
Radius, r (10^{-13} cm)	6	7.9

d.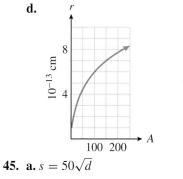

45. a. $s = 50\sqrt{d}$ **b.** 30 cm/sec

c.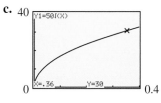

47. a. $r = 12.1\sqrt{P}$ **b.** 94 ft/sec

c.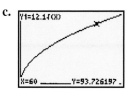

49. a. 287; 343 **b.** 2015; 2058

c. The membership grows rapidly at first but is growing less rapidly with time.

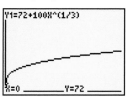

51. a. I **b.** III **c.** II **d.** none

53. a. The graphs of $x^{1/n}$ become closer and closer to horizontal when n increases (for $x < 1$).

b. 10, 4.64, 3.16, 2.51

c. 1.58, 1.05, 1.005; the values decrease towards 1.

55. The graphs of y_1 and y_2 are symmetric about $y_3 = x$.

57. The graphs of y_1 and y_2 are symmetric about $y_3 = x$.

59. a. 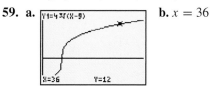 **b.** $x = 36$

61. a. $x^{1/2}$ **b.** $\left(x^{1/2}\right)^{1/2}$

c.
$$\sqrt{\sqrt{x}} = \left(x^{1/2}\right)^{1/2} \quad \text{By definition of fractional exponents.}$$
$$= x^{1/4} \quad \text{By the third law of exponents.}$$
$$= \sqrt[4]{x} \quad \text{By definition of fractional exponents.}$$

63. $\dfrac{1}{4}x^{1/2} - 2x^{-1/2} + \dfrac{1}{\sqrt{2}}x$

65. $3x^{-1/3} - \dfrac{1}{2}$ **67.** $x^{0.5} + x^{-0.25} - x^0$

HOMEWORK 3.4

1. a. 27 **b.** 25 **c.** 125

3. a. $\dfrac{1}{64}$ **b.** $\dfrac{1}{16}$ **c.** $\dfrac{1}{256}$

5. a. $\sqrt[5]{x^4}$ **b.** $\dfrac{1}{\sqrt[6]{b^5}}$ **c.** $\dfrac{1}{\sqrt[3]{(pq)^2}}$

7. a. $3\sqrt[5]{x^2}$ **b.** $\dfrac{4}{\sqrt[3]{z^4}}$ **c.** $-2\sqrt[4]{xy^3}$

9. a. $x^{2/3}$ **b.** $2a^{1/5}b^{3/5}$ **c.** $-4mp^{-7/6}$

11. a. $(ab)^{2/3}$ **b.** $8x^{-3/4}$ **c.** $\frac{1}{3}RT^{-1/2}K^{-5/2}$

13. a. 8 **b.** -81 **c.** $2y^3$

15. a. $-a^4b^8$ **b.** $2x^3y^9$ **c.** $-3a^2b^3$

17. a. 7.931 **b.** 10.903 **c.** 0.090 **d.** 35.142

19. a.

t	5	10	15	20
$I(t)$	131	199	254	302

Range: $[0, 302]$

b. ≈ 19.812 or about 20 days

c.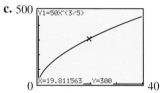

21. All the graphs are increasing and concave up. For $x > 1$, each graph increases more quickly than the previous one.

23. a. $V = L^3$, $A = 6L^2$ **b.** $L = V^{1/3}$, $L = \left(\dfrac{A}{6}\right)^{1/2}$

c. $A = 6V^{2/3}$

d. $\frac{A}{V} = \frac{6}{L}$. As L increases, the surface-to-volume ratio decreases.

25. a. **b.** 7114.32

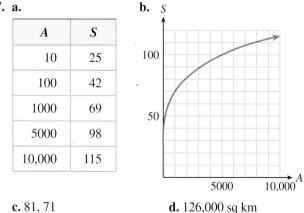

27. a.

A	S
10	25
100	42
1000	69
5000	98
10,000	115

b.

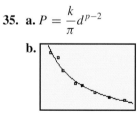

c. 81, 71 **d.** 126,000 sq km

29. a. Home range size: II, lung volume: III, brain mass: I, respiration rate: IV

b. If $p > 1$, the graph is increasing and concave up. If $0 < p < 1$, the graph is increasing and concave down. If $p < 0$, the graph is decreasing and concave up.

31. a. *Tricosanthes* is the snake gourd and *Lagenaria* is the bottle gourd. *Tricosanthes* is thinner and *Lagenaria* is fatter.

b. $a \approx 9.5$ **c.** $a \approx 2$ **d.** Yes

33. a.

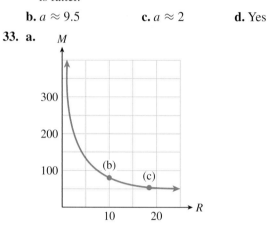

b. ≈ 78.5 or about 79 species **c.** $18.4°C$

d. $f(9) \approx 85$, $f(10) \approx 79$, $f(19) \approx 49$, $f(20) \approx 47$; from $9°C$ to $10°C$ has the greater decrease corresponding to the steeper slope. If the temperature range is 9, there will be approximately 85 species. If the temperature range is 10, there will be approximately 79 species. If the temperature range is 19, there will be approximately 49 species. If the temperature range is 20, there will be approximately 47 species.

35. a. $P = \dfrac{k}{\pi}d^{p-2}$

b.

The power function is a good fit on this interval.

c. 1.3

37. a. $4a^2$ **b.** $9b^{5/3}$ **39. a.** $4w^{3/2}$ **b.** $3z^2$

41. a. $\dfrac{1}{2k^{1/4}}$ **b.** $\dfrac{4}{3h^{1/3}}$

43. a. Wren: 15 days, greylag goose: 28 days

b. $\dfrac{I(m) \cdot W(m)}{m} = 0.18m^{-0.041}$

c. Because $m^{-0.041}$ is close to m^0, the fraction lost is close to 0.18.

45. $x = 64$ **47.** $x = \dfrac{1}{243}$ **49.** $x \approx 2.466$

51. a. $p = \sqrt{Ka^3}$ or $(Ka^3)^{1/2}$ **b.** 1.88 years

53. $\dfrac{13}{3}$ **55.** 0.655   **57.** $2x^{3/2} - 2x$

59. $\dfrac{1}{2}y^{1/3} + \dfrac{3}{2}y^{-7/6}$ **61.** $2x^{1/2} - x^{1/4} - 1$

63. $a^{3/2} - 4a^{3/4} + 4$ **65.** $x(x^{1/2} + 1)$

67. $\dfrac{y - 1}{y^{1/4}}$ **69.** $\dfrac{a^{2/3} + a^{1/3} - 1}{a^{1/3}}$

HOMEWORK 3.5

1. a. $R = f(x, y) = 129x + 240y$

  b. $f(24, 12) = 5976$ dollars is the maximum revenue.

3. a. $f\left(4, \dfrac{3}{2}\right) = \dfrac{25}{12} \approx 2.1$ inches

  b. No: We do not have $r = kL^2$ for any constant k.

  c. When $L = 2r$ and $h = r$, $f(L, h) = r$.

5. a. 16,220 sq cm   **b.** Height   **c.** 4.1%

7. a. $C = f(s, w)$

  b. $f(4.5, 160) = 110$, so someone walking 4.5 mph and weighing 160 pounds burns 110 calories per mile.

  c. $s > 4.5$. A person who weighs 160 pounds must walk faster than 4.5 mph in order to burn more than 110 calories per mile.

  d. 7 mph

  e. Find the row with your walking speed in the left column and move along that row until you are in the column with your weight at the top. The value in that row and column is the number of calories you burn per mile.

9. a. When is $f(r, 20) \le 800$? $r \le 7\%$; When is $f(r, 30) \le 800$? $r \le 9\%$

  b. Reducing interest rate by 5%

  c. No   **d.** No

  e. 30-year loan

11. a. Direct variation: In each row, $E = km$ for some constant k that depends on the row.

  b. Inverse variation: In each column, $E = \dfrac{c}{g}$ for some constant c that depends on the column.

  c. $E = \dfrac{m}{g}$ miles/gallon

d.

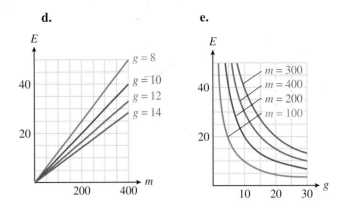

e.

13. a. In each row, $R = kp$ for some constant k that depends on the row.

  b. In each column, $R = cd^2$ for some constant c that depends on the column.

  c. $R = 1.57d^2 p$

  d. 588.75 pounds: When we keep p constant and double d, R is multiplied by a factor of 4. So the value at $d = 2\frac{1}{2}$, $p = 60$ should be 4 times the value at $d = 1\frac{1}{4}$, $p = 60$.

15. a. $a = \dfrac{2d}{t^2}$

  b. Mercedes-Benz: 19.05 ft/sec^2, Porsche: 19.73 ft/sec^2, Dodge: 20.47 ft/sec^2, Saleen: 23.59 ft/sec^2, Ford: 25.66 ft/sec^2

  c.

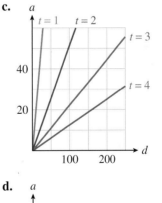

  d.

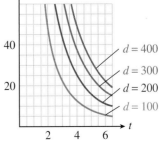

17. **a.** $L = \dfrac{3.2v^3}{R}$

   b. Increased by 72.8%

   c. Decreased by $16\frac{2}{3}\%$

19. **a.**

Temperature (°C)	Pressure (atmospheres)							
	50	100	150	200	250	300	350	400
350	25	38	46	53	58	62	66	68
400	16	26	33	38	45	48	53	56
450	9	17	23	28	32	37	40	43
500	6	11	16	20	23	27	29	32
550	4	8	11	14	17	19	22	24

   b. The ammonia yield decreases.

   c.

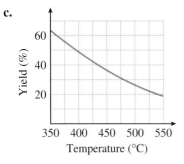

CHAPTER 3 REVIEW PROBLEMS

1. **a.** $d = 1.75t^2$ **b.** 63 cm

3. 480 bottles

5. **a.** $w = \dfrac{k}{r^2}$

   b.

   c. $3960\sqrt{3} \approx 6860$ miles

7. $y = 1.2x^2$ **9.** $y = \dfrac{20}{x}$

11. **a.** $\dfrac{1}{81}$ **b.** $\dfrac{1}{64}$

13. **a.** $\dfrac{1}{243m^5}$ **b.** $\dfrac{-7}{y^8}$

15. **a.** $\dfrac{2}{c^3}$ **b.** $\dfrac{99}{z^2}$

17. **a.** $25\sqrt{m}$ **b.** $\dfrac{8}{\sqrt[3]{n}}$

19. **a.** $\dfrac{1}{\sqrt[4]{27q^3}}$ **b.** $7\sqrt{u^3v^3}$

21. **a.** $2x^{2/3}$ **b.** $\dfrac{1}{4}x^{1/4}$

23. **a.** $6b^{-3/4}$ **b.** $\dfrac{-1}{3}b^{-1/3}$

25.

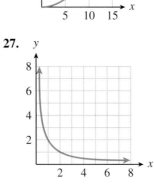

27.

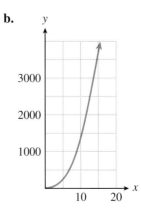

29. $f(x) = \dfrac{2}{3}x^{-4}$

31. **a.**

x	16	$\dfrac{1}{4}$	3	100
$Q(x)$	4096	$\dfrac{1}{8}$	$4\sqrt{3^5} \approx 62.35$	400,000

   b.

33. a.

x	0	1	5	10	20	50	70	100
$f(x)$	0	1	1.62	2.00	2.46	3.23	3.58	3.98

b. $f(x)$

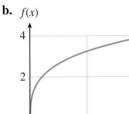

35. 112 kg

37. a. $M(t)$

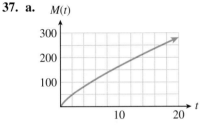

b. 283.7 or ≈ 284 **c.** 2051

39. a. It is the cost of producing the first ship.

b. $C = \dfrac{12}{\sqrt[8]{x}}$ million

c. About \$11 million; about 8.3%

d. About 8.3%

41. $t = 10$ **43.** $x = 7$

45. $x = 5$ **47.** $x = 75$

49. $y = 29{,}524$ **51.** $g = \dfrac{2v}{t^2}$

53. $p = \pm 2\sqrt{R^2 - R}$ **55.** $49t^2$

57. $\dfrac{k^7}{64}$ **59.** $8a^2$

61. a. 132.6 km   **b.** w

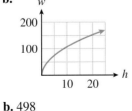

63. a. 480   **b.** 498

65. a. \$450

b. $t = 8$: It costs \$864 to insulate a ceiling with 8 cm of insulation over an area of 600 square meters.

c. $C = 0.72A$ **d.** $C = 18T$

e. $C = 0.18AT$ **f.** \$1440

67. a. $N = \dfrac{k}{d^2 E^3}$, where N is number of people, d is distance in miles from the road, E is the elevation gain, and k is the constant of variation.

b. $k \approx 0.01$ **c.** 3

HOMEWORK 4.1

1. a. \$28   **b.** \$31.36

3. It is 99% of what it was 2 years ago.

5. a. $P = 1200 + 150t$; 1650

b. $P = 1200 \cdot 1.5^t$; 4050

7. a. $V = 18{,}000 - 2000t$; \$8000

b. $V = 18{,}000 \cdot 0.8^t$; \$5898.24

9. A: 20%;   B: 2%;   C: 7.5%;

     D: 100%;   E: 115%

11. a. $P = 20{,}000 \cdot 2.5^{t/6}$

b. P (thousands)

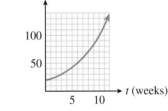

c. 36,840 bees; 424,128 bees

13. a. $A = 4000 \cdot 1.08^t$

b. A

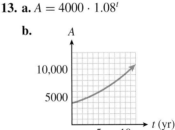

c. \$4665.60; \$8635.70

15. a. $P = 200{,}000 \cdot 1.05^t$

b. P (thousands)

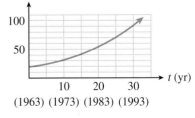

(1963) (1973) (1983) (1993)

c. \$359,171; \$746,691

17. a. $P = 250{,}000 \cdot 0.75^{t/2}$

b. P (thousands)

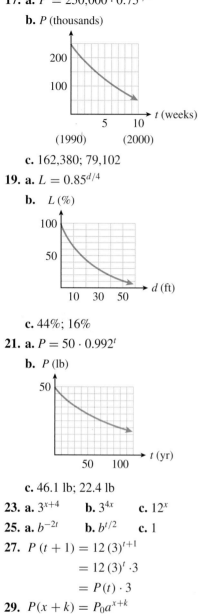

c. 162,380; 79,102

19. a. $L = 0.85^{d/4}$

b. L (%)

c. 44%; 16%

21. a. $P = 50 \cdot 0.992^t$

b. P (lb)

c. 46.1 lb; 22.4 lb

23. a. 3^{x+4} **b.** 3^{4x} **c.** 12^x

25. a. b^{-2t} **b.** $b^{t/2}$ **c.** 1

27. $P(t+1) = 12(3)^{t+1}$
$$= 12(3)^t \cdot 3$$
$$= P(t) \cdot 3$$

29. $P(x+k) = P_0 a^{x+k}$
$$= P_0 a^x \cdot a^k$$
$$= P(x) \cdot a^k$$

31. a. In the expression $2 \cdot 3^t$, only the 3 is raised to a power t, and the result is doubled, but if both the 2 and the 3 were raised to the power t, the result would be 6^t.

b.

t	0	1	2
$P(t)$	2	6	18
$Q(t)$	1	6	36

33. 4 **35.** 1.2

37. $r \approx 0.14$ or $r \approx -2.14$ **39.** $r \approx 0.04$ or $r \approx 1.96$

41. a. $P(t) = 1{,}545{,}387 b^t$

b. Growth factor 1.049; Percent rate of growth 4.9%

c. 2,493,401

43. a. 365 **b.** $N(t) = 365(0.356)^t$

c. 400

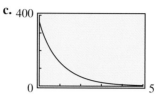

d. 0.03. Therefore, none.

45. a.

t	0	1	2	3	4
P	8	12	18	27	40.5

The growth factor is 1.5.

47. a.

x	0	1	2	3	4
Q	20	24	28.8	34.56	41.47

The growth factor is 1.2.

49. a.

w	0	1	2	3	4
N	120	96	76.8	61.44	49.15

The decay factor is 0.8.

51. a.

t	0	1	2	3	4
C	10	8	6.4	5.12	4.10

The decay factor is 0.8.

53. a.

n	0	1	2	3	4
B	200	220	242	266.2	292.82

The growth factor is 1.1.

55. a. Initial value 4, growth factor $2^{1/3}$

b. $f(x) = 4 \cdot 2^{x/3}$

57. a. Initial value 80, decay factor $\frac{1}{2}$

b. $f(x) = 80 \cdot \left(\frac{1}{2}\right)^x$

59. 84.6%, 55.8%

61. No, an increase of 48% in 6 years corresponds to a growth factor of $1.48^{1/6} \approx 1.0675$, or an annual growth rate of about 6.75%.

63. a. $P(t) = 16{,}986{,}335(1 + r)^t$ **b.** 2.07%

65. a. 3.53% **b.** 3.53% **c.** No **d.** 3.53%

67. a. 39; 1.045 **b.** 36; 1.047 **c.** Species B

69. a.

t	0	2	4	6	8
$L(t)$	3	6	9	12	15

$L(t) = 3 + 1.5t$

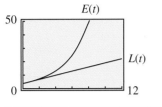

b.

t	0	2	4	6	8
$E(t)$	3	6	12	24	48

$E(t) = 3 \cdot 2^{t/2}$

71. a. 244 tigers per year

b. 0.97; 3%

c. Linear: 3067; Exponential: 4170

HOMEWORK 4.2

1. a. 26; increasing **b.** 1.2; decreasing

c. 75; decreasing **d.** $\frac{2}{3}$; increasing

3.

x	-3	-2	-1	0	1	2	3
$f(x) = 3^x$	$\frac{1}{27}$	$\frac{1}{9}$	$\frac{1}{3}$	1	3	9	27
$g(x) = \left(\frac{1}{3}\right)^x$	27	9	3	1	$\frac{1}{3}$	$\frac{1}{9}$	$\frac{1}{27}$

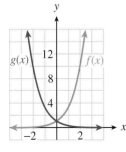

The two graphs are reflections of each other across the y-axis. f is increasing, g is decreasing. f has the negative x-axis as an asymptote, and g has the positive x-axis as its asymptote.

5.

t	-3	-2	-1	0	1	2	3
$h(t) = 4^{-t}$	64	16	4	1	$\frac{1}{4}$	$\frac{1}{16}$	$\frac{1}{64}$
$q(t) = -4^t$	$\frac{-1}{64}$	$\frac{-1}{16}$	$\frac{-1}{4}$	-1	-4	-16	-64

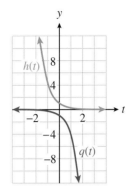

The graphs are reflections of each other across the origin. Both are decreasing, but h has the negative x-axis as an asymptote, and q has the positive x-axis as its asymptote.

7. a. I **b.** IV **c.** III **d.** II

9. a.

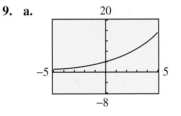

b. [1.08, 14.85]

11.

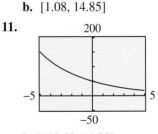

b. [152.59, 16.38]

13. Because they are defined by equivalent expressions, (b), (c), and (d) have identical graphs

15. a. To evaluate f we subtract 1 from the input before evaluating the exponential function; to evaluate g we subtract 1 from the output of the exponential function.

b.

x	$y = 2^x$	$f(x)$	$g(x)$
-2	$\frac{1}{4}$	$\frac{1}{8}$	$\frac{-3}{4}$
-1	$\frac{1}{2}$	$\frac{1}{4}$	$\frac{-1}{2}$
0	1	$\frac{1}{2}$	0
1	2	1	1
2	4	2	3

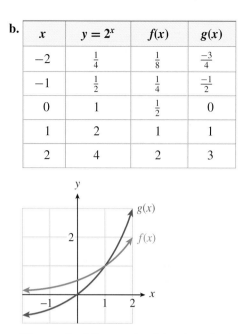

c. The graph of f is translated 1 unit to the right; the graph of g is shifted 1 unit down.

17. a. To evaluate f we take the negative of the output of the exponential function; to evaluate g we take the negative of the input.

b.

x	$y = 3^x$	$f(x)$	$g(x)$
-2	$\frac{1}{9}$	$\frac{-1}{9}$	9
-1	$\frac{1}{3}$	$\frac{-1}{3}$	3
0	1	-1	1
1	3	-3	$\frac{1}{3}$
2	9	-9	$\frac{1}{9}$

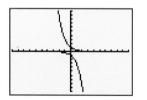

c. The graph of f is reflected about the x-axis; the graph of g is reflected about the y-axis.

19. a. $3(5^{a+2})$ is not equivalent to $9 \cdot 3(5^a)$.

b. $3(5^{2a})$ is not equivalent to $2 \cdot 3(5^a)$.

21. a. $8^w - 8^z$ is not equivalent to 8^{w-z}.

b. 8^{-x} is equivalent to $\frac{1}{8^x}$.

23. a. $P_0 = 300$

b.

x	0	1	2
$f(x)$	300	600	1200

c. $b = 2$ **d.** $f(x) = 300(2)^x$

25. a. $S_0 = 150$ **b.** $b \approx 0.55$ **c.** $S(d) = 150(0.55)^d$

27. $\frac{2}{3}$ **29.** $\frac{-1}{4}$ **31.** $\frac{1}{7}$ **33.** $\frac{-5}{4}$ **35.** ± 2

37. a. $N(t) = 26(2)^{t/6}$

b. 500,000   **c.** 72 days later

39. a. $V(t) = 700(0.7)^{t/2}$

b. 800   **c.** 4 yr

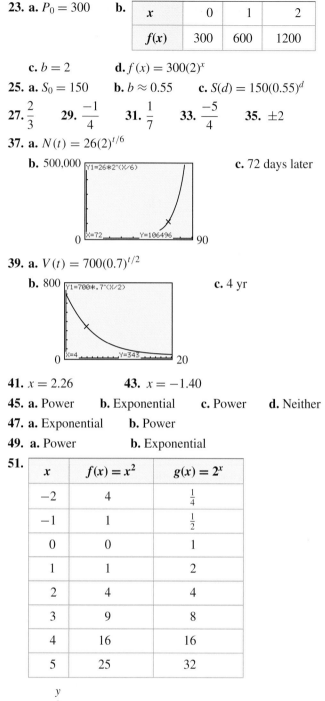

41. $x = 2.26$ **43.** $x = -1.40$

45. a. Power   **b.** Exponential   **c.** Power   **d.** Neither

47. a. Exponential   **b.** Power

49. a. Power   **b.** Exponential

51.

x	$f(x) = x^2$	$g(x) = 2^x$
-2	4	$\frac{1}{4}$
-1	1	$\frac{1}{2}$
0	0	1
1	1	2
2	4	4
3	9	8
4	16	16
5	25	32

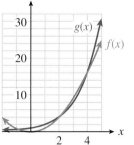

a. Range of f: $[0, \infty)$; Range of g: $(0, \infty)$

b. 3   **c.** $-0.7667, 2, 4$

d. $(-0.7667, 2)$ and $(4, \infty)$ **e.** g

53. a. $y = 3^x - 4$,

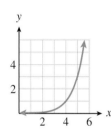

Domain: $(-\infty, \infty)$; range: $(-4, \infty)$, x-intercept $(1.26, 0)$; y-intercept $(0, -3)$; horizontal asymptote $y = -4$

b. $y = 3^{x-4}$,

Domain: $(-\infty, \infty)$; range: $(0, \infty)$; no x-intercept; y-intercept $(0, 0.012)$; the x-axis is the horizontal asymptote.

c. $y = -4 \cdot 3^x$,

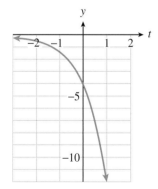

Domain: $(-\infty, \infty)$; range: $(-\infty, 0)$; no x-intercept; y-intercept $(0, -4)$; the x-axis is the horizontal asymptote.

55. a. $y = -6^t$,

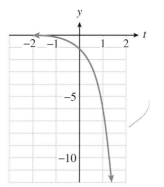

Domain: $(-\infty, \infty)$; range: $(-\infty, 0)$; no x-intercept; y-intercept $(0, 1)$; the x-axis is the horizontal asymptote.

b. $y = 6^{-t}$,

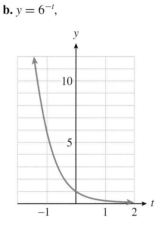

Domain: $(-\infty, \infty)$; range: $(0, \infty)$; no x-intercept; y-intercept $(0, 1)$; the x-axis is the horizontal asymptote.

c. $y = -6^{-t}$,

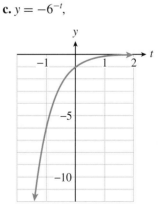

Domain: $(-\infty, \infty)$; range: $(-\infty, 0)$; no x-intercept; y-intercept $(0, -1)$; the x-axis is the horizontal asymptote.

57. a. $y = 2^{x-3}$, (0.0125)

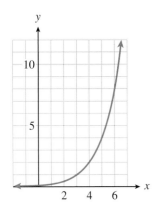

Domain: $(-\infty, \infty)$; range $(0, \infty)$; no x-intercept; y-intercept $(0, 0.125)$; the x-axis is the horizontal asymptote.

b. $y = 2^{x-3} + 4$, (0, 4.125)

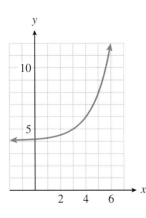

Domain: $(-\infty, \infty)$; range $(0, \infty)$; no x-intercept; y-intercept $(0, 4.125)$; horizontal asymptote: $y = 4$

61. a. The graph of $y = 2^x$ has been reflected about the y-axis and shifted up 2 units.

b. $y = 2^{-x} + 2$

63. a. The graph of $y = 2^x$ has been reflected about the x-axis and shifted up 10 units.

b. $y = -2^x + 10$

65. a. I   **b.** III   **c.** II

67. a.

t	3.5	4	8	10	15
$f(t)$	128	154.75	184.05	150.93	103.96

b.

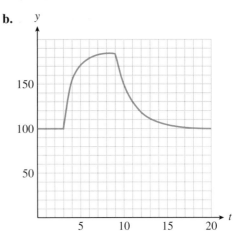

c. The initial walking corresponds to the time interval from 0 to 3 minutes; the volunteer's heart rate was 100 beats per minute during the first 3 minutes. The jogging corresponds to the increasing portion of the graph, with the interval from 3 to 4 minutes, and an increase in the heart rate to about 155 beats per minutes. From 4 to 9 minutes, the volunteer maintained the higher speed, and the heart rate rose to about 185 beats per minute. The decreasing portion, from 9 to 20 minutes, corresponds to the cool down walking period; the heart rate at this decreased rapidly and leveled off to about 100 beats per minute.

HOMEWORK 4.3

1. a. 2   **b.** 5   **3. a.** $\dfrac{1}{2}$ **b.** -1

5. a. 1   **b.** 0   **7. a.** 5   **b.** 6

9. a. -1 **b.** -3

11. $\log_{10} 1024 = 10$ **13.** $\log_{10} 5 \approx 0.699$

15. $\log_t 16 = \dfrac{3}{2}$ **17.** $\log_{0.8} M = 1.2$

19. $\log_x(W - 3) = 5t$ **21.** $\log_3(2N_0) = -0.2t$

23. a. $\log_4(2.5)$ **b.** 0.7

25. a. $\log_{10} 0.003$ **b.** -2.5

27. a. $0 < \log_{10} 7 < 1$ **b.** 0.85

29. a. $3 < \log_3 67.9 < 4$ **b.** 3.84

31. a. 0.7348   **b.** 1.7348   **c.** 2.7348

   d. 3.7348 When the input to the common logarithm is multiplied by 10, the output is increased by 1.

33. a. 0.3010   **b.** 0.6021   **c.** 0.9031

   d. 1.2041 When the input to the common logarithm is doubled, the output is increased by about 0.3010.

35. -0.23 **37.** 2.53   **39.** 0.77

41. -0.68 **43.** 3.63

45. $2 \cdot 5^x \neq 10^x$; the first step should be to divide both sides of the equation by 2; $x = \log 424$.

47. $\frac{10^{4x}}{4} \neq 10^x$; the first step should be to write $4x = \log 20$; $x = \frac{\log 20}{4}$.

49. a. 33,855,812

b. 38,515,295; 41,080,265; 43,816,051

c. 2002 **d.** 2012

e.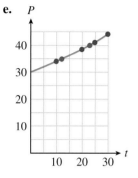

51. a. 85.5

b. Decreasing; range: [5.4, 1355.2]

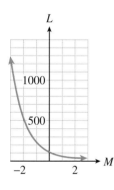

c. 1.45 **d.** $\frac{1}{100}$

e. $10^{0.4} \approx 2.5119$ **f.** 2.15×10^{-6} to 855,067

53. 9.60 in **55.** 1.91 mi **57.** 3.34 mi **59.** 1

61. 0 **63.** 1 **65.** 0

HOMEWORK 4.4

1. a. 10^8 **b.** 2; 6; 8; $2 + 6 = 8$

3. a. b^3 **b.** 8; 5; 3; $8 - 5 = 3$

5. a. 10^{15} **b.** 15; 3; $15 = 3 \cdot 5$

7. a. $\log_b 2 + \log_b x$ **b.** $\log_b x - \log_b 2$

9. a. $\log_3 3 + 4 \log_3 x$ **b.** $\frac{1}{t} \log_5 1.1$

11. a. $\frac{1}{2} + \frac{1}{2} \log_b x$ **b.** $\frac{1}{3} \log_3 (x^2 + 1)$

13. a. $\log P_0 + t \log (1 - m)$ **b.** $4t[\log_4 (4 + r) - 4t]$

15. a. $\log_b 4$ **b.** $\log_4 x^2 y^3$

17. a. $\log 2x^{5/2}$ **b.** $\log (t - 4)$

19. a. $\log \frac{1}{27}$ **b.** $\log_6 (2w^2)$

21. a. 1.7917 **b.** -0.9163

23. a. 2.1972 **b.** 1.9560

25. 2.8074 **27.** 0.8928 **29.** ± 1.3977

31. 0.8109 **33.** 0.2736 **35.** -12.4864

37. a. $S(t) = S_0(1.09)t$ **b.** 4.7 hours

39. a. $C(t) = 0.7(0.80)^t$ **b.** After 2.5 hours

c.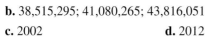

41. a. $J(t) = 1,041,000 \cdot 1.0182^t$ **b.** In 2040

43. a. $S(t) = S_0 \cdot 0.9527^t$ **b.** 28.61 hours

45. a. 5 **b.** 6 **c.** 5; (a) and (c) are equal.

47. a. 6 **b.** 9 **c.** 4; a, c

49. a. $\log 24 \approx 1.38$ **b.** $\log 240 \approx 2.38$

c. $\log 230 \approx 2.36$; None are equal.

51. a. $\log 60 \approx 1.78$ **b.** $\log 5 \approx 0.70$

c. $\frac{\log_{10} 75}{\log_{10} 15} \approx 1.59$; None are equal.

53. 12.9% **55.** About 11 years

57. a. $A = 1000\left(1 + \frac{0.12}{n}\right)^{5n}$

b. A increases.

c. 16; 31; 553

d. Increasing, concave down, asymptotically approaching A ≈ 1822.12

59. $k = \frac{1}{t} \frac{\log(N/N_0)}{\log a}$

61. $\frac{1}{k} \log\left(\frac{A}{A_0} + 1\right)$ **63.** $\frac{\log(w/p)}{\log v}$

65. a. $x = b^m$, $y = b^n$ **b.** $\log_b(b^m \cdot b^n)$

    c. $\log_b(b^m \cdot b^n) = \log_b b^{m+n}$ **d.** $\log_b b^{m+n} = m + n$

    e. $\log_b b^{m+n} = (\log_b x) + (\log_b y)$

67. a. $x = b^m$ **b.** $\log_b (b^m)^k$

    c. $\log_b (b^m)^k = \log_b b^{mk}$ **d.** $\log_b b^{mk} = mk$

    e. $\log_b b^{mk} = (\log_b x) \cdot k$

HOMEWORK 4.5

1. $A(x) = 0.14(50)^{x/3}$ **3.** $f(x) = \dfrac{65{,}536}{729}\left(\dfrac{3}{4}\right)^x$

5. $M(x) = 62{,}500(0.2)^x$ **7.** $s(x) = \dfrac{1}{135}(9)^x$

9. $y = \dfrac{4}{3}(3)^{x/4}$ **11.** $y = 50(2)^{-x/4}$

13. a. $y = 2.6 - 1.3x$ **b.** $y = 2.6(0.5)^x$

c.

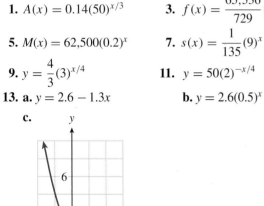

15. a. $y = -36 - 16x$ **b.** $y = \dfrac{12}{5}(5)^{-x/3}$

c.

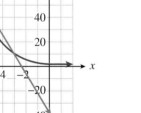

17. a. $y = 2.5 + 0.875x$ **b.** $y = 1.5(2)^{x/2}$

c.

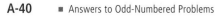

19. a. $P = P_0(1.052)^t$; t is the number of years since 1990.

    b. $\dfrac{\log 2}{\log 1.052} \approx 13.7$ years

    c.

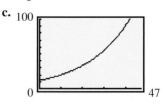

21. a. $GDP = 1.028^t$ million pounds

    b. $\dfrac{\log 2}{\log 1.028} \approx 25.1$ years   **c.** 50.2 years

    d. *GDP*

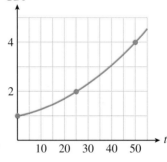

23. a. $\dfrac{\log 0.5}{\log 0.946} \approx 12.5$ hours   **b.** 25 hours

    c.

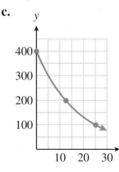

25. a. $\dfrac{\log 0.5}{\log 0.844} \approx 4.1$ hours   **b.** 8.2 hours

    c.

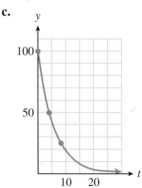

27. a. $P = 2000(2)^{t/5}$ **b.** 14.87%

29. a. $D = D_0\left(\dfrac{1}{2}\right)^{t/18}$ **b.** 3.78%

31. a. $A = A_0\left(\dfrac{1}{2}\right)^{t/1620}$ **b.** 0.043%

33. a. $P = P_0(2)^{t/25}$ **b.** 2.81%

35. a. $ab^D = 2 \cdot ab^0 = 2a$ **b.** $b^D = 2$

     c. $f(t + D) = ab^{t+D} = a \cdot b^t \cdot b^D = ab^t \cdot 2 = 2f(t)$

     d. For any value of t, after D units of time, the new value of f is 2 times the old value.

37. a. $ab^R = \dfrac{1}{3} \cdot ab^0 = \dfrac{1}{3}a$ **b.** $b^R = \dfrac{1}{3}$

     c. $g(t + R) = ab^{t+R} = a \cdot b^t \cdot b^R = ab^t \cdot \dfrac{1}{3} = \dfrac{1}{3}g(t)$

     d. For any value of t, after R units of time, the new value of g is $\dfrac{1}{3}$ times the old value.

39. a. $A = A_0\left(\dfrac{1}{2}\right)^{t/5730}$ **b.** 760 years old

41. a. $A = A_0\left(\dfrac{1}{2}\right)^{t/432}$ **b.** 220 years

43. ≈ 27 years; ≈ 29 years

45. \$446.43; \$243.23

47. a. $N(t) = 2200(2)^{t/1.5}$

     b. The given model has a smaller growth factor, 1.356, than $2^{1/1.5} \approx 1.59$.

     c.

Name of chip	Year	Moore's law	$N(t)$	Actual number
Pentium IV	2000	2,306,867,200	20,427,413	42,000,000
Pentium M (Banias)	2003	9,227,468,800	50,932,200	77,000,000
Pentium M (Dothan)	2004	14,647,693,680	69,064,063	140,000,000

     d. About 2.3 years

CHAPTER 4 REVIEW PROBLEMS

1. a. $D = 8(1.5)^{t/5}$ **b.** 18; 44

3. a. $M = 100(0.85)^t$ **b.** 52.2 mg; 19.7 mg

5. $16n^{2x+10}$ **7.** $\dfrac{1}{m^{x+2}}$

9. $g(t) = 16(0.85)^t$ **11.** $f(x) = 500\left(\dfrac{1}{5}\right)^x$

13. 4.8% loss   **15.** 6%

17 a.

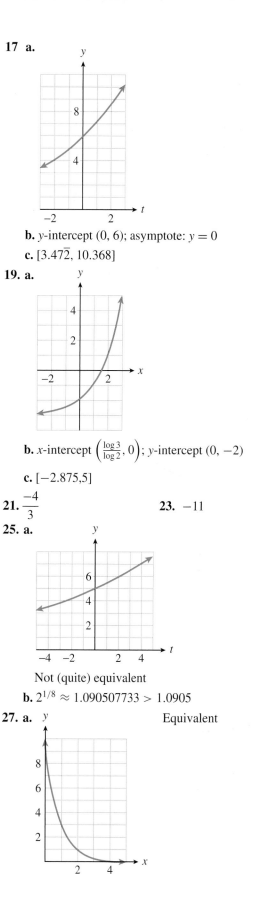

     b. y-intercept $(0, 6)$; asymptote: $y = 0$

     c. $[3.47\overline{2}, 10.368]$

19. a.

     b. x-intercept $\left(\dfrac{\log 3}{\log 2}, 0\right)$; y-intercept $(0, -2)$

     c. $[-2.875, 5]$

21. $\dfrac{-4}{3}$ **23.** -11

25. a.

Not (quite) equivalent

     b. $2^{1/8} \approx 1.090507733 > 1.0905$

27. a.   Equivalent

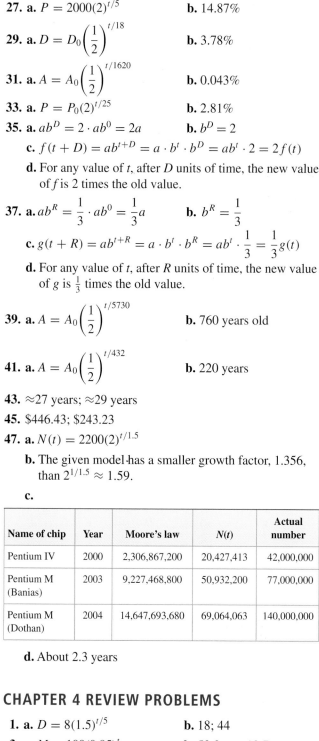

b. $\left(\dfrac{1}{3}\right)^{x-2} = \left(\dfrac{1}{3}\right)^{x} \cdot \left(\dfrac{1}{3}\right)^{-2} = \left(\dfrac{1}{3}\right)^{x} \cdot 9$

29. a. $y = 4 + 2^{x+1}$

b. Shift the graph of f 1 unit left, 4 units up.

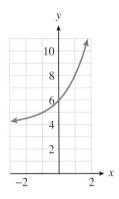

31. a. $y = 6 - 3 \cdot 2^{x}$

b. Expand vertically by 3, reflect about x-axis, shift 6 units up.

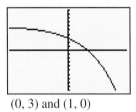

$(0, 3)$ and $(1, 0)$

33.

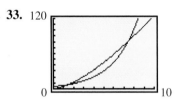

g eventually grows faster.

35. $2^{1.5} \approx 2.83;\ 2.25$

37. $M = M_0(2)^{t/10}$, where M is the organic content, M_0 is the organic content at $0°$ C, and t is the temperature in $°$ Celsius.

39. 4   **41.** -1 **43.** -3 **45.** $\log_{0.3}(x + 1) = -2$

47. $\dfrac{\log 5.1}{1.3} \approx 0.5443$ **49.** $\dfrac{\log(2.9/3)}{-0.7} \approx 0.21$

51. $\log_b x + \dfrac{1}{3}\log_b y - 2\log_b z$

53. $\dfrac{4}{3}\log x - \dfrac{1}{3}\log y$ **55.** $\log \sqrt[3]{\dfrac{x}{y^2}}$

57. $\log \dfrac{1}{8}$ **59.** $\dfrac{\log 63}{\log 3} \approx 3.77$

61. $\dfrac{\log 50}{-0.3 \log 6} \approx -7.278$ **63.** $\dfrac{\log(N/N_0)}{k}$

65. a. 238   **b.** 2010

67. a. $C = 90(1.06)^{t}$ **b.** \$94.48   **c.** 5 years

69. a. 7.4 years   **b.** 6.1%

71. $f(x) \approx 1600(1.035)^{x}$ **73.** $g(x) \approx 600(0.075)^{x}$

75. a. $\dfrac{\log 2}{\log 1.001} \approx 693$ years   **b.** 105 years

77. 17%   **79.** \$2192.78

81. a.

Day	1	2	3	$\cdots$	t	$\cdots$	30
Wage (¢)	2	4	8		2^{t}		2^{30}

b. $W(t) = 2^{t}$ cents

c. \$327.68; \$10,737,418.24

HOMEWORK 5.1

1. a.

x	-1	0	1	2
$f(x)$	0	1	-2	-1

y	0	1	-2	-1
$f^{-1}(y)$	-1	0	1	2

b. $f^{-1}(1) = 0$ **c.** $f^{-1}(-1) = 2$

3. a.

x	-1	0	1	2
$f(x)$	-1	1	3	11

y	-1	1	3	11
$f^{-1}(y)$	-1	0	1	2

b. $f^{-1}(1) = 0$ **c.** $f^{-1}(3) = 1$

5. a. $f(60) \approx 38$. The car that left the 60-foot skid marks was traveling at 38 mph.

b. $f^{-1}(60) \approx 150$. The car traveling at 60 mph left 150-foot skid marks

7. a. (60 hours, 78 grams)

b. $f^{-1}(90) \approx 19$, so that the vampire bat's weight has dropped to 90 grams about 19 hours after its last meal.

9. a. $g(0.05) = 0.28$. At 5% interest, \$1 earns \$0.28 interest in 5 years.

b. 8.45%

c. $g^{-1}(I) = (I+1)^{1/5} - 1$

d. $g^{-1}(0.50) \approx 0.0845$

11. a. $f(0.5) \approx 62.9$. At an altitude of 0.5 miles, you can see 62.9 miles to the horizon.

b. 0.0126 mile, or 66.7 feet   **c.** $h = f^{-1}(d) = \dfrac{d^2}{7920}$

d. $f^{-1}(10) = 0.0126$

13. a. $h^{-1}(3) \approx -4$ **b.** $h^{-1}(x) = h^{-1}(3) = 5 - x^2 = -4$

15. a. $f^{-1}(y) = \sqrt[3]{y} + 2$

b. $f^{-1}(f(4)) = f^{-1}(8) = 4$

c. $f(f^{-1}(-8)) = f(0) = -8$

d.

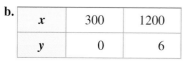

17. 6   **19.** $\dfrac{2}{9}$ **21.** 4

23. a.

x	0	6
y	300	1200

b.

x	300	1200
y	0	6

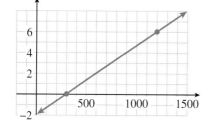

25. a.

x	0	1	2
y	5	20	100

b.

x	5	20	100
y	0	1	2

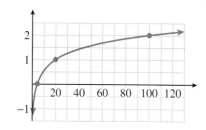

27. a. $f^{-1}(x) = \dfrac{x+6}{2}$

b.

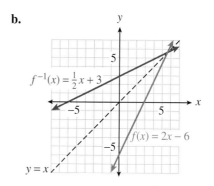

29. a. $f^{-1}(x) = \sqrt[3]{x-1}$

b.

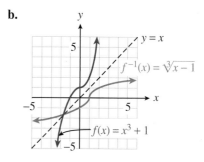

31. a. $f^{-1}(x) = \dfrac{1}{x} + 1$

b.

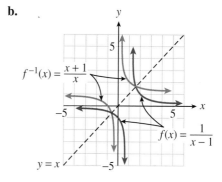

33. a. Domain: $(-\infty, 4]$; Range: $[0, \infty)$

b. $g^{-1}(x) = 4 - x^2$

c. Domain: $[0, \infty)$; Range: $(-\infty, 4]$

d.

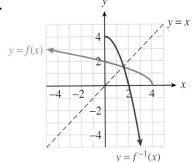

35. a and d

37. a

39. a

41. a and b

43. **a.** $f(x) = 4 + 2x$; IV

    b. $f(x) = 2 - \dfrac{x}{2}$; III

    c. $f(x) = -4 - 2x$; I

    d. $f(x) = \dfrac{x}{2}$; II

45. **a.** III   **b.** II   **c.** I

HOMEWORK 5.2

1. a.

x	-1	0	1	2
2^x	$\frac{1}{2}$	1	2	4

x	$\frac{1}{2}$	1	2	4
$\log_2 x$	-1	0	1	2

b.

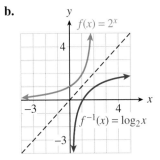

3. a.

x	-2	-1	0	1
$\left(\frac{1}{3}\right)^x$	9	3	1	$\frac{1}{3}$

x	9	3	1	$\frac{1}{3}$
$\log_{1/3} x$	-2	-1	0	1

b.

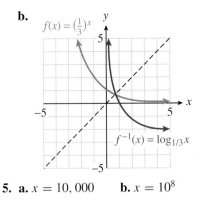

5. a. $x = 10,000$ **b.** $x = 10^8$

7. $0 < x < 0.01$

9. a. $\log 100{,}322 \approx 5.001$

    b. $\log 693 \approx 2.841$

11. a. $\log(-7)$ is undefined.

    b. $6 \log 28 \approx 8.683$

13. a. 15.614 **b.** 0.419

15. a. 81 **b.** 4

    c. Definition of logarithm base 3

    d. 1.8 **e.** a

17. a. 2^8 **b.** -2

19. a. $2k$ **b.** x^3 **c.** $\sqrt{x}$ **d.** $2m$

21. a. $(9, \infty)$ **b.** $f^{-1}(x) = 3^{x-4} + 9$

23. a. $f^{-1}(x) = \log_4(100 - x) - 2$

    b. $f^{-1}(f(1)) = f^{-1}(36) = \log_4(64) - 2 = 1$

    c. $f\left(f^{-1}(84)\right) = f(0) = 100 - 4^2 = 84$

25. a. IV   **b.** I   **c.** II   **d.** III

27 a and b.

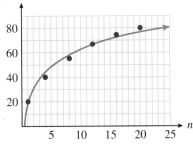

The graph resembles a logarithmic function. The (translated) log function is close to the points but appears too steep at first and not steep enough after $n = 15$. Overall, it is a good fit.

    c. f grows (more and more slowly) without bound. f will eventually exceed 100 per cent, but no one can forget more than 100% of what is learned.

29. a. $10^{1.41} \approx 25.704$ **b.** $10^{-1.69} \approx 0.020417$

    c. $10^{0.52} \approx 3.3113$

31. $16^w = 256$ **33.** $b^{-2} = 9$ **35.** $10^{-2.3} = A$

37. $u^w = v$ **39.** $b = 2$ **41.** $b = 100$

43. $x = 11.$ **45.** $x = 7^{2/3}$ **47.** $x = 4$

49. $x = 11$ **51.** $x = 3$

53. No solution **55.** $A = k(10^{t/T} - 1)$

57. $s = \dfrac{b^{N/N_0}}{k}$ **59.** $H = (H_0)^{kM^2}$

61. a. II **b.** VI **c.** III **d.** V **e.** I **f.** IV

63. a.

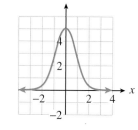

No inverse function

b.

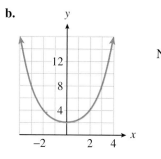

No inverse function

65.

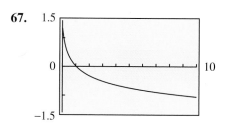

The functions are equal.

67.

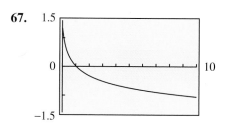

The functions are equal.

69. a.

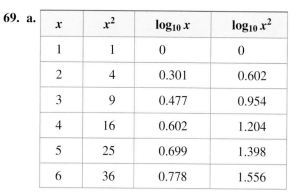

x	x^2	$\log_{10} x$	$\log_{10} x^2$
1	1	0	0
2	4	0.301	0.602
3	9	0.477	0.954
4	16	0.602	1.204
5	25	0.699	1.398
6	36	0.778	1.556

b. $\log_{10} x^2 = 2 \log_{10} x$

71.

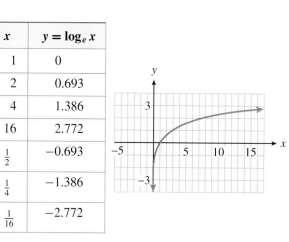

x	$y = \log_e x$
1	0
2	0.693
4	1.386
16	2.772
$\frac{1}{2}$	-0.693
$\frac{1}{4}$	-1.386
$\frac{1}{16}$	-2.772

HOMEWORK 5.3

1.

x	-10	-5	0	5	10	15	20
$f(x)$	0.135	0.368	1	2.718	7.389	20.086	54.598

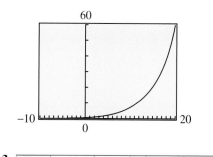

3.

x	-10	-5	0	5	10	15	20
$f(x)$	20.086	4.482	1	0.223	0.05	0.011	0.00248

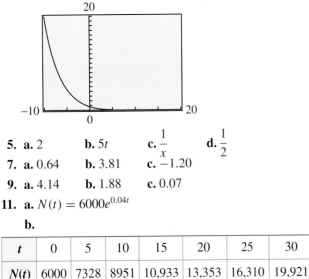

5. a. 2   **b.** $5t$ **c.** $\dfrac{1}{x}$ **d.** $\dfrac{1}{2}$

7. a. 0.64   **b.** 3.81   **c.** -1.20

9. a. 4.14   **b.** 1.88   **c.** 0.07

11. a. $N(t) = 6000e^{0.04t}$

b.

t	0	5	10	15	20	25	30
$N(t)$	6000	7328	8951	10,933	13,353	16,310	19,921

c.   **d.** 15,670

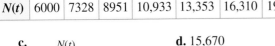

e. 70.3 hrs

13. a.

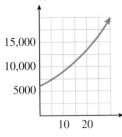

b. 941.8 lumens   **c.** 2.2 cm

15. $P(t) = 20\left(e^{0.4}\right)^t \approx 20 \cdot 1.492^t$; increasing; initial value 20

17. $P(t) = 6500\left(e^{-2.5}\right)^t \approx 6500 \cdot 0.082^t$; decreasing; initial value 6500

19. a.

x	0	0.5	1	1.5	2	2.5
e^x	1	1.6487	2.7183	4.4817	7.3891	12.1825

b. Each ratio is $e^{0.5} \approx 1.6487$: Increasing x-values by a constant $\Delta x = 0.5$ corresponds to multiplying the y-values of the exponential function by a constant factor of $e^{\Delta x}$.

21. a.

x	0	0.6931	1.3863	2.0794	2.7726	3.4657	4.1589
e^x	1	2	4	8	16	32	64

b. Each difference in x-values is approximately $\ln 2 \approx 0.6931$: Increasing x-values by a constant $\Delta x = \ln 2$ corresponds to multiplying the y-values of the exponential function by a constant factor of $e^{\Delta x} = e^{\ln 2} = 2$. That is, each function value is approximately equal to double the previous one.

23. 0.8277   **25.** -2.9720

27. 1.6451   **29.** -3.0713

31. $t = \dfrac{1}{k} \ln y$ **33.** $t = \ln\left(\dfrac{k}{k-y}\right)$

35. $k = e^{T/T_0} - 10$

37. a.

n	0.39	3.9	39	390
$\ln n$	-0.942	1.361	3.664	5.966

b. Each difference in function values is approximately $\ln 10 \approx 2.303$: Multiplying x-values by a constant factor of 10 corresponds to adding a constant value of $\ln 10$ to the y-values of the natural log function.

39. a.

n	2	4	8	16
$\ln n$	0.693	1.386	2.079	2.773

b. Each quotient equals k, where $n = 2^k$. Because $\ln n = \ln 2^k = k \cdot \ln 2, k = \dfrac{\ln n}{\ln 2}$.

41. a. $N(t) = 100e^{(\ln 2)t} \approx 100e^{0.6931t}$

b.

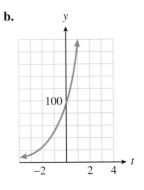

43. a. $N(t) = 1200e^{(\ln 0.6)t} \approx 100e^{-0.5108t}$

b.

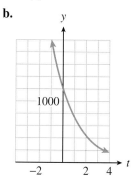

d–e.

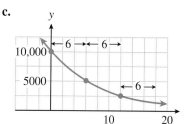

45. a. $N(t) = 10e^{(\ln 1.15)t} \approx 10e^{0.1398t}$

b.

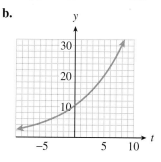

47. a. 20,000

b. $\left(\dfrac{35,000}{20,000}\right)^{1/10} \approx e^{0.056}$

c. $P(t) = 20,000e^{0.056t}$ **d.** 107,188

49. a. $\left(\dfrac{385}{500}\right)^{1/2} \approx e^{-0.1307}$ **b.** $N(t) = 500e^{-0.1307t}$

c. 135.3 mg

51. a. $A(t) = 500e^{0.095t}$ **b.** 7.3 years

c. 7.3 years

53. a. 6 hours   **b.** 6 hours

c.

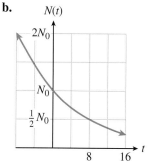

55. a. $\frac{1}{2}N_0, \frac{1}{4}N_0, \frac{1}{16}N_0$

b.

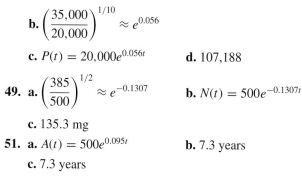

c. $N(t) = N_0e^{-0.0866t}$

57.

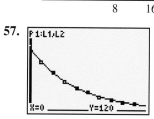

a. $y = 116(0.975)^t$

b. $G(t) = 116e^{-0.025t}$

c. 28 minutes

HOMEWORK 5.4

1. a.

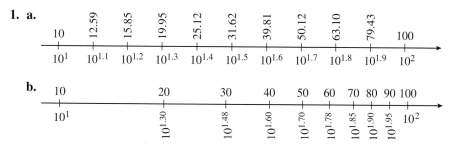

b.

3.

5. 1.58, 6.31, 15.8, 63.1   **7.** 1, 80, 350, 1600, 7000, 4×10^7

9.

11. Proxima Centauri: 15.5; Barnard: 13.2; Sirius: 1.4; Vega: 0.6; Arcturus: -0.4; Antares: -4.7; Betelgeuse: -7.2

13. a. 1   **b.** 0.5012

   c. 0.1259   **d.** 0.01

   e. 0.000079   **f.** 3.2×10^{-7}

   g. 2×10^{-8} **h.** 8×10^{-10}

15. a. $10^{1.75} \approx 56.2341$

   b. $10^{(\log 600)/2} \approx 24.4949$

17. $10^{3.4} \approx 2512$

19. A: $a \approx 45$, $p \approx 7.4\%$; B: $a \approx 400$, $p \approx 15\%$; C: $a \approx 6000$, $p \approx 50\%$; D: $a \approx 13000$, $p \approx 45\%$

21. 3.2   **23.** 0.0126   **25.** 100

27. 6,309,573 watts per square meter

29. 1000   **31.** 12.6   **33.** 100

35. $\approx$25,000   **37.** 4.7   **39.** 53

CHAPTER 5 REVIEW PROBLEMS

1.

y	-1	1	3	11
$x = f^{-1}(y)$	-1	0	1	2

3.

y	0	$\frac{-1}{3}$	-1	-3
$w = f^{-1}(y)$	-1	0	1	2

5. $P^{-1}(350) = 40$, $P^{-1}(100) = 0$

7. a. $f^{-1}(x) = x - 4$

b.

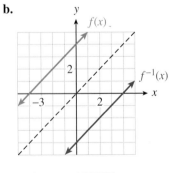

9. a. $f^{-1}(x) = \sqrt[3]{x+1}$

b.

11. a. $f^{-1}(x) = \dfrac{1}{x-2}$

b.

13. 0

15. a. $f^{-1}(300) = 200$: \$200,000 in advertising results in \$300,000 in revenue.

   b. $f(A) = 250$ or $A = f^{-1}(250)$

17. $10^z = 0.001$ **19.** $2^{x-2} = 3$ **21.** $b^3 = 3x + 1$

23. $n^{p-1} = q$ **25.** $6n$ **27.** $2x + 6$

29. -1 **31.** $\dfrac{1}{2}$ **33.** 4

35. $\dfrac{-15}{8}$ **37.** $\dfrac{9}{4}$ **39.** 3

41. $x \approx 1.548$ **43.** $x \approx 411.58$ **45.** $x \approx 2.286$

47. $\sqrt{x}$ **49.** $k - 3$

51. a. $P = 7{,}894{,}862e^{-0.011t}$ **b.** 1.095%

53. a. \$1419.07   **b.** 13.9 years   **c.** $t = 20\ln\left(\dfrac{A}{1000}\right)$

55. $t = \dfrac{-1}{k}\ln\left(\dfrac{y-6}{12}\right)$

57. $M = N^{Qt}$ **59.** $P(t) = 750(1.3771)^t$

61. $N(t) = 600e^{-0.9163t}$

63.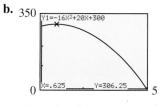

65. Order 3: 17,000; Order 4: 5000; Order 8: 40; Order 9: 11

67. 5×10^{-7}　　**69.** 3160

HOMEWORK 6.1

1. a.

t	0	0.5	1	1.5	2	2.5	3	3.5	4	4.5	5
h	300	306	34	294	276	250	216	174	124	66	0

b.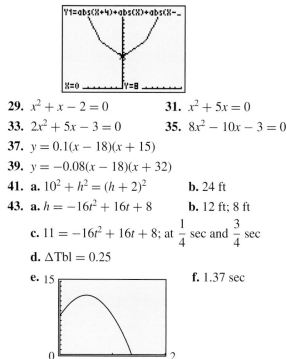

c. 306.25 ft at 0.625 sec　　**d.** 1.25 sec　　**e.** 5 sec

3. $\dfrac{-5}{2}, 2$　　**5.** $0, \dfrac{-10}{3}$　　**7.** $\dfrac{-3}{4}, -8$　　**9.** 4

11. $\dfrac{1}{2}, -3$　　**13.** $0, 3$　　**15.** 1　　**17.** $\dfrac{1}{2}, 1$

19. $2, 3$　　**21.** $-1, 2$　　**23.** $-3, 6$

25 and 27. The 3 graphs have the same x-intercepts. In general, the graph of $y = ax^2 + bx + c$ has the same x-intercepts as the graph of $y = k(ax^2 + bx + c)$.

29. $x^2 + x - 2 = 0$　　**31.** $x^2 + 5x = 0$

33. $2x^2 + 5x - 3 = 0$　　**35.** $8x^2 - 10x - 3 = 0$

37. $y = 0.1(x - 18)(x + 15)$

39. $y = -0.08(x - 18)(x + 32)$

41. a. $10^2 + h^2 = (h + 2)^2$　　**b.** 24 ft

43. a. $h = -16t^2 + 16t + 8$　　**b.** 12 ft; 8 ft

c. $11 = -16t^2 + 16t + 8$; at $\dfrac{1}{4}$ sec and $\dfrac{3}{4}$ sec

d. $\Delta\text{Tbl} = 0.25$

e. 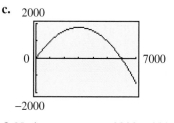　　**f.** 1.37 sec

45. a.

Width	Length	Area
10	170	1700
20	160	3200
30	150	4500
40	140	5600
50	130	6500
60	120	7200
70	110	7700
80	100	8000

b. $l = 180 - x$, $A = 180x - x^2$; 80 yd by 100 yd

c. $180x - x^2 = 8000$, 80 yd by 100 yd, or 100 yd by 80 yd. There are two solutions because the pasture can be oriented in two directions.

47. a. $l = x - 4$, $w = x - 4$, $h = 2$, $V = 2(x - 4)^2$

b.

x	4	5	6	7	8	9	10
V	0	2	8	18	32	50	72

c. As x increases, V increases.

d. 9 inches by 9 inches.

e. $2(x - 4)^2 = 50$, $x = 9$

49. a.

x	0	500	1000	1500	2000	2500	3000	3500
I	0	550	1000	1350	1600	1750	1800	1750

x	4000	4500	5000	5500	6000	6500	7000
I	1600	1350	1000	550	0	-650	-1400

b. 1600, 1000, −1400

c.

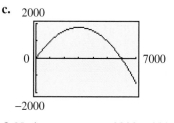

d. No increase　　**e.** 3000; 1800

51. ± 1 **53.** $\sqrt[3]{-3/4}, 1$ **55.** $-27, 1$ **57.** $\log 2, \log 3$

59. $1, 2$ **61.** $\dfrac{-1}{6}, 1$

63. **a.** $A = \dfrac{1}{2}(x^2 - y^2)$ **b.** $A = \dfrac{1}{2}(x - y)(x + y)$
   c. 18 sq ft

HOMEWORK 6.2

1. **a.** $(x + 4)^2$ **b.** $\left(x - \dfrac{7}{2}\right)^2$ **c.** $\left(x + \dfrac{3}{4}\right)^2$

3. 1 **5.** $-4, -5$ **7.** $\dfrac{-3 \pm \sqrt{21}}{2}$ **9.** $-1 \pm \sqrt{\dfrac{5}{2}}$

11. $\dfrac{-4}{3}, 1$ **13.** $\dfrac{1 \pm \sqrt{13}}{4}$ **15.** $-1, \dfrac{4}{3}$ **17.** $-2, \dfrac{2}{5}$

19. $-1 \pm \sqrt{1 - c}$ **21.** $\dfrac{-b \pm \sqrt{b^2 - 4}}{2}$

23. $\dfrac{-1 \pm \sqrt{4a + 1}}{a}$

25. **a.** $A = (x + y)^2$ **b.** $A = x^2 + 2xy + y^2$
   c. x^2, xy, xy, y^2

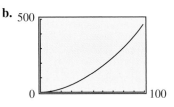

27. $1.618, -0.618$ **29.** $1.449, -3.449$

31. $1.695, -0.295$ **33.** $1.434, 0.232$

35. $-5.894, 39.740$

37. **a.**

s	10	20	30	40	50	60	70	80	90	100
d	9	27	53	87	129	180	239	307	383	467

b. 500

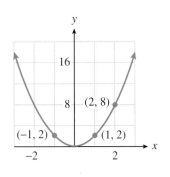

c. $\dfrac{s^2}{24} + \dfrac{s}{2} = 50$; 29.16 mph

39. **a.**

t	0	5	10	15	20	25
h	11,000	10,520	9240	7160	4280	600

b. 12,000

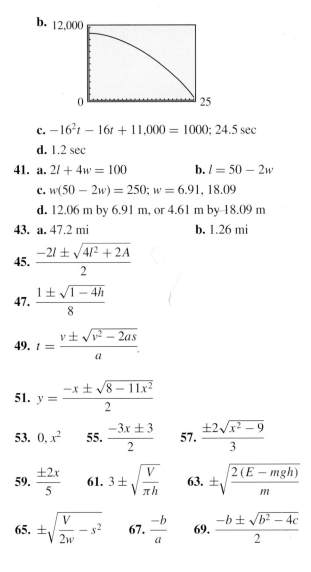

c. $-16^2 t - 16t + 11{,}000 = 1000$; 24.5 sec

d. 1.2 sec

41. **a.** $2l + 4w = 100$ **b.** $l = 50 - 2w$
   c. $w(50 - 2w) = 250$; $w = 6.91, 18.09$
   d. 12.06 m by 6.91 m, or 4.61 m by 18.09 m

43. **a.** 47.2 mi   **b.** 1.26 mi

45. $\dfrac{-2l \pm \sqrt{4l^2 + 2A}}{2}$

47. $\dfrac{1 \pm \sqrt{1 - 4h}}{8}$

49. $t = \dfrac{v \pm \sqrt{v^2 - 2as}}{a}$

51. $y = \dfrac{-x \pm \sqrt{8 - 11x^2}}{2}$

53. $0, x^2$ **55.** $\dfrac{-3x \pm 3}{2}$ **57.** $\dfrac{\pm 2\sqrt{x^2 - 9}}{3}$

59. $\dfrac{\pm 2x}{5}$ **61.** $3 \pm \sqrt{\dfrac{V}{\pi h}}$ **63.** $\pm\sqrt{\dfrac{2(E - mgh)}{m}}$

65. $\pm\sqrt{\dfrac{V}{2w} - s^2}$ **67.** $\dfrac{-b}{a}$ **69.** $\dfrac{-b \pm \sqrt{b^2 - 4c}}{2}$

HOMEWORK 6.3

1. **a.** The parabola opens up, twice as steep as the standard parabola.

b. The parabola is the standard parabola shifted 2 units up.

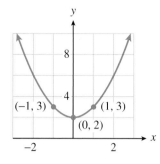

c. The parabola is the standard parabola shifted 2 units left.

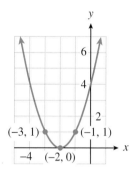

d. The parabola is the standard parabola shifted 2 units down.

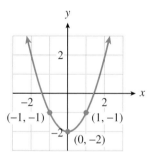

3. a. Vertex $(0, -16)$; x-intercepts $(\pm 4, 0)$

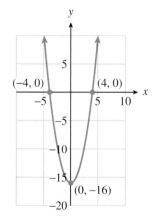

b. Vertex $(0, 16)$; x-intercepts $(\pm 4, 0)$

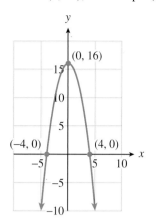

c. Vertex $(8, 64)$; x-intercepts $(0, 0)$ and $(16, 0)$

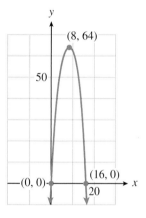

d. Vertex $(8, -64)$; x-intercepts $(0, 0)$ and $(16, 0)$

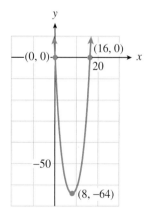

5. a. Vertex $(-1, -3)$; x-intercepts $(0, 0)$ and $(-2, 0)$

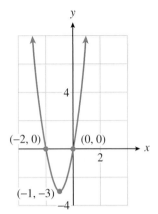

b. Vertex $(1, -3)$; x-intercepts $(0, 0)$ and $(2, 0)$

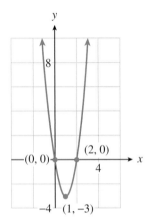

c. Vertex $(0, 6)$; no x-intercepts

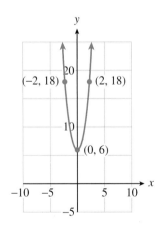

d. Vertex $(0, -6)$; x-intercepts $(\pm\sqrt{2}, 0)$

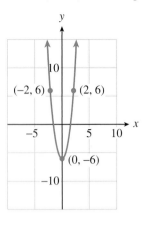

7. a. II **b.** IV **c.** I **d.** III **e.** VI **f.** V

9. a. $(2000, 400)$. The largest annual increase in biomass, 400 tons, occurs when the biomass is 2000 tons.

b.
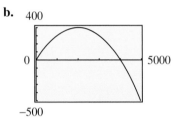

c. $4000 < x \le 5000$; When there are too many fish, there will not be enough food to support all of them.

11. a.

x	1	2	3	4	5
y	1.6	1.2	0.8	0.4	0
A	1.6	2.4	2.4	1.6	0

b. $A = x(2 - 0.4x)$ or $A = 2x - 0.4x^2$

c.
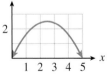

d. The maximum number of young marmots, on average, is 2.5; the optimal number of female marmots is 2.5.

13. a.
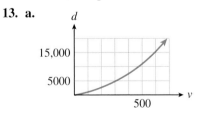

Vertex: $(-6, -1.5)$; Horizontal intercepts $(-12, 0)$ and $(0, 0)$. The point $(0, 0)$ means that no distance is required to stop a plane that is not moving.

b. 594 ft/sec

15. a. $\left(\dfrac{3}{2}, \dfrac{17}{4}\right)$, maximum

b. $\left(\dfrac{2}{3}, \dfrac{1}{9}\right)$, minimum

c. $(-4.5, 18.5)$, maximum

17. a. x-intercepts: $\left(\dfrac{-1}{2}, 0\right)$ and $(4, 0)$; y-intercept: $(0, 4)$; vertex: $\left(\dfrac{7}{4}, \dfrac{81}{8}\right)$

b.

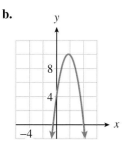

19. a. x-intercepts: $(-2, 0)$ and $(1, 0)$; y-intercept: $(0, -1.2)$; vertex: $(-0.5, -1.35)$

b.

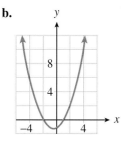

21. a. No x-intercepts; y-intercept: $(0, 7)$; Vertex: $(-2, 3)$

b.

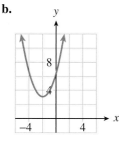

23. a. x-intercepts: $(-1 \pm \sqrt{2}, 0)$; y-intercept: $(0, -1)$; Vertex: $(-1, -2)$

b.

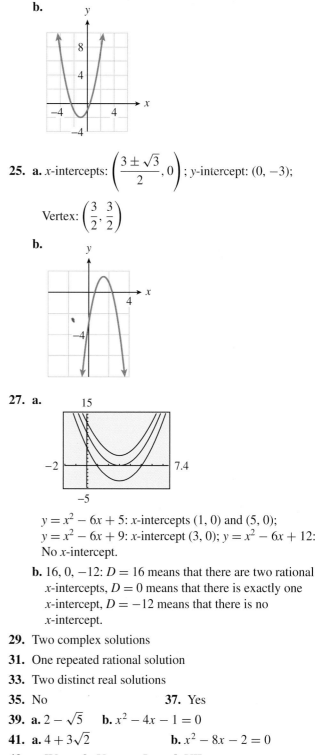

25. a. x-intercepts: $\left(\dfrac{3 \pm \sqrt{3}}{2}, 0\right)$; y-intercept: $(0, -3)$;

Vertex: $\left(\dfrac{3}{2}, \dfrac{3}{2}\right)$

b.

27. a.

$y = x^2 - 6x + 5$: x-intercepts $(1, 0)$ and $(5, 0)$; $y = x^2 - 6x + 9$: x-intercept $(3, 0)$; $y = x^2 - 6x + 12$: No x-intercept.

b. $16, 0, -12$: $D = 16$ means that there are two rational x-intercepts, $D = 0$ means that there is exactly one x-intercept, $D = -12$ means that there is no x-intercept.

29. Two complex solutions

31. One repeated rational solution

33. Two distinct real solutions

35. No   **37.** Yes

39. a. $2 - \sqrt{5}$ **b.** $x^2 - 4x - 1 = 0$

41. a. $4 + 3\sqrt{2}$ **b.** $x^2 - 8x - 2 = 0$

43. a. IV   **b.** V   **c.** I   **d.** VII

45. a. $y = x^2 + x - 6$; $x = \dfrac{-1}{2}$

b. $y = 2x^2 + 2x - 12$; $x = \dfrac{-1}{2}$

47. a.

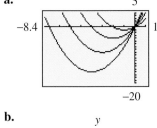

b.

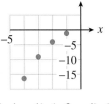

$(-1, -1), (-2, -4), (-3, -9), (-4, -16)$

c. $y = -x^2$

d. The vertex of $y = x^2 + 2kx$ is $(-k, -k^2)$

49. a.

t	0	0.5	1.0	1.5	2.0	2.5	3.0	3.5
x	0	6.075	11.5	16.275	20.400	23.875	26.700	28.875
y	0	7.44	12.48	15.12	15.36	13.2	8.64	1.68

b.

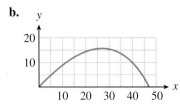

c. $y \approx 15.4$ m **d.** $x \approx 30$ m **e.** 3.6 sec

f. $x \approx 29.2$ m **g.** $y \approx 15.55$ m

HOMEWORK 6.4

1. a.

No. of price increases	Price of room	No. of rooms rented	Total revenue
0	20	60	1200
1	22	57	1254
2	24	54	1296
3	26	51	1326
4	28	48	1344
5	30	45	1350
6	32	42	1344
7	34	39	1326
8	36	36	1296
10	40	30	1200
12	44	24	1056
16	52	12	624
20	60	0	0

b. Price of a room: $20 + 2x$; Rooms rented: $60 - 3x$; Revenue: $1200 + 60x - 6x^2$

d. 20

e.

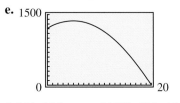

f. $24; $36 **g.** $1350; $30; 45 rooms

3. a. (For example) 10 m by 20 m with area 200 sq m; or 15 m by 15 m, area 225 sq m

b. $30 - x$ **c.** $30x - x^2$

5. 3 sec, 144 ft **7.** 100 baskets, $2000

9. a. Length: $50 - w$; Area: $50w - w^2$ **b.** 625 sq in

11. a. $300w - 2w^2$ **b.** 11,250 sq yd

13. a. Number of people: $16 + x$; Price per person: $2400 - 100x$; Total revenue: $38,400 + 800x - 100x^2$

b. 20

15. $a = 0.9$; $I = $865.80

17.

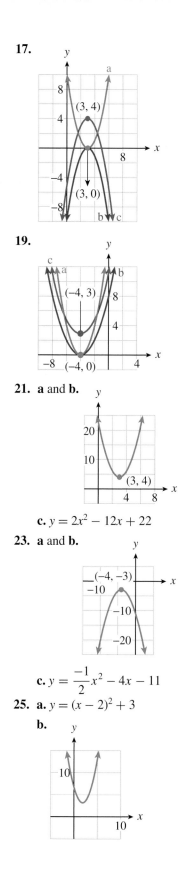

19.

21. a and **b.**

c. $y = 2x^2 - 12x + 22$

23. a and **b.**

c. $y = \dfrac{-1}{2}x^2 - 4x - 11$

25. a. $y = (x - 2)^2 + 3$

b.

27. a. $y = 3(x + 1)^2 - 5$

b.

29. a. $y = -2(x + 2)^2 + 11$

b.

31. No solution:

One solution:

Two solutions:

33. $(-1, 12), (4, 7)$ **35.** $(-2, 7)$ **37.** No solution

39. $(-2, -5), (5, 16)$ **41.** $(1, 4)$ **43.** $(3, 1)$

45. a.

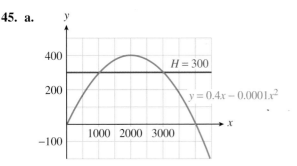

b. Larger, by 75 tons. Smaller, by 125 tons.

c. 1000 tons and 3000 tons

d. The fish population will decrease each year until it is completely depleted.

47. a.

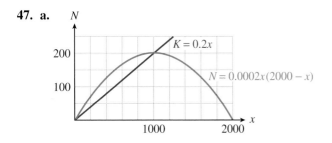

b. $K > N$. The population will decrease by 48 bears.

c. The population will increase by 18 bears.

d. 1000

e. Populations between 0 and 1000 will increase; populations over 1000 will decrease.

f. 1000 (unless the population is 0)

g. 500 (unless the population is 0)

49. a. (200, 2600), (1400, 18,200)

b.

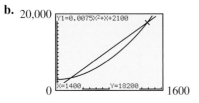

c. $x = 800$

51. a. (60, 39,000), (340, 221,000)

b.

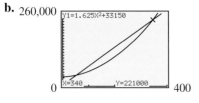

c. $x = 200$

53. a.

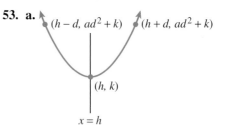

b–c. $ad^2 + k$

d. The two points on the parabola that are the same horizontal distance from the line $x = h$ have the same y-coordinate, so they are symmetric about that line.

HOMEWORK 6.5

1. a–b.

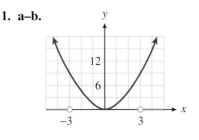

c. $x > 3$ or $x < -3$; It omits the solutions $x < -3$.

3. a–b.

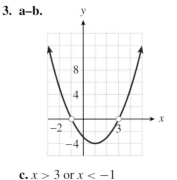

c. $x > 3$ or $x < -1$

5. a. $-12, 15$ **b.** $x < -12$ or $x > 15$

7. a. $0.3, 0.5$ **b.** $0.3 \le x \le 0.5$

9. a. $(-\infty, -3) \cup (6, \infty)$ **b.** $(-3, 6)$

   c. $[-2, 5]$ **d.** $(-\infty, -2] \cup [5, \infty)$

11. a. $(-4, 4)$ **b.** $(-\infty, -4) \cup (4, \infty)$

    c. $(-\infty, -3] \cup [3, \infty)$ **d.** $[-3, 3]$

13. $(-\infty, -2) \cup (3, \infty)$ **15.** $[0, 4]$

17. $(-\infty, -1) \cup (6, \infty)$ **19.** $(-4.8, 6.2)$

21. $(-\infty, -7.2] \cup [0.6, \infty)$ **23.** $(-\infty, 5.2) \cup (8.8, \infty)$

25. $x < -3.5$ or $x > 3.5$

27. All x

29. $-10.6 < x < 145.6$

31. $(-3, 4)$

33. $[-7, 4]$ **35.** $\left(-\infty, \dfrac{-1}{2}\right) \cup (4, \infty)$

37. $(-8, 8)$ **39.** $(-2.24, 2.24)$

41. $(-\infty, 0.4] \cup [6, \infty)$ **43.** $(-\infty, -0.67) \cup (0.5, \infty)$

45. $(-\infty, 0.27] \cup [3.73, \infty)$

47. All m

49. No solution

51. a. $320t - 16t^2 > 1024$; $4 < t < 16$: Between 4 and 16 seconds

b.

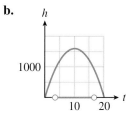

53. a. $-0.02x^2 + 14x + 1600 < 2800$; $[0, 100) \cup (600, 700]$: Either less than 100 or more than 600 shears.

b.

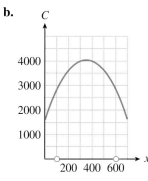

55. a. $p(1200 - 30p) > 9000$; $10 < p < 30$: Between $10 and $30

b.

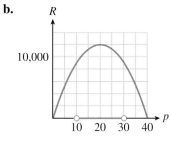

57. a. $500\pi < 20\pi r^2 < 2880\pi$; $5 < r < 12$; The radius must be between 5 and 12 ft.

b.

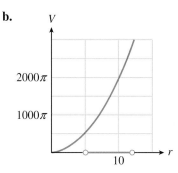

59. a. Size of group: $20 + x$; Price per person: $600 - 10x$

b. $I = (20 + x)(600 - 10x)$ **c.** $16,000$; $20

d. At least 35 people, but not more than 45.

e.

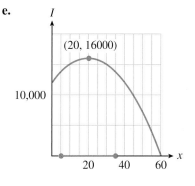

HOMEWORK 6.6

1. $a = -2, b = 3, c = -4$ **3.** $a = 1, b = -4, c = 7$

5. $a = 3, b = 1, c = -2$. The equation for the parabola is $y = 3x^2 + x - 2$

7. a. $P = -0.16x^2 + 7.4x - 71$

b. 14%. It predicts that 14% of the 25-year old population use marijuana on a regular basis.

c.

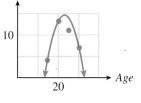

9. a. $C = 0.75t^2 - 1.85t + 56.2$ **b.** 65.7 lb

c.

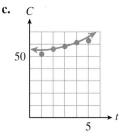

11. $D = \dfrac{1}{2}n^2 - \dfrac{3}{2}n$

13. a. $y = a(x + 2)^2 + 6$ **b.** 3

15. a. $y = \dfrac{3}{4}x^2 - 3$ **b.** $y = ax^2 - 3$ for any $a < 0$

17. $y = -2(x - 30)^2 + 280$ **19.** $y = x^2 - 9$

21. $y = -2x^2$ **23.** $y = x^2 - 2x - 15$

25. $y = x^2 - 4x + 5$

27. a. $y = \dfrac{-1}{40}(x - 80)^2 + 164$

   b. 160.99 ft

29. a. Vertex: $(\frac{1991}{2}, 79)$; y-intercept: $(0, 297)$

   b. $y = 0.00022(x - 995.5)^2 + 79$

31. a. 8 m   **b.** $y = \dfrac{x^2}{32}$ **c.** 3.125 m

33. a. $h = 8.24t + 38.89$ **b.** 162.5 m

c.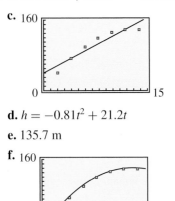

   d. $h = -0.81t^2 + 21.2t$

   e. 135.7 m

   f.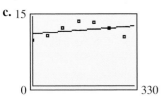

   g. Quadratic: Gravity will slow the projectile, giving the graph a concave down shape.

35. a. $y = -0.587t^2 + 7.329t - 2.538$

   b. The predicted peak was in 2000, near the end of March. The model predicts 7 deaths for 2005.

37. a. $y = 0.0051t + 11.325$ **b.** 13.2 hr

   c.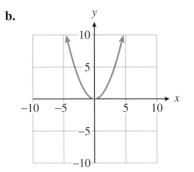

   d. $y = -0.00016t^2 + 0.053t + 9.319$

   e. 7.4 hr

f.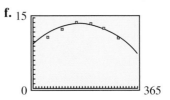

g. 9.8 hr; Neither model is appropriate. The quadratic model estimate is too low, and the linear model is too high.

CHAPTER 6 REVIEW PROBLEMS

1. $0, \dfrac{-5}{2}$ **3.** $-1, 2$ **5.** $-2, 3$

7. $4x^2 - 29x - 24 = 0$ **9.** $y = (x - 3)(x + 2.4)$

11. $1, 2$ **13.** $-1, \dfrac{1}{4}$

15. $2 \pm \sqrt{10}$ **17.** $\dfrac{3 \pm \sqrt{3}}{2}$

19. $1, 2$ **21.** $2 \pm \sqrt{2}$

23. $\pm\sqrt{\dfrac{2K}{m}}$ **25.** $\dfrac{3 \pm \sqrt{9 - 3h}}{3}$

27. 9

29. 10 ft by 18 ft or 12 ft by 15 ft

31. 1 sec

33. a. $h = 100t - 2.8t^2$ **b.**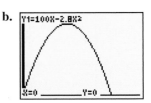

   c. 893 ft

   d. $15\frac{5}{7}$ sec on the way up and 20 sec on the way down

35. A_1 is the area of a square minus the area of two triangles:

$$x^2 - 2\left(\dfrac{1}{2}y \cdot y\right) = x^2 - y^2$$

37. a. Vertex and intercepts are all $(0, 0)$.

   b.

39. a. Vertex $\left(\frac{9}{2}, \frac{-81}{4}\right)$; x-intercepts $(9, 0)$ and $(0, 0)$; y-intercept$(0, 0)$

b.

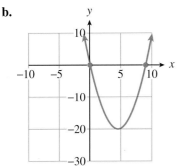

41. a. Vertex: $\left(\frac{-1}{2}, \frac{-25}{4}\right)$; x-intercepts $(-3, 0)$ and $(2, 0)$; y-intercept $(0, -6)$

b.

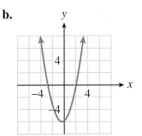

43. a. Vertex: $\left(\frac{-1}{4}, \frac{65}{8}\right)$; x-intercepts $\left(\frac{-1\pm\sqrt{65}}{4}, 0\right)$; y-intercept $(0, 8)$

b.

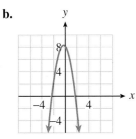

45. a. Vertex: $\left(\frac{1}{2}, \frac{-37}{4}\right)$; x-intercepts $\left(\frac{1\pm\sqrt{37}}{2}, 0\right)$; y-intercept $(0, -9)$

b.

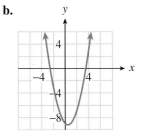

47. Two **49.** One rational solution

51. No real solutions

53. a. 45; $410

b.

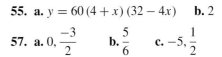

55. a. $y = 60 (4 + x) (32 - 4x)$ **b.** 2

57. a. $0, \dfrac{-3}{2}$ **b.** $\dfrac{5}{6}$ **c.** $-5, \dfrac{1}{2}$

d.

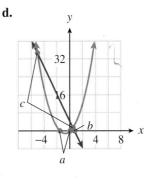

59. a. 0, 1 **b.** None **c.** $-1, 3$

d.

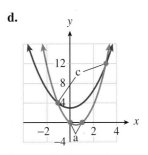

61. $(-\infty, -2) \cup (3, \infty)$

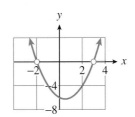

63. $\left[-1, \dfrac{3}{2}\right]$

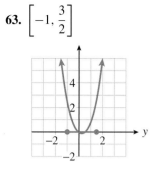

65. $[-2, 2]$

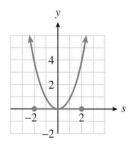

67. a. $R = p\left(220 - \dfrac{1}{4}p\right)$

  b. Between $4.00 and $4.80

69. $(1, 3), (-1, 3)$

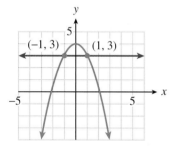

71. $(-1, -4), (5, 20)$

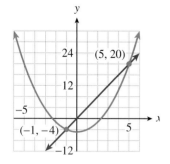

73. $(-9, 155), (5, 15)$

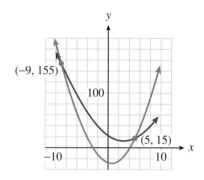

75. $(-1, 2), (3, 0)$

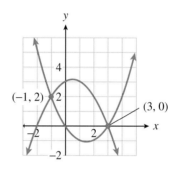

77. $a = 1, b = -1, c = -6$

79. $p(x) = \dfrac{-1}{2}x^2 - 4x + 10$

81. $y = 0.2(x - 15)^2 - 6$

83. a. $f(x) = (x - 12)^2 - 100$

  b.

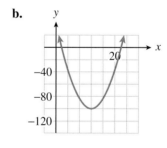

85. a. $y = \dfrac{1}{3}(x + 3)^2 - 2$

  b.

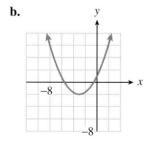

87. a. $h = 36.98t + 5.17$ **b.** 116.1 m, 153.1 m

c.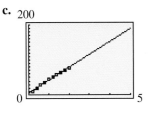

d. $h = -4.858t^2 + 47.67t + 0.89$

e. 100.2 m, 113.9 m

f.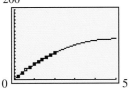

g. Quadratic: Gravity will slow the cannonball, giving the graph a concave down shape.

HOMEWORK 7.1

1. $12x^3 - 5x^2 - 8x + 4$ **3.** $x^3 - 6x^2 + 11x - 6$

5. $6a^4 - 5a^3 - 5a^2 + 5a - 1$ **7.** $y^4 + 5y^3 - 20y - 16$

9. $6 + x + 5x^2$ **11.** $4 - 7x^2 - 8x^4$

13. $0x^2$ **15.** $-8x^3$ **17. a.** 4   **b.** 5   **c.** 7

19. $(x + y)^3 = (x + y)(x + y)^2$
$$= (x + y)(x^2 + 2xy + y^2)$$
$$= x^3 + 2x^2y + xy^2 + x^2y + 2xy^2 + y^3$$
$$= x^3 + 3x^3y + 3xy^3 + y^3$$

21. $(x + y)(x^2 - xy + y^2)$
$$= x^3 - x^2y + xy^2 + x^2y - xy^2 + y^3$$
$$= x^3 + y^3$$

23. a. The formula begins with x^3 and ends with y^3. As you proceed from term to term, the exponents on x decrease while the exponents on y increase, and on each term the sum of the exponents is 3. The coefficients of the two middle terms are both 3.

    b. The formula is the same as for $(x - y)^3$, except that the terms alternate in sign.

25. $1 + 6z + 12z^2 + 8z^3$

27. $1 - 15\sqrt{t} + 75t - 125t\sqrt{t}$

29. $x^3 - 1$ **31.** $8x^3 + 1$ **33.** $27a^3 - 8b^3$

35. $(x + 3)(x^2 - 3x + 9)$

37. $(a - 2b)(a^2 + 2ab + 4b^2)$

39. $(xy^2 - 1)(x^2y^4 + xy^2 + 1)$

41. $(3a + 4b)(9a^2 - 12ab + 16b^2)$

43. $(5ab - 1)(25a^2b^2 + 5ab + 1)$

45. $(4t^3 + w^2)(16t^6 - 4t^3w^2 + w^4)$

47. a. $\left(6 - \dfrac{5}{4}\pi\right)x^2$ **b.** ≈ 132.67 square inches

49. a. $\dfrac{2}{3}\pi r^3 + \pi r^2 h$ **b.** $V(r) = \dfrac{14}{3}\pi r^3$

51. a. $500(1 + r)^2$; $500(1 + r)^3$; $500(1 + r)^4$

    b. $500r^2 + 1000r + 500$;
      $500r^3 + 1500r^2 + 1500r + 500$;
      $500r^4 + 2000r^3 + 3000r^2 + 2000r + 500$

53. a. Length: $16 - 2x$; Width: $12 - 2x$; Height: x

    b. $V = x(16 - 2x)(12 - 2x)$

    c. Real numbers between 0 and 6

    d.

x	1	2	3	4	5
V	140	192	180	128	60

    e. 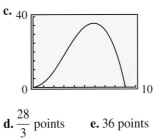   **f.** 2.26 in, 194.07 cu in

55. a. 0, 9   **b.** $0 \le x \le 9$; $R \ge 0$ for these values

    c.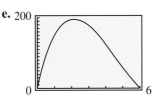

    d. $\dfrac{28}{3}$ points   **e.** 36 points

    f. 3 ml or 8.2 ml; But, if 3 ml work, then there is no need to use 8.2 ml of this medication.

57. a.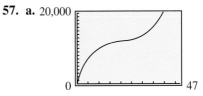

    b. 900; 11,145; 15,078   **c.** 1341; 171; 627

    d. Between 1990 and 1991

59. a.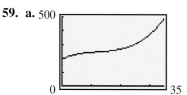

b. The graph is concave down until about $x = 12.5$ and is concave up afterwards. The cost is growing at the slowest rate at the inflection point at about $x = 12.5$, or 1250 students.

c. About 2738

61. a. 20 cm   **b.** 100 cm

63. a. $763.10 < H(t) < 864$

b.

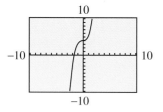

c. 864 min   **d.** 859.8 min

e. Within 34 days of the summer solstice

f. More than 66 days from the summer solstice

HOMEWORK 7.2

1. a. The end behavior is the same as for the basic cubic because the lead coefficient is positive.

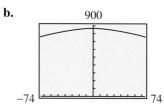

b. There is one x-intercept, no turning points, one inflection point.

3. a. The end behavior is opposite to the basic cubic (the graph starts in the upper left and extends to the lower right) because the lead coefficient is negative.

b. There is one x-intercept, no turning points, one inflection point.

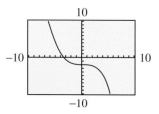

5. a. The end behavior is the same as for the basic cubic because the lead coefficient is positive.

b. There are three x-intercepts, two turning points, one inflection point.

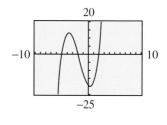

7. a. The end behavior is the same as for the basic cubic because the lead coefficient is positive.

b. There are three x-intercepts, two turning points, one inflection point.

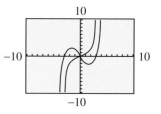

9. a.

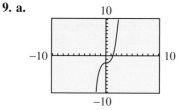

b.

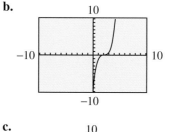

c.

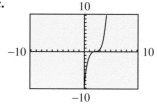

(b) and (c) are the same.

11. a. The end behavior is the same as for the basic quartic because the lead coefficient is positive.

b. There are two *x*-intercepts, one turning point, no inflection point.

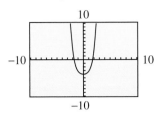

13. a. The end behavior is opposite to the basic quartic (the graph starts in the lower left and ends in the lower right) because the lead coefficient is negative.

b. There are no *x*-intercepts, three turning points, two inflection points.

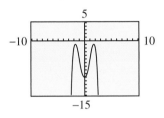

15. a. The end behavior is the same as for the basic quartic because the lead coefficient is positive.

b. There are two *x*-intercepts, one turning point, two inflection points.

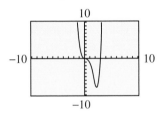

17. a. The end behavior is opposite to the basic quartic (the graph starts in the lower left and ends in the lower right) because the lead coefficient is negative.

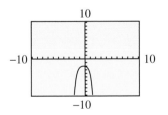

b. There are no *x*-intercepts, one turning point, two inflection points.

19. The graph of a cubic polynomial with a positive lead coefficient will have the same end behavior as the basic cubic, and a cubic with a negative lead coefficient will have the opposite end behavior. Each graph of a cubic

polynomial has one, two, or three *x*-intercepts, it has two, one or no turning point, and it has exactly one inflection point.

21. a.

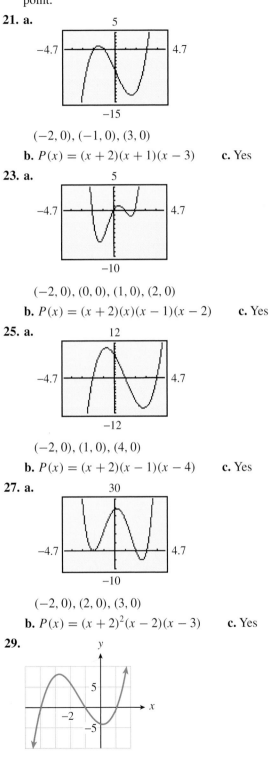

$(-2, 0), (-1, 0), (3, 0)$

b. $P(x) = (x + 2)(x + 1)(x - 3)$ **c.** Yes

23. a.

$(-2, 0), (0, 0), (1, 0), (2, 0)$

b. $P(x) = (x + 2)(x)(x - 1)(x - 2)$ **c.** Yes

25. a.

$(-2, 0), (1, 0), (4, 0)$

b. $P(x) = (x + 2)(x - 1)(x - 4)$ **c.** Yes

27. a.

$(-2, 0), (2, 0), (3, 0)$

b. $P(x) = (x + 2)^2(x - 2)(x - 3)$ **c.** Yes

29.

31.

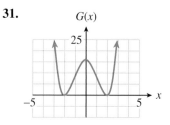

33.

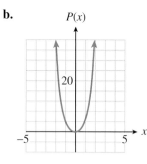

35.

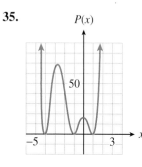

37. a. 0 (multiplicity 2)

b.

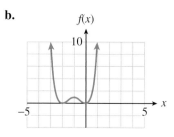

39. a. 0 (multiplicity 2), 2 (multiplicity 2)

b.

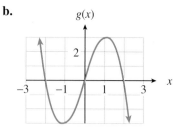

41. a. 0, ±2

b.

43. a. $\pm\sqrt{2}, \pm\sqrt{8}$

b.

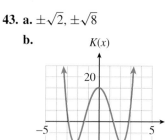

45. a. ±1, −3 (multiplicity 2)

b.

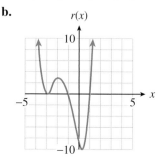

47. $P(x) = (x + 2)(x − 1)(x − 4)$

49. $P(x) = (x + 3)^2(x − 2)$ **51.** $P(x) = (x + 2)(x − 2)^3$

53. a. $y = x^3 − 4x + 3$; The graph of $y = f(x)$ shifted 3 units up.

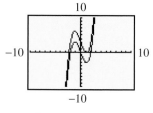

b. $y = x^3 − 4x − 5$; The graph of $y = f(x)$ shifted 5 units down.

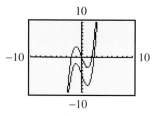

c. $y = (x − 2)^3 − 4(x − 2)$; The graph of $y = f(x)$ shifted 2 units right.

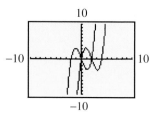

d. $y = (x+3)^3 - 4(x+3)$; The graph of $y = f(x)$ shifted 3 units left.

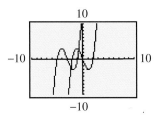

55. a. $y = x^4 - 4x^2 + 6$; The graph of $y = f(x)$ shifted 6 units up.

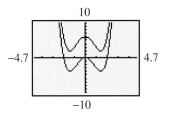

b. $y = x^4 - 4x^2 - 2$; The graph of $y = f(x)$ shifted 2 units down.

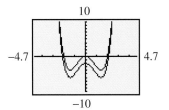

c. $y = (x-1)^4 - 4(x-1)^2$; The graph of $y = f(x)$ shifted 1 unit right.

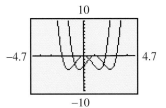

d. $y = (x+2)^4 - 4(x+2)^2$; The graph of $y = f(x)$ shifted 2 units left.

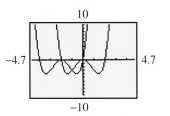

57. $q(x) = 2x^2 + 4x - 7$; $r(x) = -32$

59. $q(x) = x^3 - 6x$; $r(x) = -6x + 5$

61. a. If $P(x)$ is a nonconstant polynomial with real coefficients and a is any real number, then there exist unique polynomials $q(x)$ and $r(x)$ such that

$$P(x) = (x-a)q(x) + r(x)$$

where $\deg r(x) < \deg (x-a)$.

b. Zero

c. $P(a) = (a-a)q(a) + r(a) = r(a)$. Because $\deg r(x) = 0$, $r(x)$ is a constant. That constant value is $P(a)$, so $P(x) = (x-a)q(x) + P(a)$.

63. a. From the remainder theorem,

$$\begin{aligned} P(x) &= (x-a)Q(x) + P(a) \\ &= (x-a)Q(x) + 0 \\ &= (x-a)Q(x) \end{aligned}$$

b. By definition of a factor, if $x - a$ is a factor of $P(x)$, then $P(x) = (x-a)q(x)$, so $P(x) = (x-a)q(x) + 0$. The uniqueness guaranteed in the remainder theorem tells us that $P(a) = 0$.

65. a. $P(1) = 0$ **b.** $\dfrac{1 \pm \sqrt{5}}{2}$

67. a. $P(-3) = 0$ **b.** 0, 2, 4

69. a. About $-1 < x < 2$

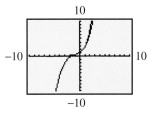

b.

x	-1	-0.5	0	0.5	1	1.5	2
$f(x)$	0.368	0.607	1	1.649	2.718	4.482	7.389
$p(x)$	0.333	0.604	1	1.646	2.667	4.188	6.333

c. 0.122

d.

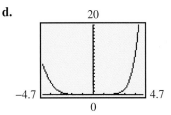

The error is relatively small for values of x between -3 and 2.5.

HOMEWORK 7.3

1. $-4 + 5i$ **3.** $-4 + i$ **5.** $\dfrac{-5}{6} - \dfrac{\sqrt{2}}{6}i$

7. $-3 \pm 2i$ **9.** $\dfrac{1}{6} \pm \dfrac{\sqrt{11}}{6}i$ **11.** $13 + 4i$

13. $-0.8 + 3.8i$ **15.** $20 + 10i$ **17.** $-17 + 34i$

19. $46 + 14i\sqrt{3}$ **21.** 52 **23.** $-2 - 2i$

25. $-1 + 4i$ **27.** $7 + 4i$ **29.** $\dfrac{-25}{29} + \dfrac{10}{29}i$

31. $\dfrac{3}{4} - \dfrac{\sqrt{3}}{4}i$ **33.** $\dfrac{-2}{3} + \dfrac{\sqrt{5}}{3}i$ **35.** i

37. a. 0 **b.** 0

39. a. 0 **b.** 0

41. a. 0 **b.** 0

43. $4z^2 + 49$ **45.** $x^2 + 6x + 10$

47. $v^2 - 8v + 17$ **49.** $x \geq 5;\ x < 5$

51. a. -1 **b.** 1 **c.** $-i$ **d.** -1

53. a. $2 - \sqrt{5}$ **b.** $x^2 - 4x - 1$

55. a. $4 + 3i$ **b.** $x^2 - 8x + 25$

57. a. 4 **b.** 5

59. $x^4 - 6x^3 + 23x^2 - 50x + 50$

61. $x^4 - 7x^3 + 20x^2 - 19x + 13$

63. The complex conjugates are reflections of each other across the real axis.

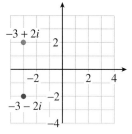

65. The complex conjugates are reflections of each other across the real axis.

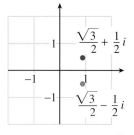

67.

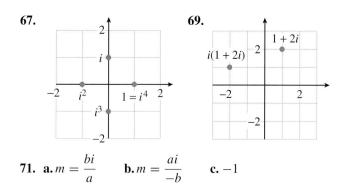

69.

71. a. $m = \dfrac{bi}{a}$ **b.** $m = \dfrac{ai}{-b}$ **c.** -1

HOMEWORK 7.4

1. a. $t = \dfrac{150}{50 - v}$

b.

v	0	5	10	15	20	25	30	35	40	45	50
t	3	3.33	3.75	4.29	5	6	7.5	10	15	30	—

The travel time increases as the headwind speed increases.

c.

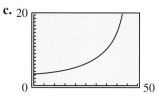

d. The horizontal asymptote is $t = 0$, and the vertical asymptote is $v = 50$, corresponding to the fact that the duck would never reach its destination if the headwind speed matched the airspeed.

3. a. $0 \leq p < 100$

b.

p	0	15	25	40	50	75	80	90	100
C	0	12.7	24	48	72	216	288	648	—

c.

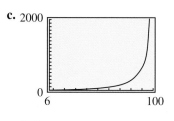

60%

d. $p > 96\%$

e. $p = 100$; As the percentage immunized approaches 100, the cost grows without bound.

5. a. $C = 8 + \dfrac{20,000}{n}$

b.

n	100	200	400	500	1000	2000	4000	5000	8000
C	208	108	58	48	28	18	13	12	10.5

c.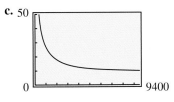

d. 2000   **e.** $n > 5000$

f. $C = 8$; As n increases, the average cost per calculator approaches \$8.

7. a. $4500 + \dfrac{3000}{x}$; $C(x) = 6x + 4500 + \dfrac{3000}{x}$

b.

x	10	20	30	40	50
C	4860	4770	4780	4815	4860

x	60	70	80	90	100
C	4910	4963	5018	5073	5130

c.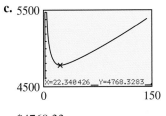

\$4768.33

d. 22; 14

e.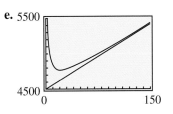

The graph of C approaches the line as an asymptote.

9. a. The surface area is $2x^2 + 4xh = 96$. Solving for h,

$$h = \frac{96 - 2x^2}{4x} = \frac{24}{x} - \frac{x}{2}.$$

b. $V = 24x - \dfrac{1}{2}x^3$

c.

x	1	2	3	4	5	6	7
h	23.5	11	6.5	4	2.3	1	−0.07
V	23.5	44	58.5	64	57.5	36	−3.5

If the base is more than 7 cm, the top and bottom alone exceed the total area allowed.

d.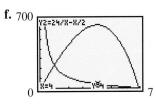

Maximum of 64 cu cm

e. 4 cm

f.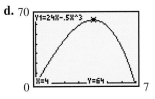

$h = 4$ cm

11. a.

v	−100	−75	−50	−25	0
P	338.15	358.92	382.41	409.19	440

v	25	50	75	100
P	475.83	518.01	568.4	629.66

b.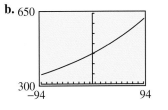

c. −20m/sec; 68 m/sec   **d.** $v > 12$ m/sec

e. $v = 332$; As v approaches 332 m per sec, the pitch increases without bound.

13.

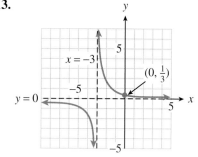

15.

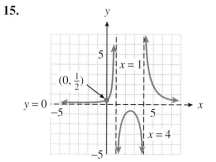

17.

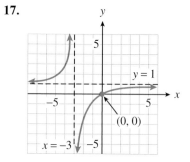

19.

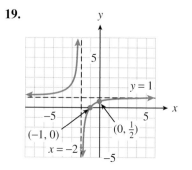

21.

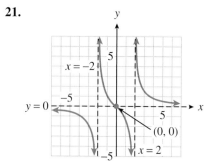

23.

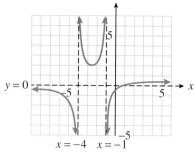

25.

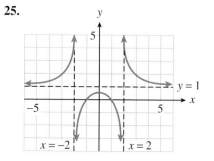

27.

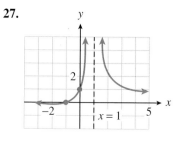

29.

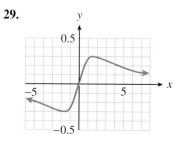

31.

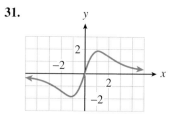

33. a. $y = \dfrac{2}{x} + 2$

b.

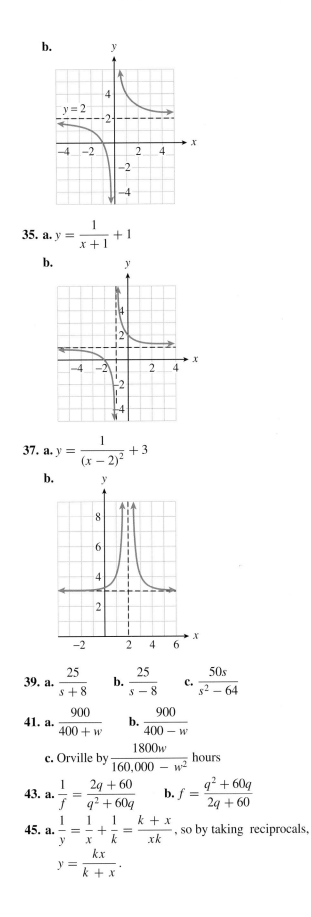

35. a. $y = \dfrac{1}{x+1} + 1$

b.

37. a. $y = \dfrac{1}{(x-2)^2} + 3$

b.

39. a. $\dfrac{25}{s+8}$ **b.** $\dfrac{25}{s-8}$ **c.** $\dfrac{50s}{s^2-64}$

41. a. $\dfrac{900}{400+w}$ **b.** $\dfrac{900}{400-w}$

   c. Orville by $\dfrac{1800w}{160{,}000-w^2}$ hours

43. a. $\dfrac{1}{f} = \dfrac{2q+60}{q^2+60q}$ **b.** $f = \dfrac{q^2+60q}{2q+60}$

45. a. $\dfrac{1}{y} = \dfrac{1}{x} + \dfrac{1}{k} = \dfrac{k+x}{xk}$, so by taking reciprocals,

   $y = \dfrac{kx}{k+x}$.

b.

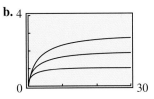

The graphs increase from the origin and approach a horizontal asymptote at $y = k$.

47. $y = \dfrac{12x}{x+20}$

49. a. V **b.** $\dfrac{V}{2}$

   c–d.

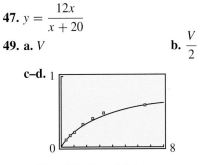

$V \approx 0.7$, $K \approx 2.2$ (many answers are possible)

51. a. $\dfrac{1}{V} = \dfrac{K}{V} \cdot \dfrac{1}{s} + \dfrac{1}{V}$; Therefore, $a = \dfrac{K}{V}$ and $b = \dfrac{1}{V}$

b.

$\frac{1}{s}$	3.0	1.5	1	0.6	0.4	0.3	0.15
$\frac{1}{v}$	12.5	7.1	5	3.3	2.6	2.2	1.7

c.

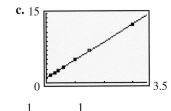

$\dfrac{1}{v} = 3.8 \cdot \dfrac{1}{s} + 1.1$

   d. $V \approx 0.89$, $K \approx 3.37$

53. a. $x \neq 2$ **b.** $x + 2$

c.

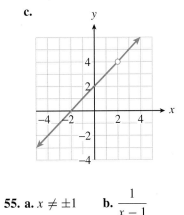

55. a. $x \neq \pm 1$ **b.** $\dfrac{1}{x-1}$

c.

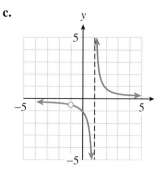

HOMEWORK 7.5

1. $\dfrac{-1}{2}$ **3.** $\dfrac{13}{8}$ **5.** $\pm\sqrt{\dfrac{15}{8}}$

7. $\dfrac{1800}{1849} \approx 0.97$ **9.** 37 ft

11. a. $t = \dfrac{150}{50-v}$

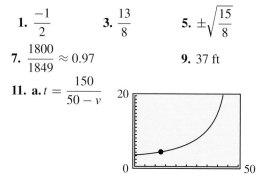

b. $4 = \dfrac{150}{50-v}$; $v = 12.5$ mph

13. $168 = \dfrac{72p}{100-p}$; $p = 70\%$

15. a.

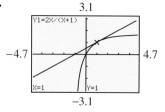

b. $x = 1$

17. a.

b. $x = \dfrac{1}{2}$

19. a.

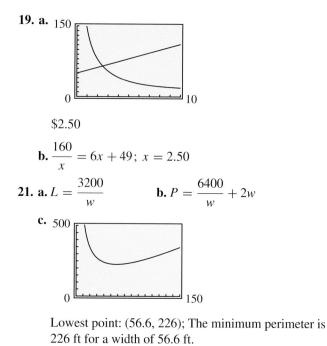

$2.50

b. $\dfrac{160}{x} = 6x + 49$; $x = 2.50$

21. a. $L = \dfrac{3200}{w}$ **b.** $P = \dfrac{6400}{w} + 2w$

c.

Lowest point: (56.6, 226); The minimum perimeter is 226 ft for a width of 56.6 ft.

d. $240 = \dfrac{6400}{w} + 2w$ **e.** 40 ft by 80 ft

23. Multiply both sides of the equation by bd and simplify.

$$\dfrac{a}{b} \cdot \dfrac{bc}{1} = \dfrac{c}{d} \cdot \dfrac{bc}{1}, \text{ so } ac = bd$$

25. 4 **27.** 40 **29.** $6187.50

31. 45 mi **33.** 889

35. a. 19,882 m **b.** 0.3% **c.** 0.00657 in

37. a. $AE = 1$, $ED = x - 1$, $CD = 1$

b. $\dfrac{1}{x} = \dfrac{x-1}{1}$ **c.** $\dfrac{1+\sqrt{5}}{2}$

39. $r = \dfrac{S-a}{S}$ **41.** $x = \dfrac{Hy}{2y-H}$

43. $d = \pm\sqrt{\dfrac{Gm_1 m_2}{F}}$ **45.** $r = \dfrac{2QI}{I+Q}$

47. $P = \dfrac{ES}{E+S}$ **49.** 5 **51.** 1

53. $\dfrac{-14}{5}$ **55.** $\dfrac{-1}{6}, \dfrac{-4}{3}$

57. a. $t_1 = \dfrac{144}{s-20}$ **b.** $t_2 = \dfrac{144}{s+20}$

c.

If the airspeed is 100 mph, then the airplane will complete the round trip in 3 hours.

d. $\dfrac{144}{s-20} + \dfrac{144}{s+20} = 3$ **e.** 100 mph

59. a. $t_1 = \dfrac{d}{r_1}, t_2 = \dfrac{d}{r_2}$

b. Total distance is $2d$; total time $\dfrac{d}{r_1} + \dfrac{d}{r_2}$.

c. $\dfrac{2d}{\dfrac{d}{r_1} + \dfrac{d}{r_2}}$ **d.** $\dfrac{2r_1 r_2}{r_1 + r_2}$ **e.** $58\dfrac{1}{3}$ mph

CHAPTER 7 REVIEW PROBLEMS

1. $2x^3 - 11x^2 + 19x - 10$

3. $t^3 + 3t^2 - 5t - 4$

5. $31x^2$

7. $-13x^3$

9. $(2x - 3z)(4x^2 + 6xz + 9z^2)$

11. $(y + 3x)(y^2 - 3xy + 9x^2)$

13. $v^3 - 30v^2 + 300v - 1000$

15. a. $\dfrac{1}{6}n^3 - \dfrac{1}{2}n^2 + \dfrac{1}{3}n$ **b.** 220   **c.** 20

17. a.

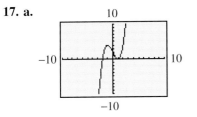

b. $[-968, 972]$

19. a. $2, -1$

b.

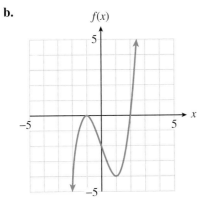

21. a. $0, 1, -3$

b.

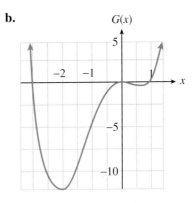

23. a. $0, 1, -1$

b.

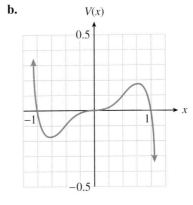

25. a. $-1, 1$

b.

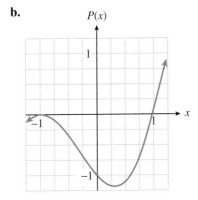

27. a. $-1, 1, \pm\sqrt{6}$

b.

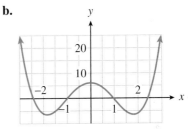

29. $x(x + 2)(x - 3)$ **31.** $x^3(x + 2)(x - 2)$

33. $x^2(x + 4)(x - 4)$

35. a. $P(-2) = 0$ **b.** $\dfrac{3 \pm \sqrt{13}}{2}$

37. $-2 \pm i\sqrt{6}$ **39.** $1 \pm \dfrac{\sqrt{6}}{3}i$

41. a. -8 **b.** -8

43. $\dfrac{11}{10} - \dfrac{13}{10}i$ **45.** $x^4 - 2x^3 + 14x^2 - 18x + 45$

47. a.

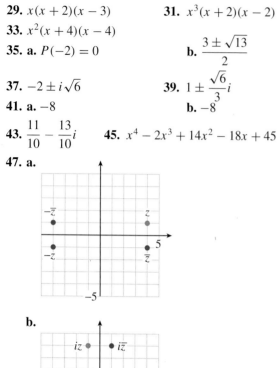

b.

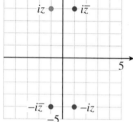

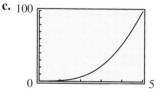

49. a. $V = \dfrac{\pi h^3}{4}$

b. 2π cm$^3 \approx 6.28$ cm^3; 16π cm$^3 \approx 50.27$ cm^3

c.

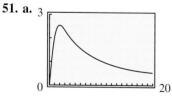

$h \approx 5$ cm

51. a.

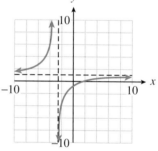

b. 338   **c.** Months 2 and 20

d. During month 6. The number of members eventually decreases to zero.

53. All numbers except $-2, 0, 2$.

55.

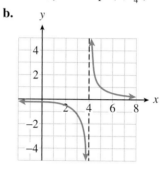

57. a. Horizontal asymptote $y = 0$; Vertical asymptote $x = 4$; y-intercept $(0, \frac{-1}{4})$

b.

59. a. Horizontal asymptote $y = 1$; Vertical asymptote $x = -3$; x-intercept $(2, 0)$; y-intercept $(0, \frac{-2}{3})$

b.

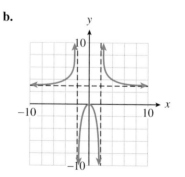

61. a. Horizontal asymptote $y = 3$; Vertical asymptotes $x = \pm 2$; x-intercept $(0, 0)$; y-intercept $(0, 0)$

b.

63. a. $y = \dfrac{-5}{x+3} + 3$

b.

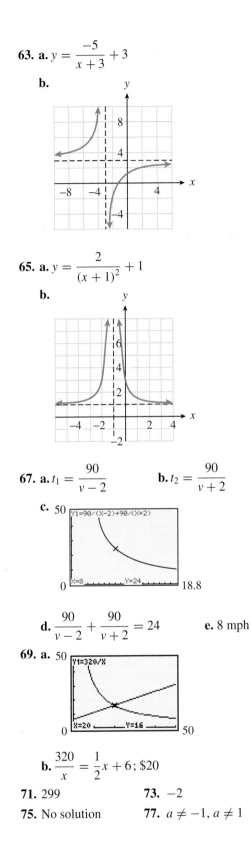

65. a. $y = \dfrac{2}{(x+1)^2} + 1$

b.

67. a. $t_1 = \dfrac{90}{v-2}$ **b.** $t_2 = \dfrac{90}{v+2}$

c. 50

d. $\dfrac{90}{v-2} + \dfrac{90}{v+2} = 24$ **e.** 8 mph

69. a. 50

b. $\dfrac{320}{x} = \dfrac{1}{2}x + 6$; $20

71. 299   **73.** -2

75. No solution   **77.** $a \neq -1, a \neq 1$

79. 0

81. $n = \dfrac{ct}{c-v}$

83. $q = \dfrac{pr}{r-p}$

HOMEWORK 8.1

1. $(3, 0)$

3. $(50, 70)$

5. $(-2, 3)$

7. $(2, 3)$

9.

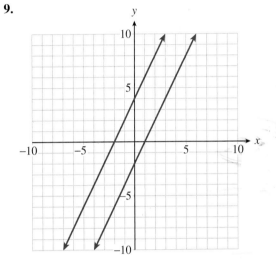

Inconsistent

11.

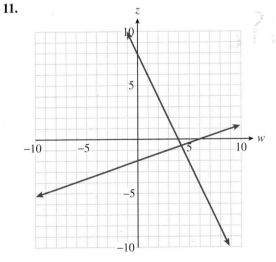

Consistent and independent

13.

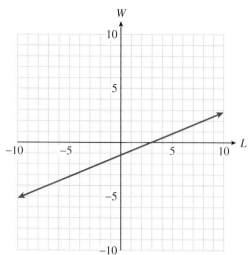

Dependent

15. Inconsistent

17. Consistent and independent

19. Dependent

21. a. $D = 10 + 0.09x$ **b.** $F = 15 + 0.05x$

   c. 125 min

23. a. $y = 50x$ **b.** $y = 2100 - 20x$

   c. 30¢ per bushel, 1500 bushels

25. a. $C = 200 + 4x$ **b.** $R = 12x$ **c.** 25 pendants

27. a.

	Number of tickets	Cost per ticket	Revenue
Adults	x	7.50	$7.50x$
Students	y	4.25	$4.25y$
Total	82	—	465.50

   b. $x + y = 82$ **c.** $7.50x + 4.25y = 465.50$

   d. 36 adults, 46 students

29. a.

	Rate	Time	Distance
P waves	5.4	x	y
S waves	3	$x + 90$	y

   b. $y = 3(x + 90)$ **c.** $y = 5.4x$

   d. (112.5, 607.5): The seismograph is 607.5 miles from the earthquake.

31. $(-4.6, 52)$ **33.** $(7.15, 4.3)$

HOMEWORK 8.2

1. $(1, 2, -1)$ **3.** $(2, -1, -1)$ **5.** $(4, 4, -3)$

7. $(2, -2, 0)$ **9.** $\left(\dfrac{3}{4}, -\dfrac{1}{2}, 0\right)$ **11.** $(-1, 1, -2)$

13. $\left(\dfrac{1}{2}, \dfrac{2}{3}, -3\right)$ **15.** $(4, -2, 2)$ **17.** $(1, 1, 0)$

19. $\left(\dfrac{1}{2}, \dfrac{-1}{2}, \dfrac{1}{3}\right)$ **21.** Inconsistent   **23.** $\left(\dfrac{1}{2}, 0, 3\right)$

25. $(-1, 3, 0)$ **27.** Inconsistent   **29.** $\left(\dfrac{1}{2}, \dfrac{1}{2}, 3\right)$

31. 60 nickels, 20 dimes, 5 quarters

33. $x = 40$ in, $y = 60$ in, $z = 55$ in

35. 0.3 cup carrots, 0.4 cup green beans, 0.3 cup cauliflower

37. 40 score only, 20 evaluations, 80 narrative report

39. 20 tennis, 15 Ping Pong, 10 squash

HOMEWORK 8.3

1. $\begin{bmatrix} -2 & 1 & | & 0 \\ -9 & 3 & | & -6 \end{bmatrix}$ **3.** $\begin{bmatrix} 1 & -3 & | & 6 \\ 0 & -2 & | & 11 \end{bmatrix}$

5. $\begin{bmatrix} 1 & 0 & -2 & | & 5 \\ 2 & 6 & -1 & | & 3 \\ 0 & -3 & 2 & | & -3 \end{bmatrix}$ **7.** $\begin{bmatrix} 1 & 2 & 1 & | & -5 \\ 0 & 4 & -2 & | & 3 \\ 0 & -9 & 2 & | & 12 \end{bmatrix}$

9. $\begin{bmatrix} 1 & -3 & | & 2 \\ 0 & 7 & | & 0 \end{bmatrix}$ **11.** $\begin{bmatrix} 2 & 6 & | & -4 \\ 4 & 0 & | & 3 \end{bmatrix}$

13. $\begin{bmatrix} 1 & -2 & 2 & | & 1 \\ 0 & 7 & -5 & | & 4 \\ 0 & -9 & -11 & | & -1 \end{bmatrix}$ **15.** $\begin{bmatrix} -1 & 4 & 3 & | & 2 \\ 3 & 0 & -5 & | & 14 \\ -3 & 0 & -3 & | & 8 \end{bmatrix}$

17. $\begin{bmatrix} -2 & 1 & -3 & | & -2 \\ 0 & 4 & -6 & | & -2 \\ 0 & 0 & -4 & | & -5 \end{bmatrix}$ **19.** $(2, 3)$

21. $(-2, 1)$ **23.** $(3, -1)$ **25.** $\left(\dfrac{-7}{3}, \dfrac{17}{3}\right)$

27. $(1, 2, 2)$ **29.** $\left(2, \dfrac{-1}{2}, \dfrac{1}{2}\right)$

31. $(-3, 1, -3)$ **33.** $\left(\dfrac{5}{4}, \dfrac{5}{2}, \dfrac{-1}{2}\right)$

35. $(3, 4, -1, 2)$ **37.** $\left(\dfrac{5}{3}, \dfrac{4}{3}, \dfrac{-1}{2}, \dfrac{-2}{3}\right)$

39. $(-5, -2, 3, 4, 2)$

41. a. $p(x) = \dfrac{-7}{6}x^3 + 7x^2 - \dfrac{65}{6}x + 7$

    b. $p(x) = \dfrac{-1}{2}x^3 + 3x^2 - \dfrac{7}{2}x + 3$

43. $p(x) = -2x^4 + x^3 - x^2 + 3x + 5$

45. a. $0 = 1$ **b.** No

47. a. $0 = 0$ **b.** $a = 9, b = 4, c = 2$

 c. Yes (infinitely many)

HOMEWORK 8.4

1.

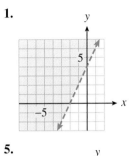

3.

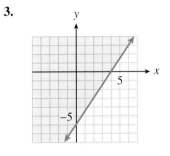

5.

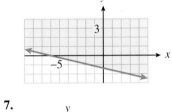

7.

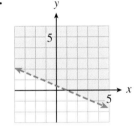

9.

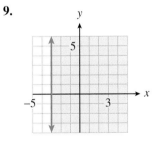

11.

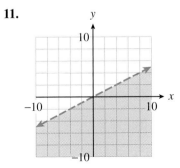

13.

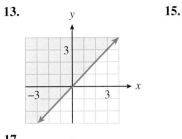

15.

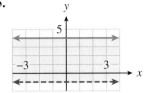

17.

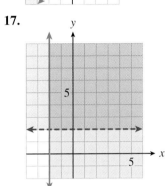

19.

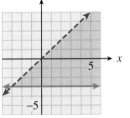

21.

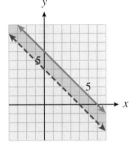

23.

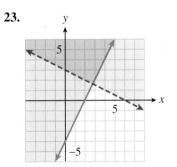

25.

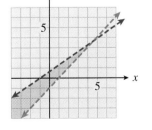

27.

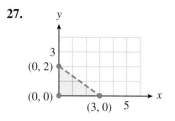

29.

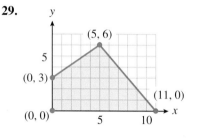

31.

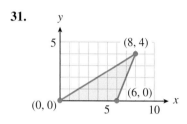

33.

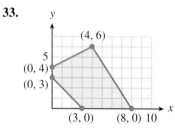

35.

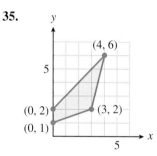

37.

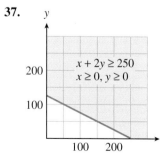

39.

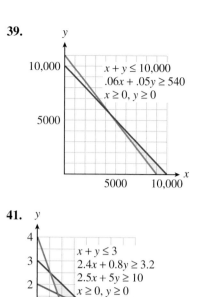

41.   *y*

4
3
2
1

$x + y \leq 3$
$2.4x + 0.8y \geq 3.2$
$2.5x + 5y \geq 10$
$x \geq 0, y \geq 0$

1   2   3   4   *x*

HOMEWORK 8.5

1. The graph of $12 = 3x + 4y$ does not intersect the set of feasible solutions.

3. (8, 2); $32   **5.** (2, 0)

7. a. $22   **b.** (8, 0)   **c.** $32

9. a. (1, 4); 7   **b.** (4, 5); 17

11. a. (0, 5); −10   **b.** (5, 0); 25

13. b. (10, 0); 0   **c.** (3, 2); 13

15. b. (0, 14); −14   **c.** (10, 0); 30

17. b. (0, 8); −160   **c.** (8, 0); 1600

19. a. $C = x + 3y$ **b.** $x \geq 0, y \geq 0, x + 2y \geq 250$
   d. 250

21. a. $P = 36x + 24y$
   b. $x \geq 0, y \geq 0, x + y \leq 180, 2x + y \leq 240$ **d.** $5040

23. a. $C = 433x + 400y$
   b. $x \geq 0, y \geq 0, 2.4x + 0.8y \geq 3.2, 2.5x + 5.2y \geq 10$
   d. 967.7 cal

25. Maximum 17.4; minimum −8.4

27. Maximum 4112; minimum 0

29. Maximum 1908; minimum 0

CHAPTER 8 REVIEW PROBLEMS

1. (−1, 2)   **3.** $\left(\dfrac{1}{2}, \dfrac{7}{2}\right)$ **5.** (12, 0)

7. Consistent and independent

9. Dependent   **11.** $(2, 0, -1)$ **13.** $(2, -5, 3)$

15. $(-2, 1, 3)$ **17.** $(3, -1)$ **19.** $(4, 1)$

21. $(-1, 0, 2)$ **23.** $(4, 3, 1, 2)$

25. $(2, -1, 5, -3, 4)$ **27.** 26

29. \$3181.82 at 8%, \$1818.18 at 13.5%

31. 5 cm, 12 cm, 13 cm

33.

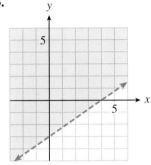

35.

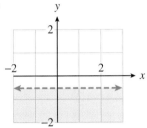

37.

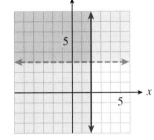

39.

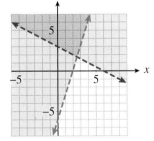

41.

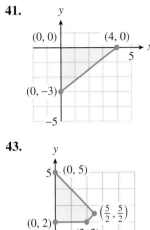

43.

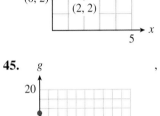

45.

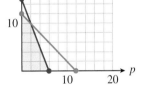

$20p + 8g \leq 120, 10p + 10g \leq 120$

47. b. $(1, 0)$; 18   **c.** $(5, 2)$; 186

49. a. $20p + 8g \leq 120, 10p + 10g \leq 120, p \geq 0, g \geq 0$

b. 2 batches peanut butter cookies; 10 batches of granola cookies; for \$125

APPENDIX A.1

1. 6   **3.** -42 **5.** 5   **7.** -2 **9.** -2

11. -60 **13.** -25 **15.** 81   **17.** -64

19. -32 **21.** 50   **23.** -2 **25.** -5

27. a. 8   **b.** -5 **c.** -0.4 **d.** 1.2

29. a. -4 **b.** $\dfrac{-1}{3}$

31. -19 **33.** 98.4   **35.** 36   **37.** -3

39. 45   **41.** -5 **43.** 100   **45.** 1080

47. 0   **49.** 72.5904

51. a. 2.84×10^2 **b.** 8.372×10^6

c. 2.4×10^{-2} **d.** 5.23×10^{-4}

53. a. 240   **b.** 6,870,000,000,000,000

c. 0.005   **d.** 0.000202

55. a. 3,000,000 **b.** 112,500

57. a. 8.062156×10^{12} **b.** $27,087.30

59. a. 7.9×10^9 km per year **b.** 250,000 meters per second

61. a. 250,000 times **b.** 8×10^{19} meters

APPENDIX A.2

1. 7 **3.** $\dfrac{31}{3}$ **5.** $\dfrac{2}{3}$ **7.** $-4.8\overline{3}$ **9.** 34.29

11. $y = \dfrac{-2}{3} - \dfrac{4x}{3}$ **13.** $y = \dfrac{x}{4} - 2$

15. $y = \dfrac{7}{2} - \dfrac{21x}{2}$ **17.** $y = \dfrac{19}{6} - \dfrac{x}{6}$

19. $y = \dfrac{-25}{4} - \dfrac{7x}{4}$ **21.** $t = \dfrac{v - k}{g}$

23. $h = \dfrac{S - 2w^2}{4w}$ **25.** $n = \dfrac{P - a + d}{d}$

27. $h = \dfrac{A - \pi r^2}{\pi r}$ **29.** $x > 3$ **31.** $x > 0$ **33.** $x \le \dfrac{6}{13}$

35. $-4 < x < \dfrac{5}{2}$ **37.** $\dfrac{4}{3} \le x \le 5$ **39.** $-10 \le x \le -6$

41. $(-5, 3]$

−5 0 3

43. $[-4, 0]$

−4 0

45. $(-6, \infty)$

−6 0

47. $(-\infty, -3) \cup [-1, \infty)$

−3 −1 0

49. $[-6, -4) \cup (-2, 0]$

−6 −4 −2 0

APPENDIX A.3

1. a. $j = a + 27$ **b.** 49

3. a. $h = \dfrac{1260}{r}$ **b.** 28 hours

5. a. $A = \pi r^2$ **b.** 78.54 sq cm

7. a. $b = 1.079p$ **b.** $528.71

---

9. a. $C = 1.97 + 0.39m$ **b.** $12.50

11. a. $c = 0.4w$ **b.** $r = 50 - 0.4w$ **c.** 47.6 lbs

13. $18

15. a. $t =$ time wife drives **b.** $45t$

   c. $16(t + 6)$ **d.** 3.3 hours

17. a. $n =$ number of copies

   b. $20,000 + 0.02n$; $17,500 + 0.025n$

   c. 500,000 copies

19. a. 145,800 **b.** 125,000

21. 7.53% **23.** 87

25. a. $x =$ amount of 6% fertilizer

   b. x; $10 - x$ **c.** $0.06x$; $0.15(10 - x)$

   d. $0.06x + 0.15(10 - x) = 0.08(10)$; $7.\overline{7}$ lbs of 6%, $2.\overline{2}$ lbs of 15%

27. a. $x =$ clerk's salary

   b. $4 \times 28,000$; $12 \times 22,000$; $30x$

   c. $4 \times 28,000 + 12 \times 22,000 + 30x$; $19,000(4 + 12 + x)$

   d. $4 \times 28,000 + 12 \times 22,000 + 30x = 19,000(4 + 12 + x)$; $16,600

APPENDIX A.4

1. a. High: 7°F; Low: −19°F

   b. Above 5°F from noon to 3 p.m.; Below −5°F from midnight to 9 a.m. and from 7 p.m. to midnight

   c. 7 a.m.: −10°F; 2 p.m.: 6°F; 10 a.m. and 5 p.m.: 0°F; 6 a.m. and 10 p.m.: −12°F

   d. Between 3 a.m. and 6 a.m.: 6°F; Between 9 a.m. and noon: 10°F; Between 6 p.m. and 9 p.m.: 9°F

   e. Increased most rapidly: 9 a.m. to noon; Decreased most rapidly: 6 p.m. to 9 p.m.

3. a. 28 mpg **b.** 50 mph

   c. Best gas mileage at 70 mph. The graph seems to be leveling off for higher speeds; any improvement in mileage probably would not be significant, and the mileage might in fact deteriorate.

   d. Road condition, weather conditions, traffic, weight in the car

5. $(-6, 6)$, $(-3, 5)$, $(0, 4)$, $(3, 3)$, $(6, 2)$

7. $(-2, 11)$, $(-1, 2)$, $(0, -1)$, $(1, 2)$, $(2, 11)$

9.

x	y
-3	5
0	$\frac{1}{2}$
$\frac{1}{3}$	0
1	-1
$\frac{5}{3}$	-2
3	-4

11.

x	y
-4	2
0	1
3	$\frac{1}{4}$
4	0
5	$\frac{-1}{4}$
8	-1

13.

x	y
-6	-3
0	-1
0	-1
3	0
6	1

15.

x	y
-4	3
-2	1
0	-1
5	-2

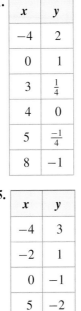

17. -4

19. $-4, 2$

21.

x	-3.2	-1.5	0.1	1.9	2.5	3
y	4	3	0.2	3.8	4	4

23.

x	-3	-2.2	-2	4	6	8
y	2	21	undefined	$0.3\overline{3}$	0.5	0.6

25. $(-3, -2), (1, 6), (-2, 0), (0, 4)$

27. $(-2, 6), (2, 6), (1, 3)$ or $(-1, 3), (0, 2)$

29. $\left(-1, -\dfrac{1}{2}\right), \left(\dfrac{1}{2}, -2\right), \left(4, \dfrac{1}{3}\right), (0, -1)$

31. $(-2, -8), \left(\dfrac{1}{2}, \dfrac{1}{8}\right), (0, 0), (-1, -1)$

APPENDIX A.5

1. $(2, -4)$ 3. $(3, 0)$ 5. $\left(\dfrac{-2}{3}, \dfrac{1}{3}\right)$ 7. $\left(-3, \dfrac{1}{2}\right)$

9. $(2, 1)$ 11. $(1, -2)$ 13. $(6, 0)$ 15. $(1, 2)$

17. $(1, 2)$

19. a.

	Principal	Interest rate	Interest
Bonds	x	0.10	0.10x
Certificate	y	0.08	0.08y
Total	x + y	----	0.10x + 0.08y

b. $x + y = 2000$

c. $0.10x + 0.08y = 184$

d. \$800 at 8%, \$1200 at 10%

21. a.

	Pounds	% Silver	Amount of silver
First alloy	x	0.45	0.45x
Second alloy	y	0.60	0.60y
Mixture	x + y	----	0.45x + 0.60y

b. $x + y = 40$ c. $0.45x + 0.60y = 0.48(40)$ d. 32 lb

23. True-false: 2 points; fill-ins: 5 points

25. a.

	Rate	Time	Distance
Detroit to Denver	x - y	4	1120
Denver to Detroit	x + y	3.5	1120

b. $4(x - y) = 1120$ c. $3.5(x + y) = 1120$

d. Airplane: 300 mph; wind: 20 mph

27. a.

	Cups	Calories per cup	Calories
Oat flakes	x	310	310x
Wheat flakes	y	290	290y
Mixture	x + y	----	310x + 290y

b. $x + y = 1$ c. $310x + 290y = 302$

d. 0.6 cup oats, 0.4 cup wheat

29. $(2.3, 1.6)$ 31. $(182, 134)$

APPENDIX A.6

1. a. b^9 b. b^{10} c. q^9 d. p^{10}

3. a. 2^9 b. 6^8 c. 2^5 d. 8^4

5. a. $216x^3$ b. $81y^4$ c. $32t^{15}$ d. $36s^4$

7. a. $\dfrac{h^8}{m^{12}}$ b. $\dfrac{n^{24}}{k^{32}}$ c. $256a^8b^{16}$ d. $-125a^3b^{24}$

9. a. $2w$ b. w^2

11. a. $-2z^2$ b. $-24z^4$

13. a. Cannot be simplified   b. $12p^5$

15. a. 3^{17} b. Cannot be simplified

17. a. $-24y^2$ b. $32z^2$

19. a. $-12x^3y^4$ b. $-50x^5y$

21. a. $\dfrac{1}{4ab^4}$ b. $\dfrac{2}{3a^3b^2}$

23. a. $\dfrac{5}{c^2}$ **b.** $\dfrac{5}{c^5}$

25. a. b^{13} **b.** b^{25}

27. a. $4x^{10}y^{14}$ **b.** $108x^7y^{10}$

29. a. $\dfrac{-8x^3}{27y^6}$ **b.** $\dfrac{x^8}{16y^4}$

31. a. $-x^4y$ **b.** x^4

APPENDIX A.7

1. Binomial; 3 **3.** Monomial; 4

5. Trinomial; 2 **7.** Trinomial; 3

9. b and c **11.** a, b, c

13. a. -1 **b.** -21 **c.** $8b^3 - 12b^2 + 2b + 1$

15. a. $\dfrac{11}{4}$ **b.** $\dfrac{1}{9}$ **c.** $w^2 - 3w + 1$

17. a. 28.0128 **b.** 126.5728

c. $3k^4 - 12k^3 + 16k^2 - 8k + 4$

19. a. 2 **b.** 96 **c.** $\dfrac{m^6}{729} - \dfrac{m^5}{243}$

21. $4xy - 8y^2$ **23.** $-12x^3 + 6x^2 - 6x$

25. $3a^4b - 2a^3b^2 - a^2b^2$ **27.** $8x^3y^7 - 4x^3y^4 - 6x^5y^5$

29. $n^2 + 10n + 16$ **31.** $r^2 + 3r - 10$

33. $2z^2 - 5z - 3$ **35.** $8r^2 + 2rs - 3s^2$

37. $6x^2 - 13xy + 6y^2$ **39.** $9t^2 - 16s^2$

41. $2a^4 - 5a^2b^2 - 3b^4$ **43.** $4xz(x + 2)$

45. $3n^2(n^2 - 2n + 4)$ **47.** $3r(5rs + 6s^2 - 1)$

49. $m^2n^2(3n^2 - 6mn + 14m)$

51. $3a^2b^2c^4(5a^2b - 4c + 2b)$ **53.** $(a + b)(a + 3)$

55. $(y - 3x)(y - 2)$ **57.** $4(x - 2)^2(-2x^2 + 4x + 1)$

59. $x(x - 5)^2(-x^2 + 5x + 1)$ **61.** $-(2n - 3m)$

63. $-2(x - 1)$ **65.** $-a(b + c)$

67. $-(-2x + y - 3z)$ **69.** $6rs - 5 + \dfrac{2}{rs}$

71. $-5s^8 + 7s^3 - \dfrac{2}{s^2}$ **73.** $2y + 5 + \dfrac{2}{2y + 1}$

75. $x^2 + 4x + 9 + \dfrac{19}{x - 2}$

77. $4z^3 - 2z^2 + 3z + 1 + \dfrac{2}{2z + 1}$

79. $x^3 + 2x^2 + 4x + 8 + \dfrac{15}{x - 2}$

APPENDIX A.8

1. $(x + 2)(x + 3)$ **3.** $(y - 3)(y - 4)$

5. $(x - 3)(x + 2)$ **7.** $(2x - 1)(x + 2)$

9. $(4x - 1)(x + 2)$ **11.** $(3y + 1)(3y - 8)$

13. $(2u + 1)(5u - 3)$ **15.** $(3x - 7)(7x + 2)$

17. $(9a + 4)(8a - 3)$ **19.** $(2x - 3)(15x - 4)$

21. $2(3t + 2)(9t - 11)$ **23.** $(x - 2a)(3x - a)$

25. $(3x - 2y)(5x + 2y)$ **27.** $(3u - 4v)(6u - 5v)$

29. $(3a + 2b)(4a - 7b)$ **31.** $(5ab - 2)(2ab - 3)$

33. $2(4xy + 1)(7xy - 2)$ **35.** $(2az - 3)(11az + 7)$

37. $x^2 + 6x + 9$ **39.** $4y^2 - 20y + 25$

41. $x^2 - 9$ **43.** $9t^2 - 16s^2$

45. $25a^2 - 20ab + 4b^2$ **47.** $64x^2z^2 + 48xz + 9$

49. $(x + 5)(x - 5)$ **51.** $(x - 12)^2$

53. $(x + 2y)(x - 2y)$ **55.** $(2x + 3)^2$

57. $(3u - 5v)^2$ **59.** $(2a + 5b)(2a - 5b)$

61. $(xy + 9)(xy - 9)$ **63.** $(3xy + 1)^2$

65. $(4xy - 1)(4xy + 1)$ **67.** $(x + 2 - y)(x + 2 + y)$

APPENDIX A.9

1. $\dfrac{-2}{d^2}$ **3.** $\dfrac{2x + 3}{3}$ **5.** $\dfrac{3a^2 - 2a}{2}$ **7.** $\dfrac{6(1 + t)}{1 - t}$

9. $y - 2$ **11.** $\dfrac{-2}{v^2 + 3v + 9}$ **13.** $\dfrac{2(x - 3)}{3}$

15. $\dfrac{y + 3x}{y - 3x}$ **17.** $\dfrac{2x - 3}{x - 1}$ **19.** $\dfrac{4z^2 + 6z + 9}{2z + 3}$

21. (b) **23.** None **25.** $\dfrac{-np^2}{2}$ **27.** $\dfrac{5}{ab}$

29. 5 **31.** $\dfrac{a(2a - 1)}{a + 4}$ **33.** $\dfrac{x - 2}{x + 1}$

35. $\dfrac{6x(x - 2)(x - 1)^2}{(x^2 - 8)(x^2 - 2x + 4)}$ **37.** $\dfrac{2}{9}$ **39.** $\dfrac{a + 1}{a - 2}$

41. $3(x^2 - xy + y^2)$ **43.** $\dfrac{x + 2}{x^2 - 1}$ **45.** $\dfrac{x^2(x - 4)}{(x + 1)}$

47. $\dfrac{x + 3}{6y}$ **49.** $\dfrac{x - 3}{2}$ **51.** $\dfrac{a + b - 5c}{6}$

53. $\dfrac{2x - 1}{2y}$ **55.** $\dfrac{-2x + 7}{x + 2y}$ **57.** $12xy^2(x + y)^2$

59. $(a + 4)(a + 1)^2$ **61.** $x(x - 1)^3$ **63.** $\dfrac{7x}{6}$

65. $\dfrac{y}{12}$ **67.** $\dfrac{7x + 1}{6x}$ **69.** $\dfrac{8x - 5}{x(x - 1)}$

71. $\dfrac{3y - 3y^2}{(y + 1)(2y - 1)}$

73. $\dfrac{y^2 - 4y + 5}{(y + 1)(2y - 3)}$

75. $\dfrac{-4}{15(x - 2)}$

77. $\dfrac{3y + 1}{y(y - 3)(y + 2)}$

79. $\dfrac{x^2 - 1}{x}$

81. $\dfrac{x^3 - 2x^2 + 2x - 2}{(x - 1)^2}$

83. $\dfrac{7}{10a + 2}$

85. $\dfrac{a}{a - 2}$

87. h

89. $\dfrac{q}{q - 1}$

91. LC

93. $\dfrac{-2(x + z)}{xz}$

95. $\dfrac{x^2 + y^2}{x^2 y^2}$

97. $\dfrac{8w - 1}{4w^2}$

99. $\dfrac{b^2 - a^2}{ab}$

101. $\dfrac{xy}{x + y}$

103. $\dfrac{x^3 + 1}{x^3}$

105. $b + a$

APPENDIX A.10

1. a. $3\sqrt{2}$ **b.** $2\sqrt[3]{3}$ **c.** $-2\sqrt[4]{4} = -2\sqrt{2}$

3. a. $100\sqrt{6}$ **b.** $10\sqrt[3]{900}$ **c.** $\dfrac{-2}{3}\sqrt[3]{5}$

5. a. $x^3\sqrt[3]{x}$ **b.** $3z\sqrt{3z}$ **c.** $2a^2\sqrt[4]{3a}$

7. a. $-6s^2$ **b.** $-7h$ **c.** $2\sqrt{4 - x^2}$

9. a. $A\sqrt[3]{8 + A^3}$ **b.** $3xy\sqrt{x}$ **c.** $\dfrac{2b\sqrt[3]{b^2}}{a^2}$

11. $5\sqrt{7}$ **13.** $\sqrt{3}$ **15.** $9\sqrt{2x}$ **17.** $-\sqrt[3]{2}$

19. $6 - 2\sqrt{5}$ **21.** $2\sqrt{3} + 2\sqrt{5}$ **23.** $2\sqrt[3]{5} - 4\sqrt[3]{3}$

25. $x - 9$ **27.** $-4 + \sqrt{6}$ **29.** $7 - 2\sqrt{10}$

31. $a - 4\sqrt{ab} + 4b$ **33.** $(1 + \sqrt{3})^2 - 2(1 + \sqrt{3}) - 3 = 0$

35. $(-3 + 3\sqrt{2})^2 + 6(-3 + 3\sqrt{2}) - 9 = 0$

37. $2\sqrt{3}$ **39.** $\dfrac{\sqrt{14x}}{6}$ **41.** $\dfrac{\sqrt{2ab}}{b}$ **43.** $\dfrac{\sqrt{6k}}{k}$

45. $-2(1 - \sqrt{3})$ **47.** $\dfrac{x(x + \sqrt{3})}{x^2 - 3}$ **49.** $\dfrac{\sqrt{6}}{2}$

51. a. $y = \sqrt{x^2} = |x|$ **b.** $y = \sqrt[3]{x^3} = x$

53. a. $2|x|$ **b.** $|x - 5|$ **c.** $|x - 3|$

55. 64 **57.** -2 **59.** $\dfrac{9}{4}$ **61.** -5

63. $\dfrac{-1}{3}$ **65.** $\dfrac{-1}{2}, 2$ **67.** 12 **69.** 5

71. 4 **73.** 5 **75.** 0 **77.** 4

79. We cannot square each term separately; we must square each side of the equation.

81. $\dfrac{\sqrt{7} + \sqrt{3}}{2}$

83. $\dfrac{6\sqrt{3} + 7\sqrt{2}}{5}$

APPENDIX A.11

1. $47°, 57°, 76°$ **3.** $30°, 60°$ **5.** $50°, 50°, 80°$ **7.** 15 cm

9. It is more than 4 and less than 16 feet long

11. 14 ft **13.** 17.1 sq ft **15.** 89.23 ft

17. a. 7.24 cu m **b.** 6.16 sq cm

19. a. 2623.86 cu m **b.** 1903.43 sq in

21. $5; \left(\dfrac{5}{2}, 3\right)$ **23.** $\sqrt{20}; (0, -2)$ **25.** $5; \left(\dfrac{1}{2}, 5\right)$

27. a. 13 miles **b.** $\dfrac{\sqrt{125}}{2} \approx 5.6$ miles

29. $7 + \sqrt{89} + \sqrt{68} \approx 24.7$ **31.** $AC = BC = \sqrt{18}$

33.

35.

37.

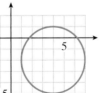

39.

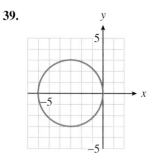

41. $(x + 2)^2 + (y - 5)^2 = 12$

43. $\left(x - \dfrac{3}{2}\right)^2 + (y + 4)^2 = \dfrac{29}{4}$

45. $(x - 2)^2 + (y - 2)^2 = 10$

APPENDIX A.12

1. Rationals   **3.** Irrationals   **5.** Integers

7. Whole numbers   **9.** Rationals   **11.** Irrationals

13. 0.375, terminates   **15.** $0.\overline{285714}$, repeats a pattern

17. 0.4375, terminates   **19.** $0.\overline{846153}$, repeats a pattern

21. 7   **23.** 3   **25.** 147   **27.** $2 \cdot 5$ **29.** 5

31. 78   **33.** 70   **35.** 70   **37.** 1300   **39.** 600

41. $<$ **43.** $>$ **45.** $<$

INDEX

U

unit price, 56
upper triangular form (of a matrix), **667**

V

Variable (dependent and independent), 19
variation (or vary), 235–237,
    240–241, 243
  constant of, 237, 239–240, 241, 243
  direct, 237, 240–241, 243
  inverse, 243
  joint, 301–302
  with a power, 240, 243

vertex, 501, 504, 542, 693, 817
  form (of a quadratic function), 517
vertical
  asymptote, 143, 601, 604–605
  compression, 160
  coordinate, 39
  intercept, *See* y-intercept.
  line test, 42–**43**
  stretch, 160
  translation, 154

W

window, 7, 11, 346, 802

X

x-intercept, 9, 11, 478, 506, 581

Y

y-intercept, 9, 11, 77, 340

Z

zero
  exponent, 256
  factor principle, **477**
  of a polynomial, **581**–583, 593
zero-factor principle, **477**

FORMULAS AND EQUATIONS

Linear Equations

General form $Ax + By = C$

Slope

$$m = \frac{\Delta y}{\Delta x} = \frac{\text{change in } y\text{-coordinate}}{\text{change in } x\text{-coordinate}}$$

$$m = \frac{y_2 - y_1}{x_2 - x_1} = \frac{f(x_2) - f(x_1)}{x_2 - x_1}$$

Slope-intercept form $\quad y = b + mx$

Point-slope form $\quad y = y_1 + m(x - x_1)$

Linear model

$\quad f(x) = (\text{starting value}) + (\text{rate of change}) \cdot x$

Absolute value $|x| = \begin{cases} x & \text{if } x \geq 0 \\ -x & \text{if } x < 0 \end{cases}$

$$|a + b| \leq |a| + |b|$$
$$|ab| = |a||b|$$

Exponential Notation

$$a^{-n} = \frac{1}{a^n}, \quad a \neq 0$$

$$a^0 = 1, \quad a \neq 0$$

$$a^{1/n} = \sqrt[n]{a}, n \text{ an integer}, n > 2$$

$$a^{m/n} = (a^{1/n})^m = (a^m)^{1/n}, a > 0, n \neq 0$$

Laws of exponents

I. $a^m \cdot a^n = a^{m+n}$

II. $\dfrac{a^m}{a^n} = a^{m-n}$

III. $(a^m)^n = a^{mn}$

IV. $(ab)^n = a^n b^n$

V. $\left(\dfrac{a}{b}\right)^n = \dfrac{a^n}{b^n}, b \neq 0$

Power Functions, $\quad f(x) = kx^p$

Direct variation $\quad y = kx^n, n > 0$

Inverse variation $\quad y = \dfrac{k}{x^n}, n > 0$

Exponential Functions, $\quad f(x) = ab^x$

Exponential growth

$$P(t) = P_0 b^t = P_0(1 + r)^t$$
$$= P_0 e^{kt}, k > 0$$
$$P(t) = P_0 2^{t/D}$$

Exponential decay

$$Q(t) = Q_0 b^t = Q_0(1 - r)^t$$
$$= Q_0 e^{kt}, k < 0$$
$$Q(t) = Q_0(0.5)^{t/H}$$

Properties of logarithms

If x, y, and $b > 0$, then

1. $\log_b(xy) = \log_b x + \log_b y$

2. $\log_b \dfrac{x}{y} = \log_b x - \log_b y$

3. $\log_b x^k = k \log_b x$

4. $\log_b b^x = x \quad$ and $\quad b^{\log_b x} = x$

Properties of natural logarithms

If $x, y > 0$, then

1. $\ln(xy) = \ln x + \ln y$

2. $\ln \dfrac{x}{y} = \ln x - \ln y$

3. $\ln x^m = m \ln x$

4. $\ln e^x = x \quad$ and $\quad e^{\ln x} = x$

Logarithmic scales

Acidity $\text{pH} = -\log_{10}[\text{H}^+]$

Decibels $D = 10 \log_{10}\left(\dfrac{I}{10^{-12}}\right)$

Richter $M = \log_{10}\left(\dfrac{A}{A_0}\right)$